MASONRY DESIGN AND DETAILING

MASONRY DESIGN AND DETAILING

For Architects, Engineers, and Contractors

CHRISTINE BEALL, NCARB, CCS

Fourth Edition

McGRAW-HILL

New York San Francisco Washington, D.C. Auckland Bogotá
Caracas Lisbon London Madrid Mexico City Milan
Montreal New Delhi San Juan Singapore
Sydney Tokyo Toronto

Library of Congress Cataloging-in-Publication Data

Beall, Christine.
Masonry design and detailing : for architects, engineers, and contractors / Christine Beall. — 4th ed.
p. cm.
Includes bibliographical references (p.) and index.
ISBN 0-07-005844-X
1. Masonry. I. Title.
TA670.B43 1997
693'.1—dc21 97-1066
CIP

McGraw-Hill
A Division of The ***McGraw-Hill*** *Companies*

5 6 7 8 9 0 KGP/KGP 9 0 2 1 0 9

ISBN 0-07-005844-X

The sponsoring editor for this book was Larry Hager, the editing supervisor was Peggy Lamb, and the production supervisor was Tina Cameron. It was set in Century Schoolbook by Ron Painter of McGraw-Hill's Professional Book Group composition unit.

Printed and bound by Quebecor/Kingsport Press.

This book is printed on acid-free paper.

To Star

CONTENTS

PREFACE TO FOURTH EDITION

This edition has been updated to reflect revisions in the 1994 edition of the *Uniform Building Code* and the 1995 edition of the Masonry Standards Joint Committee (MSJC) *Building Code Requirements for Masonry Structures (ACI 530/ASCE 5/TMS 402)* and *Specifications for Masonry Structures (ACI 530.1/ASCE 6/TMS 602)*. The text has also been revised to include new ASTM standards and to change references to revised standards. The information on mortar selection has been expanded to better explain selection criteria and methods of specifying, and a new chapter has been added on quality assurance and quality control. The QA/QC chapter is intended to guide readers through the intricacies of specifying a quality assurance program for masonry, choosing the proper industry standards and ASTM quality control tests, and identifying appropriate levels of field observation or structural inspection. Information has also been added on new masonry products such as segmental retaining walls, EPDM and rubberized asphalt flashings, "mortar cement," and seismic veneer anchors.

Christine Beall, NCARB, CCS
McQueeney, Texas

PREFACE TO THIRD EDITION

There are several major changes in this edition. First of all, the text has been expanded significantly to cover new topics and discuss existing ones in greater detail. Along with this new text, many new drawings, tables, and photographs have been added. New coverage includes discussion of the Masonry Standards Joint Committee Building Code Requirements for Masonry Structures ACI 530/ASCE 5/TMS 402 and its accompanying Specifications. Revisions have also been made to update code requirements for the 1991 Uniform Building Code, 1990 National Building Code, and 1991 Standard Building Code. Much of the detailed information that has been added first appeared in articles which I have written for *Magazine of Masonry Construction*, *The Construction Specifier*, *Architecture*, and *Architectural Record*.

The text has also been reorganized to consolidate topics on structural design and to create separate chapters on movement and moisture control, paving and fireplaces, and cleaning and restoration. This new arrangement should make information easier to find.

Christine Beall, NCARB
Austin, Texas

PREFACE TO SECOND EDITION

This handbook addresses a broad range of aesthetic, technical, and environmental considerations. In addition to the engineering aspects of design, technical information on energy and sound control, maintenance, life-cycle costing, and workmanship is included. My goal has been to assemble and correlate existing industry information into a single, concise, and complete reference aimed at architects and other design professionals.

Two major sources of information have been publications of the Brick Institute of America and the National Concrete Association, especially the BIA "Technical Notes" series and the NCMA "TEK Bulletins." A bibliography of detailed sources is given at the back of this book as well as a list of national and regional masonry organizations through which design assistance can be obtained.

I would like to thank Excy Johnston, AIA, for preparing the sketches in Figs. 1-1 and 2-1; Gregg Borchelt, P.E., of the Masonry Institute of Houston/Galveston for his technical editing of Chapters 10, 11, and 12 and for the preparation of sample problems to accompany the text; Bernie Beall for typing the rough draft; and Kathy Cogburn for putting the text on computer disk.

Photographs and many technical illustrations, charts, graphs, and tables have been provided through the courtesy of the Brick Institute of America, the National Concrete Masonry Association, the American Concrete Institute, the Masonry Institute of America, and the Portland Cement Association.

Christine Beall, AIA
Austin, Texas

INTRODUCTION

1

HISTORY AND DEVELOPMENT OF MASONRY TECHNOLOGY

The unwritten record of history is preserved in buildings—in temples, fortresses, sanctuaries, and cities constructed of brick and stone. Early efforts to build permanent shelter were limited to the materials at hand. The trees of a primeval forest, the clay and mud of a river valley, the rocks, caves and cliffs of a mountain range afforded only primitive opportunity for protection, security, and defense, and few examples survive. But the stone and brick of skeletal architectural remains date as far back as the temples of Ur built in 3000 B.C., the early walls of Jericho of 8000 B.C., and the vaulted tombs at Mycenae of the fourteenth century B.C. It was the permanence and durability of the masonry which safeguarded this prehistoric record of achievements, and preserved through centuries of war and natural disaster the traces of human development from cave dweller to city builder. Indeed, the history of civilization is the history of its architecture, and the history of architecture is the history of masonry.

1.1 DEVELOPMENT

Stone is the oldest, most abundant, and perhaps the most important raw building material of prehistoric and civilized peoples. Stone formed their defense in walls, towers, and embattlements. They lived in buildings of stone, worshiped in stone temples, and built roads and bridges of stone. Builders began to form and shape stone when tools had been invented that were hard enough to trim and smooth the irregular lumps and broken surfaces. Stone building was then freed from the limitations of monolithic slab structures like those at Stonehenge and progressed through the shaped and fitted blocks of the Egyptians to the intricately carved columns and entablatures of the Greeks and Romans.

Brick is the oldest *man-made* building material, invented almost 10,000 years ago. Its simplicity, strength, and durability led to extensive use, and gave it a dominant place in history alongside stone.

Rubble stone and mud bricks were small, easily handled materials that could be stacked and shaped to form enclosures of simple or complex design. Hand-shaped, sun-dried bricks, reinforced with such diverse materials as straw and dung, were so effective that kiln-fired bricks did not appear until the third millennium B.C., long after the art of pottery had demonstrated the effects of high temperatures on clay. Some of the oldest bricks in the world, taken from archaeological digs at the site of ancient Jericho, resemble long loaves of bread with a bold pattern of Neolithic thumbprint impressions on their rounded tops (*see Fig. 1-1*). The use of wooden molds did not replace such hand-forming techniques until the early Bronze Age, around 3000 B.C.

Perhaps the most important innovations in the evolution of architecture were the development of masonry arches and domes. Throughout history, the arch was the primary means of overcoming the span limitations of single blocks of stone or lengths of timber, making it possible to bridge spaces once thought too great. Early forms only approximated true "arching" action and were generally false, corbeled arches. True arches carry their loads in simple compression to each abutment, and as long as the joints are roughly aligned at right angles to the compressive stress, the precise curve of the arch is not critical.

The excavation of ruins in Babylonia exposed a masonry arch believed to have been built around 1400 B.C. Arch construction reached a high level of refinement under the Romans, and later developments were limited primarily to the adaptation of different shapes. Islamic and Gothic arches led to the design of groined vaults, and eventually to the high point of cathedral architecture and masonry construction in the thirteenth century.

Simple dome forms may actually have preceded the true arch because, like the corbeled arch, they could be built with successive horizontal rings of masonry, and required no centering. These domes were seen as circular walls gradually closing in on themselves rather than as rings of vertical arches. Barrel vaults were built as early as the thirteenth century B.C., and could also be constructed without centering if one end of the vault was closed off.

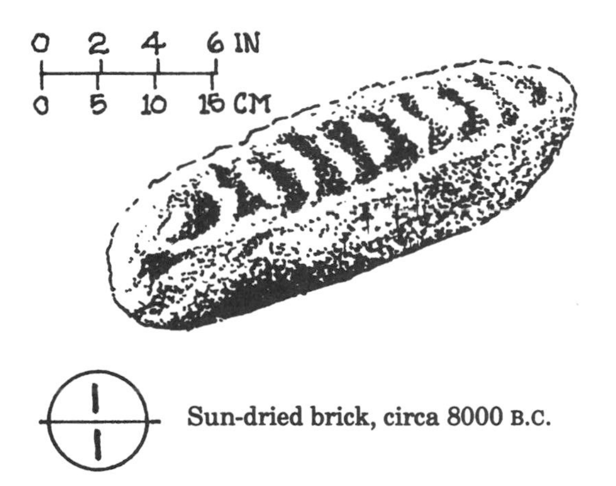

Sun-dried brick, circa 8000 B.C.

Initial exploitation of the true dome form took place from the middle of the first century A.D. to the early second century, under the reigns of Nero and Hadrian. The brick dome of the Pantheon in Rome exerts tremendous outward thrusts counteracted only by the massive brick walls encircling its perimeter. Later refinements included the masonry squinch and pendentive, which were instrumental in the construction of the dome of the Florence Cathedral, and buttressing by means of half domes at the sides, as in the Church of Hagia Sophia in Constantinople.

1.2 DECLINE

Renaissance architecture produced few significant innovations in structural building practices, since designs were based primarily on the classical forms of earlier eras. The forward thrust of structural achievements in masonry essentially died during this period of "enlightenment," and masonry structures remained at an arrested level of development.

With the onslaught of the Industrial Revolution, emphasis shifted to iron, steel, and concrete construction. The invention of portland cement in 1824, refinements in iron production in the early nineteenth century, and the development of the Bessemer furnace in 1854 turned the creative focus of architecture away from masonry.

By the early twentieth century, the demand was for high-rise construction, and the technology of stone and masonry building had not kept pace with the developments of other structural systems. The Chicago School had pioneered the use of iron and steel skeleton frames, and masonry was relegated to secondary usage as facings, infill, and fireproofing. The Monadnock Building in Chicago (1891) is generally cited as the last great building in the "ancient tradition" of masonry architecture (*see Fig. 1-2*). Its 16-story *unreinforced* loadbearing walls were required by code to be several feet thick at the base, making it seem unsuited to the demands of a modern industrialized society. Except for the revivalist periods following the 1893 World's Columbian Exposition and the "mercantile classicism" which prevailed for some time, a general shift in technological innovation took place, and skeleton frame construction began to replace loadbearing masonry.

During this period, only Antonio Gaudi's unique Spanish architecture showed innovation in masonry structural design (*see Fig. 1-3*). His "structural rationalism" was based on economy and efficiency of form, using ancient Catalan vaulting techniques, parabolic arches, and inclined piers to bring the supporting masonry under compression. His work also included vaulting with hyperbolic paraboloids and warped "helicoidal" surfaces for greater structural strength. Gaudi, however, was the exception in a world bent on developing lightweight, high-rise building techniques for the twentieth century.

At the time, most considered both concrete and masonry construction to be unsophisticated systems with no tensile strength. Very soon, however, the introduction of iron and steel reinforcement brought concrete a step forward. While concrete technology developed rapidly into complex steel-reinforced systems, masonry research was virtually nonexistent, and the widespread application of this new reinforcing technique to masonry never occurred.

The first reinforced concrete building, the Eddystone Lighthouse (1774), was actually constructed of both concrete *and* stone, but the use of iron or steel as reinforcing was soon limited almost entirely to concrete. A few reinforced brick masonry structures were built in the early to mid-nineteenth century, but these experiments had been abandoned by about

Monadnock Building, Chicago, 1891. Burnham and Root, architects. (*Photo courtesy of the School of Architecture Slide Library, the University of Texas at Austin.*)

1880. Reinforced masonry design was at that time intuitive or empirical rather than rationally determined, and rapid advances in concrete engineering quickly outpaced what was seen as an outmoded, inefficient, and uneconomical system. Even by the time the Monadnock Building was constructed, building codes still recognized lateral resistance of masonry walls only in terms of mass, and this did indeed make the system expensive and uneconomical.

(A)

(B)

(C)

(D)

Gaudi's innovative masonry structures: (A) Inclined brick column, Colonia Guell Chapel; (B) warped masonry roof, schools of the Sagrada Familia Church; (C) thin masonry arch ribs, Casa Mila; and (D) arch detail, Bell Esguard. (*Photos courtesy of the School of Architecture Slide Library, the University of Texas at Austin.*)

1.3 REVIVAL

In the early 1920s, economic difficulties in India convinced officials that alternatives to concrete and steel structural systems had to be found. Extensive research began into the structural performance of reinforced masonry which led not only to new systems of low-cost construction, but also to the first basic understanding of the structural behavior of masonry. It was not until the late 1940s, however, that European engineers and architects began serious studies of masonry bearing wall designs—almost 100 years after the same research had begun on concrete bearing walls.

By that time, manufacturers were producing brick with compressive strengths in excess of 8000 psi, and portland cement mortars had strengths as high as 2500 psi. Extensive testing of some 1500 wall sections generated the laboratory data needed to develop a rational design method for masonry. These studies produced the first reliable, mathematical analysis of a very old material, freed engineers for the first time from the constraints of empirical design, and allowed formulation of rational structural theories. It was found that no new techniques of analysis were required, but merely the application of accepted engineering principles already being used on other systems.

The development of recommended practices in masonry design and construction in the United States took place during the decade of the 1950s, and resulted in publication of the first "engineered masonry" building code in 1966. Continued research throughout the following two decades brought about refinements in testing methods and design procedures, and led to the adoption of engineered masonry structural systems by all of the major building codes in the United States. Laboratory and field tests have also identified and defined the physical properties of masonry and verified its excellent performance in fire control, sound attenuation, and thermal resistance.

Masonry construction today includes not only quarried stone and clay brick, but a host of other manufactured products as well. Concrete block, cast stone, structural clay tile, terra cotta, glass block, mortar, grout, and metal accessories are all a part of the mason's trade. In various definitions of masonry, this group of materials is often expanded to include concrete, stucco, or precast concrete. However, the most conventional application of the term "masonry" is limited to relatively small building units of natural or manufactured stone, clay, concrete, or glass that are assembled by hand using mortar, dry-stacking, or mechanical connectors.

1.4 CONTEMPORARY MASONRY

Contemporary masonry may take one of several forms. Structurally, it may be divided into loadbearing and non-loadbearing construction. Walls may be of single- or multi-wythe design. They may also be solid masonry, solid walls of hollow units, or cavity walls. Finally, masonry may be reinforced, partially reinforced, or unreinforced, and either empirically or analytically designed. Loadbearing masonry supports its own weight as well as the dead and live loads of the structure, and all lateral wind and seismic forces. Non-loadbearing masonry also resists lateral loads, and veneers may support their own weight for the full height of the structure or be wholly supported by the structure at each floor. Solid masonry is built of solid units or fully grouted hollow units in multiple wythes with the collar joint between wythes filled with mortar or grout. Solid walls of hollow units have open cores in the units, but grouted collar joints. Cavity walls have two or more wythes of solid or hollow units separated by an open collar joint or cavity at least 2 in. wide (*see Fig. 1-4*).

Empirical designs are based on arbitrary limits of height and wall

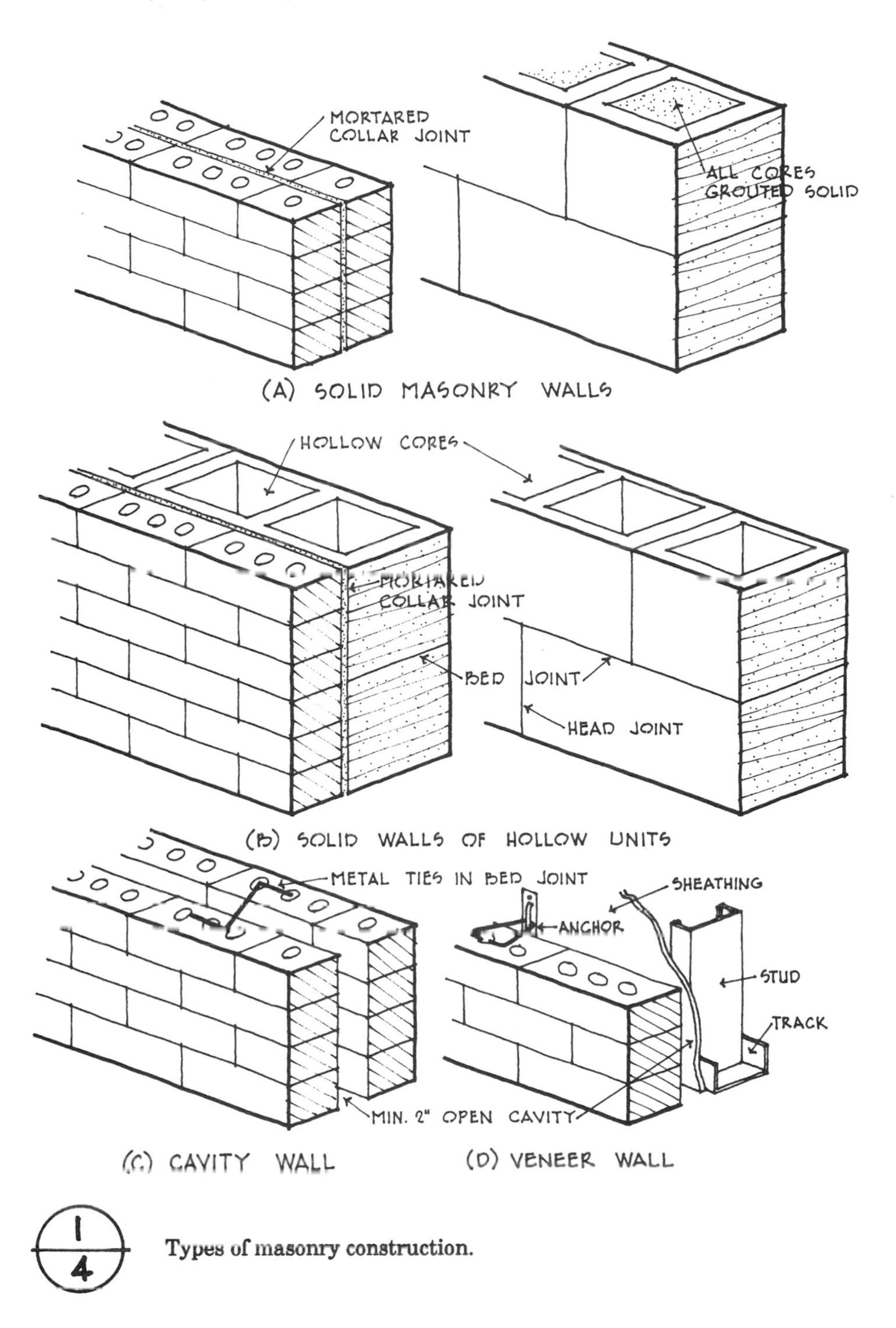

1-4 **Types of masonry construction.**

thickness. Engineered designs, however, are based on rational analysis of the loads and the strength of the materials used in the structure. Standard calculations are used to determine the actual compressive, tensile, and shear stresses, and the masonry designed to resist these forces. Unreinforced masonry is still sometimes designed by empirical methods, but is applicable only to low-rise structures with modest loads. Unreinforced masonry is strong in compression, and small lateral loads and overturning moments are resisted by the weight of the wall. Shear and flexural stresses are resisted only by the bond between mortar and units. Where lateral loads are higher, flexural strength can be increased by solidly grouting reinforcing steel into hollow unit cores or wall cavities wherever design analysis indicates that tensile stress is developed. The cured grout binds the masonry and the steel together to act as a single load-resisting element.

Contemporary masonry is very different from the traditional construction of earlier centuries. Its structural capabilities are still being explored as continuing research provides a better understanding of masonry structural behavior. Contemporary masonry buildings have thinner, lighter-weight, more efficient structural systems and veneers than in the past, and structures designed in compliance with current code requirements perform well, even in cases of significant seismic activity and extreme fire exposure.

1.5 COMMON CONCERNS

Although there is continuing structural research aimed at making masonry systems stronger, more efficient, and more economical, many of the concerns commonly expressed by both design professionals and contractors are related to weather resistance. Moisture penetration and durability, in fact, seem to be more significant day-to-day issues for most people than structural performance. Building codes, which have traditionally provided minimum performance requirements only for structural and life safety issues, are now beginning to address water penetration, weather resistance, and durability issues for masonry as well as other building systems.

Contemporary masonry walls are more water-permeable than traditional masonry walls because of their relative thinness, and more brittle because of the portland cement that is now used in masonry mortar. As with any material or system used to form the building envelope, the movement of moisture into and through the envelope has a significant effect on the performance of masonry walls. Contemporary masonry systems are designed not as a barrier to water penetration, but as drainage walls in which penetrated moisture is collected on flashing membranes and expelled through a series of weepholes. Higher-performance wall systems for extreme weather exposures can be designed as pressure-equalized rain screens, but at a higher cost than drainage walls. Design, workmanship, and materials are all important to the performance of masonry drainage and rain screen walls:

- Mortar joints must be full
- Mortar must be compatible with and well bonded to the units
- Drainage cavity must be kept free of mortar droppings
- Appropriate flashing material must be selected for the expected service life of the building
- Flashing details must provide protection for all conditions
- Flashing must be properly installed
- Weepholes must be properly sized and spaced
- Weepholes must provide rapid drainage of penetrated moisture

With adequate provision for moisture drainage, masonry wall systems can provide long-term performance with little required maintenance. The chapters which follow discuss materials, design and workmanship with an eye toward achieving durability and weather resistance as well as adequate structural performance in masonry systems.

2 RAW MATERIALS AND MANUFACTURING PROCESSES

The quality and characteristics of masonry products are directly and exclusively determined by the raw materials and methods of manufacture used in their production. A basic introduction to this aspect of masonry will aid in understanding the finished products and how they may best be used in specific design applications.

2.1 CLAY MASONRY

Clay, the raw material from which brick, structural clay tile, and terra cotta are made, is the most plentiful natural substance used in the production of any building product. Clay is the end product of the chemical alteration over long periods of time of the less stable minerals in rock. This chemical weathering produces minute particles that are two-dimensional or flake-shaped. The unique plastic characteristics of clay soils are a result of the enormous amount of surface area inherent in this particle size and shape. The natural affinity of clay soils and moisture result in cohesiveness and plasticity from the surface tension of very thin layers of water between each of these minute particles. It is this plasticity which facilitates the molding and shaping of moist clay into usable shapes.

For the architect, the importance of understanding clay characteristics and methods of manufacture is their relationship to finished appearance and physical properties. Color depends, first, on the composition of the raw material and the quantitative presence of metallic oxides. Second, it is an indication of the degree of burning to which the clay has been subjected. Lighter-colored units (called salmons) for a given clay are normally associated with underburning. They may also be indicative of high porosity and absorption along with decreased strength, durability, and resistance to abrasion. On the other hand, the very dark-colored units (called clinkers) produced from the *same* clay result from overburning. This indicates that the units have been pressed and burned to a very high compressive strength and abrasion resistance, with greatly reduced absorption and increased resistance to freezing and thawing.

Most of the brick used in building construction falls between the extremes of salmon and clinker brick. Since clay composition is the primary determinant of brick color, lightness or darkness cannot be used as an absolute indicator of physical properties for brick made from different raw materials. It can, however, assist generally in the evaluation and selection of brick to meet specific design or exposure requirements.

2.1.1 Clay Composition

Clays are basically compounds of silica and alumina with varying amounts of metallic oxides and other minor ingredients and impurities. Metallic oxides act as fluxes to promote fusion at lower temperatures, influence the range of temperatures in which the material vitrifies, and give burned clay the necessary strength for structural purposes. The varying amounts of iron, calcium, and magnesium oxides also influence the color of fired clay.

Clays may be classified as either calcareous or non-calcareous. While both are hydrous aluminum silicates, the calcareous clays contain around 15% calcium carbonate, which produces a yellowish color when fired. The non-calcareous clays are influenced by feldspar and iron oxide. The oxide may range from 2 to 25% of the composition, causing the clay to burn from a buff to a pink or red color as the amount increases.

Any lime that is present in a clay must be finely crushed to eliminate large lumps. Lime becomes calcined in the burning process and later slakes or combines with water when exposed to the weather, so that any sizable fragments will expand and possibly chip or spall the brick.

2.1.2 Clay Types

There are three different types of clay which, although they are similar in chemical composition, have different physical characteristics. Surface clays, shales, and fire clays are common throughout the world, and result from slight variations in the weathering process.

Surface clay occurs quite close to the earth's surface, and has a high oxide content, ranging from 10 to 25%. Surface clays are the most accessible and easily mined, and therefore the least expensive.

Shale is a metamorphic form of clay hardened and layered under natural geologic conditions. It is very dense and harder to remove from the ground than other clays, and as a result, is more costly. Like surface clay, shale contains a relatively high percentage of oxide fluxes.

Fire clay is formed at greater depths than either surface clay or shale. It generally has fewer impurities, more uniform chemical and physical properties, and only 2 to 10% oxides. The lower percentage of oxide fluxes gives fire clay a much higher softening point than surface clay and shale, and the ability to withstand very high temperatures. This refractory quality makes fire clay best suited to producing brick and tile for furnaces, fireplaces, flue liners, ovens, and chimney stacks. The low oxide content also causes the clay to burn to a very light brown or light buff color, approaching white.

Clay is well suited to the manufacture of masonry products. It is plastic when mixed with water, and easily molded or formed into the desired shapes; it has sufficient tensile strength to maintain those shapes after the dies or molds are removed; and its particles are ceramically fused at high temperatures.

2.1.3 Material Preparation

Brick plants commonly mine from several clay pits at a time. Since the raw clay is not always uniform in quality and composition, two or more clays from different pits or from remote locations within the same pit are blended to minimize much of the natural variation in chemical composition and physical properties. Blending produces a higher degree of product uniformity, helps control the color of the units, and permits some latitude in providing raw material suitable for specific types of brick or special product requirements. The clay is first washed to remove stones, soil, or excessive sand, then crushed into smaller pieces and finally ground to a powdered mix. Particle size is carefully controlled so that only the finer material is taken to storage bins or directly to the forming machine or pug mill for tempering and molding.

2.1.4 Manufacturing

After preparation of the raw clay, the manufacture of fired brick is completed in four additional stages: *forming, drying, burning,* and *drawing and storage* (*see Fig. 2-1*). The basic process is always the same, and differences occur only in the molding techniques. In ancient as well as more recent history, brick was exclusively hand-made. Since brick-making machines were invented in the late nineteenth century, however, most of the structural clay products manufactured in the United States are machine made by one of three forming methods: *stiff-mud, soft-mud,* or *dry-press.*

2.1.5 Forming

The first step in each forming method is tempering, where the clay is thoroughly mixed with a measured amount of water. The amount of water and the desired plasticity vary according to the forming method to be used.

The *stiff-mud extrusion method is used for more than 80% of the brick manufactured in the United States.* A minimum amount of water, generally 12 to 15% moisture by weight, is mixed with the dry clay to make it plastic. After thorough mixing in a pug mill, the tempered clay goes through a de-airing process which increases the workability and plasticity of the clay and produces units with greater strength. The clay is then forced through a steel die in a continuous extrusion of the desired size and shape, and at the same time, is cored to reduce weight and to facilitate drying and burning. Automatic cutting machines using thin wires attached to a circular steel frame cut the extruded clay into pieces (*see Fig. 2-2*). Since the clay will shrink as it is dried and burned, die sizes and cutter wire spacing must be carefully calculated to compensate. Texturing attachments may be affixed to roughen, score, scratch, or otherwise alter the smooth skin of the brick column as it emerges from the die (*see Fig. 2-3*). After cutting, a clay slurry of contrasting color or texture may also be applied to the brick surface to produce different aesthetic effects.

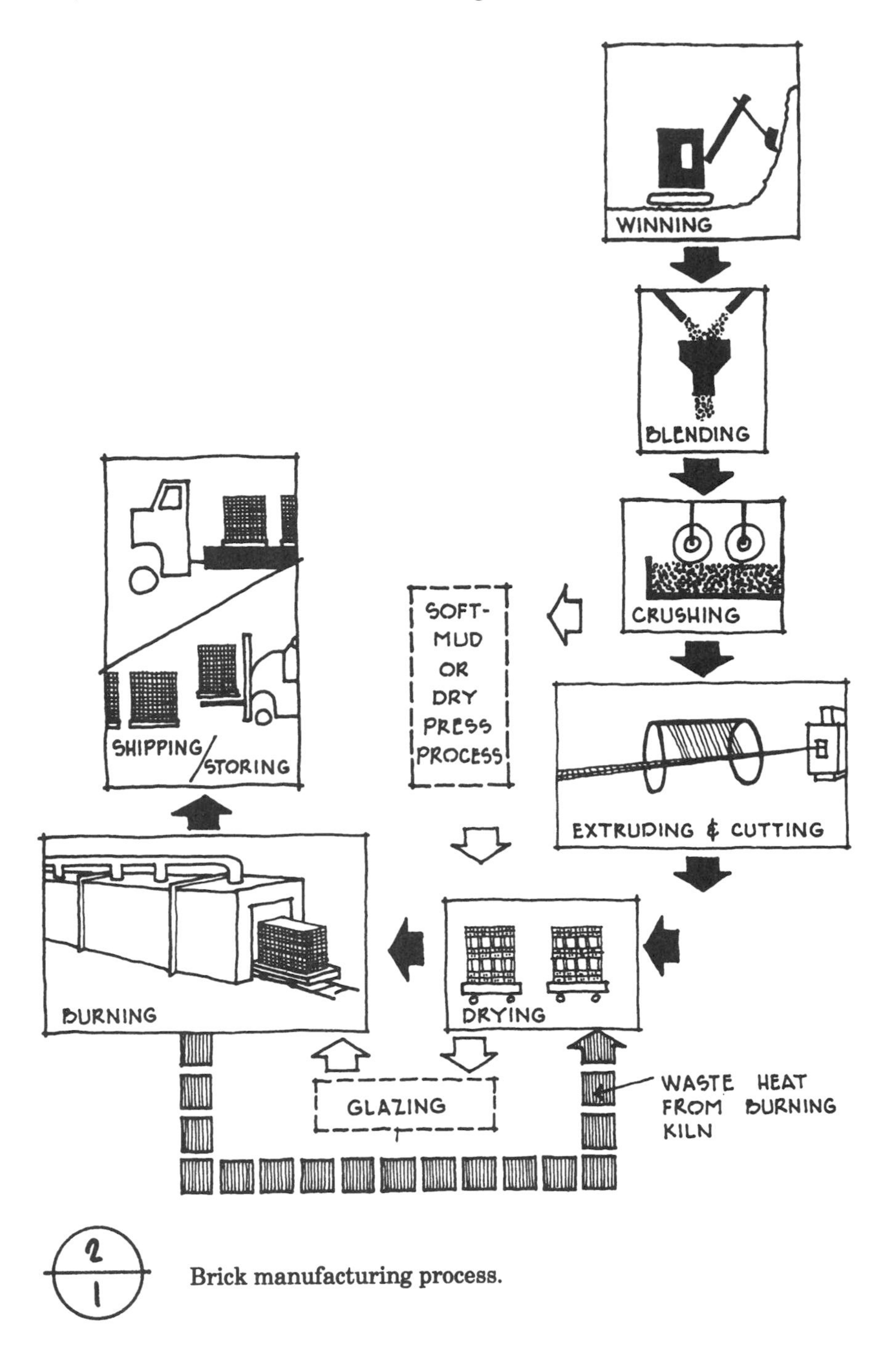

2-1 **Brick manufacturing process.**

A conveyor belt moves the "green" or wet brick past inspectors, who remove imperfect units and return them to the pug mill. Satisfactory units are moved from the conveyor to dryer cars and stacked in a prescribed pattern to allow free flow of air and kiln gases for burning. The stiff-mud process produces the hardest and most dense of the machine-made bricks, and also delivers the highest volume of production.

The *soft-mud method* of production is the oldest, and was used exclusively up until the nineteenth century (*see Fig. 2-4*). All hand-made brick are formed by this process even today. Only a few manufacturers still pro-

Wire-cutting extruded, stiff-mud brick. (*Photo courtesy BIA.*)

duce genuine hand-made brick, but demand is increasing as more historic restoration projects are undertaken.

Automated machinery can accomplish soft-mud molding more uniformly and efficiently than hand work, and is now widely used. The soft-mud process is particularly suitable for clays which contain too much natural water for the extrusion method. The clay is tempered to a 20 to 30% moisture content (about twice that of the stiff-mud clays) and then pressed into wooden molds by hand or machine to form standard or special shapes. To prevent the clay from sticking, the molds are lubricated with sand or water. The resulting "sand-struck" or "water-struck" brick has a unique appearance characterized by either a rough, sandy surface, or a relatively smooth surface with only slight texture variations from the individual molds (*see Fig. 2-3*). In addition to having an attractive rustic appearance, soft-mud units are more economical to install because less precision is required, and bricklayers can usually achieve a higher daily production. Manufacturers often simulate the look of hand-made brick by tumbling and roughening extruded brick.

The mortar bedding surfaces of sand-struck or sand-molded brick must be brushed clean of loose sand particles so that mortar bond is not adversely affected. Even if sand is not actually applied to the bed surfaces in the manufacturing process, stray particles along the edge of a unit can inhibit the critical mortar-to-unit bond at the weathering face of a wall, creating an unwanted increase in moisture penetration.

The *dry-press method,* although it produces the most accurately formed units, is used for less than one-half of one percent of U.S.-made brick. Clays

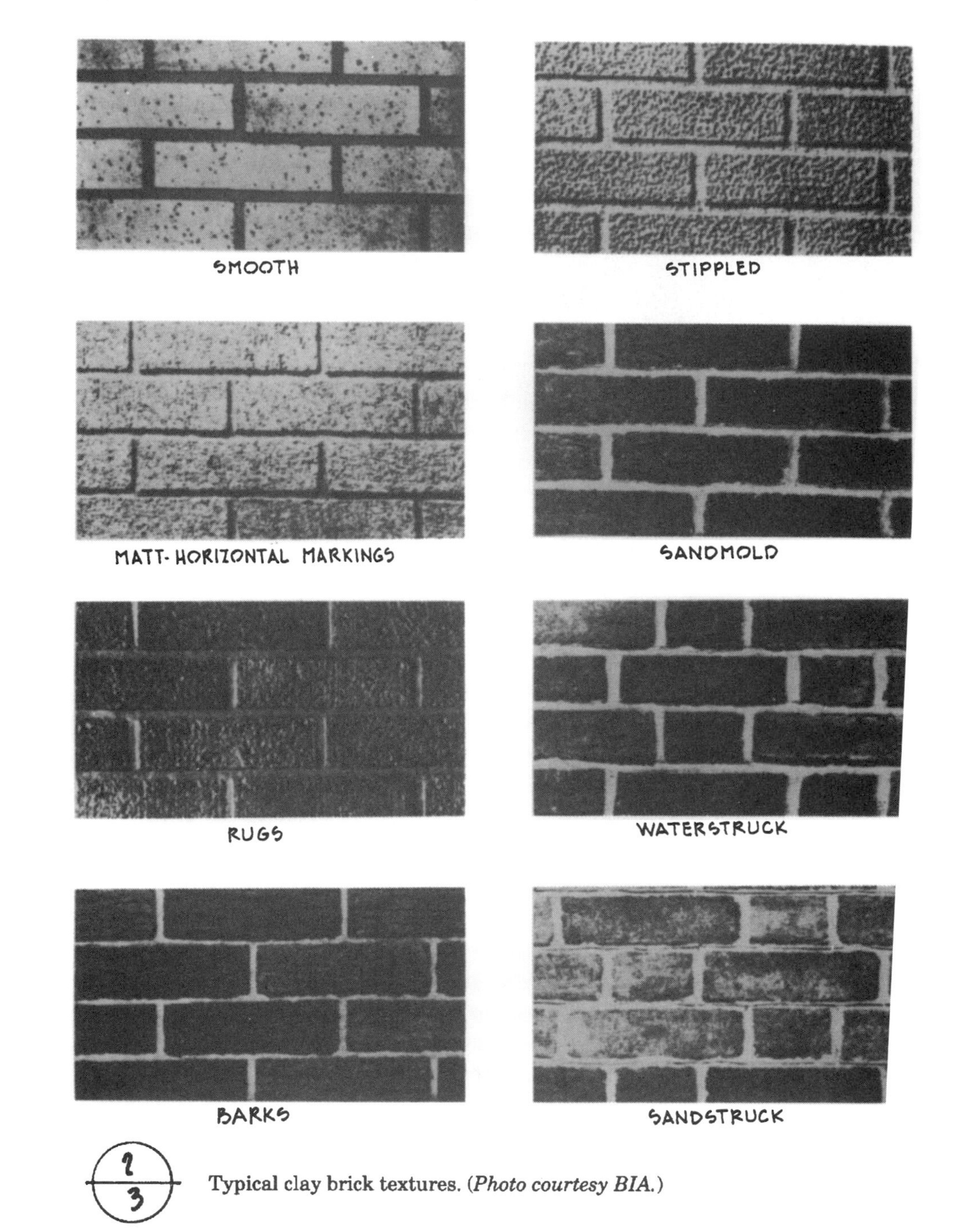

2/3 Typical clay brick textures. (*Photo courtesy BIA.*)

of very low natural plasticity are required, usually with moisture contents of 10% or less. The relatively dry mix is pressed into steel molds by hydraulic plungers exerting a force of 500 to 1500 psi to form the unit.

2.1.6 Drying

Green clay units coming from the molding or cutting machines may contain 10 to 30% free moisture, depending on the forming process used. Before

Artist's engraving of a colonial brick-making operation. (*Courtesy BIA.*)

burning can begin, most of this excess must be evaporated. The open sheds once used for natural air drying were affected by weather conditions, and the evaporation process took anywhere from 7 days to 6 weeks. Today, brick plants use separate dryer kilns or chambers supplied with waste heat from the exhaust of the firing kilns. Drying time takes only 24 to 48 hours, depending on the original moisture content. Drying temperatures range from 100 to 400°F, but must be carefully regulated, along with humidity, to prevent sudden changes which could crack or warp the units.

2.1.7 Glazing

Glazing is a highly specialized, carefully controlled procedure used in the production of decorative brick. *High-fired ceramic glazes* are the most widely used. The glaze is a blend of clays, ceramic frit, fluxes, and base metals sprayed on the units before burning, and then subjected to normal firing temperatures to fuse it to the clay body. Glazes with a higher flux content will burn to a glossy finish, while more refractory mixes produce a matte glaze. After the basic glass material is prepared, ceramic pigments are used to stain it to the desired color. Cobalt, vanadium, chrome, tin, nickel, alumina, and other metals are used singly or in combinations to produce standard, custom or color-matched blues, greens, ochres, pinks, lavenders, buffs, grays, and blacks. Color consistency is easier to maintain with high-gloss glazes, both within batches and between kiln runs.

Low-fired glazes are for colors which cannot be produced at high firing temperatures, such as bright red, bright yellow, burgundy, and orange. If fired too hot, bright red, for instance, will craze or burn transparent because the cadmium and lead in the glaze are unstable at high temperatures. The glaze is applied after the brick has been burned to maturity, and then requires a second firing at lower temperatures of 1300 to 1800°F. Low-fired glazes are much more expensive because of the two-step process.

Clay coat glazes (sometimes called slip glazes) produce a dull, nonreflective, vitreously applied surface in softer tones than ceramic glazes. *Salt glazes* are produced by applying a vapor of sodium-iron silicate to the brick while it is at maximum firing temperature. The transparent finish shows the natural color of the fired brick under a lustrous gloss.

Producing some ceramic glazes leaves contaminants in the kiln which can affect the next batch of brick. The residue from ceramic glazes is also classified by the Environmental Protection Agency as hazardous waste which must be recovered for reuse or disposal.

2.1.8 Burning

After excess moisture has been evaporated from the clay units and desired glazes, if any, have been applied, the bricks are ready for burning. This is one of the most specialized and critical steps in the manufacture of clay products. Burning is accomplished by controlled firing in a kiln to achieve ceramic fusion of the clay particles and hardening of the brick. Since so many of the properties of brick and clay tile depend on the method and control of firing, the development over the years of more sophisticated kilns has been instrumental in improving the quality and durability of clay masonry.

Originally, bricks were cured by sun drying. This permitted hardening by evaporation, but did not achieve the chemical fusion necessary for high strength. High-temperature kiln firing of clay brick was done as early as 3500 B.C. Early scove kilns heated by wood fires were eventually replaced by beehive kilns. The heat source was originally at the bottom of the kiln, and could not be controlled effectively, so uneven firing resulted in hard-burned "clinker" brick nearest the fire and soft, underburned "salmon" brick at the top of the kiln. Salmon brick were sometimes used in unexposed locations such as filler courses in multi-wythe walls, but clinker brick were usually discarded. Builders in colonial Williamsburg, Virginia, however, were fond of clinker brick and often used shiny, black, overburned units as headers to create checkerboard patterns with ordinary red brick. Tudor style homes of the early 1900s also used clinker brick in the same way. Some manufacturers still produce and sell clinkers for use, not only in restoring or renovating old buildings which used clinkers originally, but in new construction as well. The dark-colored, warped or twisted shapes provide textures which are unusual in brick walls (*see Fig. 2-5*).

Beehive kilns were later heated by more precisely controlled gas and oil fires in separate fireboxes. Heat was circulated by a system of ducts

Clinker brick adds texture and color to walls. (*Photo courtesy The Brickyard.*)

from both the bottom and the top of the kiln which resulted in more uniform firing of the brick. However, the excessive time required for burning in a "periodic" kiln of this nature yields only a limited quantity of bricks.

Most plants now use continuous straight-line tunnel kilns, with sophisticated computer equipment for precisely controlled firing temperatures (*see Fig. 2-6*). The clayware which was stacked on flat rail cars for drying, is moved into the first stage of the tunnel kiln, where it travels through various temperature zones. A European manufacturer has recently patented a "rotary circular kiln" that can reportedly save up to 30% on fuel consumption. Brick move through the kiln on a hydraulically controlled turntable. The system can capture and re-use 70 to 75% of the waste heat compared to only about 45% for tunnel kilns.

Burning consists essentially of subjecting brick units to gradually increasing temperatures until fusion chemically alters the structure of the clay. The burning process consists of six phases which are accomplished in the dryer kiln and in the preheating, firing, and cooling chambers of the burning kiln. The drying and evaporating of excess moisture is often called the water-smoking stage. This initial preheat may be done in separate dryers or, if high-fired glazes will not be added, in the forward section of the burning kiln. This exposure to relatively low temperatures of up to 400°F begins the gradual, controlled heating process. Dehydration, or removal of the remaining trapped moisture, requires anywhere from 300 to 1800°F, oxidation from 1000 to 1800°F, and vitrification 1600 to 2400°F. It is only within this final temperature range that the silicates in the clay melt and fill the voids between the more refractory materials binding and cementing them together to form a strong, dense, hard-burned brick. The actual time and exact temperatures required throughout these phases vary according

2/6 Tunnel kiln.

to the fusing characteristics and moisture content of the particular clay. Near the end of the vitrification phase, a reducing atmosphere may be created in which there is insufficient oxygen for complete combustion. This variation in the process is called flashing, and is intended to produce different hues and shadings from the natural clay colors. For example, if the clay has a high iron oxide content, an oxygen-rich fire will produce a red brick. If the same clay is fired in a reducing atmosphere with low oxygen, the brick will be more purple.

The final step in the firing of brick masonry is the cooling process. In a tunnel kiln, this normally requires up to 48 hours, as the temperatures must be reduced carefully and gradually to avoid cracking and checking of the brick.

2.1.9 Drawing and Storage

Removing brick from the kiln is called drawing. The loaded flatcars leave the cooling chamber and are placed in a holding area until the bricks reach room temperature. They are then sorted as necessary for size, chippage and warpage tolerances, bound into "cubes" equaling 500 standard size bricks, and either moved to storage yards or loaded directly onto trucks or rail cars for shipment.

2.2 CONCRETE MASONRY

The development of modular concrete masonry was a logical outgrowth of the discovery of portland cement, and was in keeping with the manufacturing trends of the Industrial Revolution. Although the first rather unsuccessful attempts produced heavy, unwieldy, and poorly adaptable units, the molding of cementitious ingredients into large blocks promised a bright new industry. With the invention and patenting of various block-making machines, unit concrete masonry began to have a noticeable effect on building and construction techniques of the late nineteenth and early twentieth centuries. Concrete masonry today is made from a relatively dry mix of cementitious materials, aggregates, water, and occasionally special admixtures. The material is molded and cured under controlled conditions to produce a strong, finished block that is suitable for use as a structural building element. Both the raw materials and the method of manufacture influence strength, appearance, and other critical properties of the block and are important in understanding the diversity and wide-ranging uses of concrete masonry products.

2.2.1 Aggregates

The aggregates in concrete block and concrete brick account for as much as 90% of their composition. The characteristics of these aggregates therefore play an important role in determining the properties of the finished unit. Aggregates may be evaluated on the basis of (1) hardness, strength, and resistance to impact and abrasion; (2) durability against freeze-thaw action; (3) uniformity in gradation of particle size; and (4) absence of foreign particles or impurities. A consistent blend of fine and coarse particle sizes is necessary to produce a mixture that is easily workable and a finished surface that is dense and impervious.

There are two categories of aggregates used in the manufacture of concrete masonry: *lightweight aggregates* and *heavyweight aggregates* (also called normal-weight). Early concrete masonry units were, for the most part, made with the same heavyweight aggregates as those used today. Well-graded sand, gravel, crushed stone, and air-cooled slag are combined with other ingredients to produce a block that is heavy, strong, and fairly low in water absorption.

Efforts to make handling easier and more efficient led to the introduction of lightweight aggregates. Pumice, cinders, expanded slag, and other natural or processed aggregates are often used, and the units are sometimes marketed under proprietary trade names. Testing and performance have proven that lightweight aggregates affect more than just weight, however. Thermal, sound, and fire resistance are also influenced, as well as color and texture. Lightweight aggregates increase the thermal and fire resistance of concrete masonry, but sound transmission ratings generally are lower because of reduced density. Moisture absorption is also generally much higher with lightweight aggregates.

In an effort to recycle materials, reduce landfill demand, and economize production, some block manufacturers are now using crushed block as a portion of the aggregate content in manufacturing new units. Broken units are crushed and blended with new aggregate to save money on raw materials and to give contractors an alternative means for disposing of construction site debris. Currently about 50 to 60% of the block produced at some manufacturing plants uses at least some recycled material, and companies are finding new ways to blend aggregates in order to use more recycled material. Some federal agencies are already requiring certain percentages of recycled-content materials in new construction projects.

Concrete masonry colors resulting from the mix of aggregate and cement may range from white, to buff or brownish tones, to dull grays. Special colors may be produced by the use of selected crushed stones or the addition of special pigments. Color variation in units is affected by several things. Aggregate gradation should be carefully controlled during manufacture, but shipping of raw materials, particularly by rail, can cause separation of fine surface material from coarse aggregate. The degree of separation and resultant dust content varies from one shipment to the next, causing a variation in the color of the block (particularly with split face units). As ambient temperatures rise during the day, moisture evaporates from the aggregate. If the moisture content is not accurately monitored, particularly in hot climates, the drier aggregate effectively changes the water-cement ratio of the mix within a single day's production. Higher water-cement ratios produce lighter-colored block. Temperature and moisture variations in the kiln affect unit color, and units loaded first may also experience a slightly longer hydration period. Units which are air-dried can be significantly affected by changes in ambient temperature and relative humidity.

Surface textures depend on the size and gradation of aggregates. Classification of surface effects is only loosely defined as "open" or "tight," with either fine, medium, or coarse texture. Although interpretation of these groups may vary, an open surface is generally characterized by numerous large voids between the aggregate particles. A tight surface has few pores or voids of the size easily penetrated by water or sound. Fine textures are smooth, and consist of small, very closely spaced granular particles. Coarse textures are large grained and rough, and medium textures are, of course, intermediate (*see Fig. 2-7*). Both coarse and medium textures provide better sound absorption than the smoother faces, and are also recommended if the units are to be plastered.

The American Society for Testing and Materials (ASTM) has developed standards to regulate quality and composition. Within the limits of the required structural properties of the masonry, the architect may select different aggregates to serve other nonstructural functions required by building type, occupant use, or aesthetics.

2.2.2 Cements

The cementitious material in concrete masonry is normally Type I, all-purpose portland cement. Type III, high-early-strength cement, is sometimes used to provide early strength and avoid distortion during the curing process. The air-entraining counterparts of these two cements (Types IA and IIIA) are sometimes used to improve the molding and off-bearing characteristics of the uncured units, and to increase resistance to weathering cycles. Air entrainment, however, does cause some strength reduction.

2.2.3 Admixtures

Admixtures marketed chiefly for use in site-cast concrete have shown few beneficial or desirable effects in the manufacture of concrete masonry. Air entrainment facilitates compaction and the close reproduction of the contours of the molds, but increased air content always results in lower compressive strengths. Calcium chloride accelerators speed the hardening or set of the units, but tend to increase shrinkage. Water-repellent admixtures are commonly used in decorative architectural block intended for exterior exposures without protective coatings. However, the bond between mortar and units (and consequently the flexural strength of the wall) will be seriously impaired unless the mortar is also treated with a chemically

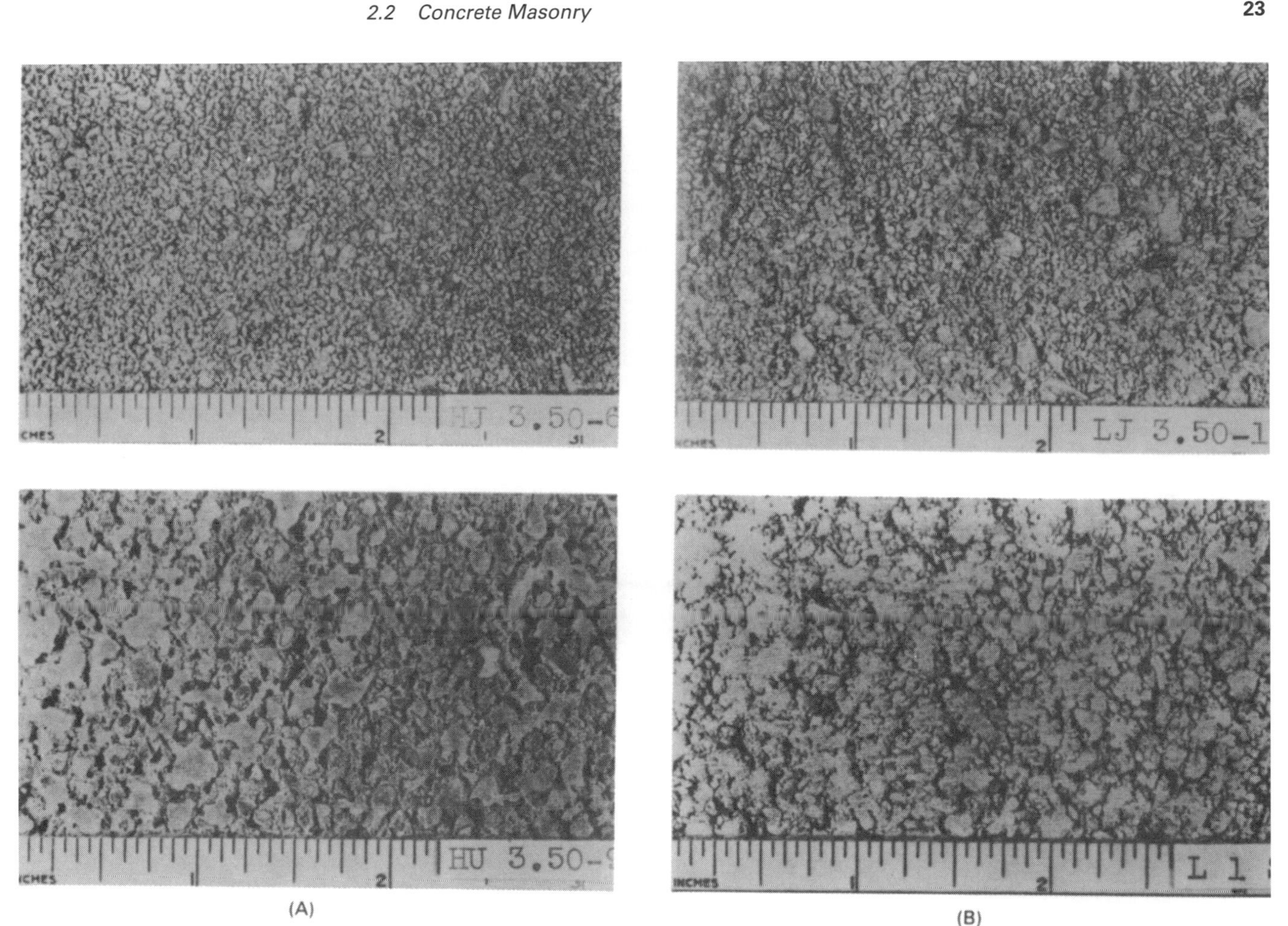

2/7 Concrete block surface textures: (A) Lightweight aggregate; (B) normal weight aggregate. (*Photo courtesy PCA.*)

compatible admixture. ASTM Standards do not permit the use of any admixtures in concrete masonry without laboratory tests or performance records which prove that the additives are in no way detrimental to the performance of the masonry.

Architectural concrete masonry units (CMUs) are sometimes treated with an integral water-repellent admixture during manufacture to resist soil accumulations and to decrease surface water absorption. Some research indicates that calcium stearate–based products are more effective in creating hydrophobic surfaces than those based on oleic/linoleic acid chemistries, and are also less likely to leach out of the masonry. An integral water-repellent which lasts the life of the masonry will provide more economical performance than a surface-applied water-repellent which must be re-applied every few years. Whenever an integral-water repellent is used in a concrete masonry product, compatibility and bond with mortar and grout must be considered because the bonding characteristics of the unit are affected. CMU products that have been treated with an integral water-repellent require mortar and grout that have compatible chemical admixtures to promote better bond.

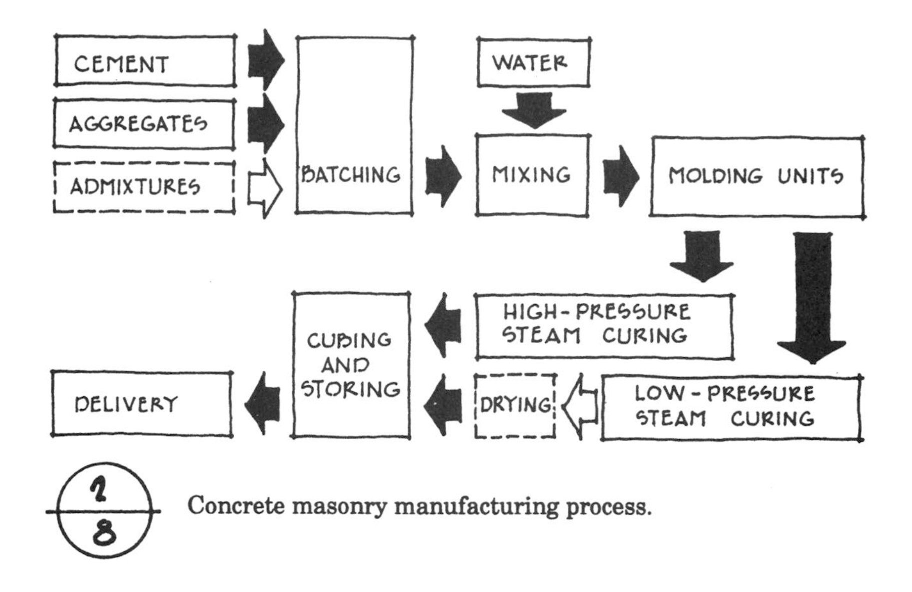

Concrete masonry manufacturing process.

Special colors can be produced by using pure mineral oxide pigments, but many factors affect color consistency. Even in natural block, color variations can be caused by the materials, processing, curing, and weathering. In integrally colored units, such variations may be magnified. Natural aggregate colors are more durable, and more easily duplicated in the event of future additions to a building.

2.2.4 Manufacturing

Concrete masonry manufacturing consists of six phases: (1) receiving and storing raw materials, (2) batching and mixing, (3) molding unit shapes, (4) curing, (5) cubing and storage, and (6) delivery of finished units (*see Fig. 2-8*).

2.2.5 Material Preparation

Materials are delivered in bulk quantities by truck or rail. Aggregates are stored separately and later blended to produce different block types. Mixes will vary depending on aggregate weight, particle characteristics, and water absorption properties. Ingredients must be carefully regulated so that consistency in texture, color, dimensional tolerances, strength, and other physical properties is strictly maintained. Batching by weight is more common than volume proportioning.

The mixes normally have a low water-cement ratio, and are classified as zero-slump concrete. Special high-strength units are made with more cement and water, but still have no slump. In the production of some slump block units, the batching is changed so that the mix will slump within controlled limits when the unit is removed from its mold. The soft roll in texture produces the appearance of a hand-made adobe.

2.2.6 Forming

Early block production consisted of hand-tamping the concrete mix into wooden molds. A team of two could turn out about 80 blocks a day. By the mid-1920s, automatic machines could produce as many as 3000 blocks a day. Today, units are molded with a combination of mechanical vibration and hydraulic pressure, and production is typically in the neighborhood of 1000 units per hour.

Custom concrete masonry units are often designed by architects. Special forms or textures can lend themselves to certain decorative motifs or to the expression of larger building shapes and masses (*see Fig. 2-9*). The economics of custom units generally requires that the volume of the order offset the expense of producing the molds.

(A)

(B)

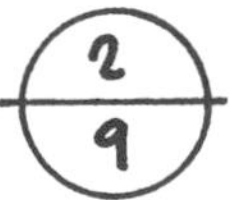

Paul Rudolph's Crawford Manor Apartments in New Haven, Connecticut (A). Customized masonry units used (B). (*Photos courtesy National Concrete Masonry Association.*)

2.2.7 Curing

Freshly molded blocks are lightly brushed to remove loose aggregate particles, then moved to a kiln or autoclave for accelerated curing.

A normal 28-day concrete curing cycle is not conducive to the mass production of unit masonry. Experiments in accelerated steam curing were conducted as early as 1908. In addition to hastening the hydration process, steam curing also increases compressive strength, helps control shrinkage, and aids in uniformity of performance and appearance. Both high-pressure and low-pressure curing are used in the industry.

Most of the block manufactured in the United States is produced by *low-pressure steam curing.* The first phase is the holding or preset period of 1 to 3 hours. The units are allowed to attain initial hardening at normal temperatures of 70 to 100°F before they are exposed to steam. During the heating period, saturated steam is injected to raise the temperature to a maximum of 190°F. The exact time duration and temperature span recommended by the American Concrete Institute (ACI) depends on the composition of the cementitious materials and the type of aggregate used. Once maximum temperature is reached, the steam is shut off and a soaking period begins. Blocks are held in the residual heat and moisture for 12 to 18 hours or until the required compressive strengths are developed. An accelerated drying period may also be used, with the temperature in the kiln raised to evaporate moisture.

The entire cycle is generally accomplished within 24 hours. Compressive strengths of 2- to 4-day-old units cured by low-pressure steam are approximately 90% of ultimate strength compared with only 40% for blocks of the same age cured by 28-day moist sprinkling. Steam-cured units are also characterized by a generally lighter color.

A variation of the low-pressure steam method adds a carbonation phase in which carbon dioxide is introduced into the drying atmosphere to cause irreversible shrinkage. Pre-shrinking decreases volume changes caused by atmospheric moisture conditions and reduces shrinkage cracking in the wall. Carbonation also increases tensile and compressive strength, hardness, and density of the block.

High-pressure steam curing improves the quality and uniformity of concrete masonry, speeds production, and lowers manufacturing costs. Curing takes place in an autoclave kiln 6 to 10 ft wide and as much as 100 ft long (*see Fig. 2-10*).

A typical high-pressure curing cycle consists of four phases: preset, temperature rise, constant temperature and pressure, and rapid pressure release. The low-heat preset period hardens the masonry sufficiently to withstand the high-pressure steam. The temperature rise period slowly brings both pressure and temperature within the autoclave to maximum levels where they remain constant for 5 to 10 hours. Temperature is actually the critical curing factor. Pressure is used as a means of controlling steam quality. Rapid pressure release or "blow-down" causes quick moisture loss from the units without shrinkage cracks. For normal-weight aggregates, the cycle produces relatively stable, air-dry blocks soon after removal from the autoclave. Lightweight blocks may require additional time to reach this same air-dry condition.

Blocks cured by high-pressure autoclaving undergo different chemical reactions from those cured at low pressure. They are more stable and less subject to volume change caused by varying moisture conditions. The improved dimensional stability reduces shrinkage cracking in completed wall assemblies.

Loading racks of fresh units into an autoclave for high-pressure steam curing. (*Photo courtesy PCA.*)

2.2.8 Surface Treatment

Concrete blocks are sometimes finished with ceramic, organic, or mineral *glazes.* These special finishes are applied after curing, and then subjected to heat treatment. The facings vary from epoxy or polyester resins to specially treated glass silica sand, colored ceramic granules, mineral glazes, and cementitious finishes. The treated surfaces are resistant to water penetration, abrasion, and cleaning compounds, and are very durable in high-traffic areas.

Surface *textures* are applied to hardened concrete blocks in a number of ways. Grinding the unit face produces a smooth, polished finish that highlights the aggregate colors (*see Fig. 2-11*). Ground faces can be supplementally treated with a wax or clear sealer. Sandblasting a block face exposes the underlying aggregate, adding color, texture, and depth. Split-face units are produced by splitting ordinary blocks lengthwise. Solid units produce a rough stone appearance, while cored units are used to make split-ribbed block (*see Fig. 2-12*).

2.2.9 Cubing and Storage

Once the masonry units have been cured and dried, and any additional surface treatments have been completed, the blocks are removed from the curing racks and assembled in "cubes." The cubes are moved to a storage yard where, depending on the curing method used, they may remain in inventory anywhere from a few days to several weeks before they are shipped to a job site.

2.3 MORTAR AND GROUT MATERIALS

Mortar may account for as little as 7% of the volume of a masonry wall, but the role that it plays and the influence that it has on performance and appearance are far greater than the proportion indicates. The selection

Concrete blocks face ground to expose natural aggregate colors. (*Photo courtesy PCA.*)

and use of various mortar ingredients directly affects the performance and bonding characteristics of masonry. It is important to be aware of the materials available and the effects they may have on the overall integrity of the masonry.

The principal components of masonry mortar and grout are cement, lime, sand, and water. Each of these constituents is essential in the performance of the mix. Cement gives the mortar strength and durability. Lime adds workability, water retentivity, and elasticity. Sand acts as a filler and contributes to economy and strength, and water imparts plasticity. To produce high-quality mortar and grout, each of the ingredients must be of the highest quality.

2.3.1 Cements

The Romans used natural pozzolans to give hydraulic setting qualities to mortar. Concrete and mortar are said to be "hydraulic" if they will set and harden in water. Natural hydraulic cements were widely manufactured in the late 18th and early 19th centuries. Natural cement rock was burned in kilns and the calcined lumps were then ground into a fine powder.

Since its discovery in the early nineteenth century, portland cement has become the most widely used material of its kind. Portland cement is a carefully controlled combination of lime, silica, alumina, and iron oxide. Although production of portland cement is a lengthy and complicated procedure, it consists principally of grinding the raw materials, blending them to the desired proportions, and burning the mix in a rotary kiln until it

Detail of a veneer wall using split-face and split-rib units to create a pattern (Avner Nager, architect).

reaches incipient fusion and forms clinkers. These hardened pellets are ground with gypsum, and the fine powder is then bagged for shipment. When mixed with water, portland cement undergoes hydration—a change in the chemical composition of the ingredients in which crystals of various complex silicates are formed, causing the mass to harden and set.

There are five types of portland cement, each with different physical and chemical characteristics. Since the properties required for mortar are significantly different from the qualities called for in concrete, not all of these types are suitable for masonry construction. For most ordinary mortars, Type I, all-purpose cement, is most widely used. In some instances, such as masonry catch basins or underground drainage structures where mortar may come in contact with sulfates in the soil, Type II portland cement can be used to resist chemical attack. A more common substitute for Type I is Type III, high-early-strength cement. This mixture attains ultimate compressive strength in a very short period of time, and generates greater heat during the hydration process. For use in cold weather construction, these properties help keep the wet mortar or grout from freezing and permit a reduction in the period of time required for protection against low temperatures.

Air-entraining portland cement, Types IA, IIA, IIIA, and so on, are made by adding selected chemicals to produce minute, well-distributed air bubbles in the hardened concrete or mortar. Increased air content improves workability, increases resistance to frost action and the scaling caused by chemical removal of snow and ice, and enhances moisture, sulfate, and abrasion resistance. Air-entrained mixes are not as strong as ordinary portland cements, and excessive air is detrimental in mortar and grout because it impairs bond to masonry units and reinforcing steel.

Air-entrained cements are used primarily in horizontal concrete applications where exposure to ponded water, ice, and snow is greatest. Entrained air produces voids in the concrete into which freezing water can expand without causing damage. Rigid masonry paving applications installed with mortared joints may also enjoy some of the benefits of air-entrained cements in resisting the expansion of freezing water. Although industry standards for masonry mortar generally limit the air content of mortar, the benefits of higher air contents in resisting freeze-thaw damage may outweigh the decrease in bond strength. Since rigid masonry paving systems are generally supported on concrete slabs, the flexural bond strength of the masonry is less important than its resistance to weathering. In these applications, lower bond strength might be tolerated in return for increased durability.

2.3.2 Lime

The term "lime" when used in reference to building materials means a burned form of lime derived from the calcination of sedimentary limestone. Powdered, hydrated lime is the most common and convenient form used today. Of the two grades of hydrated lime covered in ASTM C207, *Standard Specification for Hydrated Lime for Masonry Purposes,* only Type S is suitable for masonry work because of its ability to develop high early plasticity and higher water retentivity, and because of limits on the unhydrated oxide content.

The mortar used in historic buildings was made with lime and sand only and did not contain any cement. Lime mortars cured very slowly, though. The invention of portland cement in the late 1800s changed the way mortar was made by substituting cement in the mix for a portion of the lime. Contemporary cement and lime mortars are now made with a higher proportion of cement than lime. Although this has reduced curing time and speeded up construction, the trade-off is that the higher the portland cement content, the stiffer the mixture is when it is wet and the more rigid the mortar when it is cured. A cement mortar without lime is stiff and unworkable, high in compressive strength, but weak in bond and other required characteristics. The continued use of lime, although reduced in proportion, has many beneficial effects in masonry mortar and grout. Lime increases water retentivity, improves workability, and makes the cured mortar or grout less brittle and less prone to shrinkage.

Lime adds plasticity to mortar, enabling the mason to spread it smoothly and fill joints completely, improving both productivity and workmanship. The plastic flow quality of lime helps mortar and grout to permeate tiny surface indentations, pores, and irregularities in the masonry units and develop a strong physical bond. Lime also improves water retention. The mortar holds its moisture longer, resisting the suction of dry, porous masonry so that sufficient water is maintained for proper curing and development of good bond. Lime has low efflorescing potential because of its relatively high chemical purity. Its slow setting quality allows retempering of a mix to replace evaporated moisture. Lime undergoes less volume change or shrinkage than other mortar ingredients. It contributes to mortar integrity and bond by providing a measure of autogenous healing, the ability to combine with moisture and carbon dioxide to reconstitute or re-knit itself if small cracks develop. Some manufacturers pre-blend portland cement and lime, and sell bagged mixes which require only the addition of sand and water at the job site.

2.3.3 Masonry Cements

Proprietary mixes of cement and workability agents or "masonry cements" are popular with masons because of their convenience and good workability. However, ASTM C91, *Standard Specification for Masonry Cement,* places no limitations on chemical composition, and the ingredients as well as the properties and performance vary widely among the many brands available. Although the exact formula is seldom disclosed by the manufacturer, masonry cements generally contain combinations of portland cement, plasticizers, and air-entraining additives. Finely ground limestone, clay, and lime hydrate are often used as plasticizers because of their ability to adsorb water and thus improve workability. Air-entraining additives protect against freeze-thaw damage and provide some additional workability. ASTM C91 limits air content to a range of 12 to 22% (*see Fig. 8-13*), and sets water retentivity at a minimum of 70%.

Like all proprietary products, different brands of masonry cements will be of different qualities. Because of the latitude permitted for ingredients and proportioning, the properties of a particular masonry cement cannot be accurately predicted solely on the basis of compliance with ASTM standards. They must be established through performance records and laboratory tests.

The 1991 Uniform Building Code (UBC) incorporated, for the first time, a new classification of masonry mortars called "mortar cement mortars." Generally, proprietary masonry cements that can produce mortars which meet the performance requirements for labeling as "mortar cements" are considered to be the higher-quality masonry cements among those on the market. The physical requirements for mortar cement are based on ASTM C91, except that maximum air content is lower, values have been added for minimum flexural bond strength, and certain harmful or deleterious materials are limited or excluded as ingredients (refer to Chapter 6).

2.3.4 Sand

Sand aggregate accounts for at least 75% of the volume of masonry mortar and grout. Manufactured sands have sharp, angular grains, while natural sands obtained from banks, pits, and river beds have particles that are

Masonry cement type	N	S	M
Fineness, residue on a No. 325 sieve, max. (%)	24	24	24
Autoclave expansion, max. (%)	1.0	1.0	1.0
Time of setting			
Initial set, min. (hr)	2.0	1.5	1.5
Final set, max. (hr)	24.0	24.0	24.0
Compressive strength (average of 3 cubes), min.			
7 days (psi)	500	1300	1800
28 days (psi)	900	2100	2900
Air content, volume (%)			
Minimum	12	12	12
Maximum	22	20	20
Water retention, flow after suction as % of original flow, min.	70	70	70

ASTM C91 requirements for masonry cements. ***(Copyright, American Society for Testing and Materials, 1916 Race Street, Philadelphia, Pa. 19103. Reprinted with permission.)***

smoother and more round. Natural sands generally produce mortars that are more workable than those made with manufactured sands.

For use in masonry mortar and grout, sand must be clean, sound, and well-graded according to requirements set by ASTM C144, *Standard Specification for Aggregate for Masonry Mortar,* or ASTM C404, *Standard Specification for Aggregates for Masonry Grout.* Sand particles should always be washed and treated to remove foreign substances. Silt can cause mortar to stick to the trowel, and can impair proper bond of the cementitious material to the sand particles. Clay and organic substances reduce mortar strength and can cause brownish stains varying in intensity from batch to batch.

The sand in masonry mortar and grout acts as a filler. The cementitious paste must completely coat each particle to lubricate the mix. Sands that have a high percentage of large grains produce voids between the particles, and will make harsh mortars with poor workability and low resistance to moisture penetration. When the sand is well proportioned of both fine and coarse grains, the smaller grains fill these voids and produce mortars that are more workable and plastic. If the percentage of fine particles is too high, more cement is required to coat the particles thoroughly, more mixing water is required to produce good workability, and the mortar will be weaker, more porous, and subject to greater volume shrinkage.

Figure 2-14 illustrates the range and distribution of particle gradation that is acceptable under ASTM C144, from the coarsest allowable gradation to the finest allowable gradation, with the ideal gradation shown in the middle. Both the coarse and fine gradations have a void content much higher than that of the ideal gradation. Many commercially available sands fall outside of ASTM gradation requirements for mortar and may have void contents even larger than those shown. Such shortcomings may be corrected by the addition of the deficient fine or coarse sands.

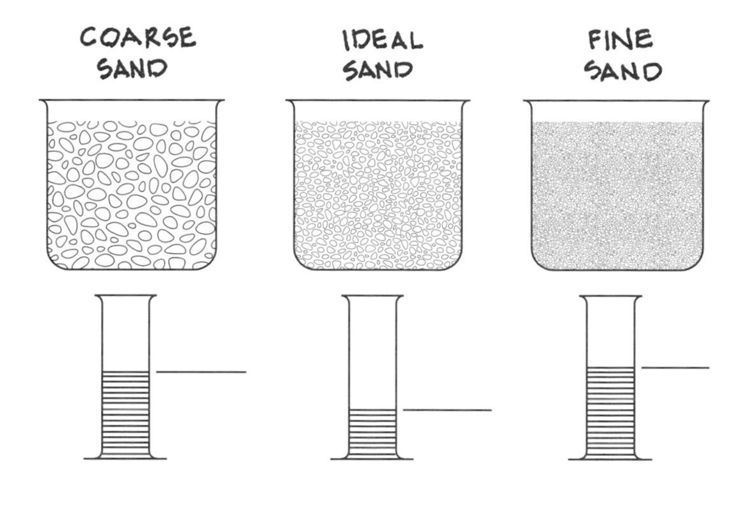

The level of liquid in the graduated cylinders at the bottom, representing voids, is less for a sand having the ideal gradation compared to coarse or fine sand. (*From Portland Cement Association, Trowel Tips "Mortar Sand," 1992.*)

When locally available mason's sand does not meet ASTM C144 gradation requirements, it can still be used if laboratory tests determine that a mortar can be produced which meets the property specification requirements of ASTM C270, *Standard Specification for Mortar for Unit Masonry.* The volume ratio of aggregate to cementitious materials may be selected and tested at various levels within the specified range of $2\frac{1}{4}$ to $3\frac{1}{2}$ times the sum of the volume of the cementitious materials. If test results show that ASTM C270 property requirements for compressive strength, air content, and water retentivity are met, the aggregate is qualified for use at the tested ratio.

2.3.5 Water

Water for masonry mortar must be clean and free of harmful amounts of acids, alkalis, and organic materials. Whether the water is drinkable is not in itself a consideration, as some drinking water contains appreciable amounts of soluble salts, such as sodium and potassium sulfate, which can contribute to efflorescence. If necessary, laboratory analysis of the water supply should be used to verify suitability.

2.3.6 Mortar Admixtures

Although admixtures are often used with some success in concrete construction, they can have adverse effects on the properties and performance of masonry mortar and grout. ASTM standards do not incorporate, nor in fact even recognize, admixtures of any kind.

A variety of proprietary materials are available which are reported by their manufacturers to increase workability or water retentivity, lower the freezing point, and accelerate or retard the set. Although they may produce some effects, they can also reduce compressive strength, impair bond, contribute to efflorescence, increase shrinkage, or corrode metal accessories and reinforcing steel. As a rule, if admixtures are used to produce or enhance some special property in the mortar, the specifications should require that laboratory tests establish the effects on strength, bond, volume change, durability, and density.

Also ask the manufacturer for test data that reports performance under field conditions. Tests done in a laboratory at 73°F do not necessarily reflect how an admixture will perform on the job site at 40°F. If relevant data are scarce, test the admixture at an independent laboratory and determine exact dosage rates with the materials which will be used at the job site. Make sure the mortar still meets ASTM specification requirements, and that the admixture does not contribute to other problems such as efflorescence or corrosion of embedded metals. Request and retain test results which support the manufacturer's claims.

Several *proprietary plasticizers* are sold to partially or wholly replace lime in masonry mortar and grout. One plasticizer used as a complete lime replacement contains, among other ingredients, natural bentonite clay as a lubricant. The water-carrying capacity of the clay gives mortar a longer board life than conventional portland-lime or masonry cement mortars.

Other types of plasticizing agents work by changing the viscosity of the mixing water and its evaporation rate, or by modifying the cement reaction rate. This increased workability can be beneficial in relatively stiff, high-compressive-strength mixes.

Air-entraining agents help hardened mortar resist freeze-thaw damage and improve the workability of wet mortar by creating minute air bubbles in the mix. In hardened mortar, freezing water expands into these bubbles instead of building up pressure which might otherwise fracture the mortar. In wet mixes, the bubbles act as a lubricant and a water reducer to increase workability and significantly lower water content. Air entrainment may be useful whenever the hardened mortar will be exposed to freeze-thaw cycles in the presence of moisture (such as paver installations). During cold weather, air entrainment may also be helpful because the lower water content of the mortar offers less potential for freezing before set.

Neutralized vinsol resins are used most widely in air-entraining admixtures, but organic acid salts, fatty acids, and hydrocarbon derivatives are also used. Although job site admixtures are available, air-entraining agents should not be added in the field, because it is difficult to obtain a consistent air content. Instead, air-entrained portland cement or masonry cements or air-entrained lime should be used, so that the batching is pre-measured. Excess air entrainment decreases both compressive strength and bond strength. ASTM C270 limits the air content of masonry mortars and prohibits the use of more than one air-entrained ingredient in a mix (see Chapter 6).

Set accelerators are sometimes used in winter construction to speed cement hydration, shorten setting time, increase 24-hour strength, and reduce the time required for cold weather protective measures. *Water-reducing accelerators* increase early strength and ultimate strength by reducing the water-cement ratio needed to produce a workable mix. Set accelerators, sometimes mistakenly referred to as "antifreeze" compounds, contain calcium chloride, calcium nitrite, calcium nitrate, calcium formate, or other aqueous solutions of organic and inorganic polymer compounds such as soluble carbonates, silicates, and flurosilicates, calcium aluminates, and triethanolamine. Accelerators are added to the mortar mixing water as a percentage of the weight of the cement.

Calcium chloride and other chloride ions contain salts which can contribute to efflorescence. Calcium chloride and, to a lesser extent, calcium nitrate also cause corrosion of embedded steel anchors and reinforcement. Non-chloride accelerators are more expensive, but less damaging to the masonry. Chlorides should be prohibited in mortar and grout which contains embedded metals such as anchors, ties, or joint reinforcement. Triethanolamine (TEA) and calcium aluminate accelerators should also be prohibited because of ultimate strength reductions and flash setting problems. Automotive antifreeze should never be used in masonry mortar or grout.

Set retarders extend the board life of fresh mortar and grout for as long as four to five hours by helping to retain water for longer periods of time. Set retarders, which contain sodium gluconate, sodium lignosulfonate, or sodium citrate, are sometimes used during hot weather to counteract the effects of rapid set and high evaporation rates. With soft, dry brick or block, set retarders are also sometimes used to counteract rapid suction and help achieve better bond. Mortar with set retarders can *not* be retempered.

Extended-life retarders slow the hydration of the cement and water to give the mortar a 12- to 72-hour board life, depending on the dosage rate. The extended workability allows the mortar to be mixed at a central batching plant where quality control can be closely maintained, and then shipped to the site in plastic tubs. The admixture has little or no effect on setting time, because the retarder is absorbed by the masonry units on contact, allowing normal cement hydration to begin. The extended life retarders used in ready-mixed mortars contain hydroxycarboxylic acids and other ingredients. Hot weather may require higher dosage rates. Most extended life retarders increase air content in the mortar slightly, so use with other air-entrained mortar ingredients should be very carefully controlled or avoided entirely.

Bond modifiers are intended to improve adhesion to smooth, dense-surfaced units such as glass block. Made of acrylic polymer latex, polyvinyl acetate, styrene butadiene rubber, or methol cellulose, bond modifiers cannot be used with air-entraining agents or air-entrained cements.

In marine environments or where deicing salts may be used, calcium nitrite *corrosion inhibitors* are used to offset the effects of chloride intrusion and prevent steel reinforcement and anchors from corroding. Corrosion inhibitors may also accelerate setting time and reduce entrained air content.

Integral water repellents reduce the water absorption of hardened mortar by as much as 60%. They are typically used in conjunction with architectural concrete masonry units that have also been treated with an integral water-repellent admixture. Stearate-, fatty acid-, or polymeric-based water repellents reduce the capillarity of the mortar while still permitting moisture vapor transmission. Using water repellent treated mortar with untreated masonry units, or vice versa, can reduce mortar-to-unit bond and the flexural strength of the wall. Reduced bond can negate the effects of the water repellent by allowing moisture to penetrate the wall freely at the joint interfaces. Mortars and block treated with integral water repellents achieve better bond and better moisture resistance only if the admixtures are chemically compatible. Wall panels should be tested both for flexural bond strength and water permeance compared to an identical but untreated wall.

Some water repellents based on fatty acids or stearates other than calcium stearate perform satisfactorily only for a limited time. Solvent migration eventually renders the treatment ineffective. Obtain manufacturer's test data on long-term performance to verify that the service life of the product is commensurate with the expected service life of the masonry.

2.3.7 Mortar Colors

Natural and synthetic *pigments* are used to color masonry mortar. Most mortar colorants are made from iron oxide pigments. Iron oxides are non-toxic, colorfast, chemically stable in mortar, and resistant to ultraviolet radiation. Iron oxides come in yellows, reds, browns, and blacks. Chromium oxides (which produce greens) and cobalt (which produces blue) also are stable in alkalis and resist ultraviolet radiation. Ultramarine blues, which are made from sulfur, sodium carbonate, and kaolin, are less stable in mortar. Carbon black and lampblack (used to make blacks and browns) are less weather-resistant than the iron oxides used to make the same colors.

Iron oxide pigments are either natural or synthetic. Natural iron oxides are made by crushing and grinding iron ore to a fine particle size. Synthetic iron oxides are made by several processes, including precipitation of iron salts, calcination of iron salts, and as a by-product in the manufacture of aniline, which is used in dyes. Synthetic iron oxides have more tinting power, so less pigment is required per unit of mortar to produce a given color. Synthetic oxides also produce brighter, cleaner colors than natural iron oxides. Natural and synthetic pigments may also be blended together.

Beyond a certain point, called the saturation point, the color intensity of the mortar does not increase in proportion to the amount of pigment added. The saturation point varies with the tinting strength of the particular pigment. Synthetic iron oxides generally are saturated at about 5% of the weight of the cement, and natural oxides at about 10%. Adding pigment beyond the saturation point produces little additional color.

When pigments are used in recommended dosages, colored mortar has not been found to adversely affect the compressive strength of the masonry, but bond strength is reduced by 3 to 5%. Colored mortar can be made at the job site from powdered or liquid pigments. Powdered pigments are used most frequently, and the majority are packaged so that one bag contains enough pigment to color one cubic foot of cementitious material (i.e., for each bag of masonry cement, portland cement, or lime, one bag of color is added). Pigment manufacturers supply charts which identify the exact number of bags of pigment required for various mortar proportions. Similarly, liquid colorants are generally packaged so that one quart of pigment is needed for each bag of cementitious material. Manufacturers can also custom-blend packaged pigment so that one bag or bottle contains enough colorant for an entire batch of mortar. Liquid pigments create less mess and blowing dust than dry powders, but they also cost more. The same pigments used to color mortars are used to produce colored concrete masonry units. Some manufacturers market colored masonry cements, mortar cements and pre-bagged portland lime mortar mixes in which pigments are pre-blended in the bag with the other ingredients.

The color of a finished mortar joint is affected by the properties of the component materials (including the sand aggregate and cement), workmanship, curing conditions, cleaning procedures, joint type, and joint tooling techniques. When colored mortar is used, it is best to evaluate and select materials from samples that closely approximate job site materials and design, and to incorporate the colored mortar into a job site sample panel before acceptance.

2.3.8 Grout Admixtures

Shrinkage-compensating admixtures (commonly called grouting aids) are the most common grout additives. Grout typically shrinks 5 to 10% after

Admixture type	*Uses*
Shrinkage compensating	Expands grout to compensate for moisture shrinkage
Set retarder	Delays set during hot weather, long transit, or time delays
Set accelerator, noncorrosive type	Accelerates set during cold weather
Corrosion inhibitor	Reduces corrosion in harsh environments
Superplasticizer	Increases slump without additional water and without strength reduction

Admixtures commonly used in masonry grout.

placement as the surrounding masonry units absorb water. To minimize volume loss, maintain good bond, and give workers more time to vibrate the grout before it stiffens, these specially blended admixtures expand the grout, retard its set, and lower the water requirements. Admixtures can also be used to accelerate set in cold weather or retard set in hot weather. Superplasticizers may also be used in hot weather to increase slump without adding water or reducing strength. All grout mixes which contain admixtures should be tested in advance of construction to assure quality. Grout mix designs which meet project requirements and ASTM guidelines can be determined in the laboratory by preconstruction testing of trial batches.

The table in *Fig. 2-15* lists the types of admixtures most commonly used in masonry grout. Air-entraining admixtures for increased freeze-thaw durability are used less frequently because the grout is normally not exposed to moisture saturation.

2.4 ENVIRONMENTAL IMPACT

Environmental issues are a growing concern in the construction industry. New terms like "green buildings," "sustainable architecture," "embodied energy," and "building ecology" have crept into the vocabularies of architects, owners, and contractors alike. Ecological issues are being driven beyond the philosophical and ideological into the mainstream of business economics. The cost of energy, the cost of raw materials and the cost of solid waste and hazardous waste disposal are directly linked to profitability in any industry. The operational efficiency of buildings and occupant productivity also have a direct effect on overhead and profit as well as health.

ASTM's Subcommittee E50.06 on Green Buildings defines that term as "building structures...that are designed, constructed, operated and demolished in an environmentally enhanced manner." That means using recycled materials wherever possible, and avoiding materials that create the clinical symptoms of "sick building syndrome." Areas of particular concern include resource efficiency, energy efficiency, pollution control, waste minimization, and indoor air quality.

The concept of green buildings and sustainable architecture are so new that guidelines are only now being developed. Generally, a building is evaluated throughout its life cycle, from construction through operation and demolition. The amount of energy consumed and the amount of waste generated at each phase, as well as the building's internal environment and its relationship to the external global environment should enter into site considerations, design decisions, and product selections.

The green building movement seeks to identify building materials that minimize environmental impacts in their creation and use, and minimize health risks to building occupants. But there is no such thing as an environmentally perfect material. Product selection for green buildings is therefore a process of evaluation and compromise, seeking the best overall solution for a given program and budget. For example, steel may have more embodied energy than wood, but steel framing is more efficient and can produce smaller structural members and longer spans. Ceramic tile is more energy-intensive than hardwood for flooring, but requires no finish coatings and no chemical cleaners for maintenance. By the same token, masonry products may require more energy to produce than some other building materials, but their performance characteristics, durability, and chemical stability usually justify similar trade-offs.

Masonry's multi-functional properties have always made it an attractive choice as a building material. From an environmental standpoint, this ability to serve more than one purpose is a particular bonus. Coatings are generally not required because most types of masonry already have a finished surface. Sound batts are not required because the masonry has inherent sound-damping capacity. Fireproofing is not required because masonry is noncombustible. And structural framing is eliminated in buildings where loadbearing systems can be used. The thermal mass of masonry can reduce the amount of insulation material required in some climates. It can also, when properly integrated with passive solar design techniques, reduce total energy consumption and reduce utility service demand through off-peak loading (see Chapter 8). Such multi-functional applications, as well as the long service life and low maintenance traditionally associated with masonry buildings, mean that the energy embodied in the materials goes further, and delivers more than many other materials.

Two of the premier examples typically cited for their environmentally responsible design are the Audubon Society Headquarters in New York and the Natural Resource Defense Council Building in Washington, both designed by the Croxton Collaborative architects. One thing the two have in common is their adaptive reuse of historic masonry buildings. The rehabilitation of historic buildings, many of which are masonry, conserves the embodied energy already invested in such structures.

Products and systems must demonstrate reduced life-cycle energy consumption, increased recycled content, and minimal waste products in manufacture, construction, use, and demolition. Such requirements may result in the introduction of mortarless interlocking masonry systems, a renewed interest in "bio-bricks," or the successful re-introduction of autoclaved cellular concrete block from Europe.

2.4.1 Resource Management, Recycled Content, and Embodied Energy

The raw materials for making clay brick are an abundant resource that is easily acquired and produces little waste. Clay mining operations are regulated by the Environmental Protection Agency, and dormant pits have been reclaimed as lakes, landfills, and nature preserves. Recycled materials are not often used in the manufacture of clay brick, but additives such as oxidized sewage sludge, incinerator ash, fly ash, waste glass, papermaking sludge, and metallurgical wastes have been incorporated with varying degrees of success. The waste materials are either burned to complete combustion at the high kiln temperatures needed to bake the brick, or encapsulated within the clay body where they cannot leach out. The primary energy cost associated with brick manufacturing is the fuel burned in the firing process. Most brick kilns now use natural gas instead of coal. This has reduced sulfur dioxide emissions and also allows more precise control of fuel consumption. Waste heat from the firing kilns is also ducted and reused to dry unfired units. When the costs of transporting brick to job sites is factored in, the embodied energy is estimated at approximately 4000 Btu per pound of brick.

The primary ingredients in concrete masonry units are the sand and aggregates, which account for as much as 90% of a unit's composition. These materials are abundant, easily extracted, and widely distributed geographically. Recycled materials such as crushed concrete or block and by-products such as blast furnace slag, cinders, and mill scale can be used for some of the aggregate. The portland cement used as the binder in concrete masonry is energy-intensive in its production, but it accounts for only about 9 to 13% of the unit. Energy consumption for cement production has decreased 25% during the last 20 years, mostly as a result of more efficient equipment and production methods. The proportion of portland cement in concrete masonry units can be reduced by substituting fly ash, which is a by-product of coal-fired power plants.

Natural stone uses less energy in its production and fabrication than other masonry materials, but its transportation costs can be significantly higher. It is not unusual for a stone to be quarried on one continent, shipped to another for fabrication, and to yet another for installation. The use of local or regional building stones greatly reduces transportation and embodied energy costs.

2.4.2 Construction Site Operations

Masonry construction is generally less hazardous to the environment than some other building systems because most of the materials used are chemically inert. Mortar mixing and stone cutting operations can generate airborne particulate wastes such as silica dust. Keeping aggregate piles covered and using water-cooled saws can reduce this hazard. Modular dimensioning of masonry can reduce job site waste by limiting construction to the use of only whole and half-size units.

Cleaning compounds, mortar admixtures, coatings, and the chemicals used to clean and maintain equipment may include potentially hazardous materials. Precautions should be taken in the disposal of such products, and runoff should be controlled to prevent the migration of chemicals into natural waterways and municipal storm sewer systems. On small cleaning projects, this may be a simple matter of temporary flashings and catch basins, but on large projects, it may become a complex task. The rinse material should be tested after cleaning a sample wall area to make sure it is safe to dispose of in the public storm sewer system.

2.4.3 Indoor Air Quality/Building Ecology

When the cost of energy went up dramatically in the 1970s, building standards began to change. Construction was tightened up to reduce or eliminate air leakage and the heat loss or heat gain associated with it. Ventilation standards also changed, reducing the number of air changes per hour which the mechanical systems delivered to increase the efficiency of heating and cooling systems. Unfortunately, these changes also led to increased concentrations of chemical air pollutants in buildings. Many building products contain substances that are known to pose health risks through continued exposure. Substances used in the manufacture of plywood, insulation, sealants, adhesives, paints, pigments, and solvents include formaldehyde and benzenes (both of which are carcinogens), as well as trichloroethylene. Synthetic carpet can emit formaldehyde, toluene, and xylene as well as methyl methacrylate, ethylbenzene, and a host of other chemicals. Even softwood framing lumber contains terpenes that continually off-gas and are of concern for sensitive individuals. Masonry products are generally inert and do not contribute to indoor air quality problems. They contain no toxins or volatile organic compounds (VOCs), do not emit any chemical pollutants as they age, and none of the natural stone that is typically used in building is known to emit radon.

PART 2

MASONRY PRODUCTS AND ACCESSORIES

3

CLAY AND CERAMIC PRODUCTS

Clay as a raw material is most valued for its ceramic characteristics. When subjected to high firing temperatures, the silicates in clay melt, fusing the particles to a density that approaches vitrification. The resulting strength and weather resistance make brick, structural clay tile, and terra cotta among the most durable of building materials.

3.1 BRICK

There are many different shapes, sizes, and types of brick. The first distinction made by ASTM standards is between building brick and facing brick, based on appearance of the unit.

Building brick (sometimes called common brick) is used primarily as a structural material or as a backing for other finishes, where strength and durability are of more importance than appearance. Under ASTM C62, *Standard Specification for Building Brick,* grading is based on physical requirements and directly related to durability and resistance to weathering (*see Fig. 3-1*).

Grade SW (Severe Weathering) is used where a high degree of resistance to frost action is required and where conditions of exposure indicate the possibility of freezing when the unit is permeated with water. Grade SW is recommended for below-grade installations in moderate and severe weathering areas, and for horizontal or other non-vertical surfaces in all weathering conditions. Grade MW (Moderate Weathering) may be used only in negligible weathering regions for vertical installations and for above-grade non-vertical installations. Grade NW (No Weathering) is permitted only for interior work where there will be no weather exposure.

Moisture enters the face of a brick by capillary action. When present in sufficient quantity and for an extended time, water will penetrate through the brick and approximate the laboratory condition defined as "permeated" (which results from 24-hour submersion in cold water). Permeation may easily occur in units exposed in parapet walls, retaining

TABLE A GRADE REQUIREMENTS FOR FACE EXPOSURES			
	Weathering index		
Exposure	*Less than 50*	*50 to 500*	*500 and greater*
In vertical surfaces			
In contact with earth	MW	SW	SW
Not in contact with earth	MW	SW	SW
In other than vertical surfaces			
In contact with earth	SW	SW	SW
Not in contact with earth	MW	SW	SW

TABLE B PHYSICAL REQUIREMENTS						
	Minimum compressive strength (brick flatwise), gross area (psi)		*Maximum water absorption by 5-hr boiling (%)*		*C/B Maximum saturation coefficient**	
Designation	*Average of 5 brick*	*Individual*	*Average of 5 brick*	*Individual*	*Average of 5 brick*	*Individual*
Grade SW	3000	2500	17.0	20.0	0.78	0.80
Grade MW	2500	2200	22.0	25.0	0.88	0.90
Grade NW	1500	1250	No limit	No limit	No limit	No limit

*The saturation coefficient is the ratio of absorption by 24-hr submersion in cold water (C) to that after 5-hr submersion in boiling water (B).

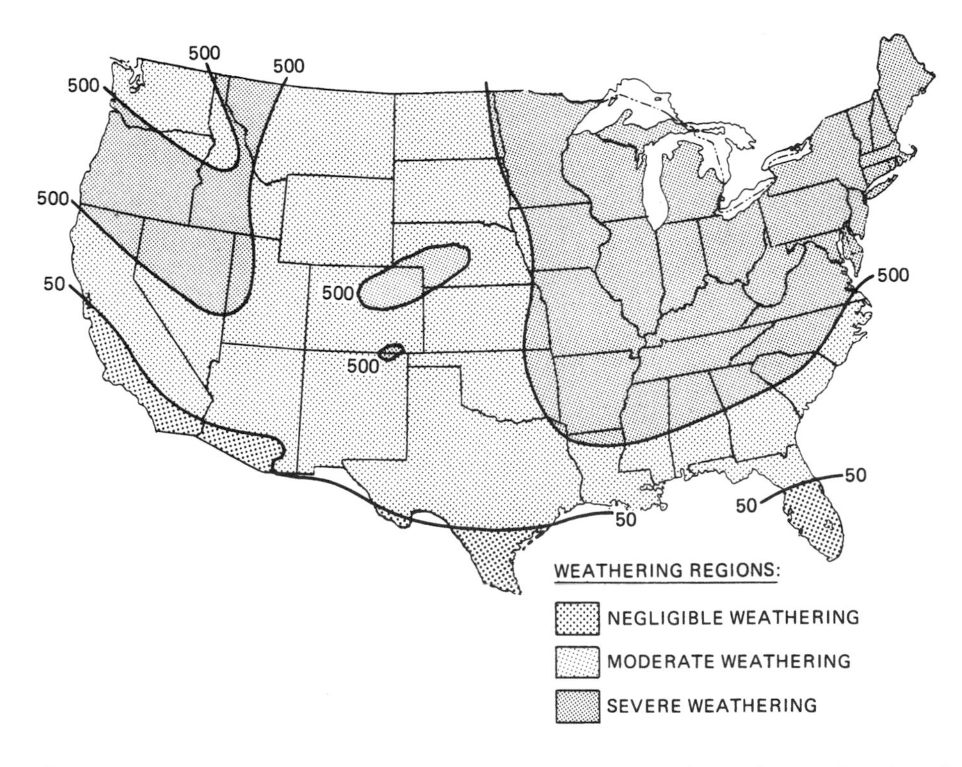

Requirements for building brick, from ASTM C62. ***(Copyright, American Society for Testing and Materials, 1916 Race Street, Philadelphia, Pa. 19103. Reprinted with permission.)***

walls, and horizontal surfaces, but is unlikely for ordinary exterior wall exposures if the brick is suitably protected at the top by copings, metal flashings, or overhanging eaves. Under most circumstances, permeation of the brick in building walls would result only from defective workmanship or faulty drainage.

Face brick is used for exposed areas where appearance is an important design criterion. These units are typically selected for specific aesthetic criteria such as color, dimensional tolerances, uniformity, surface texture, and limits on the amount of cracks and defects. ASTM C216, *Standard Specification for Facing Brick,* covers Grades SW and MW, which correspond to the same physical and environmental requirements as those for building brick. Within each of these grades, face brick may be produced in three specific appearance Types. Type FBS (Standard) is for general use. Type FBX (Select) is for use in exposed applications such as stack bond patterns where a high degree of mechanical perfection and minimum size variation are required. Type FBA (Architectural) is manufactured with characteristic architectural effects, such as distinctive irregularity in size and texture of the individual units to simulate historic brick (*see Fig. 3-2*). Extruded, stiff-mud brick may be produced in any of the three types by progressively increasing the amount of texturing and roughening the units receive after leaving the die. Dry-press brick normally falls well within the strict tolerances required for Type FBX, but is not widely used because of higher production costs and the higher labor costs associated with laying up such precision units. Soft-mud brick, including hand-molded units, are rustic in appearance and meet the specifications only for Type FBA. Both labor economy and distinctive appearance make FBA brick very popular in residential and light commercial construction. All three types meet the same requirements for strength and durability.

ASTM C216 formerly made reference to color range in describing brick Types, but the reference has been dropped in recent editions. Color range used to be associated with size variation when the older kiln types were used. Dark colors indicated hard-burned brick that experienced greater shrinkage during firing than lighter-colored, soft-burned brick. Brick Types FBS, FBX, and FBA differ only in appearance as related to degree of precision and uniformity in size tolerance. Types FBS, FBX, and FBA are all available in a wide range of colors as well as in both weathering Grades, SW and MW.

The allowable size tolerances for brick Types FBS and FBX have also been modified in recent editions of ASTM C216, to tighten the allowable size variation within a job lot. The changes are meant to reflect the actual variations in the majority of brick manufactured in the United States, and is based on a survey conducted by the Brick Institute of America.

Used brick are sometimes specified by architects because of their weathered appearance and broad color range. In many instances, these specimens are not totally in compliance with accepted standards of durability for exposed usage. Sources for salvaged masonry are generally buildings at least 30 to 40 years old, constructed of solid masonry walls with hard-burned brick on the exterior and inferior "salmon" brick as backup. Since the color differences used in originally sorting and selecting the brick become obscured with exposure and contact with mortar, salmon brick may inadvertently be used for an exterior exposure, where they can undergo rapid and excessive deterioration. Building code requirements may vary regarding the use of salvaged brick, and should be consulted prior to their selection and specification.

TABLE A TOLERANCES ON CHIPPAGE

Type	Percentage Allowed*	Chippage (in.) in from Edge	Chippage (in.) in from Corner	Percentage allowed*	Chippage (in.) in from Edge	Chippage (in.) in from Corner
FBX	5 or less	1/8–1/4	1/4–3/8	95–100	0–1/8	0–1/4
FBS, smooth†	10 or less	1/4–5/16	3/8–1/2	90–100	0–1/4	0–3/8
FBS, rough‡	15 or less	5/16–7/16	1/2–3/4	85–100	0–5/16	0–1/2
FBA§						

*Percentage of exposed brick allowed in wall with chips measured listed dimensions in from edge or corner.
†Smooth texture is the unbroken natural die finish.
‡Rough texture is the finish produced when the face is sanded, combed, scratched, or scarified or the die skin on the face is entirely broken by mechanical means such as wire cutting or wire brushing.
§To meet designated sample or as specified by purchaser, but not more restrictive than for Type FBS, rough.

TABLE B TOLERANCES ON DIMENSIONS

Specified dimension or average brick size in job lot sample (in.)	Maximum permissible variation (in.) plus/minus				
	Column A (for specified dimension)		Column B (for average brick size in job lot sample)*		
	Type FBX	Type FBS	Type FBX	Type FBS, Smooth†	Type FBS, Rough‡
3 and under	1/16	3/32	1/16	1/16	3/32
Over 3 to 4 incl.	3/32	1/8	1/16	3/32	1/8
Over 4 to 6 incl.	1/8	3/16	3/32	3/32	3/16
Over 6 to 8 incl.	5/32	1/4	3/32	1/8	1/4
Over 8 to 12 incl.	7/32	5/16	1/8	3/16	5/16
Over 12 to 16 incl.	9/32	3/8	3/16	1/4	3/8

*Lot size shall be determined by agreement between purchaser and seller. If not specified, lot size shall be understood to include all brick of one size and color in the job order.
†Type FBS Smooth units have relatively fine texture and smooth edges, including wire cut surfaces. This definition relates to dimensional tolerances only.
‡Type FBS Rough units have textured, rounded, or tumbled edges or faces. This definition relates to dimensional tolerances only.

TABLE C TOLERANCES ON DISTORTION

Maximum dimension, (in.)	Maximum permissible distortion (in.)	
	Type FBX	Type FBS
8 and under	1/16	3/32
Over 8 to 12 incl.	3/32	1/8
Over 12 to 16 incl.	1/8	5/32

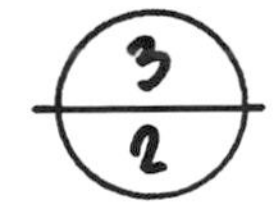

ASTM C216 brick chippage, dimension, and distortion tolerances. (*Copyright, American Society for Testing and Materials, 1916 Race Street, Philadelphia, Pa. 19103. Reprinted with permission.*)

Imported Mexican brick give a distinctive, handcrafted quality to masonry. They also lack uniformity in conformance with U.S. durability standards. Officials at the Brick Institute of Texas estimate that as much as 85% of the imported brick sold in that state each year are found to be substandard in water absorption, weathering, and compression tests. The abbreviated burning period and low firing temperatures typical of some Mexican brick plants produce units that are extremely soft and porous, causing severe maintenance problems even in relatively dry climates (*see Fig. 3-3*).

Under-burned *salmon brick* are not acceptable under any building code for use in areas exposed to weather. Some codes, however, do permit unburned clay products such as *adobe* brick. Unless they are protected by a surface coating such as plaster, however, these sun-dried brick are susceptible to severe moisture damage or disintegration. Commercially available adobe brick treated with emulsified asphalt have been tested and approved for use by some local authorities. Traditional blends of clay with straw or fiber reinforcing that have not been treated or certified must either be used in a completely sheltered location or receive a protective plaster or stucco coating.

Over-burned *clinker brick* were more common when coal-fired periodic kilns were used. The units currently produced are made for special aesthetic effects, and should not be used in structural masonry or severe weathering exposures unless the masonry assemblage is tested for flexural strength and water permeance.

3.1.1 Sizes and Shapes

Masonry unit sizes and shapes have proliferated over the last 5000 years to meet various regional standards and design requirements throughout the world. Even within the United States, unit dimensions may vary from one area to the next and be further confused by different names for the same size of unit. At one time, there were only three commonly used brick sizes: "standard," Norman, and Roman. Industry demand has increased that number substantially. Brick is now available in thicknesses or bed depths ranging from 3 to 12 in., heights from 2 to 8 in., and lengths of up to 16 in. Production includes both non-modular and modular sizes conforming to the 4-in. grid system of structural and material coordination. Some typical units are illustrated in *Fig. 3-4*. *Figure 3-5* lists several of the modular sizes, their recommended joint thicknesses, and coursing heights.

For clarity in specifying brick, units should be identified first by dimensions, then by name, and actual dimensions should be used, listed width × height × length. Nominal dimensions may vary from actual sizes by the thickness of mortar joint with which the unit was designed to be used. Firebrick, however, are laid without true mortar beds, and sizes given should always be actual dimensions. Mortar joint thicknesses are determined by the type and quality of the unit. In general, glazed brick are laid with a ¼-in. joint, face brick with a ⅜- or ½-in. joint, and building brick with a ½-in. joint.

The bricks in *Fig. 3-4* show a variety of core designs. Although they are typical of commercially available products, the corings vary with the manufacturer, and are not necessarily typical for or limited to the particular size with which they are shown. These design modifications have been developed over the years to facilitate, among other things, ease of forming,

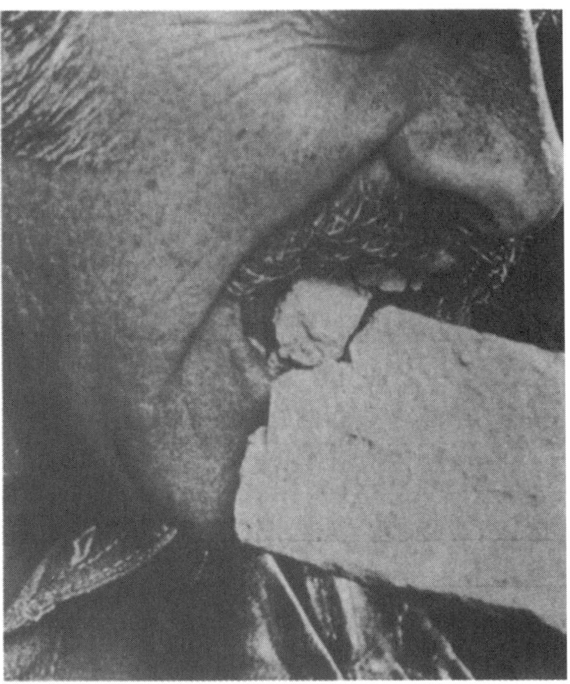

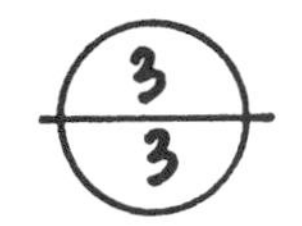

Substandard brick. (*Photos courtesy Brick Institute of Texas.*)

ease of handling, and improved grip and mortar bond. The oldest pattern is an indentation or "frog" producible only by dry-press or soft-mud processes. Originally conceived as a scheme for reducing the weight of a solid unit, this depression provided a space for identification by early craftsmen, who would write the name of the reigning monarch during the time of construction. This practice has since aided archaeologists in dating ancient buildings. Still in use today, the "frog" is now often stamped with the name of the brick manufacturer or date of production. Extruded brick are made with a series of holes cored through the unit which, for "solid

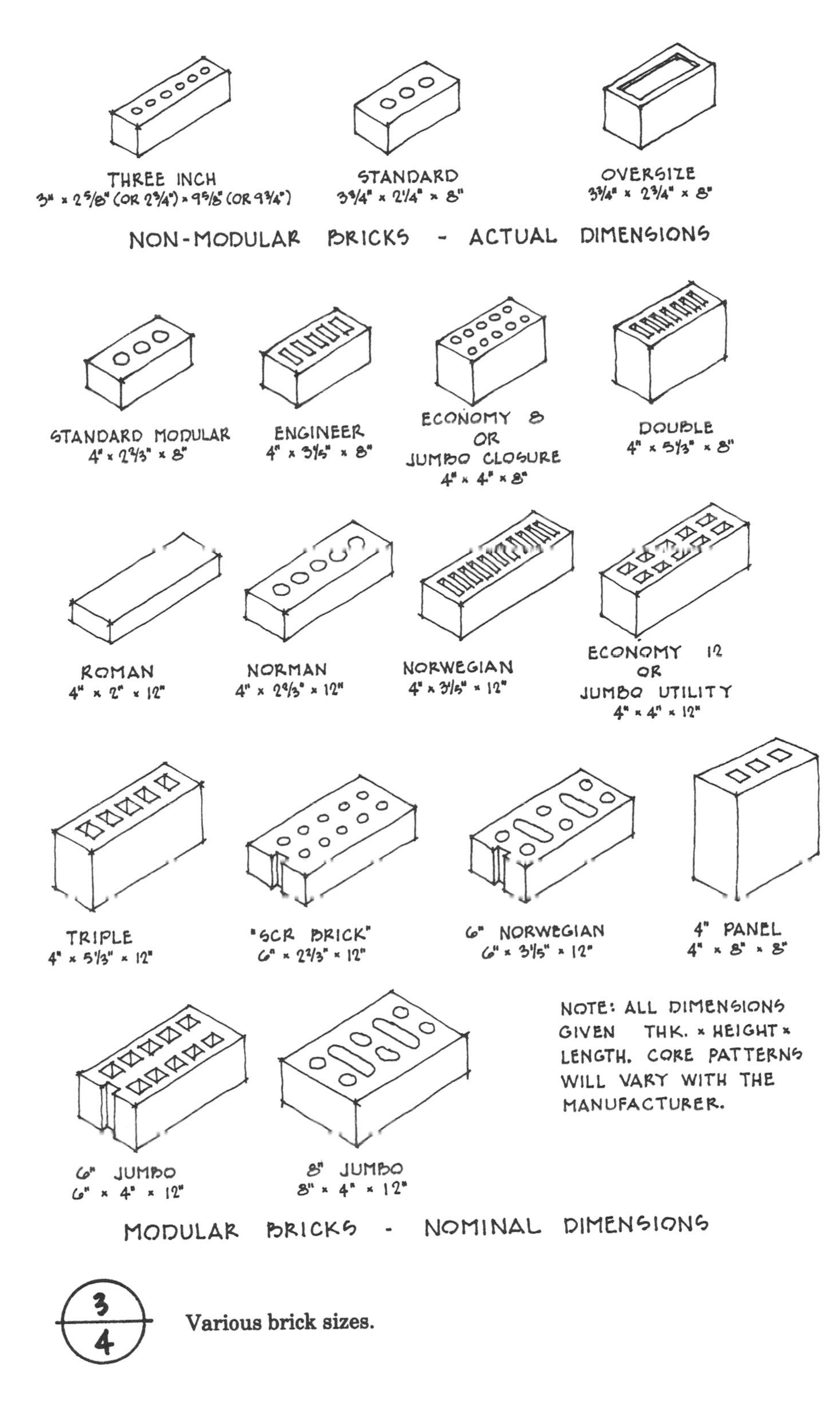

3/4 **Various brick sizes.**

Unit designation	Sizes: Nominal dimensions (in.) T	H	L	Joint thickness (in.)	Manufactured dimensions (in.) T	H	L	Modular coursing (in.)
Standard Modular	4	$2\frac{2}{3}$	8	$\frac{3}{8}$	$3\frac{5}{8}$	$2\frac{1}{4}$	$7\frac{5}{8}$	3C = 8
				$\frac{1}{2}$	$3\frac{1}{2}$	$2\frac{1}{4}$	$7\frac{1}{2}$	
Engineer	4	$3\frac{1}{5}$	8	$\frac{3}{8}$	$3\frac{5}{8}$	$2\frac{13}{16}$	$7\frac{5}{8}$	5C = 16
				$\frac{1}{2}$	$3\frac{1}{2}$	$2\frac{11}{16}$	$7\frac{1}{2}$	
Economy 8 or Jumbo Closure	4	4	8	$\frac{3}{8}$	$3\frac{5}{8}$	$3\frac{5}{8}$	$7\frac{5}{8}$	1C = 4
				$\frac{1}{2}$	$3\frac{1}{2}$	$3\frac{1}{2}$	$7\frac{1}{2}$	
Double	4	$5\frac{1}{3}$	8	$\frac{3}{8}$	$3\frac{5}{8}$	$4\frac{15}{16}$	$7\frac{5}{8}$	3C = 16
				$\frac{1}{2}$	$3\frac{1}{2}$	$4\frac{13}{16}$	$7\frac{1}{2}$	
Roman	4	2	12	$\frac{3}{8}$	$3\frac{5}{8}$	$1\frac{5}{8}$	$11\frac{5}{8}$	2C = 4
				$\frac{1}{2}$	$3\frac{1}{2}$	$2\frac{1}{4}$	$11\frac{1}{2}$	
Norman	4	$2\frac{2}{3}$	12	$\frac{3}{8}$	$3\frac{5}{8}$	$2\frac{1}{4}$	$11\frac{5}{8}$	3C = 8
				$\frac{1}{2}$	$3\frac{1}{2}$	$2\frac{1}{4}$	$11\frac{1}{2}$	
Norwegian	4	$3\frac{1}{5}$	12	$\frac{3}{8}$	$3\frac{5}{8}$	$2\frac{13}{16}$	$11\frac{5}{8}$	5C = 16
				$\frac{1}{2}$	$3\frac{1}{2}$	$2\frac{11}{16}$	$11\frac{1}{2}$	
Economy 12 or Jumbo Utility	4	4	12	$\frac{3}{8}$	$3\frac{5}{8}$	$3\frac{5}{8}$	$11\frac{5}{8}$	1C = 4
				$\frac{1}{2}$	$3\frac{1}{2}$	$3\frac{1}{2}$	$11\frac{1}{2}$	
Triple	4	$5\frac{1}{3}$	12	$\frac{3}{8}$	$3\frac{5}{8}$	$4\frac{15}{16}$	$11\frac{5}{8}$	3C = 16
				$\frac{1}{2}$	$3\frac{1}{2}$	$4\frac{13}{16}$	$11\frac{1}{2}$	
6-in. Norwegian	6	$3\frac{1}{5}$	12	$\frac{3}{8}$	$5\frac{5}{8}$	$2\frac{13}{16}$	$11\frac{5}{8}$	5C = 16
				$\frac{1}{2}$	$5\frac{1}{2}$	$2\frac{11}{16}$	$11\frac{1}{2}$	
6-in. Jumbo	6	4	12	$\frac{3}{8}$	$5\frac{5}{8}$	$3\frac{5}{8}$	$11\frac{5}{8}$	1C = 4
				$\frac{1}{2}$	$5\frac{1}{2}$	$3\frac{1}{2}$	$11\frac{1}{2}$	
8-in. Jumbo	8	4	12	$\frac{3}{8}$	$7\frac{5}{8}$	$3\frac{5}{8}$	$11\frac{5}{8}$	1C = 4
				$\frac{1}{2}$	$7\frac{1}{2}$	$3\frac{1}{2}$	$11\frac{1}{2}$	

Modular brick sizes and coursing heights. (*From Brick Institute of America,* Technical Note 10B, *BIA, Reston, Va.*)

brick" as defined by ASTM, may not exceed 25% of the area in the bearing plane. In addition to the cores, a $\frac{3}{4} \times \frac{3}{4}$-in. notch may be cut in one end of 6-in. brick to serve as a jamb unit. Roman brick are made in double form and broken into two units on the job site, leaving a rough, exposed edge.

The trend in development of different brick sizes has been toward modular coordination and toward slightly larger dimensions. Most contemporary masonry products, including clay tile and concrete block, are designed for connection at 8- or 16-in. course heights. For example, two courses of concrete block with mortar joints will equal 16 in. vertically, while three, five, or six courses of various size brick, and two, three, or four courses of clay tile equal the same height. This permits horizontal mechanical connection between the facing and backup elements of a multi-wythe wall. "Standard" brick produced before 1946–1947, when modular coordination was adopted, had actual heights of $2\frac{1}{4}$ in. (designed to lay up three courses to 8 in.). This size is still widely available so that, in renovation or restoration work, coursing heights can be effectively matched.

One of the first oversize brick units was introduced by the Brick Institute of America (BIA). The SCR brick was developed for use in single-wythe, 6-in. loadbearing walls. Larger brick sizes have also increased labor production. Although a mason can lay fewer of the large units in a day, the square footage of wall area completed is greater, less mortar is required, and projects are completed faster.

In addition to the common rectangular cut, brick may be formed in many special shapes for specific job requirements. Some of the more commonly used items include square and hexagonal pavers, bullnose and stair tread units, caps, sills, special corner brick, and wedges for arch construction (*see Fig. 3-6*). Unique custom shapes may be available on request from some manufacturers, but can be expensive to produce depending on the size of the order. Job-cut shapes must often be made for corners or other locations where a full brick length may not fit. These job-cut units are called half or bat, three-quarter closure, quarter closure, queen closure, king closure, and split.

The most unusual examples of customized masonry are sculptured pieces handcrafted from the green clayware before firing. The unburned units are firm enough to allow the artist to work freely without damage to the brick body, but sufficiently soft for carving, scraping, and cutting. After execution of the design, the units are returned to the plant for firing and the relief is permanently set in the brick face (*see Fig. 3-7*).

3.1.2 Hollow Brick

One of the traditional distinctions made between different clay masonry products is based on the definition of brick as "solid" (core area of less than 25% and clay tile as "hollow" (more than 25% cored area). However, during the 1970s, new standards were developed for "hollow brick" with a greater core area than that previously permitted for brick, but less than that allowed for tile.

The trend toward larger unit sizes led to production of jumbo brick in $8 \times 4 \times 12$-in. dimensions as early as the 1920s. In the southeastern United States, this prompted experimentation with greater coring as an effective means of reducing the weight and production costs of such large units. Originally made and marketed under a number of different proprietary names and specifications, these hollow brick are now classified in ASTM C652, *Standard Specification for Hollow Brick* (*see Fig. 3-8*). Sometimes

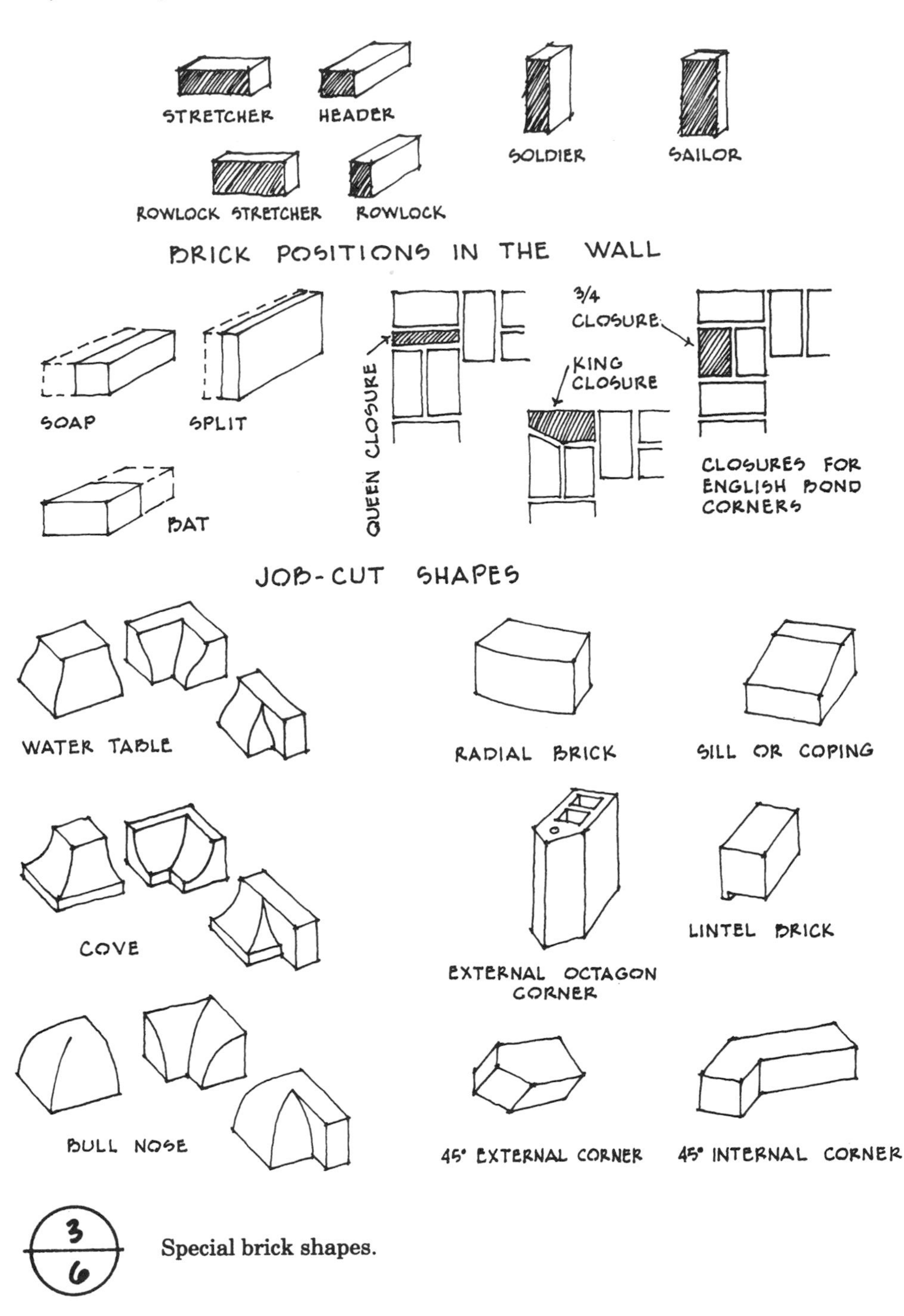

Special brick shapes.

referred to as through-the-wall units, hollow brick may be laid with opposite faces exposed. They offer considerable economy in speed and construction of masonry walls while maintaining the aesthetic appeal of conventional multi-wythe systems.

ASTM C652 covers hollow brick with core areas between 25% and 40% (Class H40V) and between 40% and 60% (Class H60V) of the gross cross-sectional area in the bearing plane. The two grades listed correspond to the same measure of durability as that used for building brick and face brick:

Sculptured brick panels at Loew's Anatole Hotel, Dallas. Deran and Shelmire, architects; Mara Smith, sculptor. (*Photo courtesy BIA.*)

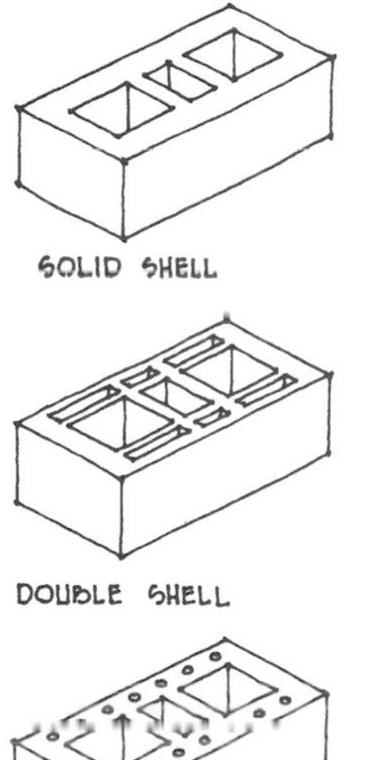

Hollow brick. (***Copyright, American Society for Testing and Materials, 1916 Race Street, Philadelphia, Pa. 19103. Reprinted with permission.***)

Grade SW (Severe Weathering) and Grade MW (Moderate Weathering). Types HBX (Select), HBS (Standard), and HBA (Architectural) are identical to face brick Types FBX, FBS, and FBA. Another type, HBB, is for general use in walls and partitions where color and texture are not a consideration and greater variation in size is permissible (as for building brick or common brick). Hollow brick are used for both interior and exterior construction in much the same way as solid brick. Sizes range from $4 \times 2\frac{1}{4} \times 12$ in. to $8 \times 4 \times 16$ in.

3.1.3 Special-Purpose Bricks

Special-purpose bricks serve many functions in architecture and industry. Refractory brick or *fire brick,* for instance, are used in furnaces, chimney stacks, fireboxes, and ovens. The fire clay from which they are made has a much higher fusing point than that of ordinary clay or shale. Once the initial kiln firing has been accomplished, fire brick are extremely resistant to high temperatures without cracking, decomposition, or distortion. Fire brick are normally heavier and softer than other units and are produced in a slightly larger size ($4\frac{1}{2} \times 2\frac{1}{2} \times 9$ in.), to be laid with a thin coating of refractory mortar in lieu of standard mortar joints. Fire clays typically burn to a white or buff color, so fire brick are usually in this color range as well.

For residential fireplace applications, ASTM C1261, *Standard Specification for Firebox Brick for Residential Fireplaces,* covers material requirements, physical properties, and fabrication tolerances. Units must be 100% solid, with no cores or frogs, must have a minimum modulus of rupture of 500 psi, and a pyrometric cone equivalent of 13. Since firebox brick are designed to be laid with very thin refractory mortar joints, the size tolerances permitted by ASTM C1261 are very restrictive.

Refractory brick of different chemical compositions are covered in a series of ASTM standards, and are graded according to fusion temperature, porosity, spalling strength, resistance to rapid temperature changes, thermal conductivity, and heat capacity. Some commonly used types of refractory brick are alumina brick, chrome brick, magnesite brick, and silica brick. The highly specialized nature of refractory design requires consultation with manufacturers to assure correlation between design needs and product specifications.

Glazed brick are fired with ceramic coatings which fuse to the clay body in the kiln and produce an impervious surface in clear or color, matte, or gloss finish. Standards for glazed brick are outlined in ASTM C126, *Standard Specification for Ceramic Glazed Structural Clay Facing Tile, Facing Brick, and Solid Masonry Units.* Requirements cover compressive strength, imperviousness, chemical resistance, crazing, and limitations on distortion and dimensional variation. Durability and weather resistance are not covered, so for exterior use, the body of the brick should be specified to conform to the requirements for ASTM C216 face brick, Grade SW, Type FBX, with the glaze in accordance with ASTM C126 standards. Glazed brick may suffer severe freeze-thaw damage in cold climates if not adequately protected from moisture permeance, and are not recommended for copings or other horizontal surfaces in any climate. Units are manufactured in Grade S (Select) and Grade SS (Select Sized, or ground edge), where a high degree of mechanical perfection, narrow color range, and minimum variation in size are required. Units may be either Type I, single-faced, or Type II, double-faced (opposite faces glazed). Type II are generally special-order items and are not widely used. Glazed brick are commonly available in "standard," oversize, Norman, and modular sizes, and in stretchers, jambs, corners, sills, and other supplementary shapes (*see Fig. 3-9*).

Most colors are fired at temperatures around 2100°F. The glaze, which is about the same consistency as thick house paint, is sprayed on the raw clay unit, and both are fired together. Some color glazes such as bright reds, primary yellows, burgundies, and oranges must be fired at lower temperatures ranging from 1300 to 1800°F. A red glaze burns clear if it gets too hot because the cadmium and lead ingredients are not stable

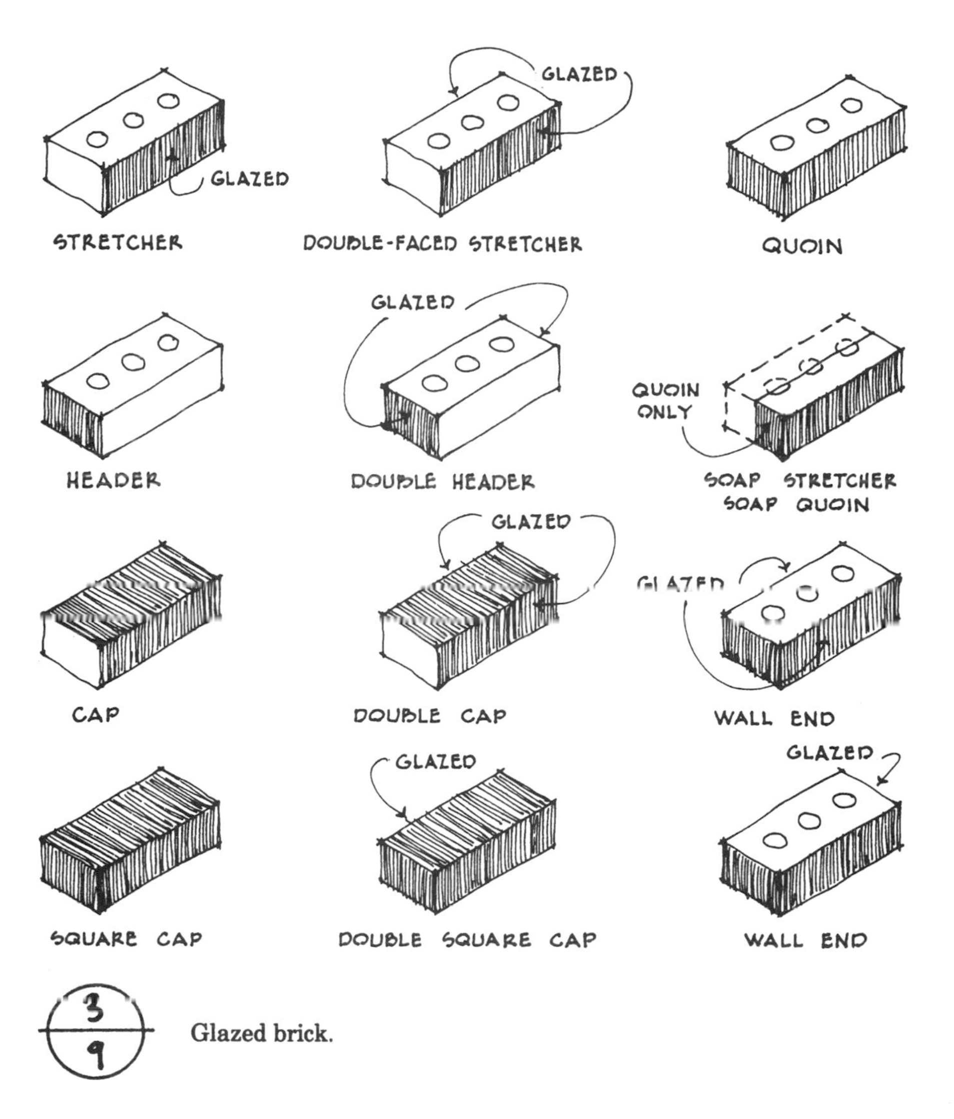

Glazed brick.

at high temperatures. This requires two firings. First the brick is fired at normal kiln temperatures, then the glaze is applied and the units are fired again at a low temperature. This two-fire process greatly increases the cost of the brick, and usually limits such colors to accents and specialty applications. Some low-fired glazes are prone to crazing because they are not as hard as high-fired glazes.

The naturally high abrasion resistance of ceramic clay products makes them very durable as paving materials. *Paving brick* are unique in color, pattern, and texture and are often specified as a wearing surface for roadways, walks, patios, drives, and interior floors. ASTM C902, *Standard Specification for Pedestrian and Light Traffic Paving Brick,* lists specific physical requirements. Three traffic uses are covered: Type I, highly abrasive; Type II, intermediate; and Type III, light traffic exposure. Three weathering classifications, SX, MX, and NX correspond to similar exposure limitations for face brick of severe, moderate, and negligible weathering. For extruded brick, Class SX requires a minimum average compressive strength of 8000 psi, a maximum average cold-water absorption of 8%, and a maximum saturation coefficient of 0.78. Compressive strength and cold water absorption for Class SX molded brick are one-half those of extruded brick at 4000 psi and 16%, respectively. Maximum saturation coefficient is the same at 0.78. Size tolerances for paving brick are governed by the intended method of installation: application PS for setting with mortar joints or in running bond or other patterns not requiring extremely close dimensional tolerances; application PX for setting without mortar joints; and application PA, characteristically nonuniform to simulate the appearance of hand-made brick.

ASTM C1272, *Standard Specification for Heavy Vehicular Paving Brick,* covers units intended for service in heavy use areas such as streets, commercial driveways and aircraft taxiways. Two Types of brick are covered. Type R (Rigid paving) is intended to be set in a mortar setting bed supported by an adequate concrete base or on an asphalt setting bed supported by an asphalt or concrete base. Type F (Flexible paving) is intended to be set in a sand setting bed with sand joints and may be installed on a flexible or rigid base. Three different applications are also covered, corresponding roughly to face brick appearance types. Application PS pavers are intended for general use. Application PX pavers are intended for use where dimensional tolerances, warpage, and chippage are limited. Application PA pavers are intended to produce characteristic architectural effects resulting from nonuniformity in size, color, and texture. Type R pavers must have a minimum average compressive strength of 8000 psi, a minimum modulus of rupture of 1200 psi, a maximum cold water absorption of 6%, and a minimum thickness of $2\frac{1}{4}$ in. Type F pavers must have a minimum average compressive strength of 10,000 psi, a minimum modulus of rupture of 1500 psi, a maximum cold water absorption of 6%, and a minimum thickness of $2\frac{5}{8}$ in.

Appearance depends largely on color, size, texture, and bond pattern. Paving brick are usually uncored and designed to be laid flat. Colors may range from reds to buffs, grays, and browns. Surface textures include smooth, velour, and rough, slip-resistant finishes. Standard or round-edge pavers are available in rectangular as well as square and hexagonal shapes (*see Fig. 3-10*).

Many industrial operations, such as foundries, steel mills, refineries, and breweries, require flooring materials that are resistant to vibration, impact, heavy vehicular traffic, thermal shock, and chemical attack. *Industrial floor brick* have been used successfully in these applications because of their dense structure, chemical stability, hardness, and "non-dusting" characteristics. Four basic types of units are described in ASTM C410, *Standard Specification for Industrial Floor Brick,* and are classified on the basis of absorption, chemical resistance, and modulus of rupture. Type T provides high resistance to thermal shock and mechanical impact, but also has relatively high absorption (10%). Type H has a lower percentage of absorption (6%), but offers only moderate resistance to chemicals and thermal shock. Type M should be used where low absorption is required (2%), with limited mechanical shock resistance and high resistance to abrasion. Type L provides minimum absorption (1%) and maximum chemical and abrasion resistance, but limited resistance to thermal and mechanical shock. Jointing material for industrial floor brick should be portland cement mortar or grout or, when required, chemical-resistant mortar.

Most well-burned clay masonry, including conventional face brick, building brick, structural clay tile, and ceramic glazed units, have excellent resistance to chemicals and chemical agents. In some installations, however, such as waste treatment facilities, dairies, chemical plants,

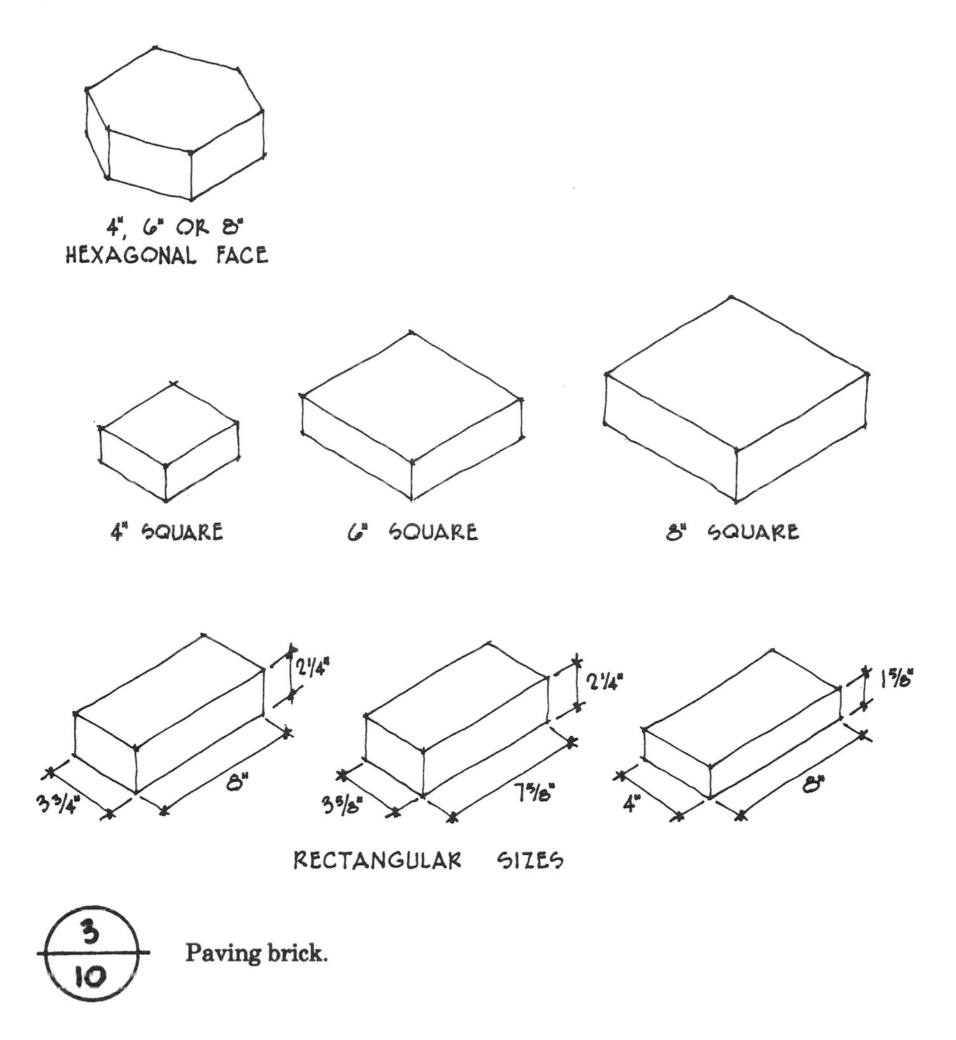

3
10
Paving brick.

refineries, and food processing plants, extraordinary resistance may be required. *Chemical-resistant brick* or acid-proof brick are machine-made, uncored, kiln-fired units made specifically for this purpose. They are strong, free of laminations, burned to vitrification to close all pores, and sufficiently rough in texture to ensure complete and intimate bond with the mortar. Conditions of temperature and acidity and the absorption rate of the unit are the primary factors governing material selection for use in corrosive environments. Determination of the nature and severity of exposure will dictate which of the three types of units covered in ASTM C279, *Standard Specification for Chemical-Resistant Masonry Units,* should be used. Type I has low absorption (6%) and high sulfuric acid resistance. Type II has lower absorption (4%) and higher acid resistance. Type III has minimum absorption (1%) and maximum acid resistance. The three types do not differ significantly in thermal shock resistance. Chemical-resistant brick performs satisfactorily in the presence of mild alkalis and all acids except hydrofluoric. In instances where strong alkalis or hydrofluoric acid and its salts are present, a special "carbon brick" is required. Chemical-resistant mortars must be used with these units to assure effective performance (see Chapter 6).

ASTM C32, *Standard Specification for Sewer and Manhole Brick,* identifies two grades for each usage. For *sewer brick,* Grades SS and SM distinguish between the amounts and velocities of abrasive materials carried. Grade SS is lower in absorption and offers greater erosion resistance. *Manhole brick* are graded on their ability to withstand freezing action rather than abrasion. Grade MS provides a high and uniform degree of resistance, while Grade MM offers only moderate and nonuniform resistance. These brick may be used in drainage structures for the conveyance of sewage, industrial wastes, and storm water, and for related structures such as manholes and catch basins.

3.1.4 Glass Block

Glass block can be used as security glazing, or to produce special daylighting effects. Glass block is considered a masonry material since it is laid up in cement mortar and uses the same type of joint reinforcement as other units. Although they are not made from clay, glass blocks do share some common characteristics with burned clay products. Both contain silicates as a primary raw ingredient and the glass units, like brick, undergo vitrification when subjected to a heat process. They are available in a variety of sizes, and in both solid and hollow form. Decorative blocks are produced in clear, reflective, or color glass with smooth, molded, fluted, etched, or rippled texture. Functionally, glass block is used to diffuse or direct light for different illuminating requirements, and provide a high level of security and energy efficiency for glazed areas. Compressive strengths range from 400 to 800 psi.

Most glass block is made of clear, colorless glass that admits the full spectrum of natural light (*see Fig. 3-11*). Hollow block with patterns pressed into the interior face partly or totally distort images, creating visual privacy. Units made with glass fiber inserts reduce glare and brightness. Other units diffuse or reflect light. Glass block can increase or reduce solar heat gain, and because of their large air cavity, hollow block have greater thermal resistance than ordinary flat glass (*see Fig. 3-12*). A partial vacuum created when the hollow units are made further improves their thermal resistance.

Solar reflective block is coated with a heat-bonded oxide which can reduce solar heat gain by as much as 80% compared to conventional ⅛-in. flat glass (*see Fig. 3-13*). Glass fiber inserts further reduce solar heat gain by about 5%, and also increase thermal resistance.

Glass block comes in nominal face sizes of 6×6-, 8×8-, and 12×12-in. square units and 4×8- and 6×8-in. rectangular units. Actual dimensions vary by manufacturer and style. Units made in the United States are ¼ in. less than nominal dimensions. Most hollow block are 3⅞ in. thick (nominal 4 in.), but some manufacturers also make thin block which measure only

Type of glass block	*% light transmitted*
Solid	80
Hollow	50–75
Diffusion	28–40
Reflective	5–20

Light transmission characteristics of glass block.

Type of glass	% light transmitted	U-value	Shading coefficient	Heat gain (Btu/hr/sq ft)
8 × 8 reflective block	20	0.51	0.25	42
8 × 8 clear block	62	0.51	0.65	140
1/8-in clear sheet glass	90	1.04	1.00	215

Thermal performance of glazing materials.

Alfred C. Glassel School of Art, Houston. Morris/Aubry, architects. *(Photo courtesy Houston Museum of Fine Arts.)*

$3\frac{1}{8}$ in. and weigh 20% less than standard units. Solid glass block are used for high-security glazing. They come in 3×8-in. rectangular units and 8×8-in. square units. Ordinary construction methods require limiting the number of courses laid at one time so that fresh mortar is not extruded from the joints by the weight of the block. There are several proprietary types of spacers which help speed construction. Unit weight is transferred directly from block to block by the spacers, allowing work to progress rapidly without waiting for substantial mortar cure to support the weight of the units. Mortar adhesion to glass block is limited.

3.2 STRUCTURAL CLAY TILE

Structural clay tile is the most recently developed of clay masonry products, first produced in this country in 1875. Up until that time, most buildings were constructed with solid loadbearing masonry walls. With the invention and mastery of cast iron and steel structural framing, a need arose for lightweight backing materials for the facing masonry used to clad these skeleton frames. Clay tile satisfied the demand and added elements of economy and fire resistance.

Structural clay tile is still produced by a limited number of manufacturers, both for new construction and for restoration/retrofit work. It is used both as structural, facing, and backup material in construction. Tile, like brick, is made of clay that is molded and then fired in a kiln to ceramic fusion. Clay tile may be used with the hollow cells either horizontal or vertical (side construction or end construction tile), for both loadbearing and non-loadbearing applications. "Structural" clay tile is distinguished from flat clay wall tile and flat clay floor tile by its ability to carry load and support its own weight. The numerous types of tile used today are classified by function. *Structural clay loadbearing wall tile* and *structural clay non-loadbearing tile* may be used in the construction of walls and partitions where a finish coat of plaster or other material will be applied or where appearance is not a primary concern. These units are the equivalent of building brick, and are considered principally utilitarian in nature. *Structural clay facing tile* and *ceramic glazed facing tile* may be loadbearing or non-loadbearing, but are distinguished from the above on the basis of finish, much the same as face brick is distinguished from building brick.

3.2.1 Loadbearing Wall Tile

ASTM C34, *Standard Specification for Structural Clay Loadbearing Wall Tile,* divides units into two grades based on compressive strength and resistance to frost action in the presence of moisture (*see Fig. 3-14*). The higher grade, LBX, is suitable for areas exposed to weathering provided they meet the same durability requirements as Grade SW, ASTM C216 face brick. Grade LB is limited to unexposed areas unless protected by at least 3 in. of stone, brick, terra cotta, or other masonry. In either case, the tile carries the structural load, the live load, and the weight of the facing material, plus its own weight. Loadbearing tile may also be used in composite wall construction with facing tile, brick, or other masonry units. In this instance, the wythes of the wall are bonded together structurally so that the tile bears an equal share of the superimposed load.

3.2.2 Non-Loadbearing Tile

ASTM C56, *Standard Specification for Structural Clay Non-Loadbearing Tile,* covers partition tile, furring tile, and fireproofing tile. The standard

Grade	*Maximum water absorption by 1-hr boiling* (%)*		*Minimum compressive strength (based on gross area)† (psi)*			
			End-construction tile		*Side-construction tile*	
	Average of 5 tests	*Individual*	*Average of 5 tests*	*Individual*	*Average of 5 tests*	*Individual*
LBX	16	19	1400	1000	700	500
LB	25	28	1000	700	700	500

Physical requirements for structural clay loadbearing wall tile, from ASTM C34. (*Copyright, American Society for Testing and Materials, 1916 Race Street, Philadelphia, Pa. 19103. Reprinted with permission.*)

includes only one grade, NB, and one physical property specification, which limits the rate of water absorption. Maximum allowable absorption is 28%, and the maximum range of absorptions for tile delivered to any one job may not exceed 12%. *Partition tile* is used to construct non-loadbearing interior partitions. *Furring tile* is used to line the inside surface of exterior walls, providing an insulating air space and a surface suitable for plastering. Partition and furring tile may be used to fireproof structural steel members, but for some applications around beams and girders, special shapes of *fireproofing tile* are required to conform to the profile of the steel. Clip and angle shapes have been devised for this purpose and, when used in conjunction with conventional rectangular tiles, provide a simple means of complete coverage (*see Fig. 3-15*).

Tile that will be plastered must have a surface texture which provides good bond between plaster and unit (*see Fig. 3-16*). ASTM C56 covers smooth (wire-cut), scored, combed, and roughened finishes.

3.2.3 Facing Tile

Facing tile combines the loadbearing capacity of ordinary clay tile with a permanent, finished surface suited for architectural applications. These natural-color unglazed tile are covered in ASTM C212, *Standard Specification for Structural Clay Facing Tile.* Two classes of tile are defined based on face shell and web thickness. "Standard" tile are general-purpose units for exterior or interior locations. "Special duty" tile have heavier webs and shells designed to increase resistance to impact and moisture penetration. Aesthetic factors are designated the same as for face brick. Type FTX (Select) tile have a smooth finish for general use in interior and exterior applications requiring minimum absorption, easy cleaning, and resistance to staining. They provide a high degree of mechanical perfection, narrow color range, and minimum variation in face dimensions. Type FTS (Standard) may have a smooth or rough texture, are suitable for interior and exterior construction where moderate absorption and moderate variation in face dimensions are permissible, and may be used where minor defects in surface finish are not objectionable. ASTM C212 lists compressive strength and absorption, and sets limits on chippage, dimensional variation, and face distortion (see *Fig. 3-17* for sizes and shapes available).

3.2.4 Ceramic Glazed Facing Tile

Most of the structural clay tile used in new construction today is glazed. Glazed units are also of loadbearing quality, but have an impervious finish in either clear or color glaze. Physical requirements are outlined in ASTM C126, which also governs glazed brick. For exposed exterior applications, the tile body should also meet the durability requirements for ASTM C652, Grade SW hollow brick units. Exterior applications should also be limited to vertical cell tile, since horizontal cells can trap moisture in the wall. If the units are frozen when wet, the glazed surface can easily spall. Grade and type classifications for glazed tile are identical to those for glazed brick. Grade S (select) units are used with comparatively narrow mortar joints. Grade SS (selected sized, or ground edge) are used where variation of face dimension must be very small. Both grades may be produced in either Type I, single-faced units, where only one face is glazed; or Type II, double-faced, where two opposite faces are glazed. ASTM C126 covers com-

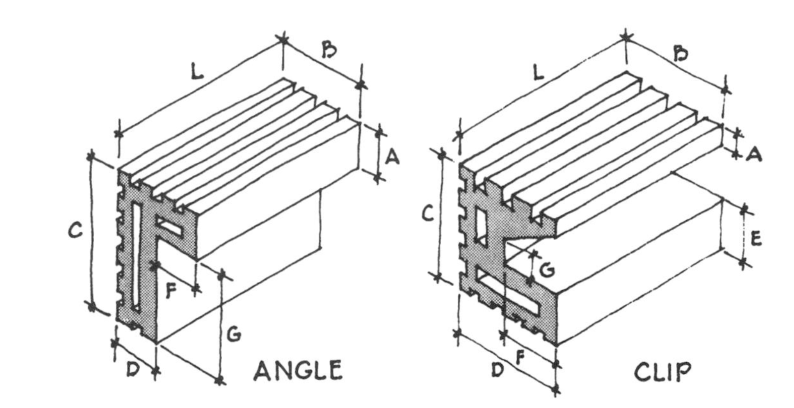

TYPICAL CLIP AND ANGLE TILE SIZES									
Unit number	*Outside measure (sq ft)*	*Weight per piece (lb)*	*A*	*B*	*C*	*D*	*E*	*F*	*G*
G20	0.65	13	$\frac{3}{4}$	$3\frac{3}{8}$	$3\frac{3}{8}$	$3\frac{3}{4}$	2	2	$\frac{3}{4}$
G25	0.70	14	$\frac{3}{4}$	$3\frac{7}{8}$	4	$4\frac{1}{2}$	2	$2\frac{1}{2}$	$\frac{13}{16}$
G30	0.75	15	$\frac{7}{8}$	$4\frac{1}{4}$	$4\frac{1}{4}$	$4\frac{3}{4}$	2	3	$\frac{7}{8}$
G35	0.80	16	$\frac{3}{4}$	$4\frac{3}{4}$	$4\frac{1}{4}$	$5\frac{1}{2}$	2	$3\frac{1}{2}$	1
G40	0.85	17	$\frac{3}{4}$	$4\frac{3}{4}$	$4\frac{1}{4}$	6	2	4	1
G45	0.90	18	$\frac{3}{4}$	$4\frac{3}{4}$	$4\frac{1}{4}$	$6\frac{1}{2}$	2	$4\frac{1}{2}$	1
G46	0.95	19	$\frac{3}{4}$	$4\frac{3}{4}$	$4\frac{3}{4}$	$6\frac{5}{8}$	2	$4\frac{5}{8}$	$1\frac{1}{2}$
G50	0.95	19	$\frac{3}{4}$	$4\frac{3}{4}$	$4\frac{1}{4}$	7	2	5	1
G55	1.00	20	$\frac{3}{4}$	$4\frac{3}{4}$	$4\frac{1}{2}$	$7\frac{1}{2}$	2	$5\frac{1}{2}$	$1\frac{1}{4}$
G60	1.05	21	1	$5\frac{1}{2}$	$4\frac{5}{8}$	8	2	6	$1\frac{3}{8}$
G61	1.05	21	1	$4\frac{1}{2}$	$4\frac{5}{8}$	8	2	6	$1\frac{3}{8}$
G70	1.16	23	1	6	$4\frac{7}{8}$	9	2	7	$1\frac{1}{2}$
G71	1.16	23	1	$4\frac{1}{2}$	$4\frac{7}{8}$	9	2	7	$1\frac{1}{2}$
G75	1.33	29	2	$7\frac{1}{2}$	$6\frac{3}{8}$	$9\frac{1}{2}$	2	$7\frac{1}{2}$	$2\frac{3}{8}$
G80	1.40	32	2	8	$6\frac{7}{8}$	10	2	8	$2\frac{7}{8}$
L23	0.60	12	2	4	5	2	—	2	3
L26	0.90	18	2	4	$8\frac{1}{2}$	2	—	2	$6\frac{1}{2}$
L43	0.65	13	2	6	5	2	—	4	3
L46	0.95	19	2	6	$8\frac{1}{2}$	2	—	4	$6\frac{1}{2}$

Clip and angle tile sizes. (*From Harry C. Plummer,* **Brick and Tile Engineering,** *Brick Institute of America, Reston, Va., 1962.*)

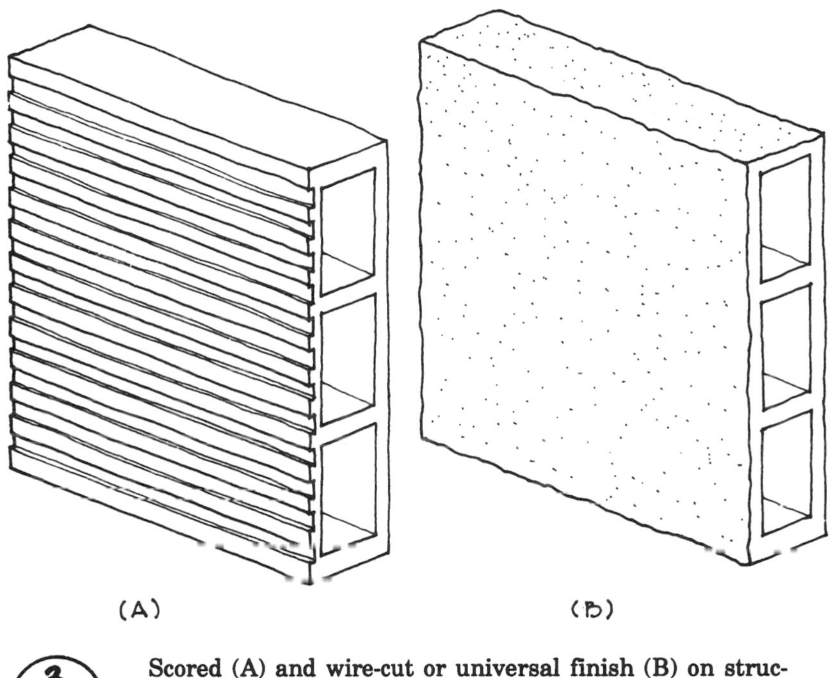

Scored (A) and wire-cut or universal finish (B) on structural clay tile units.

pressive strength, absorption rate, number of cells, shell and web thickness, dimensional tolerances, and properties of the ceramic finish, including imperviousness, chemical resistance, and crazing.

The shapes of all structural tile units are controlled by the dies through which the plastic clay is extruded. The relative ease with which various designs can be produced led to the development of a large number of sizes and patterns. Through a process of standardization, this number has been reduced to only the most economical and useful units. Development and acceptance of the criteria for modular coordination encouraged refinement aimed at correlation with other manufactured masonry products. Structural clay tile are designed for use with $\frac{1}{4}$-, $\frac{3}{8}$-, or $\frac{1}{2}$-in. mortar joints, while facing tile uses only $\frac{1}{4}$-in. joints. Nominal dimensions, as for brick, include this thickness and are multiples of the 4-in. module or fractions of a multiple of that module (i.e., three courses of $5\frac{1}{3}$-in.-high tiles equals 16 in.).

The nomenclature of shape number can be bewildering because of the many possible combinations, but the system is really fairly simple (*see Fig. 3-18*). The prefix is an alphanumeric designation of length, height, and coring; followed by numbers denoting horizontal and vertical axis conditions (such as cove base, bullnose, or stretcher), and bed depth; and a letter suffix denoting return and reveal, back surface finish, and right- or left-handed unit when required.

Because of the glazed surface, a larger variety of special shapes is required to facilitate door and window openings, headers, corners, and so on. In addition to full-size stretcher units, shapes include half-lengths, half-heights, corner and jamb units, as well as sills, caps, lintels, cove bases, and coved internal corners (*see Fig. 3-19*). Some manufacturers prepare shop drawings from the architectural plans to show actual tile shapes and locations. If the project is laid out with modular dimensions, very few

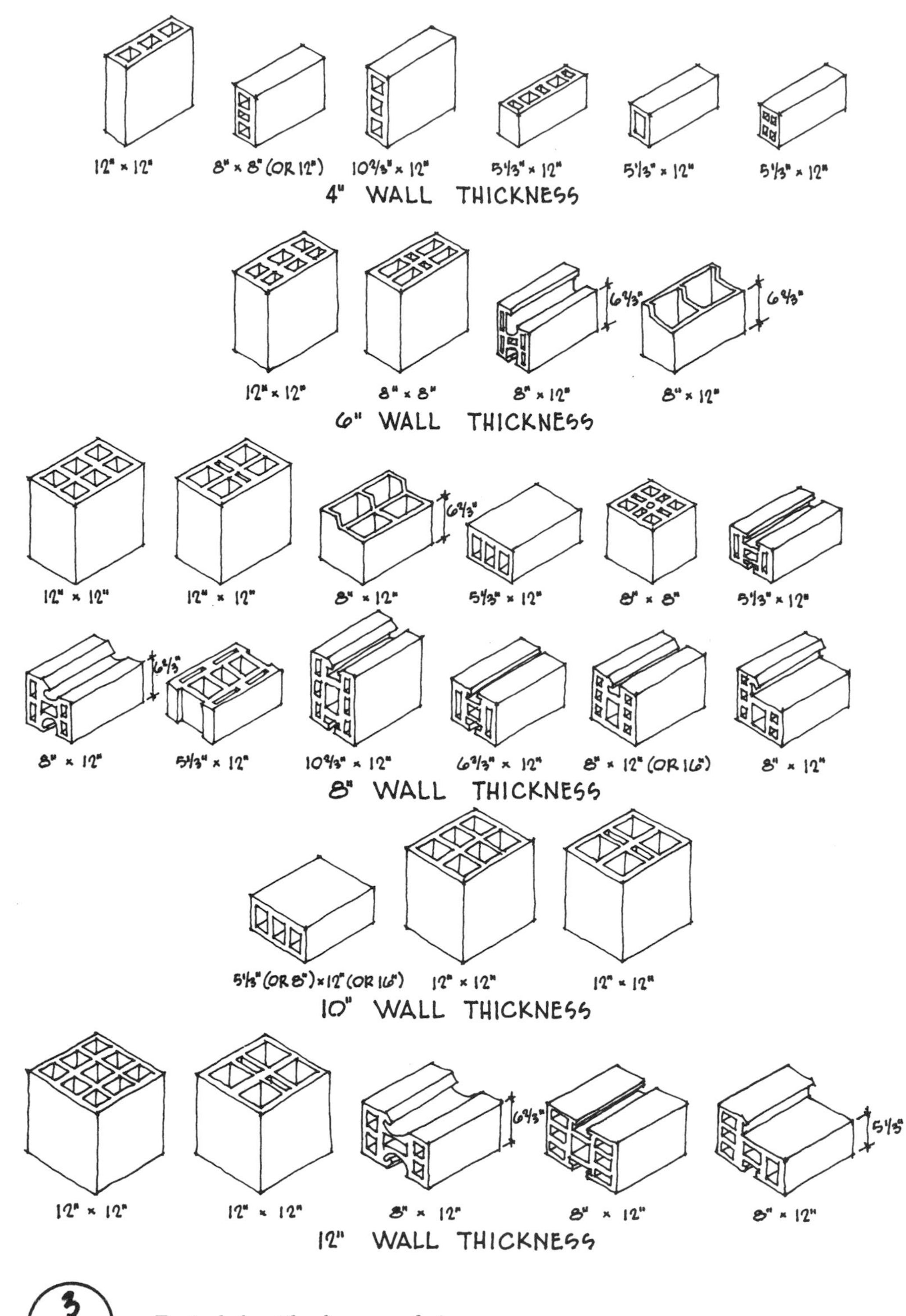

Typical clay tile shapes and sizes.

AVAILABLE SIZES		
Series	*Nominal face dimensions, inches*	*Nominal thickness, inches*
PREFIX 6T	$5\frac{1}{3} \times 12$	2,4,6,8
4D	$5\frac{1}{3} \times 8$	2,4,6,8
4S	$2\frac{2}{3} \times 8$	2,4
4W	8×8	2,4,6,8
8W	8×16	2,4,6,8

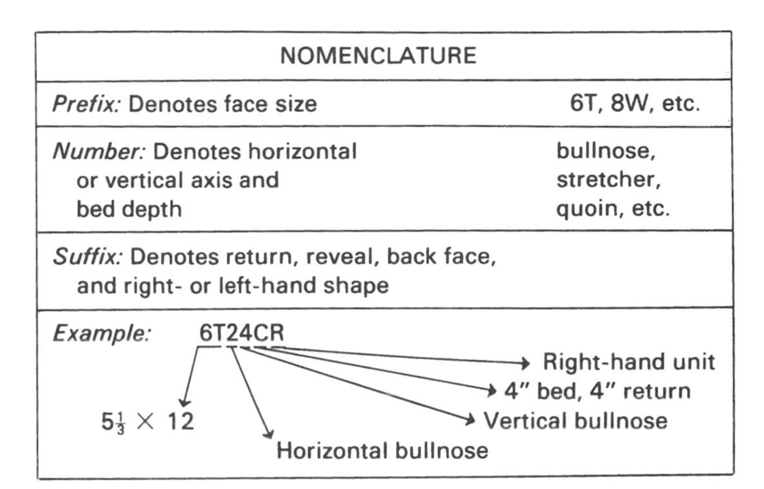

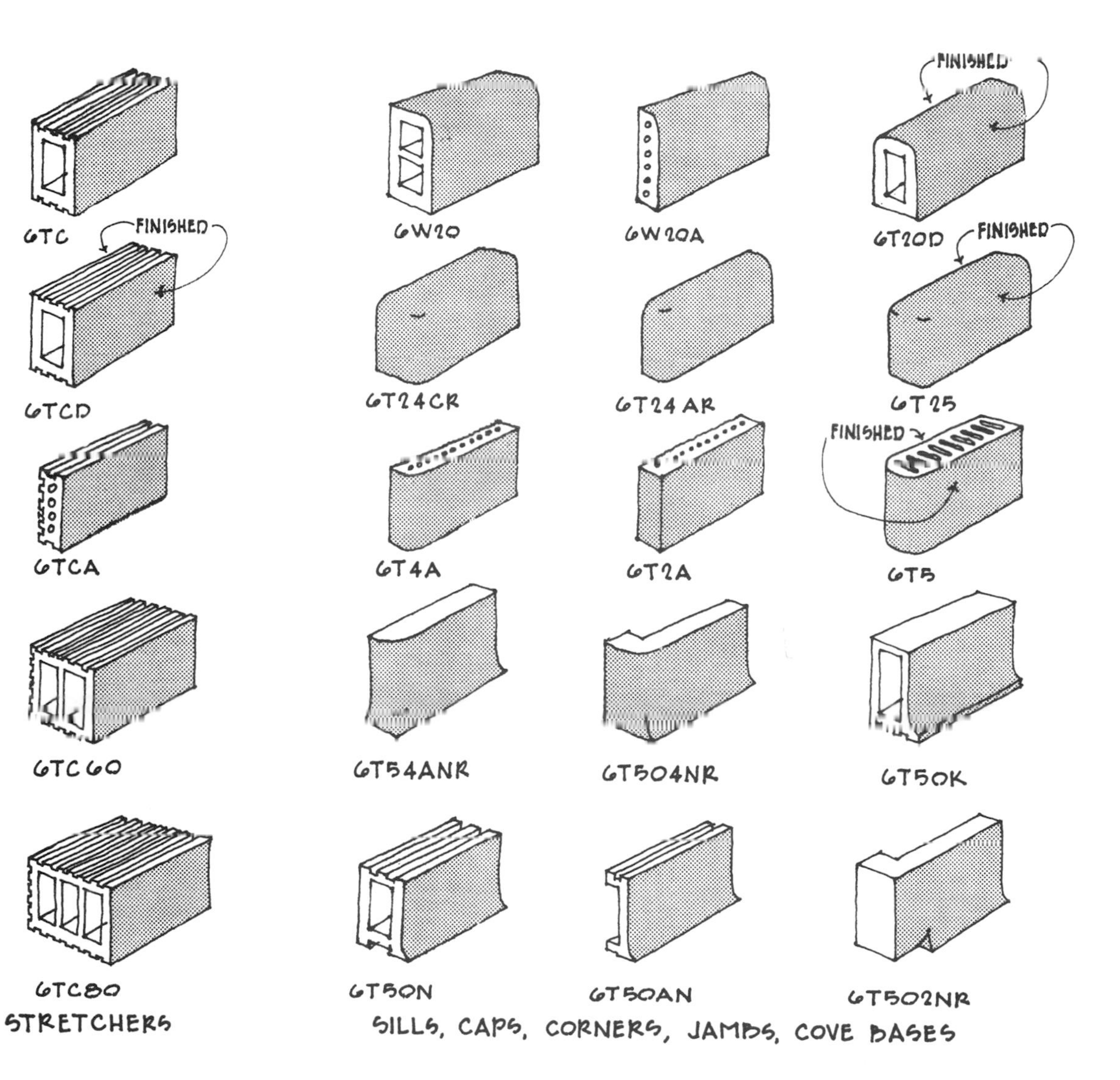

Glazed facing tile shapes.

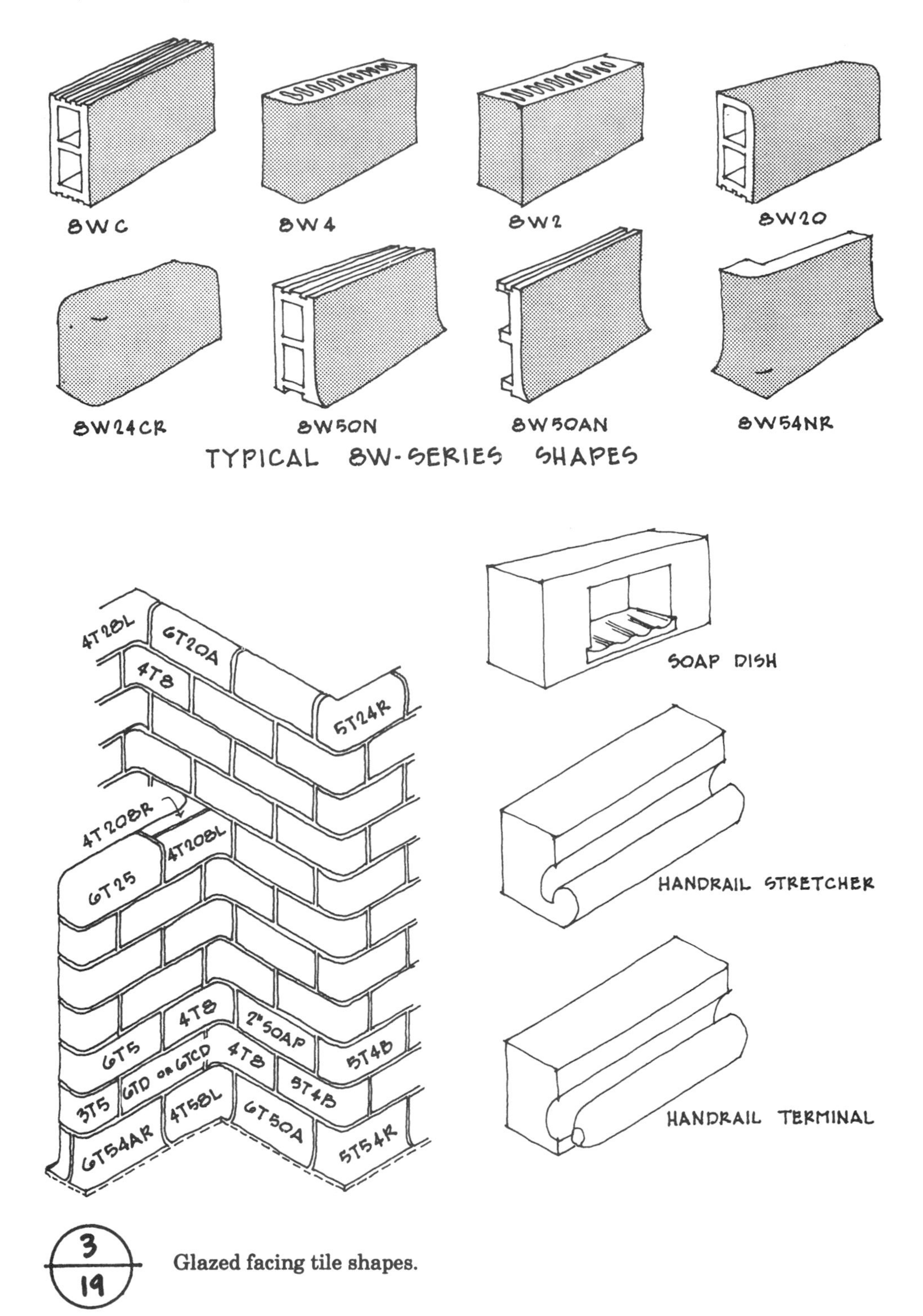

3/19 Glazed facing tile shapes.

(if any) extraordinary special shapes will be required and job-site cutting and waste will be minimum.

Structural glazed tile has long had a place in commercial kitchens, bottling and food processing plants, schools, and hospitals because of its durable surface and low maintenance requirements. But more and more architects are turning to this material for use in correctional facilities and in high-traffic public buildings such as airports, shopping malls, and sporting arenas. Unlike glazed brick, glazed structural clay tile allows single-wythe construction of walls and partitions glazed on both sides.

Glazed tile is available in traditional pastels and in bold colors such as fire engine red and cobalt blue. It also comes with either smooth or textured surfaces. The 8×8 modular and 8×16 face sizes are most popular today because they course easily with other types of masonry. Scored 8×16 units are available which simulate the look of 8×8 stack bond. Like glazed brick, glazed structural tile is impervious to stains, resistant to fading and crazing, and unaffected by many chemicals including hydrochloric acid and caustic cleaning solutions. Its abrasion resistance is greater than that of ordinary steel when rated on the Mohs Hardness Scale. As long as the mortar is designed to resist the same abuse expected of the units, a structural glazed tile wall will last the life of the building with no maintenance other than washing. Even when concrete masonry is used for walls, a structural glazed cove base provides better resistance to the abuse of floor cleaning equipment and traffic than ordinary block.

For applications requiring extremely sanitary conditions, and for high-abuse areas, joints can be raked out and pointed with epoxy mortar. Walls can then be hosed down, scrubbed, or steam-cleaned without damaging the mortar and without allowing moisture to enter the wall. After the joints are raked, the setting mortar should cure for 24 hours before pointing with the epoxy mortar.

3.2.5 Screen Tile

Clay masonry solar screens have always found wide acceptance whether constructed of screen tile or of standard units ordinarily used for other purposes. Screen tile are available in a variety of shapes and patterns and in all colors of glazed and unglazed clay masonry (*see Fig. 3-20*). Lighter colors, because of greater reflectivity, provide brighter interiors. Darker colors absorb more of the sun's heat and light and give greater protection from its harsh rays.

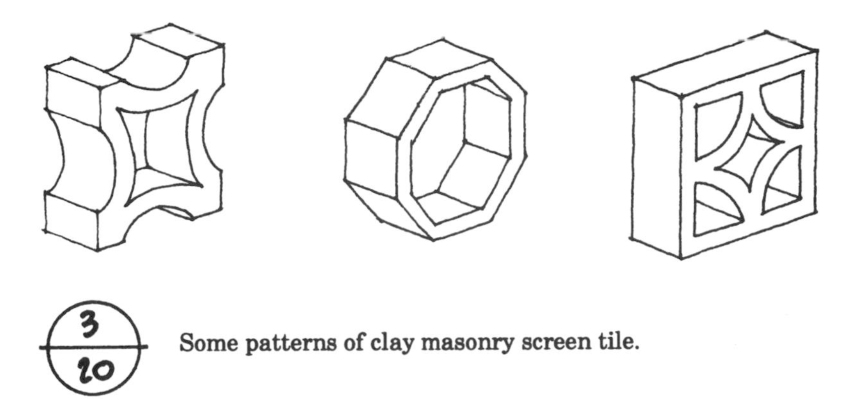

3/20 **Some patterns of clay masonry screen tile.**

Screen tile are covered by ASTM C530, *Standard Specification for Structural Clay Non-Loadbearing Screen Tile.* Grades SE, ME, and NE correspond to the Severe Weather, Moderate Weather, and No Weather exposure durability ratings of other clay masonry products. Only two aesthetic types are included, though, Type STX (Select) and Type STA (Architectural).

3.3 TERRA COTTA

Architectural terra cotta has been used as a decorative veneer for centuries. The name itself, which means "baked earth," dates from Roman antiquity. Hand-molded slabs with either plain or sculptured surfaces are still produced in the traditional manner, and extruded units are mechanically fabricated with smooth-ground, beveled, scored, scratched, or fluted surfaces. Both the hand- and machine-made types may be glazed in clear, monochrome, or polychrome colors and in matte, satin, or gloss finishes.

Sometimes referred to as ceramic veneer, architectural terra cotta is an enriched clay mixture fired at high temperatures to a hardness and density not obtainable with brick. Glazed units are durable and weather-resistant, and provide an almost infinite range of colors which will retain their sharpness and clarity for the life of the product.

Architectural terra cotta was originally used as a loadbearing element in multi-wythe walls, but in the late nineteenth and early twentieth centuries it gained popularity as a cladding material, particularly for structural frame buildings. It was lightweight, relatively inexpensive, and particularly adaptable to rich ornamental detailing. Architectural terra cotta figured prominently in the work of H. H. Richardson, Cass Gilbert, Louis Sullivan, and Daniel Burnham, among others, and was a key element in such architectural idioms as the Chicago School, the Gothic and Romanesque Revival movements and the Beaux Arts style. Though architectural terra cotta is one of the most prevalent materials from this period, many are unaware of its presence because it frequently masqueraded as stone. Building owners and architects alike are often surprised to discover that what they presumed to be a granite, limestone, or brownstone facade is actually glazed terra cotta.

Today, terra cotta is produced for both historic restoration and new construction. Flat field panels, the basic units of veneer systems, must have scored or dovetailed backs to form a key with the mortar. Accessory pieces such as copings, sills, and projecting courses are also die-formed, and may be simple or elaborate in profile. *Figure 3-21* shows the stock shapes used to create the cornice and base courses on the Best Products corporate headquarters building. The backs and webs of units, as shown, are often cut at the factory or broken out at the job site to accommodate mechanical fasteners. Decorative balusters and hand-molded ornamental shapes are also available. Reproduction pieces can be made by taking a plaster cast of existing features and then hand-packing wet clay into a mold made from the cast.

There are no ASTM standards for terra cotta, but units should meet the minimum requirements of the *Standard Specifications for Ceramic Veneer* and *Standard Methods for Sampling and Testing Ceramic Veneer* published by the Architectural Terra Cotta Institute (1961). Glazes, however, are covered by ASTM C126, *Standard Specifications for Ceramic Glazed Structural Clay Facing Tile, Facing Brick and Solid Masonry Units.*

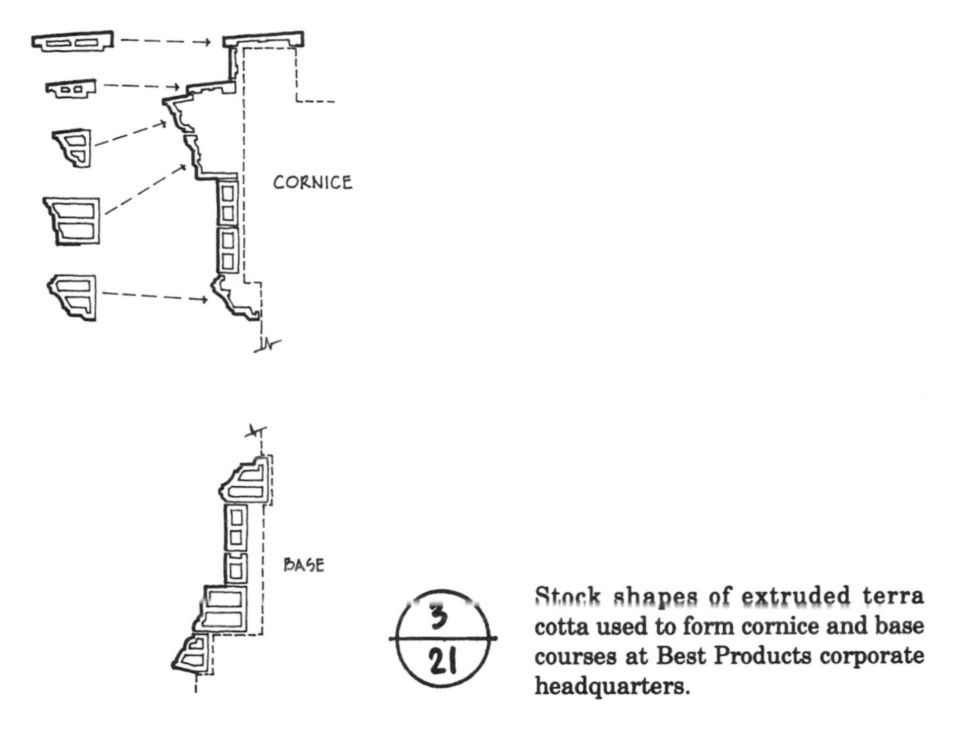

Stock shapes of extruded terra cotta used to form cornice and base courses at Best Products corporate headquarters.

3.4 PROPERTIES AND CHARACTERISTICS OF FIRED CLAY PRODUCTS

Physical properties and characteristics of masonry units are important to the architect only insofar as they affect performance and appearance of the finished wall or structure. The major building codes in the United States rely primarily on ASTM standards and requirements of the American National Standards Institute (ANSI) for minimum property specifications. These deal mainly with compressive strength, absorption, and saturation coefficients as indicators of acceptable performance (*see Fig. 3-22*).

3.4.1 Compressive Strength

The compressive strengths of brick and tile are usually based on gross area. Extruded brick generally have higher compressive strength and lower absorption than those produced by the soft-mud or dry-press processes. For a given clay and method of manufacture, higher compressive strength and lower absorption are also associated with higher burning temperatures. The minimum compressive strength values listed in *Fig. 3-22* are substantially exceeded by most manufacturers. Actual compressive strengths of clay masonry units are usually higher than those of ordinary structural concrete. For standard run brick, strengths typically range from 1500 to 22,500 psi, with the majority of units produced being in excess of 4500 psi (*see Fig. 3-23*).

3.4.2 Transverse Strength

The transverse strength of a brick acting as a beam supported at both ends is called the *modulus of rupture*. Tests at the National Institute of Standards and Technology (NIST) indicate minimum values for well-burned brick to be

Unit	Grade	Minimum compressive strength, gross area (psi): Average of 5 tests	Minimum compressive strength, gross area (psi): Individual unit	Maximum water absorption by 5-hr boiling (%): Average of 5 tests	Maximum water absorption by 5-hr boiling (%): Individual unit	C/B Maximum saturation coefficient: Average of 5 tests	C/B Maximum saturation coefficient: Individual unit
Building brick (ASTM C62)	SW	3000	2500	17.0	20.0	0.78	0.80
	MW	2500	2200	22.0	25.0	0.88	0.90
	NW	1500	1250	No limit		No limit	
Facing brick (ASTM C216)	SW	3000	2500	17.0	20.0	0.78	0.80
	MW	2500	2200	22.0	25.0	0.88	0.90
Hollow brick (ASTM C652)	SW	3000	2500	17.0	20.0	0.78	0.80
	MW	2500	2200	22.0	25.0	0.88	0.90

Unit	Grade	End construction Av.	End construction Ind.	Side construction Av.	Side construction Ind.	By 1-hr boiling, % (Av.)	By 1-hr boiling, % (Ind.)
Structural clay load bearing tile (ASTM C34)	LBX	1400	1000	700	500	16.0	19.0
	LB	1000	700	700	500	25.0	28.0
Structural clay facing tile (ASTM C212)	FTX					9.0	11.0
	FTS					16.0	19.0
	Standard	1400	1000	700	500		
	Special-duty	2500	2000	1200	1000		

Minimum physical requirements of clay masonry products.

in excess of 500 psi, with a maximum average of 2890 psi. There is no general rule, however, for converting values of compressive strength to transverse strength, or vice versa. Tensile and shearing properties of brick have not been widely tested, but data from NIST indicate that *tensile strength* normally falls between 30 and 40% of the modulus of rupture and *shear values* from 30 to 45% of the net compressive strength. Tensile strength of structural clay tile is quite low and usually will not exceed 10% of the compressive strength. The *modulus of elasticity* for brick ranges from 1,400,000 to 5,000,000.

3.4.3 Absorption

The weight of burned clay products ranges from 100 to 500 lb/cu ft. Variations may generally be attributed to the process used in manufacturing and burning. Increased density and weight result from fine grinding of

Compressive strength, flatwise		Modulus of rupture	
Range (psi)	Percentage of production within range	Range (psi)	Percentage of production within range
21,001–22,500	0.46	2101–3450	6.95
19,501–21,000	0.69	1951–2100	3.00
18,001–19,501	0.46	1801–1950	2.74
16,501–18,000	2.04	1651–1800	7.57
15,001–16,500	1.49	1501–1650	8.34
13,501–15,000	3.71	1351–1500	5.34
12,001–13,500	4.76	1201–1350	7.12
10,501–12,000	7.78	1051–1200	10.55
9,001–10,500	8.61	901–1050	10.44
7,501–9,000	11.92	751–900	13.60
6,001–7,500	15.47	601–750	11.74
4,501–6,000	16.81	451–600	7.52
3,001–4,500	17.97	301–450	4.35
1,501–3,000	7.46	151–300	0.37
0–1,500	0.36	0–150	0.37
Total percent	99.99		100.00

Actual compressive strengths of brick produced in the United States. (*From Brick Institute of America, Principles of Brick Masonry, BIA, Reston, Va., 1973.*)

raw materials, uniform mixing, pressure exerted on the clay as it is extruded, de-airing, and hard or complete burning. The extrusion process produces very dense brick and tile characterized by high strength and a small percentage of voids. Since properties of absorption are also affected by the method of manufacture and degree of burning, these factors indicate fairly close relationships among total absorption, weight, density, and compressive strengths. With few exceptions, hard-burned units are highest in strength and density and lowest in absorption.

The absorption of a brick or clay tile is defined as the weight of water taken up by the unit under given laboratory test conditions, and is expressed as a percentage of the dry weight of the unit. Since highly absorptive brick exposed to weathering can cause a buildup of damaging moisture in the wall, ASTM standards limit face brick absorption to 17% for Grade SW and 22% for Grade MW units. Most brick produced in the United States have absorption rates of only 4 to 10% (*see Fig. 3-24*).

An important property of brick which critically affects bond strength is the *initial rate of absorption* (IRA), or *suction*. High-suction brick absorb water from the mortar too quickly, impairing bond, proper hydration, and curing. Laboratory tests and field experience indicate that maximum strength and minimum water penetration occur with units having initial rates of absorption between 5 and 25 grams/minute/30 sq in. at the time they are laid. (The requirement is based on the area of the bed surface of the brick being approximately 30 sq in. Figures reported for brick of other sizes must be adjusted in accordance with the test procedure described in ASTM C67, *Standard Test Methods of Sampling and Testing Brick and Structural Clay Tile.*) Brick with high suction rates should be thoroughly wetted prior to installation, then allowed to "surface dry." Since suction

Designation	Absorption by total immersion: 24-hr cold, C (%)	Absorption by total immersion: 5-hr boil, B (%)	Saturation coefficient, C/B (%)	Saturation by partial immersion, flat, 1 min (g)
A	1.9	3.5	0.53	10
B	9.4	13.4	0.70	33
C	14.6	16.9	0.86	112
D	11.3	15.1	0.74	25
E	10.2	14.7	0.69	38
F	13.8	18.7	0.74	42
G	5.4	7.8	0.69	12
H	3.3	6.0	0.54	6
J	9.3	13.5	0.68	31
K	6.8	13.4	0.54	27
L	3.5	7.6	0.44	9
M	1.6	1.9	0.73	3
N	7.4	8.5	0.87	16
O	3.6	6.6	0.55	20
P	1.5	2.8	0.51	6
R1	4.2	6.5	0.64	7
R2	2.1	4.7	0.45	3
R3	4.9	6.6	0.73	5
U	5.3	9.3	0.56	30
V	8.98	15.07	0.59	32
W	4.02	9.82	0.38	4
X	9.11	15.83	0.57	35

Absorption properties of various U.S. manufactured brick. (*From Brick Institute of America,* **Pocket Guide to Brick Construction,** ***BIA, Reston, Va.***)

can be controlled by this means, it is not covered in ASTM requirements. It should however, be included in project specifications. Refer to Chapter 15 for field testing of IRA.

Saturation coefficient, or *C/B ratio,* is a measure of the relationship of two aspects of water absorption: the amount freely or easily absorbed and the amount absorbed under pressure. The former (C) is determined by the 24-hour cold water absorption test, and the latter (B) by the 5-hour boil absorption test. For Grade SW brick, the ratio must be 0.78 or less to meet ASTM standards. The C/B ratio determines the volume of open pore space remaining after free absorption has taken place. This is important under severe weathering conditions when a unit has taken in water which must have room to expand if frozen in order to avoid damage to the clay body. The theory does not apply to hollow masonry units or to certain types of de-aired products. In those cases, strength and absorption alone are used as measures of resistance to frost action.

Underburned bricks are easily scratched or scored with a coin, cut with a knife, or broken by hand. ***(Photos courtesy Brick Institute of Texas.)***

3.4.4 Durability

The durability of clay masonry usually refers to its ability to withstand freezing in the presence of moisture, since this is the most severe test to which it is subjected. Compressive strength, absorption, and saturation coefficient are evaluated together as indicators of freeze-thaw resistance since a value cannot be assigned specifically for this characteristic.

Resistance to wear and abrasion are important aspects of durability for brick paving, and for the lining of structures which will carry sewage, industrial waste, and so on. Abrasion resistance is closely associated with the degree of burning, and ranges from underburned salmon brick at the low end to vitrified shale and fire clay at the high end. The stronger the unit, and the lower the absorption, the greater the abrasion resistance will be. In salvaged brick or imported brick, underburned units are easily detected without sophisticated laboratory equipment or procedures. Extremely soft units are easily scratched or scored with a coin, cut with a knife, or even broken by hand (*see Fig. 3-25*). Brick conforming to ASTM standards, however, are high-quality products with proven records of performance in service.

3.4.5 Expansion Coefficients

Clay masonry products are among the most dimensionally stable of building materials. The *coefficient of thermal expansion* ranges from 0.0000025 in./°F for fire clay units to 0.0000036 in./°F for surface clay and shale units. This minute thermal expansion and contraction is reversible. Moisture expansion, however, is not reversible. Fired brick are at their smallest dimension when leaving the kiln. All natural moisture and the water added for forming and extrusion are evaporated during the firing process. Once fired, clay products begin to re-hydrate by absorbing atmospheric moisture, causing irreversible expansion of the units. Test results have assigned a value of 0.02 to 0.07% for the *coefficient of moisture expansion*. Both vertical and horizontal expansion joints must be provided in the masonry to permit this movement. Severe problems can develop when clay masonry expansion is restrained, particularly by concrete elements which have an opposing potential for shrinkage. Jointing details must provide flexible anchorage to accommodate such differential movement. (Refer to Chapter 9.)

3.4.6 Fire and Thermal Resistance

Masonry fire resistance and thermal performance are both determined by mass. The characteristics of the individual units are not considered, but ratings are established for finished wall assemblies. Detailed analysis of these properties is covered in Chapter 8.

3.4.7 Acoustical Characteristics

The density of clay masonry determines its acoustical characteristics. Although sound absorption is almost negligible, the heavy mass provides excellent resistance to the transmission of sound through walls. This suggests best use as partitions or sound barriers between areas of different occupancy. Where higher absorption is required in addition to sound isolation, special acoustical units are used. Acoustical tile was developed to offer 60 to 65% absorption. The unit is a structural facing tile with a perforated face shell. The adjacent cell(s) are factory-filled with a fibrous glass pad. The perforations may be round or slotted and arranged in random or uniform patterns. The tile itself is of loadbearing quality, may be glazed or unglazed, and otherwise exhibits the same properties and characteristics of structural clay facing tile manufactured in accordance with ASTM C212 or C126.

3.4.8 Colors and Textures

Brick and tile are available in an almost unlimited variety of colors and textures. They may be standard items or custom units produced for unique project requirements. Natural clay colors can be altered or augmented by the introduction of various minerals in the mix, and further enhanced by application of a clear, lustrous glaze. Ceramic glazed finishes range from the bright primary colors through the more subtle earth tones in solid, mottled, or blended shades. Glossy, matte, and satin finishes, as well as applied textures, add other aesthetic options (*see Fig. 2-3*).

3.5 ADOBE MASONRY

Adobe masonry is constructed of large, sun-dried bricks made from clay, sand, silt, and water with additives sometimes used as stabilizers. There are no industry standards for reliable soil selection. The National Parks Service performed tests on a number of historic and contemporary adobe structures and found a wide range of soil types, clay contents, and particle sizes. The Uniform Building Code (UBC) Standard 21-9, *Unburned Clay Masonry Units and Standard Methods of Sampling and Testing Unburned Clay Masonry Units,* requires soil with not less than 25% nor more than 45% of material passing a No. 200 mesh sieve, and containing sufficient clay to bind the particles together. Soil can be tested for approximate composition. Place a soil sample in a jar and then fill the jar with water. Shake the mixture and allow it to settle until the water at the top is clear. The resulting bands of coarser aggregates at the bottom of the jar, sand, silt, and clay on top will indicate the approximate proportions of the constituent materials. Another field test for clay content and plasticity is the rope test. Mix a sample of soil with a small amount of water to make a stiff lump of mud. Roll the mud by hand into a rope-like shape. The rope should bend easily without breaking if the soil composition is suitable for making adobes.

Soils more often contain too much rather than too little clay, and can be modified by adding sand, straw, hay, or other vegetable fibers. This tempering process helps minimize the shrinkage cracking which can be caused by using soils with too much clay. Any sand that is added should be sharp, angular sand rather than rounded bank run particles. The proper proportions are usually determined by trial and error and tested by making sample bricks. There is generally a broad tolerance range on sand/clay proportions which produce good quality adobe bricks.

Adobe mixtures are seldom specified. Most often, test bricks are made, dried, and checked. A good mix of clay, sand, silt, and water should be easy to hand-mix by shovel or hoe, should not fall apart when turned into a small mound, should slip free of forms, and should not warp, curl, or crack as the brick dries. The dried brick should not chip or break off at corners when moved, and should be able to withstand 10 to 15 minutes of light to moderate rain with little or no erosion or washing. When broken in half, a unit should exhibit uniform color throughout. Expansive clays such as montmorillinite should be avoided because or their shrink/swell potential, minimized by adding straw to the mix. Kaolin clays are non-expansive.

Traditional additives or stabilizers for adobe have ranged from straw to horsehair, grass, and pine needles. These materials were used primarily as tensile reinforcement to resist shrinkage cracking. More recently, other chemical materials have been used to increase moisture resistance. Emulsified asphalt is most commonly used, mixed with the adobe at a rate of 5 to 8% by weight. The asphalt emulsion coats the clay particles to reduce natural moisture absorption. Portland cement is also sometimes added to adobe soils. This will increase the compressive strength of the bricks, but will not improve moisture resistance. Lime should not be used as a soil additive.

3.5.1 Manufacturing Adobe

Adobe bricks are usually rectangular in shape with a length that is twice the width. Sizes range from 4 in. high×9×18 in. to 6 in. high×12×24 in., but the latter units are very heavy. In New Mexico, the majority of adobe bricks are cast at 4 in. high×10×14 in. and weigh about 30 lb each. The maximum practical weight for handling by a single person is about 40 lb.

Adobe manufacturing can range from small, on-site operations to fully or partially mechanized commercial operations (*see Fig. 3-26*). A two-person crew can turn out 300 to 400 bricks per day by hand. A fully mechanized operation may produce as many as 20,000 bricks per day using a pug mill for mixing, self-propelled molding and screeding machines, and permanent steel molds.

Molds are most commonly made of wood, but can also be of metal. Wood molds must be oiled or wetted to prevent sticking. All molds must be cleaned frequently to remove dried mud. The molds are laid on flat or leveled ground which can be covered with sand to prevent the bricks from sticking to the ground. The consistency of the mud mix may be either damp or liquid, depending on the method of mixing and the number of molds available. Liquid mixes generally produce stronger and more dense

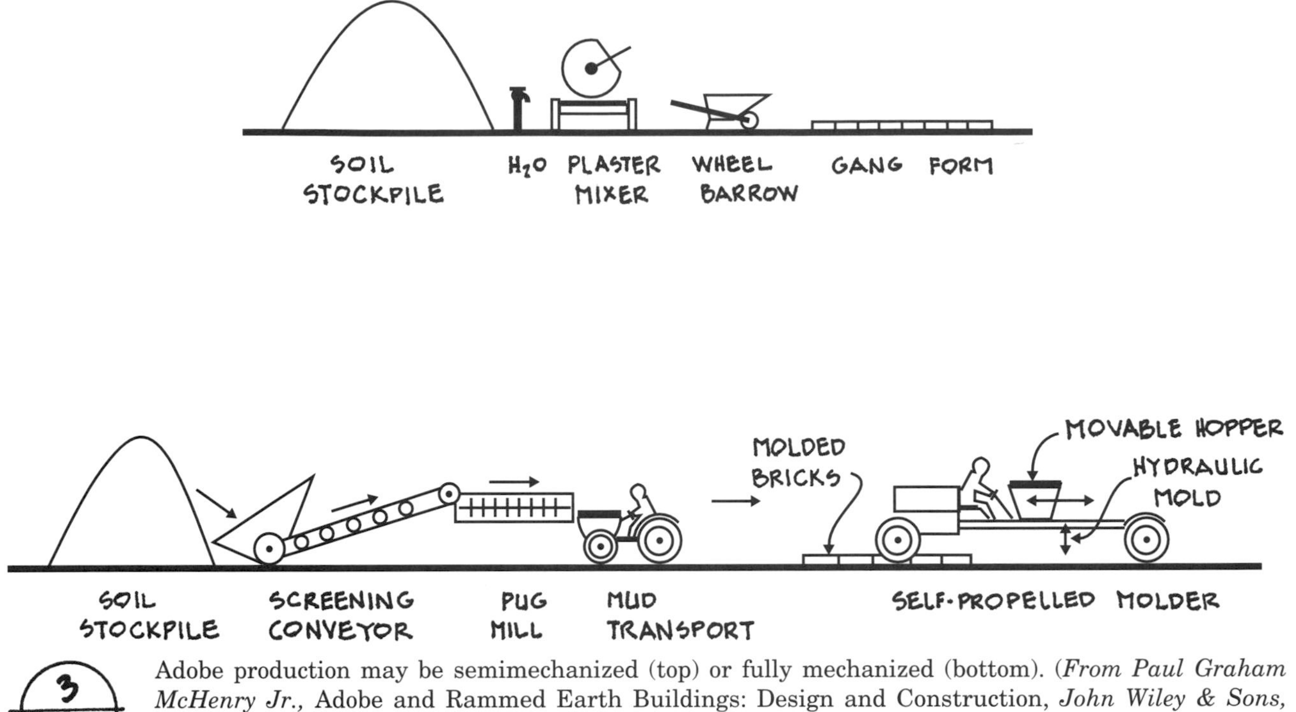

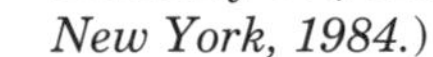

Adobe production may be semimechanized (top) or fully mechanized (bottom). (*From Paul Graham McHenry Jr.,* Adobe and Rammed Earth Buildings: Design and Construction, *John Wiley & Sons, New York, 1984.*)

bricks, but since initial set takes longer, it will be necessary to use large gang molds in order to produce in sufficient quantities (*see Fig. 3-27*). Molds can be removed from damp, stiff mixes more quickly and reused right away. After the mold is filled, the top is leveled and screeded. Stiff mixes may require tamping to assure complete filling of the mold corners.

After the molds have been removed, the bricks must remain flat until they are dry enough to handle. This initial drying may take anywhere from 2 to 3 days in the summer or several weeks in the winter, depending on the moisture content of the mix. During this period, the units should be protected from rain by covering as necessary with tarps or plastic sheeting. The covers will retard the drying and curing process, though, so they should be removed as soon as possible. When the bricks are dry enough to be handled without breaking, they are set on edge to expose both large surfaces for better drying. At this stage, any cleaning or trimming that is necessary can be performed. The length of the drying and curing period will vary depending on the size and thickness of the unit and the weather conditions. Freezing temperatures during the drying period can destroy the units, and many U.S. companies suspend production during the winter months. After the bricks have dried for at least 7 days, they are stacked on edge two units deep and three or four courses high. The top of the stack should be protected from rain, but the sides left open. With such protection, the units can remain in storage in this manner for extended periods of time.

3.5.2 Physical Properties and Characteristics

Finished bricks can be simply tested in a couple of different ways. The point of a knife blade should not penetrate the surface of a fully dried

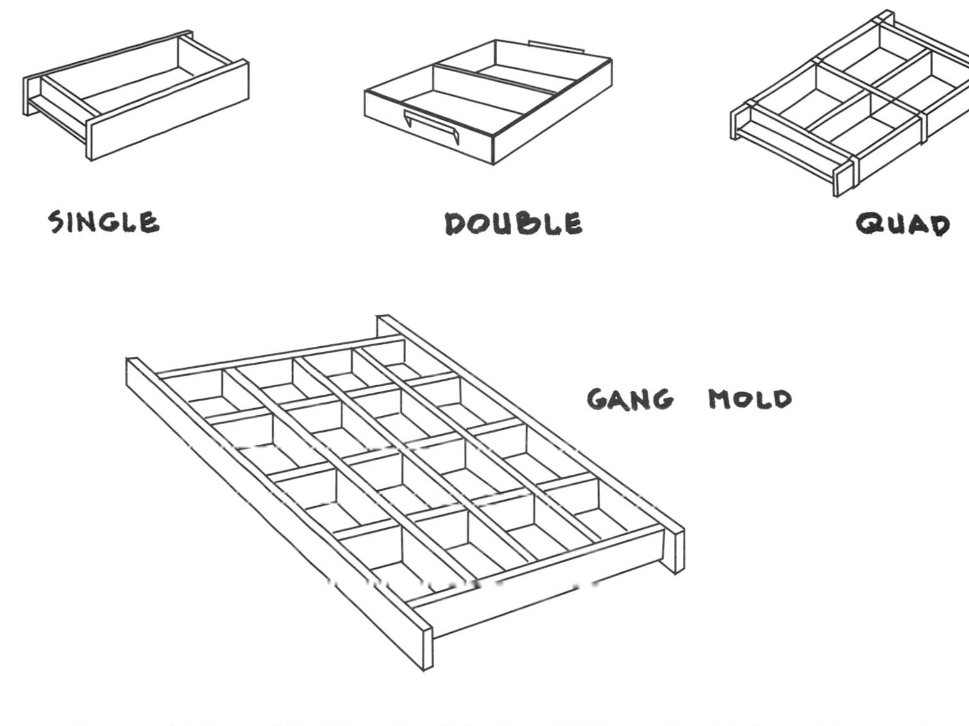

Adobe molds. (*From Paul Graham McHenry Jr.,* Adobe and Rammed Earth Buildings: Design and Construction, *John Wiley & Sons, New York, 1984.*)

Unit type	Minimum net area compressive strength		Water absorption*	Moisture content*	Minimum modulus of rupture	
	Average of five tests	Individual unit			Average of five tests	Individual unit
Unburned clay masonry	300	250	2.5%	4%	50	35

*Based on percentage of dry weight.

UBC Standard 21-1 Physical Requirements for Unburned Clay Masonry Units. (*From UBC Standards, International Conference of Building Officials, 1994.*)

brick more than about ⅛ in. If the unit appears dry on the surface but is still damp on the inside, the knife blade will penetrate deeply. A finished brick can also be dropped on its corner from a height of about 3 ft. A thoroughly dried and cured brick of good quality will suffer little or no damage other than minor chipping at the corner. If the unit is not totally dry, it will shatter. If the soil was not properly and thoroughly mixed to homogeneity, it will split along the planes of weakness.

UBC Standard 21-1, *Building Brick, Facing Brick and Hollow Brick (Made From Clay or Shale)*, requires that finished adobe brick have the minimum physical properties indicated in *Fig. 3-28.*

3.5.3 Adobe Mortar

Although adobe can be laid with an ASTM C270 Type S portland cement-lime mortar, it is more common to use an adobe mortar made from the same soil mix as the units. When laid with adobe mortar, a wall that is subjected to cracking stresses will crack monolithically through the units. When laid with a traditional masonry mortar, the same wall will crack at the mortar joints. Maximum aggregate size in adobe mortar should be $\frac{1}{4}$ in., and the mortar should be thoroughly soaked and mixed to prevent clay balling.

4

CEMENTITIOUS MASONRY UNITS

Cementitious masonry units are hardened by chemical reactions rather than by ceramic fusion. This group includes concrete brick and block as well as sand-lime brick and cast stone. The majority of these units are classified as solid, having less than 25% core area in relation to the gross cross section in the bearing plane. Concrete block, however, typically has 40 to 50% coring and is thus defined as hollow.

4.1 CONCRETE BRICK

Concrete brick are produced from a controlled mixture of portland cement and aggregates in sizes, colors, and proportions similar to clay brick. They are governed by the requirements of ASTM C55, *Standard Specification for Concrete Building Brick,* and can be loadbearing or non-loadbearing. Aggregates include gravel, crushed stone, cinders, burned clay, and blast-furnace slag, producing both normal-weight and lightweight units. Coring or "frogging" may be used to reduce weight and improve mechanical bond.

Grading is based on strength and resistance to weathering. Grade N provides high strength and maximum resistance to moisture penetration and frost action. Grade S has only moderate strength and resistance to frost action and moisture penetration. Each grade may be produced as Type I, moisture-controlled, or Type II, non-moisture-controlled, designated as N-I, N-II, S-I, or S-II. For moisture-controlled units, the standards limit the moisture content at the time of construction according to the relative humidity conditions typical of the project site's geographic location. These requirements establish predictable shrinkage characteristics so that proper allowance can be made for control joints in the structure (*see Fig. 4-1*).

Concrete mixes may be altered to produce a slight roll or slump when forms are removed, creating a unit similar in appearance to adobe brick. Color is achieved by adding natural or synthetic iron oxides, chromium oxides, or other pigments to the mix, just as in colored mortar (refer to Chapter 2). ASTM standards do not include color, texture, weight classifi-

TABLE A MOISTURE CONTENT REQUIREMENTS FOR TYPE I CONCRETE BRICK			
	Moisture content, max. percent of total absorption (average of 3 concrete brick)		
	Humidity conditions at job site or point of use*		
Linear shrinkage (%)	*Humid*	*Inter-mediate*	*Arid*
0.03 or less	45	40	35
From 0.03 to 0.045	40	35	30
0.045 to 0.065, max.	35	30	25

*Arid, average annual relative humidity less than 50%; intermediate, average annual relative humidity 50 to 75%; humid, average annual relative humidity above 75%.

TABLE B STRENGTH AND ABSORPTION REQUIREMENTS					
	Compressive Strength, min. (psi) (concrete brick tested flatwise)		*Water absorption, max. (average of 3 brick) with oven-dry weight of concrete (lb/ft^3)*		
	Average gross area		*Weight classification*		
Grade	*Average of 3 concrete brick*	*Individual concrete brick*	*Lightweight (less than 105)*	*Medium weight (less than 125 to 105)*	*Normal weight (125 or more)*
N-I N-II	3500	3000	15	13	10
S-I S-II	2500	2000	18	15	13

Requirements for concrete brick, from ASTM C55. (*Copyright, American Society for Testing and Materials, 1916 Race Street, Philadelphia, Pa. 19103. Reprinted with permission.*)

cation, or other special features. These properties must be covered separately in the project specifications.

4.2 SAND-LIME BRICK

Calcium silicate brick, or sand-lime brick, are made with sand or other siliceous material and 5 to 10% hydrated lime, then steam-cured in high-pressure autoclaves at 400°F for up to 8 hours. In the autoclave, the lime reacts chemically with the silica to form hydrated calcium silicate, a strong and durable cementing agent that binds the sand particles together.

Calcium silicate brick are widely used in industrial countries where suitable siliceous sands are more readily available than clay. In the United States, calcium silicate brick has been produced since the early 1900s. The units have a natural near-white color with a slight yellow, gray, or pink tint depending on the color of the sand used. With the addition of natural or synthetic pigments, dark earth tones, reds, blacks, and light pastel colors can be produced, including blues and greens. Two colors can be blended for a swirled mixture, or units can be dipped in acid after hardening to intensify their color. Unit surfaces are smooth and uniform—the finer the sand particles, the smoother the surface. Texture is produced by sandblasting, mechanical brushing, or adding flint aggregates to the mix. Splitting hardened units produces a natural rockface finish.

Sand-lime brick are used extensively in Europe, Russia, Australia, the Middle East, Mexico, and the United Kingdom. Most U.S. building codes permit their use in the same manner as clay brick for both loadbearing and non-loadbearing applications. ASTM C73, *Standard Specification for Calcium Silicate Face Brick (Sand-Lime Brick)*, includes grading standards identical to those for clay face brick for severe weathering (Grade SW) and moderate weathering (Grade MW). Compressive strength minimums are 4500 and 2500 psi, respectively, and absorption rates are 10 and 13%, respectively. Strength and hardness are increased as carbon dioxide in the air slowly converts the calcium silicate to calcium carbonate.

Alternate wetting and drying, and repeated freeze-thaw cycles have little effect on calcium silicate brick, and efflorescence is not a problem because the raw materials do not ordinarily contain soluble sulfates or other salts. Sand-lime brick is also resistant to attack when in contact with soils containing high levels of sulfates. As with all limestone-based products, the sulfur dioxide in heavily polluted air affects the brick after 20 to 30 years. And like cement-based products, calcium silicate brick are not resistant to acids or repeated exposure to saturated salt water solutions.

At 100 to 300 lb/cu ft, density is similar to that of a medium-density clay brick. Using expanded aggregates reduces density to about 80 lb/cu ft, and coring or perforations can reduce overall weight. Both sound transmission and fire resistance are similar to those of clay brick, but ordinary calcium silicate units can also be used in flues, chimney stacks, and other locations requiring moderate refractories.

Calcium silicate units are produced in modular brick sizes as well as larger blocks measuring from 8×8×16 in. to face sizes of 12×24 or 24×36 in. with a bed thickness of 4 to 12 in. The larger block units are widely used in some European countries to increase labor production. Single or multi-wythe walls can be erected using small, transportable electric cranes and mechanical grippers. The units are self-aligning, either by tongue-and-groove slots or by mechanical plugs, and vertical reinforcing steel can be incorporated through core holes and end grooves.

4.3 CAST STONE

Cast stone is most widely used as an accessory for masonry construction in the form of lintels, sills, copings, and so on. Some manufacturers also produce simulated stone products designed for use as facing materials. The shape of the mold used for casting will determine the appearance of the unit. Any shape which can be carved in natural stone can probably also be formed in cast stone.

Cast stone is a highly refined architectural precast concrete which is manufactured to resemble natural cut and dressed building stone. Unlike "simulated stone" produced in random sizes as cleft-face quarried stone, cast stone exhibits the same finish as a good grade of limestone or brownstone which has been cut or honed.

The French made lintels and door trim out of cast stone as early as the twelfth century. Today, cast stone is made of a carefully proportioned mix containing natural gravel, washed and graded sand, and crushed and graded stone such as granite, marble, quartz, or limestone meeting requirements of ASTM C33, *Standard Specification for Concrete Aggregates.* White portland cement usually is used to produce light colors and color consistency, although gray cement and color pigments are sometimes blended with the white cement. Because a rich cement-aggregate ratio of 1:3 is normally used, cast stone properly cured in a warm, moist environment is dense, relatively impermeable to moisture, and has a fine-grained, natural texture. Cast stone is relatively heavy at 144 lb/cu ft, and at a minimum of 7000 psi, compressive strength is higher than ordinary cast-in-place concrete. Maximum water absorption should be 6%.

Cast stone is made by two methods. In the wet cast method, stone is cast in much the same way as other architectural precast concrete. If most of the stone surfaces are to be flat, the concrete is vibrated with external vibrators. If the stone is highly ornamental, it is vibrated with internal vibrators. The stone is then cured in the mold until the next day when it is stripped. In the vibrant dry-tamp method of manufacture, a pneumatic machine is used to ram and vibrate moist, zero-slump concrete against rigid formwork. When the concrete is densely compacted, it is removed from the form and left to cure overnight.

To ensure that the stone undergoes little change in appearance due to weathering, the outer surface of mortar is removed to expose the fine aggregates. Hydrochloric acid is used to etch the surface because it produces the most brilliant colors and leaves a surface that stays clean. Sandblasting and chemical retarders which are normally used to finish architectural precast are not used on cast stone because they dull the aggregates and cause the loss of fine detail.

Two industry test standards are available which can be used for laboratory investigation of physical characteristics. ASTM C1194, *Compressive Strength of Architectural Cast Stone,* and ASTM C1195, *Absorption of Architectural Cast Stone,* provide standardized test methods for evaluating strength and absorption characteristics for comparison of different materials.

Although some cast stone manufacturers produce and stock standard items of architectural trim such as balusters, door pediments, and balcony rails, cast stone is more often custom-designed and -fabricated for each project. For greatest economy, design shapes should be tailored to the fabrication process. Projections should be slightly angled rather than flat to facilitate removal of the molds (*see Fig. 4-2a*). The length of projections also should not exceed their thickness (*see Fig. 4-2b*). Transitions from a finished to an unfinished surface should be sharp angles rather than feathered edges (*see Fig. 4-2c*). L- and U-shaped sections are vulnerable to breakage during fabrication and shipment. Returns should be formed instead by butt-jointing or mitering two pieces (*see Fig. 4-2d*).

4.4 CELLULAR CONCRETE BLOCK

Cellular concrete block has excellent insulating, sound-damping, and fire-resistive properties, but is relatively low in compressive strength (*see Fig. 4-3*) and must be covered with stucco or other weather-resistant treatments. The units weigh only one-fourth to one-third as much as normal concrete block, but not because of lightweight aggregates. The mix contains portland cement, lime, sand or fly ash, and aluminum powder, with water added to form a slurry. Large steel vats are used as molds. A chemical reaction releases hydrogen gas and generates heat, which causes the concrete to expand and set in cellular form. Smaller units are wire-cut or

4/2 Guidelines for designing cast stone shapes. (*From "Working with Cast Stone," Ken Hooker,*) Aberdeen's Magazine of Masonry Construction, *October 1995.*)

TABLE A COMPRESSIVE STRENGTH	
Density (lb/cu ft)	Compressive strength (psi)
19	72–232
25	189–464
31	290–725
37	435–870

TABLE B THERMAL RESISTANCE	
Density (lb/cu ft)	R-value per inch of thickness
25	1.66
31	1.14

TABLE C FIRE RESISTANCE	
Wall thickness, (in.)	Fire rating (hours)
6	6
4	4
2.5	2
2	1

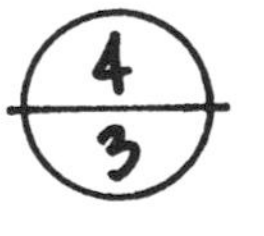

Properties of cellular concrete block. (*Source: North American Cellular Concrete Company.*)

saw-cut from the large forms and curing is completed under steam pressure in autoclave kilns. The material is sometimes referred to as autoclaved cellular concrete.

Cellular concrete block is laid up in mortar, typically with joints that are only ⅛ in. thick. It can be used for bearing walls in low-rise construction, for interior partition walls, as lightweight fireproofing for steel structural frames, and as acoustical partitions. Cellular concrete block can be cut or sawed with ordinary woodworking tools and is also nailable. Although it is a popular material in most parts of the world, cellular concrete block has never caught on in the United States. Two companies are currently re-introducing the material in the United States, one using sand aggregate and the other using fly ash. One of the manufacturers also plans to offer higher strength block as well as roof, floor, and wall panels.

4.5 CONCRETE BLOCK

Of the cementitious masonry products marketed in this country, concrete block is the most familiar and most widely used. Aggregates determine the weight of the block and give different characteristics to the units. Lightweight aggregates reduce the weight by as much as 20 to 45% with little or no sacrifice in strength. Specifications for aggregates are covered in ASTM C33, *Standard Specification for Concrete Aggregates,* and ASTM C331, *Standard Specification for Lightweight Aggregates for Concrete Masonry Units.* Weight classifications are based on density of the concrete and are

Classi-fication	Aggregate	Unit weight of concrete (pcf)	Average weight of 8×8×16 unit (lb)
Normal weight	Sand and gravel aggregate	135	44
	Crushed stone and sand aggregate	135	40
Medium weight	Air-cooled slag	120	35
Light weight	Coal cinders	95	28
	Expanded slag	95	28
	Scoria	95	28
	Expanded clay, shale, and slate	85	25
	Pumice	75	22

Weight variations of concrete with different aggregates.

subdivided as follows: normal-weight units are those whose concrete mix weighs more than 125 lb/cu ft; medium weight units, between 105 and 125 lb/cu ft; and lightweight units, less than 105 lb/cu ft.

Some of the more commonly used aggregates are listed in *Fig. 4-4* along with the concrete unit weight and weight classifications. Exact individual unit weights depend on the coring design of the block and the percentage of solid volume and voids. An ordinary 8×8×16-in. unit weighs 40 to 50 lb when made from the more dense aggregates, and 25 to 35 lb when made from the lighter aggregates. Manufacturers can supply information regarding exact weight of their products, or the figures may be calculated if the percent of solid volume is known. Both heavy and lightweight block can be used in any type of construction, but lightweight units have higher fire, thermal, and sound resistance. Choice of unit will depend largely on local availability and project design requirements. Two kinds of concrete block are recognized—ASTM C90, *Standard Specification for Loadbearing Concrete Masonry Units,* and ASTM C129, *Standard Specification for Non-Loadbearing Concrete Masonry Units.*

Units defined as solid must have a minimum of 75% net solid area. Although the industry has standardized exterior dimensions of modular units, no such standardization exists for the number, size, or configuration of cores. Coring design and percent of solid volume vary depending on the unit size, the equipment, and the methods of the individual manufacturers. For structural reasons, ASTM standards for loadbearing units specify minimum face shell and web thickness, but these stipulations do not apply for non-loadbearing units. Although minimum face shell and web thickness will not necessarily correspond to actual dimensions for all units, they can be used to estimate properties for preliminary design (*see Fig. 4-5*).

4.5.1 Coring

Block is produced in two-core and three-core designs and with smooth or flanged ends (*see Fig. 4-5*). Two-core designs offer several advantages, including a weight reduction of approximately 10%, and larger cores for

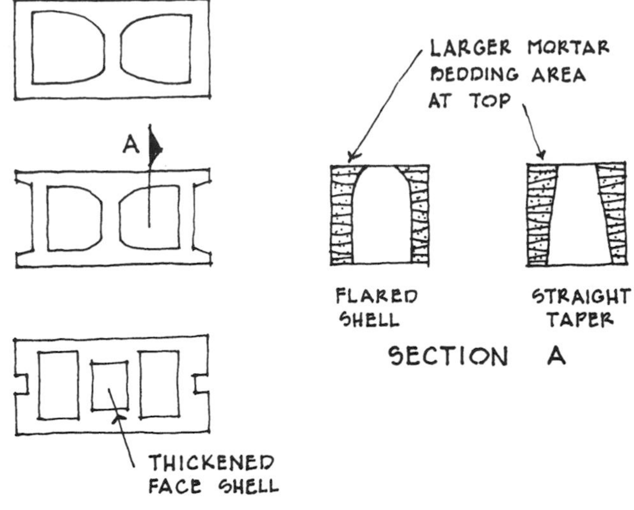

(A)

		Web thickness (WT)	
Nominal width, W, of units (in.)	Face-shell thickness (FST), min. (in.)*	Webs,* min. (in.)	Equivalent web thickness, min. (in./lin ft)†
3 and 4	$\frac{3}{4}$	$\frac{3}{4}$	$1\frac{5}{8}$
6	1	1	$2\frac{1}{4}$
8	$1\frac{1}{4}$	1	$2\frac{1}{4}$
10	$1\frac{3}{8}$ $1\frac{1}{4}$‡	$1\frac{1}{8}$	$2\frac{1}{2}$
12	$1\frac{1}{2}$ $1\frac{1}{4}$‡	$1\frac{1}{8}$	$2\frac{1}{2}$

*Average of measurements on 3 units taken at the thinnest point.

†Sum of the measured thickness of all webs in the unit, multiplied by 12, and divided by the length of the unit.

‡This face-shell thickness (FST) is applicable where allowable design load is reduced in proportion to the reduction in thickness from basic face-shell thicknesses shown.

(B)

Concrete block coring: (A) Various concrete block core patterns; (B) table of minimum thickness of face shells and webs (ASTM C90 units only). (*From National Concrete Masonry Association,* TEK Bulletin 36, *NCMA, Herndon, Va.*)

the placement of vertical reinforcing steel and conduit. In addition, the thickened area of the face shell at the center web increases tensile strength and helps to reduce cracking from drying shrinkage and temperature changes. Accurate vertical alignment of both two-core and three-core designs is important in grouted and reinforced construction. End designs of block may be smooth or flanged, and some also have a mortar key or groove for control joints and jamb units. Smooth face ends must be used for corner construction, piers, pilasters, and so on. The cores of hollow

Width (in.)	Gross volume [cu in. (cu ft)]	Minimum thicknesses		Three-core units		Two-core units	
		Shell (in.)	Web (in.)	Percent solid volume	Equivalent solid thickness (in.)	Percent solid volume	Equivalent solid thickness (in.)
$3\frac{5}{8}$	432 (0.25)	0.75	0.75	63	2.28	64	2.32
		1.00	1.00	73	2.66	73	2.66
$5\frac{5}{8}$	670 (0.388)	1.00	1.00	59	3.32	57	3.21
		1.12	1.00	63	3.54	61	3.43
		1.25	1.00	66	3.71	64	3.60
		1.37	1.12	70	3.94	68	3.82
$7\frac{5}{8}$	908 (0.526)	1.25	1.00	56	4.27	53	4.04
		1.37	1.12	60	4.57	57	4.35
		1.50	1.12	62	4.73	59	4.50
$9\frac{5}{8}$	1145 (0.664)	1.25	1.12	53	5.10	48	4.62
		1.37	1.12	55	5.29	51	4.91
		1.50	1.25	58	5.58	54	5.20
$11\frac{5}{8}$	1395 (0.803)	1.25	1.12	49	5.70	44	5.12
		1.37	1.12	51	5.93	46	5.35
		1.50	1.25	54	6.28	49	5.70
		1.75	1.25	57	6.63	52	6.05

Volume characteristics of typical concrete masonry units (based on 7⅝ in. height × 15⅝ in. length). (*From National Concrete Masonry Association,* TEK Bulletin 2A, *NCMA, Herndon, Va.*)

units are usually tapered, with the face-shell thickness wider at the top than at the bottom of the unit. This facilitates form removal, provides a larger bedding area for mortar, and a better grip for the mason. Minimum thickness required by ASTM standards refers to the narrowest cross section, not an average thickness of top and bottom. Since compressive strengths of hollow units are established on the basis of gross area, and fire-resistance ratings on equivalent solid thickness, these details of unit design become important in determining actual ratings for a particular unit (*see Fig. 4-6*).

Hollow concrete masonry units (CMUs) are more widely used than solid units because of reduced weight, ease of handling, and lower cost. Most hollow blocks have core areas of 40 to 50%, leaving a net solid volume of 50 to 60%. Some concrete brick manufacturers have begun to capitalize on this economy by producing a cored "through-the-wall" unit that has an increased thicknesses of 8 in., but maintains the typical face dimensions of brick. They may be classified as either solid or hollow depending on the percentage of voids created.

4.5.2 Grading

Durability under freeze-thaw conditions is taken as a measure of the absorption characteristics of the unit. Loadbearing concrete masonry units were formerly graded in the same manner as concrete brick. Grade N units provided higher strength and resistance to frost action, and Grade S provided only moderate strength and resistance to freezing. The separate grading requirements of ASTM C90 have been eliminated, and all load-bearing units must now have the higher requirements formerly classified as Grade N (*see Fig. 4-7*). ASTM C129 does not list absorption requirements for non-loadbearing units.

4.5.3 Moisture Content

For both solid and hollow, loadbearing and non-loadbearing block, a Type I designation represents "moisture-controlled" units, and Type II, "non-moisture-controlled" units (*see Fig. 4-7*). Since concrete shrinks with water loss, limits on unit moisture content were established to control the potential for shrinkage cracking. When moist units are built into a wall and shrinkage is restrained, tensile and shearing stresses develop which may cause cracking. Allowable moisture content depends not only on the relative humidity of the project location, but on the shrinkage characteristics of the particular block. In order to equalize the expected drying shrinkage, units made from dense materials with lower shrinkage characteristics are allowed higher moisture contents than are lighter-weight units with higher shrinkage characteristics. The table in *Fig. 4-8* lists the average linear shrinkage for various aggregate types and the corresponding ASTM moisture values permitted.

Type I block that are unprotected during storage at the job site will absorb rainwater, and can no longer be considered as Type I block. When Type I block are not available, or when protection cannot be provided during storage and transport, control joint spacing should be adjusted as for a Type II block (refer to Chapters 10 and 15). All concrete masonry units (Type I and Type II) should be stored off the ground and covered with weatherproof coverings to prevent wetting and increased shrinkage. Partially completed walls should also be covered at the top when work is not in progress so that rain and snow do not cause excessive wetting of the masonry.

4.6 UNIT TYPES

Concrete masonry units are governed by the same modular standards as clay masonry products. The basic concrete block size is derived from its relationship to modular brick. A nominal 8×8×16-in. block is the equivalent of two modular bricks in width and length, and three brick courses in height. Horizontal ties may be placed at 8- or 16-in. vertical intervals with either brick or structural clay tile facing. These are nominal dimensions that include allowance for a standard $^3/_8$-in. mortar joint. Concrete brick dimensions are the same as clay brick, but fewer sizes are generally available. Some variation in face size of standard concrete block stretcher units has been introduced to increase productivity on the job. Both the 12-in.-high×16-in.-long and the 8-in.-high×24-in.-long units have 50% larger face area. To compensate for the additional size, lighter-weight aggregates are used to yield an 8-in.-thick unit weighing only 33 lb (less than a normal-weight 8×8×16-in. block). Each of the larger units can be laid as easily as a standard block, but covers 50% more wall area. These oversize units, however, are not typical. Size variation in most concrete block is limited to 2-in. incremental widths of 4 to 12 in., with a standard face size of 8×16 in. (*see Fig. 4-9*). Half-lengths and half-heights are available for special conditions at openings, corners, and so on. A number of special shapes have been developed for specific structural functions, such as lintel blocks, sash blocks, pilaster units, and control joint blocks

TABLE A MOISTURE CONTENT REQUIREMENTS FOR HOLLOW LOADBEARING, SOLID LOADBEARING, AND HOLLOW NON-LOADBEARING UNITS (ASTM C90 AND C129).			
	Moisture content, max. percent of total absorption (average of 3 units)		
	*Humidity conditions at job site or point of use**		
Linear shrinkage (%)	*Humid†*	*Interme-diate‡*	*Arid§*
0.03 or less	45	40	35
From 0.03 to 0.045	40	35	30
0.045 to 0.065, max.	35	30	25

*See map in Fig. 9-4.
†Average annual relative humidity above 75%.
‡Average annual relative humidity 50 to 75%.
§Average annual relative humidity less than 50%.

TABLE B STRENGTH REQUIREMENTS FOR HOLLOW NON-LOADBEARING UNITS (ASTM C129).	
	Compressive strength (average net area), min. (psi)
Average of 3 units	600
Individual unit	500

TABLE C STRENGTH AND ABSORPTION REQUIREMENTS FOR HOLLOW LOADBEARING UNITS (ASTM C90).					
Compressive strength, min. (psi)		*Water absorption, max. (average of 3 units) with oven-dry weight of concrete (lb/ft³)*			
Average net area		*Weight classification (lb/ft³)*			
		Light weight			
Average of 3 units	*Individual unit*	*Less than 85*	*Less than 105*	*Medium weight (105 to less than 125)*	*Normal weight (125 or more)*
1900	1700	—	18	15	13

Note: To prevent water penetration, protective coating should be applied on the exterior face of basement walls and when required on the face of exterior walls above grade.

Requirements for concrete masonry, from ASTM C90 and C129. *(Copyright, American Society for Testing and Materials, 1916 Race Street, Philadelphia. Pa. 19103. Reprinted with permission.)*

Aggregate type	Steam curing method	Average Linear shrinkage* (%)	Maximum moisture content, as percent of total absorption (average of 3 units) for average humidity conditions at job site†‡		
			Humid	Inter-mediate	Arid
Normal weight	Autoclave	0.019	45	40	35
	low-pressure	0.027	45	40	35
Light weight	Autoclave	0.023	45	40	35
	low-pressure	0.042	40	35	30
Pumice	Autoclave	0.039	40	35	30
	low-pressure	0.063	35	30	25

*As determined by test, ASTM C426.
†From Table A, Fig. 4-7
‡See map in Fig. 9-4.

Actual shrinkage and permissible moisture content for various aggregate units.

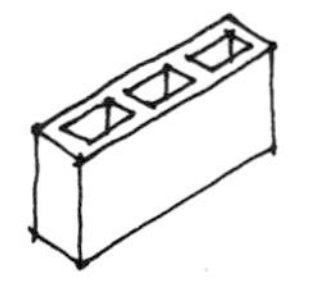
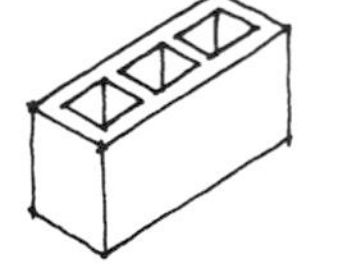
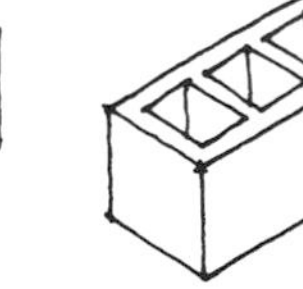
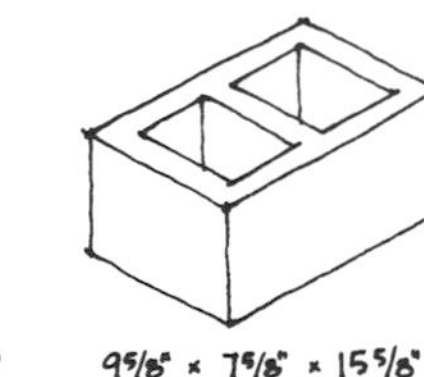
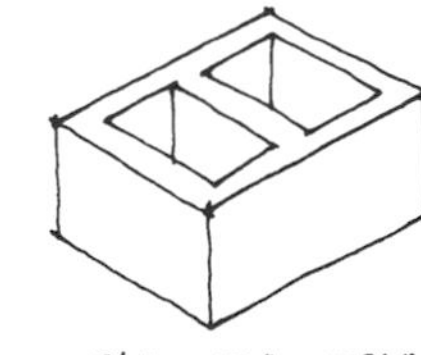
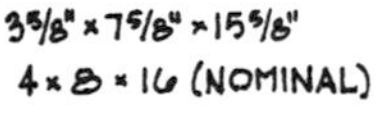
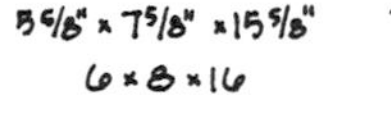

Standard-size stretcher units.

(*see Fig. 4-10*). Terminology is not fully standardized, and availability will vary, but most manufacturers produce and stock at least some of the more commonly used special items. In the absence of such shapes, however, standard units can be field-cut to accommodate many functions.

Innovations in special concrete block shapes are increasing the versatility of concrete masonry. The Innovative Design Research (IDR) Division of the National Concrete Masonry Association (NCMA) has developed several new shapes for special applications such as footings, roof ballast, retaining walls, floors, thermal mass for wood stud walls, and blocks that accommodate horizontal wiring and plumbing runs.

Used instead of poured footings, *footer block* are laid dry, without mortar, on a solid base of gravel or sand. Adjacent blocks interlock, and slots in the top accept steel reinforcement (*see Fig. 4-11*). Mortar is placed on top of the blocks to provide a level bed for the foundation wall. When laid on soil with a 2000-psf bearing capacity, the 16-in.-wide footer blocks used for sup-

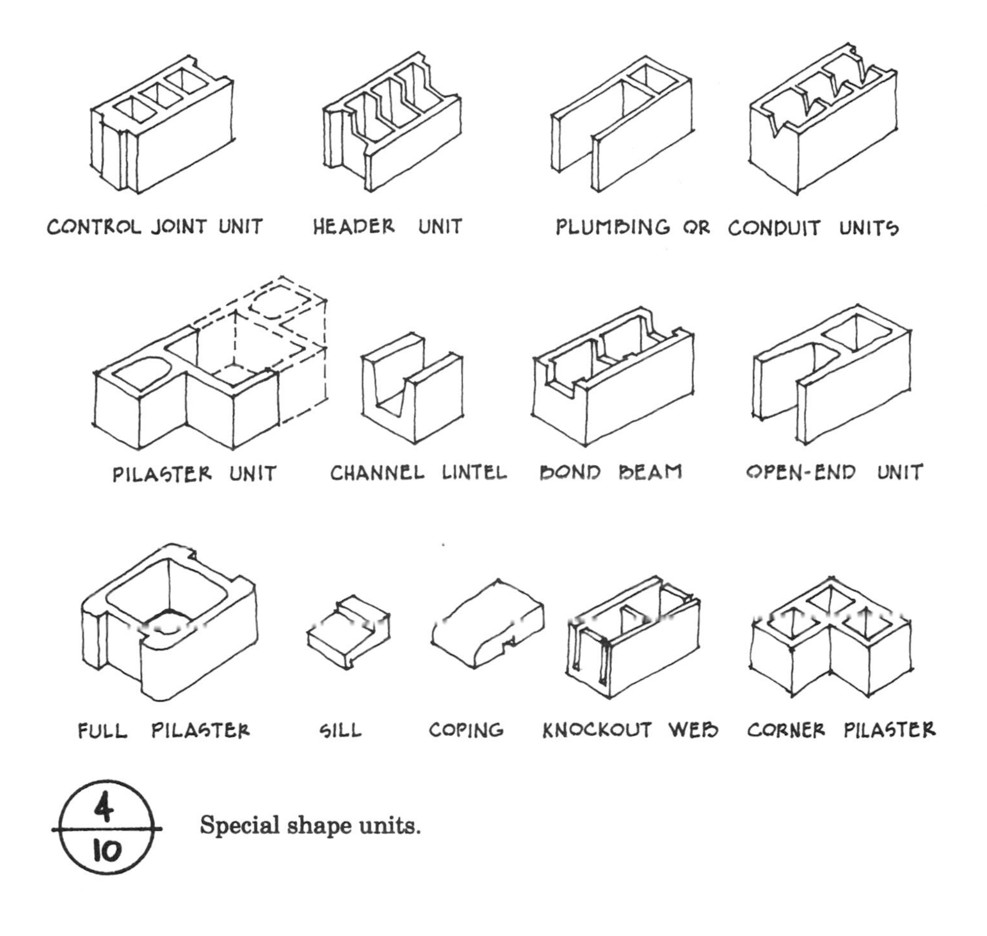

4-10 Special shape units.

porting 8-in. block walls can carry 2000 lb/lin ft, while the 20-in.-wide units can carry 3300 lb/lin ft. By spacing the footers about 1/4 -in. apart, they also may help vent radon gas from below the slab.

Ballast units designed for single-ply roof applications can resist wind speeds in excess of 120 mph. The blocks can be interlocked with optional mechanical connectors, but even without them, the patterned surface of the units reportedly disrupts wind flow and minimizes uplift pressures at corners and perimeters. These pavers weigh less than gravel ballast, and built-in drainage channels prevent water ponding and promote rapid drying. *Biaxial units* permit installation of wiring and plumbing in single-wythe walls. Each web of the block has a round cutout so that wires, conduit, and pipes can be passed longitudinally through the wall. To insert outlets and switches, some block are made with modular openings in the face shell for special multi-use utility insert boxes. Specially shaped *heat soak units* are designed to fit between wall studs in ordinary wood frame construction to absorb solar heat and reduce indoor temperature fluctuations.

Other proprietary specialty units include form blocks for pouring grouted and reinforced walls, flashing blocks, angled units for making 45° corners and intersections, and special block for laying curved walls. Another proprietary design incorporates channels in the block webs to accommodate reinforcing bars and hold them in place without the need for spacers. Still others offer cornice, sill, and water table units, inspection blocks for grouting, angled keystone block for arches, and others (*see Fig. 4-11*). Specialty block are usually patented designs and may not be available in all areas.

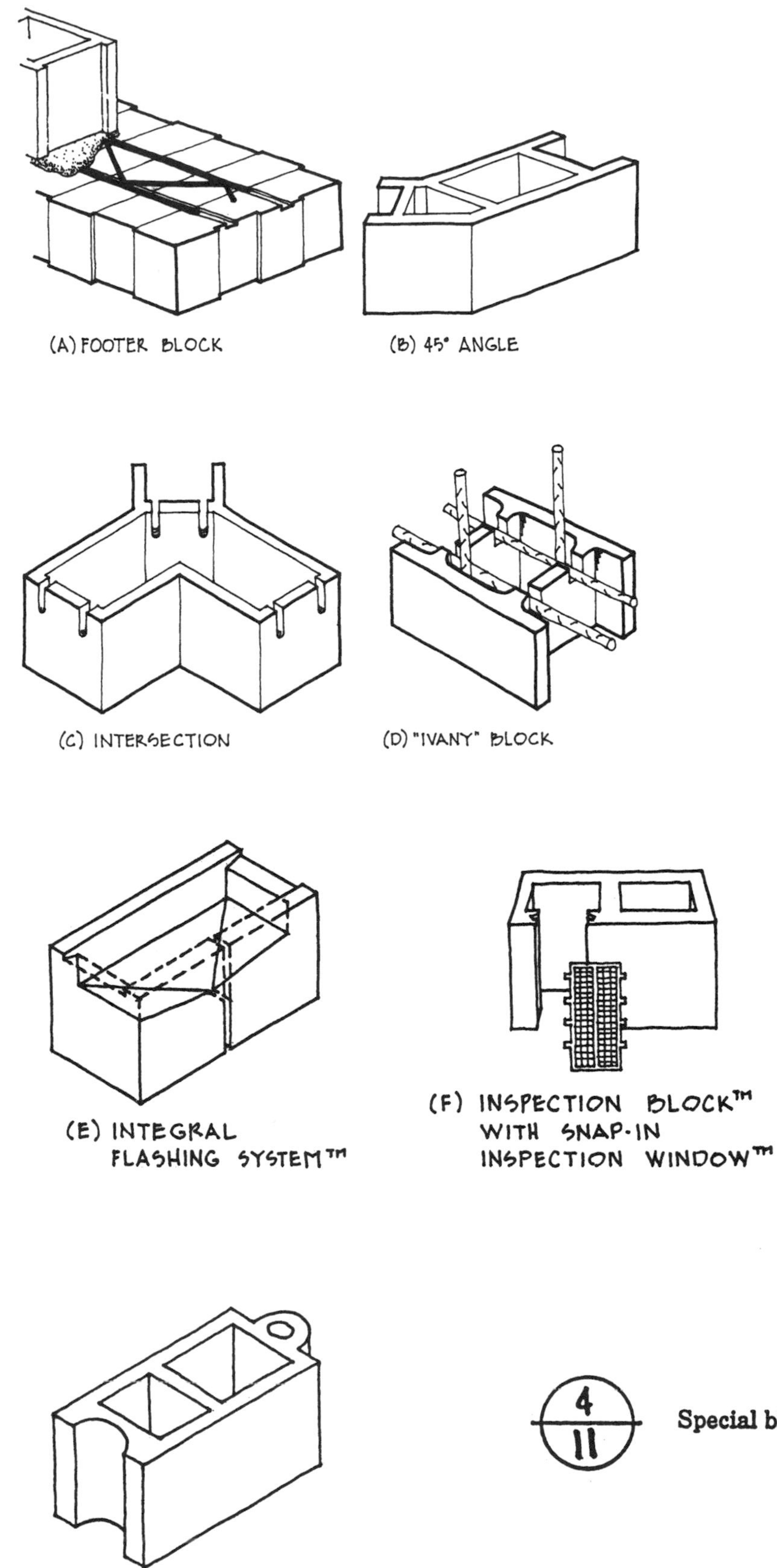

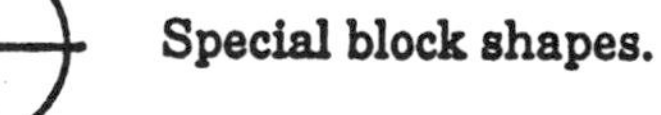

Special block shapes.

4.6.1 Decorative Units

Many decorative effects can be achieved through various CMU surface treatments. Perforated screen block are available in several patterns and can be used as sun screens, ornamental partitions, and exterior sound baffles for damping low-frequency airborne noise (*see Fig. 4-12*). Ribbed, grooved, and fluted faces laid in various bonding patterns can create effects scarcely resembling ordinary concrete block masonry. Split units simulate the effect of rough quarried stone. Surface grinding gives a smooth, burnished appearance that enhances the color of the natural aggregate. Slump block units made with warm-tone cement and light-color aggregates give an irregular, hand-made adobe brick look. Customized designs have been created by many architects in shapes, patterns, and sculptured faces particularly suited to their projects, some with loadbearing capability and others strictly for veneer (*see Figs. 4-13 through 4-16*).

Since the selection of custom units over standard block is based primarily on appearance, the physical properties necessary for satisfactory performance may sometimes be taken for granted. Although a standard is currently being developed, at the present time there is no ASTM specification covering decorative units for exposure in exterior walls without protective coatings or facing materials. Most manufacturers have developed their own recommended specifications which are more stringent than those for conventional CMUs. Stricter requirements are essential to assure durability and resistance to water penetration when units are exposed to weather without surface protection. In the absence of specific minimum standards, NCMA studies have shown that the strength and absorption

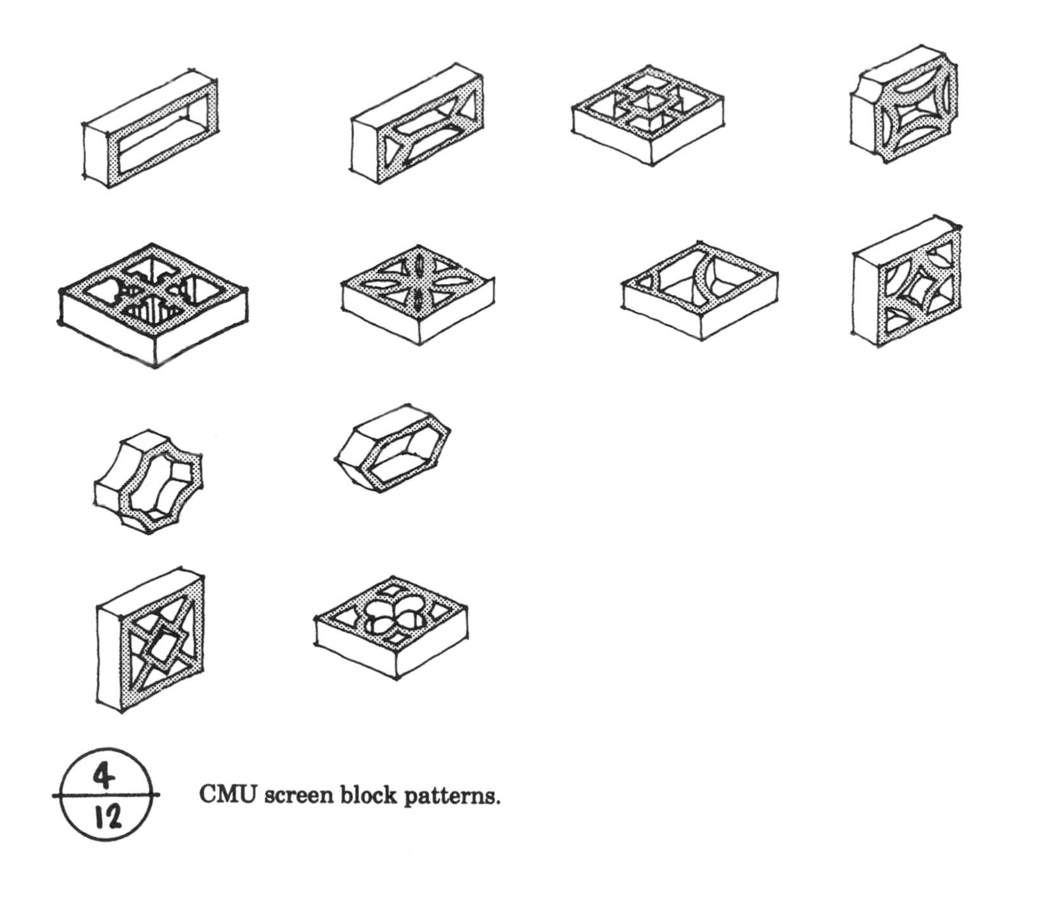

4/12 CMU screen block patterns.

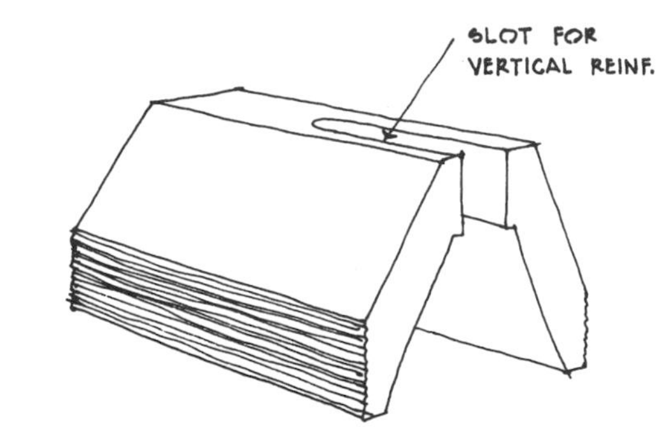

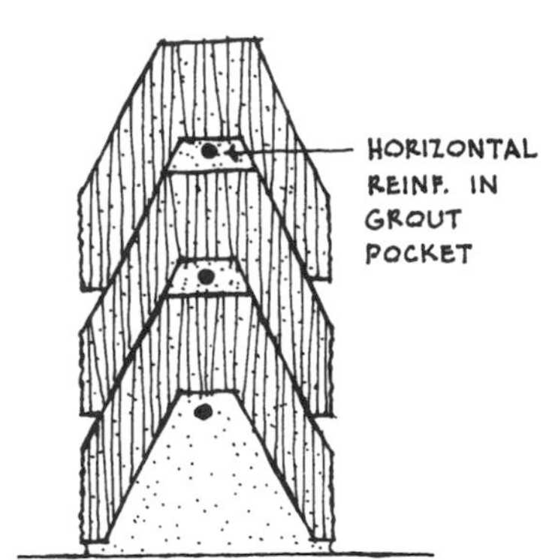

CUSTOM UNIT DESIGNED BY EDWARD HARDIN, ARCHITECT

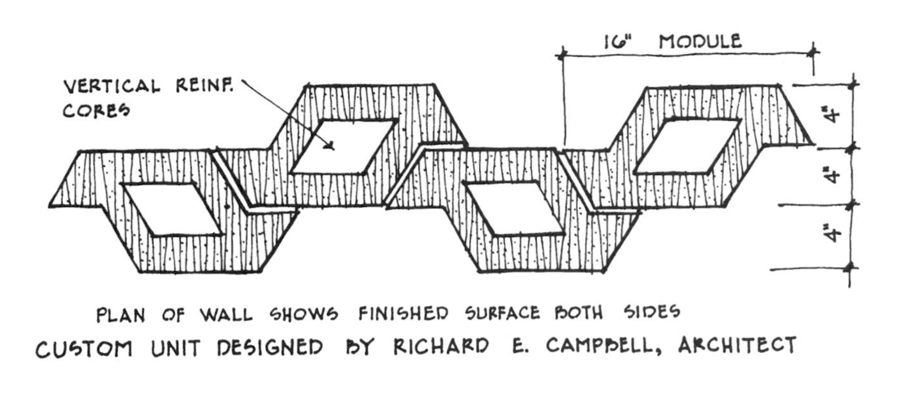

CUSTOM UNIT DESIGNED BY RICHARD E. CAMPBELL, ARCHITECT

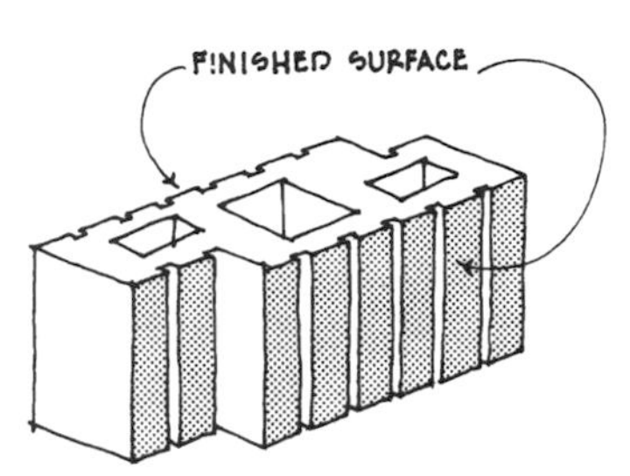

CUSTOM UNIT DESIGNED BY CHARLES W. DUEMMLING, ARCHITECT

Custom concrete block. *(Courtesy National Concrete Masonry Association.)*

requirements for normal-weight, Grade N concrete brick (ASTM C55) are applicable for decorative facing units where high strength and resistance to weathering are desired.

4.6.2 Glazed Units

Glazed surfaces may be applied to concrete brick or block as well as to sand-lime brick. Surfaces may consist of epoxy, polyester, ceramic, porcelainized, or mineral glazes, or cementitious finishes. All applied surfaces must meet the requirements of ASTM C744, *Standard Specification for Prefaced Concrete and Calcium Silicate Masonry Units,* in tests of imperviousness, abrasion, stain-resistance, chemical-resistance, and fire-resistance as well as crazing and adhesion of facing material to unit. Grade N, Type I, hollow loadbearing units are used and a thermosetting, resinous coating combined with specially treated silica sand, pigments, and/or ceramic colored granules is applied. The minimum requirements for both strength and abrasion are lower for glazed cementitious and concrete products than for glazed clay masonry units.

Custom 12-rib split-face block, Coldspring New Town, Baltimore. Moshe Safdie, architect. (*Photo courtesy National Concrete Masonry Association.*)

(A)

(B)

Split-rib and round-fluted blocks. (A) Split-rib block: Guilderland Reinsurance Bldg., Delmar, N.Y., Howard Geyer Assoc., architects. (B) Round-fluted block: Polytarp Office Bldg., Toronto, Canada, Boigon and Armstrong, architects. (*Photos courtesy National Concrete Masonry Association.*)

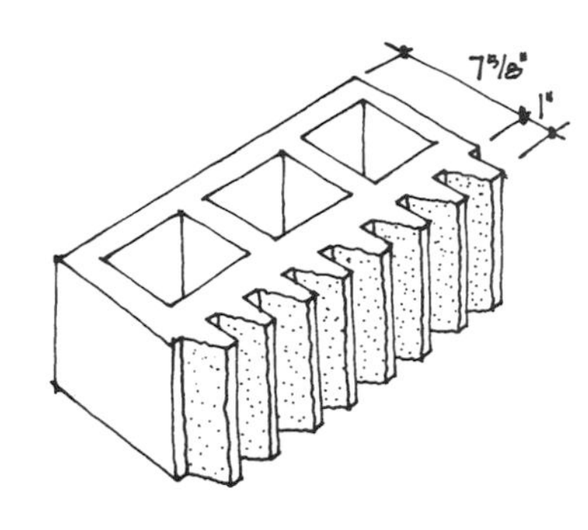

SPLIT-SAWTOOTH

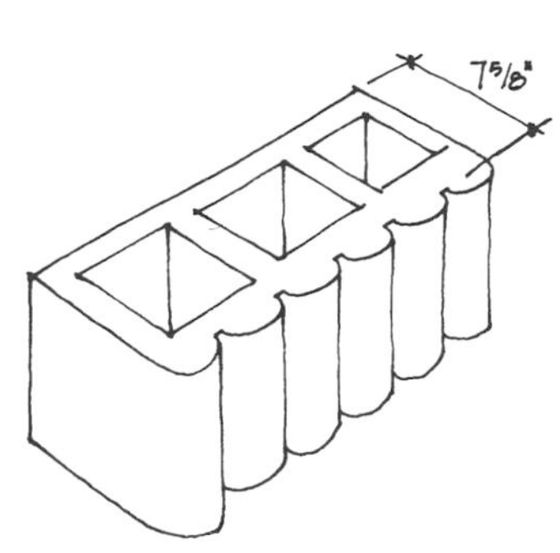

CURVED FLUTE

Two custom face designs by Paul Rudolph, architect. (*Photos courtesy National Concrete Masonry Association.*)

4.6.3 Paving Units

Two kinds of concrete masonry paving units are available for roadway and parking area surfacing. Solid interlocking units in a number of patterns provide a continuous topping over standard sand and gravel base materials. Open grid blocks permit grass to grow through the perforations while stabilizing the soil, protecting vegetation, and supporting vehicular traffic (*see Fig. 4-17*). Densely compacted units of 5000- to 10,000-psi compressive strength have a high resistance to moisture penetration and great durability in severe weathering conditions. CMU paving systems permit percolation of rainwater back into the soil despite the relative imperviousness of the unit itself. Prevention of excessive runoff is an important environmental consideration for standard installations as well as those where erosion and drainage of surrounding areas may be a problem. Grid units are generally 15 to 18 in. wide, about 24 in. long, and from $4\frac{1}{2}$ to 6 in. deep. Solid units are usually of proprietary design, and sizes and shapes will vary among manufacturers. Thicknesses range from $2\frac{1}{2}$ to $5\frac{1}{2}$ in., depending on

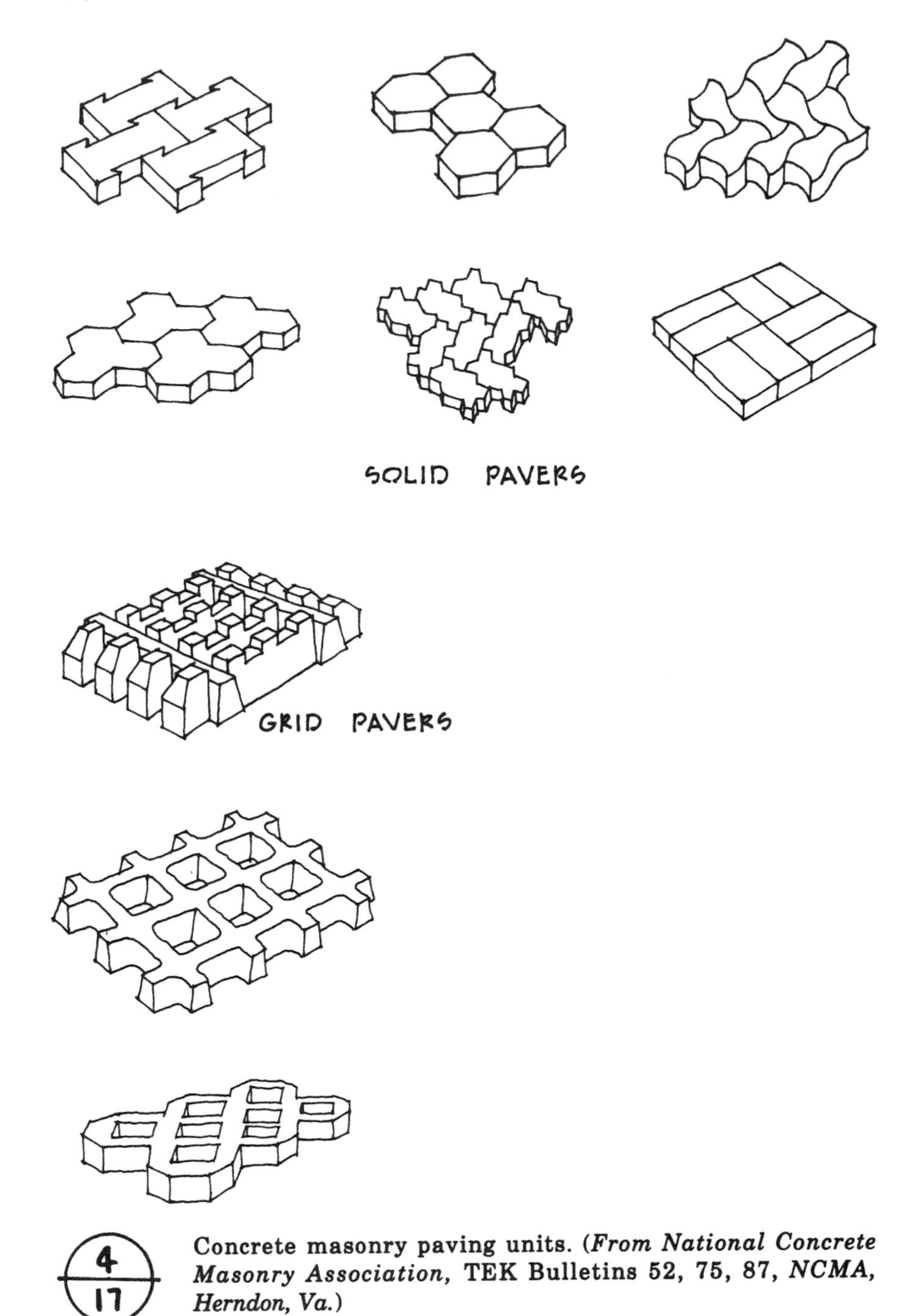

Figure 4-17 Concrete masonry paving units. (*From National Concrete Masonry Association,* **TEK Bulletins 52, 75, 87,** ***NCMA,*** *Herndon, Va.*)

the type of service and traffic load anticipated. Heavy-duty performance can be provided for industrial areas and roadways when speeds do not exceed about 40 mph.

ASTM C936, *Standard Specification for Solid Concrete Interlocking Paving Units,* governs abrasion resistance and resistance to freeze-thaw, limits absorption to 5%, and sets minimum compressive strength at 8000 psi. Tensile splitting strength is thought to be a better indicator of paver performance than simple compressive strength, and recommended values are currently being developed by NCMA for incorporation into the standard.

4.6.4 Segmental Retaining Wall Units

One of the newest developments in the concrete masonry industry is the dry-stacked, interlocking concrete block retaining wall. Sometimes referred to as segmental retaining walls (SRW), a variety of proprietary units and systems are available (*see Fig. 4-18*). The systems are designed to step back slightly in each course toward the embankment. Some units interlock simply by their shape, while others use pins or dowels to connect successive courses. The units use high compressive strengths and low absorption characteristics to resist spalling and freeze/thaw damage. Because they are dry-stacked without mortar, segmental retaining wall systems are simple and fast to install. The open joints in SRWs allow free drainage of soil moisture, and the stepped-back designs reduce overturning stresses. (Refer to Chapter 13 for design and installation requirements.)

4.7 PROPERTIES AND CHARACTERISTICS

CMU physical properties and characteristics fall into a number of structural, aesthetic, and functional categories. The two basic aspects, strength and absorption, have the greatest influence on overall performance. Compressive strength varies with the type and gradation of the aggregate, the water-cement ratio, and the degree of compaction achieved in molding. In general, the lighter-weight aggregates produce slightly lower strength values and have increased rates of absorption (*see Fig. 4-19*).

4.7.1 Unit Strength

Aggregate size and gradation as well as the amount of mixing water affect compaction and consolidation, and are important determinants of strength. Reducing unfilled voids between particles by 1% with extra compaction may increase block strength by as much as 5%. Higher compressive strengths are generally associated with wetter mixes, but manufacturers must individually determine optimum water proportions to obtain a balance among moldability, handling, breakage, and strength. For special applications, higher-strength units may be obtained from the same aggregates by careful design of the concrete mix and slower curing, increasing net strength ratings to as much as 4000 psi.

Other CMU structural values can be estimated from compressive strength. *Tensile strength* generally ranges from 3 to 5% of net compressive strength, *flexural strength* from 7 to 10%, and the *modulus of elasticity* from 150 to 600 times the value in compression. For engineering calculations in reinforced masonry construction, exact figures must be computed, but for general design purposes, these rules of thumb give a fairly accurate idea of the properties and capabilities of the block or concrete brick being considered.

4.7.2 Absorption

Water absorption characteristics are an indication of durability in resistance to freeze-thaw cycles. Highly absorptive units, if frozen when permeated with water, can be fractured by the expanding ice crystals. A drier unit can accommodate some expansion into empty pore areas without damage. Minimum ASTM requirements differentiate between unit weights because of the effect of aggregate characteristics on this property. Absorption values are measured in pounds of water per cubic foot of con-

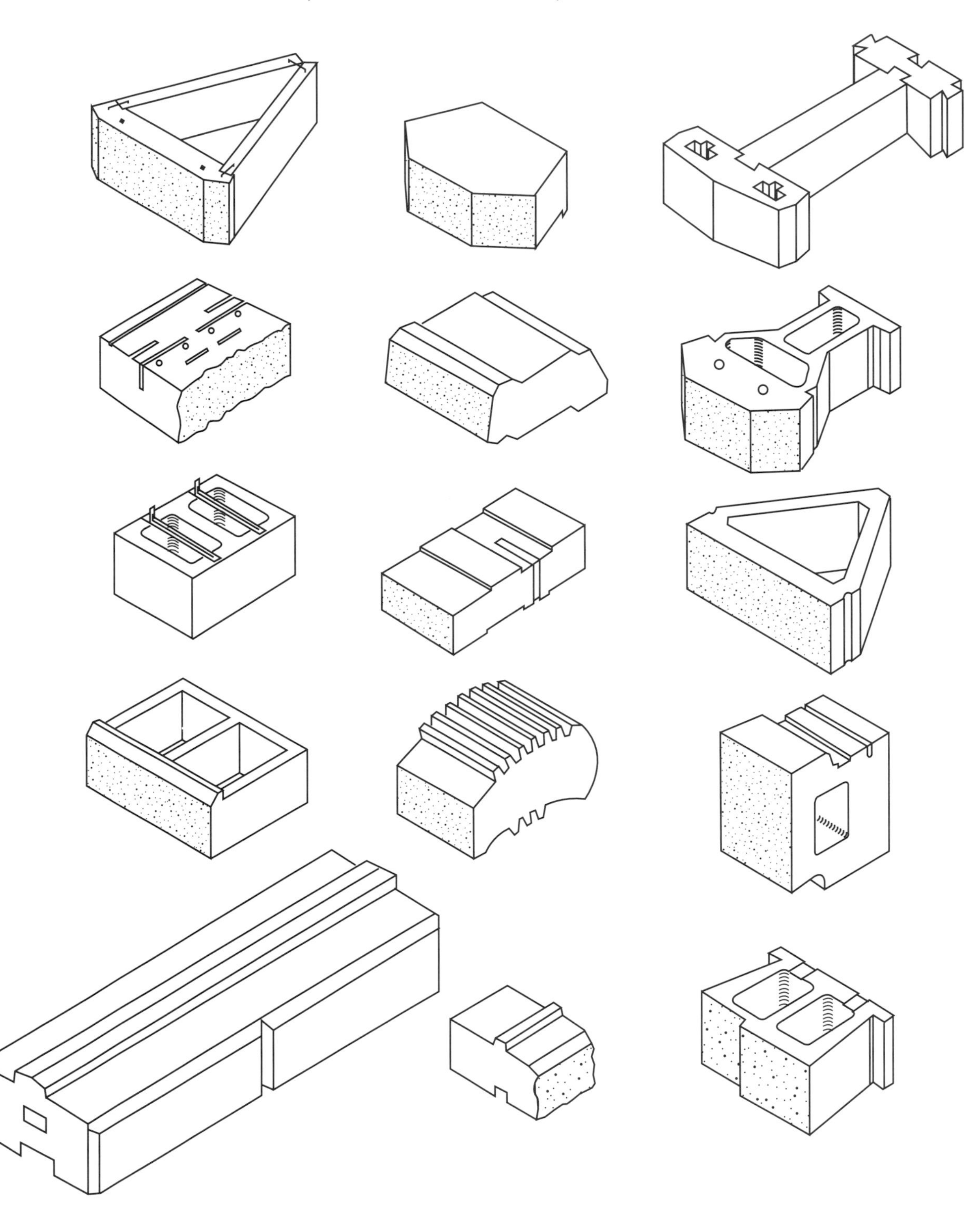

A number of proprietary CMU retaining wall systems are available. (*From National Concrete Masonry Association,* Design Manual for Segmental Retaining Walls, *NCMA, Herndon, Va., 1993.*)

Classification	*Aggregate*	*Compressive strength (net area) (psi)*	*Water absorption (lb/cu ft of concrete)*	*Thermal expansion coefficient (per °F) × 10^{-6}*
Normal weight	Sand and gravel	2200–3400	7–10	5.0
	Crushed stone	2000–3400	8–12	5.0
Medium weight	Air-cooled slag	2000–2800	10–15	4.6
Light weight	Coal cinders	1300–1800	12–18	2.5
	Expanded slag	1300–2200	12–16	4.0
	Scoria	1300–2200	12–16	4.0
	Expanded clay, shale, and slate	1800–2800	12–15	4.5
	Pumice	1300–1700	13–18	4.0

Effects of aggregate on concrete masonry properties.

crete. They may range from as little as 4 or 5 lb/cu ft for heavy sand and gravel materials to 20 lb/cu ft for the most porous, lightweight aggregates.

Porosity influences other properties, such as thermal insulation and sound absorption. Increases in these characteristics are often accompanied by an undesirable increase in moisture absorption as well. Pore structure varies for different aggregates and material types and has varying influence on these values and their relationships to one another. Relatively large interconnected pores readily absorb air and sound as well as water, and offer less resistance to damage from freezing. Unconnected or closed pores such as those in structural grade expanded aggregate offer good insulating qualities, and reduced absorption of water and sound. A high initial rate of water absorption, or suction, adversely affects the bond between mortar and unit just as it does in clay masonry. Unlike brick, however, concrete products may not be pre-wetted at the job site to control suction because of the moisture shrinkage inherent to concrete. Prewetting concrete masonry units could cause excessive shrinkage cracking in the wall. Suction can be controlled only through proper product specification by ASTM standards, and through the use of highly water retentive mortars to ensure the integrity of the bond.

Architectural block are sometimes treated with an integral water repellent to resist soil accumulations and to decrease surface water absorption. Some research indicates that calcium stearate–based products are more effective in creating hydrophobic surfaces than those based on oleic/linoleic acid chemistries, and are also less likely to leach out of the masonry. Whenever an integral water repellent is used in a concrete masonry product, compatibility and bond with mortar and grout must be considered because the bonding characteristics of the unit are affected. In general, a CMU product that has been treated with an integral water repellent requires use of mortar and grout that have compatible chemical admixtures to promote better bond.

4.7.3 Volume Changes

Volume changes in concrete masonry are caused by several things. Moisture shrinkage can be the most damaging because initial cycles of wetting and drying cause permanent shrinkage. Aged material has more ability to expand and contract in reversible movement. Controlling the initial moisture content at the time of construction is essential in preventing excessive shrinkage cracks in the finished wall. The different manufacturing techniques described in Chapter 2 bear significantly on this characteristic because of the variations in curing and drying methods. For a given aggregate, shrinkage tendencies due to moisture change can be reduced by as much as half by using high-pressure autoclave curing methods as opposed to low-pressure steam curing.

Small dimensional variations may occur due to changes in temperature. These changes, however, are fully reversible, and the units return to their original size after being heated and cooled through the same temperature range. Coefficients of thermal expansion (*see Fig. 4-19*) vary with different aggregates and are generally greater than values for clay masonry. As a result, provisions must be made for flexible anchorage and pressure-relieving control joints.

Volume changes are also caused by a natural chemical reaction called carbonation. Cured concrete absorbs carbon dioxide from the air, causing irreversible shrinkage. Preliminary tests indicate that, under certain conditions, the magnitude of this change may nearly equal that of moisture shrinkage. Carbonation stages added to the normal manufacturing process can eliminate many field problems by effectively "preshrinking" the masonry and producing a more dimensionally stable unit.

4.7.4 Fire, Sound, and Heat Resistance

Fire resistance, thermal insulation, and acoustical characteristics are all related to the density of the product. *Fire-resistance* ratings are based on the rate of heat transmission through the unit and the rate of temperature rise on the opposite face rather than on structural failure because no such failure occurs. Ratings are calculated on the equivalent solid thickness of the unit exclusive of voids. For some aggregates and core designs, maximum 4-hour ratings can be obtained with 8-in. hollow units. *Thermal insulation* characteristics vary with aggregate type and density. Exact values may be easily determined from basic information. (Insulating qualities based on engineering calculations are discussed in Chapter 8.)

Acoustical characteristics may be subdivided into two categories: (1) sound absorption and reflectance, which depend primarily on surface texture; and (2) sound transmission, which is a function of density and mass. Normal-weight or heavyweight units have higher resistance to sound transmission. They will produce walls with higher sound transmission class (STC) ratings than those of lightweight units because of their resistance to diaphragm action. Sound absorption is higher for coarse, open-textured surfaces with large pores. Sound reflectance is greater for tighter, closer grained, or painted surfaces with few, if any, open pores. CMUs can absorb from 18 to 68% of the sound striking the face of the wall, with lightweight units having the higher values. Specially designed block with slotted face shells provide high absorption by permitting sound waves to enter the cores, where their energy is absorbed by fiber inserts or dissipated through internal reverberation. Noise problems, particularly of middle- and high-frequency sounds, can often be controlled by these units, but they are proprietary products and may not be available in all locations.

4.7.5 Colors

CMU colors may be altered through the use of different aggregates, cements, or the integral mixing of natural or synthetic pigments (refer to Chapter 2). Pearl grays, buffs, tans, or even whites can easily be produced, offering great versatility within the generic product group. Penetrating stains may also be applied to the finished wall to achieve a uniform color.

5
NATURAL STONE

The earth's hard crust has undergone many changes throughout the millennia of geologic history. The stress and strains, the wearing away by atmospheric forces, by rain, wind, and heat, have produced a great variety of stones differing widely in appearance, but sharing some similarities of composition. All stone is made up of one or more minerals of specific crystalline structure and definable chemical makeup. No two blocks of stone, however, even if quarried side by side, are identical in internal structure or physical and chemical composition.

5.1 GEOLOGICAL CHARACTERISTICS

As a natural, inorganic substance, stone can be categorized by form and geological origin. *Igneous rock* is formed by the solidifying and cooling of molten material lying deep within the earth or thrust to its surface by volcanic action. Granite is the only major building stone of this origin. *Sedimentary rock* such as sandstone, shale, and limestone is formed by waterborne deposits of minerals produced from the weathering and destruction of igneous rock. The jointed and stratified character of the formation makes it generally weaker than igneous rock. *Metamorphic rock* is either igneous or sedimentary material whose structure has been changed by the action of extreme heat or pressure. Marble, quartzite, and slate are all metamorphically formed.

Stone may also be classified by mineral composition. Building stone generally contains as the major constituent (1) silica, (2) silicates, or (3) calcareous materials. The primary silica mineral is quartz, the most abundant mineral on the earth's surface, and the principal component of granite. Silicate minerals include feldspar, hornblende, mica, and serpentine. Feldspar may combine with lime or potash to produce red, pink, or clear crystals. Hornblende, combining often with lime or iron, appears green, brown, or black. Mica, with iron or potash, produces clear crystals. Serpentine, in combination with lime, is generally green or yellow in color. The most common silicate building stone is also called serpentine after this mineral. Calcareous minerals include carbonates of lime and magnesia, such as calcite and dolomite, forming limestone, travertine, and marble.

5.2 PROPERTIES

Prior to the twentieth century, stone was the predominant material used in major building construction. It was not only the structural material, but also the exterior and interior finish, and often the flooring and roofing as well. The term "masonry" at one time referred exclusively to stonework, and the "architects" of medieval castles and cathedrals were actually stone masons. Because of its massive weight and the resulting foundation requirements, stone is seldom used today as a structural element in contemporary architecture. It is, however, still widely used as a facing or veneer; in retaining walls, steps, walks, paths, and roads; as a floor finish; and is enjoying renewed popularity as a roofing material.

Despite their abundant variety, relatively few types of stone are suitable as building materials. In addition to accessibility and ease of quarrying, the stone must also satisfy the requirements of *strength, hardness, workability, porosity, durability,* and *appearance.*

The strength of a stone depends on its structure, the hardness of its particles, and the manner in which those particles are interlocked or cemented together. Generally, the denser and more durable stones are also stronger, but this is not always true. A minimum *compressive strength* of 5000 psi is considered adequate for building purposes, and the stones most often used are many times stronger in compression than required by the loads imposed on them. Failures from bending or uneven settlement are not uncommon, however, since stone is much stronger in compression than in flexure or shear. Stones of the same type may vary widely in strength, those from one quarry being stronger or weaker than those from another. Thus the average crushing strength of any type of stone may be misleading because of the wide variation in test results produced by stones within the same classification. The table in *Fig. 5-1* illustrates the ranges typical for several major types of stone. In modern building construction, *shearing strength* in stone is not nearly so important as compressive strength. The allowable unit stress of stone in shear should not be taken at more than one-fourth the allowable compressive unit stress. In *tension,* a safe working stress for stone masonry with portland cement mortar is 15 psi.

Hardness of stone is critically important only in horizontal planes such as flooring and paving, but hardness does have a direct influence on workability. Characteristics may vary from soft sandstone, which is easily scratched, to some stones which are harder than steel. Both strength and hardness are proportional to silica content. *Workability* in this instance refers to the ease with which a stone may be sawed, shaped, dressed, or carved, and will directly affect the cost of production. Workability decreases as the percentage of siliceous materials increases. Limestone, for instance, which contains little silica, is easily cut, drilled, and processed. Granite, however, which consists largely of quartz, is the most difficult stone to cut and finish.

Porosity, the percentage of void content, affects the stone's absorption of moisture, thus influencing its ability to withstand frost action and repeated freeze-thaw cycles. Pore spaces are usually continuous and often form microscopic cracks of irregular shape. The method of stone formation, and the speed of cooling of the molten material, influence the degree and structure of these voids because of compaction and the possibility of trapped gases. Thus, sedimentary rock, formed in layers without high levels of pressure, is more porous than rock of igneous or metamorphic origin. Closely linked to this characteristic are grain and texture, which influence the ease with which stones may be split, and for ornamental purposes contribute to aesthetic effects as much as color.

Stone type	Absorption by weight, max. (%) ASTM C97	Density, min. (lb/cu ft) ASTM C97	Compressive strength, min. (psi) ASTM C170	Modulus of rupture, min. (psi) ASTM C99	Abrasion resistance, min. hardness ASTM C241	Flexural strength, min. (psi) ASTM C880	Acid resistance, max. (in.) ASTM C217	Thermal expansion coefficient ($10^{-6}/°F$)	Modulus of elasticity (psi)	Ultimate shear strength (psi)	Ultimate tensile strength (psi)
Marble ASTM C503											
I. Calcite	0.75	162	7,500	1000	10	1000	N/A	3.69–12.30	1,970,000–	1638–4812	50–2300
II. Dolomite	0.75	175	7,500	1000	10	1000	N/A				
III. Serpentine	0.75	168	7,500	1000	10	1000	N/A		14,850,000		
IV. Travertine	0.75	144	7,500	1000	10	1000	N/A				
Limestone ASTM C568											
I. Low density	12	110	1,800	400	10	—	N/A	2.4–3.0	3,300,000–	900–1800	300–715
II. Medium density	7.5	135	4,000	500	10	—	N/A				
III. High density	3	160	8,000	1000	10	—	N/A		5,400,000		
Granite ASTM C615	0.40	160	19,000	1500	—	—	N/A	6.3–9.0	5,700,000–8,200,000	2000–4800	600–1000
Sandstone ASTM C616											
I. Sandstone	20	140	2,000	300	8	—	N/A	5.0–12.0	1,900,000–7,700,000	300–3000	280–500
II. Quartzite sandstone	3	150	10,000	1000	8	—	N/A				
III. Quartzite (bluestone)	1	160	20,000	2000	8	—	N/A				
Slate ASTM C629	ASTM C121			ASTM C120							
I. Exterior	0.25	—	—	Across grain, 9000	8	—	0.015	9.4–12.0	9,800,000–18,000,000	2000–3600	3000–4300
II. Interior	0.24	—	—	Along grain, 7200	8	—	0.025				

5/1 Properties of building stone.

Durability of stone, or its resistance to wear and weathering, is also considered roughly analogous to silica content. This is perhaps the most important characteristic of stone because it affects the life span of a structure. The stones traditionally selected for building construction have exhibited almost immeasurable durability compared to other building materials.

5.3 PRODUCTION

Stone is quarried from its natural bed by various techniques, depending on the nature of the rock. The most basic, and the oldest, method is drilling and splitting. With stratified material such as sandstone and limestone, the process is facilitated by natural cleavage planes, but also limited in the thickness of stone which can be produced. Holes are drilled close together along the face of the rock, and plugs and wedges are then driven in with sufficient pressure to split the rock between holes. For stratified rock, holes are drilled only on the face perpendicular to the bed, but nonstratified material must be drilled both vertically and horizontally. Channeling machines are often used on sandstone, limestone, and marble, but cannot be used with granite or other very hard stone. Wire saws are now used by most stone producers to cut a smoother surface, reduce the required mill finishing, and to subdivide large blocks of stone for easier transport, handling, and finishing.

The first stones cut from the quarry are large, with rough, irregular faces (*see Fig. 5-2*). These monolithic pieces are cut or split to the required rough size, then dressed at the mill with power saws and/or hand tools. Finished stone surface textures may vary from a rough rock face to a more refined hand-tooled or machine-tooled finish. For thin facings of marble or granite, gang saws cut several slabs from a block of stone at the same time. Although the sawing is a slow process, the surface it produces is so even that much work is saved in later dressing and polishing. Other saws, such as chat saws, shot saws, and diamond saws, are used to cut rough blocks of stone to required dimensions. Each type of saw produces a different surface texture.

In the 1970s the Italian stone industry developed new technology that enabled them to produce thin-sliced marble and granite panels which were light enough to clad high-rise buildings and inexpensive enough to dress the lobbies of speculative office buildings. Diamond-studded cables were devised to cut large blocks of stone from the quarry with little waste, and large diamond-tipped blades were ganged together to cut the slabs into ¾-in., 1¼-in., and 1½-in. thicknesses. Ultra-thin marble and granite tiles could also be cut in ¼- and ⅜-in. thicknesses and then gang-ground and polished with large multi-headed machines. While the cost of other cladding materials went up, the price of stone came down because of this new capability of producing more surface for less cost. Between 1980 and 1985, the use of travertine in the United States increased 600%, marble 625%, and granite an astonishing 1735%.

For exterior use, most building codes require a minimum ¾-in.-thick stone for low-rise buildings, and 1¼-in. thickness for high-rises. The use of veneers less than 2 in. thick is still relatively new compared to the long history of stone masonry, and much is still being learned about their in-service behavior and long-term performance.

In addition to sawed finishes, stone may also be dressed with hand or machine tools. Planing machines prepare a surface for hammered finishes,

Quarrying stone. (*Photo courtesy Georgia Marble Co.*)

for polished finishes, and for honed or rubbed finishes. A carborundum machine, used in place of a planer, will produce a very smooth finish. Honing is accomplished by rubbing the stone surface with an abrasive such as silicon carbide or sand after it has been planed, while a water spray is used to control dust. Larger surfaces are done by machine, smaller surfaces and moldings by hand. Polished surfaces require repeated rubbing with increasingly finer abrasives until the final stage, which is done with felt and a fine polish-

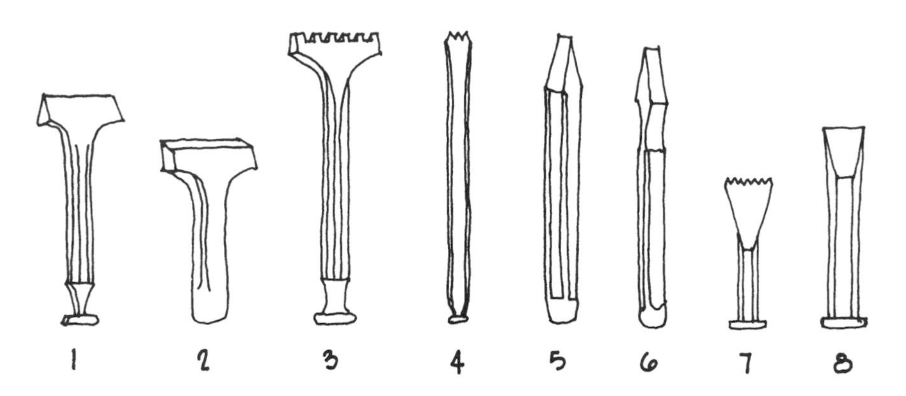

1. 2-3" WIDE DROVE CHISEL
2. 3½-4½" WIDE BOASTER OR BOLSTER TOOL
3. 19TH CENTURY TOOTH CHISEL
4. 16TH CENTURY ITALIAN TOOTH CHISEL
5. 19TH CENTURY NARROW CHISEL
6. SPLITTING CHISEL
7. 1¾", 7-TOOTH CHISEL
8. 1½" CHISEL

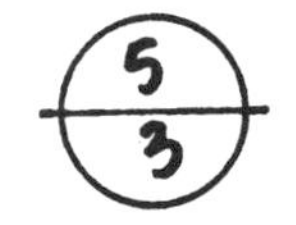

Traditional stone chisels. (*From Harley J. McKee,* **Introduction to Early American Masonry,** *National Trust for Historic Preservation and Columbia University, Preservation Press, Washington, D.C., 1973.*)

ing material. Only granite, marble, and some very dense limestones will take and hold a high polish. Power-driven lathes have been developed for turning columns, balusters, and other members that are round in section.

Hand-tooling is the oldest method of stone dressing. Working with pick, hammer, and chisels (*see Fig. 5-3*), the mason dressed each successive face of the stone, giving it the desired finish and texture. The drawings in *Fig. 5-4* illustrate the various steps in dressing the face, beds, and joints of a rough stone. Other hand-applied finishes include the bush-hammered, patent-hammered, pick-pointed, crandalled, and peen-hammered surface (*see Figs. 5-5 and 5-6*). Many of these finishes are now applied with pneumatic rather than hand tools, resulting in a more uniform surface. Ornate carving is still done by hand, both for new construction and for restoration and rehabilitation projects, although it is sometimes aided by pneumatic chisels.

Another finishing technique which produces a roughened surface is called flame cutting or thermal finishing. A natural gas or oxyacetylene flame is passed over a polished surface that has been wetted. The water that has been absorbed by the stone changes to steam and breaks off the surface, leaving an irregular finish. This finish can be selectively applied to portions of a stone surface to provide contrast.

A polished finish, by providing some measure of sealing of the stone pores, helps protect the surface of the veneer from deterioration by atmospheric weathering agents. A thermal finish, frequently used on granite, reduces the effective thickness by about ⅛ in. Bush-hammered and other similar surface finishes also reduce the effective thickness. For 1¼-in.

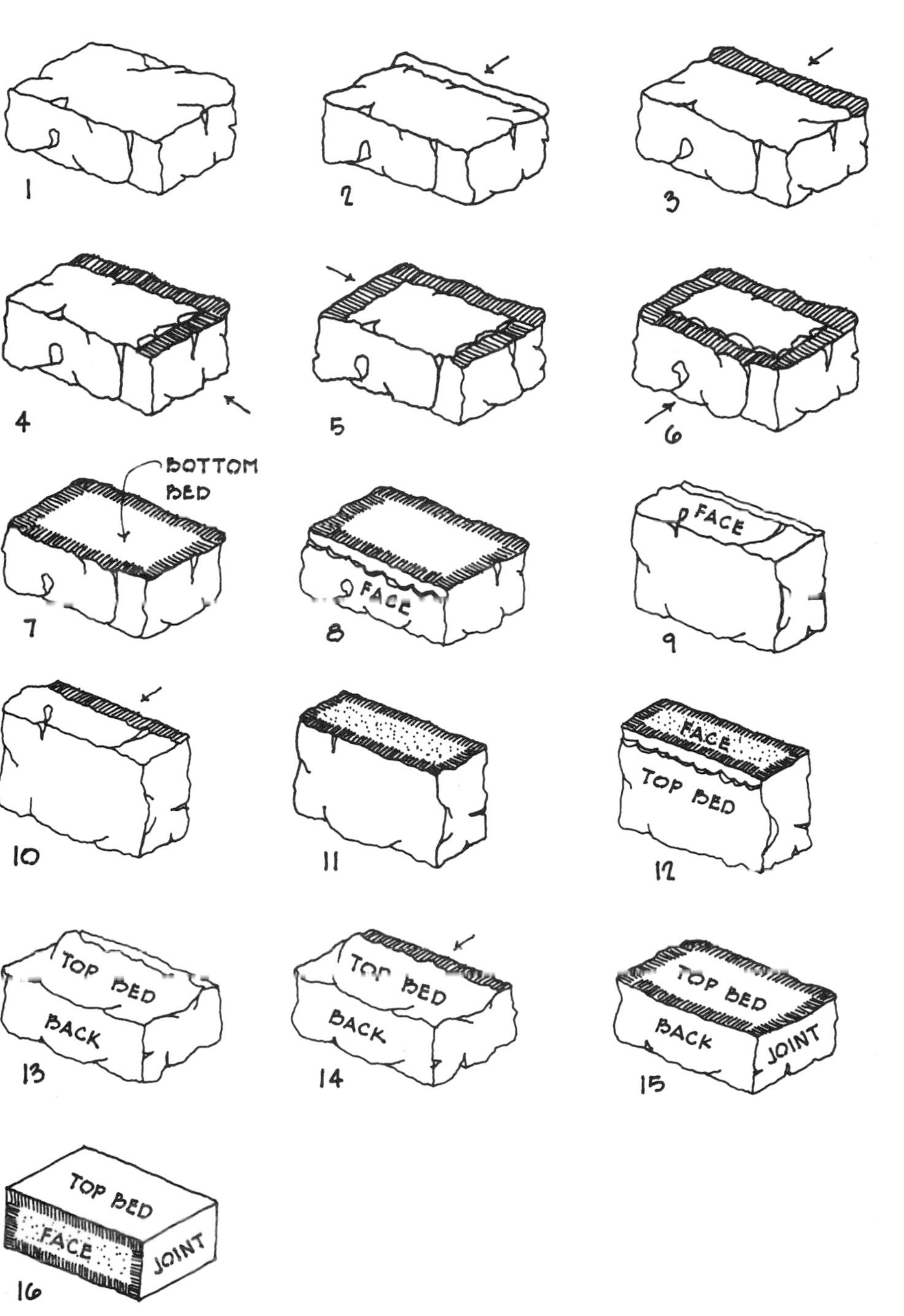

Various steps in hand-dressing the face, beds, and joints of a rough stone. (*From Harley J. McKee,* Introduction to Early American Masonry, *National Trust for Historic Preservation and Columbia University, Preservation Press, Washington, D.C., 1973.*)

veneers, a reduction in thickness of ⅛ in. reduces bending strength by 20 to 30% and increases the theoretical elastic deflection under wind loads by 37%.

Stone is used for masonry construction in many forms and is available commercially as (1) rubble stone, (2) flagstone, (3) dimension stone, (4) thin veneers, and (5) tile. *Rubble* includes rough fieldstone and irregular stone fragments with at least one good face. The stone may be either broken into

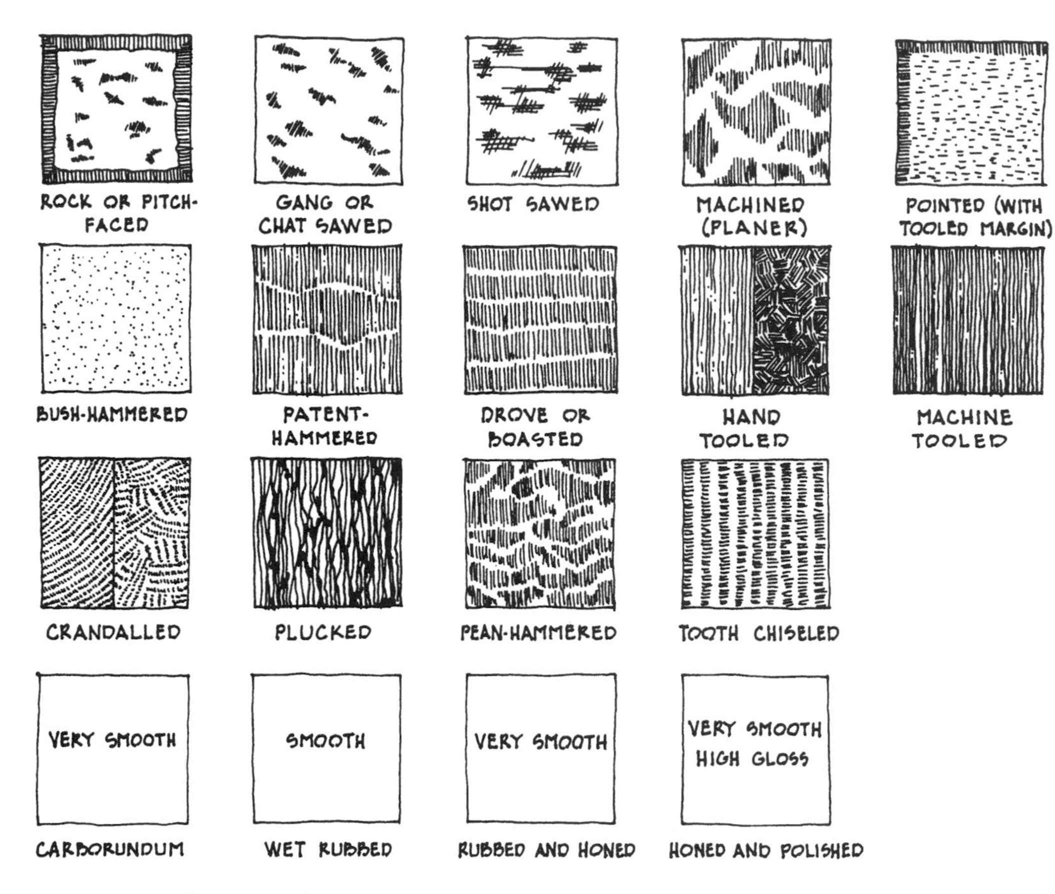

5/5 Stone surface finishes. (*From Charles G. Ramsey and Harold S. Sleeper,* Architectural Graphic Standards, *6th ed., ed. Joseph N. Boaz. Copyright © 1970 by John Wiley & Sons, Inc. Reprinted by permission of John Wiley & Sons, Inc.*)

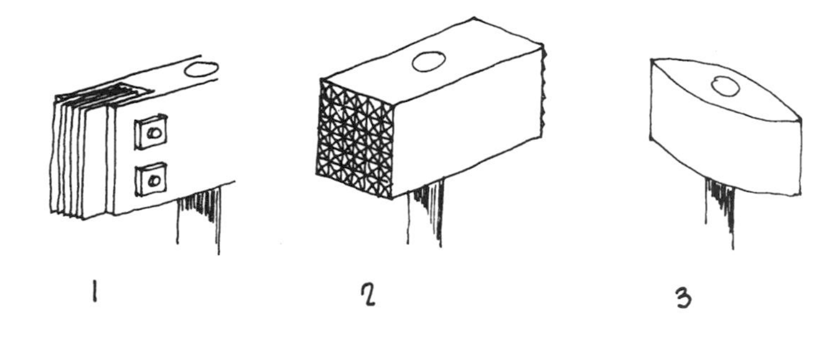

1. PATENT HAMMER (MAY HAVE 4 TO 8 BLADES)
2. BUSH HAMMER
3. PEEN HAMMER

Stone hammers. (*From Harley J. McKee,* Introduction to Early American Masonry, *National Trust for Historic Preservation and Columbia University, Preservation Press, Washington, D.C., 1973.*)

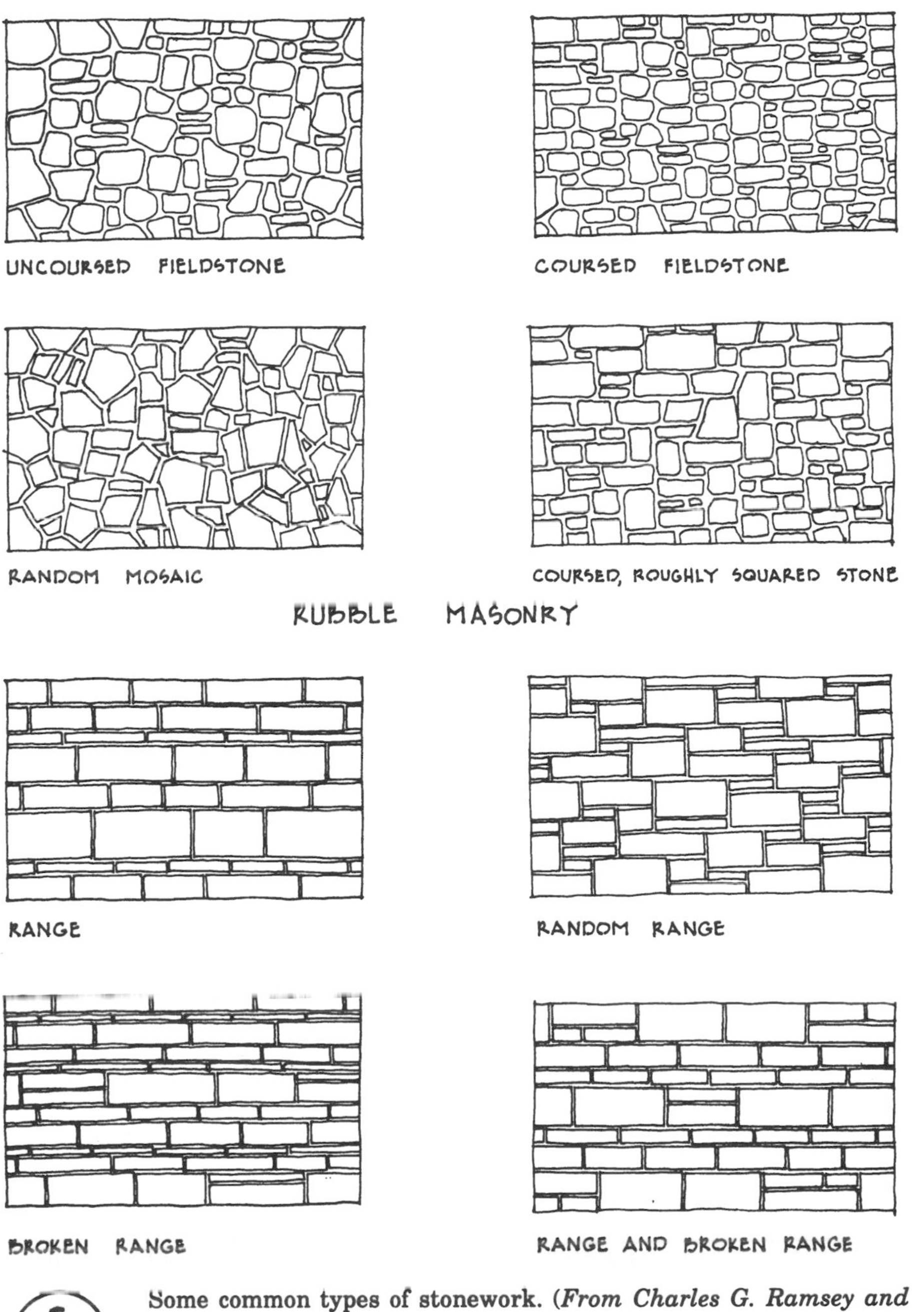

Some common types of stonework. *(From Charles G. Ramsey and Harold S. Sleeper,* **Architectural Graphic Standards,** *6th ed., ed. Joseph N. Boaz. Copyright © 1970 by John Wiley & Sons, Inc. Reprinted by permission of John Wiley & Sons, Inc.)*

suitable sizes, or roughly cut to size with a hammer. Some common types of stonework, as shown in *Fig. 5-7,* include rough fieldstone, random mosaic, coursed fieldstone, and cut-stone ashlar of various patterns. *Flagstone* consists of thin slabs from ½ to 2 in. thick in either squared or irregular shapes. Surfaces may be slightly rough, smooth, or polished. Flagstone is used on the exterior for walks, paths, and terraces, and on the interior as stair treads, flooring, coping, sills, and so on. *Dimension stone* (or cut stone) is delivered from stone mills cut and dressed to a specific size, squared to dimension each way, and to a specific thickness. Surface treatments include a rough or natural split-face, smooth, slightly textured, or polished finishes. Ashlar is a type of flat-faced dimension stone, generally in small squares or rectangles, with sawed or dressed beds and joints. Dimension stone is used for interior and exterior surface veneers, prefabricated panels, bearing walls, toilet partitions, arch stones, flooring, copings, stair treads, sills, and so on.

Thin stone veneers are a type of dimension stone, cut to a thickness less than 2 in. Unlike conventionally set dimension stone, which is mechanically anchored to a backing system at the project site, thin stone may be anchored directly to precast concrete panels, to glass fiber reinforced concrete (GFRC) panels, or to prefabricated steel truss panels. Thin stone may also be incorporated into stick-built or unitized metal curtainwall systems. *Stone tile* is generally limited to interior surfaces as wall and floor finish systems.

5.4 BUILDING STONE

Some of the natural stones that satisfy the requirements of building construction are granite, limestone, sandstone, slate, and marble (*see Fig. 5-1*). Many others, such as quartzite and serpentine, are used locally or regionally, but to a much lesser extent.

5.4.1 Granite

Granite has been used as a building material almost since the inception of man-made structures. Because of its hardness, it was first used with exposed, hand-split faces. As tools and implements were improved, the shapes of the stone became more sophisticated. With the development of modern technology and improved methods of sawing, finishing, and polishing, granite was more readily available in the construction market and more competitive with the cost of other, softer stones.

Granite is an igneous rock composed primarily of quartz, feldspar, mica, and hornblende. Colors vary depending on the amount and type of secondary minerals. Feldspar produces red, pink, brown, buff, gray, and cream colors, while hornblende and mica produce dark green or black. Granite is classified as fine-, medium-, or coarse-grained. It is very hard, strong, and durable, and is noted for its hard-wearing qualities. Compressive strength may range from 7700 to 60,000 psi, but ASTM C615, *Standard Specification for Granite Building Stone,* requires a minimum of 19,000 psi for acceptable performance in building construction (*see Fig. 5-1*). While the hardness of the stone lends itself to a highly polished surface, it also makes sawing and cutting very difficult. Granite is often used for flooring, paneling, veneer, column facings, stair treads, flagstones, or in landscape applications. Carving or lettering on granite, which was formerly done by hand or pneumatic tools, is now done by sandblasting, and can achieve a high degree of precision.

For granite, the National Building Granite Quarries Association recommends a maximum variation in the dimensions of any individual piece of stone of one quarter of the specified bed and joint width. Variations from true plane or flat surfaces on polished, honed, and fine-rubbed finishes at the bed and joint arris lines may not exceed $^3/_{64}$ in. or one-sixth the specified joint width, whichever is greater. For other types of finishes, the maximum variation cannot exceed one-quarter the specified joint width. Variations from true plane on other parts of the face surface are based of the type of finish (*see Fig. 5-8*).

Type of finish	Variations from true plane on parts of face surfaces other than bed and joint arris lines (in.)
Polished, honed or fine rubbed	3/64
Rubbed or fine stippled sand blasted	1/16
Shot ground, eight- and six-cut	3/32
Four-cut and sawn	1/8
Thermal and coarse stippled sand blasted	3/16

Variations from true plane for granite building stone. (*From* Specifications for Architectural Granite, *National Building Granite Quarries Association Inc., West Chelmsford, Mass., 1986.*)

5.4.2 Limestone

Limestone is a sedimentary rock which is durable, easily worked, and widely distributed throughout the earth's crust. It consists chiefly of calcium carbonate deposited by chemical precipitation or by the accumulation of shells and other calcareous remnants of animals and plants. Very few limestones consist wholly of calcium carbonate. Many contain magnesium carbonates in varying proportions, sand or clay, carbonaceous matter, or iron oxides, which may color the stone. The most "pure" form is *crystalline* limestone, in which calcium carbonate crystals predominate, producing a fairly uniform white or light gray stone of smooth texture. It is highest in strength and lowest in absorption of the various types of limestone. *Dolomitic* limestone contains between 10 and 45% magnesium carbonate, is somewhat crystalline in form, and has a greater variety of texture. *Oolitic* limestone consists largely of small, spherical calcium carbonate grains cemented together with calcite from shells, shell fragments, and the skeletons of other marine organisms. It is distinctly noncrystalline in character, has no cleavage planes, and is very uniform in composition and structure.

The compressive strength of limestone varies from 1800 to 28,000 psi depending on the silica content, and the stone has approximately the same strength in all directions. ASTM C568, *Standard Specification for Limestone Building Stone,* classifies limestone in three categories: I (low-density); II (medium-density); and III (high-density), with minimum required compressive strengths of 1800, 4000, and 8000 psi, respectively. Limestone is much softer, is more porous, and has a higher absorption capacity than granite, but is a very attractive and widely used building stone. Although soft when first taken from the ground, limestone weathers hard on exposure. Its durability is greatest in drier climates, as evidenced by the remains of Egyptian and Mayan monuments.

Impurities affect the color of limestone. Iron oxides produce reddish or yellowish tones while organic materials such as peat give a gray tint. Limestone textures are graded as A, statuary; B, select; C, standard; D, rustic; E, variegated; and F, old Gothic. Grades A, B, C, and D come in buff or gray, and vary in grain from fine to coarse. Grade E is a mixture of buff and gray, and is of unselected grain size. Grade F is a mixture of D and E and includes stone with seams and markings.

When quarried, limestone contains groundwater (commonly called quarry sap) which includes varying amounts of organic and chemical matter. Gray stone generally contains more natural moisture than buff-colored stone. As the quarry sap dries and stabilizes, the stone lightens in color and is said to "season." Buff stone does not normally require seasoning beyond the 60 to 90 days it takes to quarry, saw, and fabricate the material. Gray stone, however, may require seasoning for as long as 6 months. If unseasoned stone is placed in the wall, it may be very uneven in color for several months, or even as long as a year. No specific action or cleaning procedure will notably improve the appearance during this period, nor can it reduce the seasoning time. Left alone to weather, the stone eventually attains its characteristic light neutral color. No water repellents or other surface treatments should be applied until after the stone is seasoned.

Limestone is used as cut stone for veneer, caps, lintels, copings, sills, and moldings, and as ashlar with either rough or finished faces. Naturally weathered or fractured fieldstone is often used as a rustic veneer on residential and low-rise commercial buildings. Veneer panels may be sliced in thicknesses ranging from 2 to 6 in. and face sizes from 3×5 ft to 5×14 ft. When the stone is set or laid with the grain running horizontally, it is said to be on its natural bed. When the grain is oriented vertically, it is said to be on edge. Fabrication tolerances for limestone are shown in *Fig. 5-9.*

Travertine is a porous limestone formed at the earth's surface through the evaporation of water from hot springs. It is characterized by small pockets or voids formed by trapped gases. This natural and unusual texturing presents an attractive decorative surface highly suited to facing materials and veneer slabs.

The denser varieties of limestone, including travertine, can be polished and for that reason are sometimes classed as marble in the trade. Indeed, the dividing line between limestone and marble is often difficult to determine.

	Length tolerance (±in.)	*Height tolerance (±in.)*	*Deviation from flat surface exposed face (±in.)*	*Critical depth (±in.)*	*Non-critical depth (±in.)*	*Deviation from square (±in.)*
Smooth machine finish	1/16	1/16	1/16	1/16	1/2	1/16
Diamond gang finish	1/16	1/16	1/4	1/8	1/2	1/16
Chat-sawed finish	1/16	1/16	1/4	1/8	1/2	1/16
Shot-sawed finish	1/16	1/16	1/2	1/4	1/2	1/16
Preassembled units	1/8	1/8	1/8	1/8	1/2	1/8
Panels over 50 sq. ft.	1/8	1/8	1/8	1/8	1/2	1/8

Note: Tolerances for deviation from flat surface, exposed face and deviation from square are measured within the length of a standard 4-ft straightedge applied at any angle on the face of the stone.

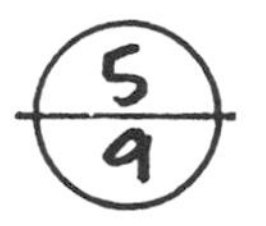

Dimensional tolerances for Indiana limestone. (*From* Indiana Limestone Handbook, *17th ed., Indiana Limestone Institute, Bedford, Ind.*)

5.4.3 Marble

Marble is a crystallized, metamorphosed form of noncrystalline limestone or dolomite. Its texture is naturally fine, permitting a highly polished surface. The great color range found in marbles is due to the presence of oxides of iron, silica, mica, graphite, serpentine, and carbonaceous matter in grains, streaks, or blotches throughout the stone. The crystalline structure of marble adds depth and luster to the colors as light penetrates a short distance and is reflected back to the surface by the deeper-lying crystals. Pure marbles are white, without the pigmentation caused by mineral oxides. Brecciated marbles are made up of angular and rounded fragments embedded in a colored paste or cementing medium.

Marble often has compressive strengths as high as 20,000 psi, and when used in dry climates or in areas protected from precipitation, the stone is quite durable. Some varieties, however, are decomposed by weathering or exposure to industrial fumes, and are suitable only for interior work. ASTM C503, *Standard Specification for Marble Building Stone (Exterior)*, covers four marble classifications, each with a minimum required compressive strength of 7500 psi: I, Calcite; II, Dolomite; III, Serpentine, and IV, Travertine. Over 200 imported and domestic marbles are available in the United States. Each has properties and characteristics that make it suitable for different types of construction.

Marbles are classified as A, B, C, or D on the basis of working qualities, uniformity, flaws, and imperfections. For exterior applications, only group A, highest-quality materials should be used. The other groups are less durable, and will require maintenance and protection. Group B marbles have less favorable working properties than Group A, and will have occasional natural faults requiring limited repair. Group C marbles have uncertain variations in working qualities; contain flaws, voids, veins, and lines of separation; and will always require some repair (known as sticking, waxing, filling, and reinforcing). Group D marbles have an even higher proportion of natural structural variations requiring repair, and have great variation in working qualities.

Marble is available as rough or finished dimension stone and as thin veneer slabs for wall and column facings, flooring, partitions, and other decorative surface work. Veneer slabs may be cut in thicknesses from ¾ to 2 in. Light transmission and translucence diminish as thickness increases. Fabrication tolerances for marble are shown in *Fig. 5-10.*

5.4.4 Slate

Slate is also a metamorphic rock, formed from argillaceous sedimentary deposits of clay and shale. Slates containing large quantities of mica are stronger and more elastic than clay slates. The texture of slate is fine and compact with very minute crystallization. It is characterized by distinct cleavage planes permitting easy splitting of the stone mass into slabs ¼ in. or more in thickness. Used in this form, slate provides an extremely durable material for flooring, roofing, sills, stair treads, and facings. ASTM C629, *Standard Specification for Slate Building Stone,* requires that Type I exterior slate have a minimum modulus of rupture of 9000 psi across the grain and 7200 psi along the grain.

Small quantities of other mineral ingredients give color to the various slates. Carbonaceous materials or iron sulfides produce dark colors such as black, blue, and gray; iron oxide produces red and purple; and chlorite pro-

Thickness, in. (cm)	Panel finished on both faces in. (mm)	Panel finished on one face in. (mm)
Thin stock, $\frac{3}{4}$–2 (2–5)	+0 (+0), $-\frac{1}{16}$ (−2)	+ $\frac{1}{8}$, $-\frac{1}{16}$ (±3)
Cubic stock, over 2 (5)	$\pm\frac{1}{16}$ (±2)	$+\frac{3}{16}$ (+5), $-\frac{1}{8}$ (−3)
Marble tile	—	$\pm\frac{1}{32}$ (±1)
	Sizes and squareness	
Thin stock	$\pm\frac{1}{16}$ (±2)	$\pm\frac{1}{16}$ (±2)
Cubic stock	—	$\pm\frac{1}{16}$ (±2)

Allowable tolerances for marble building stone. (*From* Dimension Stone, *Vol. III, Marble Institute of America, 1987.*)

duces green tints. "Select" slate is uniform in color and more costly than "ribbon" slate, which contains stripes of darker colors.

5.4.5 Sandstone

Sandstone is a sedimentary rock formed of sand or quartz grains. Its hardness and durability depend primarily on the type of cementing agent present. If cemented with silica and hardened under pressure, the stone is light in color, strong, and durable. If the cementing medium is largely iron oxide, the stone is red or brown, and is softer and more easily cut. Lime and clay are less durable binders subject to disintegration by natural weathering. ASTM C616, *Standard Specification for Sandstone Building Stone,* recognizes three classifications of sandstone. Type I, sandstone, is characterized by a minimum of 60% free silica content; Type II, quartzite sandstone, by 90% free silica; and Type III, quartzite, by 95% free silica content. As a reflection of these varying compositions, minimum compressive strengths are 2000 psi, 10,000 psi, and 20,000 psi respectively. Absorption characteristics also differ significantly, ranging from 20% for Type I, to 3% for Type II, and 1% for Type III. When first taken from the ground, sandstone contains large quantities of water, which make it easy to cut. When the moisture evaporates, the stone becomes considerably harder.

Sandstones vary in color from buff, pink, and crimson to greenish brown, cream, and blue-gray. It is traces of minor ingredients such as feldspar or mica which produce the range of colors. Both fine and coarse textures are found, some of which are highly porous and therefore low in durability. The structure of sandstone lends itself to textured finishes, and to cutting and tooling for ashlar and dimension stone in veneers, moldings, sills, and copings. Sandstone is also used in rubble masonry as fieldstone. Flagstone or bluestone is a form of sandstone split into thin slabs for flagging.

5.5 SELECTING STONE

Stone for building construction is judged on the basis of (1) appearance, (2) durability, (3) strength, (4) economy, and (5) ease of maintenance. Design and aesthetics will determine the suitability of the color, texture, aging characteristics, and general qualities of the stone for the type of building

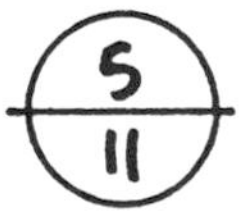

Fieldstone veneers.

under consideration. Colors may range from dull to brilliant hues, and from warm to cool tones. Textures may vary from coarse or rough to fine and dense. Some stones, such as marble and granite, are typically used on structures of grandeur and importance, whereas others, such as rubble sandstone, are more often seen on smaller, more picturesque buildings (*see Figs. 5-11 and 5-12*). Limestones are generally considered in the broader range of commercial, institutional, and utilitarian applications. Some stones, such as granite, will soften very slowly in tone and outline, and will retain a sharp edge and hard contour indefinitely. Others mellow in tone and outline, becoming softer in shape without losing their sense of strength and durability. Elaborately carved ornaments and lettered panels require stones of fine grain to produce and preserve the detail of the artist's design.

The compressive strength of stone was of great importance when

Veneer slabs and coursed dimension stone.

large buildings were constructed of loadbearing stone walls and foundations. Today however, stone is more often used as a thin veneer over steel, concrete, or unit masonry structures, or as loadbearing elements only in low-rise structures. In these applications, the compressive loads are generally small, and nearly all of the commonly used building stones are of sufficient strength to maintain structural integrity.

In terms of practicality and long-term cost, durability is the most important consideration in selecting building stones. Suitability will depend not only on the characteristics of the stone, but also on local environmental and climatic conditions. Frost is the most active agent in the destruction of stone. In warm, dry climates, almost any stone may be used with good results. Stones of the same general type such as limestone, sandstone, and marble differ greatly in durability based on softness and porosity. Soft, porous stones are more liable to absorb water and to flake or disintegrate in heavy frosts, and may not be suitable in the colder and more moist northern climates.

Weathering of stone is the combined chemical decomposition and physical disintegration of the material. The thinner the stone is cut, the more susceptible it is to weathering. Marble naturally has a lower fatigue endurance than other stones, and there are a number of variables which affect its strength and stiffness. Certain environmental conditions will weaken marble over time, causing panels to fracture, crumble, or bow.

Most stone used for exterior building construction is relatively volume stable, returning to its original dimensions after undergoing thermal expansion and contraction through a range of temperatures. Some fine-grained, uniformly-textured, relatively pure marbles, however, retain small incremental volume increases after each heating cycle. Marble is actually composed of layers of crystals, and repeated thermal and moisture cycles tend to make these crystals loosen and slide apart. The marble becomes less dense when it expands during heating, but does not return to its prior state during the cooling cycle. This irreversible expansion is called *hysteresis.* In relatively thick veneers, the greater expansion on the exposed exterior surface is restrained or accommodated by the unaffected interior mass. In thin veneers, however, dilation of the surface region can easily overcome the restraint of the inner layers, causing a dishing effect because the greatest expansion is across the diagonal axis.

Expansion of the exterior face of marble panels increases the porosity of the stone and its vulnerability to attack by atmospheric acids and cyclic freezing. Thermal finishes, in addition to reducing the effective thickness of marble and granite panels, also cause micro-fracturing of the stone. The micro-cracks, in turn, permit moisture absorption to depths of at least ¼ in., which can result in physical degradation if the stone freezes while it is saturated.

Some soft marbles can be easily "granulated," even by light impact forces such as pelting wind and rain. In addition to environmental problems, marble may bow naturally after it is quarried, and the thinner it is cut, the greater the tendency. Each time the thickness of marble is halved, the stresses are quadrupled. Marble can be a very non-uniform and unpredictable material, and pre-construction testing is critical to assure adequate performance.

Limestone and marble are both vulnerable to attack by sulfurous and sulfuric acids, and to a lesser extent, by carbonic acid and ammonium salts. Rainwater is a weak carbonic acid that dissolves the calcite or lime component, causing stones to flake, crumble, and eventually disintegrate. Sulfur-based acids form gypsum which is eventually washed from the stone matrix. Urban environments that produce stronger acid rain also produce accelerated disintegration. Chloride ions, such as those derived from deicing salts like sodium chloride or calcium chloride do not chemically react directly with stone. However, chloride can cause physical distress from the forces of crystal growth caused by calcium chloride salts precipitating from solutions within the stone, and by osmotic forces created by cyclic wetting. Porosity-permeability relationships and macro- and micro-fracturing influence these types of chemical weathering. Permeability is of increased significance in thin veneers. It is likely that water will penetrate thin stone veneers in greater amounts and at faster rates than would normally be expected.

Polished marble is not recommended for commercial floors. Polished finishes wear off rapidly, becoming dull and showing traffic patterns. Honed finishes are less slippery, require less maintenance, and look better with wear, becoming more polished from normal foot traffic. Granite is normally a better choice for polished floors. Porous stones require commercial sealers to protect them from stains. Food, grease, and sugared drinks readily penetrate porous stone faces, leaving unsightly stains that are difficult, if not impossible, to remove. Sealers not only protect these floors, but also enhance their natural colors.

Abrasion resistance of the stone must also be considered. If two or more varieties of stone are used, the abrasion resistance should be approximately the same, or uneven wear will result. Only stones highly resistant to wear should be used on stair treads.

Polished marble is also a very poor choice for bar and table tops. Acidic fruit juices, sugared drinks, and cola products can etch polished marble finishes, leaving spots and rings. Honed marble makes good bar and table tops, but polished granite is virtually impervious to damage from drink and food spills.

The costs of various stones will depend on the proximity of the quarry to the building site, the abundance of the material, and its workability. In general, stone from a local source will be less expensive than those which must be imported; stone produced on a large scale less expensive than scarce varieties; and stone quarried and dressed with ease less expensive than those requiring excessive time and labor.

6

MORTAR AND GROUT

Although mortar may account for as little as 7% of the volume of a masonry wall, it influences performance far more than the percentage indicates. Aesthetically, mortar adds color and texture to the masonry. Functionally, it binds the individual units together, seals against air and moisture penetration, and bonds with anchors, ties, and reinforcing to structurally join the building components. For engineered construction and loadbearing applications, mortar strength and performance are as critical as unit strength and workmanship.

6.1 MORTAR PROPERTIES

During this century, portland cement has become the principal ingredient of most mortar and grout mixes. Since it is also the principal ingredient of concrete, many designers assume that the methods and materials used to produce strong, durable concrete also apply to mortar and grout technology. However, both laboratory tests and performance records disprove this assumption. The most important physical property of concrete is compressive strength, but compressive strength is only one of several properties important to mortar and grout, such as bond strength and durability. These qualities are influenced by three distinct sets of properties which interact to affect overall performance: (1) *properties of the plastic mortar* include workability, water retentivity, initial flow, and flow after suction; (2) *properties of the hardened mortar* are bond strength, durability, and extensibility, as well as compressive strength; and (3) *mortar/unit assembly properties*.

6.1.1 Workability

Workability significantly influences most other mortar characteristics. Workability is not precisely definable in quantitative terms because there are no definitive tests or standards for measurement. Workability is recognized as a complex rheological property including adhesion, cohesion, density, flowability, plasticity, and viscosity which no single test method can measure. A "workable" mortar has a smooth, plastic consistency, is easily spread with a trowel, and readily adheres to vertical surfaces. Well-graded, smooth aggregates enhance workability as do lime, air entrainment, and proper amounts of mixing water. The lime imparts plasticity and increases the water-carrying capacity of the mix. Air entrainment introduces minute bubbles which act as lubricants in promoting flow of the mortar particles, but maximum air content is limited in mortars to minimize the reduction of bond strength. When structural reinforcement is incorporated in the mortar, cement-lime mixes are limited to 12% air content, and masonry cement mixes to 18%. Unlike concrete, mortar requires a *maximum amount of water* for workability, and retempering to replace moisture lost to evaporation should be permitted.

Variations in unit materials and in environmental conditions affect optimum mortar consistency and workability. Mortar for heavier units must be more dense to prevent uneven settling after unit placement or excessive squeezing of mortar from the joints. Warmer summer temperatures require a softer, wetter mix to compensate for evaporation. Although workability is easily recognized by the mason, the difficulty in defining this property precludes a statement of minimum requirements in mortar specifications.

6.1.2 Retentivity and Flow

Other mortar characteristics that influence general performance, such as aggregate grading, water retentivity, and flow, can be accurately measured by laboratory tests and are included in ASTM Standards. *Water retentivity* allows mortar to resist the loss of mixing water by evaporation and the suction of dry masonry units to maintain moisture for proper cement hydration. It is the mortar's ability to retain its plasticity so that the mason can carefully align and level the units without breaking the bond between mortar and unit.

Highly absorptive clay units may be prewetted at the job site, but concrete products may not be moistened, thus requiring that the mortar itself resist water loss. Conversely, if low-absorption units are used with a highly retentive mortar, they may "float." Less retentive mortars may also "bleed" moisture, creating a thin layer of water between mortar and unit which can substantially reduce bond strength (*see Fig. 6-1*). Water retentivity generally increases as the proportion of lime in the mix increases. High suction units, especially if laid in hot or dry weather, should be used with a mortar that has high water retentivity (i.e., a higher proportion of lime). Low suction units, especially if laid in cold or wet weather, should be used with a mortar that has low water retentivity (i.e., a lower proportion of lime). ASTM C91, *Standard Specification for Masonry Cements,* includes a water retention test which simulates the action of absorptive masonry units.

Under laboratory conditions, water retention is measured by flow tests, and is expressed as the percentage of *flow after suction* to *initial flow*. The flow test is similar to a concrete slump test, but is performed on a "flow table" that is rapidly vibrated up and down for several seconds. Suction is applied by vacuum pressure to simulate the absorption of the masonry units, and the mortar is tested a second time on the flow table.

Although they accurately predict the water retention characteristics of mortar, laboratory values differ significantly from field requirements. Construction mortars need initial flow values on the order of 130 to 150%, while laboratory mortars are required to have an initial flow of only 105 to 115%. The amount of mixing water required to produce good workability,

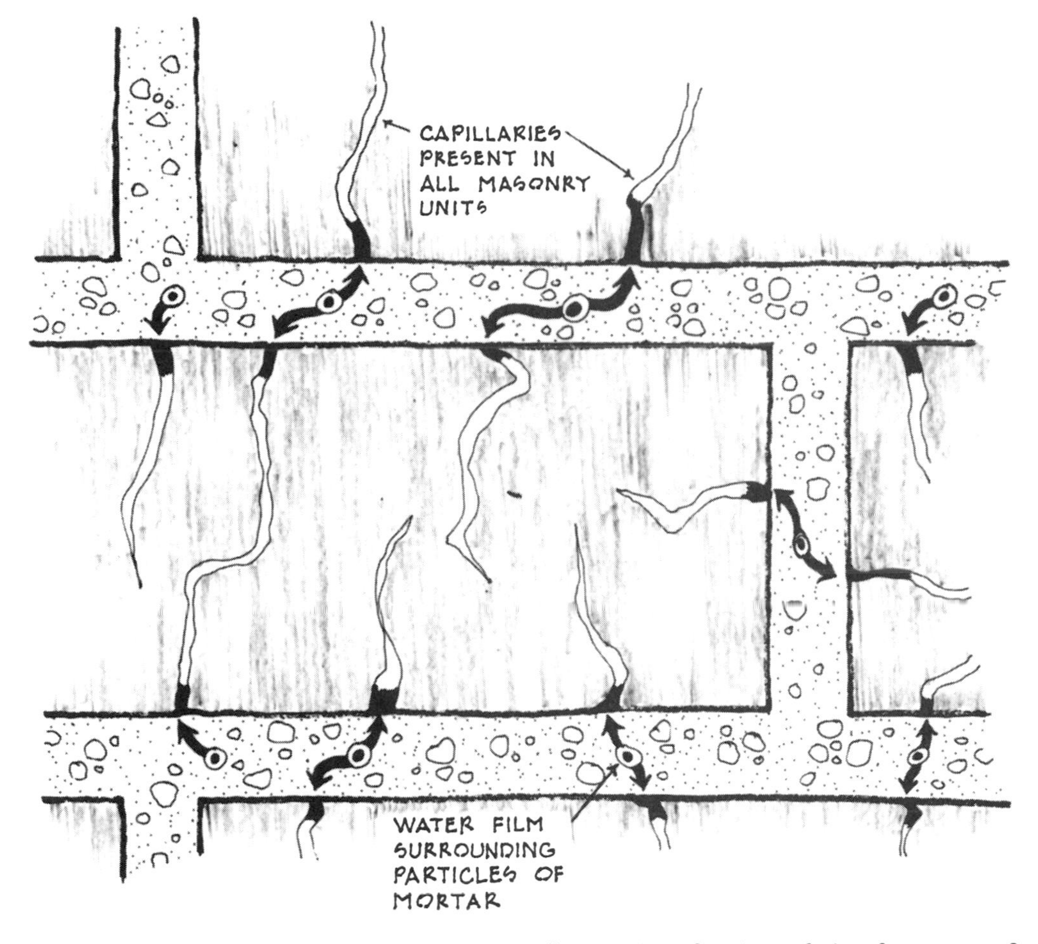

6/1 **Exaggerated section showing capillary action of moisture being drawn out of mortar by dry units.** (*Courtesy Acme Brick Co., Fort Worth, Tex.*)

proper flow, and water retention are quickly and accurately adjusted by experienced masons. Results produced from masonry assemblages prepared in the field reliably duplicate the standards set by laboratory researchers. Dry mixes lose too much water to the masonry units and will not cure properly. Excessively wet mixes cause units to float, and will decrease bond strength. The "proper" amount of mixing water is universally agreed upon as the maximum compatible with "workability," and workability is best judged by the mason. Project specifications should not dictate water/cement ratios for masonry mortar or grout.

Mortar is subject to water loss by evaporation, particularly on hot, dry days. *Retempering* (the addition of mixing water to compensate for evaporation) is acceptable practice in masonry construction. Since highest bond strengths are obtained with moist mixes having good flow values, a partially dried and stiffened mortar is less effective if the evaporated water is not replaced. Mortar which has begun to harden as a result of cement hydration, however, should be discarded. Since it is difficult to determine by either sight or touch whether mortar stiffening is due to evaporation or hydration, it is customary to determine the suitability of mortar on the basis of the time elapsed after initial mixing. Evaporative drying is related to both time and temperature. When ambient temperatures are above 80°F, mortar may be safely retempered as needed during the first $1\frac{1}{2}$ to 2 hours after mixing. When temperatures are below 80°F, mortar may be retempered for $2\frac{1}{2}$ to 3 hours after mixing before it should be discarded. ASTM C270, *Standard Specification for Mortar for Unit Masonry,* requires that all mortar be used within $2\frac{1}{2}$ hours without reference to weather conditions, and permits retempering as frequently as needed within this time period. Tests have shown that the decrease in compressive strength is minimal if retempering occurs within recommended limits, and that it is much more beneficial to the performance of the masonry to maximize workability and bond by replacing evaporated moisture.

6.1.3 Bond Strength

For the majority of masonry construction, the single most important property of mortar is bond strength and integrity. For durability, weather resistance, and resistance to loads, it is critical that this bond be strong and complete. The term mortar bond refers to a property that includes

- Extent of bond or area of contact between unit and mortar
- Bond strength or adhesion of the mortar to the units

Bond strength can be tested as tensile bond or flexural bond. The mechanical bond between the mortar and the individual bricks, blocks, or stones unifies the assembly for integral structural performance, provides resistance to tensile and flexural stress, and resists the penetration of moisture. The strength and extent of the bond are affected by many variables of material and workmanship. Complete and intimate contact between the mortar and the unit is essential, and the mortar must have sufficient flow and workability to spread easily and wet the contact surfaces. The masonry units must have surface irregularities to provide mechanical bond, and sufficient absorption to draw the wet mortar into these irregularities (*see Fig. 6-2*). The moisture content, absorption, pore structure, and surface characteristics of the units, the water retention of the mortar, and curing conditions such as temperature, relative humidity, and wind combine to influence the completeness and integrity of

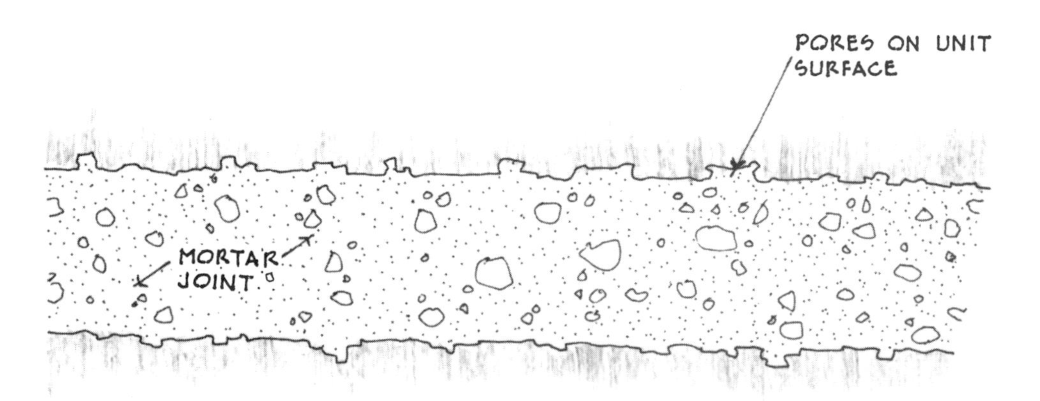

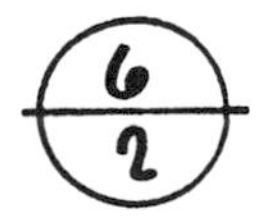

Enlargement showing the increased mechanical bond of mortar to a porous masonry unit surface.

the mortar-to-unit bond. Voids at the mortar-to-unit interface offer little resistance to water infiltration and facilitate subsequent disintegration and failure if repeated freezing and thawing occurs.

Investigations have shown that bond strength derives primarily from the mechanical interlocking of cement hydration crystals formed in the unit pores and on its surface. Higher bond strengths result if the extent of bond is good and the network of hydration products is complete. Although a certain amount of unit suction is desirable to increase the depth of penetration of the mortar paste, excessive suction reduces the amount of water available for hydration at the unit surface. *Moist curing* of masonry after construction assures complete hydration of the cement and improves mortar bond to high-suction brick and to dry, absorptive concrete masonry units (see Chapter 15). Clay brick with low initial rates of absorption and nonabsorptive glass block provide little or no suction of the mortar paste into surface pores. These types of units require a relatively stiff, low-water-content mortar.

Unit texture also affects bond. Coarse concrete masonry units and the wire-cut surfaces of extruded clay brick produce a better mechanical bond than molded brick or the die-formed surfaces of extruded brick. Smooth glass block and smooth stone surfaces provide little or no mechanical bond with the mortar. Loose sand particles, dirt, coatings and other contaminants also adversely affect mortar bond.

All other factors being equal, mortar bond strength increases slightly as compressive strength increases, although the relationship has no direct proportions. Mortar with a laboratory compressive strength of 2500 psi develops tensile bond strength on the order of 50 to 100 psi. Although higher cement ratios in the mix increase both compressive and bond strength, high cement–low lime mortars are stiff and do not readily penetrate porous unit surfaces. This leaves voids and gaps which disrupt the bond and decrease bond strength. Increasing air content, or adding air-entraining ingredients, lowers both compressive and bond strength because the air voids decrease surface contact area and bearing area.

Workmanship is a critical factor in bonding. Full mortar joints must assure complete coverage of all contact surfaces, and maximum extent of bond is necessary to reduce water penetration. Once a unit has been placed and leveled, additional movement will break or seriously weaken the bond. Mortars with high water retention allow more time for placing units before evaporation or unit suction alters the plasticity and flow of the mortar. Laboratory tests show that tapping the unit to level increases bond strength 50 to 100% over hand pressure alone.

Because of the many variables involved, it is difficult to develop laboratory tests of bond strength that produce consistent results. In addition to the properties of the mortar, bond strength testing is highly sensitive to unit properties, fabrication procedures, curing environment, and testing technique. Flexural bond strength is presently measured by ASTM C952, *Standard Test Method for Bond Strength of Mortar to Masonry Units* (the crossed brick couplet test), ASTM E518, *Standard Test Method for Flexural Bond Strength of Masonry,* and ASTM C1072, *Standard Test Method for Measurement of Masonry Flexural Bond Strength* (the bond wrench test). A simple field test to check extent of bond can be made by lifting a unit from its fresh mortar bed to determine if the mortar has fully adhered to all bedding surfaces. Good extent of bond is indicated if the mortar sticks to the masonry unit and shows no air pockets or dry areas.

6.1.4 Compressive Strength

Masonry compressive strength depends on both the unit and the mortar. As with concrete, the strength of mortar is determined by the cement content and the water/cement ratio of the mix. Since water content is adjusted to achieve proper workability and flow, and since bond strength is ultimately of more importance, higher compressive values are sometimes sacrificed to increase or alter other characteristics. For loadbearing construction, building codes generally provide minimum allowable working stresses, and required compressive strengths may easily be calculated using accepted engineering methods. Strengths of standard mortar mixes may be as high as 5000 psi, but need not exceed either the requirements of the construction or the strength of the units themselves. Although compressive strength is less important than bond, simple and reliable testing procedures make it a widely accepted basis for comparing mortars. Basically, compressive strength increases with the proportion of cement in the mix and decreases as the lime content is maximized. Increases in air entrainment, sand, or mixing water beyond normal requirements also reduce compressive strength values.

For veneer construction and for two- and three-story loadbearing construction, mortar compressive strength is rarely a critical design factor because both the mortar and the masonry are usually much stronger in compression than necessary. Compressive strength is important in engineered, loadbearing construction, but structural failure due to compressive loading is rare. More critical properties such as flexural bond strength are usually given higher priority.

Although the compressive strength of masonry can be increased by using a stronger mortar, the improvement is not proportional. Tests indicate that wall strengths increase only about 10% when mortar strength increases 130%. There are incentives other than economy which dictate using mortar with only the minimum required compressive strength. An unnecessarily hard, strong mortar will restrain concrete unit shrinkage and increase the amount of wall cracking as a result of the stress. A weaker mortar with higher lime content is more flexible, permits greater movement, and more satisfactory performance as long as minimum requirements are met.

6.1.5 Extensibility and Volume Change

Two other important properties of hardened mortar are extensibility and volume change. *Volume changes* in mortar can result from the curing process, from cycles of wetting and drying, temperature change, or unsound ingredients which chemically expand. Available data indicate that expansion and contraction of masonry construction due to differential thermal volume change between units and mortar does not have a noticeable effect on performance. However, total volume change from different causes can sometimes be significant. Stronger mortars that are rich in cement can show substantial shrinkage when exposed to alternate moist-dry conditions. Shrinkage during curing and hardening is greatest with high-water-content mortars. Volume changes caused by unsound ingredients such as reactive chemical compounds can cause disintegration of the masonry.

It is commonly believed that mortar shrinkage is significant, and that it is a primary cause of wall leaks. Research indicates, however, that maximum shrinkage across a mortar joint is minute, and is not in itself a cause of leakage. The most common leakage of masonry walls is through voids at the mortar-to-unit interface, where watertightness depends on a combination of good materials, workmanship, and design. The elastic properties of mortar, in fact, often counteract both temperature and moisture shrinkage. *Extensibility* is defined as the amount per unit length that a specimen will elongate (creep), or the maximum unit tensile strain before rupture. Extensibility is sufficiently high in mortar so that when it is combined with the added plasticity which lime imparts to the hardened mix, slight movement can be accommodated without joints opening. For maximum resiliency (such as that required in chimney construction), mortar should be mixed with the highest lime content compatible with design requirements.

6.1.6 Durability

Durability is a measure of resistance to age and weathering, and particularly to repeated freeze-thaw cycles. Mortars with high compressive strength can be very durable, but a number of factors other than strength affect mortar durability. Ingredients, workmanship, volume change, elasticity, and the proper design and placement of expansion and control joints all influence durability and determine the maintenance characteristics of the construction. Although harsh environmental conditions and unsound ingredients can contribute to mortar deterioration, the most destructive factor is expansion of moisture in the wall by freezing. The bubbles introduced by air entrainment absorb the expansive forces of freezing water and provide good assurance against damage, but they also decrease both the compressive and bond strength of the mortar. Masonry cement mortars usually contain entrained air, and cement-lime mortars can be modified by using either air-entrained portland cement or air-entrained hydrated lime (ASTM C207, Type SA). The best defense against freeze-thaw destruction is the elimination of moisture leaks at the joints with high quality mortar ingredients and good bond, and the use of details which permit differential movement and provide adequate protection at the top of the wall and at penetrations.

Air-entrained cements are used in the concrete industry to provide resistance to freeze-thaw deterioration in horizontal applications where exposure to ponded water, ice, and snow is greatest. Entrained air produces voids in the concrete into which freezing water can expand without causing damage. Rigid masonry paving applications installed with mortared joints may also enjoy some of the benefits of air-entrained cements in resisting the expansion of freezing water. Although industry standards for masonry mortar generally limit the air content of mortar to 12, 14, or 18% depending on the mix, the benefits of higher air contents in resisting freeze-thaw damage in paving applications may be greater than the detrimental effects on bond strength. Rigid masonry paving systems are generally supported on concrete slabs, so the flexural strength of the masonry is less important than its resistance to weathering. Lower bond strength could probably be tolerated in such applications in return for increased durability.

6.1.7 Efflorescence and Calcium Carbonate Stains

Efflorescence is the white powdery deposit on exposed masonry surfaces caused by the leaching of soluble salts from within. If the units and the mortar ingredients contain no soluble salts such as sodium or potassium sulfate, and if insufficient moisture is present to effect leaching, efflorescence cannot occur. To minimize the possible contribution of mortar ingredients to efflorescence, specify portland cements with low alkali content, clean washed sand, and clean mixing water.

Unlike efflorescence, *calcium carbonate stains* are hard encrustations which can be removed only with acid cleaners. Calcium hydroxide is present in masonry mortar as part of the hydrated lime in cement-lime mortars, and as a by-product of the portland cement hydration process itself. Portland cement will produce about 12 to 20% of its weight in calcium hydroxide at complete hydration. Calcium hydroxide is only slightly soluble in water, but when large quantities of water enter the wall through construction defects, extended saturation of the mortar (1) prolongs the hydration process producing a maximum amount of calcium hydroxide; and (2) provides sufficient moisture to leach the calcium hydroxide to the surface. When it reacts with carbon dioxide in the air, the calcium hydroxide forms a concentrated calcium carbonate buildup, usually appearing as white streaks from the mortar joints. The existence of calcium hydroxide in cement-based mortar systems cannot be avoided. Preventing saturation of the wall both during and after construction, however, will eliminate the mechanism needed to form the liquid solution and carry it to the masonry surface.

6.2 MORTAR CLASSIFICATION

Egyptian builders of the twenty-seventh century B.C. first invented masonry mortar, when a mixture of burned gypsum and sand was used in the construction of the Great Pyramid at Giza. Greek and Roman builders later added or substituted lime or crushed volcanic materials, but it was not until the nineteenth century development of portland cement that mortar became a high-strength structural component with compressive values comparable to the masonry units it bonded together.

6.2.1 Clay Mortars

Clay is one of the oldest materials used in masonry mortar. It has been used historically with sun-dried brick, burned brick, and stone. In North America, clay mortar was often used because of its low cost, but it was also a substitute in some regions where lime was difficult to obtain. Although it is susceptible to deterioration from moisture, clay mortar has long been used in arid climates, and also in humid climates for interior work and for exterior work which can be protected from the rain. Interior chimneys were commonly constructed with clay mortar up to the roof line, and one nineteenth century specification permitted stone walls to be laid with clay mortar except for the outside 3 in. of walls above ground, and the inside 3 in. of cellar walls, which were to be pointed with lime mortar.

Ground fire clay is still used in mortars where a mild refractory quality is desired. Clay is also used as a proprietary plasticizer for mortar, and the Romans used ground clay from low-fired brick to impart pozzolanic properties to lime-sand mortars.

6.2.2 Lime-Sand Mortars

Mortars consisting of lime, sand, and water were the most common type used until the late 19th century. *Lime-sand mortars* have low compressive strength and slow setting characteristics, but offer good workability, high water retention, excellent bond, and long-term durability even in severe climates.

Lime-sand mortars cure and develop strength through a process called *carbonation*. The lime (calcium hydroxide) must combine with carbon dioxide in the air, so curing of the full joint depth occurs very slowly, over a period of months or years, and at variable rates. In the past, slower methods of construction could accommodate this gradual hardening, but modern building techniques and faster-paced production have virtually eliminated the use of lime-sand mortars except in historic restoration projects. Lime-sand mortars, however, were sufficiently flexible to accommodate slight movements caused by the uneven settlement of foundations, walls, piers, and arches. The slow curing permitted a gradual adjustment over long periods of time, and accounts for the greater elasticity of historical masonry compared to contemporary construction.

Hydraulic limes, made from limestone with clay impurities, require less water in slaking and less sand in mortar than pure lime. Hydraulic lime mortars were used extensively in civil construction during the nineteenth century, and particularly in the construction of canals, piers, and bridges. The distinction between hydraulic lime and "natural cement" is almost arbitrary. One natural cement product manufactured in the early nineteenth century, in fact, was called "artificial hydraulic lime." Natural cement rock was burned in kilns similar to those used for producing lime, and the calcined lumps ground into a fine powder in various patented processes.

Hydraulic lime or natural cement mortars were used in areas where greater strength was required and where the masonry was subject to continuous soil or moisture exposure. Volume shrinkage is high and workability often poor, so natural cement was sometimes used simply as an additive to lime-sand mortars to increase compressive strength.

6.2.3 Portland Cement–Lime Mortars

Since the latter part of the nineteenth century, portland cement has largely replaced hydraulic limes and natural cements in masonry mortars. Occasionally, portland cement is used with sand and water only in what is called a *straight cement-sand mortar*. Mixed in proportions of 1 part cement to 3 parts sand, these mortars harden quickly and consistently, exhibit high compressive strengths, and offer good resistance to freeze-thaw cycles, but are stiff and unworkable, and have low water retention and poor bond.

Portland cement, which proved to be more stable and consistent in quality than natural cement, was first used as an additive in lime-sand mortars to provide greater compressive strength and promote faster setting. As the speed of building construction increased and portland cement gained wider acceptance, the proportion was increased until it accounted for as much as 80% of the cementitious ingredients.

Cement-lime mortars represent a compromise in the attempt to take advantage of the desirable properties of both lime-sand and straight cement-sand mortars. Workability, water retentivity and compressive strength can be varied over a wide range of values by varying the proportions of cement and lime in the mix. Improvements in one property, however, are usually gained only at the expense of another. As workability and water retentivity increase with higher lime contents, for instance, compressive strength decreases. Cement-lime mortars have a high sand-carrying capacity and generally require relatively high water contents, which is beneficial in satisfying the moisture demands of unit absorption and cement hydration. During cold weather construction, however, cement-lime mortars may be more susceptible to early-age freezing because of this high moisture content. During hot weather construction, in dry conditions, or when highly absorptive units are used, cement-lime mortars generally perform better as their lime content increases. Board life is also extended with high-lime mortars.

Scanning electron microscopy has shown that cement-lime mortars can produce tight mechanical bond with a continuous structure of hydration products and a low incidence of micro-cracks at the mortar-to-unit interface (*see Fig. 6-3*). Small voids at the interface, whether caused by drying shrinkage of the cement, or by water or air bubbles in the mix, are often filled as the masonry ages by carbonation of the lime in the mortar. This process, known as *autogenous healing,* occurs when carbon dioxide reacts with the calcium hydroxide of the lime to form calcium carbonate. It is the same process of carbonation by which lime-sand mortars cure.

Cement-sand mortars gain about 75% of their ultimate strength in 10 to 14 days. With cement-lime mortars, ultimate strength development takes much longer, so small initial building movements can often be absorbed without breaking the bond between mortar and unit. Even after full cure, the extensibility of cement-lime mortars provides some elasticity to accommodate limited thermal and moisture movement in the masonry without cracking. Lime-rich mixes accommodate such movements more readily than the stronger and more rigid cement-rich mixes.

6.2.4 Masonry Cement Mortars

Proprietary masonry cements are widely used and are popular with masons because of their convenience, consistency, and economy (refer to Chapter 2). The first masonry cements were mixtures of portland cement and lime, preblended and prebagged to simplify job-site mixing operations and to increase batch-to-batch consistency. Other plasticizers such as ground clay, limestone, and air-entrained cement were soon substituted for lime. Masonry cements generally contain one or more of the following materials:

- Portland cement or blended hydraulic cement
- A plasticizing material such as finely ground limestone, hydrated lime, or certain clays or shales
- Air entraining agents
- Sometimes water-repelling agents

White and colored masonry cements containing mineral oxide pigments are available in many areas.

Air-entraining agents contribute to mortar workability by introducing millions of tiny air bubbles which act as lubricants in the mix. While the voids created by these bubbles usually reduce bond strength and increase water permeability, they also increase freeze-thaw durability by providing interstitial spaces which accommodate the expansion of ice crystals without damage to the structure of the mortar. To provide effective freeze-thaw resistance, the air content in masonry cement mortars ranges from 12 to 22%, compared to only 3 to 10% typically found in cast-in-place concrete mixes. ASTM standards limit the air content of masonry cement mortars which will contain structural reinforcement to a maximum of 18%. Air-entrained masonry cement mortars can provide a needed measure of protection against freeze-thaw deterioration in rigid masonry paving. The trade-off of reduced bond strength can usually be tolerated in paving applications because flexural stresses are carried by the supporting slab.

Masonry cement mortars generally require less mixing water to produce good workability than cement-lime mortars. The lower water content

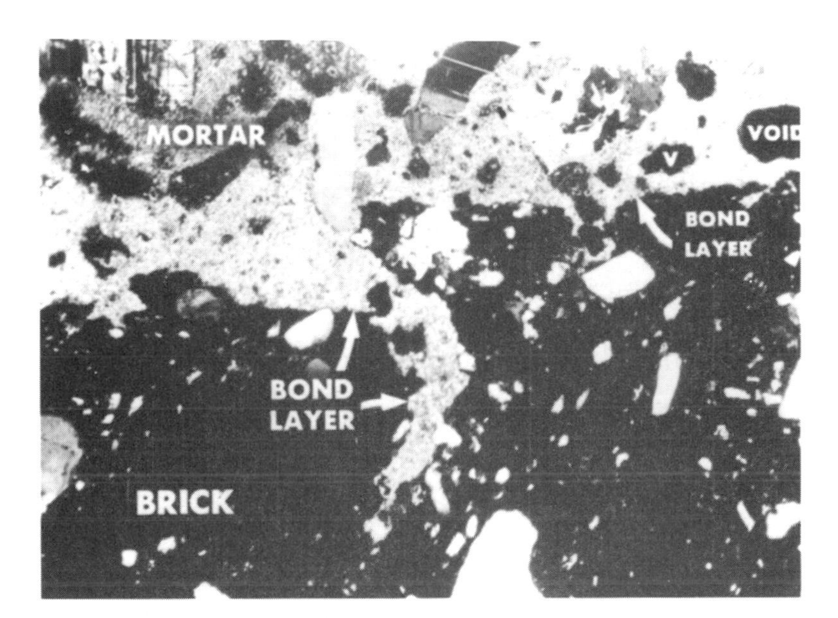

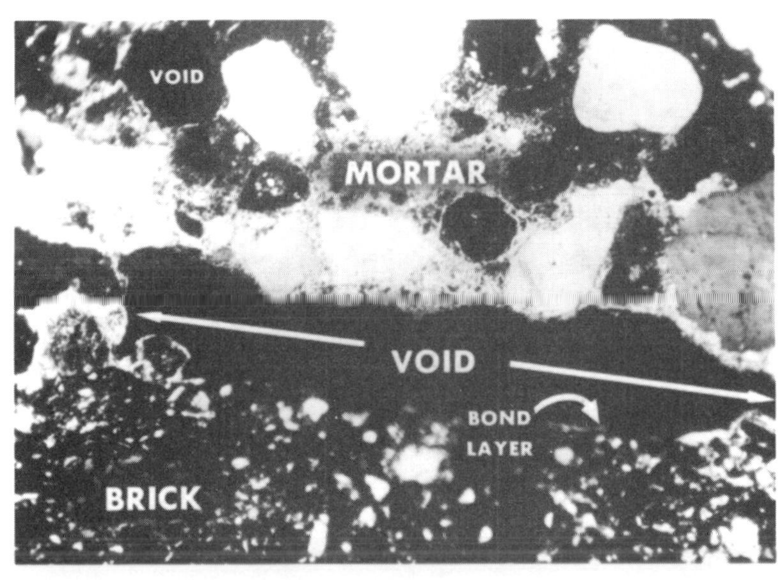

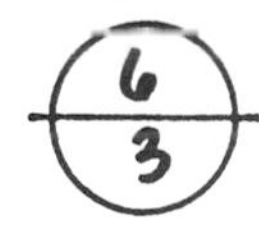

Microscopic view of mortar bond line. (*Photos courtesy National Lime Association.*)

is advantageous during winter construction, and also reduces volume shrinkage and the potential for cracking in the wall. It also means that less water is available for cement hydration. In hot, dry weather and with highly absorptive units, loss of mixing water to evaporation or suction can be sufficient to stop the hydration process and impair the bond between unit and mortar. Such *dry-outs* can be avoided by moist curing the masonry, or by re-hydrating the wall with a water fog spray (see Chapter 15).

Masonry cement mortars are less alkaline than cement-lime mortars. While this reduces the hazards of workers receiving burns to the skin, it also means the mortar will carbonate more rapidly. Carbonation is the process of chemical weathering in which the calcium hydroxide in hydrated portland cement reacts with atmospheric carbon dioxide to form calcium carbonate. Mortar that is carbonated is no longer alkaline, and no longer provides corrosion protection for embedded metal ties and reinforcing. Porosity affects the surface depth of carbonation. Porous mortars take carbon dioxide deeper into the joint to activate the process. Cracks or leaks in the construction will also increase carbonation, and water in sufficient quantity may contribute to calcium carbonate stains on the surface of the masonry.

6.2.5 "Mortar-Cement" Mortars

A relatively new classification of masonry mortar is called *mortar cement mortars.* The physical requirements for mortar cement were first developed and published in Uniform Building Code (UBC) Standard 21-14 (see Appendix B), and are based on ASTM C91, *Standard Specification for Masonry Cement* (refer to Chapter 2), except that maximum air content is lower, and values have been added for minimum flexural bond strength (*see Fig. 6-4*). Air content was limited based on the reduction in bond strength which it causes. Mortars with low flexural bond strength can crack under lateral loading, allowing water to penetrate and corrode reinforcing steel. Values for minimum flexural bond strength were established by testing cement-lime mortars and concrete brick (which develop lower bond strength than clay brick) in a UBC standard bond wrench test similar to ASTM C1072. UBC Standard 21-14 also limits or excludes certain harmful or deleterious materials as mortar cement ingredients. Since UBC 21-14 was published, ASTM C1329, *Standard Specification for Mortar Cement,* has been approved, indicative of the increased availability and use of mortar cements across the United States. UBC Standard 21-14 formed the basis for development of ASTM C1329, and the two standards are virtually identical in their requirements except for the metric conversion of some numbers.

Mortar cement type	*N**	*S*†	*M*†
Fineness, residue on a No. 325 sieve, max. (%)	24	24	24
Autoclave expansion, max. (%)	1.0	1.0	1.0
Time of setting, Gillmore method			
Initial set, min. (hr)	2.0	1.5	1.5
Final set, max. (hr)	24.0	24.0	24.0
Compressive strength (average of 3 cubes), min.			
7 days (psi)	500	1300	1800
28 days (psi)	900	2100	2900
Flexural bond strength, 28 days, min. (psi)	71	104	116
Air content of mortar			
Min. % volume	8	8	8
Max. % volume	16	14	14
Water retention, min. (%)	70	70	70

*Type N permitted in seismic zones 0, 1, and 2.

†Types S and M permitted in seismic zones 0, 1, 2, 3, and 4.

Uniform Building Code minimum physical requirements for mortar cements and mortar cement mortars (*based on* Uniform Building Code Standard No. 24-19, 1994 edition). ASTM C1329 requirements are virtually identical.

UBC Standard 21-14 and ASTM C1329 essentially sort out masonry cements with high flexural bond strength capabilities from those which can only provide lower bond strengths. The mortar cements which meet UBC 21-14 and ASTM C1329 are capable of producing mortars with flexural bond strengths equivalent to those of portland cement-lime mortars under identical laboratory test conditions. When high flexural bond strengths are required on a project and it is also desirable to use a masonry cement for its advantageous properties, a mortar cement conforming to UBC21-14 or ASTM C1329 should be specified.

6.2.6 The Portland-Lime Mortar versus Masonry Cement Mortar Controversy

For years there has been controversy over the relative merits of mortars made with portland cement and lime versus masonry cement. The preponderance of industry literature advocates the use of portland cement–lime mortars, and architects and engineers usually have a greater level of confidence in their performance. On the other hand, masons tend to prefer masonry cements because of their excellent workability, batch consistency, and easy mixing. In a survey conducted by *Aberdeen's Magazine of Masonry Construction* (February 1991, Vol. 4, No. 2), it was reported that the responding architects specified portland cement–lime mortars about 80% of the time on both commercial and residential projects. Responding masonry contractors indicated that they use masonry cement mortars nearly 70% of the time on residential projects and only about 50% of the time on commercial projects. For water leakage, bond strength and durability, both the contractors and the architects preferred portland cement–lime mortars.

Historically, portland cement–lime mortars have exhibited higher flexural strengths than masonry cement mortars. Higher flexural strengths not only increase resistance to lateral wall loads, but to moisture penetration as well. It is difficult to assess the scientific data objectively. Most laboratory test studies that have been performed have usually been sponsored by either the lime industry or the masonry cement industry, and the studies can easily be designed to emphasize the strong points of either mortar. In Grimm's *Conventional Masonry Mortar: A Review of the Literature* (published by the University of Texas at Arlington's Construction Research Center), conflicting research reports are numerous. As with any proprietary product, there are high-quality masonry cements and poor-quality ones. The selection or acceptance of a particular brand of cement should be based on its performance history and on independent laboratory verification of conformance to ASTM standards.

Masonry cements are more widely used than portland cement–lime for masonry mortars, and the vast majority of projects which incorporate them perform quite satisfactorily. On projects which have experienced flexural bond failures or excessive moisture penetration, the culprit is seldom found to be attributable solely to the use of masonry cement rather than portland cement and lime in the mortar. Usually, there are other defects which contribute more to the problems such as poor workmanship, inadequate flashing details, or low-strength backing walls. Both portland cement–lime mortars and masonry cement mortars allow water penetration through masonry walls. The amount of water entering the wall is generally higher with masonry cement mortars, but when workmanship is poor, joints are unfilled, and flashing and weeps are not functional, either type of mortar can produce a leaky wall. There are no industry standards or guidelines identifying varying amounts of water penetration that are either acceptable or unacceptable. A wall system with well-designed and properly installed flashing and weeps will tolerate a much greater volume of water penetration without damage to the wall, the building, or its contents than one without such safeguards. Ultimately, the workmanship and the flashing and weephole drainage system will determine the success or failure of most masonry installations (refer to Chapter 9).

Both masonry cement mortars and portland cement–lime mortars are capable of providing what the industry considers adequate flexural bond strength when they are designed and mixed in accordance with ASTM C270, *Standard Specification for Mortar for Unit Masonry.* If specific performance characteristics need to be enhanced for a particular application, laboratory design mixes should be based on unit/mortar compatibility and testing for the desired properties.

6.3 MORTAR TYPES

ASTM C270, *Standard Specification for Mortar for Unit Masonry,* outlines requirements for five different mortar types, designated as M, S, N, O, and K. Prior to 1954, mortar types were designated A-1, A-2, B, C, and D, but it was found that A-1 carried the connotation of "best" and that many designers consistently specified this type thinking it was somehow better than the others for all applications. To dispel this misunderstanding, the new, arbitrary letter designations were assigned so that no single mortar type could inadvertently be perceived as best for all purposes. No single mortar type is universally suited to all applications. Variations in proportioning the mix will always enhance one or more properties at the expense of others.

6.3.1 Type M Mortar

Each of the five basic mortar types has certain applications to which it is particularly suited and for which it may be recommended. Type M, for instance, is a high-compressive-strength mix recommended for both reinforced and unreinforced masonry which may be subject to high compressive loads.

6.3.2 Type S Mortar

Type S mortar produces tensile bond values which approach the maximum obtainable with portland cement–lime mortar. It is recommended for structures subject to normal compressive loads but which require flexural bond strength for high lateral loads from soil pressures, high winds, or earthquakes. Type S should also be used where mortar adhesion is the sole bonding agent between facing and backing, such as the application of adhesion-type terra cotta veneer. Because of its excellent durability, Type S mortar is also recommended for structures at or below grade and in contact with the soil, such as foundations, retaining walls, pavements, sewers, and manholes.

6.3.3 Type N Mortar

Type N is a good general-purpose mortar for use in above-grade masonry. It is recommended for exterior masonry veneers and for interior and exterior loadbearing walls. This "medium strength" mortar represents the best compromise among compressive and flexural strength, workability, and economy and is, in fact, recommended for most masonry applications.

6.3.4 Type O Mortar

Type O is a "high-lime," low-compressive-strength mortar. It is recommended for interior and exterior non-loadbearing walls and veneers which will not be subject to freezing in the presence of moisture. Type O mortar is often used in one- and two-story residential work and is a favorite of masons because of its excellent workability and economical cost.

6.3.5 Type K Mortar

Type K mortar has a very low compressive strength and a correspondingly low tensile bond strength. It is seldom used in new construction, and is recommended in ASTM C270 only for tuckpointing historic buildings constructed originally with lime-sand mortar (refer to Chapter 15).

6.3.6 Choosing the Right Mortar Type

The Appendix to ASTM C270 contains non-mandatory guidelines on the selection and use of masonry mortars which are summarized later in *Fig. 6-5*. To obtain optimum bond, use a mortar with properties compatible with those of the masonry units which will be used. To increase tensile bond in general:

- Increase the cement-to-lime ratio of the mortar within the limits established by ASTM C270.
- Keep air content within the limits established by ASTM C270.
- Use mortars with appropriate water retentivity for the absorption characteristics of the unit.
- Mix mortar with the maximum water content compatible with workability.

Location	Building segment	Mortar type	
		Recommended	Alternative
Exterior, above grade	Loadbearing wall	N	S or M
	Non-loadbearing wall	O*	N or S
	Parapet wall	N	S
Exterior, at or below grade	Foundation wall, retaining wall, manholes, sewers, pavements, walks, and patios	S†	M or N†
Interior	Loadbearing walls	N	S or M
	Non-loadbearing partitions	O*	N

*Type O mortar is recommended for use where the masonry is unlikely to be frozen when saturated and unlikely to be subjected to high winds or other significant lateral loads. Type N or S should be used in other cases.

†Masonry exposed to weather in a nominally horizontal surface is extremely vulnerable to weathering. Mortar for such masonry should be selected with due caution.

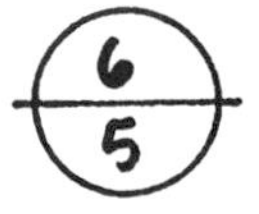

Recommended mortar uses. (*Based on* ASTM C270, Appendix XI.)

- Allow retempering of the mortar within recommended time limits.
- Use clay masonry units with moderate initial rates of absorption.
- Bond mortar to a rough surface rather than an extruded die skin.
- Minimize the time between spreading mortar and placing masonry units.
- Apply pressure in forming the mortar joint.
- Do not subsequently disturb units that have been placed.
- Moist-cure the masonry (refer to Chapter 15).

There are also several basic rules of thumb. Use mortar with the lowest compressive strength that meets structural requirements, because the lower the compressive strength, the more flexible the mortar in accommodating movements in the wall. In areas exposed to significant freeze-thaw cycling, and in particular for horizontal applications in those areas, specify mortars with a higher cement content or entrained air. For low-suction clay masonry units, use mortars with a lower lime content, and for high-suction clay masonry units, use mortars with a higher lime content.

For most projects, a Type N mortar is not only adequate in compressive and bond strength, it is the best choice for the compromise among various properties. On multi-story projects where higher wind loads at upper stories increase lateral loads, a Type S mortar will provide higher flexural bond strengths regardless of whether it is made from a masonry cement or from a portland cement and lime mix. The unnecessary specification of a Type S mortar when a Type N is adequate in strength sacrifices workability in the wet mortar and a degree of elasticity in the finished wall.

6.3.7 Proportion versus Property Method of Specifying Mortar

Conformance with ASTM C270 may be based either on volume proportions or on minimum property requirements (*see Fig. 6-6*). The *proportion specification* prescribes by volume the proportions of cementitious materials and aggregate for each mortar type. The *property specifications* are based on minimum compressive strength, minimum water retention, and maximum air content of laboratory-prepared samples made with a specified ratio of job site sand.

The proportion requirements are conservative and, for cement-lime mortars, will generally yield compressive strengths 2 to 3 times higher than the minimums given in the property specification (*see Fig. 6-7*). Conversely, the minimum compressive strengths required by the property specification generally can be achieved with a smaller proportion of cement and lime than that prescribed under the proportion specification. The property specifications encourage preconstruction testing of sample mortar cubes for a mix design to gain the economic advantage of meeting strength requirements at lower cost. On larger projects, the savings in mortar costs will more than offset the cost of the laboratory testing. Since it is generally recommended to use the mortar type with the minimum necessary compressive strength, specifying mortar by the property requirement method assures that the mortar is not any stronger in compression than it needs to be. On smaller projects where the volume of mortar is much less, using the proportion specification saves the cost of laboratory-mix designs and provides a high factor of safety in attaining adequate mortar strengths. However, it will usually yield mortars with higher compressive strengths than needed at the sacrifice of other properties.

TABLE A PROPORTION SPECIFICATION REQUIREMENTS

Mortar	Type	Proportions by volume (cementitious materials)					Aggregate ratio (measured in damp, loose condition)
		Portland cement or blended cement	Masonry cement			Hydrated lime or lime putty	
			M	S	N		
Cement–lime	M	1	—	—	—	¼	Not less than 2¼ and not more than 3 times the sum of the separate volumes of cementitious materials
	S	1	—	—	—	Over ¼ to 1½	
	N	1	—	—	—	Over ½ to 1¼	
	O	1	—	—	—	Over 1¼ to 2½	
Masonry cement	M	1	—	—	1	—	
	M	—	1	—	—	—	
	S	½	—	—	1	—	
	S	—	—	1	—	—	
	N	—	—	—	1	—	
	O	—	—	—	1	—	

Note: Two air-entraining materials shall not be combined in mortar.

TABLE B PROPERTY SPECIFICATION REQUIREMENTS (LABORATORY SPECIMENS ONLY)

Mortar	Type	Average compressive strength at 28 days, min. (psi)	Water retention, min. (%)	Air content, max. (%)	Aggregate ratio (measured in damp, loose condition)
Cement–lime	M	2500	75	12	Not less than 2¼ and not more than 3 times the sum of the separate volumes of cementitious materials
	S	1800	75	12	
	N	750	75	14*	
	O	350	75	14*	
Masonry cement	M	2500	75	—†	
	S	1800	75	—†	
	N	750	75	—†	
	O	350	75	—†	

*When structural reinforcement is incorporated in cement–lime mortar, the maximum air content shall be 12%.

†When structural reinforcement is incorporated in masonry cement mortar, the maximum air content shall be 18%.

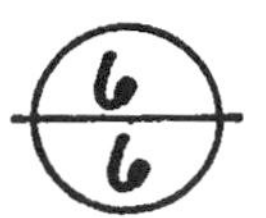

Mortar requirements by strength or by volume proportions, from ASTM C270. (*Copyright, American Society for Testing and Materials, 1916 Race Street, Philadelphia, Pa. 19103. Reprinted with permission.*)

Mortar type by proportions		Actual tested strengths in brick wallettes	
Type	Proportions	Compressive (psi)	Tensile bending (psi)
M	$1:\frac{1}{4}:3\frac{1}{2}$	3600	65
S	$1:\frac{1}{2}:4\frac{1}{2}$	3200	72
N	1:1:6	2800*	59
N	(Premixed masonry cement)	1500*	14
O	1:2:9	1600	20
K	1:3:12	Not tested	

*Note the difference between actual tested strength of Type N mortar mixed with portland cement and lime and that made with a premixed masonry cement and no added lime.

Actual strengths of mortars mixed by proportions. (*Courtesy Acme Brick Co., Fort Worth, Tex.*)

Material	Weight (lb/cu ft)
Portland cement	94
Blended cement	Weight printed on bag
Hydrated lime	40
Lime putty	80
Sand, damp and loose	80 lb of dry sand

Weights of mortar materials, from ASTM C270. (*Copyright, American Society for Testing and Materials, 1916 Race Street, Philadelphia, Pa. 19103. Reprinted with permission.*)

If ASTM C270 is referenced in project specifications without indication as to whether the property or proportion method should be used, the proportion method always governs. The volume proportions used in ASTM C270 are based on weights per cubic foot of materials as listed in *Fig. 6-8,* and proportions may be calculated on full and half bag measures using these equivalents.

The property specifications in ASTM C270 are for laboratory-prepared samples only, and the values will not correlate with those obtained from field samples tested under ASTM C780, *Standard Test Method for Preconstruction and Construction Evaluation of Mortars for Plain and Reinforced Unit Masonry.* Laboratory samples are made with a very low water-cement ratio because the molds used to form the mortar cubes are of nonabsorbent metal. Mortars mixed at the job site are made with much higher water-cement ratios because the units are absorptive and will immediately extract much of the mixing water from the mortar paste. Field-sampled mortars therefore typically yield a much lower compressive strength than the laboratory-prepared mortar because of the difference in water content. In order to compare apples to apples, the same testing procedure must be used. If the project will require field sampling of mortar during construction for laboratory testing, ASTM C780 must be used both to set the preconstruction benchmark and to perform the construction phase testing. Results from ASTM C780 tests cannot be compared to results from ASTM C270 tests or to the minimum property requirements listed in ASTM C270.

Until recently, there was no standardized test for hardened masonry mortar. ASTM C1324, *Standard Test Method for Examination and Analysis of Hardened Masonry Mortar,* now provides a standardized procedure for the petrographic and chemical analyses of hardened mortar samples to determine the proportions of ingredients used in the mix. The petrographic analysis is based on similar methods used to examine hardened concrete using a petrographic microscope and a stereoscopic low power microscope, as well as x-ray diffractometry and scanning electron microscopy. The standard also includes methods for chemical analysis. The interpretation and calculation of chemical test results depend on results of the petrographic analysis and are not intended to be used alone. The chemical data and the petrographic analysis together are intended to determine mortar composition as represented by the proportion specifications in Table 1 of ASTM C270 as Types M, S, N and O. Failure of a tested mortar specimen to comply with the proportion requirements of ASTM C270, however, does not necessarily mean that the mortar is not in compliance. Even though the proportions are different, the mortar may still meet the ASTM C270 property requirements. As yet, there is no standardized test to determine the compliance of hardened mortar samples with the property requirements of ASTM C270. Samples removed from a wall can be tested for compressive strength, but there is no correlation between these test results and the compressive strength requirements of ASTM C270.

6.4 SPECIALTY MORTARS

In determining the requirements for mortar performance, two very specialized areas demand detailed project analysis. Refractory mortars and chemical-resistant mortars are used primarily in industrial applications where exposure to extreme heat or toxic chemicals requires extraordinary mortar performance. Refractory mortars are also used in residential and commercial fireplaces.

6.4.1 Refractory Mortars

Refractory mortars may range from residential fireplace installations to extremely high heat industrial boiler incinerators or steel pouring pits. Refractory mortars are made primarily from fire clay, with calcium aluminate or sodium silicate as a binder. Mortar joints for refractory mortars should not exceed ¼ in. The fire bricks are often dipped to get a thin mortar coating, with no conventional mortar bed laid. Exposure to heat in the firebox, smoke chamber, and flue ceramically fuses the mortar and seals the joints against heat penetration. For residential and commercial fireplaces, use a medium-duty mortar as determined by ASTM C199, *Pier Test for Refractory Mortar.* Manufacturers or suppliers should be consulted regarding design details and performance characteristics for special applications.

6.4.2 Chemical-Resistant Mortars

The field of chemical-resistant mortars is highly specialized and complex in nature. Durability depends very heavily on proper mortar selection.

Even with the use of chemical-resistant brick or structural clay tile, mortar may still be attacked by acids or alkalis, causing joint disintegration and loosening of the masonry units. There are few chemicals which do not attack regular portland cement mortars. Consequently, it is necessary to develop chemical resistance by means of admixtures or surface treatments. Special cements or coatings are available which will withstand almost all service conditions, but different types react differently with various chemicals. The success of any particular treatment depends on local conditions, type and concentration of the chemical solution, temperatures, wear, vibration, type of subsurface, and workmanship. Joints should be as narrow as possible to minimize the exposed area and reduce the quantity of special material required. The selection of the optimum material for a particular installation must include the consideration of mechanical and physical properties as well as chemical-resistant characteristics.

Several special types are available, including sulfur mortars, silicate mortars, phenolic resin mortars, and furan, polyester, and epoxy resin mortars. The properties and capabilities may be altered by changing the formulations. For specific installations, full use should be made of available standards and test procedures (*see Fig. 6-9*), and the engineering

Standard number	*Title*
ASTM C395	Chemical-Resistant Resin Mortars
ASTM C279	Chemical-Resistant Masonry Units
ASTM C466	Chemically Setting Silicate and Silica Chemical-Resistant Mortars
ASTM C287	Chemical-Resistant Sulfur Mortar
ASTM C413	Absorption and Apparent Porosity of Chemical-Resistant Mortars
ASTM C321	Bond Strength of Chemical-Resistant Mortars
ASTM C608	Brittle Ring Tensile Strength of Chemical Setting Silicate and Silica Chemical-Resistant Mortars
ASTM C267	Chemical Resistance of Mortars
ASTM C579	Compressive Strength of Chemical-Resistant Mortars
ASTM C396	Compressive Strength of Chemically Setting Silicate and Silica Chemical-Resistant Mortars
ASTM C306	Compressive Strength of Chemical-Resistant Resin Mortars
ASTM C580	Flexural Strength and Modulus of Elasticity of Chemical-Resistant Mortars
ASTM C531	Shrinkage and Coefficient of Thermal Expansion of Chemical-Resistant Mortars
ASTM C307	Tensile Strength of Chemical-Resistant Resin Mortars
ASTM C414	Working and Setting Times of Chemical-Resistant Silicate and Silica Mortars
ASTM C308	Working and Setting Times of Chemical-Resistant Resin Mortars
ASTM C386	Use of Chemical-Resistant Sulfur Mortars
ASTM C397	Use of Chemically Setting Chemical-Resistant Silicate and Silica Mortars
ASTM C398	Use of Hydraulic Cement Mortars in Chemical-Resistant Masonry
ASTM C399	Use of Chemical-Resistant Resin Mortars

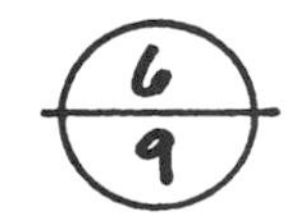

Specifications, tests, and practice standards for chemical-resistant mortars. (*From Brick Institute of America,* Technical Note 32, *BIA, Reston, Va.*)

advice, services, and recommendations of manufacturing specialists in this field should be solicited.

6.5 GROUT

Grout is a fluid mixture of cementitious material and aggregate with enough water added to allow the mix to be poured or pumped into masonry cores and cavities without segregation (*see Fig. 6-10*). ASTM C476, *Standard Specification for Grout for Masonry,* covers both fine and coarse mixtures based on aggregate size and grading.

Selection of a fine or coarse grout is based on the size of the core or cavity as well as the height of the lift to be grouted. (Some building codes and standards have different requirements for the relationship of maximum aggregate size to clear opening, so for specific projects the governing code should always be checked.) In accordance with ASTM C404, *Standard Specification for Aggregates for Masonry Grout,* if the maximum aggregate size is $\frac{3}{8}$ in. or larger, the grout is classified as *coarse.* If the maximum aggregate size is less than $\frac{3}{8}$ in., it is classified as *fine.* The smaller the grout space, the smaller the maximum aggregate size allowed. Although ASTM C404 limits the maximum aggregate size to $\frac{3}{8}$ in., some engineers allow up to $\frac{3}{4}$ in. aggregate for grouting large voids such as columns and pilasters. The larger aggregate takes up more volume, reduces grout shrinkage, and requires less cement for equivalent strength. The table in *Fig. 6-11* shows the recommended grout type for various grout spaces from the MSJC, *Specifications for Masonry Structures.*

Grout is an essential element of reinforced masonry construction. It must bond the masonry units and the steel together so that they perform integrally in resisting superimposed loads. In unreinforced loadbearing construction, unit cores are sometimes grouted to give added strength, and in non-loadbearing construction, to increase fire resistance. The fluid consistency of grout is important in determining compressive strength, in assuring that the mix will pour or pump easily and without segregation, and that it will flow around reinforcing bars and into corners and recesses without voids. ASTM C476 specifies grout proportions by volume, but does not indicate minimum strength or slump limits. Optimum water content, consistency, and slump will depend on the absorption rate of the units as well as job-site temperature and humidity conditions. Performance records indicate a minimum slump of 8 in. is necessary for units with low absorption, and as much as 10 in. for units with high absorption.

Depending on the amount of mixing water used, the mix proportions in *Fig. 6-12* will normally produce grouts with laboratory compressive strengths in non-absorptive molds of up to 2500 psi at 28 days. However, actual field compressive strength is usually higher because mixing water is absorbed immediately by the units, thus reducing the water/cement ratio and increasing the strength. The water absorbed by the units is

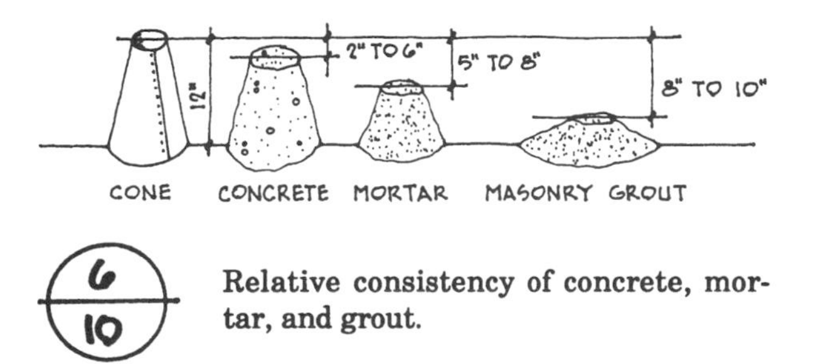

6-10 **Relative consistency of concrete, mortar, and grout.**

Grout type	Grout pour height (ft)	Min. width of grout space between wythes (in.)	Min. dimensions for grouting cells of hollow units (in. × in.)
Fine	1	¾	1½ × 2
	5	2	2 × 3
	12	2½	2½ × 3
	24	3	3 × 3
Coarse	1	1½	1½ × 3
	5	2	2½ × 3
	12	2½	3 × 3
	24	3	3 × 4

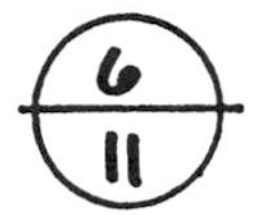

Grout space requirements. [*From Masonry Standards Joint Committee (MSJC),* **Specifications for Masonry Structures,** *ACI 530.1/ASCE 6/TMS 602.*]

Type	Grout proportions by volume			
	Parts by volume of portland cement or blended cement	Parts by volume of hydrated lime or lime putty	Aggregate, measured in a damp, loose condition	
			Fine	Coarse
Fine grout	1	0 to 1/10	2¼ to 3 times the sum of the volumes of the cementitious materials	—
Coarse grout	1	0 to 1/10	2¼ to 3 times the sum of the volumes of the cementitious materials	1 to 2 times the sum of the volumes of the cementitious materials

Grout requirements, from ASTM C476. (*Copyright, American Society for Testing and Materials, 1916 Race Street, Philadelphia, Pa. 19103. Reprinted with permission.*)

retained for a period of time, thus also providing a moist condition for optimum curing of the grout. Grout which might have a predicted 28-day strength of only several hundred psi based on the water/cement ratio may prove to have an actual strength of 3000 to 4000 psi or higher when actual core samples are tested because of unit absorption, and the resulting moist curing conditions. Unit absorption is affected not only by the characteristics of the brick or block, but also by the size of the cavity as well. The greater the surface area, the more water will be absorbed, so water content and slump limits should be adjusted accordingly.

7

MASONRY ACCESSORIES

Accessory items are important and integral components of masonry construction. Steel lintels, shelf angles, horizontal joint reinforcement, metal anchors, ties, fasteners, flashing materials, and other accessories must be of the highest quality to equal the quality of the masonry units themselves.

7.1 METALS AND CORROSION

Steel, which is most frequently used for fabrication of masonry accessories, requires protective coatings to isolate the metal from the corrosive effects of wet mortar. Several nonferrous metals are also used for masonry accessories. *Copper* and copper alloys are essentially immune to the corrosive action of wet concrete and mortar. Because of this immunity, copper can be safely embedded in fresh mortar even under saturated conditions. Galvanic corrosion will occur, however, if copper and steel items are either connected or in close proximity to one another. The presence of soluble chlorides will also cause copper to corrode.

Aluminum is also attacked by fresh portland cement mortar and produces the same expansive pressures. Galvanic corrosion also occurs if aluminum and steel are embedded in the mortar in contact with one another. If aluminum is to be used in reinforced masonry, it should be electrically insulated by a permanent coating of bituminous paint, alkali-resistant lacquer, or zinc chromate paint. If the coating is not kept intact, chlorides can greatly accelerate corrosion.

Most metal connectors used in masonry construction are of steel wire, sheet steel, or structural steel. Steel wire for reinforcement and connectors is cold-drawn wire made from low carbon steel rods (ASTM A82, *Standard Specification for Steel Wire, Plain, for Concrete Reinforcement*). It is less ductile than conventional hot-rolled structural steel and has a less well-defined yield point. Stainless steel wire is nickel-chromium stainless steel (ASTM A580, *Standard Specification for Stainless and Heat-Resisting Steel Wire*) that is annealed in the manufacturing process and, as a result, has a yield stress more consistent with structural steel. Annealed nickel-chromium stainless steels are austenitic and nonmagnetic. The table in *Fig. 7-1* lists properties of steel wire used in masonry.

Wire size					Tensile strength (lb)	
ASTM A82 designation	Gauge no.	Diameter (in.)	Area (sq in.)	Weight (lb/ft)	Yield	Ultimate
W1.1	11	0.1205	0.0114	0.0387	797	909
W1.7	9	0.1483	0.0173	0.0587	1210	1380
W2.1	8	0.1620	0.0206	0.0700	1442	1648
W2.8	3⁄16 in.	0.1875	0.0277	0.1250	1940	2220
W4.9	¼ in.	0.2500	0.0491	0.1667	3430	3935

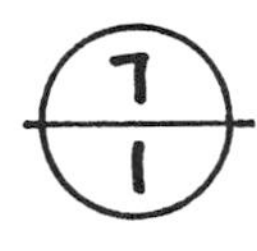

Common wire sizes used in masonry.

Sheet metal anchors are made from either cold-rolled carbon steel (ASTM A366, *Standard Specification for Steel, Sheet, Carbon, Cold-Rolled, Commercial Quality*) or stainless steel (ASTM A167, *Standard Specification for Stainless and Heat-Resisting Chromium-Nickel Steel Plate, Sheet, and Strip,* Type 304). The table in *Fig. 7-2* lists sheet metal thicknesses and standard gauges. Steel reinforcing bars may conform to a number of ASTM standards depending on the strength desired. Structural steel used for lintels, shelf angles, or strap anchors should conform to ASTM A36, *Standard Specification for Structural Steel.*

Corrosion of metals occurs from weathering, direct chemical attack, and galvanic action. Since most metals used in masonry construction are concealed within the masonry, exterior weathering is generally not a concern. However, corrosion may be caused by prolonged exposure to moisture

Gauge	Thickness (in.)
8	0.1681
10	0.1382
12	0.1084
14	0.0785
16	0.0635
18	0.0516
20	0.0396
22	0.0336
24	0.0276
26	0.0217
28	0.0187
30	0.0157
32	0.0134

Standard sheet metal gauges.

which condenses within a wall section or in open cavities or collar joints; water which penetrates the exterior face shell of single-wythe walls or the exterior wythe of cavity or veneer walls; or atmospheric humidity in excess of 75% in hollow cores and cavities. Direct chemical attack can be caused by chlorides, and set-accelerating admixtures which contain calcium chloride should not be used in masonry mortar. Deep carbonation of mortar caused by carbon dioxide intrusion through cracks or voids at the mortar-to-unit interface may also accelerate corrosion of metal anchors, ties, or reinforcement embedded in the mortar. Some metal corrosion in masonry is caused by galvanic action. Galvanic action causes corrosion between dissimilar metals in the presence of an electrolyte (such as water).

All steel used in masonry, with the exception of reinforcing bars and wire fabric, should be galvanized or stainless steel. Although zinc is also susceptible to corrosive attack, it is used in the galvanizing process to provide both a barrier coating to isolate the steel from corrosive elements and a sacrificial anodic coating that is consumed to protect the base steel at uncoated areas such as scratches and cut ends. Although corroded metal occupies a greater volume than the original material and exerts expansive pressures around the embedded item, the film of zinc used to galvanize masonry accessories is so thin that the pressure is insufficient to crack the masonry. If the masonry is absorbing excessive moisture because of design or construction defects, however, corrosion of the steel may continue and the expansive pressures increase substantially over time. As this "rust jacking" continues, the masonry is cracked, allowing even more moisture to enter the wall.

Masonry accessories in exterior walls and interior walls exposed to relative humidities of 50% or higher should be hot-dip galvanized after fabrication in accordance with ASTM A153, *Standard Specification for Zinc Coating (Hot-Dip) on Iron and Steel Hardware,* Class B. Mill galvanizing and electro-galvanizing do not provide protection at sheared edges, wire ends, shop welds, penetrations, and so on. For interior walls exposed to lower humidity, joint reinforcement can be zinc-coated in accordance with ASTM A641, *Standard Specification for Zinc Coated (Galvanized) Carbon Steel Wire.* The life expectancy of the corrosion protection afforded by galvanizing is directly proportional to its thickness (*see Fig. 7-3*). Stainless steel accessories are less susceptible to corrosion and provide greater long term durability for masonry construction. Stainless steel will provide the highest corrosion protection in severe exposures, and should conform to Series 300, ASTM A167, *Standard Specification for Stainless Steel and Heat-Resisting Chromium-Nickel Steel Plate, Sheet, and Strip.* ASTM Committee C15 on Manufactured Masonry Units is in the process of developing a standard guide for corrosion protection of embedded metals in masonry. The standard is intended to establish minimum acceptable levels of corrosion protection for ties, anchors, fasteners, and inserts based on exposure conditions and perhaps even to a driving rain index (refer to Chapter 9).

The degree of galvanic corrosion which can occur between dissimilar metals depends on the intimacy of contact, the type of electrolyte, and the voltage developed between the two metals. An electric current is conducted through the electrolyte, corroding one metal (the anode) and plating the other (the cathode). The greater the potential difference between the two metals, the more severe the corrosion (*see Fig. 7-4*). The metal that is higher in the galvanic series table is subject to corrosion by metals lower in the

TABLE A LIFE EXPECTANCY OF GALVANIZED COATINGS (ADAPTED FROM AMERICAN GALVANIZERS ASSOC.)

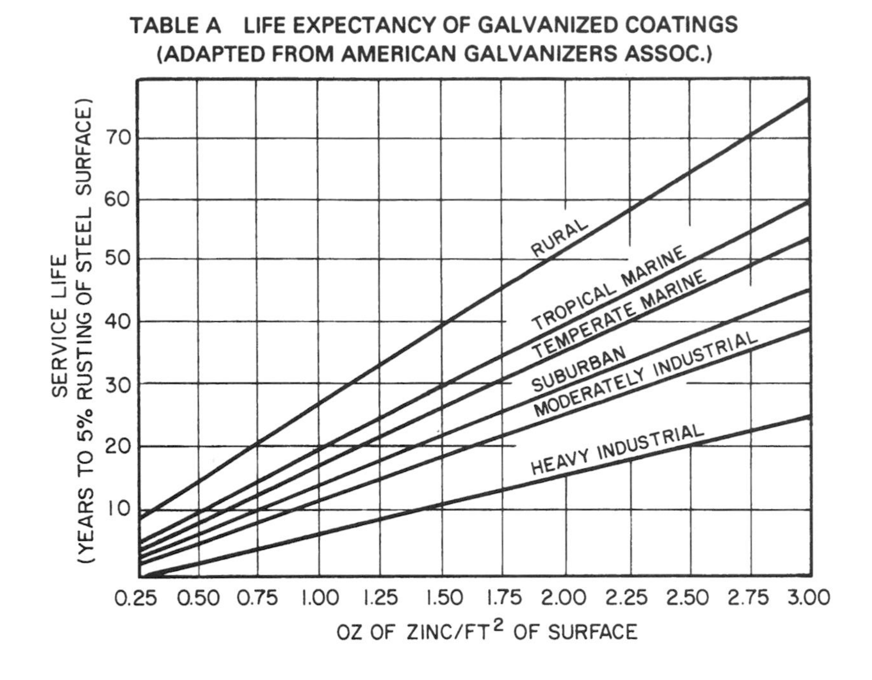

TABLE B LIFE EXPECTANCY OF GALVANIZED CAVITY WALL TIES* (ADAPTED FROM CLAYFORD T. GRIMM)

Probability of occurrence (%)	*Corrosion rate (10^{-4} oz. zinc/sq ft/yr)*	*Life expectancy (yr)*			
		ASTM A153, class B2		*ASTM A153, class B1*	
		Minimum	*Average*	*Minimum*	*Average*
5	2415	5.2	6.2	7.5	8.3
10	1791	7.0	8.4	10.1	11.2
20	1075	11.6	14.0	16.7	18.6
25	875	14.3	17.1	20.6	22.9
33	656	19.1	22.9	27.4	30.5
50	393	31.8	38.2	45.8	50.9

*Data taken in climatic areas with a driving rain index of 2.5 to 5.0 (see Chap. 9).

Life expectancy of galvanized coatings.

series. The density of the corrosion current, or the size of the current relative to the anode surface, is also important. If a fastener's surface is small compared to the metal to be fastened, its current density will be high and therefore subject to rapid corrosion. Therefore, as a general rule, a fastener in a given environment should be lower in the galvanic series table than the material to be fastened.

To protect against galvanic corrosion when dissimilar metals are used, isolation can be provided by an electrical insulator such as neoprene or asphalt-impregnated felt.

Metal or alloy	
Magnesium	Anode (+) least noble
Zinc	
Aluminum, 5052 alloy	
Aluminum, 6061 alloy	
Cadmium	
Aluminum, 2024-T4 alloy	
Iron or carbon steel	
4–6% chromium steel	
Ferritic stainless steel, 400 series (active)	
Austenitic stainless steel, 18-8 series (active)	
Lead	
Tin	
Nickel (active)	
Brass	
Copper	
Bronze	
Monel	
Silver solder	
Nickel (passive)	
Ferritic stainless steel (passive)	
Austinitic stainless steel (passive)	
Silver	
Titanium	
Graphite	
Gold	
Platinum	Cathode (−) most noble

Note: The farther apart two metals are in the galvanic series, the greater the corrosion of the less noble material.

Galvanic series of metals.

7.2 HORIZONTAL JOINT REINFORCEMENT

Horizontal joint reinforcement is used to control shrinkage cracking in concrete masonry unit (CMU) walls. It can also be used to tie the wythes of multi-wythe walls together, to bond intersecting walls, and assure maximum flexural wall strength against lateral loads. The basic types of joint reinforcement available are shown in *Fig. 7-5.* Some designs are better for certain applications than others:

- In single-wythe walls, two-wire ladder or truss type reinforcement is most appropriate. Under most circumstances, the ladder type provides adequate restraint against shrinkage cracking. The truss type is stronger and provides about 35% more area of steel, but the ladder type

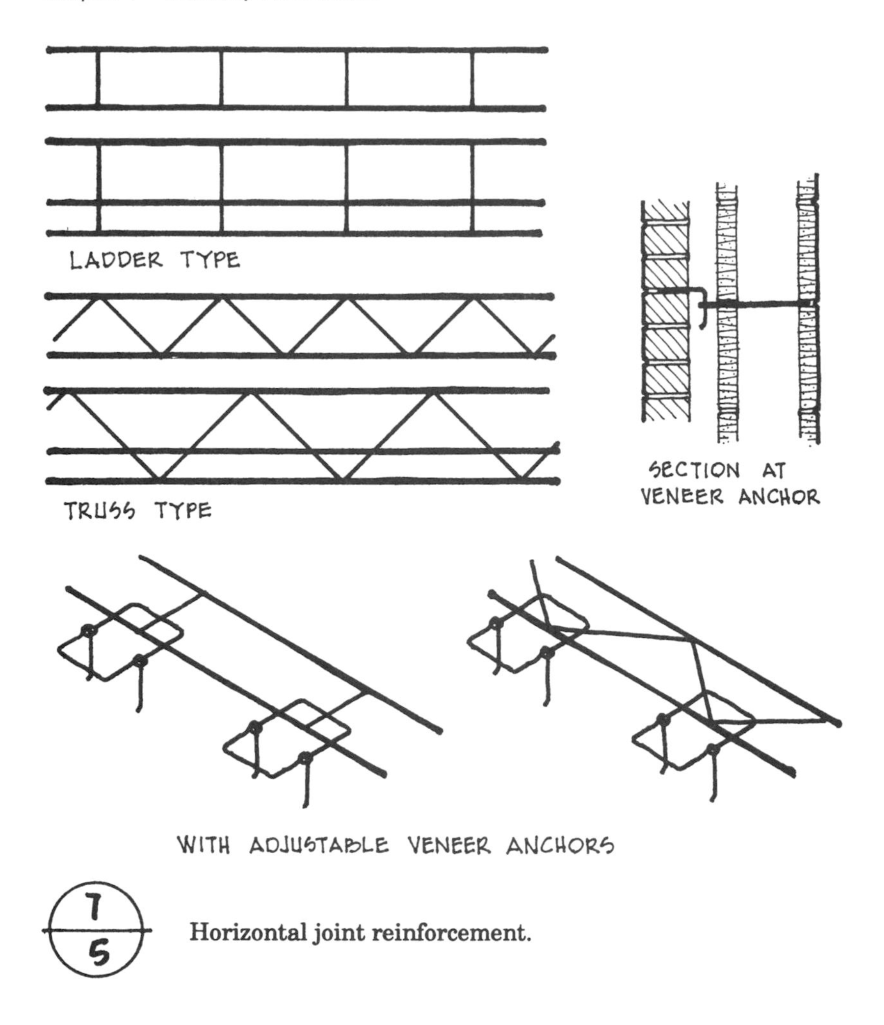

Horizontal joint reinforcement.

generally interferes less with vertical bar placement in structurally reinforced walls.

- For multi-wythe walls in which the backing and facing wythes are of the same type of masonry, three-wire joint reinforcement of either the truss- or ladder-type design is suitable. If the wythes are laid up at different times, however, the three-wire design makes installation awkward. Three-wire truss-type reinforcing should never be used when insulation is installed in the cavity between wythes because it is too stiff to allow for differential thermal movement between the backing and facing wythes.
- For walls in which the backing and facing wythes are laid at different times, or walls which combine clay and concrete masonry in the facing and backing wythes, joint reinforcement with adjustable ties allows differential movement between wythes and facilitates the installation of the outer wythe after the backing wythe is already in place. The adjustable ties may be either a tab or hook-and-eye design. Joint reinforcement with adjustable ties should not be used with concrete masonry facing wythes over concrete masonry backing wythes. The concrete masonry facing requires shrinkage restraint which is not provided by the intermittent ties. For concrete masonry facings over concrete masonry backing, three-wire joint reinforcement is more appropriate.

- For uninsulated cavity walls of block and brick where the backing and facing wythes are laid at the same time, truss- or ladder-type reinforcement with fixed welded-wire tab ties can be used. It is less expensive than reinforcement with adjustable ties, but also allows less differential movement. If the cavity is insulated, the tabs restrain differential thermal movement between the backing and facing wythes. Tab-type reinforcement also does not provide shrinkage restraint for concrete masonry facing wythes.
- For projects in seismically active areas, joint reinforcement with seismic anchors is available from several manufacturers.

Figure 7-6 summarizes the general recommendations for using various types of joint reinforcement in various applications.

Horizontal joint reinforcement is usually made of galvanized steel wire. Spacing of the welded lateral ties should not exceed 16 in. for deformed wire or 6 in. for smooth wire. If used as structural reinforcing, the longitudinal chords *must* be of deformed wire. Joint reinforcement should conform to the requirements of ASTM A951, *Standard Specification for Joint Reinforcement for Masonry*. For exterior walls and for interior walls exposed to a relative humidity of 75% or higher, joint reinforcement should be hot-dip galvanized after fabrication in accordance with ASTM A153, Class B2. For interior walls exposed to lower humidity, joint reinforcement can be zinc-coated in accordance with ASTM A641. Stainless steel joint reinforcement will provide the highest corrosion protection in severe exposures, and should conform to ASTM A167, Series 300.

Joint reinforcement is available in several wire diameters, and in standard lengths of 10 to 12 ft. Longitudinal wires are available in standard 9 gauge (W1.7) and extra heavy $^{3}/_{16}$ in. (W2.8). Standard 9-gauge wire provides better fit and more practical constructability in $^{3}/_{8}$-in. mortar joints. With extra heavy $^{3}/_{16}$-in. wire, there is little room for construction tolerances, and a Type M or Type S mortar is required to develop full bond strength with the steel. Heavy-gauge joint reinforcement should be used only when there is compelling engineering rationale. Cross wires are typically either 9 or 12 gauge. Fabricated joint reinforcement widths are approximately $1^{5}/_{8}$ in. less than the actual wall thickness to assure adequate mortar coverage. The mortar cover at the exterior wall face should be at least $^{5}/_{8}$ in. Any wire which is not covered by mortar is high in tensile strength, but weak in shear. For maximum effectiveness in structural assemblages, collar joints between wythes should be filled with grout.

7.3 CONNECTORS

There are three different types of masonry connectors. *Anchors* attach masonry to a structural support such as an intersecting wall, a floor, a beam, or a column. This type of connector includes anchor bolts and veneer anchors used to attach masonry veneers to backing walls of non-masonry construction. *Ties* connect multiple wythes of masonry together in cavity wall or composite wall construction. *Fasteners* attach other building elements or accessories to masonry.

7.3.1 Ties

While joint reinforcement can provide longitudinal strength in addition to lateral connection between wythes, individual *corrugated* or *wire ties* func-

Wall configuration	*2-Wire ladder*	*2-Wire truss*	*3-wire ladder*	*3-wire truss*	*2-wire ladder or truss with adjustable ties*	*2-wire ladder or truss with fixed tab ties*	*2-wire ladder or truss with seismic ties*
Single-wythe CMU ▪ With vertical reinforcing steel	●						
Single-wythe CMU ▪ Without vertical reinforcing steel		●					
Multi-wythe ▪ Insulated Cavity ▪ Both wythes laid at same time ▪ Backing and facing wythes both CMU			●				
Multi-wythe ▪ Uninsulated cavity ▪ Both wythes laid at same time ▪ Backing and facing wythes both CMU			●	●			
Multi-wythe ▪ Wythes laid at different times ▪ Backing wythe CMU facing wythe clay masonry				●			
Multi-wythe ▪ Uninsulated cavity ▪ Both wythes laid at same time ▪ Backing wythe CMU facing wythe clay masonry						●	
Multi-wythe ▪ Both wythes laid at same time ▪ Backing and facing wythes both CMU ▪ Seismic performance Category C			●	●			
Multi-wythe ▪ Wythes laid at different times ▪ Backing wythe CMU facing wythe clay masonry ▪ Seismic performance Category C							●

Joint reinforcement selection guide. (*Adapted from Mario Catani, "Selecting the Right Joint Reinforcement for the Job,"* The Magazine of Masonry Construction, *January 1995.*)

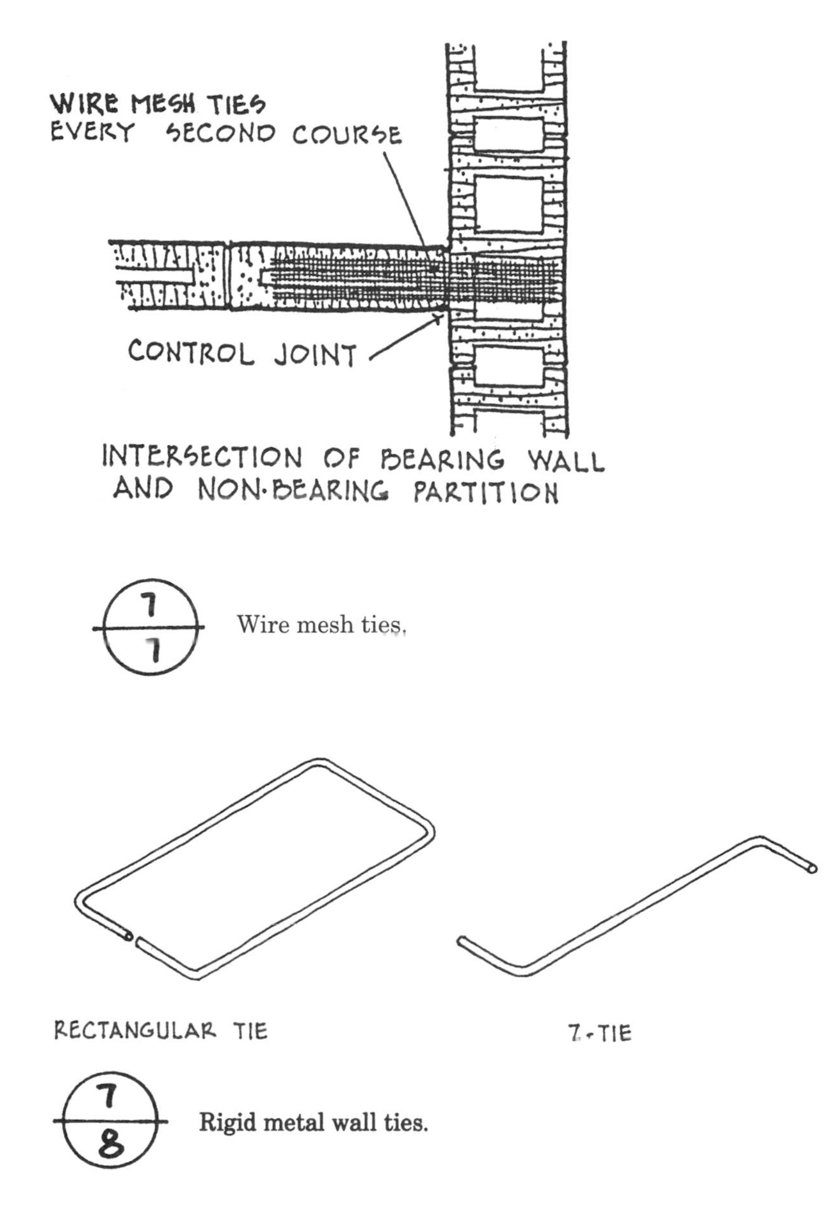

Wire mesh ties.

Rigid metal wall ties.

tion only in the lateral direction, providing intermittent rather than continuous connection. There are several shapes and configurations, different wire gauges, and various sizes to suit the wall thickness. *Woven wire mesh* is sometimes used to connect intersecting masonry walls when no load transfer is desired. This is a soft connection and requires the installation of control joints at the wall intersection (*see Fig. 7-7*). Wire ties should be used in open-cavity walls and grouted multi-wythe walls. Wire ties may be rigid for laying in bed joints at the same height, or adjustable for laying in bed joints at different levels (*see Figs. 7-8 and 7-9*). Adjustable ties also permit differential expansion and contraction between backing and facing wythes of cavity walls. This is particularly important for connecting between clay and concrete masonry because the thermal and moisture movement characteristics of the materials are so different. Crimped ties which form a water drip in the cavity are not recommended because the deformation reduces their strength in transferring lateral loads. Crimped ties, in fact, are prohibited under some building codes. Drips are incorpo-

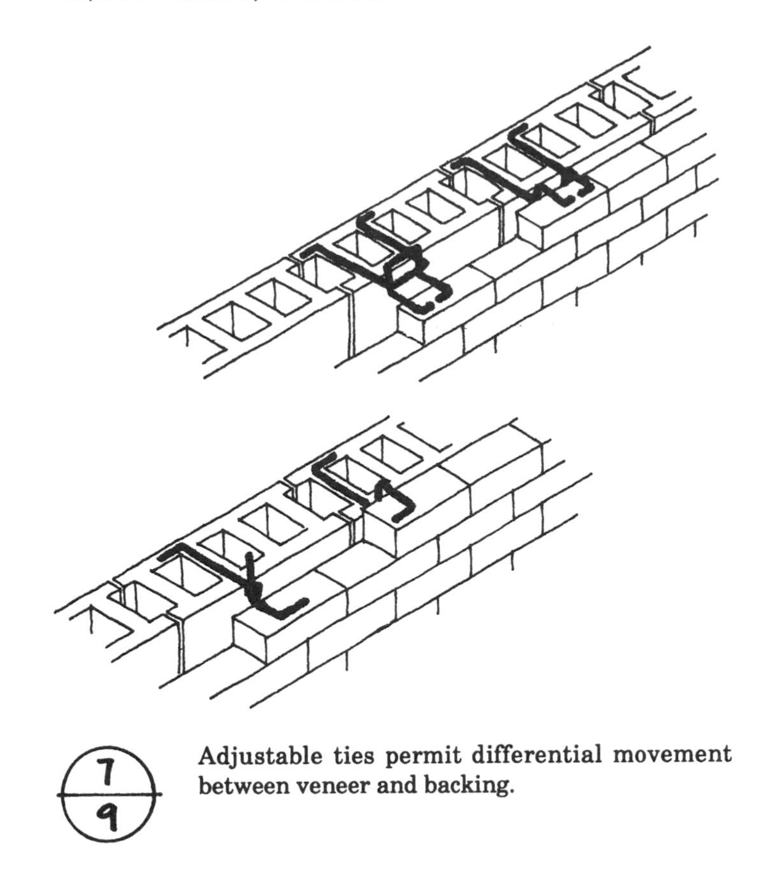

Adjustable ties permit differential movement between veneer and backing.

rated by some manufacturers by installing a plastic ring at the midsection of the wire.

Many building codes prescribe maximum tie spacing. Ties should be staggered so that no two alternate courses form a continuous vertical line, and ties should always be placed in the mortar bed rather than laid directly on the masonry unit. Structural requirements of metal wall ties can be calculated by rational design methods. Particularly in the case of adjustable ties in loadbearing construction, it is recommended that engineering analysis be used to assure adequate strength and proper performance. Adjustable ties for cavity walls should be structurally designed for each different condition of wind load, tie configuration, dimension, size, location, stiffness, embedment, modulus of elasticity of masonry, moment of inertia of each cavity wall wythe, and difference in level of connected joints.

Wire ties may be rectangular or Z-shaped in lengths of 4, 6, or 8 in. (*see Fig. 7-8*). Z-ties should have at least a 2-in. 90° leg at each end. Rectangular ties should have a minimum width of 2 in. and welded ends if the width is less than 3 in. Either type may be used for solid masonry (core area less than 25%), but Z-ties are less expensive. Only rectangular ties should be used in ungrouted walls of hollow masonry. Corrugated steel ties should have 0.3- to 0.5-in. wavelength, 0.06- to 0.10-in. amplitude, ⅞-in. width, and minimum 22-gauge thickness. Corrugated ties should be long enough to reach the outer face shell mortar bed of hollow units or the center of the mortar bed for solid units. Wire mesh ties should be formed of unwelded, woven wire, 16 gauge or heavier. A minimum width of 4 in. is required and a ½×½-in. or finer mesh. Lengths may be field-cut for convenience, and butt joints are acceptable.

Although metal ties are typically made of several materials, highest performance results from the following:

- Stainless steel, ASTM A167, Series 300
- Carbon steel, hot-dip galvanized after fabrication in accordance with ASTM A153, Class B2
  - Steel plate, headed and bent bar ties, ASTM A36
  - Sheet metal, ASTM A366
  - Wire mesh, ASTM A185
  - Wire ties, ASTM A82

7.3.2 Anchors

Masonry wall anchors provide connections which can resist compressive, tensile, and shear stresses. Rigid anchors resist all three types of loading, but flexible anchors, which may be needed to permit differential movement, do not resist shear.

Galvanized steel bolts and strips are typically used as *rigid anchors* (*see Fig. 7-10*). Hooked strips are used to anchor intersecting walls. They are typically $1\frac{1}{2} \times \frac{1}{4}$ in. in cross section and at least 24 in. long. A 1-in. 90° leg at each end is embedded in a grouted core or mortar filled joint. Bolt anchors can be of several diameters and lengths. They are threaded at one end, and have a 1-in. 90° leg at the other for embedding in a grouted core or mortar joint.

Flexible metal anchors may be of either wire or sheet metal. The configuration of the wire to be embedded in the mortar joint should be as shown in *Fig. 7-11*, or should have the equivalent pullout strength. This triangular or hooked leg may be looped through a rod or notched plate welded to a steel structural member; through a sheet metal strip dovetailed to fit a slot in a concrete structural member or screwed to a metal stud; or through a metal eye cast into the concrete, attached to joint reinforcement, or hooked to a metal stud. Flexible anchors allow the wire end to remain perpendicular to the wall when differential movement changes the vertical alignment of the backing and facing walls. This provides better performance than corrugated anchors which must bend out of plane to accommodate such movements.

Corrugated sheet metal veneer anchors should meet the same physical

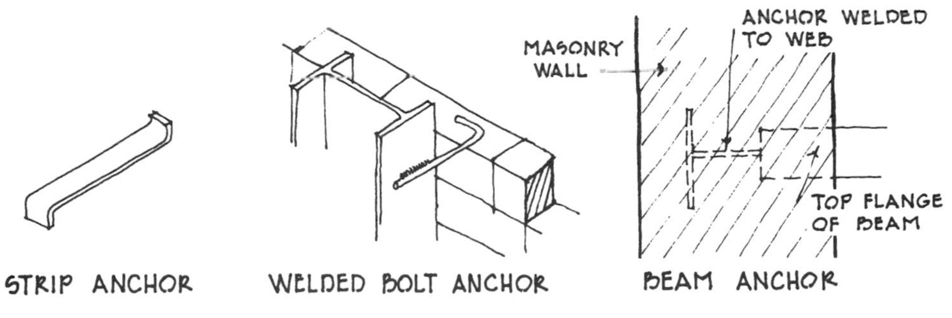

Rigid anchors. (*From National Concrete Masonry Association,* **TEK Bulletin 21A,** *NCMA, Herndon, Va.*)

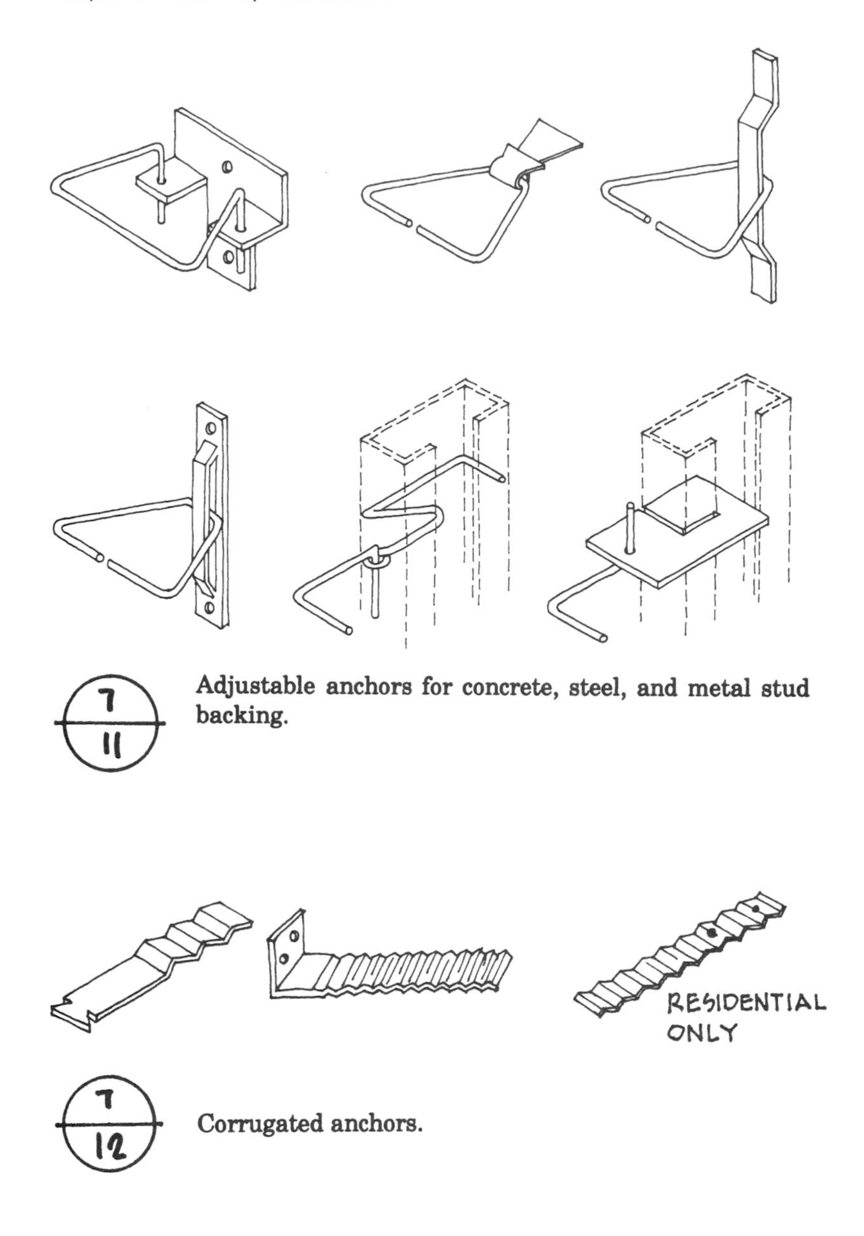

Adjustable anchors for concrete, steel, and metal stud backing.

Corrugated anchors.

requirements as corrugated ties (0.3- to 0.5-in. wavelength, 0.06- to 0.10-in. amplitude, ⅞-in. width, and 22-gauge thickness). These anchors may be used only with solid or solidly grouted units where the distance between the veneer and supporting frame is 1 in. or less. One end of the anchor is nailed or screwed directly to a wood or steel stud, and the other end is embedded in a mortar joint (*see Fig. 7-12*). Performance is greatly reduced if the attaching nail or screw is not located exactly at the bend, so these anchors should be used only for one- and two-story residential type construction. Corrugated dovetail anchors are fabricated to fit a dovetailed slot in a concrete structural frame.

Some building codes require special anchorage of masonry veneers in seismic areas. In response to this requirement, some manufacturers now produce specially designed *seismic anchors* which consist of a single or double continuous reinforcing wire attached to a plate for connection to different types of backing walls (*see Fig. 7-13*).

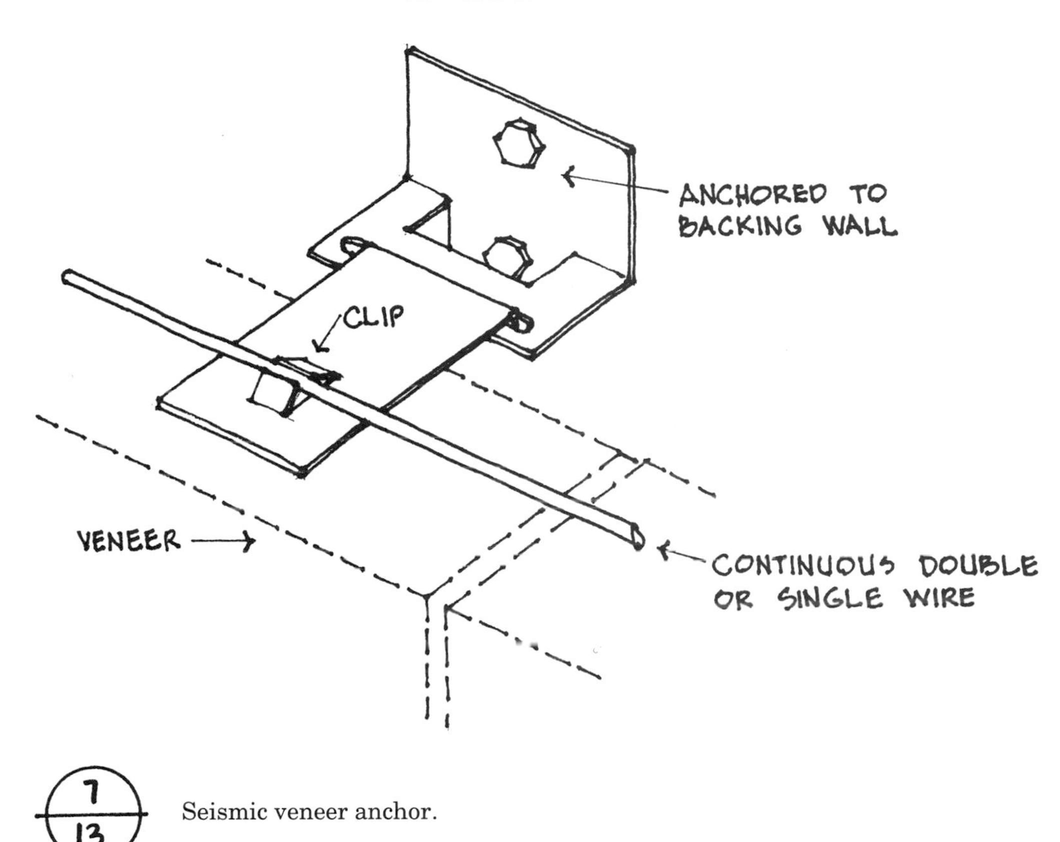

7/13 Seismic veneer anchor.

Several types of proprietary anchors have also been introduced for seismic retrofitting of unreinforced masonry and for re-anchoring masonry veneer. Retrofit veneer anchors are designed to

- Provide anchors in areas where they were not installed in the original construction
- Replace failed existing anchors
- Replace failed existing header bond units
- Upgrade older wall systems to current code, including seismic retrofitting of older buildings
- Attach new veneers over existing facades

The three general types are a mechanical expansion system, a screw system, and an epoxy adhesive system (*see Fig. 7-14*). Seismic retrofit anchors are designed to anchor existing masonry walls to existing floor and roof diaphragms for combined action under load. Seismic forces can thus be transferred from walls perpendicular to acceleration, to walls parallel to acceleration which are more capable of dissipating the force.

Figure 7-15 shows some typical anchors used to attach stone veneer to various structural frames. ASTM C1242, *Standard Guide for Design, Selection, and Installation of Exterior Dimension Stone Anchors and Anchoring Systems,* provides recommended guidelines for these complex anchoring systems. The standard defines several different generic types of anchors and discusses the design principles which must be considered in

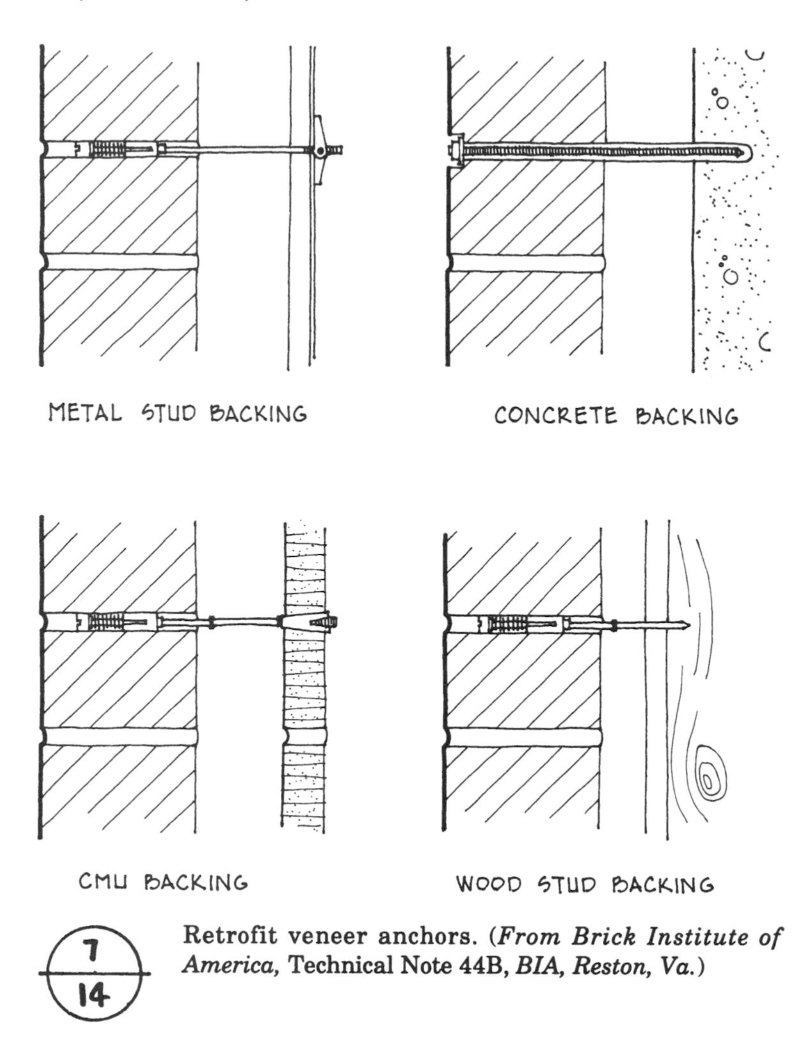

Retrofit veneer anchors. (*From Brick Institute of America,* **Technical Note 44B,** *BIA, Reston, Va.*)

resisting both lateral and gravity loads. An appendix also provides information on safety factors.

Stone anchors are almost exclusively made of stainless steel (ASTM A167, Type 304) to minimize corrosion and staining. If properly protected from moisture and from galvanic action, metal components that are not in direct contact with the stone can sometimes be made of galvanized steel, painted or epoxy coated steel, or aluminum. Copper, brass, and bronze will stain and should not be used in stone anchoring systems. Metal anchors for clay and concrete masonry are fabricated of the same materials as those used for horizontal joint reinforcement and for ties.

7.3.3 Fasteners

Attaching fixtures or dissimilar materials to masonry requires some type of fastener. Most plugs, nailing blocks, furring strips, and so on, can be installed by the mason as the work proceeds. There are a variety of products and methods from which to choose, depending largely on the kind of fixture or material to be attached and the type of masonry involved (*see Fig. 7-16*).

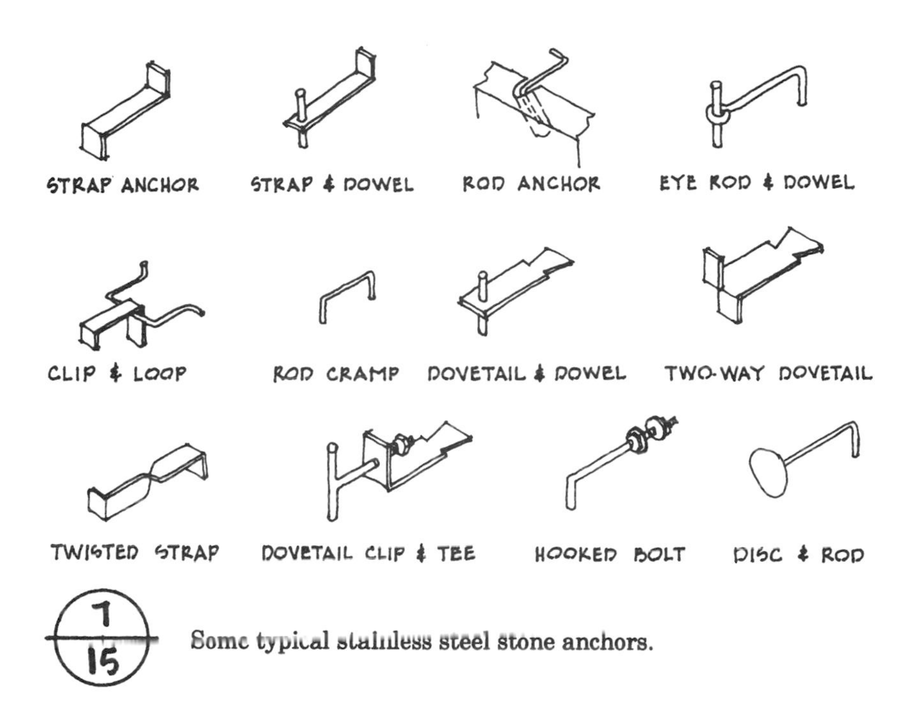

Some typical stainless steel stone anchors.

The most common method of attaching wood trim items such as baseboards or chair rails is placing *wood nailing blocks* in the vertical joints as the mason builds the wall. These blocks should be of seasoned softwood creosoted to prevent shrinkage and rot. They should never be placed in horizontal joints. *Galvanized metal nailing plugs,* with or without fiberboard inserts, provide better construction and are easily set into the joints during construction. *Toggle bolts* and *double-threaded fasteners* can be used only with hollow masonry units, and are installed after the wall is completed. *Wood plugs with threaded hooks* can be used with either solid or hollow masonry. The plug may be built into the wall or driven into a hole drilled after construction. *Plastic or fiber plugs* can also be used with solid or hollow units. They are placed in holes drilled into either the mortar joints or face shells of the masonry. *Expansion shields* and *wedge-type bolts* may be used with solid or grouted masonry. Newer attachment methods include pins or fasteners rammed or driven into solid masonry with a power tool or gun, and direct adhesive or mastic application.

Wood furring strips can be attached using nailing blocks, metal wall plugs, or direct nailing into mortar joints with case-hardened "cut nails" (wedge-shaped) or spiral threaded masonry nails. Special anchor nails may be adhesively applied to the wall, or porous clay nailing blocks inserted into the bonding pattern (*see Fig. 7-17*). Metal furring strips are attached to the wall by tie wires built into the mortar joints or by special clips designed for this purpose.

7.4 MOVEMENT JOINT FILLERS

Concrete masonry moisture shrinkage and clay masonry moisture expansion, along with reversible thermal movement, are accommodated through special jointing techniques which allow movement without damage to the wall. Control joints for concrete masonry are designed as stress-relieving contraction points, and must extend completely through the masonry wythe. Preformed rubber or PVC shear keys transfer lateral loads across the joint

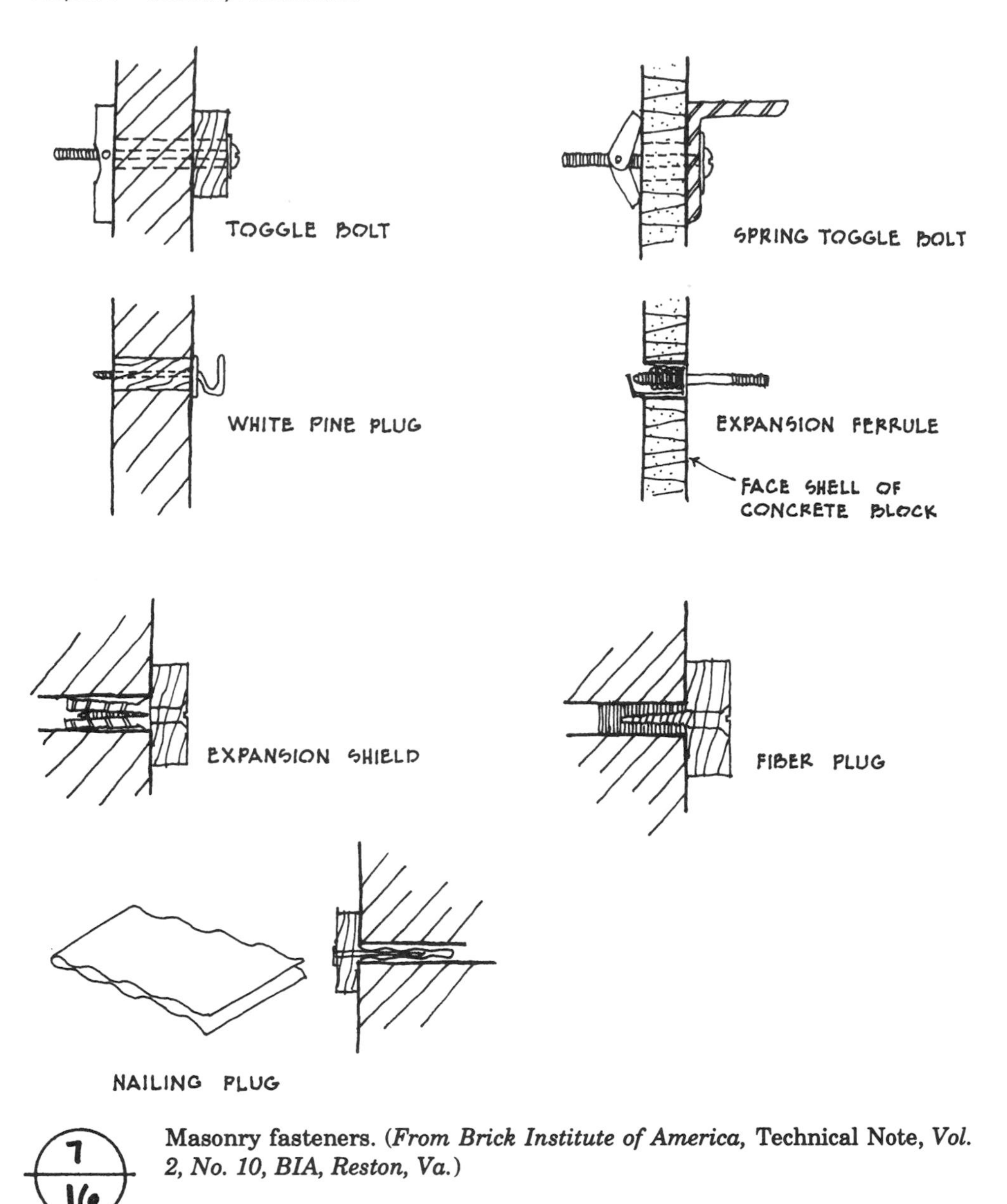

7-16 **Masonry fasteners.** (*From Brick Institute of America,* **Technical Note,** *Vol. 2, No. 10, BIA, Reston, Va.*)

while allowing it to open as the masonry shrinks (*see Fig. 7-18*), and should have a high durometer hardness. Softer materials such as neoprene rubber sponge are used for expansion joints in clay masonry walls, where brick masonry expansion will compress the filler as the joint closes. Expansion joint fillers are used only to keep mortar out of the joints during construction, and should have a compressibility at least equal to that of the sealant which will be used.

7.5 FLASHING MATERIALS

Masonry construction must include sheet flashing to divert penetrated moisture back to the exterior. Although they may be used in different situations, all flashing materials must be impervious to moisture and resistant to corrosion, abrasion, and puncture. In addition, they must be

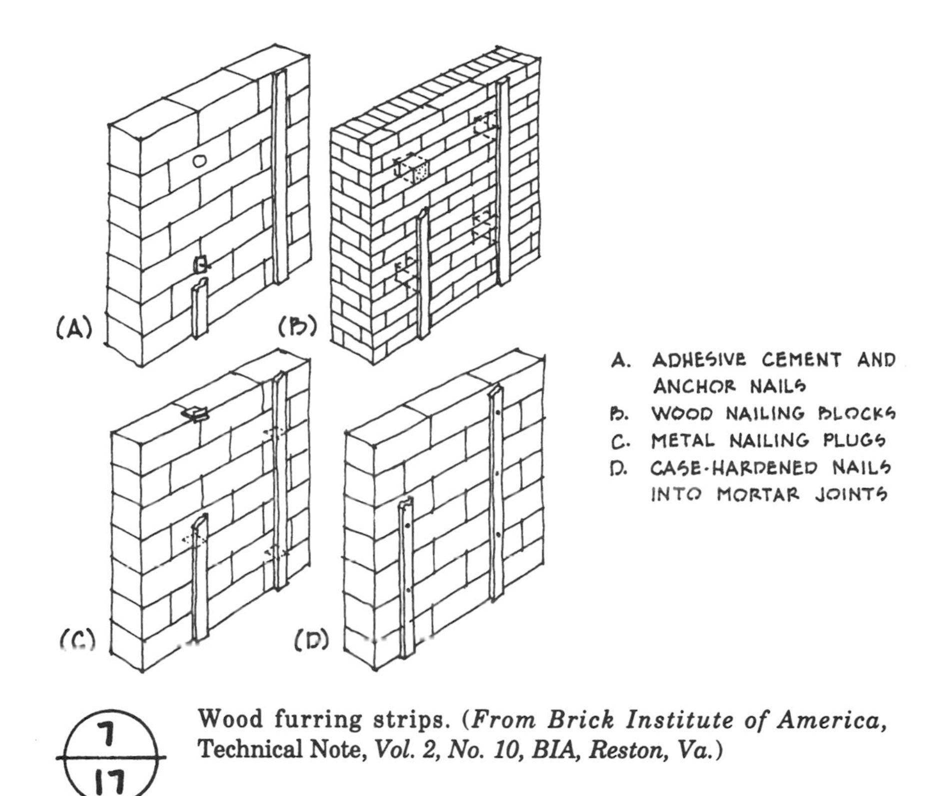

7/17 **Wood furring strips.** (*From Brick Institute of America,* **Technical Note,** *Vol. 2, No. 10, BIA, Reston, Va.*)

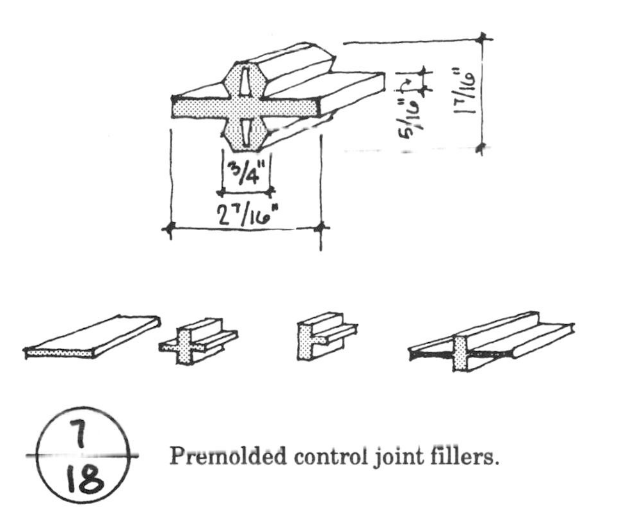

7/18 **Premolded control joint fillers.**

able to take and retain an applied shape to ensure proper performance after installation.

Stainless steel flashings are highly resistant to corrosion, and provide the highest long-term durability. *Copper sheet* resists ordinary corrosive action, provides an excellent moisture barrier, and is easily shaped. Copper flashing can stain light-colored masonry, though, unless it is coated with lead or other protective material. Copper sheet or lead-coated copper sheet should be a minimum 16-oz weight. Both stainless steel and copper flashing should also be "sawtoothed" or "dovetailed" in section to provide a mechanical bond with the mortar. *Galvanized steel* is used in both residential and commercial construction, but is subject to corrosive attack from wet mortar unless covered with a bituminous coating. Exterior exposures require a 26-gauge thickness, and concealed installations require 28-gauge. A minimum 28-gauge (0.015-in.) thickness is recommended. *Aluminum,* of course, is subject to corrosive damage from wet mortar and should not be used.

Copper is commonly used in *combination flashings* of 3-, 5-, or 7-oz copper sheet, and coatings of bitumen, kraft paper, bituminous-saturated cotton fabrics, or glass fiber fabrics. Combination flashings provide adequate protection at lower cost by allowing thinner metal sections. These coated metals are suitable only for concealed installations.

Plastic sheet flashings of PVC membrane may also be used in concealed locations, but may deteriorate with ultraviolet exposure. There is little long-term durability data on plastic flashing, but performance history does indicate that thickness should be at least 30 mils to avoid punctures during installation. The flashing must also be compatible with alkaline mortars and with elastomeric joint sealants. Prefabricated corners and end dams facilitate installation, and are sometimes used in combination with compatible metal flashing (*see Fig. 7-19*).

Most recently, EPDM (ethylene propylene diene terpolymer) rubber flashing and rubberized asphalt flashing materials have been introduced in the masonry industry. EPDM flashing should be a minimum of 45 mils in thickness, and uncured strips must be used to form corners. Like EPDM roofing membranes, this rubber flashing material is seamed with a proprietary adhesive which requires careful cleaning and priming of the mating surfaces. Rubberized (or polymer-modified) asphalt flashing has enjoyed ready acceptance from design professionals and masons alike. The rubberized asphalt is self-adhering and self-healing of small punctures. Once the workers become accustomed to handling the material, it installs quickly

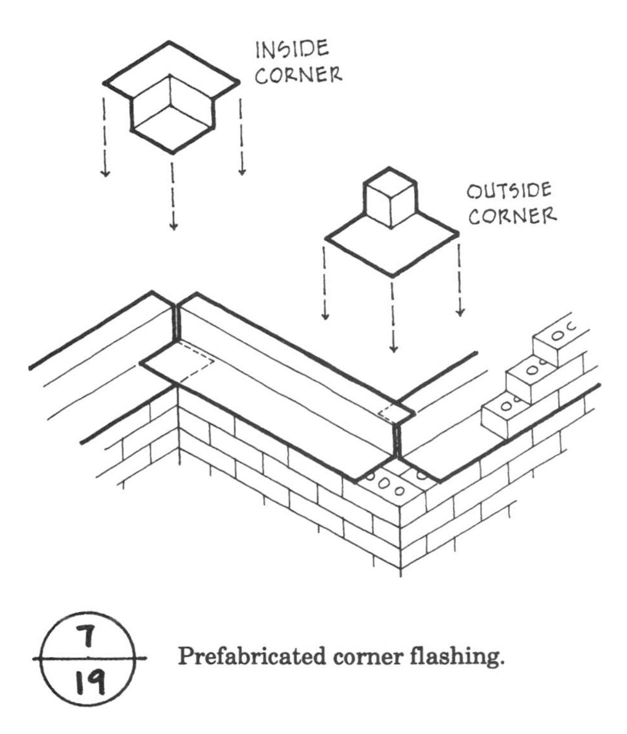

7/19 **Prefabricated corner flashing.**

and easily, and is relatively forgiving of uneven substrates. Rubberized asphalt flashing cannot tolerate ultraviolet exposure. When it is necessary or desirable to extend the flashing material beyond the face of the wall, rubberized asphalt membranes must be used in conjunction with a separate metal edge.

7.6 WEEPHOLE ACCESSORIES

Cavity walls and veneers are designed to drain moisture. Without effective weepholes in the course above flashings, walls collect moisture and hold it like a reservoir. The most common type of weephole is the open head joint which provides the largest open area, and thus the most effective evaporation and drainage. Mortar is left out of head joints every 24 in., leaving open channels that are $\frac{3}{8}$ in. wide×course height×veneer depth. The primary drawback to open joint weeps is appearance. A dark shadow is created at each opening, particularly with light-colored units and mortar. The openings are so large, in fact, that building maintenance crews all too often caulk the weepholes shut, mistakenly thinking they are the source of leaks. Some products camouflage the open joints, but still allow them to work properly. One is a vinyl or aluminum cover with louver type slots. Another is a plastic grid $\frac{3}{8}$ in. wide×course height×veneer depth less a $\frac{1}{8}$-in. recess (*see Fig. 7-20*). Both types disguise the openings and still permit drainage and evaporation.

Hollow plastic or metal tubes are also used to form weepholes. The most common ones are $\frac{1}{4}$ or $\frac{3}{8}$ in. in diameter by $3\frac{1}{2}$ to 4 in. long. Manufacturers recommend installing them at an angle in the mortar of the head joints, spaced 16 in. apart. The slight angle allows for a very small amount of mortar droppings in the cavity. The closer spacing is required because less water can drain through the tube, and less air can enter the wall, making drainage and evaporation much slower.

Tube-type weepholes are less conspicuous in the finished wall than open joints, but they also have some problems. If the installed angle is too steep, water at the bottom of the cavity cannot drain effectively. If the angle is flatter or the mortar droppings are deeper than allowed for, the

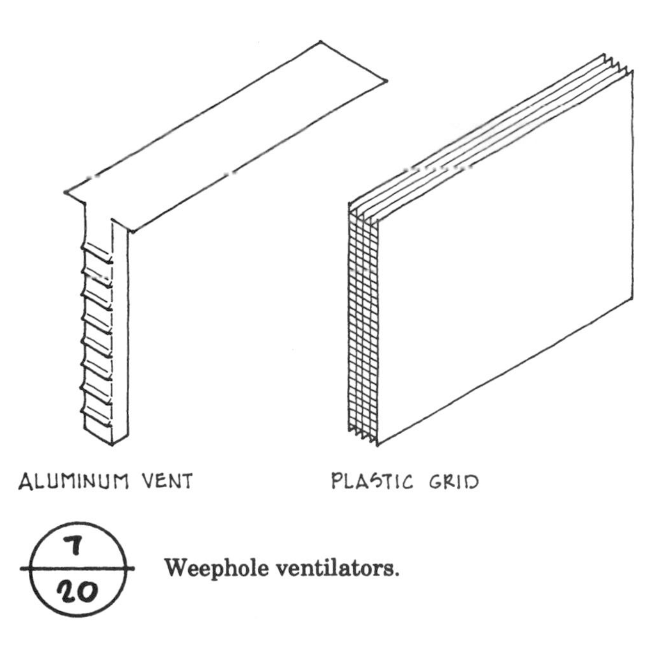

7-20 **Weephole ventilators.**

tube becomes blocked. Some contractors put a shallow layer of gravel in the bottom of the cavity to promote drainage and keep mortar droppings away from the tubes, but a piece of gravel can also lodge in or against the small opening and block it.

Some manufacturers make larger, rectangular tubes which measure $\frac{3}{8} \times 1\frac{1}{2} \times 3\frac{1}{2}$ in. Since the opening is much larger, blockage problems are reduced and drainage and evaporation rates are increased, even when the tubes are spaced 24 in. on center. With larger openings, though, the weepholes are more noticeable.

Cotton wicks are used to form another type of weep system. A $\frac{1}{4}$ to $\frac{3}{8}$-in. diameter rope is installed in the joints at 16 in. on center. The rope should be 8 to 10 in. long, and extend through the veneer face and up into the cavity well above the height of any mortar droppings. Moisture in the cavity is absorbed by the cotton material and wicked to the outside face of the wall where it evaporates. This is a slower process than open weeps, and nylon or hemp rope will not perform well. The cotton will be wet throughout its service life, and eventually will rot, leaving an open drainage hole. Using cotton wicks, however, assures that drainage is not inadvertently blocked by mortar. Wicks are also inconspicuous in the wall.

Another alternative is oiled rods or ropes which are mortared into the joints at 16 in. on center and then removed when the mortar has set. The rods function much the same as plastic tubes, and share some of the same disadvantages. The $\frac{3}{8}$-in.-diameter rods used are generally $3\frac{1}{2}$ to 4 in. long, oiled slightly to prevent mortar bond, and extended through the veneer thickness to the core or cavity. The opening left after removal is a full $\frac{3}{8}$ in, since the thickness of the tube shell is eliminated, but the hole is still small and easily blocked by mortar droppings. To avoid this, the rods can be left in place until the full story or panel height of the wall above is completed. The oiled rope technique is similar to that of the wick system where an unobstructed drainage path is provided. After the wall is completed to story height, the rope can be removed. It should be 10 to 12 in. long to allow adequate height in the cavity and to provide a handle for removal. By removing the rope instead of using it as a wick, the hole provides more rapid evaporation at the outset of construction, and its size is less noticeable than open head joint weeps.

Keeping the cavity clean during construction is essential to the performance of weepholes. A minimum cavity width of 2 in. gives the mason room to work. Techniques for placing mortar beds and removing mortar droppings are discussed in Chapter 15.

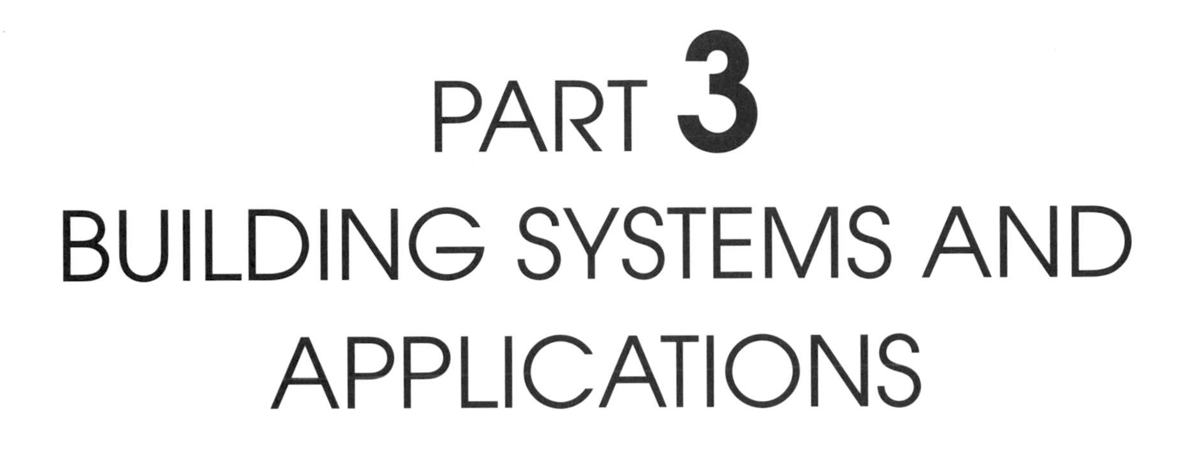

PART 3
BUILDING SYSTEMS AND APPLICATIONS

8 WALL TYPES AND PROPERTIES

Masonry can be used structurally or as veneer. Masonry walls may be single or multi-wythe, solid or hollow, grouted or ungrouted, reinforced, partially reinforced, or unreinforced, depending on the structural requirements of the design.

Masonry is non-combustible and, in its various forms, can be used as both structural and protective elements in fire-resistive construction. Masonry is durable against wear and abrasion, and most types weather well without protective coatings. The mass and density of masonry also provide efficient thermal and acoustical resistance. Although masonry is one of the most durable of building materials, masonry construction is durable only if the component materials are of equally high quality and the detailing adequately provides for movement and weather resistance. Almost any masonry material or combination of materials can be used to satisfy functional requirements, including fire, sound and thermal separation criteria. Specific material types, however, are generally selected on the basis of aesthetic criteria such as color, texture, and scale.

8.1 WALL TYPES

For *single-wythe* loadbearing applications, vertical reinforcing can be placed in the hollow cores, and horizontal steel in bond beam units. Where walls are also required to accommodate electrical conduit or plumbing piping, *multi-wythe* walls with an open or grouted cavity are more appropriate. The backing and facing wythe may be of the same or of dissimilar materials. If only the backing wythe is designed as a loadbearing element, the facing wythe actually functions as a *veneer,* and must be mechanically attached to the backing with metal anchors.

Multi-wythe walls of solid units and solidly grouted walls of hollow units are more resistant to moisture penetration than single-wythe walls of hollow units. Exterior exposures can be designed as *cavity walls* with an open separation of at least 2 in. between the facing and backing wythes.

The open cavity, when it is properly fitted with flashing and weepholes, functions as a drainage system for moisture which penetrates from the exterior or is condensed within the wall section. Single-wythe walls of hollow units must also be designed with a system of flashing and weepholes to divert collected moisture to the outside.

Loadbearing masonry is a viable and economical structural system for many building types of either low-, medium- or high-rise design. It is strong in compression, but requires the incorporation of reinforcing steel to resist tensile and flexural stresses. Repetitive, compartmentalized plans for hotels, multi-family housing, nursing homes, and other occupancies are particularly suited to the linear orientation of loadbearing walls and the characteristic fire resistance of masonry. Office buildings, schools, manufacturing facilities, and other occupancies requiring large open spaces might combine a loadbearing masonry wall system at the core and perimeter, with interior columns of steel, concrete or masonry. Loadbearing masonry exterior walls should also be considered in lieu of frame and veneer systems whenever the selected veneer is a masonry material with structural capability such as brick, concrete block, or bedded stone. Structural masonry is discussed at length in Chapters 11 and 12.

Masonry veneer can be constructed with adhesive or mechanical bond, over a variety of structural frame types and backing walls. Veneer applications of masonry are appropriate when the appearance of a masonry structure is desired but a loadbearing wall design is not considered appropriate. Masonry veneers may be used on buildings of wood, steel, or concrete structural frames, as well as on loadbearing masonry buildings. Brick, concrete block, stone, and terra cotta are the most commonly used veneer materials. Thin veneers may be adhesively attached with mortar over a solid backing, but codes limit the weight, size, and thickness of units. Veneers attached with metal anchors are more common, particularly in commercial applications. In skeleton frame construction, both brick and block veneers can be designed as reinforced curtainwalls spanning vertically or horizontally between supports. Codes generally permit the waiver of intermediate support requirements when such special design techniques are approved by the building official. When applied in this manner, masonry veneers may be constructed to 100 ft or more in height without shelf angles. More typically, however, masonry veneers are designed empirically as panel walls supported at each floor level. Masonry veneers are discussed in detail in Chapters 9 and 10.

8.2 SINGLE-WYTHE WALLS

Within the restrictions of height-to-thickness ratios prescribed by the model building codes (see Chapter 12), walls may be built with a single unit thickness of clay, concrete, or glass masonry. Single-wythe walls of hollow units provide the options of grouting the core areas for greater mass and stability, or adding steel reinforcement for additional strength. Grouted, reinforced concrete block, and hollow brick walls of a single 8-in. thickness can be used in high-rise loadbearing structures.

8.2.1 Structural Clay Tile

Hollow structural clay tile can be used in single-wythe construction of interior walls and partitions, and in some instances of exterior walls (*see Fig. 8-1*). Grading classifications (LB and LBX for loadbearing, and NB for non-

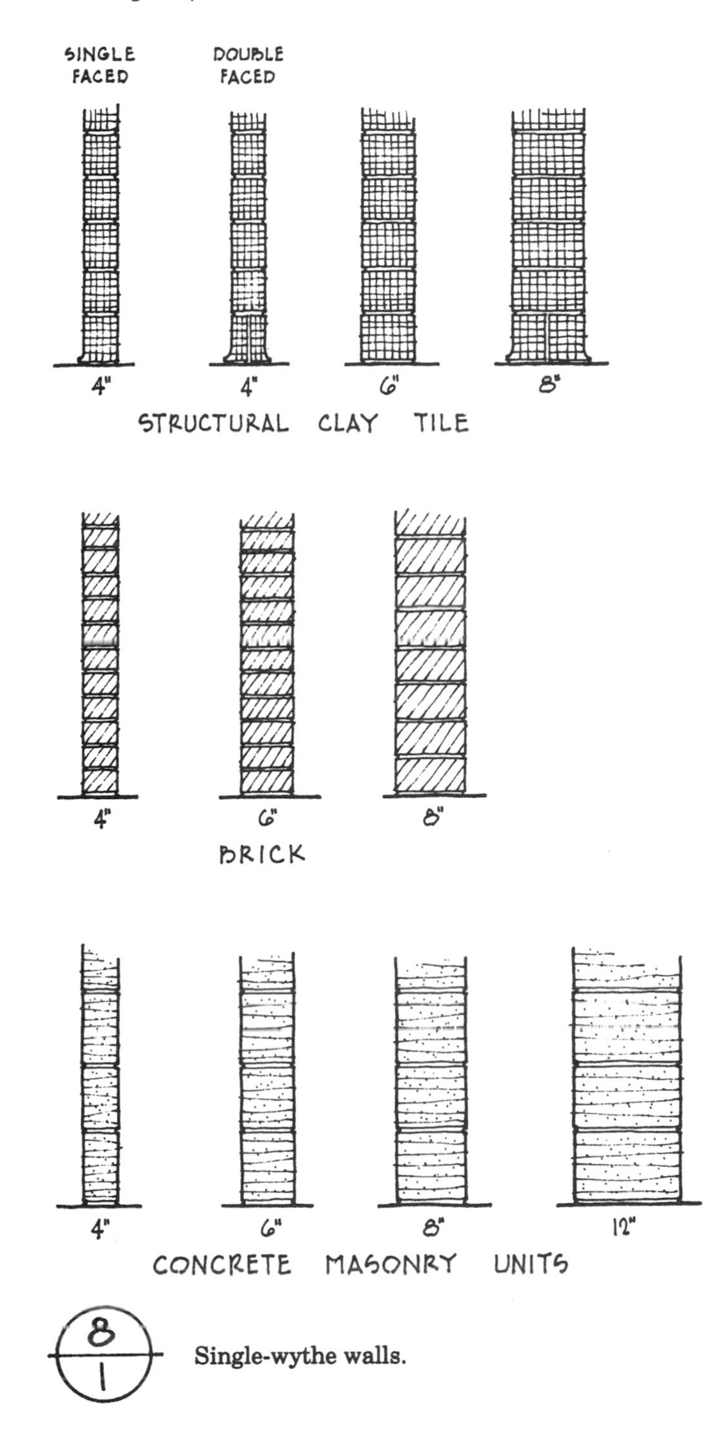

Single-wythe walls.

loadbearing) will determine the type of unit selected. Facing tile and Type II glazed tile provide a finished surface on both faces of a single-wythe wall, with only one unit thickness for simplified construction. Standard structural tile designed to receive plaster applications can also be used in through-the-wall applications of one wythe. Type I glazed units are designed for finished exposure on only one side where the other wall face will be concealed or will receive a plaster finish.

8.2.2 Brick and Block

Hollow clay brick and solid units with a 6-in. bed depth are often used in single-wythe construction, and in some instances, codes permit 4-in. walls (*see Fig. 8-1*). Hollow brick sections are usually at least 8 in. thick. Hollow concrete blocks have decorative finishes on only one side. The opposite wall face must receive paint, plaster, gypsum board, or other materials if exposed to view. Single wythes of masonry are widely used for veneer construction over wood frame, steel, concrete, or masonry structural backing. Single-wythe walls may also be loadbearing elements, or interior non-loadbearing partitions. Brick walls may of course be exposed on both sides without further finishing.

8.2.3 Glass Block

Glass block masonry is used for high-security glazing, and for glazed areas requiring light control and/or heat-gain reductions. The units are used only in single-wythe construction, and do not have loadbearing capabilities.

8.3 MULTI-WYTHE WALLS

For larger horizontal or vertical spans between lateral supports or stiffeners, or for greater resistance to fire, sound, and heat transmission, wall thicknesses are increased by adding additional wythes of masonry of the same type unit or of a different material. These wall types may be divided into (1) solid masonry walls, (2) solid walls of hollow units, and (3) cavity walls. A solid masonry wall is built of solid units, or of fully grouted hollow units laid contiguously with the collar joint between wythes filled with mortar or grout (*see Fig. 1-4*). A solid wall of hollow units is built of hollow clay or concrete masonry with mortared or grouted collar joints, but with core areas left void. A cavity wall consists of two or more wythes of hollow or solid units separated by an open collar joint or air space of at least 2 in. between two adjacent wythes.

8.3.1 Solid Masonry Walls

Solid masonry walls have been used in building construction throughout history. Strength, stability, and insulating value all depended on mass, and code requirements for empirically based, unreinforced bearing walls prescribed substantial thicknesses. The Monadnock Building in Chicago, completed in 1891, is 16 stories high with unreinforced loadbearing brick walls ranging in thickness from 12 in. at the top to more than 4 ft at the ground. At that time, wall wythes were bonded together with masonry headers as shown in *Fig. 8-2*. In 8-in. walls header courses extend the full width of the wall section, and moisture penetration from exterior to interior is facilitated, causing leakage and possible damage to the masonry. In most contemporary construction, masonry headers have been replaced by metal wall ties grouted solidly in the bed joints to tie multiple wythes together. Today, solid masonry walls do not generally exceed 12 in. in thickness except under special conditions or circumstances.

Multi-wythe walls of hollow tile, brick, or concrete units are often solidly grouted to increase strength or fire resistance. When constructed in this manner, they are considered solid masonry walls. Composite walls of different materials such as brick/concrete block or concrete block/stone are also considered solid if both the collar joint between wythes and the cores of the

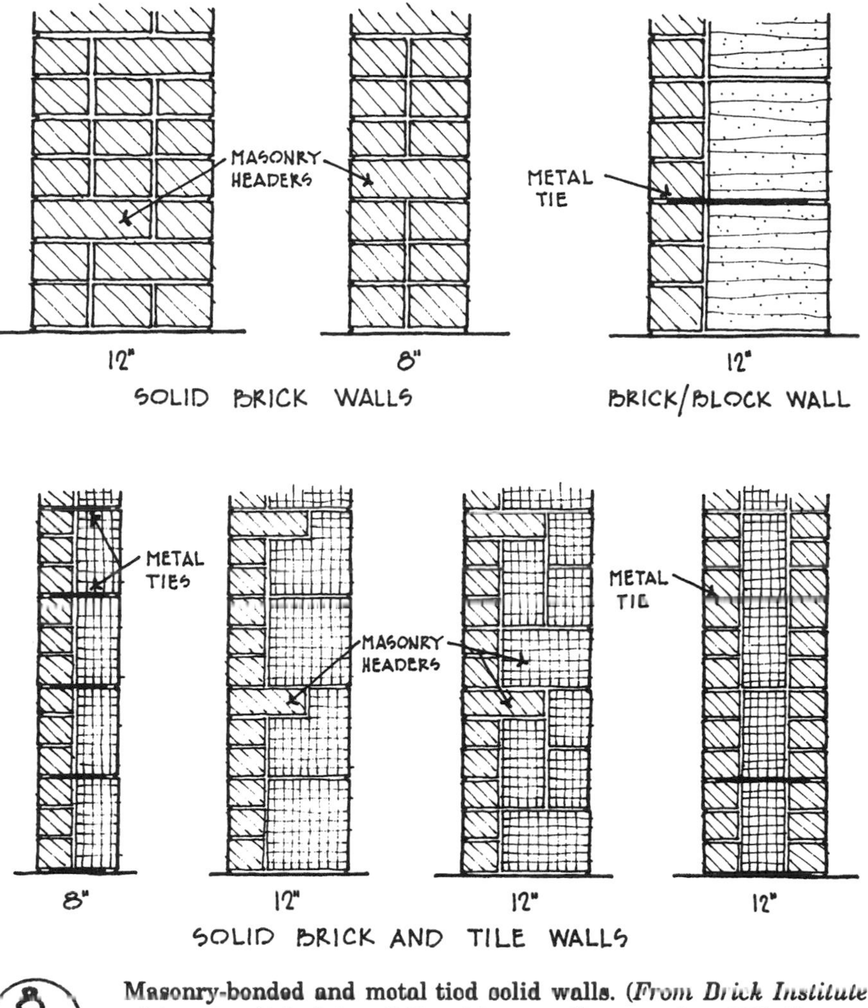

8-2 **Masonry-bonded and metal tied solid walls.** (*From Brick Institute of America,* **Principles of Clay Masonry Construction,** *BIA, Reston, Va., 1973.*)

hollow units are filled with mortar or grout. Typical material combinations include brick and tile, brick and block, brick and stone, and block and stone.

8.3.2 Solid Walls of Hollow Units

Solid walls of hollow units consist of at least two wythes of material with no cavity or open collar joint. Construction may be of hollow clay tile, hollow CMUs, or hollow brick, or of stone or solid brick units in combination with any of these. If either wythe contains hollow masonry with open cores, the wall is considered to be of hollow units (*see Fig. 8-3*). Bonding can be with either masonry headers or metal wall ties.

8.3.3 Cavity Walls

Cavity walls consist of two or more wythes of masonry units separated by an air space at least 2 in. wide. The wythes may be brick, clay tile, concrete block, or stone, anchored to one another with metal ties which span the open

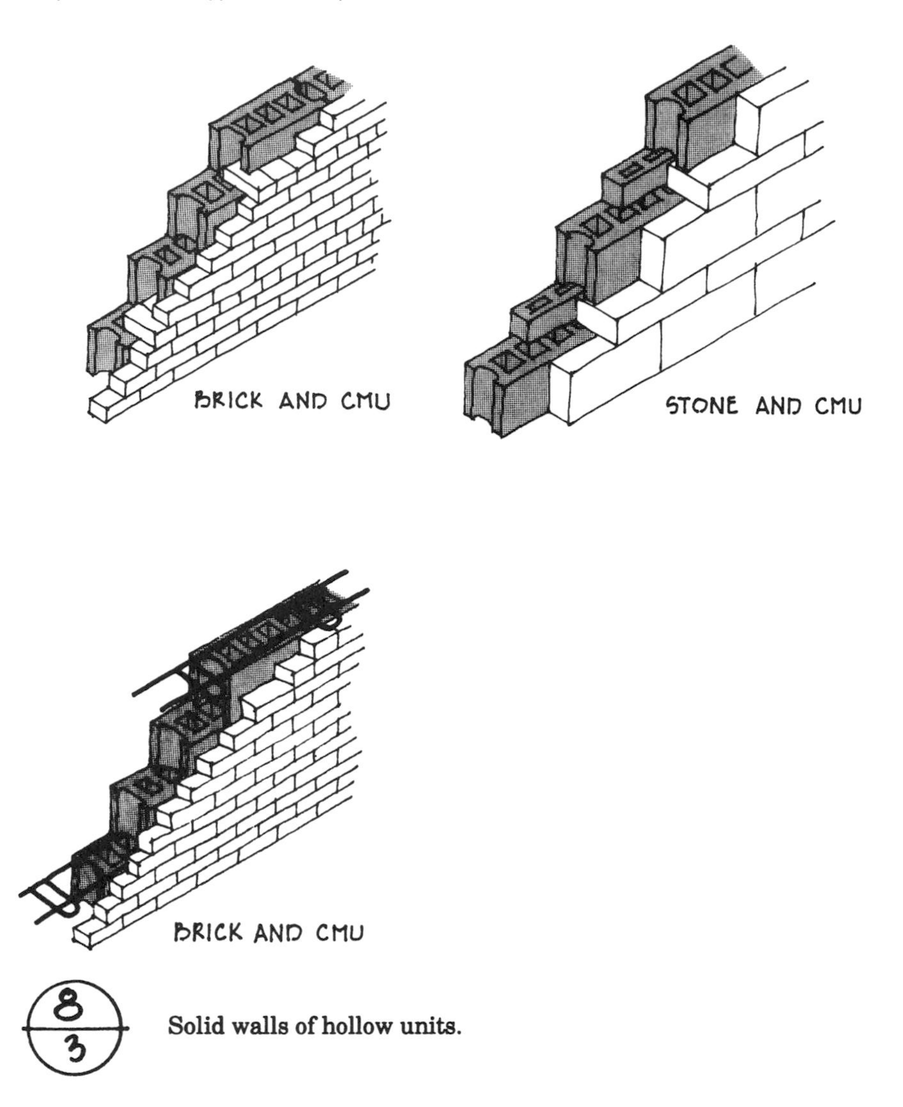

8-3 Solid walls of hollow units.

collar joint (*see Fig. 8-4*). The two wythes together can be designed as a unified loadbearing element, or the exterior wythe may be a nonstructural veneer. One of the major advantages of cavity wall construction is greater resistance to rain penetration resulting from the complete separation of the inner and outer wythes. This separation also increases thermal resistance by providing a dead air space, and allows room for additional insulating materials if desired. The open cavity, when it is properly fitted with a system of flashing and weepholes, provides drainage for moisture which may penetrate the exterior or form as condensation within the cavity.

Both sides of a cavity wall must act together in resisting wind loads and other lateral forces. The metal ties transfer these loads in tension and compression, and must be solidly bedded in the mortar joints in order to perform properly. Crimped ties with a water drip in the center are not recommended because the weakened plane created can cause buckling of the tie and ineffective load transfer.

Cavity walls can prevent the formation of condensation on interior surfaces, so that plaster and other finish materials may be directly applied without furring. Insulation may be added in the wall cavity, including

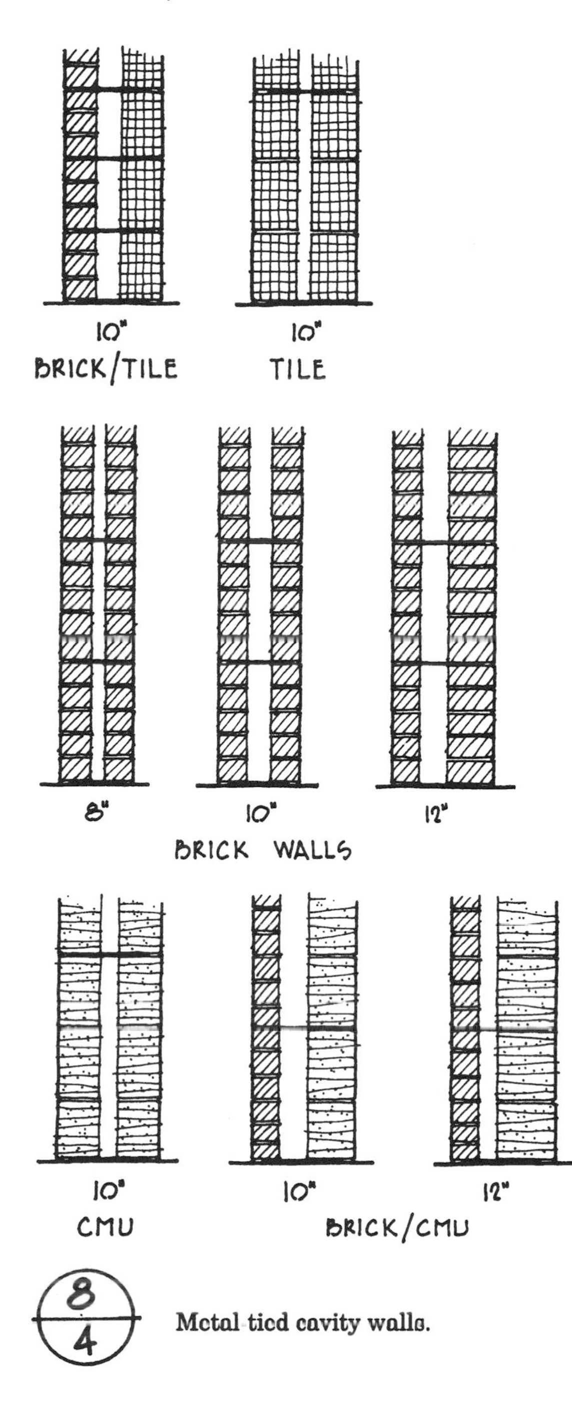

Metal tied cavity walls.

water-repellent vermiculite, silicone-treated perlite, or rigid boards. A vapor barrier or dampproof coating is usually required on the cavity face of the inner wythe (refer to Chapter 10).

In multi-wythe construction, a distinction is made between walls in which both the facing and the backup are loadbearing, and those in which only one wythe will carry the superimposed load. *Composite walls* are bonded with either masonry headers or metal ties so that both wythes act as a single element in resisting loads. In a *veneered wall,* the facing is attached but not structurally bonded, so that only the backup material is loadbearing. Clay

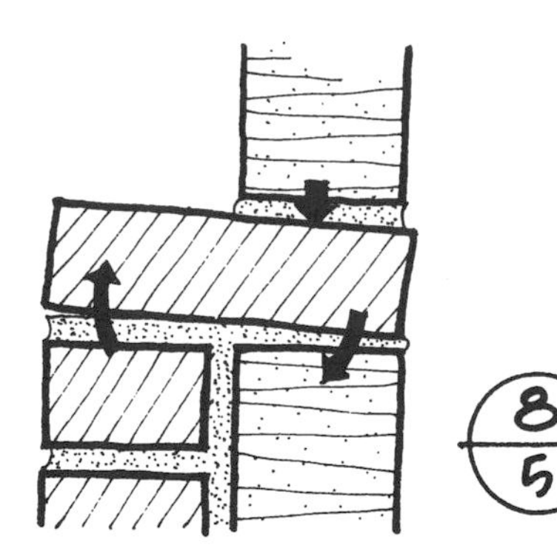

Rotation of brick from uneven settlement of heavy backup wythe. (*From Brick Institute of America,* Principles of Clay Masonry Construction, *BIA, Reston, Va., 1973.*)

brick and hollow tile, or brick and concrete block, are often used in this manner, with each material selected for optimum performance and economy.

8.3.4 Metal Wall Ties

Metal wall ties are generally recommended instead of masonry headers, for two reasons. Masonry headers increase the possibility of moisture penetration not only because of the continuity of the header through an 8-in. wall section, but also because of the difficulty of constructing some types of walls without breaking the mortar bond. If the mason is working from the exterior of a building, the backup units are naturally laid before the facing units. Brick headers are placed on top of the backup as the brick coursing reaches the proper height. With hollow units, this header is supported by two relatively thin strips of mortar. The weight of the next course of backup units can cause uneven settlement and joint separations through which moisture can penetrate (*see Fig. 8-5*). Metal wall ties eliminate this problem and simplify construction. They also provide the necessary flexibility for differential movement between the inner and outer wythes of masonry. Clay and concrete have different thermal and moisture expansion coefficients. Masonry bonding impairs movement in the wall and can create an eccentric load on the headers, rupturing the mortar bond at the exterior face.

8.4 FIRE RESISTANCE CHARACTERISTICS

Building fires are a serious hazard to life and property, and fire safety in construction is therefore a primary consideration of every building code authority. According to the National Fire Protection Association (NFPA), construction deficiencies are a major factor in large-loss fire experiences. NFPA records show that combustible construction is the predominant cause of conflagrations, particularly in areas of closely built wood-frame structures, which includes single-family and most low-rise multifamily residential buildings. Restricting the spread of fire is critical in reducing fire deaths and property loss. The overwhelming majority of U.S. fires are in residential buildings—apartments, hotels, and dwellings. Multifamily occupancies are particularly vulnerable because of the lack of physical separation between living units. In 1990, more than 95,500 structural fires in multifamily dwellings were reported to NFPA. In those fires, 680 people died, 4975 more were injured, and $623,000,000 worth of property was destroyed.

While most industrialized countries require a 2-hour fire wall between units, the United States does not yet do so, which results in greater loss of life and property each year. "One-hour" construction made of combustible materials and electronic detection and suppression systems provide a false sense of security based on unrealistic fire ratings and a reliance on poorly maintained, seldom-tested fire alarm and sprinkler equipment.

Fire regulations are concerned primarily with the safety of occupants, the safety of fire fighters, the integrity of the structure, and the reduction of damage. Construction must (1) limit the spread of fire within a building, (2) prevent fire spread to adjacent buildings, (3) maintain the integrity of occupant evacuation routes, and (4) allow for attack by fire services. The overall risk is reduced when non-combustible construction is used to construct or protect structural elements, and to divide a building into compartments for the containment of fire. Non-combustible masonry and concrete construction provide the highest level of protection through fire wall containment and structural integrity.

The degree of fire protection offered by masonry construction was recognized long ago. In 1212 A.D. an ordinance was issued by royal proclamation requiring that all alehouses in London be built of masonry. After the great fire of 1666, which destroyed most of London, King Charles II decreed that the walls of all new buildings must be of masonry. Modern masonry construction has an excellent performance record in fire containment, but non-combustible construction is not required in low-rise multifamily buildings, and standard fire ratings are misleading about the relative fire safety of different types of construction. The difference between "fire-resistive" construction of combustible and non-combustible materials is like the difference between a water *repellent* coating and a water*proof* coating. Only one keeps water out.

8.4.1 Fire Tests

Fire properties of building materials are divided into two basic categories: *combustibility* and *fire resistance*. Masonry is classified as noncombustible. Fire resistance ratings are based on standard ASTM, NFPA, or National Institute of Standards and Technology (NIST) fire endurance tests. Under these fire test standards, walls, floors, roofs, columns, and beams are tested in a furnace under controlled laboratory conditions. For walls, one end of the furnace is sealed with the actual construction assembly being tested so that one side of the wall is exposed to the fire.

Specimens are subjected to controlled heat applied by standard time-temperature curve for a maximum of 8 hours and 2300°F. Wall assemblies must also undergo a hose stream test for impact, erosion, and thermal shock. Throughout the tests, columns and bearing walls are loaded to develop full design stresses. Within 24 hours after the testing is complete, bearing walls must also safely sustain twice their normal superimposed load to simulate, for instance, a roof collapse. Fire resistance ratings, generally in 1- or ½-hour increments, are assigned according to the elapsed time at which the test is terminated. The test is terminated when any one of three possible end-point criteria is reached: (1) an average temperature rise of 250°F or a maximum rise of 350°F is measured on the unexposed side of the wall; (2) heat, flame, or gases escape to the unexposed side, igniting cotton waste samples; or (3) failure under the design load occurs (loadbearing construction only). The first two points concern only the containment of fire spread through the wall or section, while the third concerns structural failure. Despite this fundamental disparity in the level of safety provided, each of the criteria carries equal weight in determining assigned fire ratings.

How safe is safe? "Protected" wood frame construction (i.e., wood framing covered with gypsum board) generally fails the fire endurance test because of structural collapse. Concrete and masonry fire ratings, on the other hand, are almost invariably based on heat transmission alone. The temperature on the unexposed side has risen 250°F while the temperature on the opposite face a few inches away was more than 2000°F. Most building materials and contents will not burn at 250°F above room temperature, and in fact, most cooking requires higher temperatures. The wood frame assembly collapses and, in doing so, allows the fire to spread through what was supposed to be a barrier. And yet the two assemblies are given an identical fire rating (*see Fig. 8-6*). The structural integrity of the masonry wall is maintained far beyond the time indicated by its fire rating. In real fires, maintaining structural integrity is critical to the safe evacuation of occupants either on their own or by rescue workers, and it is critical in maintaining access to the fire for fire fighters and equipment. It also means that property loss is restricted to superficial damage which can be readily repaired at far less expense than rebuilding a structure whose identical fire rating was based on structural failure.

There are other discrepancies in standard fire tests which also affect the accuracy and credibility of the results. Furnace temperatures must be maintained at certain levels according to the elapsed time. As a result, the amount of fuel required for the test fire depends to some extent on properties of the test specimen. If the specimen itself burns, as it does in wood frame construction, it contributes to furnace temperature and reduces the amount of fuel needed to sustain the time-temperature curve conditions (*see Fig. 8-7*). In real fires, this means that combustible assemblies add to the fuel and therefore increase the intensity of the fire. If, on the other hand, the test specimen absorbs and stores heat from the furnace, as is the case with concrete and masonry, more fuel is required to maintain the test

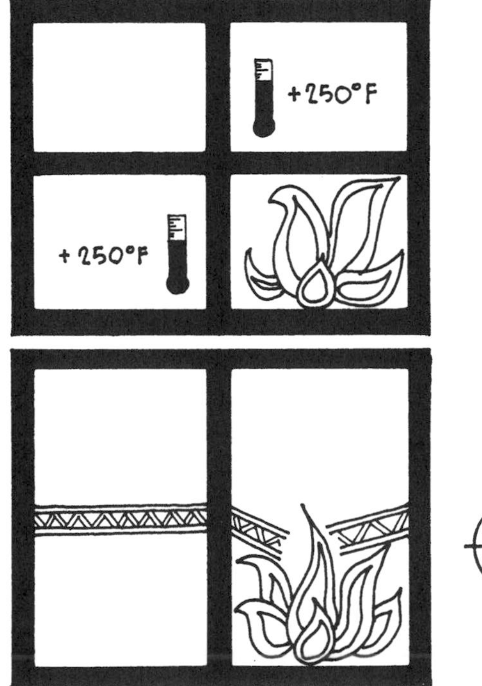

Structural collapse vs. temperature rise criteria in determining fire resistance ratings. (*From Portland Cement Association,* Fire Protection Planning Reports, *PCA, 5420 Old Orchard Rd., Skokie, Ill.*)

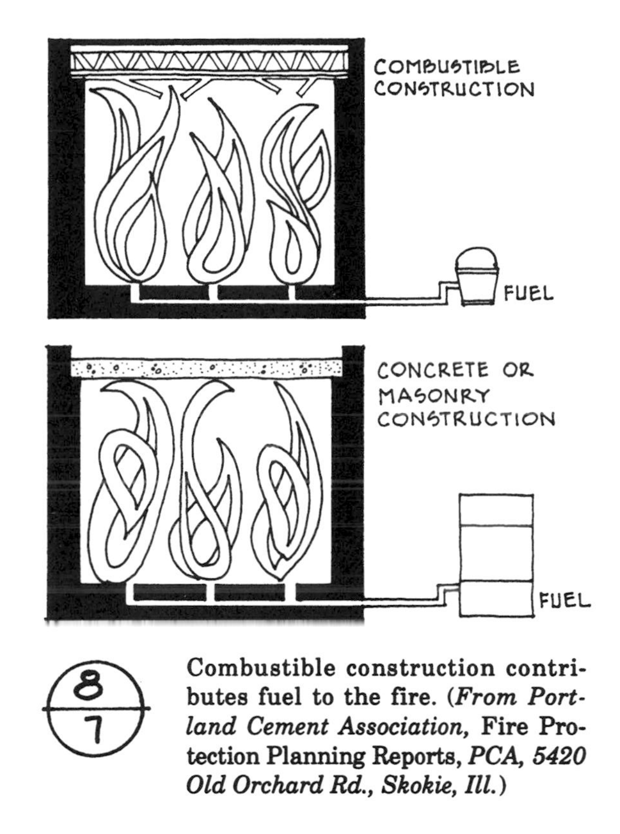

8/7 **Combustible construction contributes fuel to the fire.** ***(From Portland Cement Association,*** **Fire Protection Planning Reports,** ***PCA, 5420 Old Orchard Rd., Skokie, Ill.)***

conditions. Although these variations in fuel consumption during the test would seem to give some indication of the relative fire endurance of the construction, they are not a recognized factor in assigning ratings.

In real building fires, heat and gas movements create positive pressures, especially in the immediate vicinity of the heat source. However, fire test standards do not specify whether the test furnace should be operated with negative or positive pressure. In the United States and Canada, almost all tests are conducted with unrealistic negative pressures in order to prevent the escape of hazardous gases into the laboratory. In Europe, however, furnaces are required to operate with positive pressure and are fitted with safety devices which force emissions out an exhaust flue. Negative pressures tend to draw cool air into the furnace through cracks and gaps that typically exist in wood frame–gypsum board construction, thus extending the endurance time of the assembly beyond what it might be in an actual fire (*see Fig. 8-8*).

Because of the way fire endurance is tested in this country, and the way in which fire ratings are assigned on the basis of these tests, the ratings for masonry walls are probably too low, and the ratings for "protected" wood frame assemblies too high. Non-combustible construction controls or prevents substantial fire development because it does not contribute fuel to the fire and, in fact, can actually reduce the intensity of the fire by absorbing and storing heat. Non-combustible construction also provides true containment: it will not support fire in concealed spaces of wall, floor, or roof assemblies; it maintains the structural integrity of the building to provide safe access and egress; and it does not produce toxic gases or contribute to smoke generation. Non-combustible construction, however, is underrated because of the evaluation system we use. Although it is not

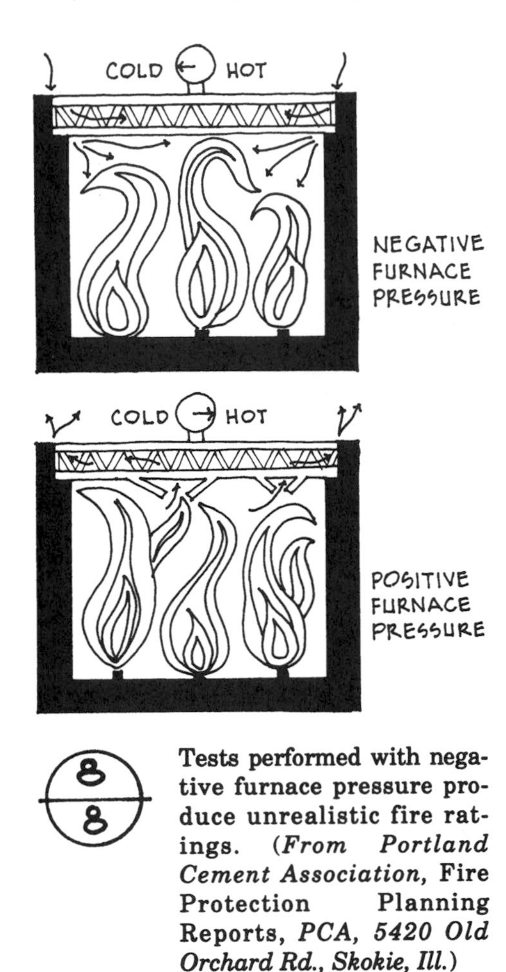

Tests performed with negative furnace pressure produce unrealistic fire ratings. (*From Portland Cement Association,* **Fire Protection Planning Reports,** *PCA, 5420 Old Orchard Rd., Skokie, Ill.*)

logical to give the same fire rating to one wall that suffers structural collapse as to another that experiences only a moderate rise in temperature, doing so perpetuates the misconception of the relative safety of various types of construction.

8.4.2 Fire Resistance Ratings

Extensive fire testing has been done on masonry walls, and ratings are listed (1) by the National Bureau of Standards (NBS) in its report BMS 92 (and not reissued since NBS became the National Institute of Science and Technology), (2) by the National Fire Protection Association in the *Fire Protection Handbook,* (3) by the Underwriters' Laboratories in its *Fire Resistance Index,* and (4) by the American Insurance Association in its publication *Fire Resistance Ratings.* The three model building codes used in the United States list fire ratings that are taken from these reports or, in some instances, refer directly to the publications as reference standards. Tables list the minimum thickness of a particular material or combination of materials required for ratings of 1, 2, 3, and 4 hours.

Ratings for brick and clay tile walls are a function of wall mass or thickness, and depend to some extent on the percent of cored area in the individual units. Units with less than 25% cored area are considered "solid," and units with more than 25% cored area are classified as hollow. An 8-in. hollow tile wall contains less mass than an 8-in. solid brick wall, and it therefore offers less resistance to fire and heat transmission. Although many fire tests on hollow clay masonry have been conducted, it would be virtually impossible to test all combinations of unit size, shape, and core area. Fire resistance ratings can be calculated based on the equivalent solid thickness (E_T) of the units and the known fire resistance characteristics of the materials. For walls of a given material and design, NBS testing showed that an increase of 50% in volume of solid material per unit area of wall surface resulted in a 100% increase in the fire resistance period. Equivalent solid thickness is the average thickness of solid material in the wall, and is calculated from the actual thickness and the percentage of solid material in the unit. E_T is found by taking the total volume of a wall unit, subtracting the volume of core or cell spaces, and dividing by the area of the exposed face of the unit, using the equation

$$E_T = \frac{V}{l \times h} \tag{8.1}$$

where E_T = equivalent thickness, in.
V = net volume (gross volume less void area), cu in.
l = length of unit, in.
h = height of unit, in.

For example, a nominal 8×8 modular face size structural clay tile unit that is 6 in. thick has a gross volume of $7\frac{5}{8} \times 7\frac{5}{8} \times 5\frac{5}{8} = 327$ cu in. A void area of 40% leaves a net volume $V = 327 - 131 = 196$ cu in. From equation (*8.1*), the equivalent thickness can be calculated:

$$E_T = \frac{196}{7.625 \times 7.625} = 3.37 \text{ in.}$$

Results of tests performed at NIST on structural clay tile partitions are listed in *Fig. 8-9,* and the appropriate E_T listed in column 1. *Figure 8-10* lists the required E_T for hollow clay brick walls from the *Standard Building Code.* If the units have less than 25% core area (i.e., solid brick), fire ratings are based on tested assemblies (*see Fig. 8-11*).

The rated wall assemblies listed by the NBS report and the American Insurance Association are illustrated in *Fig. 8-12.* These are typical of the ratings found in most model building codes.

The fire resistance of concrete masonry is a function of aggregate type and unit thickness. Walls and partitions of 1- to 4-hour ratings are governed by code requirements for actual or equivalent thickness computed on the percent of void area in the unit. Increasing the wall thickness or filling the cores with grout increases the rating. Units with less than 25% cored area are considered solid, and the actual thickness of the brick or block is used for calculations. Units with more than 25% coring are classified as hollow, and the equivalent thickness of solid material must first be computed in order to determine the fire rating. Since core size and design will vary, manufacturers' data are normally used to establish exact figures (*see Fig. 8-13*).

Concrete masonry aggregates have a significant effect on the fire resistance characteristics of the units. Lightweight aggregates such as pumice, expanded slag, clay, or shale offer greater resistance to the transfer of heat in a fire because of their increased air content. Units made with these materials require less thickness to achieve the same fire rating as a heavyweight

Wall or partition assembly, minimum nominal thickness	*Members framed into wall or partition*	
	Combustible (minutes)	*None or non-combustible (minutes)*
Hollow clay tile		
8-in. unit; 2 cells in wall thickness, 40% solid	45	75
8-in. unit; 2 cells in wall thickness, 43% solid	45	90
8-in. unit; 2 cells in wall thickness, 46% solid	60	105
8-in. unit; 2 cells in wall thickness, 49% solid	75	120
8-in. unit; 3 or 4 cells in wall thickness, 40% solid	45	105
8-in. unit; 3 or 4 cells in wall thickness, 43% solid	45	120
8-in. unit; 3 or 4 cells in wall thickness, 48% solid	60	150
8-in. unit; 3 or 4 cells in wall thickness, 53% solid	75	180
12-in. unit; 3 cells in wall thickness, 40% solid	120	150
12-in. unit; 3 cells in wall thickness, 45% solid	150	180
12-in. unit; 3 cells in wall thickness, 49% solid	180	210
12-in. wall; 2 units with 3 or 4 cells in wall thickness, 40% solid	120	210
12-in. wall; 2 units with 3 or 4 cells in wall thickness, 45% solid	150	240
12-in. wall; 2 units with 3 or 4 cells in wall thickness, 53% solid	180	240
16-in. wall; 2 or 3 units with 4 or 5 cells in wall thickness, 40% solid	240	240
Structural clay tile		
4-in. unit; 1 cell in wall thickness, 40% solid*†		75
6-in. unit; 1 cell in wall thickness, 30% solid*†		120
6-in. unit; 2 cells in wall thickness, 45% solid†		60
4-in. unit; 1 cell in wall thickness, 40% solid†‡		75
6-in. unit; 1 cell in wall thickness, 30% solid†‡		120
Hollow structural clay tile		
8-in. unit; 2 cells in wall thickness, 40% solid	45	75
8-in. unit; 2 cells in wall thickness, 49% solid	75	120
8-in. unit; 3 or 4 cells in wall thickness, 53% solid	75	180
8-in. unit; 2 cells in wall thickness, 46% solid	60	105
12-in. unit; 3 cells in wall thickness, 40% solid	120	150
12-in. wall; 2 units, with 3 cells in wall thickness, 40% solid	120	210
12-in. wall; 2 units with 3 or 4 cells in wall thickness, 45% solid	150	240
12-in. unit; 3 cells in wall thickness, 45% solid	150	180
12-in. unit; 3 cells in wall thickness, 49% solid	180	210
16-in. wall; 2 units with 4 cells in wall thickness, 43% solid	240	240
16-in. wall; 2 or 3 units with 4 or 5 cells in wall thickness, 40% solid	240	240

*Ratings are for dense hard-burned clay or shale.

†Cells filled with tile, stone, slag, cinders, or sand mixed with mortar.

‡Ratings are for medium-burned clay tile.

National Bureau of Standards (NBS) fire resistance periods for non-loadbearing and loadbearing clay tile masonry walls for units which comply with the requirements of ASTM C34, C56, C212, or C530. (*From Brick Institute of America,* Technical Note 16B, *BIA, Reston, Va.*)

Type of material	Fire resistance period (minutes)			
	60	120	180	240
Brick of clay or shale, unfilled	2.3	3.4	4.3	5.0
Brick of clay or shale, grouted or filled with perlite, vermiculite, or expanded shale aggregate	3.0	4.4	5.5	6.6

Notes:

1. Equivalent thickness is the average thickness of solid material in the wall, found by taking the total volume of the hollow wall unit, subtracting the volume of core or cell spaces, and dividing by the area of the exposed face of the unit.
2. Refer to building codes for additional details.

Minimum equivalent thickness (in.) of ASTM C652 hollow brick for various fire resistance ratings. (*From Brick Institute of America,* Technical Note 16B, *BIA, Reston, Va.*)

aggregate unit. The table in *Fig. 8-14* is taken from the *Standard Building Code.* It lists aggregate types and unit thicknesses which will satisfy specific fire rating requirements.

The fire resistance of units or wall assemblies which have not been tested can be calculated using the equation

$$R = (R_1^{0.59} + R_2^{0.59} + \cdots + R_n^{0.59} + as)^{1.7} \quad (8.2)$$

where R = calculated fire resistance of the assembly, hours
R_1, R_2, R_n = fire rating of the individual wythes, hours
as = coefficient for continuous air space

For continuous air spaces ½ to 3½ in. wide, the coefficient assigned by NBS is 0.3. The equation can be used to calculate the resistance of masonry cavity walls, composite walls which combine clay and concrete masonry, and grouted walls. For single-wythe or multi-wythe grouted walls, the grout is considered as one layer of a multi-layered assembly, and is rated on the basis of equivalent thickness of siliceous aggregate from the table in *Fig. 8-14.* The ratings of the unit or units and the rating of the grout are the values used for R_1, R_2, and R_3 in the equation, and the air space, if any, is as. For example, a 10 in. cavity wall with 4 in. brick, 2 in. open cavity, and 4 in. brick would be calculated

$$R = (1.25^{0.59} + 1.25^{0.59} + 0.03)^{1.7} = 4.15 \text{ hours (round off to 4 hours)}$$

A limestone aggregate concrete block with an E_T of 4.2 in. is rated 2 hours (from *Fig. 8-14*). If the cores of the block are grouted with a sand and gravel aggregate portland cement grout, the E_T of the grout is 7.626 in. −4.2 in. = 3.4 in. The fire rating for the grout thickness is 1 hour (from *Fig. 8-14*). Therefore

$$R = (2.0^{0.59} + 1.0^{0.59})^{1.7} = 4.78 \text{ hours (round off to 4 hours)}$$

In both instances, the whole is greater than the simple sum of the parts because of the increase in mass per unit of surface area.

Wall or partition assembly, minimum nominal thickness	Members framed into wall or partition	
	Combustible (minutes)	None or non-combustible (minutes)
Clay or shale, solid		
4-in. brick		75
6-in. brick		153
8-in. brick	120	240
12-in. brick	240	
Clay or shale, hollow		
8-in. brick, 71% solid	120	180
12-in. brick, 64% solid		240
8-in. brick, 60% solid, cells filled with loose fill insulation		240
Clay or shale, rolok		
8-in. hollow rolok	60	150
12-in. hollow rolok	180	240
8-in. hollow rolok bak		240
Cavity walls, clay or shale		
8-in. wall; two 3-in. (actual) brick wythes separated by 2-in. air space; masonry joint reinforcement spaced 16 in. o.c. vertically		180
9-in. wall; two nominal 4-in. wythes separated by 2-in. air space; ¼-in. metal ties for each 3 sq ft of wall area	60*	240
Clay or shale brick, metal furring channels		
5-in. wall, 4-in. nominal brick (75% solid) backed with a hat-shaped metal furring channel ¾ in. thick formed from 0.021-in. sheet metal attached to brick wall on 24-in. centers with approved fasteners; and ½-in. Type X gypsum board attached to the metal furring strips with 1-in.-long Type S screws spaced 8 in. on centers		120
Hollow clay tile, brick facing		
8-in. wall; 4-in. units (40% solid)† plus 4-in. solid brick	60	210
12-in. wall; 8-in. units (40% solid)† plus 4-in. solid brick	120	240

*A 9-in. wall has a 120 min rating if the hollow spaces near combustible members are filled with fire resistant materials for the full thickness of the wall and for at least 4 in. above and below and between the combustible members.

†Units shall comply with the requirements of ASTM C34.

National Bureau of Standards (NBS) fire resistance periods for non-loadbearing and loadbearing clay masonry walls for units which comply with the requirements of ASTM C62, C126, C216, or C652. (*From Brick Institute of America,* Technical Note 16B, *BIA, Reston, Va.*)

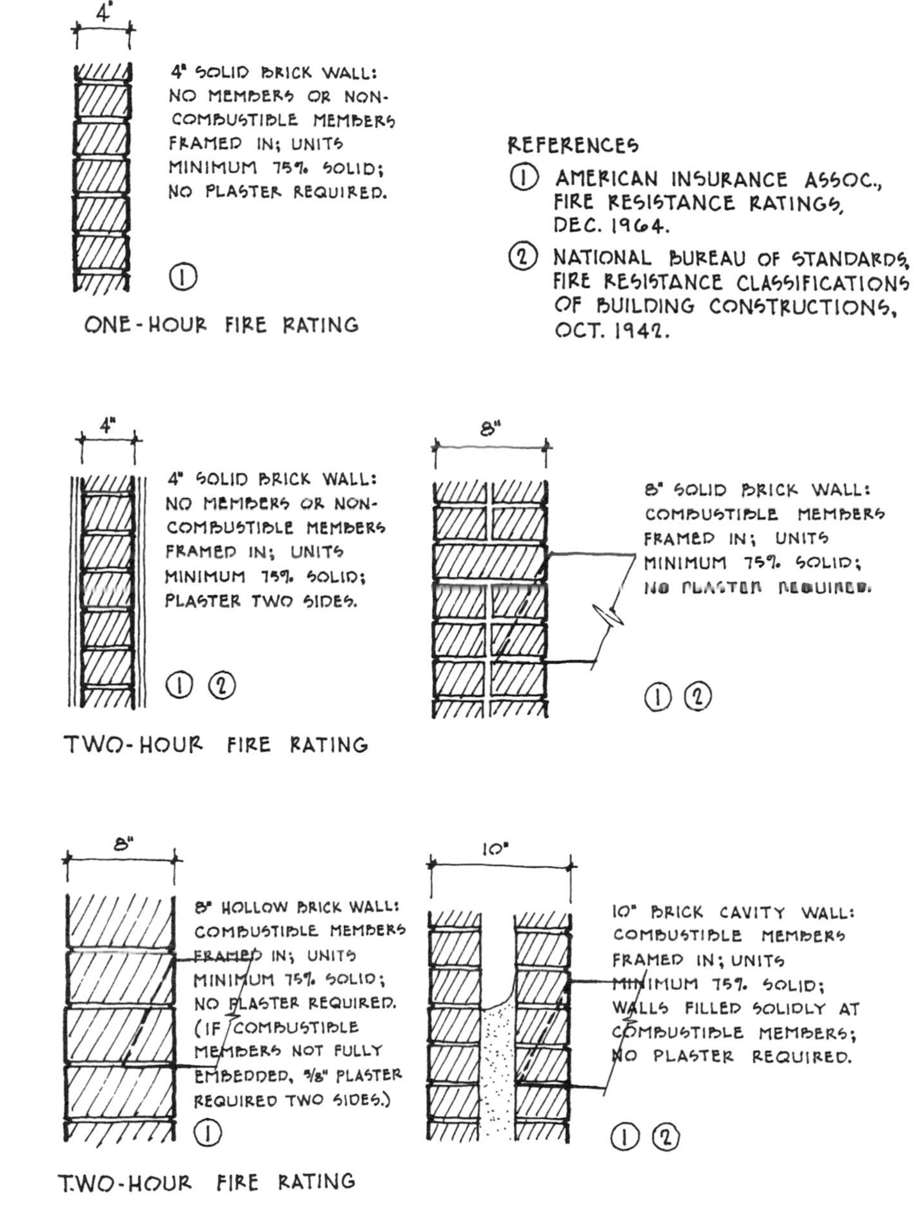

Fire resistance ratings. (*From Brick Institute of America,* **Technical Note 16 Rev.,** ***BIA, Reston, Va.***)

The application of plaster to one or both sides of a clay or concrete masonry wall increases the fire rating of the assembly. For portland cement plaster, the plaster thickness may be added to the actual thickness of solid units or to the equivalent thickness of hollow units in determining the rating. For gypsum plaster, a coefficient is added to equation (*8.2*), $R = (R_1^{0.59} + R_2^{0.59} + \cdots + R_n^{0.59} + as + pl)^{1.7}$, where pl is the thickness coefficient of sanded gypsum plaster from *Fig. 8-15.*

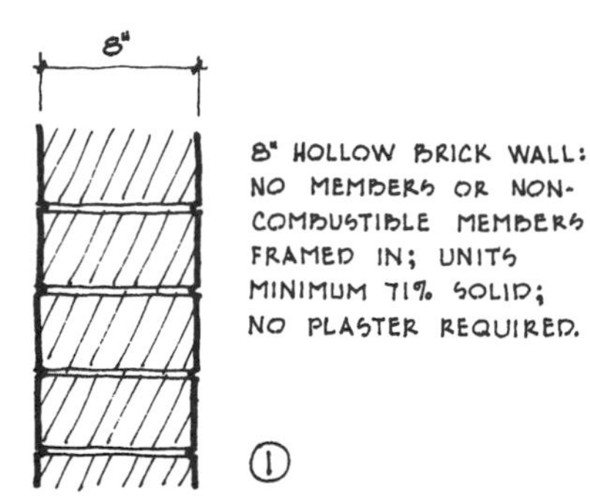

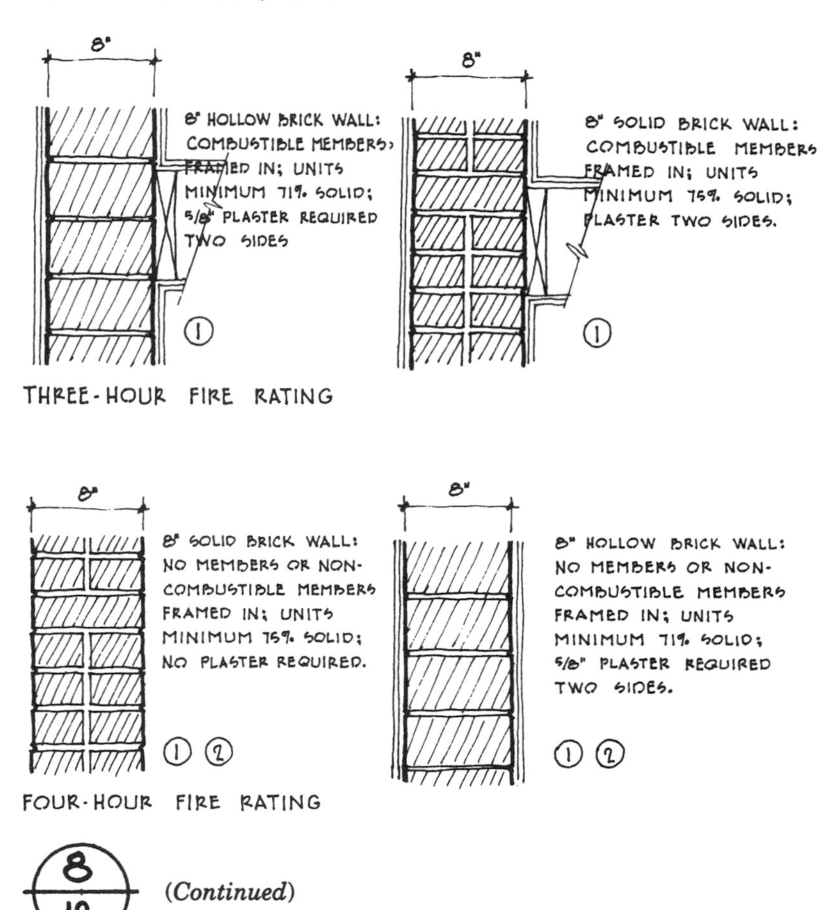

(Continued)

8.4.3 UL Ratings

Underwriters' Laboratories (UL) design numbers apply only to a specific proprietary product or assembly manufactured by a specific manufacturer or manufacturers. The fire resistance ratings of clay and concrete masonry, on the other hand, are generic. They apply to all products made from the same types of raw materials. Consequently, UL identifies masonry products by their classification, rather than by design numbers. For example, Class B-4 concrete masonry units have a 4-hour rating, Class C-3 concrete masonry units have a 3-hour rating, and Class D-2 units have a 2-hour rating. The *UL*

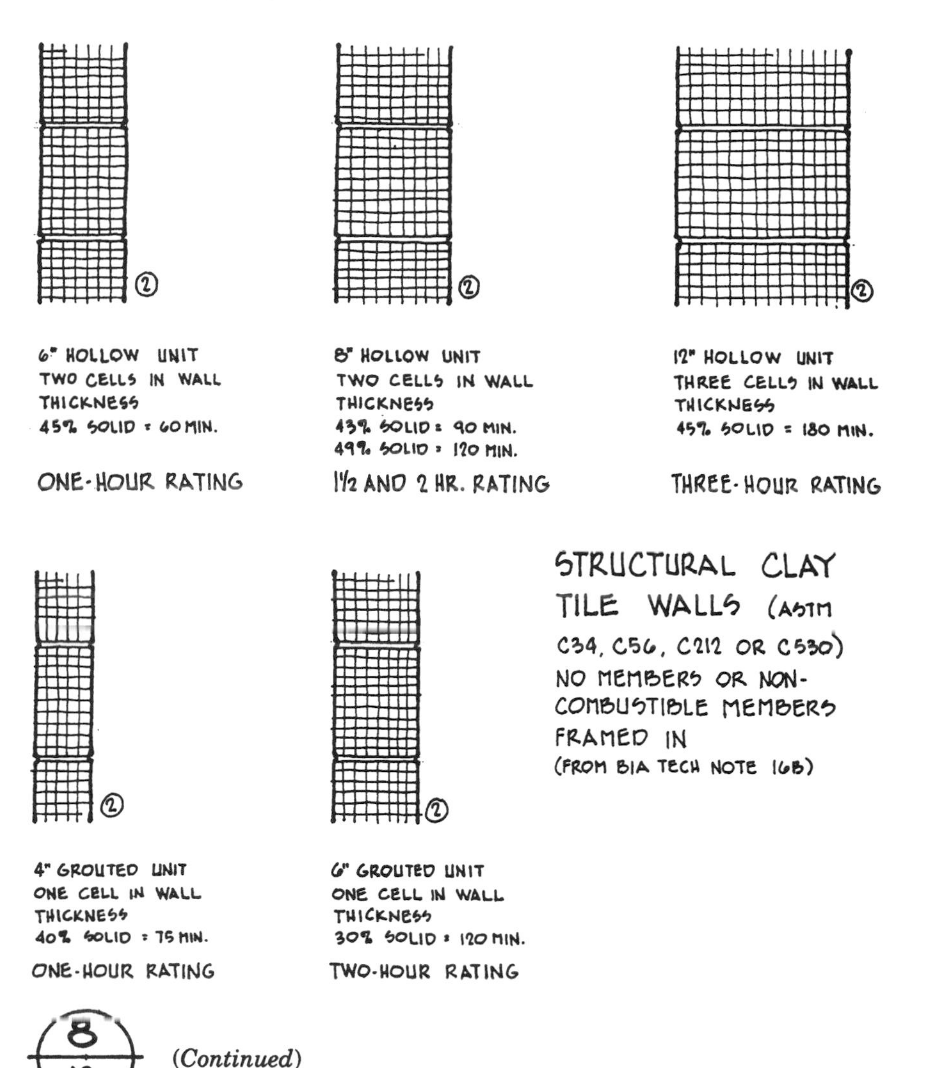

(*Continued*)

Fire Resistance Directory then lists CMU manufacturers who are eligible to issue a UL certificate for one or more of these classifications. The Directory also gives UL numbers for seven tested masonry wall assemblies. All of these assemblies were tested so that a specific manufacturer could show that a particular product (mortar mix or insulation insert, for example) could be added to or substituted in a "standard" masonry assembly and still achieve the same fire rating. For three of the substitutes, the "standard" wall assembly is also shown for comparison. In these cases, the assigned UL design number can be applied to masonry construction as follows.

- UL Design No. U901—a 4-hour rated wall assembly (loadbearing or nonloadbearing) constructed of:
  - Nominal 8-in. Class B-4 concrete block
  - ASTM C270, Type M or S portland cement–lime mortar
  - ⅜-in. joints, full mortar bedding, running bond pattern

  With certain types of loose fill insulation, Class C-3 and D-2 concrete blocks also provide a 4-hour assembly.

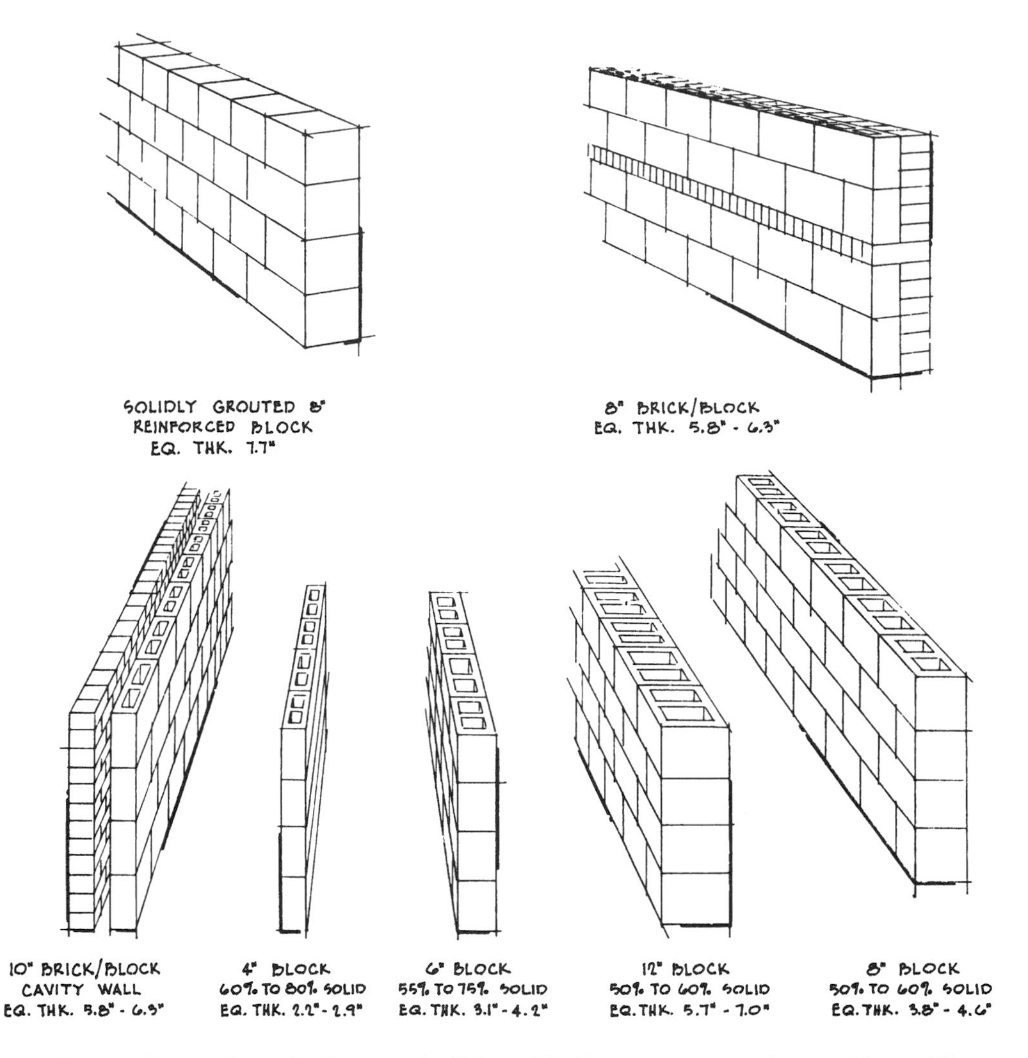

Equivalent thickness of solid and hollow masonry unit partitions. (*From National Concrete Masonry Association,* TEK Bulletin 6, *NCMA, Herndon, Va.*)

- UL Design No. U904—a 3-hour rated wall assembly (loadbearing or non-loadbearing) constructed of:
  - Nominal 8-in. Class C-3 concrete block;
  - ASTM C270, Type M or S portland cement–lime mortar;
  - $\frac{3}{8}$-in. joints, full mortar bedding, running bond pattern.

  With certain types of loose fill insulation, add 1-hour to the block classification.
- UL Design No. U905—a 2-hour rated wall assembly (loadbearing or non-loadbearing) constructed of:
  - Nominal 8-in. Class D-2 concrete block;
  - ASTM C270, Type M or S portland cement–lime mortar;
  - $\frac{3}{8}$-in. joints, full mortar bedding, running bond pattern.

  In all three assemblies, a $\frac{3}{4}$-in. plaster or stucco coating on one side adds $\frac{1}{2}$ hour to the wall rating.

Type of aggregate	Fire resistance periods (minutes)														
	30	45	60	75	90	105	120	135	150	165	180	195	210	225	240
Pumice or expanded slag	1.5	1.9	2.1	2.5	2.7	3.0	3.2	3.4	3.6	3.8	4.0	4.2	4.4	4.5	4.7
Expanded shale, clay, or slate	1.8	2.2	2.6	2.9	3.3	3.4	3.6	3.8	4.0	4.2	4.4	4.6	4.8	4.9	5.1
Limestone, cinders, or unexpanded slag	1.9	2.3	2.7	3.1	3.4	3.7	4.0	4.3	4.5	4.8	5.0	5.2	5.5	5.7	5.9
Calcareous gravel	2.0	2.4	2.8	3.2	3.6	3.9	4.2	4.5	4.8	5.0	5.3	5.5	5.8	6.0	6.2
Siliceous gravel	2.1	2.6	3.0	3.5	3.9	4.2	4.5	4.8	5.1	5.4	5.7	6.0	6.2	6.5	6.7

Notes:

1. Equivalent thickness is the average thickness of the solid material in the wall. It is found by taking the total volume of a wall unit, subtracting the volume of core spaces, and dividing this by the area of the exposed face of the unit.
2. Values between those shown in the table can be determined by direct interpolation.
3. Where combustible members are framed into the wall, the thickness of solid material between the end of each member and the opposite face of the wall, or between members set in from opposite sides, shall not be less than 93% of the thickness shown in the table.
4. Units shall comply with the requirements of ASTM C55, C73, C90, or C145.

Minimum equivalent thickness (in inches) of loadbearing and non-loadbearing concrete masonry walls. (*From* Standard Building Code, *1991 edition, Standard Building Code Congress, Birmingham, Ala.*)

Thickness of plaster (in.)	One side	Two sides
½	0.30	0.60
⅝	0.37	0.75
¾	0.45	0.90

Values used are for 1:3 sanded gypsum plaster.

Coefficients for plaster (*pl*). (*From Brick Institute of America,* Technical Note 16B, *BIA, Reston, Va.*)

The remainder of the masonry wall assemblies "listed" in the UL Directory are too proprietary to apply to masonry construction in general. The UL numbers for these other assemblies are not appropriate if any of the component materials vary from the specific brand or type of products identified, including such items as veneer anchors or lime. For masonry, a better way to note construction drawings than listing UL Design Numbers is to reference the building code. For example, designate a certain masonry wall as *1994 Uniform Building Code, Table 43B, Item No....*; or *1994 Standard Building Code, Appendix B, Table....* In the specifications, concrete masonry units may also be designated as UL Class B-4, C-3, or D-2 as applicable to the design, but there are no similar classifications for clay brick.

8.4.4 Steel Fireproofing

Steel frame construction is vulnerable to fire damage and must be protected from heat and flame. Structural clay tile, brick, and concrete block can all be used to fireproof steel columns and beams. Hollow structural clay tile

U.B.C. TABLE NO. 43-A—MINIMUM PROTECTION OF STRUCTURAL PARTS BASED ON TIME PERIODS FOR VARIOUS NONCOMBUSTIBLE INSULATING MATERIALS

Structural parts to be protected	*Item number*	*Insulating material used*	*Minimum thickness of insulating material for following fire-resistive periods (in.)*			
			4 hr	*3 hr*	*2 hr*	*1 hr*
Steel columns and all members of primary trusses	1-2.1	Clay or shale brick with brick and mortar fill	$3\frac{3}{4}$			$2\frac{1}{4}$
	1-3.1	4-in. hollow clay tile in two 2-in. layers; $\frac{1}{2}$-in. mortar between tile and column; $\frac{3}{8}$-in. metal mesh (wire diameter = 0.046 in.) in horizontal joints; tile fill	4			
	1-3.2	2-in. hollow clay tile; $\frac{3}{4}$-in. mortar between tile and column; $\frac{3}{8}$-in. metal mesh (0.046-in. wire diameter) in horizontal joints; limestone concrete fill; plastered with $\frac{3}{4}$-in. gypsum plaster	3			
	1-3.3	2-in. hollow clay tile with outside wire ties (0.08-in. diameter) at each course of tile or $\frac{3}{8}$-in. metal mesh (0.046-in. diameter wire) in horizontal joints; limestone or traprock concrete fill extending 1 in. outside column on all sides			3	
	1-3.4	2-in. hollow clay tile with outside wire ties (0.08-in. diameter) at each course of tile with or without concrete fill; $\frac{3}{4}$-in. mortar between tile and column				2

Uniform Building Code column protection requirements. (*Reproduced from the* **Uniform Building Code**, *1991 edition, copyright 1991, with permission of the publisher, the International Conference of Building Officials.*)

units were originally manufactured for this purpose in the late nineteenth century. They offer effective and relatively lightweight protection. Fire test results from the National Bureau of Standards form the basis of modern code requirements for protection of steel structural elements. The table in *Fig. 8-16* is taken from the *Uniform Building Code* to show protective masonry coverings that are acceptable for various fire ratings. By comparison, the table in *Fig. 8-17* shows fire resistance ratings for steel stud walls with brick veneer on one side and a plaster finish on the other. The masonry generally provides 2 to $2\frac{1}{2}$ times the protection of the plaster.

8.4.5 Construction Classifications

Various types of construction are classified according to the degree of fire resistance they offer. Building codes distinguish at least four types: (1) Fire-Resistive construction, (2) Non-combustible construction, (3) Ordinary or Exterior Protected construction, and (4) Wood Frame construction. Within each of these classifications, specific fire ratings are required for various building components, such as interior and exterior bearing walls, columns and beams, floors, roofs, partitions, and elevator shafts. Although the nomenclature and number of classifications differs among the codes, the table in *Fig. 8-18* compares the fire ratings required for these construction types. (Detailed exceptions and specific occupancy requirements will

Wall or partition assembly	Fire resistance period	
	Plaster side exposed (minutes)	Brick faced side exposed (minutes)
Steel studs faced outer side with ½-in. wood fiberboard sheathing next to studs, ¾-in. air space formed with ¾-in. × 1⅝-in. wood strips placed over the fiberboard and secured to the studs; metal or wire lath nailed to such strips, 3¾-in. brick veneer held in place by filling ¾-in. air space between the brick and lath with mortar. Inside facing of studs: ¾-in. unsanded gypsum plaster on metal or wire lath attached to 5/16-in. wood strips secured to edges of the studs.	90	240
Steel studs faced outer side with 1-in. insulation board sheathing attached to studs, 1-in. air space, and 3¾-in. brick veneer attached to steel frame with metal ties every 5th course. Inside facing of studs: ⅞-in. sanded gypsum plaster (1:2 mix) applied on metal or wire lath attached directly to the studs.	90	240
Same as above except use ⅞-in. vermiculite–gypsum plaster or 1-in. sanded gypsum plaster (1:2 mix) applied to metal or wire.	120	240
Steel studs faced outer side with ½-in. gypsum sheathing board, attached to studs, and 3¾-in. brick veneer attached to steel frame with metal ties every 5th course. Inside facing of studs: ½-in. sanded gypsum plaster (1:2 mix) applied to ½-in. perforated gypsum lath securely attached to studs and having strips of metal lath 3-in. wide applied to all horizontal joints of gypsum lath.	120	240

Fire resistance ratings for steel frame/brick veneer walls. (*From Brick Institute of America,* Technical Note 16B, *BIA, Reston, Va.*)

reduce or increase ratings for some elements of the structure, so local codes should be consulted to verify exact requirements.)

Required fire resistance is based on fire hazard classification for different occupancies (i.e., residential, mercantile, industrial, and so on). High-hazard occupancies and occupancies where life safety protection is critical, such as hospitals, high-rise buildings, and large assembly areas, require the highest protection to assure control of spreading fires. Because of the fire-resistant characteristics of masonry structures, they easily satisfy the requirements for Fire-Resistive or Type I construction (*see Fig. 8-19*). "Fire-Resistive" construction is defined by UBC as construction that resists the spread of fire, and "Non-combustible" materials are defined as those of which no part will ignite and burn when subjected to fire.

8.4.6 Compartmentation

A key element in fire control is compartmentation of a building to contain fire and smoke. Codes require that a building be subdivided by fire walls into areas related in size to the danger and severity of fire hazard involved. Fire walls must be constructed of non-combustible materials, have a minimum fire rating of 4 hours, and have sufficient structural stability under fire conditions to allow collapse of construction on either side without collapse of the

Building element	*Fire-Resistive*			*Non-combustive*			*Ordinary*			*Wood Frame*		
	BBC	*SBC*	*UBC*	*BBC*	*SBC*	*UBC*	*BBC*	*SBC*	*UBC*	*BBC*	*SBC*	*UBC*
Exterior bearing walls	4[1]	4	4[3,4]	2[1]	2[5]	1	2	3[6]	4[3]	1	1	1
Interior bearing walls	4[1,7]	4[7]	3[4]	2[1,8]	1	1	1	1[9]	1	1	1	1
Exterior non-bearing walls	2[10]	3[11]	4[4,12]	1½[13]	2[14,15]	1[16]	2[17]	3[18]	4[19]	1[20]	1[21]	1
Shaft enclosures (elevators and stairways)	2[22]	2[23]	2	2[22]	2[23]	1	2[22]	2[23,24]	1	1[25]	2[23,24]	1
Partitions	N[26]	1[2,27]	1	N[26]	1[27]	1	N[26]	1[27]	1	N[26]	1[27]	1
Columns[28]	4[1]	4[2,29]	3	2[1]	1	1	1	1	1	1	1	1
Beams	3[1]	4[2,29]	3[4]	1½[31]	1	1	1	1	1	1	1	1
Floors	3[1]	3[2,29]	2	1½[31]	1[30]	1	1	1[30]	1	1	1	1
Roofs	2	1½	2[4]	1	1	1	1	1	1	1	1	1

Notes: BBC—1984 ***Basic/National Building Code*** (Building Officials and Code Administrators, International); SBC—1985 ***Standard Building Code*** (Southern Building Code Congress, International); UBC—1985 ***Uniform Building Code*** (International Conference of Building Officials).

N—No fire rating required.

[1]In business and residential occupancies over 75 ft in height, rating may be reduced by 1 hr when a complete, automatic fire suppression system is installed throughout the building and only when: a) exterior bearing walls have a horizontal separation of 6 ft or greater; b) interior bearing walls and columns support more than one floor; or c) the building has floor construction which includes beams.

[2]In business and residential occupancies over 75 ft in height, ratings of partitions, columns, trusses, girders, beams, and floors may be reduced by 1 hr but no component shall be less than 1 hr when a complete automatic fire suppression system is installed throughout the building.

[3]In office buildings, apartment buildings, etc., rating may be 2 hr where openings are permitted.

[4]The fire rating may be reduced by 1 hr for interior bearing walls, exterior bearing and non-bearing walls, roofs, and beams supporting roofs provided they do not frame into columns when a complete automatic fire suppression system is installed throughout the building. This reduction applies only to buildings over 75 ft in height and shall not apply to exterior bearing and non-bearing walls whose fire rating has already been reduced by Footnote 3.

[5]Rating may be 1 hr where horizontal separation is 3 ft or greater.

[6]Rating may be 2 hr where horizontal separation is over 3 to 20 ft; rating may be 1 hr where horizontal separation is 20 ft or greater.

[7]Rating may be reduced to 3 hr when interior bearing walls support one floor only or support roof only.

[8]Rating may be reduced to 1½ hr when interior bearing walls support one floor only or support roof only.

[9]The use of combustible construction for interior bearing partitions shall be limited to the support of not more than two floors and a roof.

[10]Rating may be 1½ hr where horizontal separation is over 11 to 30 ft; no rating is required where horizontal separation is 30 ft or greater.

[11]Rating may be 2 hr where horizontal separation is over 3 to 20 ft; rating may be 1 hr where horizontal separation is over 20 to 30 ft; non-combustible building materials or exterior grade fire-retardant treated wood are permitted when horizontal separation is 30 ft or greater.

[12]Non-bearing wall fronting on public ways, or yards having a width of at least 40 ft may be of unprotected non-combustible construction. In other than hazardous occupancies, rating may be 1-hr fire-resistive, non-combustible construction where unprotected openings are permitted and 2 hr fire-resistive, non-combustible construction where protected openings are required.

[13]Rating may be 1 hr where horizontal separation is over 11 to 30 ft; no rating is required where horizontal separation is 30 ft or greater.

[14]Rating may be 1 hr where horizontal separation is over 3 to 20 ft; non-combustible building materials are required when horizontal separation is 20 ft or greater. Exterior grade fire-retardant treated wood may be used when horizontal separation is 30 ft or greater.

[15]Exterior walls of this type construction are required to be 2 hr when buildings are more than one story and greater than 2000 sq ft in area; also when wall facings are within 15 ft of common property line.

Required fire resistance ratings for various types of construction (in hours). (*From Brick Institute of America*, Technical Note 16A, *BIA, Reston, Va.*)

[16]Non-bearing walls fronting on public ways or yards having a width of at least 40 ft may be of unprotected non-combustible construction.

[17]Rating may be 1½ hr where horizontal separation is over 11 to 30 ft; no rating is required where horizontal separation is 30 ft or greater.

[18]Rating may be over 2 hr where horizontal separation is over 3 to 20 ft; 1 hr over 20 to 30 ft; and non-combustible building materials or exterior grade fire-retardant treated wood with horizontal separation of 30 ft or greater.

[19](a) Non-bearing walls fronting on public ways or yards having a width of at least 40 ft may be unprotected when entirely of non-combustible materials. (b) In other than hazardous and institutional occupancies, walls may be non-combustible 1-hr fire resistive where unprotected openings are permitted and non-combustible 2-hr fire resistive when protected openings are required. (c) Approved fire-retardant treated wood framing may be used within the assembly of exterior walls as permitted by (a) and (b), provided the required fire resistance is maintained and outer and inner faces of wall are non-combustible.

[20]No rating is required where horizontal separation is 30 ft or greater.

[21]No rating is required where horizontal separation is 20 ft or greater.

[22]Exit and shaft enclosures connecting three floor levels or less shall have a rating of not less than 1 hr; shaft enclosures must be of non-combustible materials.

[23]Exit and shaft enclosures connecting less than 4 stories shall have a rating of not less than 1 hr; except all exits and stairways in assembly and hazardous occupancies must be 2 hr; shaft enclosures must be of non-combustible materials.

[24]Same as Footnote 23, except that shaft enclosures may be of combustible materials.

[25]Fire enclosures of exits and stairways shall be of 2 hr; elevator shafts may be of combustible materials.

[26]Rating pertains only to non-bearing partitions; separation of tenant spaces, dwelling unit separations, and exit access corridors shall have a rating of 1 hr; in all occupancies other than apartment buildings and dormitories, etc., exit access corridors serving 30 or fewer occupants may be non-rated.

[27]Separation of tenant spaces and exit access corridors shall have a rating of 1 hr; separation between townhouses shall not be less than 2 hr.

[28]Applicable to columns and beams supporting loads of more than one floor or roof.

[29]Same as Footnote 2 except pertains to assembly, business, educational, factory, and residential occupancies only.

[30]For business and mercantile occupancies, when 5 or more stories in height, a 2-hr fire-resistant floor shall be required over the basement.

[31]Same as Footnote 1, but in no case shall the rating be less than 1 hr.

(Continued)

Compartmented concrete or masonry construction prevents the spread of fire. (*Photo courtesy PCA.*)

wall. Masonry fire walls may be designed as continuously reinforced cantilevered sections. They are self-supporting without dependence on connections to adjacent structural framing. For additional lateral stability, free-standing cantilever walls may be stiffened by integral masonry pilasters with vertical reinforcing steel (*see Fig. 8-20*). Double fire walls can also be used, so that if the building frame on one side collapses, half the wall can be pulled

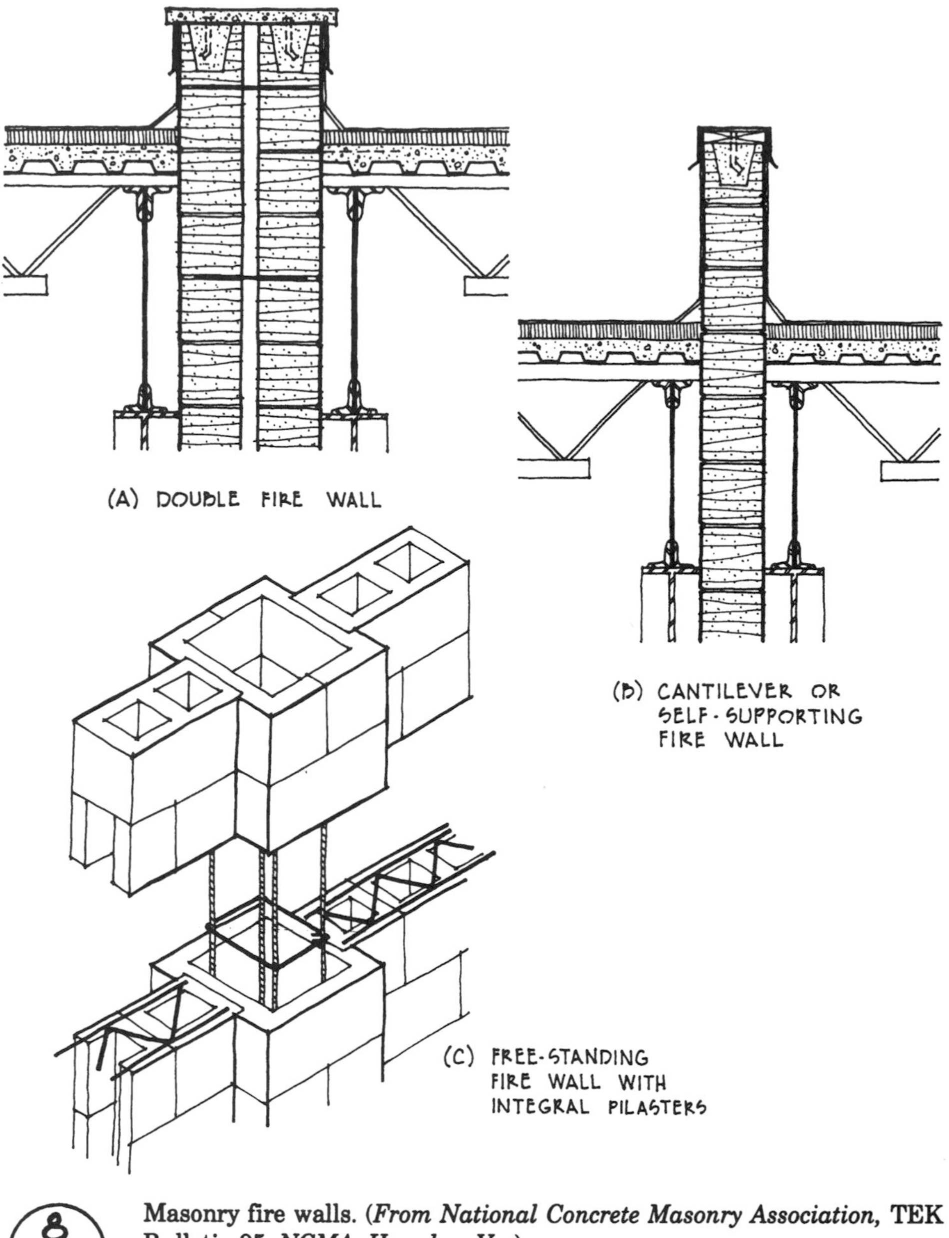

Masonry fire walls. (*From National Concrete Masonry Association,* TEK Bulletin 95, *NCMA, Herndon, Va.*)

over while the other half still protects adjacent areas. Masonry walls also provide an airtight barrier against the spread of smoke and toxic gases.

Fire walls are not extensively used in low-rise multifamily units. Low-rise multifamily buildings (apartments, nursing homes, motels, condominiums) are constructed to essentially the same fire safety standards as single-family dwellings despite the significantly increased risk posed by the proximity of adjacent units and the vulnerability of occupants to the actions of their neighbors. Code requirements are much less stringent for two- and three-story residential occupancies than for high-rise buildings, and fires can quickly consume several adjacent units of combustible construction (*see Fig. 8-21*). Multifamily dwellings are all too frequently built of wood frame and gypsum board to achieve only a "1-hour" rating. This can be provided

Spread of fire through a building of combustible construction. (*Photo courtesy PCA.*)

by simple 2×4 construction, 16 in. on center, with one layer of $\frac{5}{8}$-in. Type X gypsum board on each side of the wall. A "2-hour" rating is achieved with a double layer of Type X gypsum board on each side. Townhouse and zero-lot-line developments, however, are often required to have 2-hour non-combustible masonry fire walls separating units and as a result, statistics show that fire losses are greatly reduced.

8.4.7 Fire Insurance Rates

There is one source to which we can look for a realistic comparison of combustible and non-combustible construction, and that is fire insurance rates. The insurance industry must literally guarantee the fire safety of buildings, and they do so for a price that is based on actual fire loss experience and the corresponding degree of risk presented by various types of construction. Studies show that throughout the country, insurance premiums for wood frame apartments can be 5 to 10 times higher than for the same apartments built of 2-hour non-combustible masonry walls with concrete floor and roof slabs. Five recent studies summarized in the *Magazine of Masonry Construction* are shown in *Fig. 8-22.* In addition to comparative insurance rates, these independent studies also show that the life cycle costs of masonry buildings are less than those of the frame buildings. No two of the studies used the same factors for insurance, mortgage rates, property taxes, depreciation, maintenance, or occupancy rates, but each found that the savings in fire insurance premiums alone, even for sprinklered buildings, offset any differences in initial construction cost. *Figure 8-23* compares insurance rates for commercial buildings of various construction.

Fire detection and fire suppression systems are valuable when they work, but they can and do fail regularly from lack of maintenance, frozen pipes, and human error. Non-combustible construction cannot fail, and this fact is recognized by the insurance industry. Putting discrepancies in test results and ratings aside, this is perhaps the best indicator of just how unequal drywall and masonry fire separations are.

Construction type	Location	Construction cost	1st year insurance cost
Wood frame	Illinois	960,000	13,211
Fire-resistive		1,032,000	1,430
Wood frame	Illinois	1,228,700	16,794
Fire-resistive		1,325,400	1,751
Wood frame	Oklahoma	406,300	6,117
Fire-resistive		438,300	586
Wood frame with wood truss roof	Texas	1,270,080	32,760
Masonry with wood truss roof		1,345,400	4,500
Frame, unsprinklered	Georgia	967,680	8,928
Frame, sprinklered		997,920	7,974
Fire-resistive, unsprinklered		1,030,579	1,577

Comparative insurance rates for wood frame and fire-resistive masonry/concrete construction. (*From* The Magazine of Masonry Construction, *November 1989.*)

8.5 THERMAL PROPERTIES

The thermal efficiency of a building material is normally judged by its resistance to heat flow. A material's *R-value* is a measure of this resistance taken under laboratory conditions with a constant temperature differential from one side to the other. This is called a *steady-state* or *static condition.*

Thermal resistance depends on the density of the material. By this measure, masonry is a poor insulator. Urethane insulation, on the other hand, has a very high resistance because it incorporates closed cells or pockets to inhibit heat transfer. The reciprocal of the R-value is the *U-value,* or the overall coefficient of heat transmission. Both values are derived from the inherent thermal *conductance* of the material, and its *conductivity* per inch of thickness. Thermal characteristics of some typical building materials are listed in *Fig. 8-24.*

Materials in which heat flow is identical in all directions are considered thermally homogeneous. Materials that are not isotropic with respect to heat transmission (such as hollow masonry units) are considered thermally heterogeneous. Thermal conductance and thermal resistance of homogeneous materials of any thickness can be calculated from the equations

$$C_x = \frac{k}{x} \tag{8.3}$$

and

$$R_x = \frac{x}{k} \tag{8.4}$$

where C = thermal conductance, Btu/(hour × °F × sq ft)
R = thermal resistance, (hour × °F × sq ft)/(Btu × in.)
k = thermal conductivity (Btu × in.)/(hour × °F × sq ft)
x = thickness of material, in.

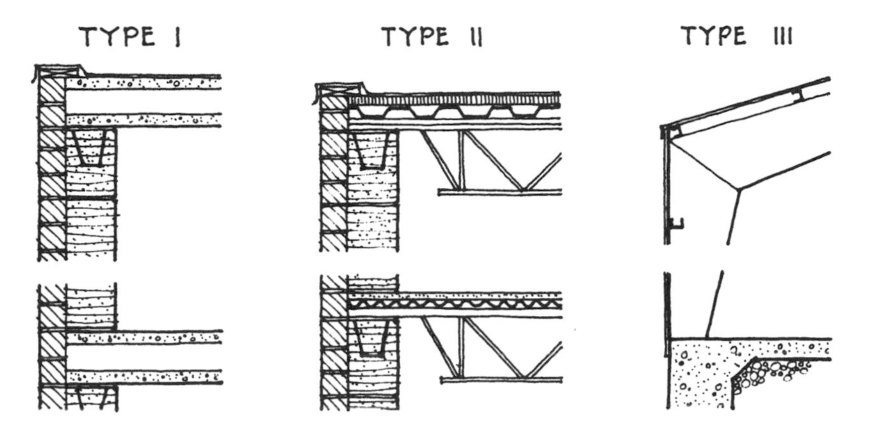

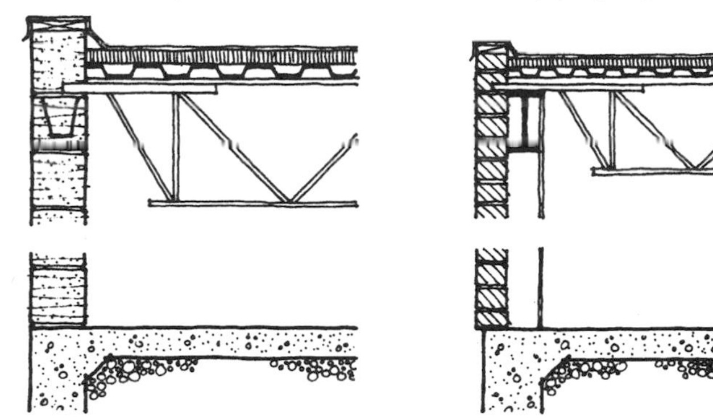

		I 12-in. masonry walls (brick faced) with roof of 2-hr fire-resistive rated		II 12-in. masonry walls (brick faced) with roof of minimum 0.22 gauge unprotected steel deck		III Preengineered steel building walls and roof of unprotected steel		IV 8-in. masonry walls (hollow concrete block) roof—minimum 0.22 gauge unprotected steel deck		V 4-in. brick veneer over steel stud, $\frac{1}{2}$-in. gypsum board each side—minimum 0.22 gauge steel deck roof	
Type of insurance	*Ins. amt. coverage*	*Rate*	*Ann. prem.*	*Rate*	*Ann. prem.*	*Rate*	*Ann. prem.*	*Rate*	*Ann. prem.*	*Rate*	*Ann. prem.*
Building	750,000	0.231	1,733	0.755	5,662	1.779	13,343	0.774	5,805	1.84	13,800
Contents	1,500,000	0.493	7,395	1.188	17,820	2.172	32,580	1.211	18,165	2.23	33,450
Extended coverage	2,250,000	0.114	2,565	0.114	2,565	0.350	7,875	0.114	2,565	0.232	5,220
Total annual premium			11,693		26,047		53,798		26,535		52,470

Note: Dimensions are 200 × 300 = 60,000 sq. ft. Divided with a standard 8-in. CMU fire wall. 1 story, no basement, 20 ft high, concrete slab floor, exterior walls maximum of 20% glass windows. Location: St. Louis County, Missouri. Class 6 Fire Protection at risk; no exposure from adjacent buildings. All insulation listed by Underwriters' Laboratories as noncombustible, having a flame spread rating under 25 and a smoke development rating under 200.

Fire insurance cost comparison for office-warehouse and storage buildings. (*From Mason Contractors Association of Greater St. Louis,* Masonry Cost Guide and Related Technical Data, *MCA of Greater St. Louis.*)

Materials description	Density (lb/cu ft)	Conductivity or conductance		Resistance (R)	
				Per inch thickness	For thickness listed
		(k)	(C)	(1/k)	(1/C)
Masonry units					
Face brick	130	9.00		0.11	
Common brick	120	5.00		0.20	
Hollow brick					
4 in. (62.9% solid)	81		1.36		0.74
6 in. (67.3% solid)	86		1.07		0.93
8 in. (61.2% solid)	78		0.94		1.06
10 in. (60.9% solid)	78		0.83		1.20
Hollow brick, vermiculite fill					
4 in. (62.9% solid)	83		0.91		1.10
6 in. (67.3% solid)	88		0.66		1.52
8 in. (61.2% solid)	80		0.52		1.92
10 in. (60.9% solid)	80		0.42		2.38
Lightweight concrete block—100-lb density concrete					
4 in.	78		0.71		1.40
6 in.	66		0.65		1.53
8 in.	60		0.57		1.75
10 in.	58		0.51		1.97
12 in.	55		0.47		2.14
Lightweight concrete block, vermiculite fill—100-lb density concrete					
4 in.	79		0.43		2.33
6 in.	68		0.27		3.72
8 in.	62		0.21		4.85
10 in.	61		0.17		5.92
12 in.	58		0.15		6.80
Building board					
⅜-in. drywall (gypsum)	50		3.10		0.32
½-in. drywall (gypsum)	50		2.25		0.45
Plywood	34	0.80		1.25	
½-in fiberboard sheathing	18		0.76		1.32
Siding					
7/16-in. hardboard	40		1.49		0.67
½-in. by 8-in. wood bevel	32		1.23		0.81
Aluminum or steel over sheathing	—		1.61		0.61

Thermal resistance of some common building materials.

Materials description	Density (lb/cu ft)	Conductivity or conductance		Resistance (R)	
				Per inch thickness	For thickness listed
		(k)	(C)	(1/k)	(1/C)
Insulating materials					
Batt or blanket					
2 to 2¾ in.					7.00
3 to 3½ in.	1.2				11.00
5½ to 6½ in.	1.2				19.00
Boards					
Expanded polystyrene					
Cut cell surface	1.8	0.25		4.00	
Smooth skin surface	1.8	0.20		5.00	
Expanded polyurethane	1.5	0.16		6.25	
Polyisocyanurate	2	0.14		7.14	
Loose fill					
Vermiculite	4–6	0.44		2.27	
Perlite	5–8	0.37		2.70	
Woods					
Hard woods	45	1.10		0.91	
Soft woods	32	0.80		1.25	
Metals					
Steel	—	312.0		0.003	
Aluminum	—	1416.0		0.0007	
Copper	—	2640.0		0.0004	
Air space					
¾ in. to 4 in., winter			1.03		0.97
¾ in. to 4 in., summer			1.16		0.86
			(h)		(1/h)
Air surfaces					
Inside—still air			1.47		0.68
Outside—15 mph wind, winter			6.00		0.17
7.5 mph wind, summer			4.00		0.25

(Continued)

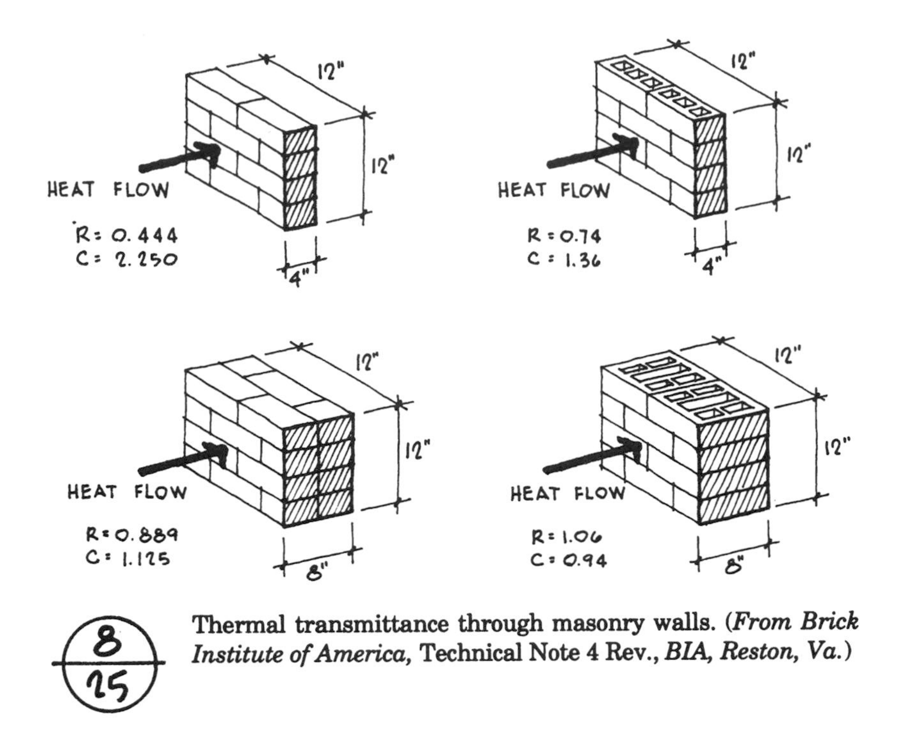

Thermal transmittance through masonry walls. (*From Brick Institute of America,* Technical Note 4 Rev., *BIA, Reston, Va.*)

Figure 8-25 shows the difference between thermal resistance (R) and thermal conductance (C) for thermally homogeneous and thermally heterogeneous masonry walls.

Whenever an opaque wall assembly is analyzed, it should include both the inside and outside air surfaces, which affect both convection and conduction of heat. The inclusion of these air surfaces makes all opaque wall assemblies "layered" construction. In computing the heat transmission coefficients of layered construction, the paths of heat flow must first be determined. If the heat flow paths are in series, the thermal resistances (R) of the layers are additive, but if the paths are in parallel, then the thermal transmittances (U) are averaged. For layered construction with paths of heat flow in series, the total thermal resistance (R) of the wall is obtained by adding the thermal resistances of each layer ($R = R_1 + R_2 + \cdots + R_n$), and the overall coefficient of heat transmission is $U = 1/R$. Average transmittances for parallel paths of heat flow are obtained using the equation

$$U_{\text{avg}} = \frac{[A_A(U_A) + A_B(U_B) + \cdots + A_n(U_n)]}{A_t} \tag{8.5}$$

or

$$U_{\text{avg}} = \frac{[1/(R_A/A_A) + 1/(R_B/A_B) + \cdots + 1/(R_n/A_n)]}{A_t} \tag{8.6}$$

where A_A, A_B, etc. = area of heat flow path, sq ft
U_A, U_B, etc. = transmission coefficients of the respective paths
R_A, R_B, etc. = thermal resistances of the respective paths
A_t = total area being considered ($A_A + A_B + \cdots + A_n$), sq ft

Such analyses are especially important when the various paths have significantly different heat flow characteristics, or when the paths involve large percentages of the total wall.

Many walls contain *thermal bridges,* which may be taken into account in different ways. The wall shown in *Fig. 8-26* has thermal bridges where the wood studs interrupt the layer of insulation. The parallel path method of calculation is used for such non-metallic bridges, where the path at the stud is Path A, and the path at the insulation is Path B. The calculations show that the average U-value is 6% higher than the U-value at the insulation.

The wall shown in *Fig. 8-27* has a thermal bridge at the metal tie. Metallic bridges are considered using the parallel zone method where a slightly larger area is assumed to be affected than just the actual area of the metal itself (Zone A). The American Society for Heating, Refrigeration and Air Conditioning Engineers (ASHRAE) *Handbook of Fundamentals* prescribes a method for determining the size and shape of each zone. In the case of a metal beam, the surface shape of Zone A would be a strip of width W centered on the beam. Since the metal tie in *Fig. 8-27* is of circular wire, Zone A is a circle of diameter W, which is calculated from the equation

$$W = m + 2d \tag{8.7}$$

where W = width or diameter of the zone, in.
m = width or diameter of the metal path, in.
d = distance from the panel surface to the metal, in. (but not less than 0.5 in.)

The larger of the two values calculated for W at each surface should be used. *Figure 8-28* shows that the effect of the metal tie is considerably less in an uninsulated cavity wall, because as the R-value of the material which the metal bridge penetrates decreases, the percent of heat loss due to thermal bridging also decreases. As the distance between the face of the wall and the edge of the metal increases, however, the area of the affected zone increases. *Figure 8-29* illustrates this phenomenon. Only the web thickness of the metal stud is considered in calculating the area of the zone. The 1⅝ in. stud flange is relatively thin compared to the wall section, and therefore does not significantly affect the average thermal performance of the system. Its distance from the exterior surface is the thickness of the masonry, plus the air space, plus the sheathing thickness.

For estimating a building's heating and cooling requirements, U-values are used in heat-loss and heat-gain calculations with specific outdoor design temperatures for winter and summer. These calculations (like the laboratory test conditions) assume a constant temperature differential between outdoor and indoor air, and do not take into account the diurnal cycles of solar radiation and air temperature. As the sun rises and sets each day, the outdoor/indoor temperature differential continually fluctuates. *The static conditions on which R- and U-values are based do not actually exist in the real world.* Building materials with heavy mass can react to temperature fluctuations, producing a dynamic thermal response which differs substantially from heat flow calculations based solely on U-values. Research indicates that the actual measured rate of heat transfer for masonry walls is 20 to 70% less than steady-state calculation methods predict.

8.5.1 Thermal Inertia

Heat transfer through solid materials is not instantaneous. The time delay involving absorption of the heat is called *thermal lag.* Although most building materials absorb at least some heat, higher density and greater mass cause slower absorption and longer retention. The speed with which a wall

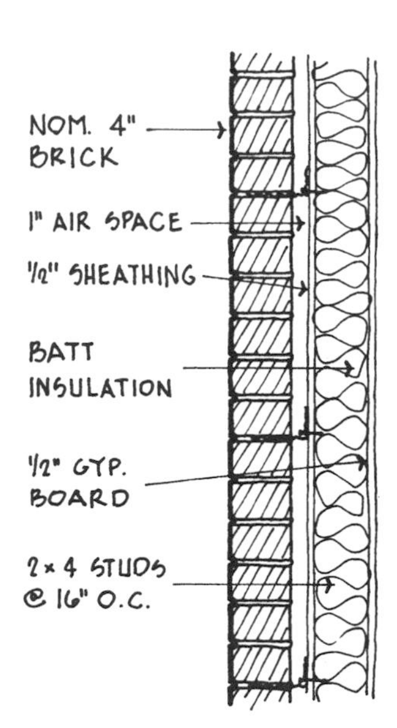

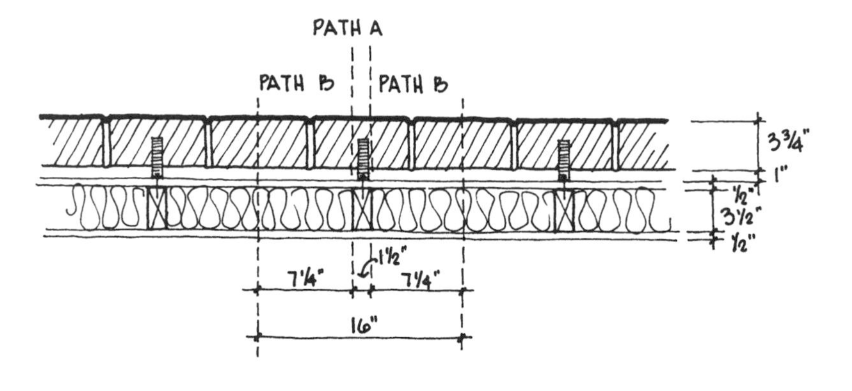

Stud spacing = 16.00 in. o.c., or 1.33 ft o.c.
Height of the section = 12.00 in. or 1.00 ft
Total area, A_t = 1.33 × 1.00 = 1.33 sq ft
Width of Path A = 1.50 in. or 0.125 ft
Area of Path A, A_A = 0.125 × 1.00 = 0.125 sq ft
Width of Path B = 16.00 − 1.50 = 14.50 in. or 1.208 ft
Area of Path B, A_B = 1.208 × 1.000 = 1.208 sq ft

Section	*C* *(Btu /(hr • °F • sq ft))*	*K* *((Btu • in.) /(hr • °F • sq ft))*	*x* *(in.)*	C_x *(Btu /(hr • °F • sq ft))*	*Path A* $1/C_x$ *((hr • °F • sq ft) /Btu)*	*Path B* $1/C_x$ *((hr • °F • sq ft) /Btu)*
Outside air surface	6.000			6.000	0.17	0.17
4-in. nominal face brick		9.000	3.75	2.400	0.42	0.42
1-in. airspace	1.030			1.030	0.97	0.97
Exterior fiberboard sheathing	0.760			0.760	1.32	1.32
2-in. × 4-in. wood stud		0.800	3.50	0.229	4.37	
3½-in. batt insulation						11.00
½-in. gypsum wall-board	2.250			2.250	0.45	0.45
Inside air surface	1.470			1.470	0.68	0.68
					$R_A = 8.38$	$R_B = 15.01$ $U_B = 0.067$

$R_A/A_A = 67.04$ $R_B/A_B = 12.43$

$1/(R_A/A_A) = 0.015$ $1/(R_B/A_B) = 0.080$

$U_{avg} = [1/(R_A/A_A) + 1(R_B/A_B)]/(A_A + A_B) = (0.015 + 0.080)/(0.125 + 1.208) = 0.071$ Btu/(hr • °F • sq ft)

$$\frac{U_{avg} - U_B}{U_B} \times 100\% = \frac{0.071 - 0.067}{0.067} \times 100\% = 6.0\%$$

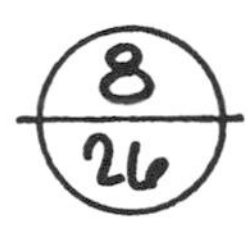

Thermal calculations for brick veneer/wood stud wall. (*From Brick Institute of America,* Technical Note 4 Rev., *BIA, Reston, Va.*)

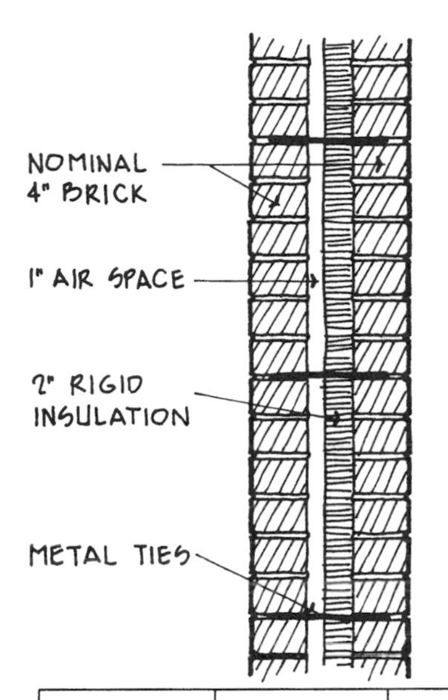

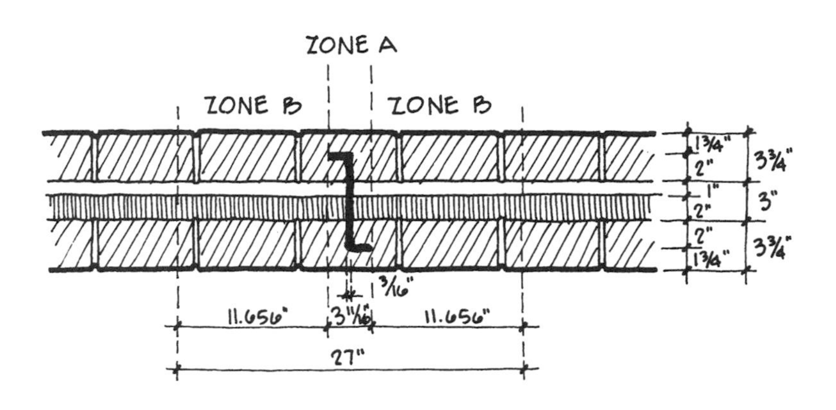

m = 0.1875 in.
d = 1.75 in.
W = 0.1875 + 2(1.75) = 3.6875 in.
Area of Zone A, $A_A = (3.6875/2)^2\pi$ = 10.6796 sq in or 0.07416 sq ft
Area of steel in Zone A = $(0.1875/2)^2\pi$ = 0.0276 sq in or 0.00019 sq ft
Nonsteel area in Zone A = 0.07416 − 0.00019 = 0.7397 sq ft
Area of Zone B, A_B = 4.50 − 0.07416 = 4.42584 sq ft

Section	C (Btu /(hr • °F • sq ft))	K ((Btu • in.) /(hr • °F • sq ft))	x (in.)	C_x (Btu /(hr • °F • sq ft))	Zone A			Zone B		
					A (sq ft)	$C_x \cdot A$ (Btu /(hr • °F))	$\frac{1}{C_x \cdot A} = \frac{R}{A}$ ((hr • °F) /Btu)	A (sq ft)	$C_x \cdot A$ (Btu /(hr • °F))	$\frac{1}{C_x \cdot A} = \frac{R}{A}$ ((hr • °F) /Btu)
Outside air surface	6.000			6.000	0.07416	0.445	2.25	4.42584	26.555	0.04
4-in. nominal face brick		9.000	3.75	2.400				4.42584	10.622	0.09
Brick		9.000	1.75	5.143	0.07416	0.381	2.62			
Brick		9.000	2.00	4.500	0.07397	0.333				
Steel		314.000	2.00	157.000	0.00019	0.030				
					Subtotal	0.363	2.75			
1-in. airspace	1.030			1.030	0.07397	0.076		4.42584	4.559	0.22
Steel		314.000	1.00	314.000	0.00019	0.060				
					Subtotal	0.136	7.35			
2-in poly-styrene rigid board insu-lation		0.250	2.00	0.125	0.07397	0.009		4.42584	0.553	1.81
Steel		314.000	2.00	157.000	0.00019	0.030				
					Subtotal	0.039	25.64			
Brick		9.000	2.00	4.500	0.07397	0.333				
Steel		314.000	2.00	157.000	0.00019	0.030				
					Subtotal	0.363	2.75			
Brick		9.000	1.75	5.143	0.07416	0.381	2.62			
4-in. nominal face brick		9.000	3.75	2.400				4.42584	10.622	0.09
Inside air sur-face	1.470			1.470	0.07416	0.109	9.17	4.42584	6.506	0.15
					R_A/A_A = 55.15 $1/(R_A/A_A)$ = 0.018			R_B/A_B = 2.40 $1/(R_B/A_B)$ = 0.417		

$U_{avg} = [1/(R_A/A_A) + 1/(R_B/A_B)]/(A_A + A_B) = (0.018 + 0.417)/(0.07416 + 4.42584) = 0.097$ Btu/(hr • °F • sq ft)

$U_B = [1/(R_B/A_B)]/A_B = 0.417/4.42584 = 0.094$ Btu/(hr • °F • sq ft)

$$\frac{U_{avg} - U_B}{U_B} \times 100\% = \frac{0.097 - 0.094}{0.094} \times 100\% = 3.19\%$$

Thermal calculations for insulated brick masonry cavity wall. (*From Brick Institute of America,* Technical Note 4 Rev., *BIA, Reston, Va.*)

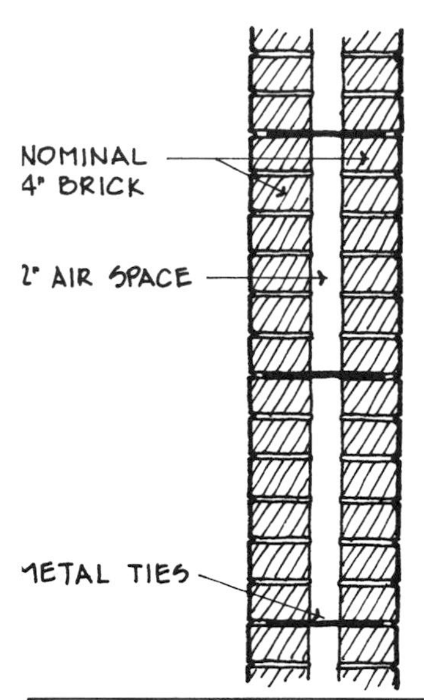

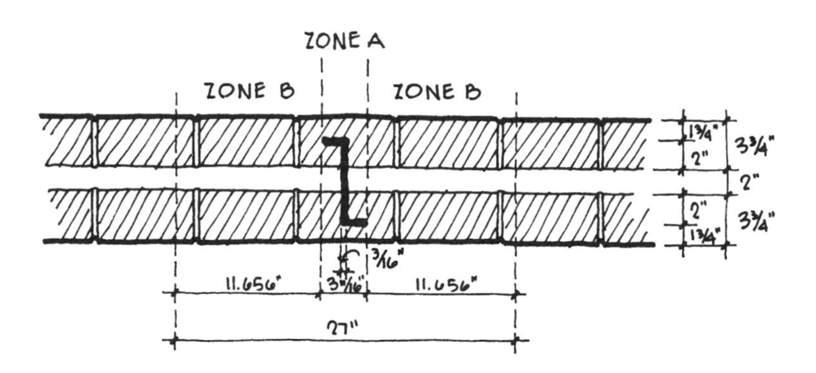

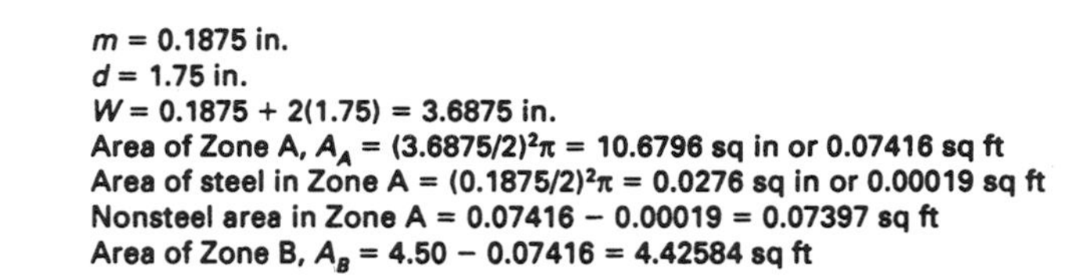

m = 0.1875 in.
d = 1.75 in.
W = 0.1875 + 2(1.75) = 3.6875 in.
Area of Zone A, $A_A = (3.6875/2)^2\pi$ = 10.6796 sq in or 0.07416 sq ft
Area of steel in Zone A = $(0.1875/2)^2\pi$ = 0.0276 sq in or 0.00019 sq ft
Nonsteel area in Zone A = 0.07416 − 0.00019 = 0.07397 sq ft
Area of Zone B, A_B = 4.50 − 0.07416 = 4.42584 sq ft

Section	C (Btu /(hr • °F • sq ft))	K ((Btu • in.) /(hr • °F • sq ft))	x (in.)	C_x (Btu /(hr • °F • sq ft))	Zone A: A (sq ft)	Zone A: $C_x \cdot A$ (Btu /(hr • °F))	Zone A: $\frac{1}{C_x \cdot A} = \frac{R}{A}$ ((hr • °F) /Btu)	Zone B: A (sq ft)	Zone B: $C_x \cdot A$ (Btu /(hr • °F))	Zone B: $\frac{1}{C_x \cdot A} = \frac{R}{A}$ ((hr • °F) /Btu)
Outside air surface	6.000			6.000	0.07416	0.445	2.25	4.42584	26.555	0.04
4-in. nominal face brick		9.000	3.75	2.400				4.42584	10.622	0.09
Brick		9.000	1.75	5.143	0.07416	0.381	2.62			
Brick		9.000	2.00	4.500	0.07397	0.333				
Steel		314.000	2.00	157.000	0.00019	0.030				
					Subtotal	0.363	2.75			
2-in. airspace	1.030			1.030	0.07397	0.076		4.42584	4.559	0.22
Steel		314.000	2.00	157.000	0.00019	0.030				
					Subtotal	0.106	9.43			
Brick		9.000	2.00	4.500	0.07397	0.333				
Steel		314.000	2.00	157.000	0.00019	0.030				
					Subtotal	0.363	2.75			
Brick		9.000	1.75	5.143	0.07416	0.381	2.62			
4-in. nominal face brick		9.000	3.75	2.400				4.42584	10.622	0.09
Inside air surface	1.470			1.470	0.07416	0.109	9.17	4.42584	6.506	0.15
					R_A/A_A = 31.59 $1/(R_A/A_A)$ = 0.032			R_B/A_B = 0.59 $1/(R_B/A_B)$ = 1.695		

$U_{avg} = [1/(R_A/A_A) + 1/(R_B/A_B)]/(A_A + A_B) = (0.032 + 1.695)/(0.07416 + 4.42584) = 0.384$ Btu/(hr • °F • sq ft)

$U_B = [1/(R_B/A_B)]/A_B = 1.695/4.42584 = 0.383$ Btu/(hr • °F • sq ft) $\qquad \frac{U_{avg} - U_B}{U_B} \times 100\% = \frac{0.384 - 0.383}{0.383} \times 100\% = 0.26\%$

Thermal calculations for uninsulated brick masonry cavity wall. (*From Brick Institute of America,* Technical Note 4 Rev., *BIA, Reston, Va.*)

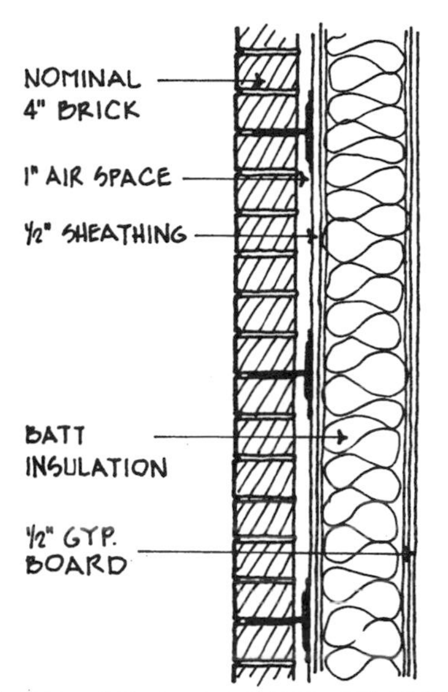

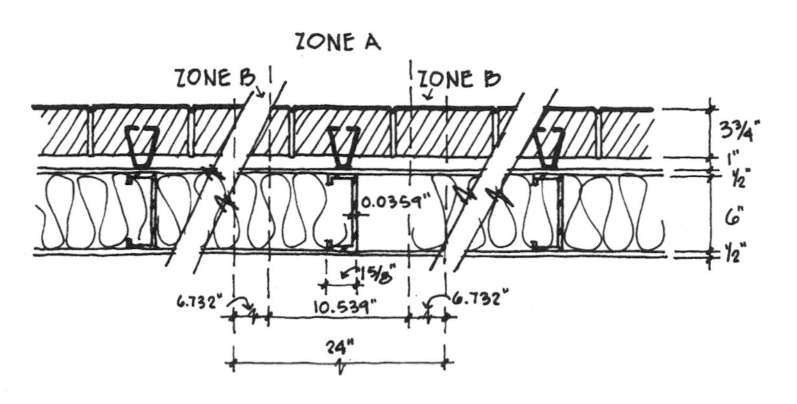

$m = 0.0359$ in.
$d = 3.75 + 1.00 + 0.50 = 5.25$ in.
$W = 0.0359 + 2(5.25) = 10.536$ in.
Area of Zone A, $A_A = 10.536 \times 12.00 = 126.43$ sq in or 0.878 sq ft
Area of steel in Zone A $= 0.0359 \times 12.00 = 0.4308$ sq in or 0.003 sq ft
Nonsteel area of Zone A $= 0.878 - 0.003 = 0.875$ sq ft
Area of Zone B, $A_B = 2.000 - 0.878 = 1.122$ sq ft

Section	C (Btu /(hr • °F • sq ft))	K ((Btu • in.) /(hr • °F • sq ft))	x (in.)	C_x (Btu /(hr • °F • sq ft))	Zone A			Zone B		
					A (sq ft)	$C_x \cdot A$ (Btu /(hr • °F))	$\frac{1}{C_x \cdot A} = \frac{R}{A}$ ((hr • °F) /Btu)	A (sq ft)	$C_x \cdot A$ (Btu /(hr • °F))	$\frac{1}{C_x \cdot A} = \frac{R}{A}$ ((hr • °F) /Btu)
Outside air surface	6.000			6.000	0.878	5.268	0.19	1.122	6.732	0.15
4-in. nominal face brick		9.000	3.75	2.400	0.878	2.107	0.47	1.122	2.693	0.37
1-in. airspace	1.030			1.030	0.878	0.904	1.11	1.122	1.156	0.87
½-in. exterior gypsum sheathing	2.250			2.250	0.878	1.975	0.51	1.122	2.525	0.40
6-in. batt insulation	0.053			0.053	0.875	0.046		1.122	0.059	16.95
Steel		314.000	6.00	52.333	0.003	0.157				
					Subtotal	0.203	4.93			
½-in. gypsum wallboard	2.250			2.250	0.878	1.976	0.51	1.122	2.525	0.40
Inside air surface	1.470			1.470	0.878	1.291	0.77	1.122	1.649	0.61
					$R_A/A_A = 8.49$ $1/(R_A/A_A) = 0.118$			$R_B/A_B = 19.75$ $1/(R_B/A_B) = 0.051$		

$U_{avg} = [1/(R_A/A_A) + 1/(R_B/A_B)]/(A_A + A_B) = (0.118 + 0.051)/(0.878 + 1.122) = 0.085$ Btu/(hr • °F • sq ft)

$U_B = [1/(R_B/A_B)]/A_B = 0.051/1.122 = 0.045$ Btu/(hr • °F • sq ft) $\quad \frac{U_{avg} - U_B}{U_B} \times 100\% = \frac{0.085 - 0.045}{0.045} \times 100\% = 88.89\%$

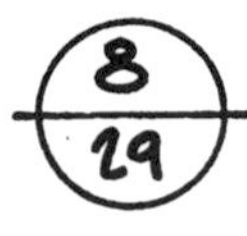

Thermal calculations for brick veneer/metal stud wall. (*From Brick Institute of America,* Technical Note 4 Rev., *BIA, Reston, Va.*)

will heat up or cool down is described as *thermal inertia,* and is dependent on wall thickness, density, specific heat, and conductivity. It is this phenomenon, in fact, which also contributes to masonry fire safety by delaying heat transfer through the walls of burning buildings.

The thermal storage properties of masonry have been used for centuries. Large, massive central fireplaces were used during the day for heating and cooking. At night, the heat stored in the fireplace shell provided radiant warmth until dawn. In the desert Southwest of the United States, thick adobe masonry walls were used, not so much for strength as for thermal stability. Buildings remained cool during the hot summer days, and heat stored in the walls was later radiated outward to the cooler night air. Until recently, however, there was no simple way of calculating this response. We now understand that the transmission of heat through building walls is a dynamic process and that any method of calculating heat loss or heat gain that assumes it is static or steady state is not an accurate measure of performance.

Heat flows from hot to cold. As temperatures rise on one side of a wall, heat begins to migrate toward the cooler side. Before heat transfer from one space to another can be achieved, the wall itself must undergo a temperature increase. The amount of thermal energy necessary to produce this increase is directly proportional to the weight of the wall. Masonry is heavy, so it can absorb and store heat and substantially retard its migration. This characteristic is called *thermal storage capacity* or *capacity insulation.* One measure of this storage capacity is the elapsed time required to achieve equilibrium between inside and outside wall surface temperatures. The midday solar radiation load on the south face of a building will not completely penetrate a 12-in. solid masonry wall for approximately 8 hours.

The effects of wall mass on heat transmission are dependent on the magnitude and duration of temperature differentials during the daily cycle. Warm climates with cool nights benefit most. Seasonal and climatic conditions with only small daily temperature differentials tend to diminish the benefits.

8.5.2 Heat-Gain Calculations

Thermal lag and capacity insulation are of considerable importance in calculating heat gain when outside temperature variations are great. During a daily cycle, walls with equal *U*-values but unequal mass will produce significantly different peak loads. The greater the storage capacity, the lower the total heat gain. Increased mass reduces actual peak loads in a building, thus requiring smaller cooling equipment. Building envelopes with more thermal storage capacity will also delay the peak load until after the hottest part of the day when solar radiation through glass areas is diminished and, in commercial buildings, after lighting, equipment, and occupant loads are reduced. This lag time decreases the total demand on cooling equipment by staggering the loads.

Steady-state heat-gain calculations do not recognize the significant benefits of thermal inertia when they employ constant indoor and outdoor design temperatures. Computer studies completed by Francisco Arumi for the Energy Research and Development Administration and the National Concrete Masonry Association (NCMA) made close comparisons between static calculations and dynamic calculations. *Figure 8-30* shows the time-temperature curves derived from each method in calculating inside room temperature.

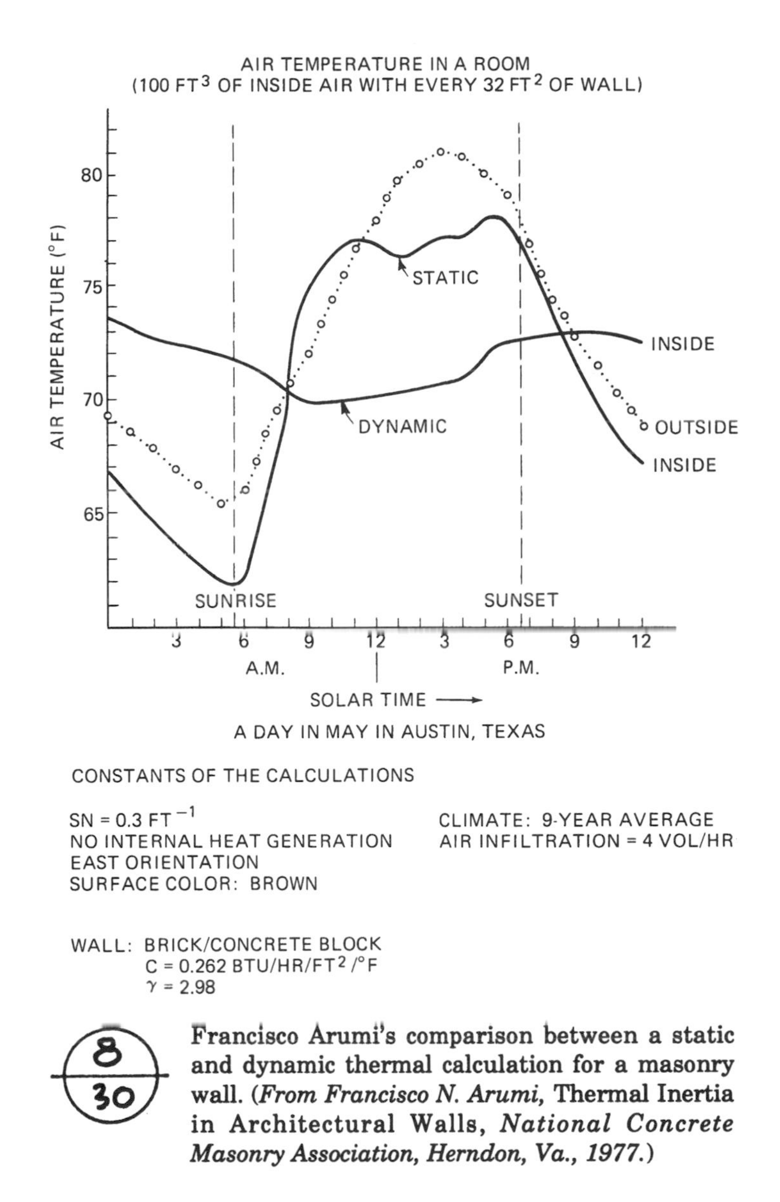

Francisco Arumi's comparison between a static and dynamic thermal calculation for a masonry wall. (*From Francisco N. Arumi,* **Thermal Inertia in Architectural Walls,** *National Concrete Masonry Association, Herndon, Va., 1977.*)

The attenuation of temperature amplitudes found with the dynamic response calculation graphically illustrates the actual effect that the thermal inertia of massive walls has on indoor comfort. Another study conducted by Mario Catani and Stanley E. Goodwin for the Portland Cement Association (PCA) and reported in the *Journal of the American Concrete Institute* shows heat-gain comparisons for several wall types (*see Fig. 8-31*). Computer analysis using dynamic response methods showed that, with *U*-values equal, the peak heat gains of the lighter-weight walls were 38 to 65% higher than for the heavy walls. In comparisons of a model building with four alternative wall types, the same results were evident. Using dynamic analysis methods, two heavy concrete walls, a concrete tilt-up wall, and a metal building wall were studied to determine peak cooling loads (*see Fig. 8-32*). Results showed that the heavier walls were far superior in performance to the lightweight sections and that, despite a *U*-value that was 33% higher than the others, the peak loads for building A were 60 to 65% less than those for the lightweight construction.

NCMA reports other cooling load tests made using NBS computer

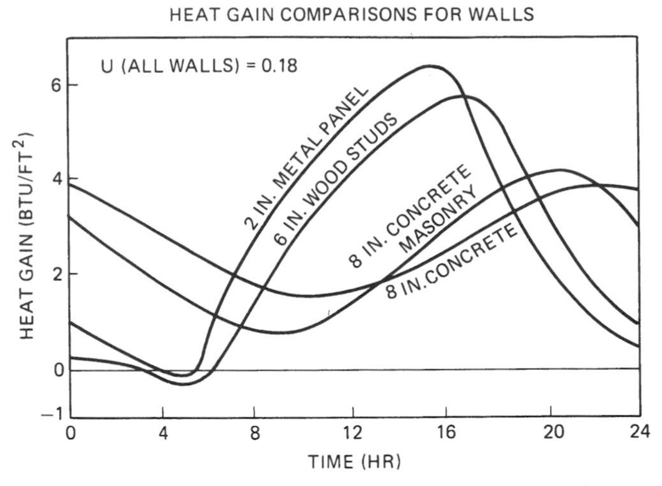

Average values of heat gain for four walls of a square building show that peak loads are increased from 38 to 65% as result of less mass.

Catani and Goodwin's heat-gain curves for various wall types. (*From Mario Catani and Stanley Goodwin, "Heavy Building Envelopes and Dynamic Thermal Response,"* ACI Journal, *February 1976.*)

programs. U-values of the walls, roof, and floor were held constant while the wall weight was varied from 10 lb/sq ft to 70 lb/sq ft in 5-lb increments. The size of the required air-conditioning equipment varied inversely with the weight of the structure. The lightest-weight walls (10 lb/sq ft) required over 35,000 Btu/hour in air conditioning. The heaviest walls (70 lb/sq ft) required less than 25,000 Btu/hour. When the data are grouped in weight categories matching those of the equivalent temperature difference graph, the relationships are easily compared (*see Fig. 8-33*).

Heat gain is known to be affected, not only by mass and density, but also by surface color and emissivity of the wall, orientation, intensity of direct and diffused solar radiation, and surface reflectivity. Because of these many factors, heat gain calculations are more complex than simple heat loss calculations. To simplify the procedure, the *sol-air temperature concept* was developed to determine simulated air temperatures which would give a rate of heat exchange equivalent to that determined by considering all of the variables.

The heat gain equation prescribed by ASHRAE includes a cooling load temperature difference (CLTD) factor (*see Fig. 8-34*) which is derived from the sol-air temperature concept and transfer function method

$$Q = UA\,(\text{CLTD}) \tag{8.8}$$

where Q = heat gain, Btu/hr
U = coefficient of thermal transmission of the opaque wall or roof assembly, Btu/(hour × °F × sq ft) (*see Fig. 8-24*)
A = opaque wall or roof surface area, sq ft
CLTD = cooling load temperature difference, °F (*see Fig. 8-34*)

The CLTD values in the table are based on the sol-air temperature for July 21 at 40° North latitude, and are valid *only* if the following conditions exist: (1) inside air temperature is 78°F; (2) maximum daily outdoor tem-

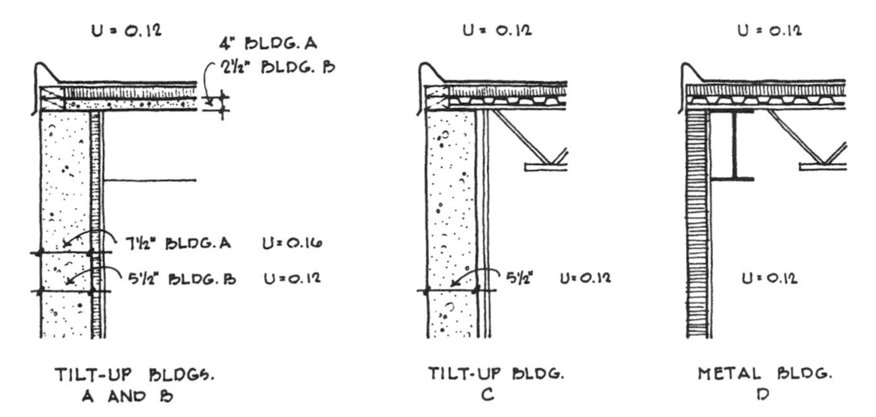

Four different building envelopes studied: two concrete, concrete and metal, and metal. Heavyweight Building A having walls with higher U-factors is used as a base for heat flow comparisons.

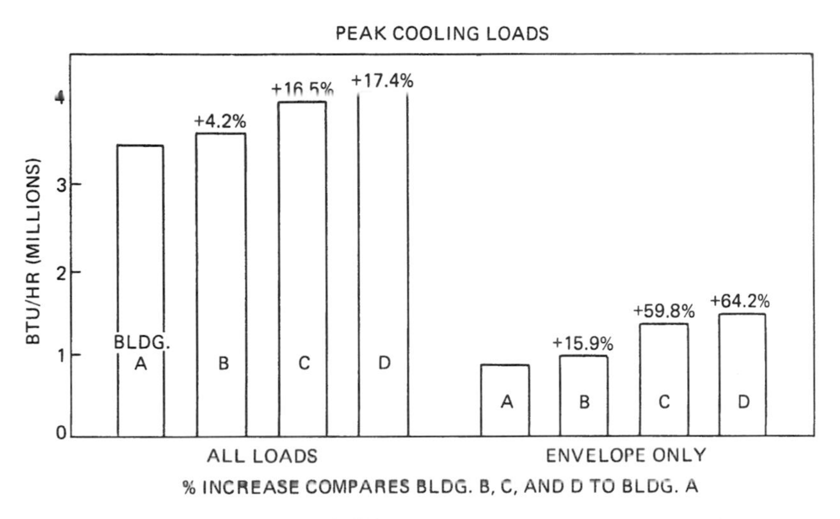

The percent increase compares Buildings B, C, and D to Building A. Because of heavier sections, peak cooling loads for Building A are reduced even though the wall U-factor is greater than for the other buildings.

Catani and Goodwin's comparison of actual heat gain for various wall types. (*From Mario Catani and Stanley Goodwin, "Heavy Building Envelopes and Dynamic Thermal Response,"* ACI Journal, *February 1976.*)

perature is 95°F; (3) average daily temperature range is 21°F; (4) outdoor daily mean temperature is 85°F; (5) building location is 40° North latitude; (6) design day is July 21; and (7) the surface color is dark. Wall construction groups are listed in *Fig. 8-35.* For wall constructions not listed, use the CLTD for a wall with similar mass (lb/sq ft), similar heat capacity [Btu/(sq ft × °F)], and similar thermal transmission or U-value.

The area of the walls for each orientation affects the hour at which the maximum cooling load occurs. The hour of maximum load in *Fig. 8-34* occurs only if the area of the four exterior walls are equal. For other conditions, it will be necessary to examine a number of hours before the maximum cooling load is determined for each of the walls of an actual building. Calculations must take into account the heat gain through the total envelope (including the roof), plus internal and latent loads, and infiltration. CLTD factors for

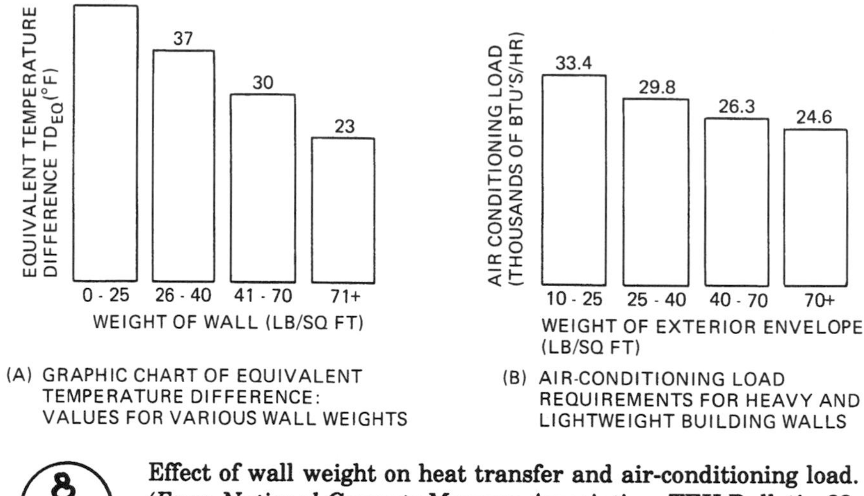

8/33 **Effect of wall weight on heat transfer and air-conditioning load. (*From National Concrete Masonry Association,* TEK Bulletin 82, *NCMA, Herndon, Va.*)**

roofs are listed in *Fig. 8-36.* If the outdoor average daily temperature for the project location (*see Fig. 8-37*) and the indoor design temperature of the building differ from those specified above, the CLTD must be corrected, as it must also be for differences in wall color, latitude and month.

$$CLTD_c = [(CLTD + LM)K] + [(78 - T_R) + (T_O - 85)] \tag{8.9}$$

where $CLTD_c$ = corrected CLTD, °F
CLTD = cooling load temperature difference, °F
LM = latitude-month correction, °F (*see Fig. 8-38*)
K = color adjustment factor
T_R = indoor design temperature, °F
T_O = outdoor average daily temperature, °F

Correction factors for latitudes and months are listed in *Fig. 8-38.* Color correction factors are

$K = 1.00$ if dark or light colored in an industrial area
$K = 0.83$ if permanently medium colored in a rural area
$K = 0.75$ if permanently light colored in a rural area

Light colors are buffs and creams; medium colors are medium blue, medium green, bright red, light brown, and so on; and dark colors are dark blue, red-brown, and green, plus dark gray and black. The color corrections for roofs are $K = 1.0$ if dark colored or light colored in an industrial area and $K = 0.5$ if permanently light colored in a rural area. The outdoor average daily temperature, if different than 85°F, can be approximated by the equation

$$T_O = t_o - \frac{t_d}{2} \tag{8.10}$$

where T_O = outdoor daily mean temperature, °F
t_o = outdoor design temperature, °F (*see Fig. 8-37*)
t_d = average daily temperature range, °F

COOLING LOAD TEMPERATURE DIFFERENCES FOR CALCULATING COOLING LOAD FROM SUNLIT WALLS*,†																													
Wall group	North latitude wall facing	Solar time (hr)																								Hr of maxi-mum CLTD	Mini-mum CLTD	Maxi-mum CLTD	Differ-ence CLTD
		1	2	3	4	5	6	7	8	9	10	11	12	13	14	15	16	17	18	19	20	21	22	23	24				
A	N	14	14	14	13	13	13	12	12	11	11	10	10	10	10	10	10	11	11	12	12	13	13	14	14	2	10	14	4
	NE	19	19	19	18	17	17	16	15	15	15	15	15	16	16	17	18	18	18	19	19	20	20	20	20	22	15	20	5
	E	24	24	23	23	22	21	20	19	19	18	19	19	20	21	22	23	24	24	25	25	25	25	25	25	22	18	25	7
	SE	24	23	23	22	21	20	20	19	18	18	18	18	18	19	20	21	22	23	23	24	24	24	24	24	22	18	24	6
	S	20	20	19	19	18	18	17	16	16	15	14	14	14	14	14	15	16	17	18	19	19	20	20	20	23	14	20	6
	SW	25	25	25	24	24	23	22	21	20	19	19	18	17	17	17	17	18	19	20	22	23	24	25	25	24	17	25	8
	W	27	27	26	26	25	24	24	23	22	21	20	19	19	18	18	18	18	19	20	22	23	25	26	26	1	18	27	9
	NW	21	21	21	20	20	19	19	18	17	16	16	15	15	14	14	14	15	15	16	17	18	19	20	21	1	14	21	7
B	N	15	14	14	13	12	11	11	10	9	9	9	8	9	9	9	10	11	12	13	14	14	15	15	15	24	8	15	7
	NE	19	18	17	16	15	14	13	12	12	13	14	15	16	17	18	19	19	20	20	21	21	21	20	20	21	12	21	9
	E	23	22	21	20	18	17	16	15	15	15	17	19	21	22	24	25	26	26	27	27	26	26	25	24	20	15	27	12
	SE	23	22	21	20	18	17	16	15	14	14	15	16	18	20	21	23	24	25	26	26	26	26	25	24	21	14	26	12
	S	21	20	19	18	17	15	14	13	12	11	11	11	11	12	14	15	17	19	20	21	22	22	22	21	23	11	22	11
	SW	27	26	25	24	22	21	19	18	16	15	14	14	13	13	14	15	17	20	22	25	27	28	28	28	24	13	28	15
	W	29	28	27	26	24	23	21	19	18	17	16	15	14	14	14	15	17	19	22	25	27	29	29	30	24	14	30	16
	NW	23	22	21	20	19	18	17	15	14	13	12	12	12	11	12	12	13	15	17	19	21	22	23	23	24	11	23	9
C	N	15	14	13	12	11	10	9	8	8	7	7	8	8	9	10	12	13	14	15	16	17	17	17	16	22	7	17	10
	NE	19	17	16	14	13	11	10	10	11	13	15	17	19	20	21	22	22	23	23	23	23	22	21	20	20	10	23	13
	E	22	21	19	17	15	14	12	12	14	16	19	22	25	27	29	29	30	30	30	29	28	27	26	24	18	12	30	18
	SE	22	21	19	17	15	14	12	12	12	13	16	19	22	24	26	28	29	29	29	29	28	27	26	24	19	12	29	17
	S	21	19	18	16	15	13	12	10	9	9	9	10	11	14	17	20	22	24	25	26	25	25	24	22	20	9	26	17
	SW	29	27	25	22	20	18	16	15	13	12	11	11	11	13	15	18	22	26	29	32	33	33	32	31	22	11	33	22
	W	31	29	27	25	22	20	18	16	14	13	12	12	12	13	14	16	20	24	29	32	35	35	35	33	22	12	35	23
	NW	25	23	21	20	18	16	14	13	11	10	10	10	10	11	12	13	15	18	22	25	27	27	27	26	22	10	27	17

**Wall construction not listed:* An actual wall construction not listed in this table (or Table 8-35) would be thermally similar to a wall in the table, if it has similar mass, lb/sq ft, similar heat capacity Btu/(sq ft • °F), and similar thermal transmission, Btu/(hr • °F • sq ft). In that case, use the *CLTD* from this table.

†*Additional insulation:* For each 7 increase in *R*-value due to insulation added to the wall structures in Table 8-35, use the *CLTD* for the wall group with the next higher letter in the alphabet.

8/34 *CLTD* for heat-gain calculations. (*From American Society of Heating, Refrigerating and Air Conditioning Engineers,* Fundamentals, *ASHRAE Handbook, 1989. Reprinted with permission.*)

COOLING LOAD TEMPERATURE DIFFERENCES FOR CALCULATING COOLING LOAD FROM SUNLIT WALLS*,†

Wall group	North latitude wall facing	Solar time (hr)																								Hr of maximum CLTD	Minimum CLTD	Maximum CLTD	Difference CLTD
		1	2	3	4	5	6	7	8	9	10	11	12	13	14	15	16	17	18	19	20	21	22	23	24				
D	N	15	13	12	10	9	7	6	6	6	6	6	7	8	10	12	13	15	17	18	19	19	19	18	16	21	6	19	13
	NE	17	15	13	11	10	8	7	8	10	14	17	20	22	23	23	24	24	25	25	24	23	22	20	18	19	7	25	18
	E	19	17	15	13	11	9	8	9	12	17	22	27	30	32	33	33	32	32	31	30	28	26	24	22	16	8	33	25
	SE	20	17	15	13	11	10	8	8	10	13	17	22	26	29	31	32	32	32	31	30	28	26	24	22	17	8	32	24
	S	19	17	15	13	11	9	8	7	6	6	7	9	12	16	20	24	27	29	29	29	27	26	24	22	19	6	29	23
	SW	28	25	22	19	16	14	12	10	9	8	8	8	10	12	16	21	27	32	36	38	38	37	34	31	21	8	38	30
	W	31	27	24	21	18	15	13	11	10	9	9	9	10	11	14	18	24	30	36	40	41	40	38	34	21	9	41	32
	NW	25	22	19	17	14	12	10	9	8	7	7	8	9	10	12	14	18	22	27	31	32	32	30	27	22	7	32	25
E	N	12	10	8	7	5	4	3	4	5	6	7	9	11	13	15	17	19	20	21	23	20	18	16	14	20	3	22	19
	NE	13	11	9	7	6	4	5	9	15	20	24	25	25	26	26	26	26	26	25	24	22	19	17	15	16	4	26	22
	E	14	12	10	8	6	5	6	11	18	26	33	36	38	37	36	34	33	32	30	28	25	22	20	17	13	5	38	33
	SE	15	12	10	8	7	5	5	8	12	19	25	31	35	37	37	36	34	33	31	28	26	23	20	17	15	5	37	32
	S	15	12	10	8	7	5	4	3	4	5	9	13	19	24	29	32	34	33	31	29	26	23	20	17	17	3	34	31
	SW	22	18	15	12	10	8	6	5	5	6	7	9	12	18	24	32	38	43	45	44	40	35	30	26	19	5	45	40
	W	25	21	17	14	11	9	7	6	6	6	7	9	11	14	20	27	36	43	49	49	45	40	34	29	20	6	49	43
	NW	20	17	14	11	9	7	6	5	5	5	6	8	10	13	16	20	26	32	37	38	36	32	28	24	20	5	38	33
F	N	8	6	5	3	2	1	2	4	6	7	9	11	14	17	19	21	22	23	24	23	29	16	13	11	19	1	23	23
	NE	9	7	5	3	2	1	5	14	23	28	30	29	28	27	27	27	27	26	24	22	19	16	13	11	11	1	30	29
	E	10	7	6	4	3	2	6	17	28	38	44	45	43	39	36	34	32	30	27	24	21	17	15	12	12	2	45	43
	SE	10	7	6	4	3	2	4	10	19	28	36	41	43	42	39	36	34	31	28	25	21	18	15	12	13	2	43	41
	S	10	8	6	4	3	2	1	1	3	7	13	20	27	34	38	39	38	35	31	26	22	18	15	12	16	1	39	38
	SW	15	11	9	6	5	3	2	2	4	5	8	11	17	26	35	44	50	53	52	45	37	28	23	18	18	2	53	48
	W	17	13	10	7	5	4	3	3	4	6	8	11	14	20	28	39	49	57	60	54	43	34	27	21	19	3	60	57
	NW	14	10	8	6	4	3	2	2	3	5	8	10	13	15	21	27	35	42	46	43	35	28	22	18	19	2	46	44

* *Wall construction not listed:* An actual wall construction not listed in this table (or Table 8-35) would be thermally similar to a wall in the table, if it has similar mass, lb/sq ft, similar heat capacity Btu/(sq ft • °F), and similar thermal transmission, Btu/(hr • °F • sq ft). In that case, use the *CLTD* from this table.

† *Additional insulation:* For each 7 increase in *R*-value due to insulation added to the wall structures in Table 8-35, use the *CLTD* for the wall group with the next higher letter in the alphabet.

8/34

(Continued)

COOLING LOAD TEMPERATURE DIFFERENCES FOR CALCULATING COOLING LOAD FROM SUNLIT WALLS*,†																													
Wall group	*North latitude wall facing*	*Solar time (hr)*																								*Hr of maxi-mum CLTD*	*Mini-mum CLTD*	*Maxi-mum CLTD*	*Differ-ence CLTD*
		1	*2*	*3*	*4*	*5*	*6*	*7*	*8*	*9*	*10*	*11*	*12*	*13*	*14*	*15*	*16*	*17*	*18*	*19*	*20*	*21*	*22*	*23*	*24*				
G	N	3	2	1	0	−1	2	7	8	9	12	15	18	21	23	24	24	25	26	22	15	11	9	7	5	18	−1	26	27
	NE	3	2	1	0	−1	9	27	36	39	35	30	26	26	27	27	26	25	22	18	14	11	9	7	5	9	−1	39	40
	E	4	2	1	0	−1	11	31	47	54	55	50	40	33	31	30	29	27	24	19	15	12	10	8	6	10	−1	55	56
	SE	4	2	1	0	−1	5	18	32	42	49	51	48	42	36	32	30	27	24	19	15	12	10	8	6	11	−1	51	52
	S	4	2	1	0	−1	0	1	5	12	22	31	39	45	46	43	37	31	25	20	15	12	10	8	5	14	−1	46	47
	SW	5	4	3	1	0	0	2	5	8	12	16	26	38	50	59	63	61	52	37	24	17	13	10	8	16	0	63	63
	W	6	5	3	2	1	1	2	5	8	11	15	19	27	41	56	67	72	67	43	29	20	15	11	8	17	1	72	71
	NW	5	3	2	1	0	0	2	5	8	11	15	18	21	27	37	47	55	55	41	25	17	13	10	7	18	0	55	55

**Wall construction not listed:* An actual wall construction not listed in this table (or Table 8-35) would be thermally similar to a wall in the table, if it has similar mass, lb/sq ft, similar heat capacity Btu/(sq ft • °F), and similar thermal transmission, Btu/(hr • °F • sq ft). In that case, use the *CLTD* from this table.

†*Additional insulation:* For each 7 increase in *R*-value due to insulation added to the wall structures in Table 8-35, use the *CLTD* for the wall group with the next higher letter in the alphabet.

8/34

(*Continued*)

Group no.	Description of construction	Weight (lb/ft)	U-value Btu/(hr • °F • sq ft)
	4-in. face brick + (Brick)		
C	Air space + 4-in. face brick	83	0.358
D	4-in. common brick	90	0.415
C	1-in. insulation or air space + 4-in. common brick	90	0.174–0.301
B	2-in. insulation + 4-in. common brick	88	0.111
B	8-in. common brick	130	0.302
A	Insulation or air space + 8-in. common brick	130	0.154–0.243
	4-in. face brick + (H. W. concrete)		
C	Air space + 2-in. concrete	94	0.350
B	2-in. insulation + 4-in. concrete	97	0.116
A	Air space or insulation + 8-in. or 12-in. concrete	143–190	0.110–0.112
	4-in face brick + (L. W. or H. W. concrete block)		
E	4-in. block	62	0.319
D	Air space or insulation + 4-in. block	62	0.153–0.246
D	8-in. block	70	0.274
C	Air space or 1-in. insulation + 6-in. or 8-in. block	73–89	0.221–0.275
B	2-in. insulation + 8-in. block	89	0.096–0.107
	4-in. face brick + (clay tile)		
D	4-in. tile	71	0.381
D	Air space + 4-in. tile	71	0.281
C	Insulation + 4-in. tile	71	0.169
C	8-in. tile	96	0.275
B	Air space or 1-in. insulation + 8-in. tile	96	0.142–0.221
A	2-in. insulation + 8-in. tile	97	0.097
	H. W. concrete wall + (finish)		
E	4-in. concrete	63	0.585
D	4-in. concrete + 1-in. or 2-in. insulation	63	0.119–0.200
C	2-in. insulation + 4-in. concrete	63	0.119
C	8-in. concrete	109	0.490
B	8-in. concrete + 1-in. or 2-in. insulation	110	0.115–0.187
A	2-in. insulation + 8-in. concrete	110	0.115
B	12-in. concrete	156	0.421
A	12-in. concrete + insulation	156	0.113
	L. W. and H. W. concrete block + (finish)		
F	4-in. block + air space/insulation	29	0.161–0.263
E	2-in. insulation + 4-in. block	29–37	0.105–0.114
E	8-in. block	47–51	0.294–0.402
D	8-in. block + air space/insulation	41–57	0.149–0.173
	Clay tile + (finish)		
F	4-in. tile	39	0.419
F	4-in. tile + air space	39	0.303
E	4-in. tile + 1-in. insulation	39	0.175
D	2-in. insulation + 4-in. tile	40	0.110
D	8-in. tile	63	0.296
C	8-in. tile + air space/1-in. insulation	63	0.151–0.231
B	2-in. insulation + 8-in. tile	63	0.099
	Metal curtain wall		
B	With/without air space + 1-in./2-in./3-in. insulation	5–6	0.091–0.230
	Frame wall		
G	1-in. to 3-in. insulation	16	0.081–0.178

Note: Where the / symbol appears in the table, it denotes that the layer on either side of the symbol may be used, i.e., an air space or 1-in. insulation.

Description of wall construction groups. (*From American Society of Heating, Refrigerating and Air Conditioning Engineers,* Fundamentals, *ASHRAE Handbook, 1989. Reprinted with permission.*)

COOLING LOAD TEMPERATURE DIFFERENCES FOR CALCULATING COOLING LOAD FROM SUNLIT WALLS*,†

Roof no.	Description of construction	Weight (lb /sq ft)	U-value (Btu /(h • sq ft • °F))	Solar time (hr)																								Hr of maxi-mum CLTD	Mini-mum CLTD	Maxi-mum CLTD	Differ-ence CLTD
				1	2	3	4	5	6	7	8	9	10	11	12	13	14	15	16	17	18	19	20	21	22	23	24				
Without Suspended Ceiling																															
1	Steel sheet with 1-in. (or 2-in.) insulation	7 (8)	0.213 (0.124)	1	−2	−3	−3	−5	−3	6	19	34	49	61	71	78	79	77	70	59	45	30	18	12	8	5	3	14	−5	79	84
2	1-in. wood with 1-in. insulation	8	0.170	6	3	0	−1	−3	−3	−2	4	14	27	39	52	62	70	74	74	70	62	51	38	28	20	14	9	16	−3	74	77
3	4-in. light-weight concrete	18	0.213	9	5	2	0	−2	−3	−3	1	9	20	32	44	55	64	70	73	71	66	57	45	34	25	18	13	16	−3	73	76
4	2-in. heavy-weight concrete with 1-in. (or 2-in.) insulation	29 (0.122)	0.206	12	8	5	3	0	−1	−1	3	11	20	30	41	51	59	65	66	66	62	54	45	36	29	22	17	16	−1	67	68
5	1-in. wood with 2-in. insulation	9	0.109	3	0	−3	−4	−5	−7	−6	−3	5	16	27	39	49	57	63	64	62	57	48	37	26	18	11	7	16	−7	64	71
6	6-in. light-weight concrete	24	0.158	22	17	13	9	6	3	1	1	3	7	15	23	33	43	51	58	62	64	62	57	50	42	35	28	18	1	64	63
7	2.5-in. wood with 1-in. insulation	13	0.130	29	24	20	16	13	10	7	6	6	9	13	20	27	34	42	48	53	55	56	54	49	44	39	34	19	6	56	50
8	8-in. light-weight concrete	31	0.126	35	30	26	22	18	14	11	9	7	7	9	13	19	25	33	39	46	50	53	54	53	49	45	40	20	7	54	47
9	4-in. heavy-weight concrete with 1-in. (or 2-in.) insulation	52 (52)	0.200 (0.120)	25	22	18	15	12	9	8	8	10	14	20	26	33	40	46	50	53	53	52	48	43	38	34	30	18	8	53	45
10	2.5-in. wood with 2-in. insulation	13	0.093	30	26	23	19	16	13	10	9	8	9	13	17	23	29	36	41	46	49	51	50	47	43	39	35	19	8	51	43

Cooling load temperature differences (*CLTD*) for calculating cooling load from flat roofs. (*From American Society of Heating, Refrigerating and Air Conditioning Engineers*, Fundamentals, *ASHRAE Handbook, 1989. Reprinted with permission.*)

COOLING LOAD TEMPERATURE DIFFERENCES FOR CALCULATING COOLING LOAD FROM SUNLIT WALLS*,†

Roof no.	Description of construction	Weight (lb /sq ft)	U-value (Btu /(h • sq ft • °F))	Solar time (hr) 1	2	3	4	5	6	7	8	9	10	11	12	13	14	15	16	17	18	19	20	21	22	23	24	Hr of maximum CLTD	Minimum CLTD	Maximum CLTD	Difference CLTD
Without Suspended Ceiling																															
11	Roof terrace system	75	0.106	34	31	28	25	22	19	16	14	13	13	15	18	22	26	31	36	40	44	45	46	45	43	40	37	20	13	46	33
12	6-in. heavy-weight concrete with 1-in. (or 2-in.) insulation	75 (75)	0.192 (0.117)	31	28	25	22	20	17	15	14	14	16	18	22	26	31	36	40	43	45	45	44	42	40	37	34	19	14	45	31
13	4-in. wood with 1-in. (or 2-in.) insulation	17 (18)	0.106 (0.078)	38	36	33	30	28	25	22	20	18	17	16	17	18	21	24	28	32	36	39	41	43	43	42	40	22	16	43	27
With Suspended Ceiling																															
1	Steel sheet with 1-in. (or 2-in.) insulation	9 (10)	0.134 (0.092)	2	0	−2	−3	−4	−4	−1	9	23	37	50	62	71	77	78	74	67	56	42	28	18	12	8	5	15	−4	78	82
2	1-in. wood with 1-in. insulation	10	0.115	20	15	11	8	5	3	2	3	7	13	21	30	40	48	55	60	62	61	58	51	44	37	30	25	17	2	62	60
3	4-in. light-weight concrete	20	0.134	19	14	10	7	4	2	0	0	4	10	19	29	39	48	56	62	65	64	61	54	46	38	30	24	17	0	65	65
4	2-in. heavy-weight concrete with 1-in. insulation	30	0.131	28	25	23	20	17	15	13	13	14	16	20	25	30	35	39	43	46	47	46	44	41	38	35	32	18	13	47	34
5	1-in. wood with 2-in. insulation	10	0.083	25	20	16	13	10	7	5	5	7	12	18	25	33	41	48	53	57	57	56	52	46	40	34	29	18	5	57	52
6	6-in. light-weight concrete	26	0.109	32	28	23	19	16	13	10	8	7	8	11	16	22	29	36	42	48	52	54	54	51	47	42	37	20	7	54	47

8/36 (Continued)

COOLING LOAD TEMPERATURE DIFFERENCES FOR CALCULATING COOLING LOAD FROM SUNLIT WALLS*,†

Roof no.	Description of construction	Weight (lb /sq ft)	U-value (Btu /(h • sq ft • °F))	1	2	3	4	5	6	7	8	9	10	11	12	13	14	15	16	17	18	19	20	21	22	23	24	Hr of maxi-mum CLTD	Mini-mum CLTD	Maxi-mum CLTD	Differ-ence CLTD
				Solar time (hr)																											
	With Suspended Ceiling																														
7	2.5-in. wood with 1-in. insulation	15	0.096	34	31	29	26	23	21	18	16	15	15	16	18	21	25	30	34	38	41	43	44	44	42	40	37	21	15	44	29
8	8-in. light-weight concrete	33	0.093	39	36	33	29	26	23	20	18	15	14	14	15	17	20	25	29	34	38	42	45	46	45	44	42	21	14	46	32
9	4-in. heavy-weight concrete with 1-in. (or 2-in.) ins.	53 (54)	0.128 (0.090)	30	29	27	26	24	22	21	20	20	21	22	24	27	29	32	34	36	38	38	38	37	36	34	33	19	20	38	18
10	2.5-in. wood with 2-in. insulation	15	0.072	35	33	30	28	26	24	22	20	18	18	18	20	22	25	28	32	35	38	40	41	41	40	39	37	21	18	41	23
11	Roof terrace system	77	0.082	30	29	28	27	26	25	24	23	22	22	22	23	23	25	26	28	29	31	32	33	33	33	33	32	22	22	33	11
12	6-in. heavy-weight concrete with 1-in. (or 2-in.) insulation	77 (77)	0.125 (0.088)	29	28	27	26	25	24	23	22	21	21	22	23	25	26	28	30	32	33	34	34	34	33	32	31	20	21	34	13
13	4-in. wood with 1-in. (or 2-in.) insulation	19 (20)	0.082 (0.064)	35	34	33	32	31	29	27	26	24	23	22	21	22	22	24	25	27	30	32	34	35	36	37	36	23	21	37	16

1. *Direct Application of Table without Adjustments:*

Values were calculated using the following conditions:

- Dark flat surface roof ("dark" for solar radiation absorption)
- Indoor temperature of 78°F
- Outdoor maximum temperature of 95°F with outdoor mean temperature of 85°F and an outdoor daily range of 21°F
- Solar radiation typical of 40 deg North latitude on July 21
- Outside surface resistance R_o = 0.333 sq ft • °F • h/Btu

(Continued)

- Without and with suspended ceiling, but no attic fans or return air ducts in suspended ceiling space
- Inside surface resistance R_i = 0.685 sq ft • °F • h/Btu

2. *Adjustments to Table Values:*

The following equation makes adjustments for deviations of design and solar conditions from those listed in Note 1 above:

$$CLTD_{corr} = [(CLTD + LM)\,K + (78 - t_R) + t_o - 85)]\,f$$

where CLTD is from this table

(a) *LM* is latitude-month correction from Table 8-38 for a horizontal surface,

(b) *K* is a color adjustment factor applied after first making month-latitude adjustments. Credit should not be taken for a light-colored roof except where permanence of light color is established by experience, as in rural areas of where there is little smoke.

K = 1.0 if dark-colored or light in an industrial area

K = 0.5 if permanently light-colored (rural area)

(c) $(78 - t_R)$ is indoor design temperature correction

(d) $(t_o - 85)$ is outdoor design temperature correction, where t_o is the average outside temperature on design day

(e) *f* is a factor for attic fan and or ducts above ceiling applied after all other adjustments have been made
f = 1.0 no attic or ducts
f = 0.75 positive ventilation

Values were calculated without and with suspended ceiling, but make no allowances for positive ventilation or return ducts through the space. If ceiling is insulated and fan is used between ceiling and roof, *CLTD* may be reduced 25% (f = 0.75). Analyze use of the suspended ceiling space for a return air plenum or with return air ducts separately.

3. *Roof Constructions Not Listed in Table:*

The *U*-values listed are only guides. The actual value of *U* as obtained from tables or as calculated for the actual roof construction should be used.

An actual roof construction not in this table would be thermally similar to a roof in the table, if it has similar mass, lb/sq ft, and similar heat capacity, Btu/sq ft • °F. In this case, use the *CLTD* from this table as corrected by Note 2 above.

Example: A flat roof without suspended ceiling has mass = 18.0 lb/sq ft, U = 0.20 Btu/h • sq ft • °F, and heat capacity = 9.5 Btu/sq ft • °F.

Use $CLTD_{uncorr}$ from Roof No. 13, to obtain $CLTD_{corr}$ and use the actual *U*-value to calculate $q/A = U\,(CLTD_{corr}) = 0.20\,(CLTD_{corr})$.

(4) *Additional Insulation:*

For each R-7 increase in *R*-value from insulation added to the roof structure, use a *CLTD* for a roof whose weight and heat capacity are approximately the same, but whose *CLTD* has a maximum value 2 hr later. If this is not possible, because a roof with longest time lag has already been selected, use an effective *CLTD* in cooling load calculation equal to 29°F.

(*Continued*)

State and city	North latitude	1% t_o*	2½% t_o*	5% t_o*	t_d
Alabama					
Birmingham	30°3′	96	94	92	21
Mobile	30°40′	95	93	91	18
Montgomery	32°20′	96	95	93	21
Alaska					
Anchorage	61°10′	71	68	66	15
Fairbanks	64°50′	82	78	75	24
Arizona					
Flagstaff	35°10′	84	82	80	31
Phoenix	33°30′	109	107	105	27
Tucson	32°10′	104	102	100	26
Yuma	32°40′	111	109	107	27
Arkansas					
Fort Smith	35°20′	101	98	95	24
Little Rock	34°40′	99	96	94	22
California					
Fresno	36°50′	102	100	97	34
Los Angeles	34°0′	83	80	77	15
Needles	34°50′	112	110	108	27
Oakland	37°40′	85	80	75	19
Sacramento	38°30′	101	98	94	36
San Diego	32°40′	83	80	78	12
San Francisco	37°40′	82	77	73	20
Colorado					
Denver	39°50′	93	91	89	28
Grand Junction	39°10′	96	94	92	29
Pueblo	38°20′	97	95	92	31
Connecticut					
Hartford	41°50′	91	88	85	22
Delaware					
Wilmington	39°40′	92	89	87	20
D.C.					
Washington	38°50′	93	91	89	18
Florida					
Jacksonville	30°30′	96	94	92	19
Miami	25°50′	91	90	89	15
Orlando	28°30′	94	93	91	17
Tampa	28°0′	92	91	90	17
Georgia					
Atlanta	33°40′	94	92	90	19
Savannah	32°10′	96	93	91	20
Hawaii					
Honolulu	21°20′	87	86	85	12
Idaho					
Boise	43°30′	96	94	91	31
Pocatello	43°0′	94	91	89	35
Illinois					
Chicago	42°0′	91	89	86	20
Springfield	39°50′	94	92	89	21
Indiana					
Evansville	38°0′	95	93	91	22
Indianapolis	39°40′	92	90	87	22
Iowa					
Des Moines	41°30′	94	91	88	23
Sioux City	42°20′	95	92	89	24
Kansas					
Topeka	39°0′	99	96	93	24
Wichita	37°40′	101	98	96	23
Kentucky					
Louisville	38°10′	95	93	90	23
Louisiana					
New Orleans	30°0′	93	92	90	16
Shreveport	32°30′	99	96	94	20
Maine					
Bangor	44°50′	86	83	80	22
Maryland					
Baltimore	39°10′	94	91	89	21
Massachusetts					
Boston	42°20′	91	88	85	16
Michigan					
Detroit	42°20′	91	88	86	20
Sault Ste. Marie	46°30′	84	81	77	23
Minnesota					
International Falls	48°30′	85	83	80	26
Minneapolis/St. Paul	44°50′	92	89	86	22
Mississippi					
Jackson	32°20′	97	95	93	21
Missouri					
Kansas City	39°10′	99	96	93	20
St. Louis	38°50′	97	94	91	21
Springfield	37°10′	96	93	91	23
Montana					
Billings	45°50′	94	91	88	31
Glendive	47°10′	95	92	89	29
Helena	46°40′	91	88	85	32
Nebraska					
Lincoln	40°50′	99	95	92	24
North Platte	41°10′	97	94	90	28
Nevada					
Las Vegas	36°10′	108	106	104	30
Reno	39°30′	95	92	90	45
New Hampshire					
Concord	43°10′	90	87	84	26
New Jersey					
Newark	40°40′	94	91	88	20
New Mexico					
Albuquerque	35°0′	96	94	92	27
Roswell	33°20′	100	98	96	33
New York					
Albany	42°50′	91	88	85	23
New York City	40°50′	92	89	87	17
Rochester	43°10′	91	88	85	22
North Carolina					
Charlotte	35°0′	95	93	91	20
Wilmington	34°20′	93	91	89	18
Winston-Salem	36°10′	94	91	89	20
North Dakota					
Bismarck	46°50′	95	91	88	27
Fargo	46°50′	92	89	85	25
Ohio					
Cincinnati	39°10′	92	90	88	21
Cleveland	41°20′	91	88	86	22
Columbus	40°0′	92	90	87	24
Toledo	41°40′	90	88	85	25
Oklahoma					
Oklahoma City	35°20′	100	97	95	23
Tulsa	36°10′	101	98	95	22
Oregon					
Medford	42°20′	98	94	91	35
Portland	45°40′	89	85	81	23
Pennsylvania					
Philadelphia	39°50′	93	90	87	21
Pittsburgh	40°30′	89	86	84	22
Rhode Island					
Providence	41°40′	89	86	83	19
South Carolina					
Charleston	32°50′	94	92	90	13
Columbia	34°0′	97	95	93	22
South Dakota					
Huron	44°30′	96	93	90	28
Rapid City	44°0′	95	92	89	28
Tennessee					
Knoxville	35°50′	94	92	90	21
Memphis	35°0′	98	95	93	21
Texas					
Amarillo	35°10′	98	95	93	26
El Paso	31°50′	100	98	96	27
Fort Worth	32°50′	101	99	97	22
Houston	29°40′	96	94	92	18
Laredo	27°30′	102	101	99	23
Utah					
Salt Lake City	40°50′	97	95	92	32
Vermont					
Burlington	44°30′	88	85	82	23
Virginia					
Richmond	37°30′	95	92	90	21
Roanoke	37°20′	93	91	88	23
Washington					
Seattle-Tacoma	47°30′	84	80	76	22
Spokane	47°40′	93	90	87	28
West Virginia					
Charleston	38°20′	92	90	87	20
Wisconsin					
LaCrosse	43°50′	91	88	85	22
Milwaukee	43°0′	90	87	84	21
Wyoming					
Casper	42°50′	92	90	87	31

*Three values of t_o are given, as some agencies require the use of different values. The designer should determine which outdoor design temperature governs on each project.

Climatological data for heat-gain calculations. (*From American Society of Heating, Refrigerating and Air Conditioning Engineers,* Fundamentals, *ASHRAE Handbook, 1989. Reprinted with permission.*)

CLTD CORRECTION FOR LATITUDE AND MONTH APPLIED TO WALLS AND ROOFS, NORTH LATITUDE

Latitude	Month	N	NNE NNW	NE NW	ENE WNW	E W	ESE WSW	SE SW	SSE SSW	S	HORIZ.
24	Dec.	−5	−7	−9	−10	−7	−3	3	9	13	−13
	Jan./Nov.	−4	−6	−8	−9	−6	−3	3	9	13	−11
	Feb./Oct.	−4	−5	−6	−6	−3	−1	3	7	10	−7
	Mar./Sept.	−3	−4	−3	−3	−1	−1	1	2	4	−3
	Apr./Aug.	−2	−1	0	−1	−1	−2	−1	−2	−3	0
	May/July	1	2	2	0	0	−3	−3	−5	−6	1
	June	3	3	3	1	0	−3	−4	−6	−6	1
32	Dec.	−5	−7	−10	−11	−8	−5	2	9	12	−17
	Jan./Nov.	−5	−7	−9	−11	−8	−4	2	9	12	−15
	Feb./Oct.	−4	−6	−7	−8	−4	−2	4	8	11	−10
	Mar./Sept.	−3	−4	−4	−4	−2	−1	3	5	7	−5
	Apr./Aug.	−2	−2	−1	−2	0	−1	0	1	1	−1
	May/July	1	1	1	0	0	−1	−1	−3	−3	1
	June	1	2	2	1	0	−2	−2	−4	−4	2
40	Dec.	−6	−8	−10	−13	−10	−7	0	7	10	−21
	Jan./Nov.	−5	−7	−10	−12	−9	−6	1	8	11	−19
	Feb./Oct.	−5	−7	−8	−9	−6	−3	3	8	12	−14
	Mar./Sept.	−4	−5	−5	−6	−3	−1	4	7	10	−8
	Apr./Aug.	−2	−3	−2	−2	0	0	2	3	4	−3
	May/July	0	0	0	0	0	0	0	0	1	1
	June	1	1	1	0	1	0	0	−1	−1	2
48	Dec.	−6	−8	−11	−14	−13	−10	−3	2	6	−25
	Jan./Nov.	−6	−8	−11	−13	−11	−8	−1	5	8	−24
	Feb./Oct.	−5	−7	−10	−11	−8	−5	1	8	11	−18
	Mar./Sept.	−4	−6	−6	−7	−4	−1	4	8	11	−11
	Apr./Aug.	−3	−3	−3	−3	−1	0	4	6	7	−5
	May/July	0	−1	0	0	1	1	3	3	4	0
	June	1	1	2	1	2	1	2	2	3	2
56	Dec.	−7	−9	−12	−16	−16	−14	−9	−5	−3	−28
	Jan./Nov.	−6	−8	−11	−15	−14	−12	−6	−1	2	−27
	Feb./Oct.	−6	−8	−10	−12	−10	−7	0	6	9	−22
	Mar./Sept.	−5	−6	−7	−8	−5	−2	4	8	12	−15
	Apr./Aug.	−3	−4	−4	−4	−1	1	5	7	9	−8
	May/July	0	0	0	0	2	2	5	6	7	−2
	June	2	1	2	1	3	3	4	5	6	1
64	Dec.	−7	−9	−12	−16	−17	−18	−16	−14	−12	−30
	Jan./Nov.	−7	−9	−12	−16	−16	−16	−13	−10	−8	−29
	Feb./Oct.	−6	−8	−11	−14	−13	−10	−4	1	4	−26
	Mar./Sept.	−5	−7	−9	−10	−7	−4	2	7	11	−20
	Apr./Aug.	−3	−4	−4	−4	−1	1	5	9	11	−11
	May/July	1	0	1	0	3	4	6	8	10	−3
	June	2	2	2	2	4	4	6	7	9	0

Notes:

1. Corrections in this table are in degrees F. The correction is applied directly to the *CLTD* for a wall as given in Table 8-36.
2. The *CLTD* correction given in this table is *not* applicable to cooling load temperature differences for conduction through glass.
3. For South latitudes, replace Jan. through Dec. by July through June.

CLTD correction factors. (*From American Society of Heating, Refrigerating and Air Conditioning Engineers,* Fundamentals, *ASHRAE Handbook, 1989. Reprinted with permission.*)

To determine the required capacity of air-conditioning equipment, use the values of CLTD for different hours of the day. For most buildings, the thermal design may be determined using the CLTD for hour 16 (4:00 pm) on July 21. For analysis of overall building thermal performance, the total building envelope must be considered (including roof, opaque walls, and glass area), plus internal, latent, ventilation, and infiltration loads. The calculation procedures given here form a basis for comparison between various opaque wall types.

In any climate where there are large fluctuations in the daily temperature cycle, the thermal inertia of masonry walls can contribute substantially to increased comfort and energy efficiency. The time lag created by delayed heat flow through the walls reduces peak cooling demands to a much greater extent than *U*-values alone indicate.

8.5.3 Heat-Loss Calculations

In northern climates where heat loss is usually more critical than heat gain, winter temperature cycles more nearly approximate static design conditions because daily temperature fluctuations are smaller. There is still, however, significant advantage to be gained by using masonry walls with thermal inertia. The methods developed by ASHRAE for measuring the dynamic thermal response of heavy construction are more complicated for heat-loss calculations than for heat gain, and require sophisticated computer programs.

The Catani and Goodwin study compared steady-state heat-loss calculations with dynamic analysis. They found that the predicted heat loss based on static conditions was 22% higher than the actual recorded loss for heavy walls, and 8% lower than the actual loss for lightweight walls. Using three different wall types with the same *U*-value, they made a direct comparison of peak heating loads. It was found that, although the effects are not as dramatic for winter conditions, peak heating load requirements decreased as the weight of the building walls increased (*see Fig. 8-39*).

Test buildings have been used to validate computer programs for dynamic heat-loss calculations by comparing them to actual measured heating loads. The National Institute of Science and Technology conducted a series of tests on a full-scale building erected in its environmental chamber where both temperature and humidity can be controlled. The study also compared maximum heat flow rates predicted by the steady-state and dynamic methods with actual measured heat flow (*see Fig. 8-40*). Steady-state calculations were an average of 52% higher than measured results.

8.5.4 The *M-Factor*

The computer programs developed by ASHRAE and NIST for dynamic heat-loss calculations are so complex that they do not easily translate into a simple equation. Researchers recognized the need for a simpler method of hand calculation that would make the concept of thermal inertia more readily usable. In response to this need, the Masonry Industry Committee sponsored a study by the engineering firm of Hankins and Anderson that resulted in development of the *M*-factor, a simplified correction factor expressing the effects of mass on heat flow.

The *M*-factor is not a new calculation procedure, but is simply used to modify steady-state calculations to account for the effect of wall mass. The

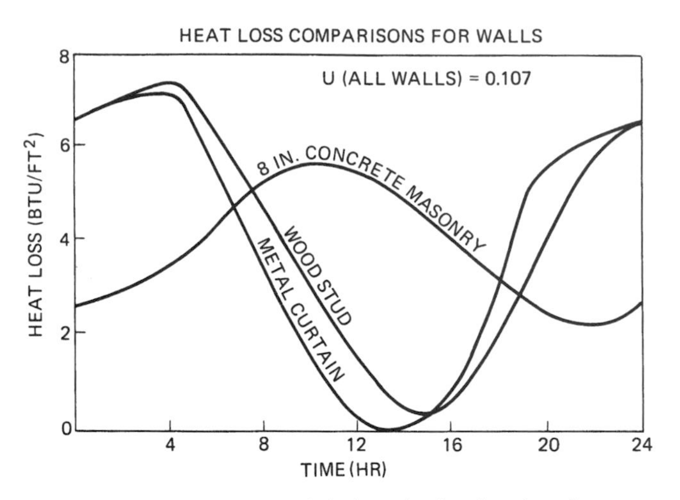

A computer program analysis shows that heat loss through walls with identical U-factors and configuration of insulation varies considerably due to mass. The effect of glass, occupants, and lights accentuates the difference.

Catani and Goodwin's heat-loss calculations for various wall types. (*From Mario Catani and Stanley Goodwin, "Heavy Building Envelopes and Dynamic Thermal Response,"* ACI Journal, *February 1976.*)

Test number	*Insulation*	*Windows*	*Maximum heat flow rate (Btu/hr)*		
			Steady-state method	*Response factor method*	*Measured heat flow*
1	None	Single glazed	15,135	11,558	11,372
2	Inside	Single glazed	4,470	2,814	2,748
3	Outside	Single glazed	4,748	3,047	2,811
4	Outside	Double glazed	4,499	2,525	2,700
5*	Outside	Double glazed	8,150	6,144	6,321

*Temperature limits changed to range of 10–70°F.

Test results of measured heat flow compared to steady-state and dynamic calculation methods. (*From National Concrete Masonry Association,* **TEK Bulletin 58,** *NCMA, Herndon, Va.*)

M-factor is a dimensionless correction factor. It is not a direct measure of the thermal storage capacity of walls. It is defined as the ratio of the cooling or heating load calculated by dynamic response methods to that computed with standard ASHRAE calculation methods.

The modifiers were plotted on a graph with variables of wall weight and number of degree-days (*see Fig. 8-41*). When the wall weight is very light, and in areas where the number of degree days is high (colder climates), the *M*-factors approach 1.0 (no correction). Ambient conditions in cold climates more closely approximate a steady-state condition, and the

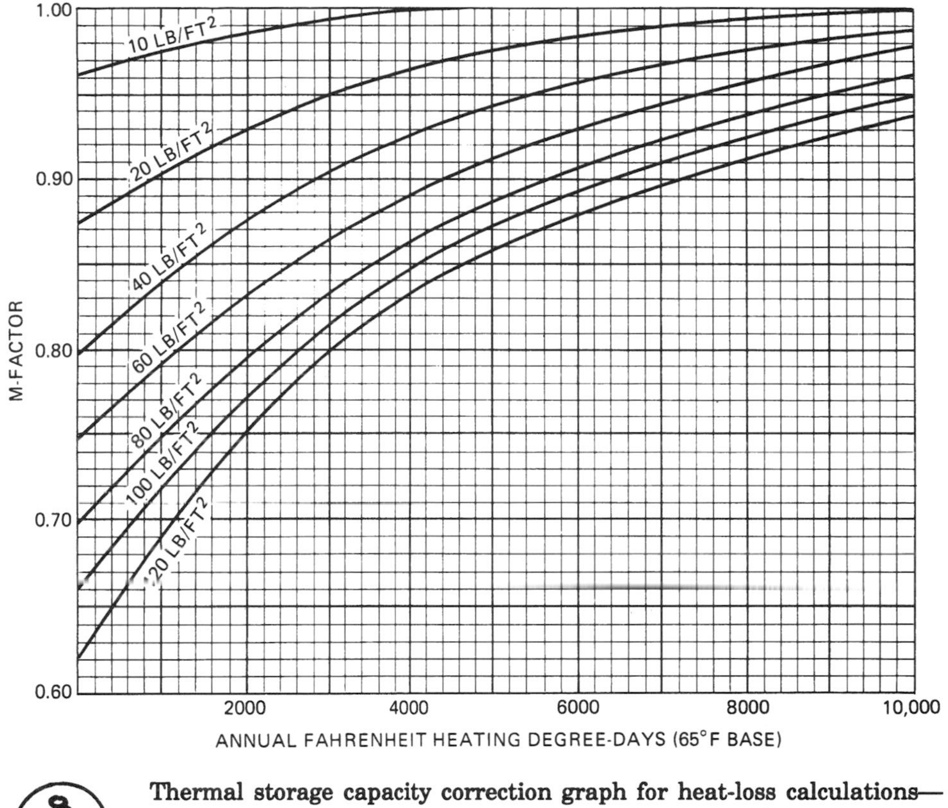

Thermal storage capacity correction graph for heat-loss calculations—*M*-factor curves. (*From Brick Institute of America,* Technical Note 4B, *BIA, Reston, Va.*)

traditional U-factor evaluation for heat loss is more accurate than for warmer regions.

The M*-factors from the curves modify only heat-loss calculations and should not be used in cooling calculations.* A study of heat-gain calculations showed that dynamic analysis results were sufficiently close to those obtained using the CLTD modifier in the ASHRAE equation that no additional correction factor was needed for cooling load calculations.

Energy conservation codes may require a minimum heat-loss rate, a minimum R-value, or a maximum U-value. M-factors adjust the normal ratings of heavy walls, taking into account the effect of thermal inertia. They may be used (1) to modify heat losses determined by steady-state calculations and (2) to determine effective U- and R-values for heavy walls.

As an example, consider a building project located in Memphis, Tennessee, with 3232 annual heating degree-days (*see Fig. 8-42*), an outdoor design temperature of 21°F, and an indoor design temperature of 72°F. Two different wall types will be considered: an insulated wood frame wall weighing 8 lb/sq ft, and a masonry wall weighing 80 lb/sq ft. From *Fig. 8-41,* the M-factor for the frame wall is 1.0 and for the masonry wall 0.84. For the purpose of this example, both walls are assumed to have steady-state U-values of 0.12.

1. *Heat-loss calculation:* Using the standard heat-loss equation with the M-factor correction, the calculation for one square foot of the wood frame wall is

City and state	Yearly total	City and state	Yearly total
Birmingham, Alabama	2,551	Omaha, Nebraska	6,612
Anchorage, Alaska	10,864	Las Vegas, Nevada	2,709
Phoenix, Arizona	1,765	Reno, Nevada	6,332
Tucson, Arizona	1,800	Concord, New Hampshire	7,383
Little Rock, Arkansas	3,219	Trenton, New Jersey	4,980
Los Angeles, California	2,061	Albuquerque, New Mexico	4,348
Sacramento, California	2,502	Albany, New York	6,875
San Diego, California	1,458	Buffalo, New York	7,062
San Francisco, California	3,015	New York, New York	4,871
Denver, Colorado	6,283	Raleigh, North Carolina	3,393
Pueblo, Colorado	5,462	Bismarck, North Dakota	8,851
Hartford, Connecticut	6,235	Cincinnati, Ohio	4,410
Wilmington, Delaware	4,930	Cleveland, Ohio	6,351
Washington, D.C.	4,224	Oklahoma City, Oklahoma	3,725
Jacksonville, Florida	1,239	Portland, Oregon	4,635
Miami, Florida	214	Philadelphia, Pennsylvania	5,144
Orlando, Florida	766	Pittsburgh, Pennsylvania	5,987
Atlanta, Georgia	2,961	Providence, Rhode Island	5,954
Savannah, Georgia	1,819	Charleston, South Carolina	2,033
Honolulu, Hawaii	0	Columbia, South Carolina	2,484
Boise, Idaho	5,809	Sioux Falls, South Dakota	7,839
Chicago, Illinois	5,882	Knoxville, Tennessee	3,494
Springfield, Illinois	5,429	Memphis, Tennessee	3,232
Indianapolis, Indiana	5,699	Nashville, Tennessee	3,578
Des Moines, Iowa	6,588	Dallas, Texas	2,363
Topeka, Kansas	5,182	El Paso, Texas	2,700
Louisville, Kentucky	4,660	Houston, Texas	1,396
Baton Rouge, Louisiana	1,560	Salt Lake City, Utah	6,052
New Orleans, Louisiana	1,385	Burlington, Vermont	8,269
Portland, Maine	7,511	Richmond, Virginia	3,865
Baltimore, Maryland	4,654	Seattle-Tacoma, Washington	5,145
Boston, Massachusetts	5,634	Charleston, West Virginia	4,476
Detroit, Michigan	6,232	Milwaukee, Wisconsin	7,635
Minneapolis, Minnesota	8,382	Cheyenne, Wyoming	7,381
Jackson, Mississippi	2,239	Montreal, Canada	7,899
Kansas City, Missouri	4,711	Quebec, Canada	8,937
St. Louis, Missouri	5,000	Toronto, Canada	6,827
Great Falls, Montana	7,750	Vancouver, Canada	5,515

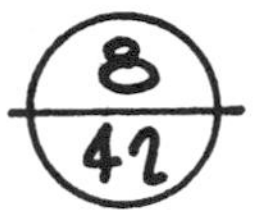

Annual degree-days for major U.S. and Canadian cities. (Data drawn from U.S. Weather Bureau information.)

$$H_L = (A)(U)(t_i - t_o)(M) \tag{8.11}$$
$$= (1.0 \text{ sq ft})(0.12)(72°\text{F} - 21°\text{F})(1.0) = 6.12 \text{ Btu/hour/sq ft}$$

and for 1 sq ft of the masonry wall,

$$H_L = (1.0 \text{ sq ft})(0.12)(72°\text{F} - 21°\text{F})(0.84) = 5.14 \text{ Btu/hour/sq ft}$$

The thermal inertia of the heavy wall accounts for an approximate 16% decrease in heat loss. This reduced figure more accurately predicts performance of the masonry wall when compared to actual measured results in field tests.

2. *Effective U-values:* The steady-state U-values for both walls are assumed to be 0.12. To find the adjusted or *effective* U-values (U_e), the formula $U_e-(U)(M)$ may be used for the frame wall:

$$U_e = (0.12)(1.0) = 0.12$$

and the masonry wall:

$$U_e = (0.12)(0.84) = 0.10$$

The better U-value for the masonry wall more accurately predicts its actual performance.

3. *Effective* R-*values:* The steady-state R-values for the two walls are assumed to be 8.33. Using the equation $R_e = R / M$, for the frame wall, the adjusted or *effective* resistance (R_e) would be

$$R_e = \frac{8.33}{1.0} = 8.33$$

and for the masonry wall,

$$R_e = \frac{8.33}{0.84} = 9.91$$

This means simply that the thermal inertia of the masonry wall increases its effective R-value to a higher resistance of 9.91. When a specified U- or R-value is required by code, the equivalent R-value for the element may be determined by

$$U_e = \frac{U}{M} \qquad \text{or} \qquad R_e = R(M) \tag{8.12}$$

where U_e = equivalent overall coefficient of thermal transmission of massive wall elements
R_e = equivalent thermal resistance of massive wall elements
M = modification factor (*see Fig. 8-41*)

The M-factor is a simple means of quantifying the effect of thermal inertia on heat-loss calculations without the aid of a computer. It permits a more accurate prediction of dynamic thermal performance than steady-state methods. The M-factors in *Fig. 8-41* are deliberately conservative. In very cold climates, they give a credit of about 10% to a heavy wall, where the more detailed computer calculations indicate a much greater actual benefit. The results of some computer calculations for various wall weights are shown in *Fig. 8-43*. The difference between the static and dynamic methods was approximately 20% for the lightweight structure, and about 30% for the heaviest wall. The relationships of heating load to wall weight determined in this and other studies appear to validate the accuracy of the M-factor concept.

8.6 ADDED INSULATION

The thermal performance of masonry walls and their resistance to heat flow can be further improved by adding insulation (*see Fig. 8-44*). In severe winter climates where diurnal temperature cycles are of minimum amplitude, the thermal inertia of brick and block walls can be complemented by the use of resistance insulation such as loose fill or rigid board materials. Hollow units can easily be insulated with loose fill or granular materials, and multi-wythe cavity walls and veneer walls over wood or metal frame construction have open collar joints for rigid

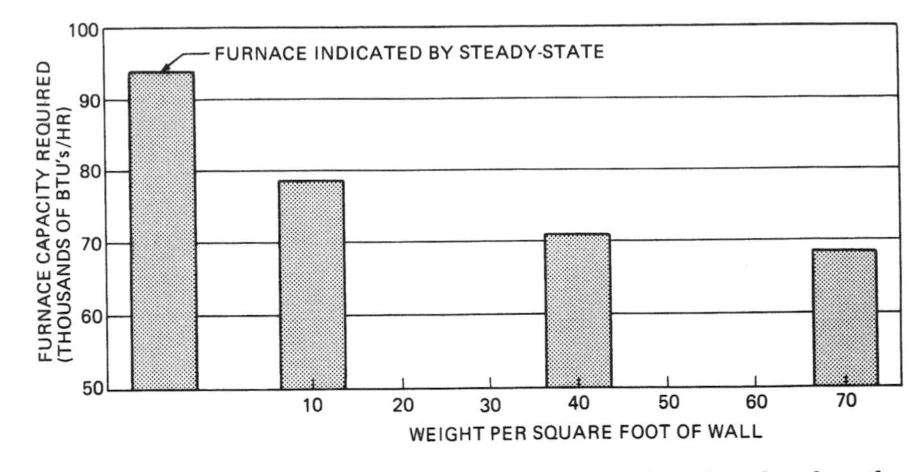

Size of furnace required for the heating load is reduced as the weight of the wall increases. (*From National Concrete Masonry Association*, TEK Bulletin 82, *NCMA, Herndon, Va.*)

Material	Density (lb/ft^3)	Thermal conductivity, k (Btu/hr/ft^2/°F)	Thermal resistance, R (per in.) (Btu/hr/ft^2/°F)	Permeability (perm-in.)	Vapor resistance, [ft^2/hr (in. Hg)]
	*Properties**				
Granular fills					
Vermiculite (expanded)	5–9	0.44	2.27	62†	0.0163†
Perlite (expanded)	5–8	0.37	2.70	N.A.‡	N.A.‡
Rigid boards					
Expanded polystyrene, extruded	3.5	0.19	5.26	1.2	0.8333
Expanded polystyrene, molded beads	0.9–1.1	0.28	3.57	2.0–5.8	0.5–0.1724
Expanded polyurethane, extruded	1.5	0.16	6.25	0.4–1.6	2.5–0.625
Perlite aggregate	11	0.38	2.63	25	0.04
Rigid urethane	2	0.16§	6.25§	2	0.50
Cellular glass	9	0.35–0.44¶	2.86–2.44¶	0	Very high
Preformed fiberglass	4–9	0.21–0.26	4.76–3.86	Very high	Very low

*Tabulated values are from varied sources. Designers should check with manufacturers and other sources for more precise values.

†Material thickness is 2.5 in.

‡N.A., not available.

§Based on *aged k*-factor.

¶From 0 to 90°F.

Insulation materials often used in masonry construction. (*From Brick Institute of America*, Technical Note 21A, *BIA, Reston, Va.*)

insulating boards. The proper selection of insulating materials for masonry walls depends on more than just thermal performance.

- The insulation must not interfere with proper cavity wall drainage.
- Thermal insulating efficiency must not be impaired by retained moisture from any source (e.g., wind-driven rain or vapor condensation within the cavity).
- Granular fill materials must be able to support their own weight without settlement, to assure that no portion of the wall is without insulation.
- Insulating materials must be inorganic, or be resistant to rot, fire, and vermin.
- Granular insulating materials must be "pourable" in lifts of at least 4 ft for practical installation.

8.6.1 Granular Fills

Two types of granular fill insulation have been tested by researchers at the Brick Institute of America and found to comply with these criteria: water-repellent-treated *vermiculite* and *perlite* fills.

Vermiculite is an inert, lightweight, insulating material made from aluminum silicate expanded into cellular granules about 15 times their original size. Perlite is a white, inert, lightweight granular insulating material made from volcanic siliceous rock expanded up to 20 times its original volume. Specifications for water-repellent treated vermiculite and perlite are published by the Vermiculite Association and the Perlite Institute, Inc. Each of these specifications contains limits on density, grading, thermal conductivity, and water repellency. Loose fill insulation should not settle more than 0.5% after placement, or a thermal bridge will be created at the top of the wall.

Cavity wall construction permits natural drainage of moisture or condensation. If insulating materials absorb excessive moisture, the cavity can no longer drain effectively, and the insulation acts as a bridge to transfer moisture across the cavity to the interior wythe. Untreated vermiculite and perlite will accumulate moisture, and suffer an accompanying decrease in thermal resistance.

Loose fill insulation is usually poured directly into the cavity from the bag or from a hopper placed on top of the wall. Pours can be made at any convenient interval, but the height of any pour should not exceed 20 ft. Rodding or tamping is not necessary and may in fact reduce the thermal resistance of the material. The insulation in the wall should be protected from weather during construction, and weepholes should be screened to prevent the granules from leaking out or from plugging the drainage path.

8.6.2 Rigid Board Insulation

Rigid board insulations are classified physically as cellular or fibrous. Cellular insulation includes polystyrenes, polyurethanes, and polyisocyanurates of open- and closed-cell construction. Fibrous insulation materials include fiberboards of wood or mineral fibers with plastic binders. To make them moisture resistant, they are sometimes impregnated with asphalt. Fibrous glass insulation is made of nonabsorbent fibers formed into boards with phenolic binders, and surfaced with asphalt-saturated, fiberglass-reinforced material.

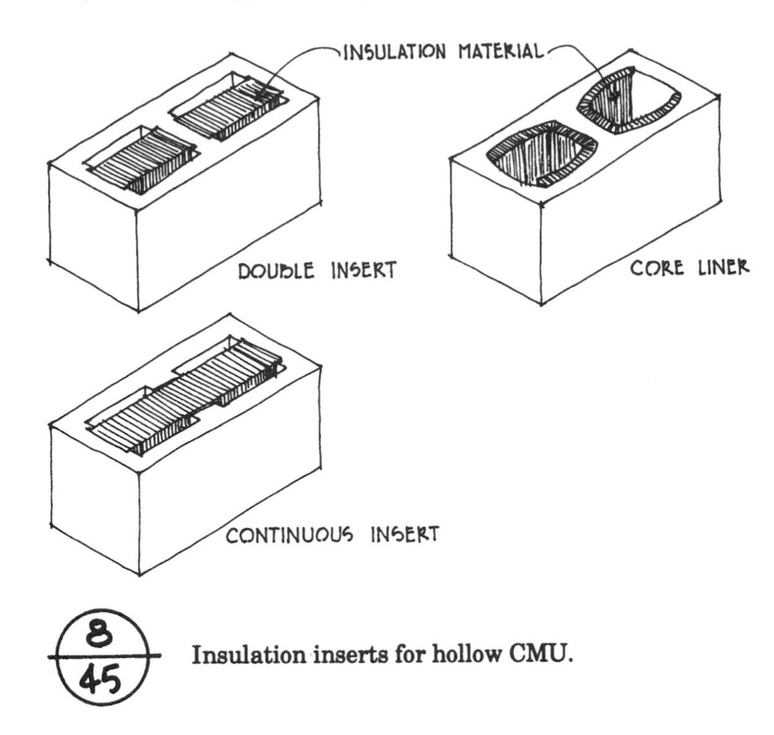

8/45 **Insulation inserts for hollow CMU.**

Water trapped in insulation can destroy its thermal insulating value. Water vapor can flow wherever air can flow between fibers, through interconnected open cells, or where a closed-cell structure breaks down. With mean wall temperatures alternating above and below freezing, ice formation can break down the walls between cells. Repeated freeze-thaw cycles can progressively destroy closed-cell insulation materials. Unimpaired cavity wall drainage is therefore important in the protection of the insulation board as well as the moisture protection of the interior spaces.

Some concrete block manufacturers produce units with built-in insulation installed at the plant prior to shipment. These inserts may be of polystyrene or polyurethane, and vary in shape and design for different proprietary products (*see Fig. 8-45*). Generally available in 6-, 8-, 10-, and 12-in. widths, these special units permit single-wythe construction of insulated walls with exposed masonry surfaces on each side, and better *U*-values than with loose fill insulation.

There are probably as many different installation procedures as there are types of rigid board insulation. Generally however, rigid insulation is installed against the cavity face of the backup wythe. A minimum of 2 in. should be left between the cavity face of the exterior wythe and the insulation board to facilitate construction and allow for drainage of the cavity. Mechanical and/or adhesive attachment as recommended by the manufacturer is used to hold the insulation in place.

8.6.3 Foams

Urea-formaldehyde foams are not recommended for use in masonry walls.

8.6.4 Vapor Retarders

Under certain conditions of design, it may be necessary to add a vapor retarder to an insulated cavity wall to retard or stop the flow of water vapor. An acceptable retarder is one with a moisture vapor permeance of less than 1 perm. Vapor retarders may be in the form of bituminous materials, continuous polyethylene films, and so on. They may be attached to the insulation as part of the fabricated product, or they may be incorporated separately in or on the wall. For greatest effectiveness, vapor retarders must be continuous and without openings or leaks through which airborne vapor might pass. *(See Chapter 9 for additional information regarding moisture control.)*

8.6.5 Insulation Location

The most effective thermal use of massive construction materials is to store and reradiate heat. This means that insulation should be on the outside of the wall. *Figure 8-46* shows that the location of the insulation within the wall section has an effect on heat flux through the wall which is not accounted for by standard *U*-value calculations. In the thermal research conducted by NIST and NCMA, the effects of variable insulation location were studied. It was found that indoor temperatures were reduced by half when insulation was placed on the outside rather than the inside of the wall, and that the thermal storage capacity of the masonry was maximized. In cavity walls, performance is improved if the insulation is placed in the cavity rather than on the inside surface.

8.7 ENERGY CONSERVATION

Masonry construction can be used in several ways with passive solar design. It can (1) provide a solar screen to shade glass areas on a facade, (2) collect

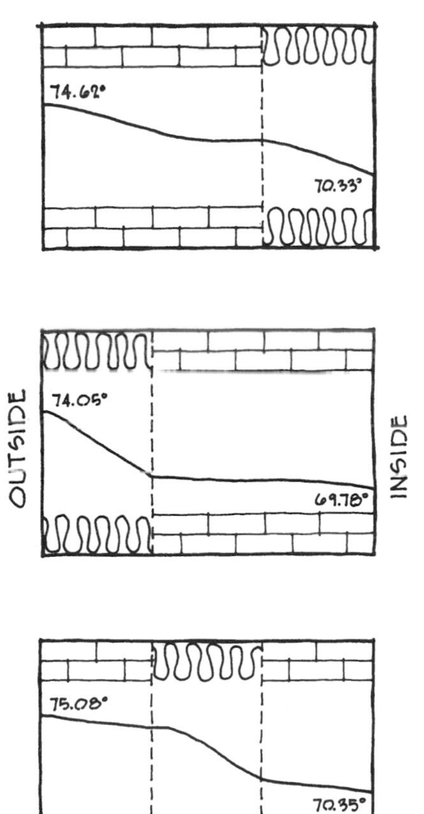

Francisco Arumi's temperature profiles for three different insulation locations. *(From* **Energy Conservation through Building Design,** *edited by Donald Watson. Copyright 1979 by McGraw-Hill, Inc. Used with the permission of the publisher.)*

and distribute solar warmth in winter, and (3) intercept excessive heat and solar radiation during the summer. In passive solar design, the buildings themselves collect, store, and distribute heat. A key element is the use of thermal mass—heavy materials which absorb and reradiate large amounts of energy. Passive measures such as cross ventilation, evaporation, exhaustion of hot air by convection, and absorption of heat by thermal mass can provide up to 100% of a building's cooling needs in summer. Masonry is particularly cost-effective in these applications because it simultaneously provides supporting structure, spatial definition, acoustical separation, fire separation, finished surfaces, and thermal storage.

Solar energy systems for buildings are divided into two categories: active and passive. Active systems use solar collectors, heat storage tanks, pumps, heat exchangers, and extensive plumbing and electrical controls. Buildings may take any form, and although building orientation is important, it need not be as critical since solar collectors can be oriented for optimum performance regardless of the building's orientation. Passive buildings, on the other hand, *must* be oriented in relation to the seasonal and daily movements of the sun to maximize heat gain in the winter and to minimize solar loads in the summer. Solar heat gain through walls, windows, roofs, skylights, and other building elements can dramatically reduce winter energy requirements. If thermal energy flow is by natural means, such as radiation, conduction, and natural convection, and if solar energy contributes a significant portion of the total heating requirement, the building is considered a passive, solar-heated structure.

Thermal mass alone does not constitute passive solar heating or cooling. Buildings must be designed as total systems in order to take advantage of masonry's thermal mass. Using climatic data for each building site, the architect or engineer must determine the optimum amount and location of thermal mass, the type of glass, orientation of windows, and the best use of shading devices, ventilation, daylighting techniques, insulation, landscaping, and efficient heating and cooling equipment. Thermal mass is only one part of passive solar design.

The National Codes and Standards Council of the Concrete and Masonry Industries has published the *Thermal Mass Handbook: Concrete and Masonry Design Provisions Using ASHRAE/IES 90.1-1989.* The handbook is intended to help design professionals take advantage of thermal mass principles in complying with the energy codes. It is an excellent design aid with in-depth coverage that is beyond the scope of this book. The handbook is available through the National Concrete Masonry Association in Herndon, Virginia.

8.7.1 Shading Devices

Solar heat gain through windows can be as much as 3 times more than heat loss because direct radiation is instantaneously transmitted to the building interior. The incident solar radiation received by a vertical surface often exceeds 200 Btu/hour/sq ft, and the annual operating cost of cooling equipment attributed to each square foot of ordinary glass is considerable.

The desirability of direct solar heat is evaluated quite differently depending on location, climate, orientation, and time of day. Hot, arid regions generally require exclusion of solar radiation to prevent overheating, excessive air-conditioning loads, glare, or deterioration of materials. In other circumstances, it may be more desirable to ensure adequate sunlight, either for heat or purely for its psychological effect.

If sun control is necessary, the most efficient means is through the use of external shading devices. ASHRAE data indicate that exterior shading devices can reduce the instantaneous rate of heat gain by as much as 85%. Different orientations require different types of shading devices. Horizontal projections or overhangs work best on southerly orientations. Vertical fins are of little value on southern exposures, where the sun is high at midday. For easterly and westerly orientations, however, vertical fins work well. Horizontal elements are of little value here because low morning and afternoon sun altitudes negate their effect. Combination horizontal/vertical egg-crate devices work well on walls facing southeast, and are particularly effective for southwest orientations. Considered by some to give the best "all-around" shading, the egg-crate patterns are most advantageous in hot climates. Their high shading ratio and low winter heat admission, however, can be undesirable in colder regions.

Clay or concrete masonry screens can be assembled in many patterns, with either standard or custom units. Their shading characteristics are all of the egg-crate type (*see Fig. 8-47*). Masonry screens can be constructed in stack bond, running bond, or split bond (where the individual units are separated horizontally and the wall contains no vertical mortar joints). Standard concrete block or clay tile can be laid with cores perpendicular to the wall surface to create screen effects, or decorative units made expressly for this purpose can be used. Solid brick can be laid in split bond to give open screen patterns of various designs. The overall texture and appearance of the wall is affected by the size and shape of the units as well as the pattern in which they are assembled. Both glazed and unglazed units are available in a variety of colors. Lighter colors provide brighter interior spaces because of greater reflectivity. Darker colors reflect less light (*see Fig. 8-48*). Depending on orientation and latitude, small screen patterns can exclude much or all of the direct sun load.

Glass block has passive solar applications too. In the winter when the sun is low on the horizon, south-facing glass block panels transmit large

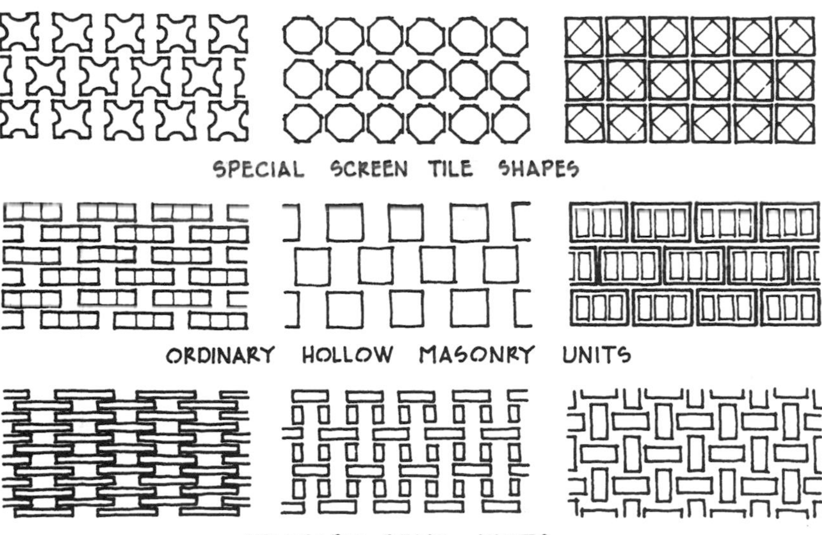

Variations of masonry solar screen designs. (*From Brick Institute of America,* Technical Note, *Vol. 11, No. 11, BIA, Reston, Va.*)

Material	*Light reflectivity (%)*
Unglazed clay masonry	
Cream manganese spot	52
Cream	50
Light buff	43
Light gray	40
Gray manganese spot	40
Golden buff	35
Red	30
Dark red	23
Ceramic glazed clay masonry	
White	83
Ivory	67
Sunlight yellow	65
White mottle	64
Coral	58
Cream glazed	51
Light gray	49
Green mottle	49
Cream mottle	49
Light green	46
Cream-tone salt glazed	44
Gray mottle	41
Ocular green	37
Tan	37
Blue	35
Buff-tone salt glazed	27
Black	5

Reflectivity of colors. (*From Brick Institute of America,* Technical Note, *Vol. 11, No. 11, BIA, Reston, Va.*)

amounts of solar energy to the interior. In the summer when the sun is high overhead, the horizontal and vertical mortar joints form an egg-crate shading device to limit heat gain.

The degree of shading provided by a masonry solar screen is a function of the shape, dimensions, and orientation of the openings. Standard sun path diagrams and shading masks can be used to compute time-shade cycles for openings of any shape, or to custom design a screen for a specific latitude and orientation. (See AIA *Architectural Graphic Standards* and Olgyay and Olgyay's *Solar Control and Shading Devices* for more detailed information on solar screens.*)

Masonry screens can be used to reduce heat gain economically when building orientation cannot be easily adjusted. They can also be retrofitted to existing buildings to substantially reduce air-conditioning loads and lower overall energy consumption.

*Major references for which complete publishing information is not given in the text can be found in the Bibliography at the end of the book.

8.7.2 Direct-Gain Solar Heating

The simplest method of solar heating is *direct gain*. If a building is constructed of lightweight materials, solar radiation will heat its low thermal mass quickly and raise inside air temperatures above comfortable levels. At night, these buildings lose their heat just as rapidly, causing temperatures to drop again. Better designs allow sunlight to strike materials of high thermal mass which can store the heat and reradiate it at a later time (*see Fig. 8-49*). Contemporary materials include poured and precast concrete as well as masonry. When these materials with high heat storage capacity are used for walls, floors, and even ceilings, performance and effi-

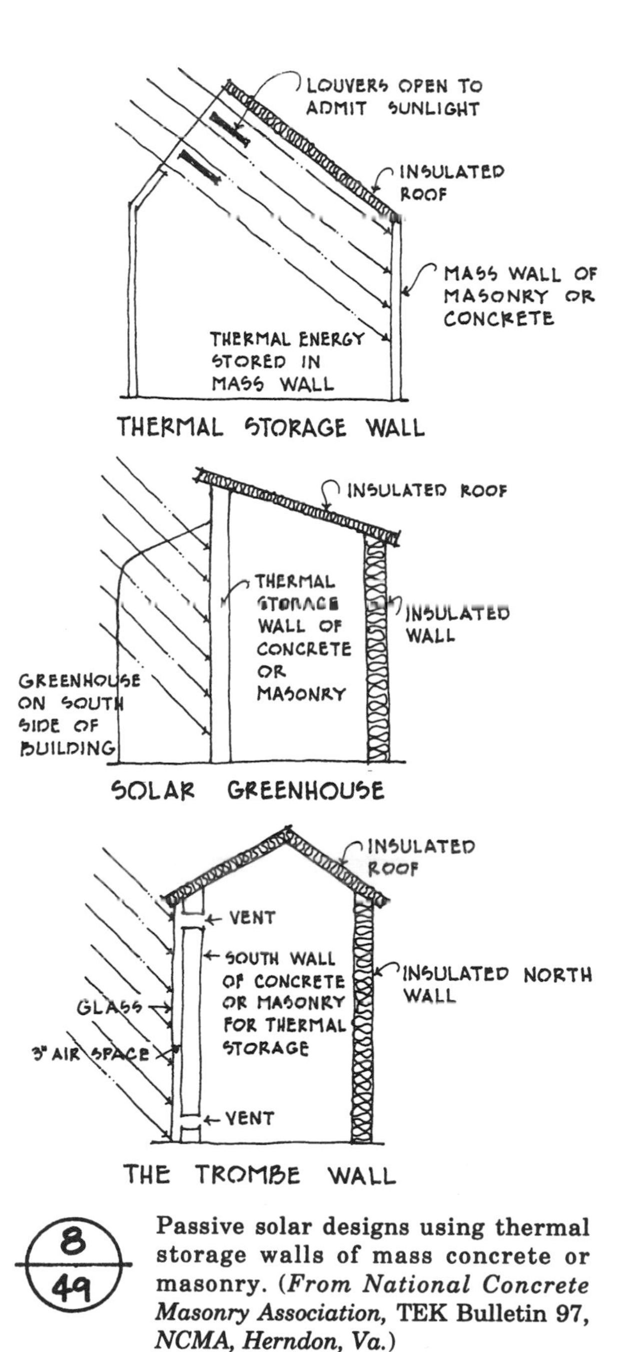

Passive solar designs using thermal storage walls of mass concrete or masonry. (*From National Concrete Masonry Association,* TEK Bulletin 97, *NCMA, Herndon, Va.*)

ciency are increased because the ratio of surface area to volume of mass is maximized.

Masonry walls in direct-gain systems can be any color, but light to medium colors are best for diffusing light over the wall. Heat distribution is generally not critical in direct-gain systems because the heat is stored in the same space in which it is used. The amount of solar heat collected and stored can be controlled by shading devices, and heat loss at night can be minimized by movable insulation. Direct gain is used primarily in mild and moderate climates.

8.7.3 Thermal Storage Walls

In regions with mild to severe winters, a thermal storage wall system provides better performance than direct gain. A loadbearing or non-loadbearing masonry or concrete wall is constructed and, leaving a 2 in. to 4 in. air space, is covered with double insulating glass to act as a collector. The masonry is heated by direct radiation, stores the heat, and then reradiates it to the interior spaces. The glass traps solar energy through the greenhouse effect. Sunlight strikes the mass wall, is converted to thermal energy, and stored. The storage mass becomes a radiant heat source, and creates natural convection currents which help to distribute the heat. Buildings are most efficient when the glass area and thermal mass are properly sized and oriented for optimum exposure, and are protected from heat loss by movable insulating panels or louvers. Efficiently designed walls may store enough heat to maintain comfortable indoor temperatures for as long as three overcast days. Thermal storage wall systems have much less temperature fluctuation than direct-gain systems, but do not usually achieve the same high initial interior temperatures.

Thermal mass walls have been built of concrete, water-filled containers, and masonry. The concept has been studied extensively in Europe and the United States during the past 25 years. Proper design requires consideration of several factors, including heat capacity and thermal conductance. There must be sufficient heat storage capacity to absorb the solar energy entering the building through windows or skylights. The conductance of the wall must be such that it can store heat for a desired length of time and then release it into the room as required. When a mass storage wall is placed directly in front of the glass, optimum thermal conductance/heat storage capacity relationships must be achieved for efficient operation. If conductance values are too high, the wall loses too much heat during the "charging period," reducing the total amount of stored heat. If the conductance is too low, the wall reaches such high temperatures during the charging period that some heat is reradiated back through the glass and lost to the exterior. Several independent studies have indicated that the optimum thickness of a masonry thermal storage wall is 10 to 18 in. Heat storage capacity should be equivalent to 30 Btu/°F/sq ft of glass, or about 150 lb of masonry for every square foot of glazing.

Storage walls should be positioned to receive maximum solar radiation by direct exposure. South-facing walls are usually most effective. Special treatment, such as a dark exterior surface, roughness, grooves, or ribs, can improve convective and radiant heat exchange. At night, insulated louvers or movable insulated panels should be closed to prevent heat loss and to permit the wall to radiate heat into the interior.

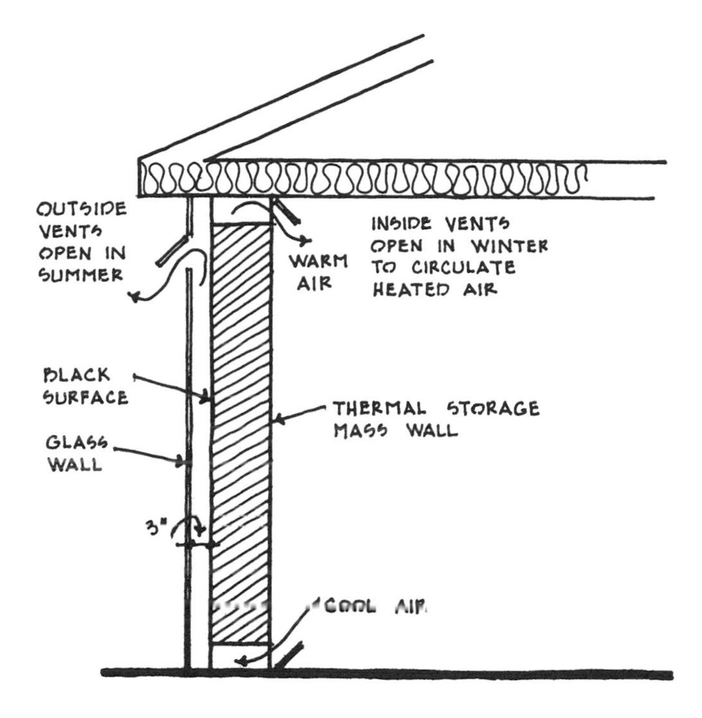

Detail of the Trombe wall. (*From S.V. Szokolay,* Solar Energy and Building, *The Architectural Press, Ltd., London, 1975.*)

8.7.4 Vented Thermal Storage Walls

The most widely used type of massive thermal storage wall was developed by Felix Trombe in a series of experimental houses built in Odeillo, France. In the Trombe wall or vented thermal storage wall, the air space between the wall and the glass enclosure is connected to the room by vents at the top and bottom of the wall. The heated air circulates into the room by thermal buoyancy currents. For summer operation, external vents are provided and the internal vents are closed (*see Fig. 8-50*). Venting the wall to the interior will reduce temperature fluctuations and increase the maximum temperature reached in the living space. Vents with automatic or manual closures should be used so that the system does not reverse itself at night and create a heat loss. If controlled vents are not installed, movable insulation is essential to prevent heat losses at night.

In the earliest Trombe wall house (1967), the wall thickness is approximately 2 ft. Computer simulations performed at the Los Alamos Scientific Laboratory tested optimum wall thickness and performance characteristics under varying outdoor conditions. It was found that although the daily fluctuations felt on the inside wall surface were considerably different for 6-, 12-, and 24-in. thicknesses, the net annual thermal contribution of the three walls was not markedly different. The 12-in. wall had the best overall performance, giving an annual solar heating contribution of 68%. Wall thickness was more important in reducing indoor temperature fluctuations to maintain a comfortable level. Calculations for several geographic locations showed that the optimum 12-in. thickness does not depend on climate.

There is also a relationship between wall thickness and thermal conductivity. For each value of conductivity, there is a wall thickness that will give a maximum annual solar energy yield. Optimum thickness decreases as thermal conductivity decreases. At the University of Texas Numerical Simulation Laboratory, Francisco Arumi conducted tests to identify the design parameters that would make a vented thermal storage wall an economical heating supplement during the central Texas winter without adding to the cooling demand during hot summer months. The results showed that for a statistically typical January day, most of the needed heat could be provided by a masonry thermal storage wall approximately equal in area to the floor area of the space to be heated if open vents were provided to allow thermocirculation of heated air. For summer operation, the data showed that the wall can be designed with variables which not only prevent additional heat transfer and increased cooling loads, but in fact reduce the required energy consumption. If the wall is shaded in summer and externally vented, it can curtail daily energy requirements by as much as 50% even without the use of additional insulating panels. This study also identified an optimum wall thickness of about 1 ft.

NCMA, however, does not recommend vented walls in residential applications in most climates, because of increased maintenance, poorly timed delivery of heat, and reverse air flow. It suggests unvented walls for residential buildings and vented walls for light industrial, commercial, and school buildings that are not occupied at night.

The best thermal performance can be obtained by combining the direct gain and thermal storage wall systems. One of the most common applications is a sun room or solarium on the south side of the building. A south-facing wall can also be designed with sections of brick or solid concrete masonry alternating with windows protected from summer radiation by overhangs or shading devices. This combination (1) permits some direct sunlight to enter and warm the interior floor and wall elements, (2) achieves higher interior temperatures than the thermal storage wall alone, (3) provides less temperature fluctuation than the direct-gain system alone, and (4) provides better distribution of natural light.

8.7.5 Hybrid Systems

Fans and blowers can be used in passive solar designs to help the natural flow of thermal energy. These mechanically assisted passive systems are often referred to as hybrid designs. One hybrid design circulates heated air by passing it through the cores of concrete block or 8 in. hollow brick to store and distribute the heat. The primary benefit is that the thermal mass can be located anywhere in the building, regardless of where the heat is collected.

One example is a floor system that uses hollow units placed on their sides with the cores aligned. The solar-heated air is blown through the cores, heating the masonry and the room above. The units should be laid on rigid insulation to prevent heat loss to the soil. Another example is a vertical plenum wall. It passes air through the hollow vertical cores which store heat during the day for later use at night. Sheet metal ducts supply and remove air from the wall.

Hollow core systems are also effective in cooling. Venting the wall at night by blowing cooler air through the cores lowers the masonry temperature so that it can absorb daytime heat from interior spaces. For commercial buildings which require cooling even in winter because of internal heat generation, outside winter air and the thermal mass of the masonry can be used to cool different zones of the building.

Determining the performance and efficiency of passive solar designs is complex. Computer programs can make the job easier by calculating solar loads, capacity of thermal mass, proper proportions of glass to storage wall areas, heating and cooling requirements, and overall thermal performance. Using software developed by the Solar Energy Research Institute (SERI), performance can be calculated for site-specific weather and solar data. Further analysis can show how combining different energy conservation techniques, passive solar design, and natural cooling strategies can improve total building performance. Although funding for solar energy research took a sharp drop in the early 1980s, a continued public awareness of environmental problems continues to spur interest. With high electric rates, diminishing nonrenewable energy sources, and problems within the nuclear power industry, the incentive is great for using masonry to capture rather than fight the sun's heat.

8.8 ACOUSTICAL PROPERTIES

Environmental comfort in multi-family housing, hotels, office buildings, and private residences can be related as much to acoustical factors as to heating and cooling. Increased technology produces more and more noise sources at the same time when human perception of the need for privacy and quiet has become acute. Interior noise sources such as furnace fans, television sets, vacuum cleaners, video games, and washing machines combine with exterior street traffic, construction equipment, power mowers, and airplanes to create high levels of obtrusive sound. Noise generated by other people is also very aggravating to residents or tenants who can overhear conversation in adjoining rooms or apartments.

For noise that cannot be either eliminated or reduced, steps can be taken to absorb the sound or prevent its transmission through walls, floors, and ceilings. Some building codes cover acoustical characteristics of construction assemblies. Clay and concrete masonry partitions have been tested and found to provide good sound insulation.

Noise is transmitted in several ways: (1) as airborne sound through open windows or doors, through cracks around doors, windows, water pipes, or conduits, or through ventilating ducts; (2) as airborne sound through walls and partitions; and (3) by vibration of the structure. Acoustical control includes absorbing the sound hitting a wall so that it will not reverberate, and preventing sound transmission through walls into adjoining spaces.

Sound absorption involves reducing the sound emanating from a source within a room by diminishing the sound level and changing its characteristics. Sound is absorbed through dissipation of the sound-wave energy. The extent of control depends on the efficiency of the room surfaces in absorbing rather than reflecting these energy waves. *Sound transmission* deals with sound traveling through barriers from one space into another. To prevent sound transmission, walls must have enough density to stop the energy waves. With insufficient mass, the sound energy will penetrate the wall and be heard beyond it.

8.8.1 Sound Ratings

There are two principal types of sound ratings: absorption and transmission loss. Sound absorption relates to the amount of airborne sound energy

	Sound absorption	*Sound transmission*
Frequency rating	SAC	STL
Class rating	NRC	STC

Acoustical ratings.

absorbed on the wall adjacent to the sound. Sound transmission loss is the total amount of airborne sound lost as it travels through a wall or floor. Each type may be identified at a particular frequency or by class (*see Fig. 8-51*). Sound absorption coefficients (SACs) and noise reduction coefficients (NRCs) are measured in sabins, sound transmission loss (STL) in decibels. In both instances, the larger the number, the better the sound insulating quality of the wall.

8.8.2 Sound Absorption

Sound is absorbed by mechanically converting it to heat. To absorb sound usefully, a material must have a certain "flow resistance"—it must create a frictional drag on the energy of sound. Sound is absorbed by porous, open-textured materials, by carpeting, furniture, draperies, or anything else in a room that resists the flow of sound and keeps it from bouncing around. If the room surfaces were capable of absorbing all sound generated within the room, they would have a *sound absorption coefficient* (SAC) of 1.0. If only 50% of it were absorbed, the coefficient would be 0.50.

The percentage of sound absorbed by a material depends not only on its surface characteristics, but also on the frequency of the sound. SAC values for most acoustical materials vary appreciably with sound frequencies. A better measure of sound absorption, which takes frequency variations into account, is the *noise reduction coefficient* (NRC), determined by averaging SAC values at different frequencies. Typical NRC values of various building materials and furnishings are given in *Fig. 8-52.*

Most materials utilized for strength and durability have low sound absorption. Masonry, wood, steel, and concrete all have low absorptions, ranging from 2 to 8%. Dense brick and heavy weight concrete block will have 1 to 3%, while lightweight block may be as high as 5%. Painting the surface effectively closes the pores of the material and reduces its absorptive capability even further. Conventional masonry products absorb little sound because of their density and their highly impervious surfaces. Specially designed structural clay tile and concrete block units combine relatively high sound absorption with low sound transmission characteristics with little or no sacrifice of strength or fire resistance. Most of these special units have a perforated face shell with the adjacent hollow cores filled at the factory with a fibrous glass pad. Perforations may be circular or slotted, uniform or variable in size, and regular or random in pattern (*see Figs. 8-52 and 8-53*). Some proprietary units have NRC ratings from 0.45 to as high as 0.85, depending on the area and arrangement of the perforations.

Sound absorption and sound reflection are directly related. If at a

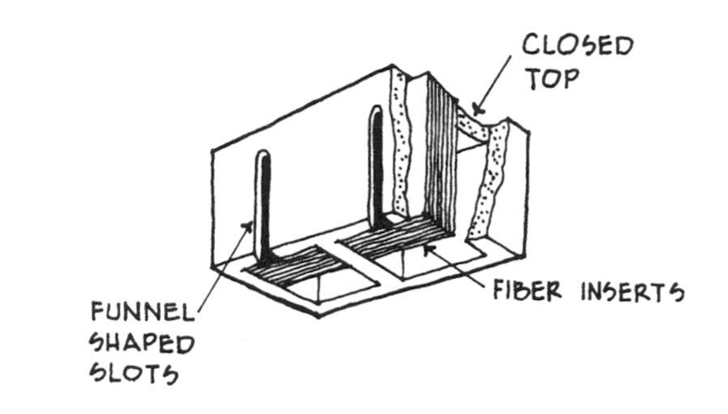

Material	*NRC*	*Material*	*NRC*
Brick, unglazed	0.04	Sound-insulated block	0.45–0.85
Carpet		Concrete floor	0.01
On concrete	0.30	V.A. tile on concrete	0.03
On pad	0.55	Wood floor	0.08
CMU, lightweight		Marble or glazed tile	0.01
Coarse texture	0.40	Single strength window glass	0.12
Medium texture	0.45	Plate glass	0.04
Fine texture	0.50	Gypsum board on 2 × 4 framing	0.07
CMU, normal weight		Gypsum board on concrete	0.03
Coarse texture	0.26	Plaster on brick or CMU	0.03
Medium texture	0.27	Wood paneling on furring	0.13
Fine texture	0.28	Draperies	
Deduct for paint		Lightweight	0.14
All types, spray		Medium weight	0.40
1 coat	−10%	Heavyweight	0.55
2 coats	−20%	Furniture (values per sq ft)	
Oil, brushed		Bed	0.80
1 coat	−20%	Sofa	0.85
2 coats	[illegible]	Wood tables, chairs, etc.	[illegible]
Latex, brushed		Leather upholstered chair	0.50
1 coat	−30%	Cloth upholstered chair	0.70
2 coats	−55%		

Noise reduction coefficients (NRC) of various building materials and furnishings. (*From Brick Institute of America,* Technical Note, *Vol. 9, No. 5, BIA, Reston, Va.*)

given frequency a particular material absorbs 75% of the incident sound, it will reflect the remaining 25%. In acoustical design, sound reflection is just as important as absorption. If too much absorption is provided, or if it is concentrated, the result will tend to "deaden" sound. Too little absorption will cause reverberation, or the persistence of sound within a room after the source has stopped. In excess, this is the principal defect associated with poor acoustics. The optimum reverberation time, which varies with room size and use, can be obtained by controlling the total sound absorption within a room. Alternating areas of reflective and absorptive materials will "liven" sound, promote greater diffusion, and provide better acoustics. Special sound-absorbing masonry units can be alternated with conventional units to achieve this effect.

Some patterns of acoustical tile (back row and right). (*Photo courtesy Stark Ceramics, Inc.*)

8.8.3 Sound Transmission

Although it is an important element in control of unwanted noise, sound absorption cannot take the place of sound insulation or the prevention of noise transmission through building elements. The NRC rating ranks wall systems only by sound absorption characteristics and does not give any indication of effectiveness in the control of sound transmission.

Sound energy is transmitted to one side of a wall by air. The impact of the successive sound waves on the wall sets it in motion like a diaphragm. Through this motion, energy is transmitted to the air on the opposite side. The amount of energy transmitted depends on the amplitude of vibration of the wall, which in turn depends on four things: (1) the frequency of the sound striking the surface, (2) the mass of the wall, (3) the stiffness of the wall, and (4) the method by which the edges of the wall are anchored. The *sound transmission loss* (STL) of a wall is a measure of its resistance to the passage of noise or sound from one side to the other. If a sound level of 80 dB is generated on one side and 30 dB measured on the other, the reduction in sound intensity is 50 dB. The wall therefore has a 50-dB STL rating. The higher the transmission loss of a wall, the better its performance as a sound barrier.

8.8.4 STC Ratings

Until the early 1960s, the most common sound rating system was the arithmetic average of STL measurements at nine different frequencies. Heavy walls have a relatively uniform STL curve and are satisfactorily

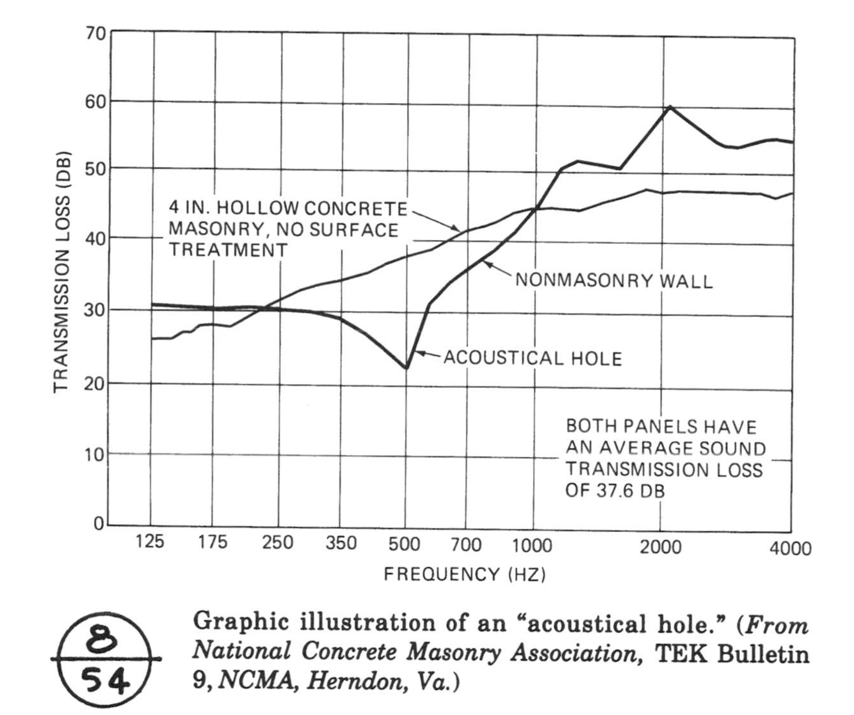

8/54 **Graphic illustration of an "acoustical hole."** (*From National Concrete Masonry Association,* **TEK Bulletin 9,** *NCMA, Herndon, Va.*)

classified by this averaging method. However, lightweight partitions often have "acoustical holes" at critical frequencies (*see Fig. 8-54*). STL averages did not identify these deficiencies, and did not accurately translate acoustical test results into useful design data. *Sound transmission class* (STC) ratings were developed to describe acoustical characteristics more accurately. STC ratings represent the overall ability of an assembly to insulate against airborne noise. They have proven more reliable in classifying the performance of both heavy- and lightweight materials over a wide range of frequencies. The higher the STC rating a wall has, the better the wall performs as a sound barrier.

For homogeneous walls, resistance to sound transmission increases with unit weight. When surfaces are impervious, sound is transmitted only through diaphragm action. The greater the inertia or resistance to vibration, the greater the ability to prevent sound transfer. The initial doubling of weight produces the greatest increase in transmission loss.

Porosity, as measured by air permeability, significantly reduces transmission loss through a wall. STC values vary inversely with porosity. Unpainted, open-textured CMU, for instance, will have lower STC values than would be expected on the basis of unit weight alone. Porosity can be reduced, and STC values increased, by sealing the wall surface. The STC value is increased by about 8% with one layer of gypsum board, 10% with two coats of paint or plaster, and 15% with two layers of gypsum board (*see Fig. 8-55*). Sealing both sides of a wall has little more effect than sealing only one side. A sealed surface not only decreases sound transmission, it also reduces sound absorption, which may not be desirable. As a general rule, leave porous surfaces unsealed in noisy areas such as stairwells or corridors, and seal them in living spaces.

Cavity walls have greater resistance to sound transmission than solid walls of equal weight. Having two wythes separated by an air space interrupts the diaphragm action and improves sound loss. Up to about 24 in.,

STC	Wall Description
	Unpainted walls
39	4 in. structural clay tile
45	8 in. structural clay tile
45	4 in. face brick
50	8 in. composite face brick and structural clay tile
50	10 in. face brick cavity wall with 2-in. air space
51	6 in. brick wall
52	8 in. solid brick wall, double wythe
59	12 in. solid brick wall
59	10 in. reinforced brick wall
49	8 in. hollow lightweight CMU
48	8 in. hollow lightweight CMU, reinforced and fully grouted
51	8 in. composite brick and hollow lightweight CMU
52	8 in. normal-weight CMU
	Painted walls, two coats latex both sides
43	4 in. hollow lightweight CMU
44	4 in. hollow normal weight CMU
48	6 in. hollow normal weight CMU
55	8 in. lightweight CMU, reinforced and fully grouted
	Plastered walls
50	4 in. brick, $\frac{1}{2}$-in. plaster one side
53	8 in. composite brick and lightweight CMU, $\frac{1}{2}$-in. plaster both sides
56	8 in. grouted, reinforced, lightweight CMU, $\frac{1}{2}$-in. plaster both sides
	Walls with gypsum board on furring strips
53	8 in. solid brick, $\frac{1}{2}$-in. gypsum board on furring strips one side
47	4 in. hollow lightweight CMU, $\frac{1}{2}$-in. gypsum board both sides
48	4 in. normal weight CMU, $\frac{1}{2}$-in. gypsum board both sides
56	8 in. composite brick and lightweight CMU, $\frac{1}{2}$-in. gypsum board both sides
56	8 in. hollow lightweight CMU, $\frac{1}{2}$-in. gypsum board both sides
49	6 in. brick, $\frac{3}{8}$-in. gypsum board over 1-in. styrofoam insulation one side
59	10 in. cavity wall brick and lightweight CMU, $\frac{1}{2}$-in. gypsum board both sides
60	8 in. grouted, reinforced lightweight CMU, $\frac{1}{2}$-in. gypsum board both sides

STC ratings of masonry walls.

the wider the air space, the more sound-efficient the wall. Cavity walls are very effective where a high transmission loss on the order of 70 to 80 dB is required. If the wythes are only an inch or so apart, the transmission loss is less because of the coupling effect of the tightly enclosed air. For maximum benefit, the walls should be further isolated from one another.

8.8.5 Code Requirements

Some building codes incorporate standards for sound transmission characteristics in buildings of residential occupancy. The standards generally specify minimum STC ratings for party wall and floor-ceiling separations between dwelling units.

Location of partition	Low background noise		High background noise	
	Bedroom adjacent to partition	Other rooms adjacent to partition	Bedroom adjacent to partition	Other rooms adjacent to partition
Living unit to living unit	50	45	45	40
Living unit to corridor	45	40	40	40
Living unit to public space (average noise)	50	50	45	45
Living unit to public space and service areas (high noise)	55	55	50	50
Bedrooms to other rooms within same living unit	45	—	40	—

FHA requirements for STC limitations. (*From National Concrete Masonry Association,* **TEK Bulletin 39,** *NCMA, Herndon, Va.*)

Party walls generally require an STC of 45 to 50. FHA minimum standards for multi-family housing are shown in *Fig. 8-56.* The required STC values range from a low of 40 to a high of 55. Requirements are lower in the city than in rural areas because higher background noise masks obtrusive sound, effectively raising the threshold of audibility.

8.9 MOVEMENT CHARACTERISTICS

All building materials expand and contract to some degree with changes in temperature. Others may also expand and contract with variations in moisture content. The thermal movement characteristics of most materials are known, and a standard coefficient can be used to calculate the expected expansion or contraction of a material for a given set of conditions. Masonry materials are relatively stable in thermal movement when compared to metals and plastics.

In addition to thermal movement, however, most masonry materials also experience moisture-related movement. Some shrinking and swelling occurs alternately through normal wetting and drying cycles, but more important are the *permanent moisture expansion* of clay masonry and the *permanent moisture shrinkage* of concrete masonry. Clay masonry begins to re-absorb moisture from the atmosphere as soon as the drying and firing process is complete, and as the moisture content increases, the units expand permanently. Concrete masonry products are moist-cured to hydrate the portland cement in the mix. Once the curing is complete, residual moisture evaporates, causing the units to shrink permanently.

The cumulative effect of reversible thermal movement and irreversible moisture movement must be accommodated in construction through the installation of expansion joints in clay masonry and control joints in concrete masonry. When clay and concrete masonry are combined, or when masonry is combined with or attached to other materials, allowance must also be made for the differential movement of the various components. Expansion, contraction, differential movement, and flexible anchorage are discussed in detail in Chapter 9.

9

MOVEMENT AND MOISTURE CONTROL

Masonry walls are relatively brittle and are characterized by thousands of linear feet of joints along which cracks can potentially open. Thermal and moisture movements and dissimilar movements between adjacent materials should always be considered, and components selected and detailed accordingly. Concrete products shrink, clay products expand, and metals expand and contract reversibly. Such movement is accommodated through flexible anchorage and the installation of control joints in concrete masonry and expansion joints in clay masonry. Coefficients of thermal expansion and moisture expansion can be used to estimate the expected movement of various materials, and movement joints sized and located accordingly. If details do not sufficiently accommodate wall movement, excessive moisture can penetrate through the resulting cracks.

9.1 MOVEMENT CHARACTERISTICS

One of the principal causes of cracking in masonry walls is differential movement. All materials expand and contract with temperature changes, but at very different rates. All materials change dimension due to stress, and some develop permanent deformations when subject to sustained loads. Clay masonry expands irreversibly with the absorption of atmospheric moisture, and concrete masonry shrinks irreversibly with loss of residual moisture from the manufacturing process. Masonry walls are much stronger than in the past because of high-strength units and portland cement mortars, but strength has come at the expense of flexibility. Using masonry as we do today with ductile steel and concrete skeleton frames requires careful consideration of the movement characteristics of each material. Cracking in the masonry can result from restraining the natural movement of the materials themselves, or from failure to allow for differential movement of adjoining or connected materials.

9.1.1 Temperature Movement

The thermal movement characteristics of most building materials are known, and a standard coefficient can be used to calculate expected movement for a given set of conditions. The table in *Fig. 9-1* shows that the potential for expansion varies from the relatively stable characteristics of clay masonry to the highly active movements of metals and plastics. Coefficients for masonry units vary with the raw material or type of aggregate used. The stress developed in a restrained element due to temperature change is equal to the *modulus of elasticity times the coefficient of thermal expansion times the mean wall temperature change.* For instance, the tensile stress in a fully restrained block wall with a thermal expansion coefficient of 4.5×10^{-6} and modulus of elasticity of 1.8×10^{6} for a temperature change of 100°F would be

$$0.0000045 \times 1{,}800{,}000 \times 100°\text{F} = 810 \text{ psi}$$

Surface temperatures must be used to calculate thermal movement because they represent greater extremes than ambient temperatures, and therefore more accurately predict actual movement. Vertical wall surface temperatures in winter are usually within a few degrees of ambient (depending on the amount of insulation present in the wall), and may safely be assumed to equal the ASHRAE winter design dry bulb temperature. Summer surface temperatures, however, are affected by solar radiation, thermal mass, and the temperature gradient through the thickness of the material. One equation used to calculate summer surface temperature taking these factors into consideration is

$$T_s = T_a + XS \tag{9.1}$$

where T_s = extreme summer surface temperature of wall, °F
T_a = extreme summer air temperature, °F (dry bulb)
X = constant for heat capacity of material (*see Fig. 9-2*)
S = solar absorption coefficient of material (*see Fig. 9-3*).

The total wall surface temperature differential (ΔT) is found by subtracting winter surface temperature from summer surface temperature:

$$\Delta T = T_s - T_w \tag{9.2}$$

where ΔT = total surface temperature differential, °F
T_s = extreme summer surface temperature of wall, °F [from equation (*9.1*)]
T_w = extreme winter surface temperature, °F (dry bulb).

The formula given in ASTM C1193, *Standard Guide for Use of Joint Sealants,* can then be used to calculate thermal movement:

$$M_t = (C_t)(\Delta T)(L) \tag{9.3}$$

where M_t = thermal movement, in.
C_t = thermal movement coefficient (*see Fig. 9-1*)
ΔT = total surface temperature differential, °F [from equation (*9.2*)]

For example, thermal movement for a clay brick panel with a thermal expansion coefficient of 3.6×10^{-6} (*from Fig. 9-1*), an estimated surface temperature differential of 145°F, and a panel length or joint spacing of 20 ft would be calculated as

$$M_t = (0.0000036)(145°\text{F})(240 \text{ in.}) = 0.2552 \text{ in. (or about } \tfrac{1}{8} \text{ in.)}$$

Material	C_t (in./in./°F) (multiply by 10^{-6})
Clay masonry	
Clay or shale brick	3.6
Fire clay brick or tile	2.5
Clay or shale tile	3.3
Concrete masonry	
Normal weight	
Sand and gravel aggregate	5.2
Crushed stone aggregate	5.2
Medium weight	
Air-cooled slag	4.6
Lightweight	
Coal cinders	3.1
Expanded slag	4.6
Expanded shale	4.3
Pumice	4.1
Stone	
Granite	2.8–6.1
Limestone	2.2–6.7
Marble	3.7–12.3
Sandstone	4.4–6.7
Slate	4.4–5.6
Travertine	3.3–5.6
Concrete	
Calcareous aggregate	5.0
Siliceous aggregate	6.0
Quartzite aggregate	7.0
Metals	
Aluminum	13.2
Steel, carbon	6.7
Steel, stainless	
301 alloy	9.1
302 alloy	9.6
304 alloy	9.6
316 alloy	8.9
410 alloy	6.1
430 alloy	5.8
Brass, 230 alloy	10.4
Bronze	10.0–11.6
Copper	9.4
Glass	4.9

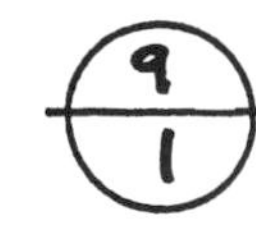

Thermal expansion coefficients of various building materials.

Material	C_t *(in./in./°F)* *(multiply by 10^{-6})*
Plaster, gypsum	
Sand aggregate	6.5–6.75
Perlite aggregate	7.3–7.35
Vermiculite aggregate	8.4–8.6
Plastic	
Acrylic sheet	41.0
Polycarbonate sheet	38.0
Wood, parallel to fiber	
Fir	2.1
Oak	2.7
Pine	3.0
Wood, perpendicular to fiber	
Fir	32.0
Oak	30.0
Pine	19.0

(*Continued*)

CONSTANT FOR HEAT CAPACITY OF MATERIAL (X)	
$X = 100$	Low-heat-capacity materials
or	
$X = 130$	Solar radiation reflected on low-heat-capacity materials
$X = 75$	High-heat-capacity materials
or	
$X = 100$	Solar radiation reflected on high-heat-capacity materials

Notes:

1. Materials such as exterior insulation and finish systems and well-insulated metal panel curtainwalls have low thermal storage capacity and therefore low heat capacity. Materials such as precast panels and masonry walls, on the other hand, have high thermal storage capacity and therefore high heat capacity.

2. If the wall surface receives reflected as well as direct radiation, use the larger constant. Reflected radiation can come from adjacent wall surfaces, roofs, and paving.

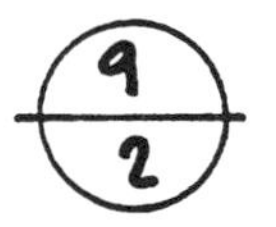

Thermal mass constant for calculating temperature movements. (*From* Building Sealants: Materials, Properties and Performance, *edited by Thomas F. O'Connor, STP 1069, ASTM, Philadelphia, Pa., 1990. Reprinted with permission.*)

Material	Coefficient (S)
Aluminum, clear finish	0.60
Aluminum, paint	0.40
Aluminum, dull	0.40–0.65
Black, nonmetallic, asphalt, slate	0.85–0.98
Mineral board, natural color	0.75
Mineral board, white	0.61
Brick, light buff (yellow)	0.50–0.70
Brick, red	0.65–0.85
Brick, white	0.25–0.50
Concrete, natural color	0.65
Copper, tarnished	0.80
Copper, patina	0.65
Galvanized steel, natural color	0.40–0.65
Galvanized steel, white	0.26
Glass, clear ¼ in.	0.15
Galss, tinted ¼ in.	0.48–0.53
Glass, reflective ¼ in.	0.60–0.83
Marble, white	0.58
Paint	
Dark red, brown, or green	0.65–0.85
Black	0.85–0.98
White	0.23–0.49
Surface color, black	0.95
Surface color, dark gray	0.80
Surface color, light gray	0.65
Surface color, white	0.45
Tinned surface	0.05
Wood, smooth	0.78

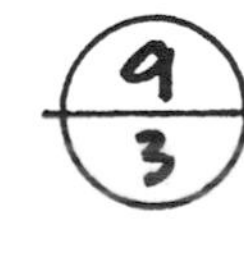

Solar absorption coefficient for calculating thermal movement. (*From* Building Seals, Sealants, Glazing and Waterproofing, *fifth volume, edited by Michael A. Lacasse, STP 1271, ASTM, Philadelphia, 1996.*)

9.1.2 Moisture Movement

Many building materials expand as moisture content increases and then contract when it decreases. In some instances, this moisture movement is almost fully reversible, but in others, the change in dimension is permanent and irreversible. Permanent moisture shrinkage affects concrete masonry products, and permanent moisture expansion affects clay masonry products.

ASTM limits the moisture content of Type I concrete masonry units depending on the unit's linear shrinkage potential and the annual average

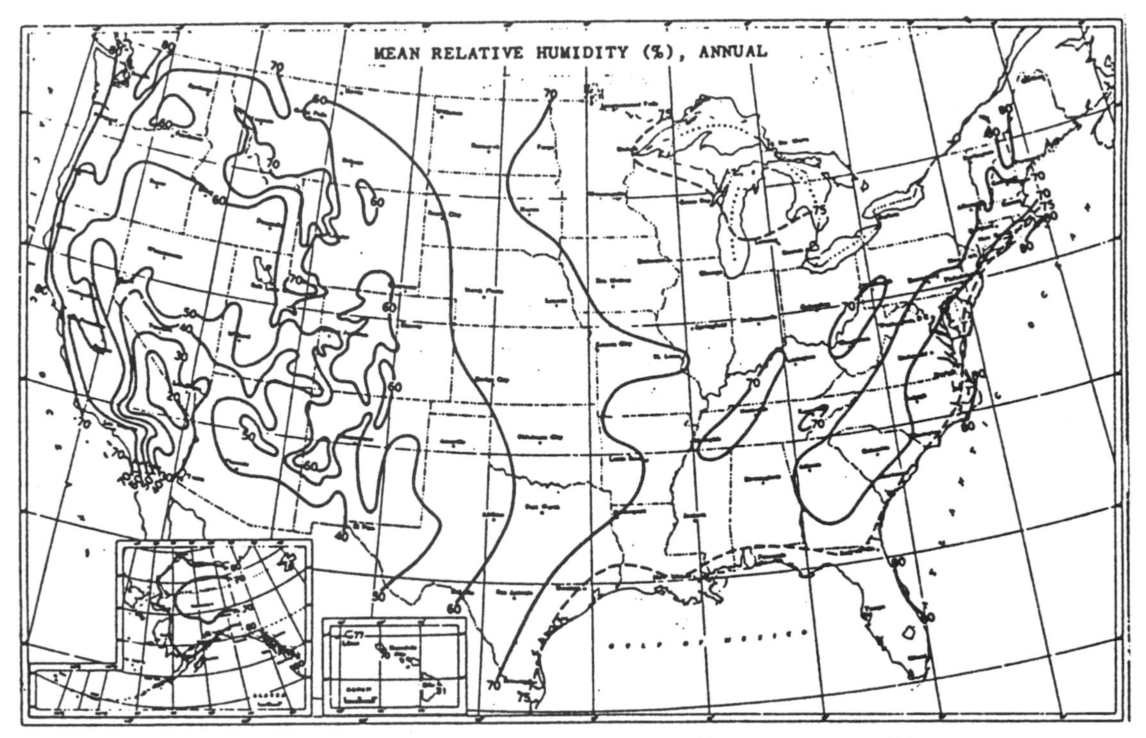

NOTE—Based on 1:30 AM and PM and 7:30 AM and PM Eastern Standard Time, observations for 20 years or more through 1964.

Linear shrinkage, (%)	Moisture content, maximum % of total absorption (average of 3 units)		
	Humidity conditions at job site or point of use		
	Humid*	Interme-diate†	Arid‡
0.03 or less	45	40	35
From 0.03 to 0.045	40	35	30
0.045 to 0.065, max	35	30	25

*Average annual relative humidity above 75%.

†Average annual relative humidity 50 to 75%.

‡Average annual relative humidity less than 50%.

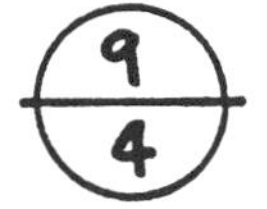

Linear shrinkage values of Type I concrete masonry units, from ASTM C90. ***(Copyright, American Society for Testing and Materials, 1916 Race Street, Philadelphia, Pa. 19103. Reprinted with permission.)***

relative humidity at the project site. The table in *Fig. 9-4* lists a range of acceptable linear shrinkage for Type I units from ASTM C90, *Standard Specification for Loadbearing Concrete Masonry Units,* and *Fig. 9-5* lists moisture movement coefficients for typical masonry products. The object is to limit residual shrinkage in the wall. Units with higher shrinkage potential must have a lower moisture content at the time of delivery than that

Material	*Movement (C_m; %)*
Concrete, gravel aggregate	−0.03 to −0.08
Concrete, limestone aggregate	−0.03 to −0.04
Concrete, lightweight aggregate	−0.03 to −0.09
Concrete block, dense aggregate	−0.02 to −0.06
Concrete block, lightweight aggregate	−0.02 to −0.06
Face brick, clay	+0.02 to +0.07

Notes:

1. If specific data for a particular unit are not available, use the maximum value given in the table.
2. Shrinkage is indicated by (−), expansion by (+).

Moisture movement coefficients. (*From* **Building Sealants: Materials, Properties and Performance,** ***edited by Thomas F. O'Connor, STP 1069, ASTM, Philadelphia, Pa., 1990. Reprinted with permission.***)

of low shrinkage units. Type I units ensure minimum moisture content and minimum moisture shrinkage, and are recommended to control wall cracking in concrete masonry. After units are delivered, it is imperative to protect on site stockpiles from rain or other wetting.

Type II concrete masonry units have greater shrinkage potential because moisture content is not limited. Walls built with Type II units may therefore require closer control joint spacing or additional joint reinforcement to control cracking, depending on the relative humidity of the project site and the service environment.

Steel reinforcement increases concrete masonry's resistance to the tensile stress of shrinkage. The most common method of shrinkage crack control is the use of horizontal joint reinforcement which distributes the stress more evenly through the wall to minimize cracking. Control joints localize cracking so that elastomeric sealants can be installed in the joint to prevent moisture penetration. Masonry that is reinforced to resist structural loads is also more resistant to shrinkage cracking.

The firing process removes virtually all moisture from clay masonry products. After they leave the kiln, reabsorption of moisture causes irreversible expansion. When combined with other types of building movement and with the conflicting shrinkage of other materials, this expansion can be significant, and must be anticipated in the design. Moisture expansion coefficients for clay face brick generally range from 0.02% to 0.07% (*see Fig. 9-5*).

9.1.3 Elastic Deformation

Shortening of axially loaded masonry walls or columns is seldom critical. More often, problems arise from elastic deformation of horizontal masonry elements such as beams and lintels. Standards limit deflection to $L/600$ or 0.3 in. under combined live and dead loads. In veneer construction, deformation of the structural frame to which the masonry is attached can also cause distress if loads are inadvertently transferred to the veneer.

9.1.4 Plastic Flow

When concrete or steel is continuously stressed, there is a gradual yielding of the material, resulting in permanent deformations equal to or greater than elastic deformation. Under sustained stress this plastic flow, or creep, continues for years, but the rate decreases with time and eventually becomes so small as to be negligible. About one-fourth of the ultimate creep takes place within the first month or so, and one-half of the ultimate creep within the first year. In clay masonry construction, the units themselves are not subject to plastic flow, but the mortar is. In concrete block construction, long-term deformations in the mortar and grout are relatively high compared to that of the unit. Joint reinforcement restrains the mortar and grout so that overall deformations of the wall are similar in magnitude to those for cast-in-place concrete. Plastic flow of the mortar in brick walls helps prevent joint separations by compensating to some degree for the moisture expansion of the units.

The creep deflection of a concrete or steel structural frame to which masonry is rigidly anchored is the most potentially damaging. Steel shelf angles and concrete ledges that sag over a period of time can exert tremendous force on the masonry below. Without soft joints below such members, the masonry can buckle or spall under the unintentional load.

9.1.5 Effects of Differential Movement

Differential movement is the primary cause of cracking in masonry walls. In a cavity wall, for example, the moisture exposure and temperature variation in the outer wythe is much greater than that of the inner wythe, especially if the wall contains insulation. Cracks can form at the external corners of brick walls because of the greater thermal and moisture expansion of the outer wythe which results. Long walls constructed without expansion joints also develop shearing stresses in areas of minimum cross section, so diagonal cracks often occur between window and door openings, usually extending from the head or sill at the jamb of the opening. When masonry walls are built on concrete foundations that extend above grade, thermal and moisture expansion of the masonry can work against the drying shrinkage of the concrete, causing extension of the masonry wall beyond the corner of the foundation or cracking of the foundation (*Fig. 9-6*). The concrete contracts with moisture loss and lower temperatures, the brick expands with moisture absorption, and cracks form near the corners. Flashing at the base of the wall serves as a bond break between the masonry and the foundation and allows independent movement without such damage.

Brick parapet walls can be particularly troublesome because, with two surfaces exposed, they are subject to temperature and moisture extremes. Differential expansion from the building wall below can cause parapets to bow, to crack horizontally at the roof line, and to overhang corners (*Fig. 9-7*). Through-wall flashing, although necessary, creates a plane of weakness at the roof line that may amplify the visual problem, but allows the differential movement to occur without physical damage to the masonry. If parapets are included in the building design, as many as twice the number of expansion joints or control joints are needed as in the wall below. The extra joints above the roof line absorb the larger movements in the parapet without excessive sliding at the flashing plane, and without

Moisture expansion of brick and drying shrinkage of concrete can cause cracking at building corners.

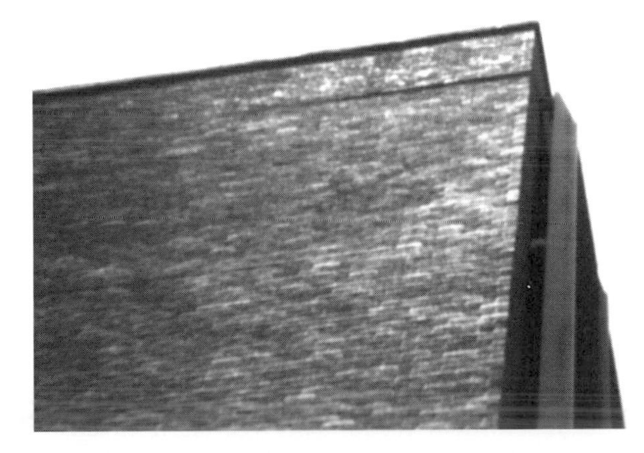

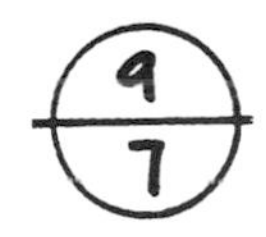

Parapets expand and contract more than the building walls below.

bowing or cracking in the parapet itself. Adding steel reinforcement also helps to counteract the tensile forces created and prevent excessive movement. The same material should be used for both wythes of multi-wythe parapets so that the back and front of the wall expand and contract at the same rate.

Floor and roof slabs poured directly on masonry bearing walls can curl from shrinkage, deflection, and plastic flow of the concrete. If the slab warps, it can rupture the masonry at the building corners and cause horizontal cracks just below the slabs. To permit flexibility, a horizontal slip plane should be installed between the slab and wall, running 12 to 15 ft back from the corners and terminating at a movement joint. This will relieve the strain at the points where movement is greatest.

Vertical shortening of steel or concrete structural frames from shrinkage or contraction can transfer excessive stress to masonry cladding. Failures are characterized by bowing, by horizontal cracks at shelf angles, by vertical cracks near corners, and by spalling of masonry units at window heads, shelf angles, and other points where stress is concentrated. Horizontal soft joints must be provided to alleviate these stresses and allow the frame to shorten without damage to the masonry (see Chapter 10 for details). Where structural steel columns are protected by masonry, the greater temperature movement of the column can be inadvertently transmitted to the masonry and cause cracking. To prevent this problem, a bond break material should isolate the mortar from the steel, and flexible anchors should be used to accommodate the differential movement.

To avoid problems of cracking and subsequent moisture leakage, differential movement between various types of masonry, and between masonry and other materials must be accommodated by flexible anchorage, and by vertical and horizontal expansion joints.

9.2 FLEXIBLE ANCHORAGE

When masonry walls are connected to steel or concrete frame buildings, differential movement must be accommodated in the anchorage of one material to another. Even if the exterior masonry veneer carries its own weight to the foundation without shelf angles or ledges, the columns or floors provide the lateral support which is required by code. Flexible connections should allow relative vertical movement without inducing stresses which could cause damage (i.e., they should resist the lateral tension and compression of wind loads, but not in-plane shear movements). Various types of flexible anchors were discussed in Chapter 7, and *Figs. 9-8 and 9-9* show ways in which they are used. Seismic anchors for the attachment of veneers are available from some manufacturers. These flexible anchors are designed to meet the requirements for Seismic Zones 3 and 4 of the *Uniform Building Code* and Seismic Performance Categories D and E of the Masonry Standards Joint Committee (MSJC) *Building Code Requirements for Masonry Structures ACI 530/ASCE 6/TMS 402* (see Chapter 12).

In loadbearing masonry construction the brick or block walls support concrete floor slabs or steel joist and metal deck floors. The methods of anchorage will vary for different conditions (see Chapter 12). With concrete block and concrete slabs, there is less concern for differential movement because of the similarity of the material characteristics. Connections may be either rigid or flexible, depending on the particular design situation. In brick masonry, however, it is more common to provide a bond break or slippage plane at the point where a concrete slab rests on the wall. Roofing felt or flashing is commonly used for this purpose, and allows each element to move independently while still providing the necessary support. The bond break may be detailed for both conditions where wall-to-slab anchorage is or is not required (*see Fig. 9-10*). Where masonry walls rest on a concrete foundation, mechanical anchorage between the two elements often is not necessary because the weight of the wall and its frictional resistance to sliding are adequate for stability. In shear wall

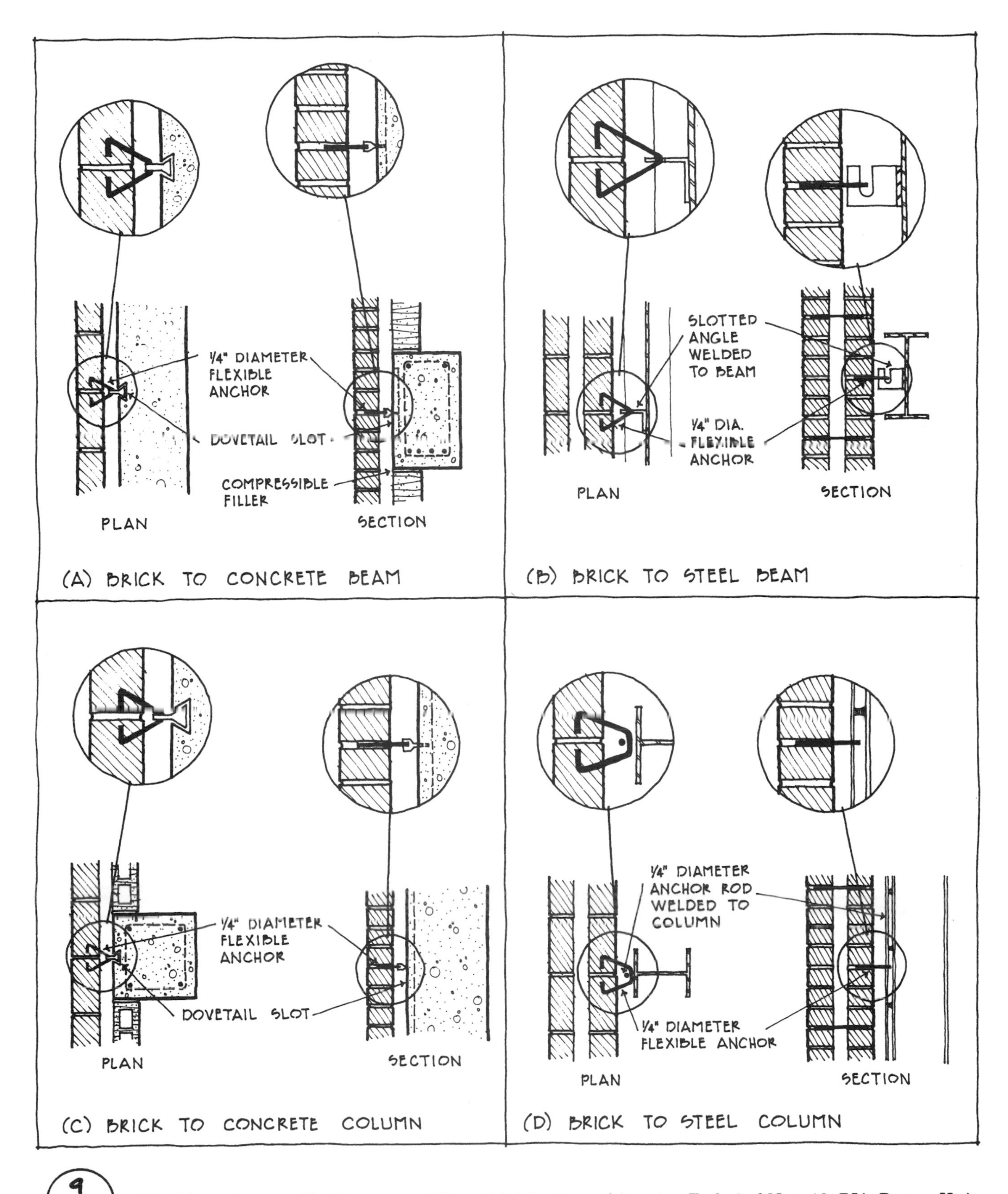

9/8 Flexible anchorage of brick masonry. (*From Brick Institute of America,* Technical Note 18, *BIA, Reston, Va.*)

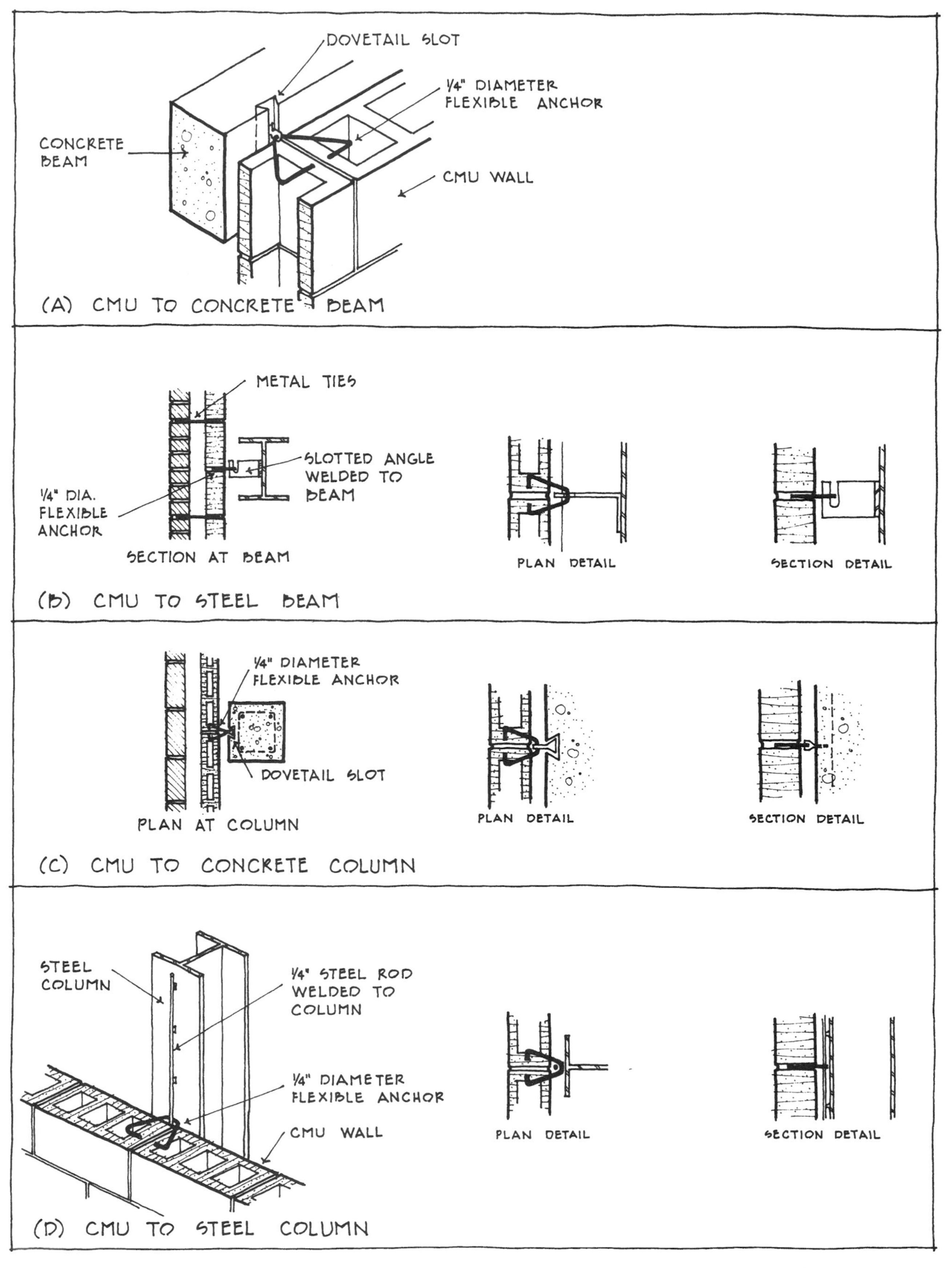

Flexible anchorage of concrete masonry.

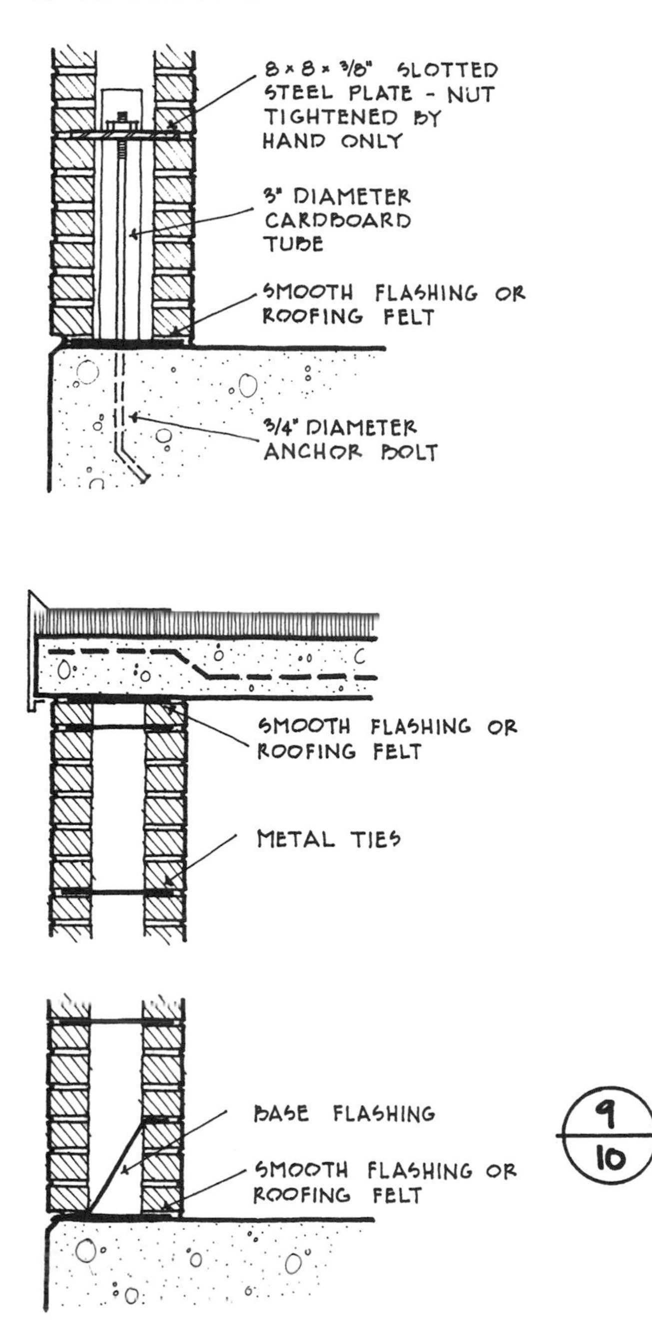

Bond breaks in wall-to-floor and wall-to-roof connections. ***(From Harry C. Plummer,*** **Brick and Tile Engineering,** ***Brick Institute of America, Reston, Va., 1962.)***

design where floor-wall connections must transfer loads through diaphragm action, anchorage must be designed as part of the engineering analysis (see Chapter 12).

9.3 MOVEMENT JOINTS

In addition to the flexible anchorage of backing and facing materials, control joints and expansion joints are used to alleviate the potential stresses caused by differential movement between materials, and by thermal and moisture movement in the masonry. *The terms control joint and expansion joint are not interchangeable.* The two types of joints are different in both function and configuration (*see Fig. 9-11*).

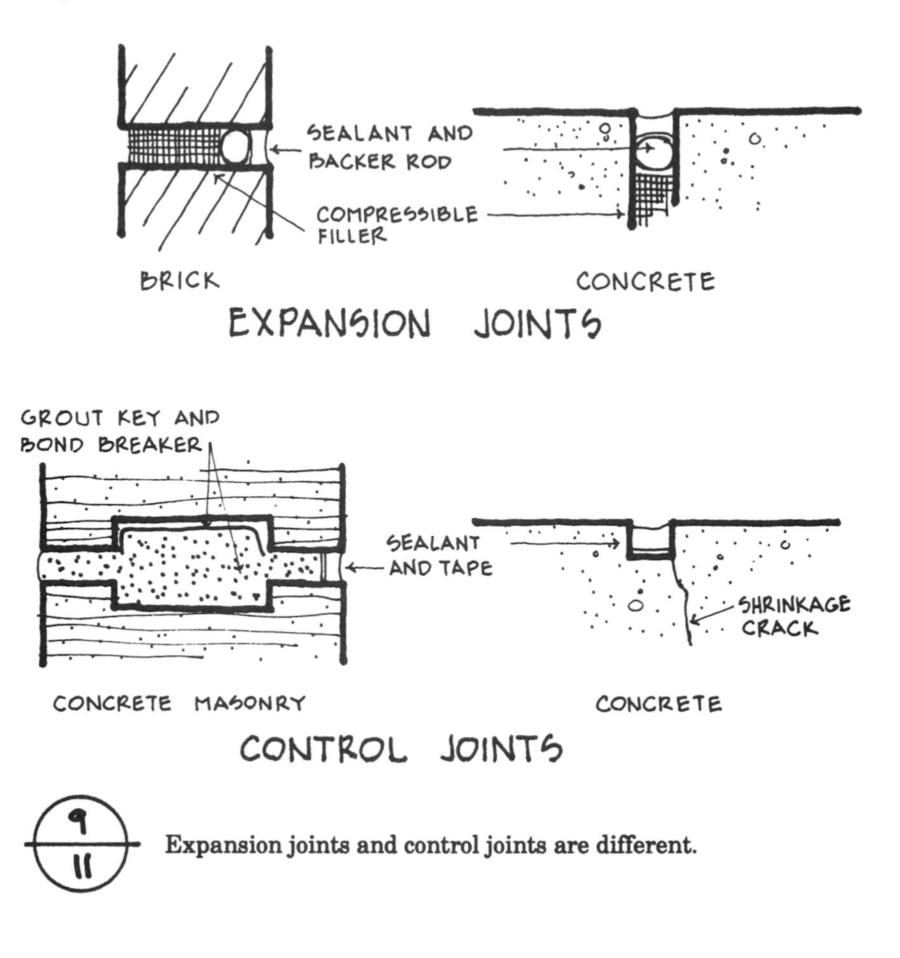

Expansion joints and control joints are different.

9.3.1 Joint Design

Control joints are continuous, weakened joints designed to accommodate the permanent shrinkage of portland cement–based products such as concrete masonry. When stress development is sufficient to cause cracks, the cracking will occur at these weakened joints rather than at random locations. Although Type I moisture-controlled units and horizontal joint reinforcement can be used to limit shrinkage cracking, strategically located control joints must also be used to eliminate random cracks and prevent the resulting moisture penetration. Cracking is not as critical in fully reinforced construction since the reinforcing steel absorbs the tensile stress.

Control joints must also provide lateral stability between adjacent wall sections. *Figure 9-12* shows several common types of joints, all of which provide a shear key for this purpose. Control joints must also be sealed against moisture leakage. The joints are first laid up in mortar just as any other vertical joint would be. After the mortar has stiffened slightly, the joints are raked out to a depth which will allow placement of a backer rod or bond-breaker tape, and a sealant joint of the proper depth. Concrete masonry shrinkage always exceeds expansion because of the initial moisture loss experienced after manufacture. So even though control joints contain hardened mortar, they can accommodate reversible thermal expansion and contraction which occur after the initial curing shrinkage.

In masonry, an expansion joint is a continuous open joint or plane designed to accommodate the permanent expansion of brick and other clay

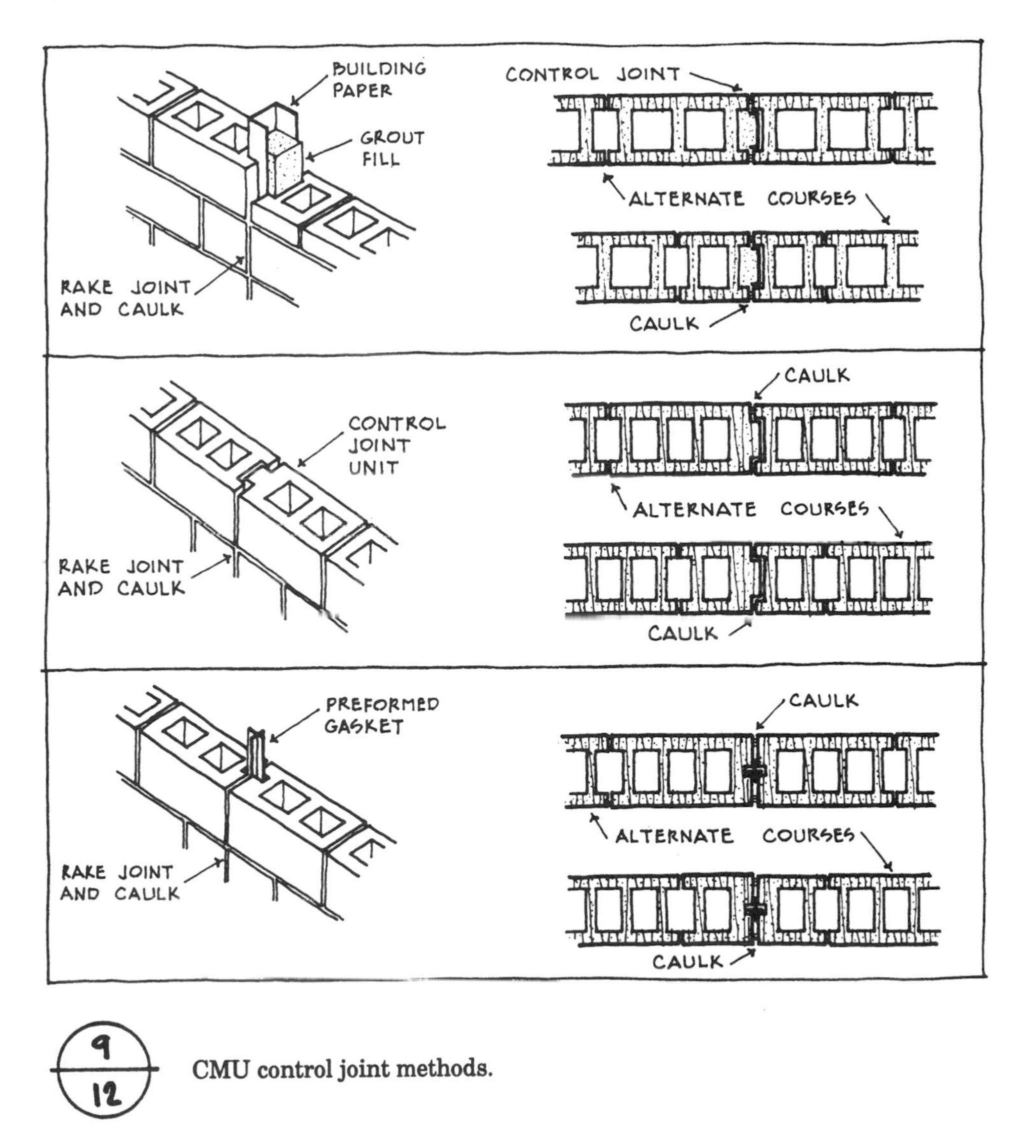

9/12 **CMU control joint methods.**

units. Brick moisture expansion always exceeds reversible thermal expansion and contraction, so the joint cannot contain mortar or other hard materials. *Figure 9-13* shows several methods of constructing vertical expansion joints. Compressible fillers may be used to keep mortar out of the joints during construction, because even small mortar bridges can cause localized spalling of the unit faces when the joint tries to close (*see Fig. 9-14*). Filler materials should be at least as compressible as the joint sealant which will be used, and the compressibility of the sealant must be considered in calculating joint width.

The required width for control joints and expansion joints can be determined by adding the widths required for thermal movement, moisture movement, and construction tolerances. If the calculated width based on an assumed joint spacing is too narrow for proper sealant function, or too wide for aesthetic reasons, the spacing can be increased or decreased and the width recalculated.

Required joint width must take into account the movement capability of the sealant itself in both extension and compression. An elastomeric

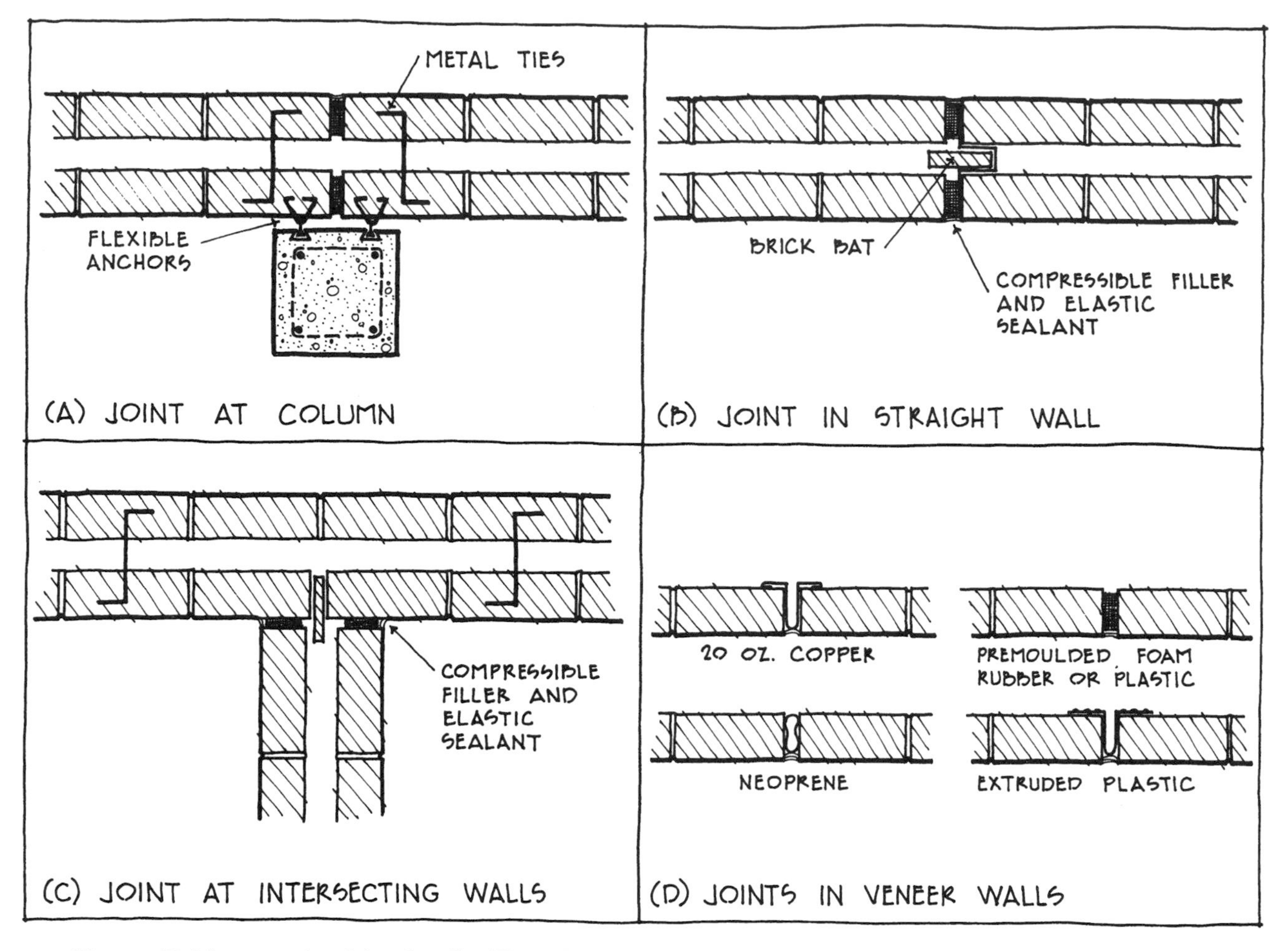

Brick expansion joint details. *(From Harry C. Plummer,* **Brick and Tile Engineering,** *Brick Institute of America, Reston, Va., 1962.)*

sealant rated ±25% can tolerate a maximum movement of +25% of the joint width when extended, and −25% of the joint width when compressed. The joint must therefore be 4 times the expected movement (100/25 = 4). A more elastic sealant rated ±50% requires a joint width twice the expected movement (100/50 = 2). A sealant reported as +100/ −50% is governed by its compressibility, so the joint still must be twice the calculated movement to allow room for the compressed thickness of the sealant itself. ASTM C1193 gives a basic formula for calculating joint width [$J_t = (100/S_m)\ (M_t)$], but to allow for imprecisions in determining surface temperatures, imperfect workmanship, and other unknowns, some researchers recommend using sealants at only a percentage of their rated movement capacity. The amount of reduction should depend on the particular circumstances of a joint design and the desired factor of safety. With this additional limitation, using the sealant at only 80% of its capacity for this example, the formula would then be

$$J_t = \frac{100}{0.8S_m}(M_t) \qquad (9.4)$$

Mortar bridges across an expansion joint can cause localized spalling when the joint tries to close.

where J_t = minimum joint width for thermal movement only, in.
S_m = sealant movement capacity
M_t = calculated thermal movement, in. [from equation (*9.3*)]

To calculate the joint width required for moisture movement in masonry, the coefficients in *Fig. 9-5* must be used in the formula

$$J_m = \frac{C_m}{100}(L) \tag{9.5}$$

where J_m = minimum joint width for moisture movement only, in.
C_m = moisture movement coefficient (*see Fig. 9-5*)
L – panel length or joint spacing, in.

Material fabrication and erection tolerances must also be considered in determining the required joint width. When unanticipated construction tolerances result in increased or decreased joint width, sealant performance is seriously affected. Narrow joints especially are a frequent cause of sealant joint failure. Although normal tolerances for different materials vary considerably (see Chapter 15), a reasonable estimate must be made of the combined or net effect on the joint. Once that allowance is determined, it must be added to the previously calculated joint width requirements in the formula

$$J = J_t + J_m + J_c \tag{9.6}$$

where J = total calculated joint width, in.
J_t = minimum joint width for thermal movement, in.
J_m = minimum joint width for moisture movement, in.
J_c = minimum joint width for construction tolerances

If the calculated joint is too wide for aesthetic considerations, the assumed spacing or panel length can be decreased and the width recalculated.

In order for the sealant to function properly, most sealant industry sources recommend that for butt joints up to ½ in wide, joint depth should be less than or equal to the width, with 2:1 a preferred ratio. Sealant depth should be constant along the length of the joint, and should never be less than ¼ in.

9.3.2 Joint Locations

The exact location of control and expansion joints will be affected by design features such as openings. The calculations for joint width and spacing apply to continuous walls with constant height and thickness. Joint locations may be adjusted, or additional joints may be required for other conditions. Openings, offsets, and intersections are the most effective locations for movement joints for both brick and concrete masonry (*see Fig. 9-15*). In brick walls, expansion joints should be located near the external corners of buildings, particularly when the masonry is resting on a concrete foundation. The shrinkage of the concrete, combined with the expansion of the brick, can cause the wall to slip beyond the edge of the foundation or to crack the concrete (*see Fig. 9-16*). The opposing push of the intersecting veneer wythes can also crack the brick itself.

Brick parapet walls experience differential movement from the walls below caused by a variation in exposures. Even in a light rain, the tops and corners of a building will always get wet, but the rest of the walls may stay dry (*see Fig. 9-17*), resulting in more wet-dry cycles at the parapet. The temperature of the building enclosure walls is also moderated by interior heat and air conditioning, so the parapet is exposed to higher and lower extremes. As a result of these differences, cracking, slippage, and separation often occur at or near the roof line. An extra expansion joint in the parapet should be located between those carried up from the building wall below. Masonry walls and parapets can be reinforced with bond beams to increase resistance to shrinkage and expansion (*see Fig. 9-18* and Chapter 10).

The joint reinforcement used to control shrinkage in concrete masonry walls affects the required location of control joints. If Type I moisture-controlled units are available, control joints may be spaced further apart than for Type II non-moisture-controlled units. If Type I units are not available, or if the units cannot be effectively protected from wetting during transport and storage, then the recommended spacing for Type II units should be used (*see Fig. 9-19*).

For both brick and concrete masonry walls, joints should always be located at points of weakness or high stress concentration such as abrupt changes in wall height; changes in wall thickness; columns and pilasters; one or both sides of windows and door openings; and coincidentally with movement joints in floors, roofs, foundations, or backing walls. In addition, joints should be located at the calculated spacing along walls or sections of walls which are not interrupted by such elements.

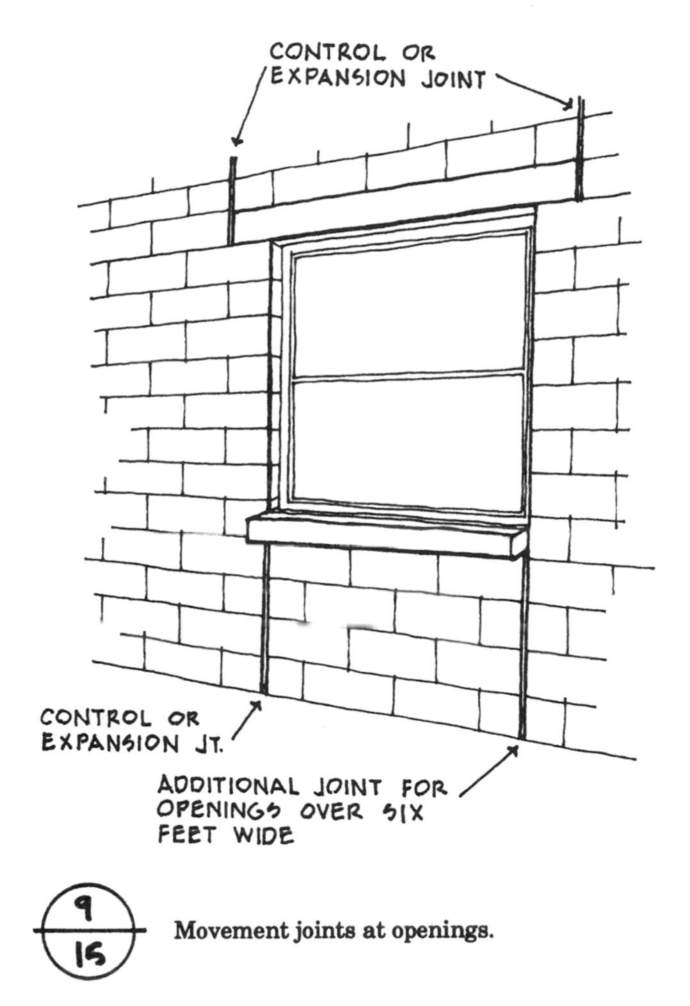

Movement joints at openings.

9.3.3 Accommodating Movement Joints in Design

Requirements for the location of movement joints in masonry are dictated by the expansion and contraction characteristics of the materials but designers can also exercise some control over joint location and the aesthetic impact of the joints themselves. The objective of movement joint placement is dividing a wall into smaller panels of masonry that can expand and contract independently of one another. The smaller the panels, the lower the cumulative stress and the less likely that cracking will occur. Wall panels that are more square than rectangular also have less stress buildup. Movement joints will be less noticeable in the appearance of a building if the exterior elevations are designed with joint locations in mind instead of placing them as an afterthought in a completed design. Just as the joint pattern in a stucco facade is part of the overall design, so too should masonry joints be a design element in masonry buildings. Joints can even be articulated with special shape units to make their visual impact stronger. Alternatively, the joints can be hidden in the shadow of a protruding pilaster while the series of pilasters articulates the panelized sections of the wall.

The location of window and door openings often governs movement joint placement because of the frequency of their occurrence. In general,

Cracking at a building corner due to brick expansion restrained by concrete shrinkage. (*Photo courtesy BIA.*)

joints should be located at one side of openings less than 6 ft wide and at both sides of openings wider than 6 ft. When the masonry above an opening is supported by a precast concrete, cast stone, or reinforced concrete masonry unit (CMU) lintel, the adjacent movement joint must be located at the ends of the lintel as shown in *Fig. 9-15*. This creates an odd-looking pattern that is not very attractive. As an alternative, movement joints can be located at the midpoint between windows. If the spacing is relatively wide (or simply as an added measure of safety), joint reinforcement can be added in the courses immediately above and below the openings to strengthen the panel (*see Fig. 9-20*). When the masonry is supported on a loose steel lintel that simply spans between the masonry on each side, special detailing can be used to avoid offsetting the joint to the end of the lintel. A piece of flashing placed under the lintel bearing area creates a slip plane so that the end of the lintel can move with the masonry over the window. With this detailing, the movement joint can then be placed adjacent to the window and run in a continuous vertical line (*see Fig. 9-21*).

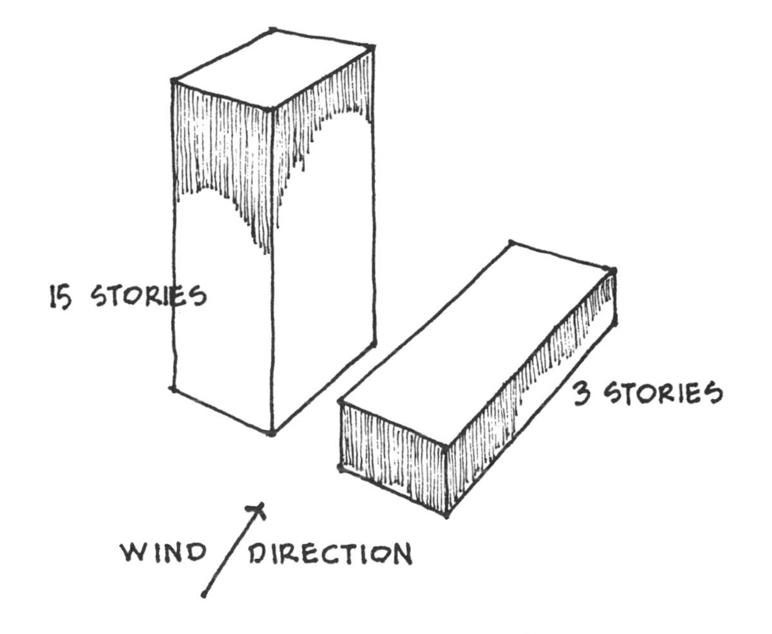

The tops and corners of a building have the greatest weather exposure. *(From R. L. Quirouette, "Rain Penetration Control,"* The Construction Specifier, *November 1994, pp. 48–56.)*

When the masonry above an opening is supported on shelf angles that are attached to the structure, a control or expansion joint can be located immediately adjacent to the opening and continue straight up the wall past the horizontal support.

Joint reinforcement can also be used to group closely spaced windows into larger panels so that the movement joints can be spread farther apart. The joints on either side of such a grouping must be sized large enough to accommodate the movement of the larger panel. In the elevation shown in *Fig. 9-22,* the two bed joints immediately above and below the groups of windows are reinforced with two-wire, truss-type joint reinforcement. Oversized movement joints can then be placed at either end of the window groupings. Because the joints are large, an offset or pilaster can be created in the wall at the joint locations to make them less noticeable. Calculating the expected movement in the masonry panels in a situation like this is very important to make sure the joints are wide enough. An extra ⅛ in. should also be added to the planned joint width to allow for construction tolerances.

Choosing the right color of sealant can also affect the appearance of movement joints. Sealant color may be selected similar to either the unit's or the mortar's color, but should be slightly darker than the unit or mortar whenever possible. Different sealant colors can blend with different bands of unit colors alternating through the height of the facade. Sand can also be rubbed into the surface of fresh sealant to remove the sheen and give it a weathered look. Project specifications should include general guidelines on the location of movement joints, but the architectural drawings should always show the location of control and expansion joints on the building elevations.

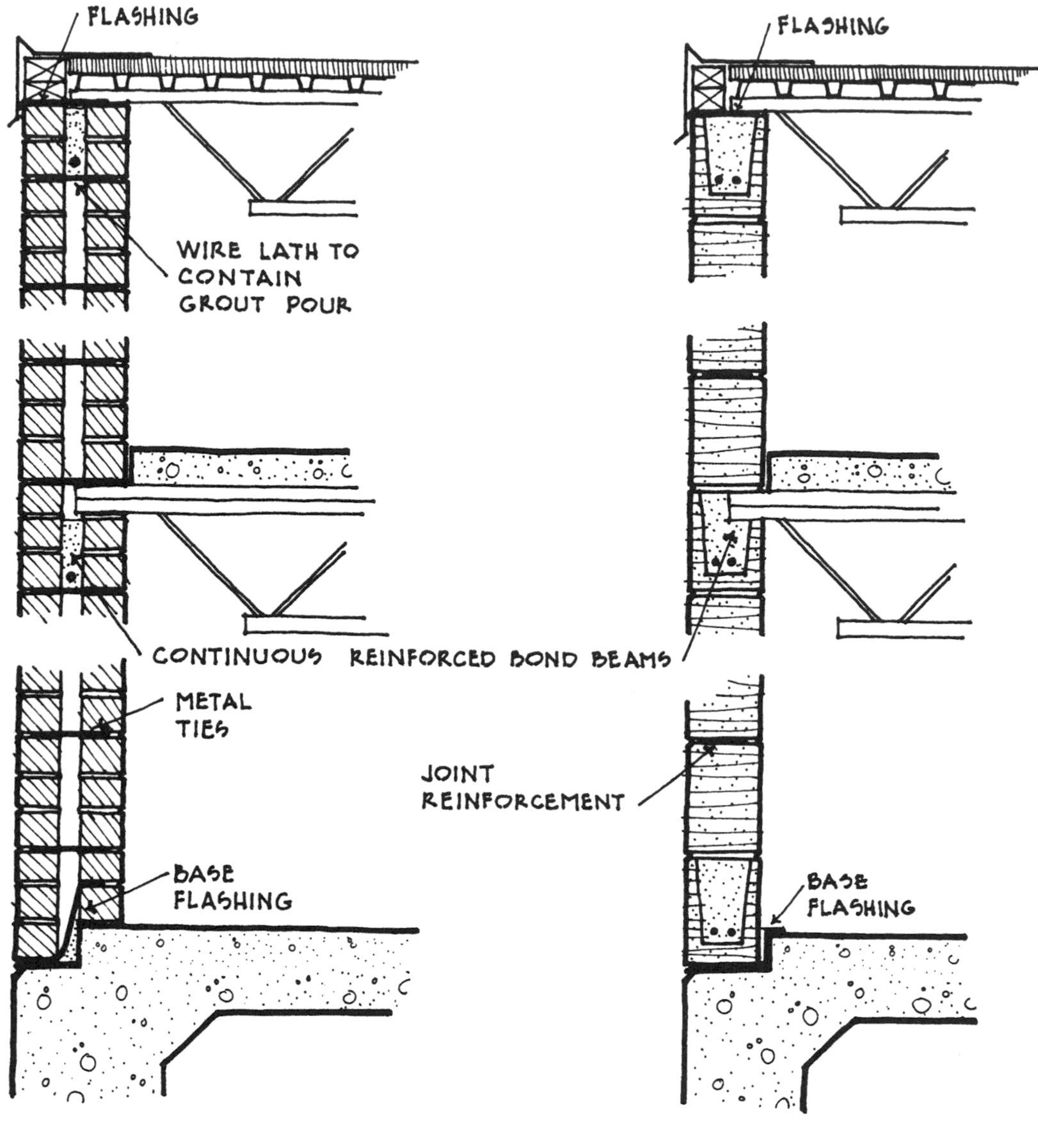

Bond beams for control of brick expansion and CMU shrinkage cracking.

9.4 MOISTURE PROTECTION

Water, which may occur as a liquid, solid, or vapor, can be a problem in any building. As wind-driven rain, it may cause decomposition or staining (*see Fig. 9-23*); as ice, sleet, snow, or hail, it is capable of physical damage; as a vapor, it may penetrate to the interior of a wall, causing decomposition or disintegration; and as surface condensation, it may be ruinous to interior finishes.

Masonry is permeable to moisture. The materials are porous, the joints are numerous, and the construction is handcrafted under diverse conditions. The key to successful performance then, is limiting the amount of moisture which enters the wall, and expediting the removal of moisture to prevent damage. The primary means of limiting moisture penetration are complete and intimate bond between units and mortar, full head and bed joints, adequate allowance for movement, and good details. The primary

Average annual relative humidity	Wall location	Vertical spacing of bed joint reinforcement, (in.)	*Maximum Recommended Control Joint Spacing, in feet*	
			ASTM C90 Type I, moisture-controlled	ASTM C90 Type II, non-moisture-controlled
Less than 50%	Exterior	None	12	6
		16	18	10
		8	24	14
	Interior	None	16.5	9
		16	24	14
		8	31.6	19
Between 50 and 75%	Exterior	None	18	12
		16	24	16
		8	30	20
	Interior	None	22.5	15
		16	30	20
		8	37.6	25
Greater than 50%	Exterior	None	24	18
		16	30	22
		8	36	26
	Interior	None	28.5	21
		16	36	26
		8	43.6	31

Control joint spacing for concrete masonry. [*From John H. Matthys, editor,* Masonry Designers' Guide Based on Building Code Requirements for Masonry Structures *(ACI 530-92/ASCE 5-92/TMS 402-92)* and Specifications for Masonry Structures *(ACI 530.1-92/ASCE 6-92/TMS 602-92). Copyright The Masonry Society and The American Concrete Institute, 1993.*)

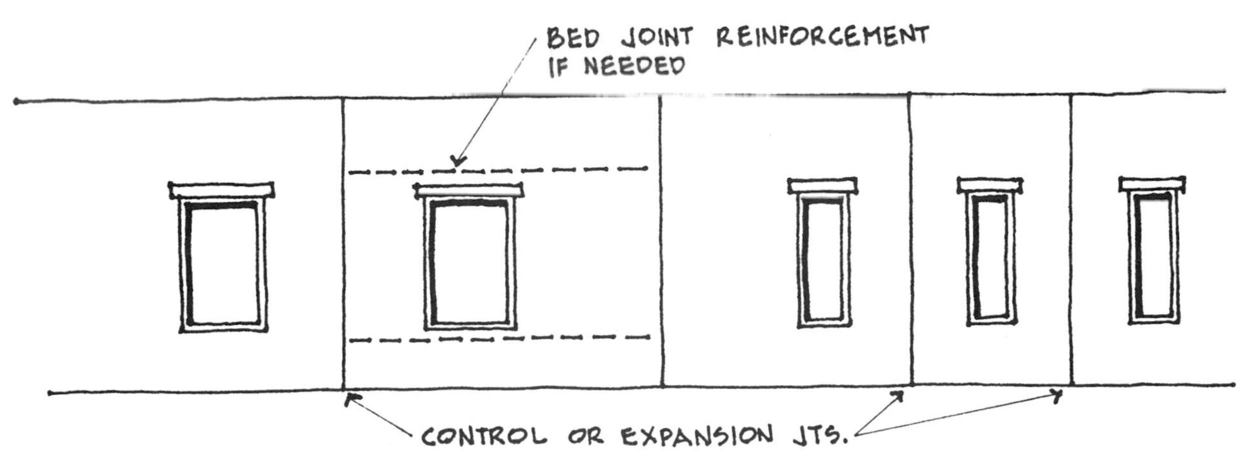

Control and expansion joints can be placed between rather than immediately adjacent to window and door openings.

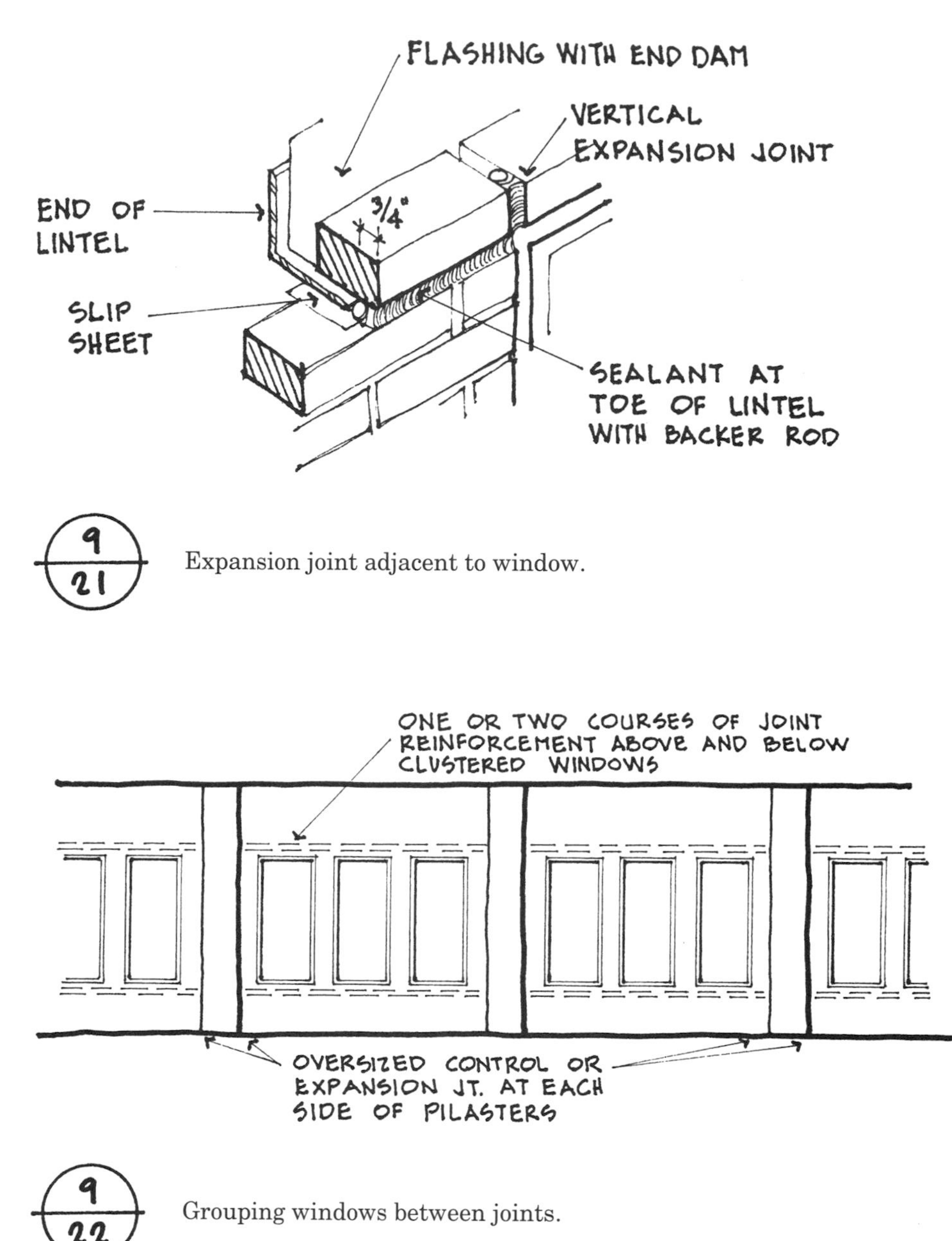

9/21 Expansion joint adjacent to window.

9/22 Grouping windows between joints.

means of removing moisture from the wall are continuous flashing, unobstructed weepholes, and good details.

Single-wythe walls are most vulnerable to moisture penetration. Solid multi-wythe walls and solidly grouted walls are more resistant to water penetration because of their increased thickness, but the term "barrier wall," which is frequently used, should not be interpreted literally. Cavity walls and veneers which have a complete separation between backing and facing provide the best protection. This *drainage wall* concept permits moisture which enters the wall or condenses within the cavity to be collected on flashing membranes and expelled through weepholes (*see Fig. 9-24*). At the base of the wall, and at any point where the cavity is interrupted such as shelf angles, floors, or openings, a layer of flashing must be installed, and with it, a row of weepholes. The cavity in a drainage wall

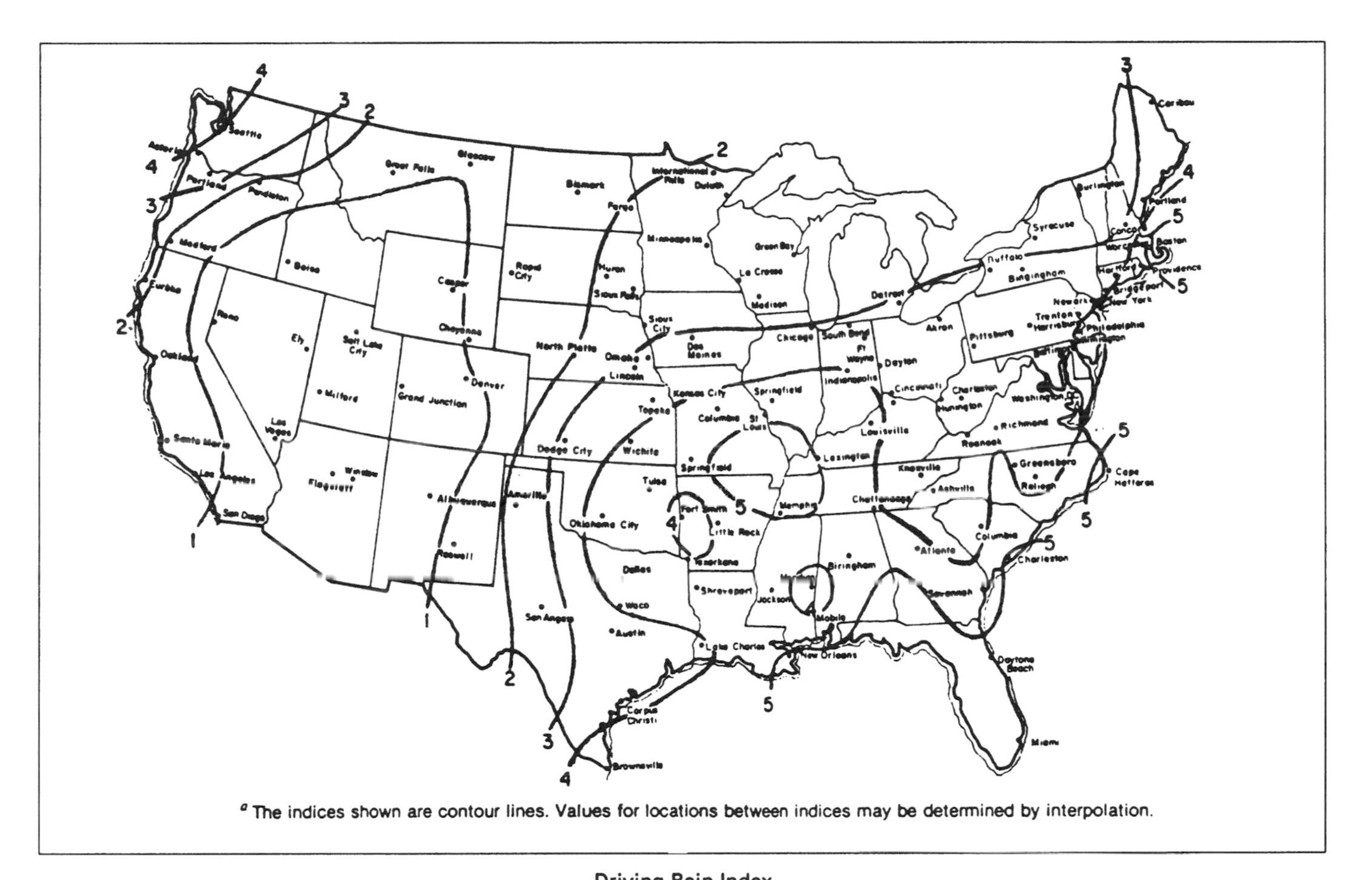

Driving Rain Index

WALL EXPOSURE TO WIND-DRIVEN RAIN*					
Driving rain index†		*Wall standing above surroundings*			
		Yes (unprotected)‡		*No (protected)‡*	
		Wall near facade edge§		*Wall near facade edge§*	
Greater than	*Less than*	*Yes*	*No*	*Yes*	*No*
0	1.5	Severe	Moderate	Sheltered	Sheltered
1.5	3.0	Severe	Moderate	Moderate	Sheltered
3.0	5.0	Severe	Severe	Severe	Moderate
5.0	—	Severe	Severe	Severe	Severe

*Exposures are for walls on buildings located more than 5 miles (8 km) from a sea, large lake, or estuary. All walls on buildings located 5 miles (8 km) or less from a sea, large lake, or estuary have a severe exposure.

†See map.

‡A wall might be considered protected where permanent buildings or terrain face the wall in all directions and have a height above the top of the wall of more than 1.2 times their individual distances from the wall or where there is a permanent solid wall overhang at the top of the wall having a width of at least 85% of the wall height.

§Near facade edge is within one-tenth of the facade width from a corner or one-tenth of the facade height from the top.

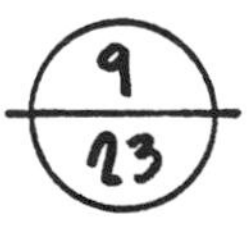

Wind-driven rain can penetrate building walls through even small hairline cracks and openings. (*From C. T. Grimm, "A Driving Rain Index for Masonry Walls," in* Masonry: Materials, Properties, and Performance, *ASTM STP 778, J. G. Borchelt, Ed., American Society for Testing and Materials, 1982. Reprinted with permission.*)

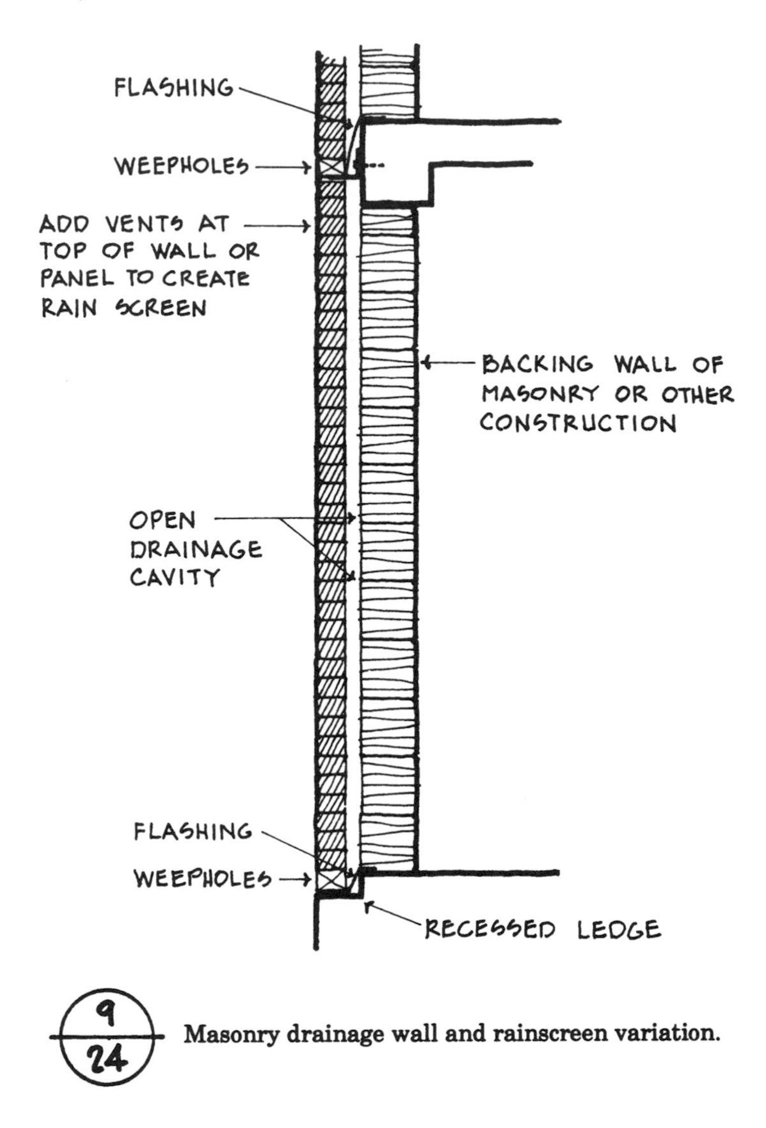

Masonry drainage wall and rainscreen variation.

should be at least 2 in. wide (exclusive of insulation or sheathing), because narrower cavities are difficult to keep clean of mortar droppings during construction.

A variation on the drainage wall concept called the *rainscreen wall* is based on equalizing the pressure between the cavity and the outside atmosphere. Blowing winds during a rain cause a low pressure condition in the cavity. In seeking a natural state of equilibrium, air moves from the high-pressure zone outside to the low-pressure zone in the cavity. With air infiltration, rainwater is carried through the wall face via any minute cracks which may exist at the mortar-to-unit interface. Under such a pressure differential, rainwater which would normally run down the face of the wall is literally driven or sucked into the building. Venting equalizes the pressure differential to eliminate the force which pushes or pulls moisture through the wall, and also promotes faster drying.

The rainscreen principle was developed in the metal curtain wall industry, and requires some special detailing for adaptation to masonry construction. To function properly as a pressure-equalized rainscreen, the wall section must include an air barrier, and the cavity must be divided into compartments. The cavity must be blocked both horizontally and vertically to prevent wind tunnel and stack effects. Without an air barrier and compartmenting, the horizontal flow of air around building corners and through the backing wall prevents pressure equalization in the wall cavity. Shelf angles in conventional masonry cavity wall and veneer construction provide compartmental barriers to the vertical flow of air, but corners require special detailing. *Figure 9-25* shows a sheet metal fin inserted into expansion joints near the building corners to form a barrier to horizontal air flow. Expanding foam sealants might also be used, if they can be adequately secured against displacement.

Each "compartment" must be properly vented so that the pressure change occurs as rapidly as possible. Rainscreen vents should be located near the top of the wall or panel section, and constructed in the same manner as open head joint weepholes. Metal or plastic inserts can be used to disguise the appearance of the void (see Chapter 7). Theoretically, if the area of the vents in the veneer is at least 3 times greater than the area of unintentional openings in the backing wall, then the air pressure in the cavity will be about the same as the outside air pressure. In practice, however, the ratio is often increased to 10:1 to provide a factor of safety that ensures equalization. Aesthetically, rainscreen vents look best if they are located at the same spacing as the weepholes in the lower course of the wall.

9.4.1 Flashing and Weepholes

An important consideration in the design of masonry walls is the proper location of flashing and weepholes. Although moisture penetration can be limited through good design and workmanship, it is virtually impossible to entirely prevent moisture from entering a masonry wall. Without proper flashing, water that does penetrate the wall cannot be diverted back to the exterior. Continuous flashing should be installed at the bottom of the wall cavity and wherever the cavity is interrupted by elements such as shelf angles and lintels. Flashing should be placed over all wall openings, at all window sills, spandrels, caps, copings, and parapet walls (*see Fig. 9-26*). Single-wythe walls also require flashing at the same coping, parapet, head, sill and base locations (*see Fig. 9-27*). Flashing in single-wythe walls can be installed using conventional brick or CMU, or using a special flashing block developed by NCMA (*see Fig. 9-28*).

The top of the vertical leg of the flashing should be installed so that water cannot flow behind it (i.e., placed in a mortar joint in the backing wythe, behind the sheathing, etc.). The flashing should also extend beyond the face of the masonry so that moisture collected on the surface cannot flow around the edge and back into the wall below. Metal flashing can be turned down and hemmed to form a drip, but plastic or rubber flashing must be extended during placement and then cut off flush with the wall after the units are laid. Bituminous flashing or metal flashing with bituminous coatings require installation with a separate metal lip (*see Fig. 9-29*) to avoid exposure to the sun which can cause bleeding, emulsification, and staining. Special flashing details at shelf angles are shown in Chapter 10. Flashing should be continuous around corners and at all horizontal terminations, the flashing should be turned up to form an end dam (*see Fig. 9-30*).

Since the purpose of flashing is to collect moisture and divert it to the outside, *weepholes must be provided wherever flashing is used.* (The sole

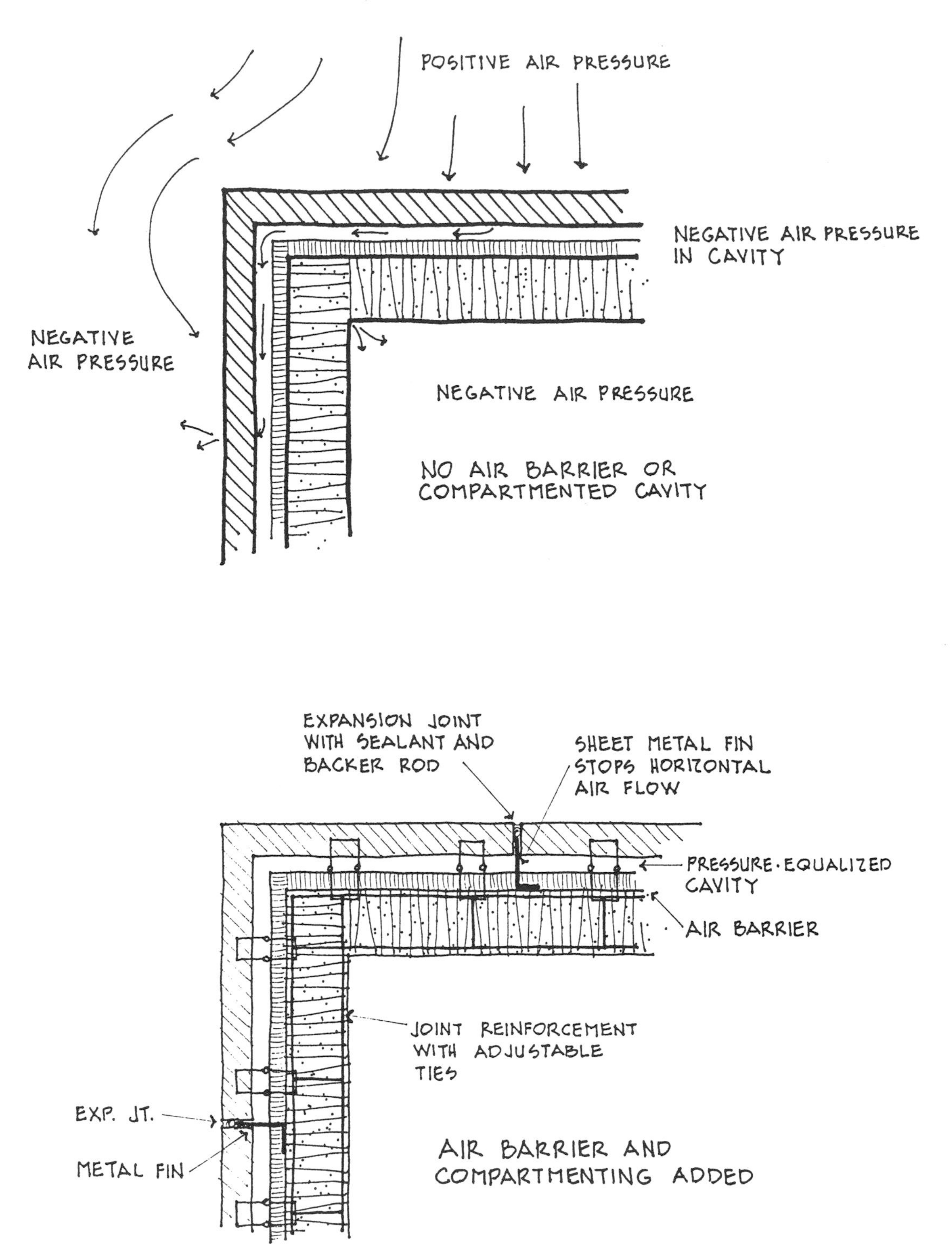

Compartmenting a masonry veneer or cavity wall for pressure equalized rain screen. (*From R. L. Quirouette, "Rain Penetration Control,"* The Construction Specifier, *November 1994, pp. 48–56.*)

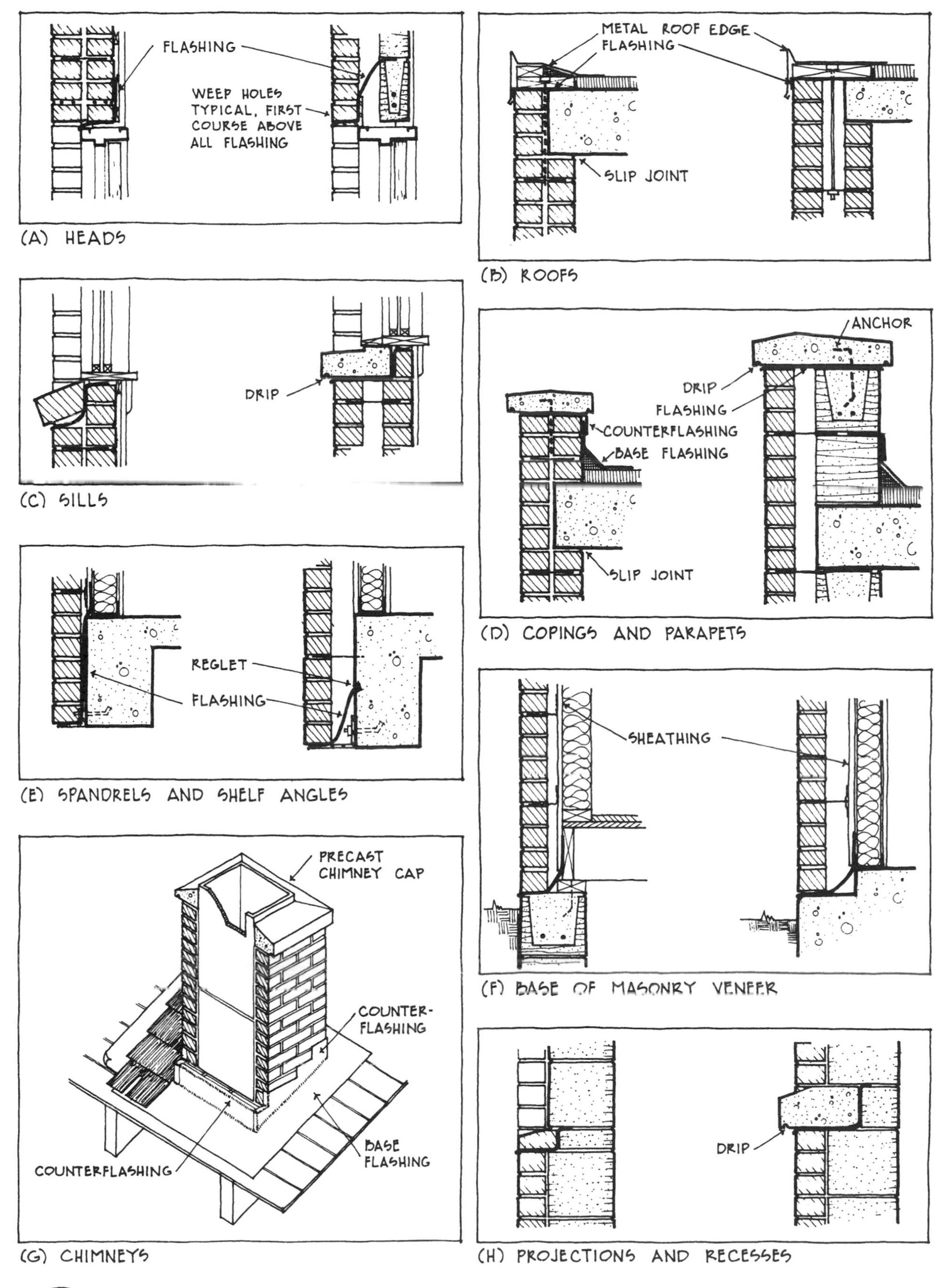

Flashing details. (*From Brick Institute of America,* Technical Note 7A, *BIA, Reston, Va.*)

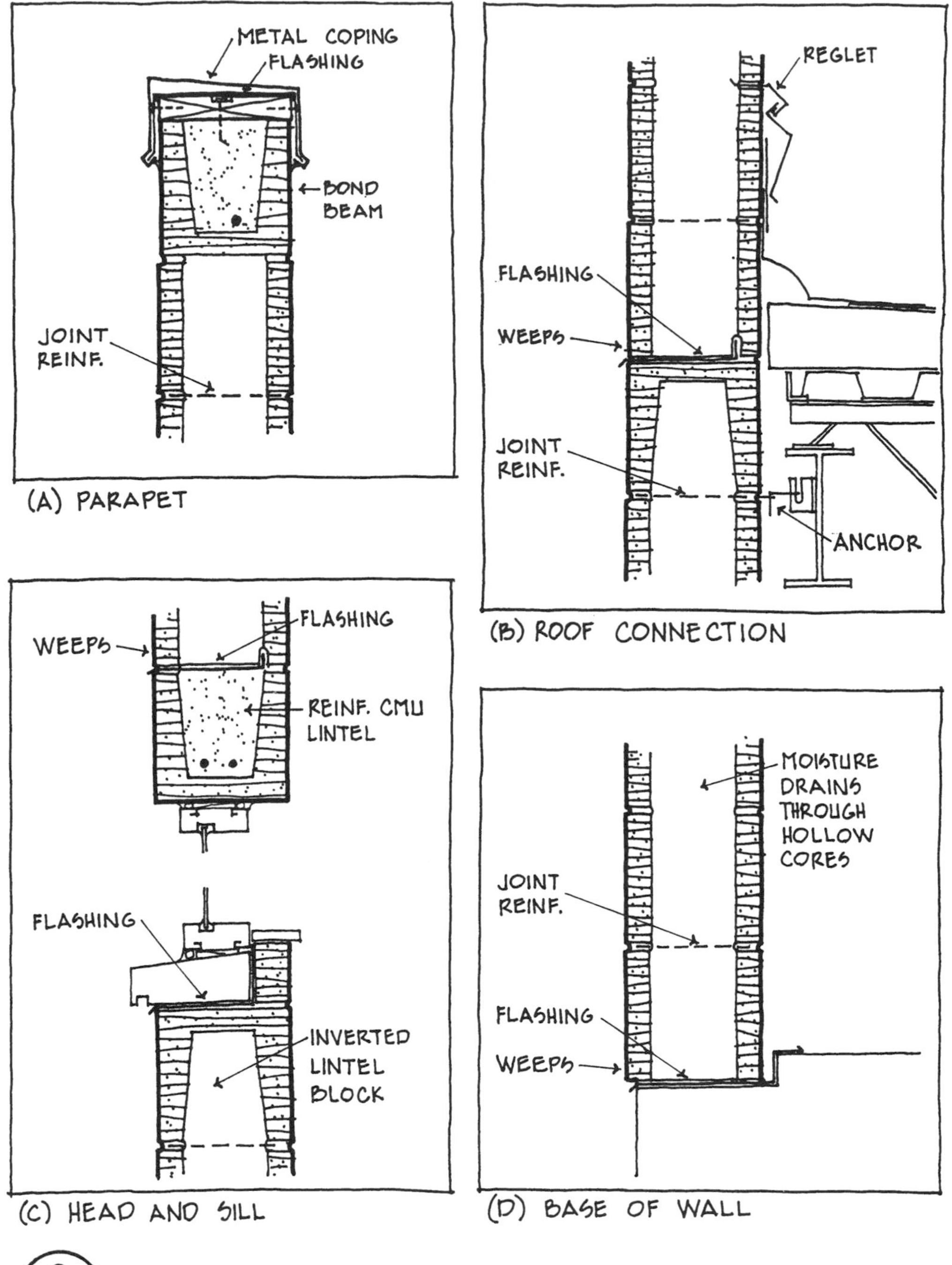

Flashing at single-wythe walls.

exception to this rule is the flashing installed directly underneath the coping, where the purpose is not to collect water, but to prevent its entrance altogether.) Weepholes are located in the head joints immediately above the flashing. Spacing of weepholes should generally be 24 in. on center for open joints, and 16 in. on center for those using wick material. Weephole tubes should not be used because the small openings are too easily obstructed by mortar droppings. In locations where units are placed directly

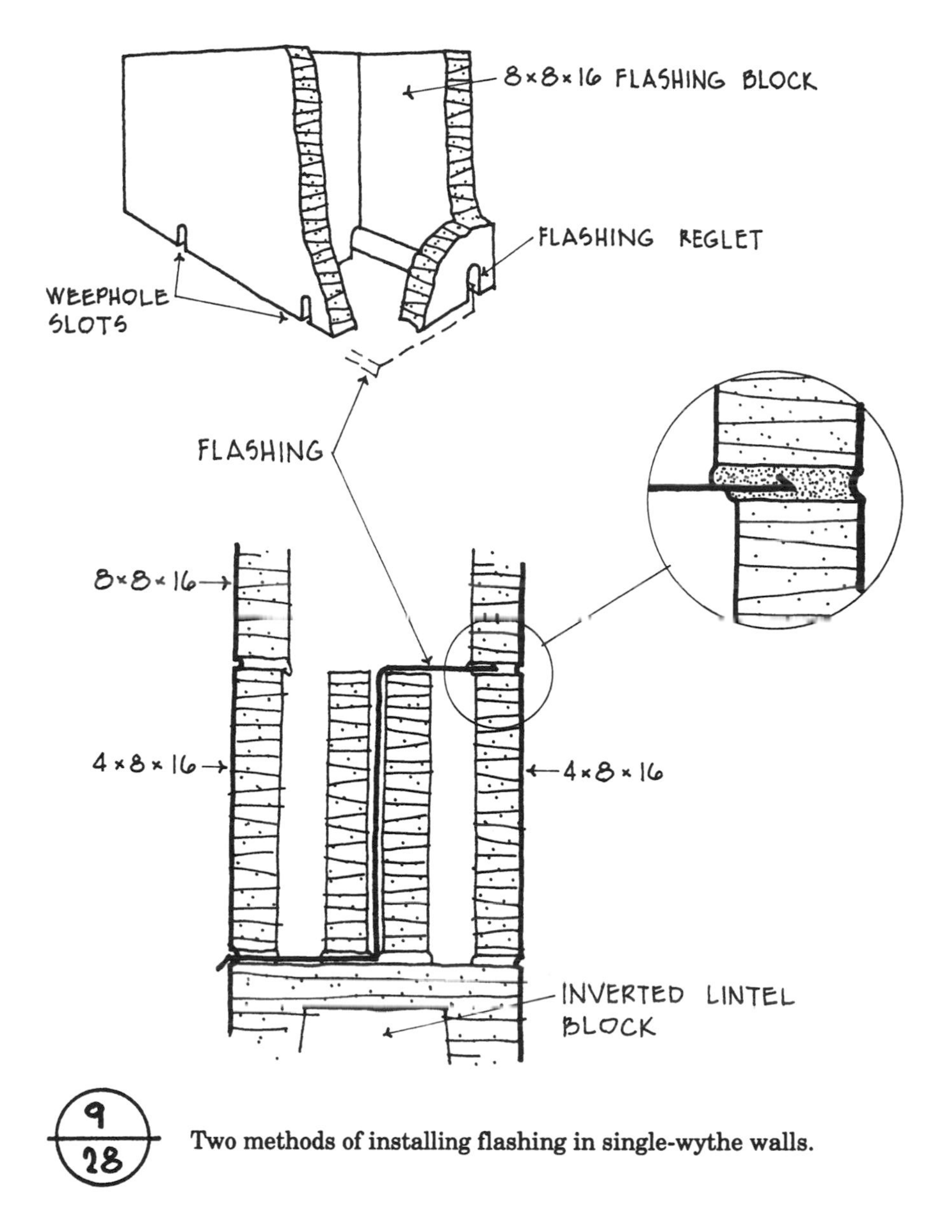

Two methods of installing flashing in single-wythe walls.

on the flashing without mortar, additional drainage is provided throughout the length of the flashing.

Flashing is generally formed from sheet metal, rubber, bituminous membranes, or composite materials selected on the basis of cost and suitability. (Chapter 7 discusses various materials and their performance.) Both installation and material costs vary, and no general recommendation can be made solely on an economic basis. It is critical, however, that only high quality materials be used since failure can lead to significant damage, and replacement is both difficult and expensive.

9.4.2 Material Selection

Proper selection of masonry units and mortar for expected weathering conditions is also an important factor. Clay brick for exterior use should conform to ASTM C216, Grade MW, or SW, depending on the severity of

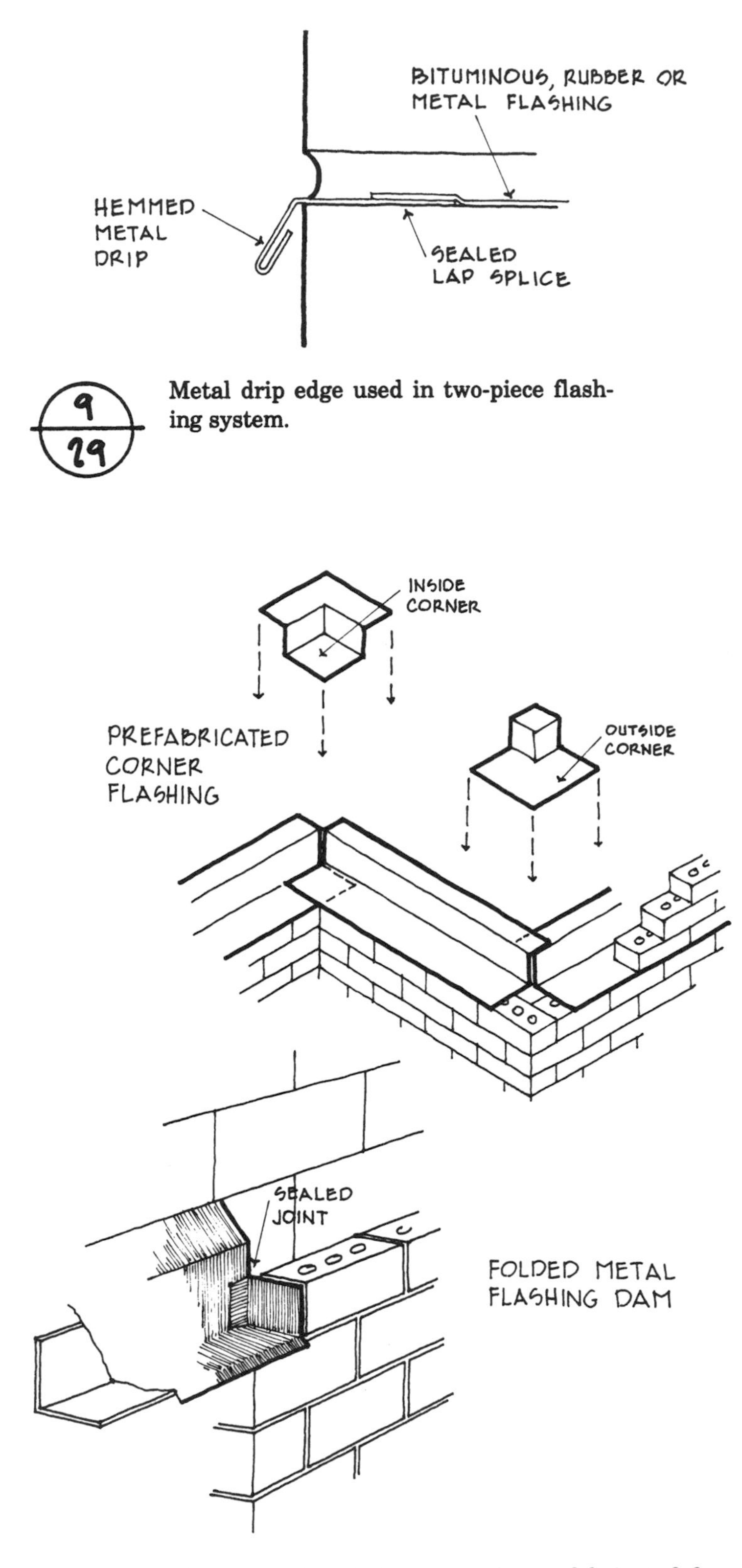

Metal drip edge used in two-piece flashing system.

Corners and end dams may be prefabricated for faster installation.

expected conditions (see Chapter 3). High-suction brick usually produce walls with poor bond. High-suction brick and porous concrete block can absorb excessive water from the mortar, thus preventing complete cement hydration at the unit surface. Mortar generally bonds best to clay masonry units with moderate initial rates of absorption (IRA) between 5 and 25 grams/minute/30 sq in. Brick with initial rates of absorption higher than 25 or 30 grams/minute should be thoroughly wetted and then allowed to surface-dry before laying. This produces better bond and more weather-resistant joints. To test units in the field for high IRA, draw a circle on the bed surface of the brick with a wax pencil, using a 25 cent coin as a guide. With a medicine dropper, place twenty drops of water inside the circle and note the time required for it to be absorbed. If the time exceeds $1\frac{1}{2}$ minutes, initial absorption is low to moderate and the unit need not be wetted. If the time is less than $1\frac{1}{2}$ minutes, initial absorption is high and the brick should be thoroughly wetted and allowed to surface-dry before installation.

Concrete masonry should be kept dry at the job site or the potential for shrinkage cracking in the wall will increase. Concrete masonry units should be Grade N, ASTM C90, C129, or C55. Type I, moisture-controlled CMUs have the lowest shrinkage potential and exhibit less tendency to form cracks in the wall which may cause eventual leakage.

Full mortar joints are essential in maintaining weather-resistant walls. Mortars should be selected on the basis of performance. Cracking or separating of bond between mortar and masonry unit invites the intrusion of water, and good bond must be maintained at all contact surfaces to prevent such leakage. Type N mortar is generally adequate for above-grade work with normal exposure. Types M and S should be used only for special conditions (see Chapter 6 for mortar Type recommendations). All head and bed joints must be fully mortared and tooled for effective weather resistance. The concave and V-joints shown in *Fig. 9-31* are most effective in excluding moisture at the surface. Steel jointing tools compress the mortar against the unit, forming a seal against water penetration at the interface. However, mortar must also be mixed with the maximum amount of water to assure good bond. The mortar mix must contain enough water to provide good workability and to assure complete hydration of the cement even after the water content has been reduced by unit suction. Optimum water content is also affected by weather conditions, so the mason should be allowed to judge the necessary amount based on workability. In hot, dry, or windy conditions, moist curing of the masonry after construction (for both clay and concrete units) can enhance bond and weather resistance by assuring proper hydration (see Chapter 15).

Masonry walls should be protected at the top by a roof overhang or roof edge flashing. When parapets are necessary to the design, they should be carefully detailed to allow expansion, contraction and differential movement, and to prevent water from infiltrating the wall (see Chapter 10). Since much of the water that penetrates masonry walls enters at the top, a roof overhang or protective fascia detail can eliminate most moisture problems. In building designs where the top of the wall is well protected, and the flashing and weepholes are properly located and detailed, decorative joint shapes other than the concave or V-tooled can also be used with less risk of moisture-related problems.

Caulking between masonry and adjacent materials completes the exterior envelope of the building. Door and window openings, intersections with dissimilar materials, penetrations, control joints, and expansion joints must all be fully and properly caulked to maintain the integrity of the sys-

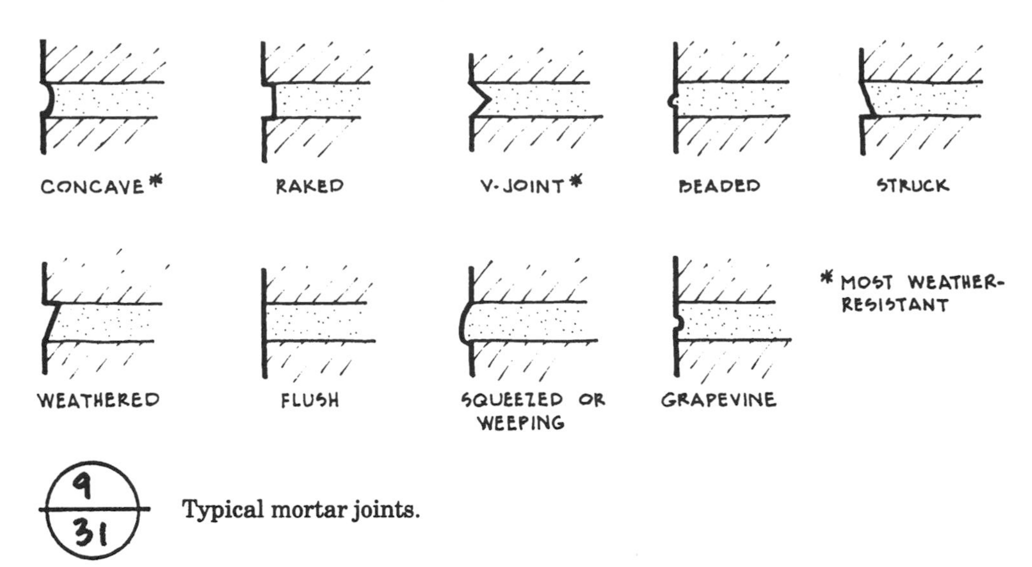

Typical mortar joints.

tem. Sealant materials must be selected for compatibility with the masonry, including adhesion, compression, extensibility, and staining characteristics. Workmanship must be of high quality, and should follow the recommendations of the Sealant, Waterproofing and Restoration Institute (SWRI). Joint surfaces should be properly cleaned and primed, backing material installed, and the sealant applied in the proper joint shape. Fillers placed in expansion joints to keep mortar out during construction may not be located at the correct depth for sealant application, or their depth may not be consistent, so they should not be used to form the back of the sealant joint. Fillers which interfere with correct installation of sealant and backer rods should be removed, and wet tooling of sealant joints should not be permitted. Masons must also ensure that cavities of double-wythe walls and veneer walls are kept clean and the weepholes free of mortar plugs.

9.4.3 Waterproofing and Dampproofing

Below-grade masonry waterproofing generally consists of a bituminous membrane or other impervious film which is resistant to water penetration even under hydrostatic pressure. In areas where soil exhibits good drainage characteristics, the membrane may actually be only a dampproof layer designed to retard moisture until the water has drained away from the building by natural gravity flow (*see Fig. 9-32*). A commonly used protective measure consists of one, or preferably two, coats of cement mortar. This method is known as parging. Although parge coats will retard leakage, wall movements may cause cracks and permit moisture penetration. Impervious membranes with some elasticity offer better assurance against leaks. These may be fluid-applied bituminous products, elastomeric sheets, bentonite clay, or any tested and approved waterproofing system (see Chapter 13).

9.4.4 Condensation and Vapor Retarders

Vapor retarders on the cavity face of backup walls in above-grade construction are often recommended in climates with high humidity. Differences in humidity between inside and outside air cause vapor flow within

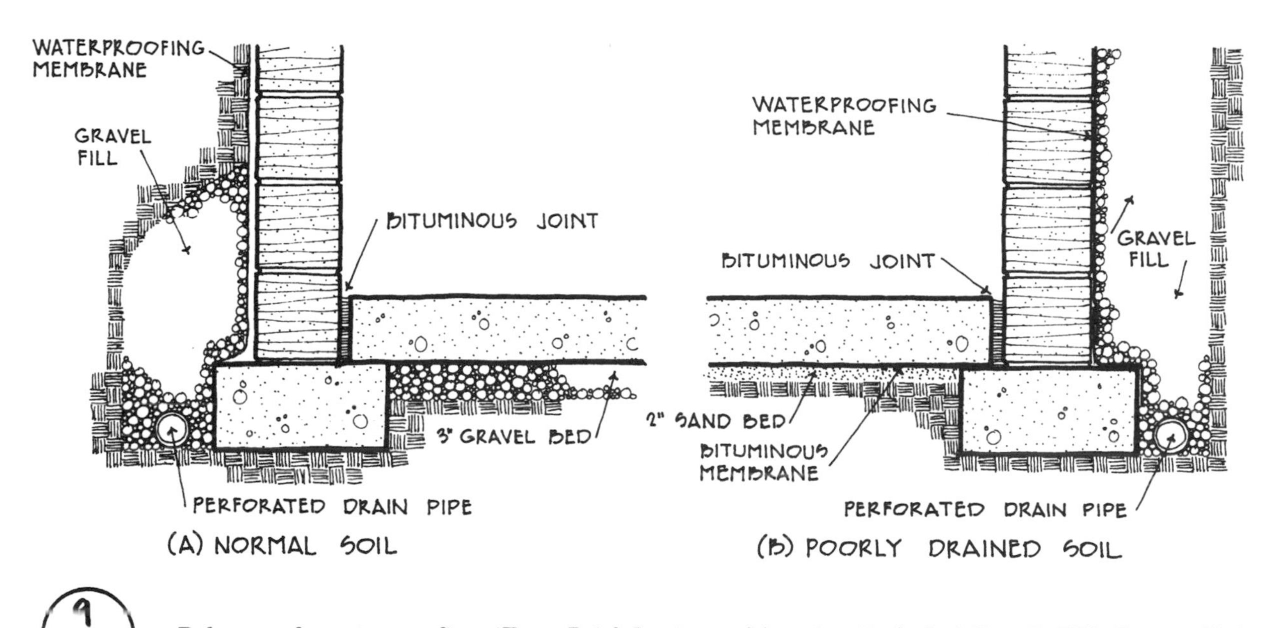

9/32 **Below-grade waterproofing.** ***(From Brick Institute of America,*** **Technical Note 7,** ***BIA, Reston, Va.)***

a wall. Unless controlled by properly placed air barriers and vapor retarders or by ventilation, the vapor may condense within the wall under certain temperature conditions.

Condensation results from saturated air. The higher the air temperature, the more water vapor the atmosphere can contain before reaching its saturation, or dew point. If warm humid air is sufficiently cooled, the water vapor it contains will condense. Condensation problems are most frequent in insulated buildings of tight construction with occupancies or heating systems that produce high humidity. The relative humidity of the air within a building is increased by cooking, bathing, washing, or other activities using water or steam. This rise in the moisture content of the air increases the vapor pressure substantially above that of the outdoor atmosphere, and tends to drive vapor outward from the building through any porous materials in the enclosure. When wall surface temperatures are substantially below air temperatures, condensation may occur on the interior wall surface. The higher the humidity, the less the temperature differential has to be for condensation to form. The table in *Fig. 9-33* lists differences in temperature between the inside air and inside wall surface for different wall types and indoor/outdoor temperature. If, for instance, an exposed 10-in. brick cavity wall separates outside air that is 50°F cooler than the inside air, the inside wall surface temperature is 11°F lower than the inside air temperature. The graph in *Fig. 9-33* shows that an 11°F temperature drop at the wall will cause surface condensation for relative humidities of 68% or more. The same wall assembly with vermiculite insulation shows a wall surface temperature differential for the same conditions of only 5°F, which would cause condensation only at humidities greater than 85.5%.

If wall surface condensation occurs, it may be eliminated by one of three methods:

- Reducing the humidity of the inside air by ventilation.
- Increasing the surface temperature of the interior wall by air movement or other means.

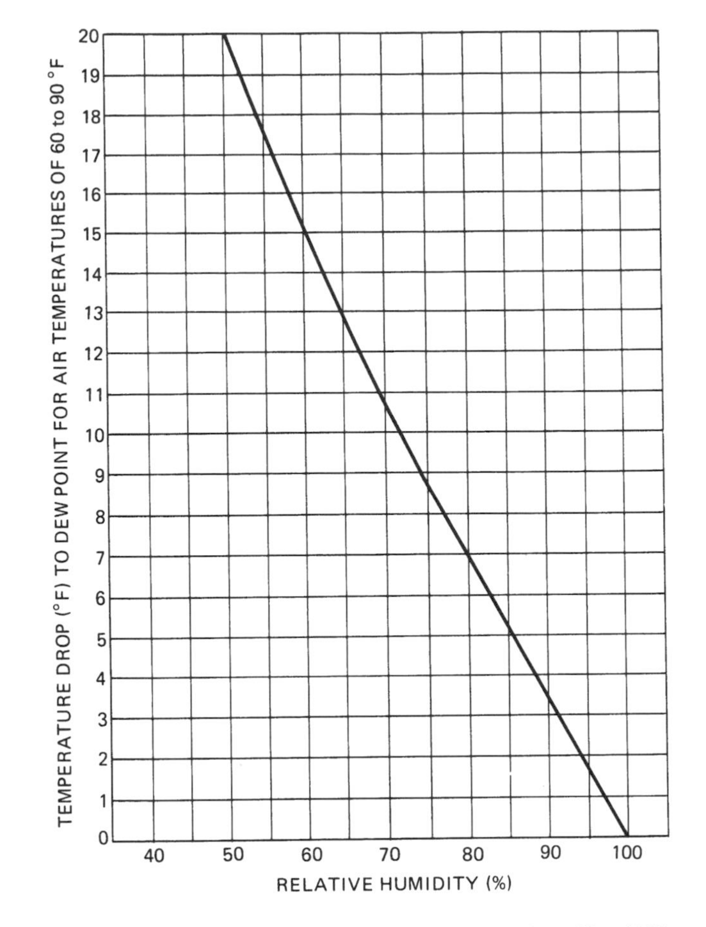

Temperature Drop Curve for Relative Humidities from 50 to 100%

DIFFERENCE IN TEMPERATURE BETWEEN INSIDE AIR AND INSIDE WALL SURFACE FOR VARIOUS TOTAL DIFFERENCES IN TEMPERATURE BETWEEN INSIDE AIR AND OUTSIDE AIR							
	Difference in temperature between inside and outside air (°F)						
Wall construction	*10*	*20*	*30*	*40*	*50*	*60*	*70*
4-in. brick and 4-in. tile, furred and plastered	2	$3\frac{1}{2}$	5	7	9	10	12
"SCR brick," furred and plastered	2	5	7	10	12	14	17
8-in. tile (3-cell), furred and plastered	$1\frac{1}{2}$	3	$4\frac{1}{2}$	6	7	$8\frac{1}{2}$	10
10-in. brick and brick cavity, exposed	$2\frac{1}{2}$	$4\frac{1}{2}$	7	9	11	$13\frac{1}{2}$	16
10-in. brick and tile cavity, plastered	$1\frac{1}{2}$	3	$4\frac{1}{2}$	6	7	$8\frac{1}{2}$	10
10-in. brick and brick cavity, vermiculite insulated, exposed	1	2	3	4	5	6	$6\frac{1}{2}$
10-in. brick and tile cavity, vermiculite insulated, plastered	1	2	$2\frac{1}{2}$	$3\frac{1}{2}$	4	5	6

Conditions affecting surface condensation. (*From Brick Institute of America,* Technical Note 7C, *BIA, Reston, Va.*)

- Increasing the thermal resistance of the wall by adding an air space or insulation (as in the example above). The table and chart can help in selecting suitable wall components for various design temperatures and occupancy conditions.

Vapor condensation *within* walls can cause extensive damage to many types of building materials. Wood framing members can warp or decay in the presence of moisture; metal can corrode; insulation can lose its effectiveness; masonry can shrink or expand, effloresce, or suffer damage from freeze-thaw cycles.

Warm air has higher saturated vapor pressures than cool air. If separated by a wall, the higher-pressure vapor will migrate through the wall toward the lower-pressure atmosphere. During the winter this flow is from inside the building toward the outside. In warm humid climates, this flow may reverse during the summer, with vapor traveling from the outside in. When vapor passes through a wall that is warm on one side and cool on the other, it may reach its dew point and condense into water within the wall. The temperature drop through a composite wall is directly proportional to the thermal resistance of the various elements. The drop in vapor pressure through the wall is in proportion to the vapor resistance of the constituent parts. The table in *Fig. 9-34* lists vapor resistances for some common building materials, and the table in *Fig. 9-35* shows the saturated vapor pressure for various temperatures.

9.4.5 Calculating Condensation

The potential for condensation can easily be calculated if indoor and outdoor design temperatures and humidities are known, as well as the thermal and vapor resistances of the component wall materials. From the table in *Fig. 9-35,* the saturated vapor pressures are determined for the indoor and outdoor temperatures, and the actual vapor pressures calculated as a percentage based on relative humidity conditions [i.e., saturated vapor pressure (SVP) × relative humidity (RH) = actual vapor pressure (AVP)]. The difference between the indoor and outdoor AVP is called the vapor pressure differential. Vapor moves from the warm higher pressure to the cool lower pressure atmosphere. During winter, this is normally an outward movement, and during summer, an inward movement.

Using the thermal resistance of each material in the wall, a temperature gradient through the section is established, and the SVP at each temperature layer is listed. Using the vapor resistance of each material, the AVP at each layer can then be determined. At any location in the wall where actual vapor pressure exceeds saturated vapor pressure, condensation will occur.

Figure 9-36 shows a sample calculation and graphic analysis for one set of conditions. In this particular case, condensation will occur at the cavity face of the exterior brick wythe. A vapor retarder on the cavity face of the interior wythe or on the interior wall surface will all but eliminate the moist vapor which reaches the cavity, and therefore reduce the risk of condensation. Every wall must be analyzed individually, because changes in materials or in temperature/humidity conditions change the location of the dew point. If condensation is expected to occur in masonry cavity or veneer walls, it should be designed to occur at the drainage cavity rather than in the masonry itself where saturation might lead to freeze-thaw damage or efflorescence.

Material	Thickness (in.)	Permeance (Perm)	Resistance* (Rep)	Permeability (Perm in.)	Resistance/in.* (Rep/in.)
Construction materials					
Concrete (1:2:4 mix)					0.31
Brick masonry	4	0.8	1.3	3.2	
Concrete block (cored, limestone aggregate)	8	2.4	0.4		
Tile masonry, glazed	4	0.12	8.3		
Plaster on metal lath	0.75	15	0.067		0.050
Gypsum wall board (plain)	0.375	50	0.020		0.050–0.020
Gypsum sheathing (asphalt impreg.)	0.5			20	
Structural insulating board (sheathing quality)				20–50	
Structural insulating board (interior, uncoated)	0.5	50–90	0.020–0.011		
Hardboard (standard)	0.125	11	0.091		
Hardboard (tempered)	0.125	5	0.2		
Built-up roofing (hot mopped)		0.0			
Wood, sugar pine				0.4–5.4†	2.5–0.19
Plywood (douglas fir, exterior glue)	0.25	0.7	1.4		
Plywood (douglas fir, interior glue)	0.25	1.9	0.53		
Acrylic, glass fiber reinforced sheet	0.056	0.12	8.3		
Polyester, glass fiber reinforced sheet	0.048	0.05	20		
Thermal insulations					
Air (still)				120	0.0083
Cellular glass				0.0	∞
Corkboard				2.1–2.6	0.48–0.38
				9.5	0.11
Mineral wood (unprotected)				116	0.0086
Expanded polyurethane (R-11 blown) board stock				0.4–1.6	2.5–0.62
Expanded polystyrene—extruded				1.2	0.83
Expanded polystyrene—bead				2.0–5.8	0.50–0.17
Phenolic foam (covering removed)				26	0.038

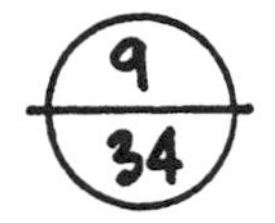

Vapor resistance of various building materials.

It is obvious that the introduction of vapor retarders within a wall assembly must be studied carefully to avoid trapping moisture in an undesirable location. Regional climatic conditions and the resulting direction of vapor flow must be analyzed and condensation points determined for both summer and winter conditions. If the flow of vapor is impeded by a highly vapor-resistant material on the warm side of the wall, the vapor cannot reach that point in the wall at which the temperature is low enough to cause condensation. Masonry cavity walls are unique in their construction and can accommodate vapor flow in either direction without retarding natural moisture drainage. A vapor retarder on the cavity face of the interior wythe will prevent the warm, moist air inside the building from reaching a

Material	Weight (lb/100 sq ft)	Permeance (Perm)			Resistance* (Rep)		
		Dry-cup	Wet-cup	Other	Dry-cup	Wet-cup	Other
Plastic and metal foils and films†							
Aluminum foil		0.001	0.0		∞		
Aluminum foil		0.00035	0.05		20		
Polyethylene		0.002	0.16		6.3		3100
Polyethylene		0.004	0.08		12.5		3100
Polyethylene		0.006	0.06		17		3100
Polyethylene		0.008	0.04		25		3100
Polyethylene		0.010	0.03		33		3100
Polyvinylchloride, unplasticized		0.002	0.68		1.5		
Polyvinylchloride, plasticized		0.004	0.8–1.4		1.3–0.72		
Polyester		0.001	0.73		1.4		
Polyester		0.0032	0.23		4.3		
Polyester		0.0076	0.08		12.5		
Building paper, felts, roofing papers§							
Duplex sheet, asphalt laminated, aluminum foil one side	0.6	0.002	0.176		500	5.8	
Saturated and coated roll roofing	65	0.05	0.24		20	4.2	
Kraft paper and asphalt laminated, reinforced 30-120-30	6.8	0.3	1.8		3.3	0.55	
Blanket thermal insulation back-up paper, asphalt coated	6.2	0.4	0.6–4.2		2.5	1.7–0.24	
Asphalt-saturated and coated vapor retarder paper	8.6	0.2–0.3	0.6		5.0–3.3	1.7	
Asphalt-saturated but not coated sheathing paper	4.4	3.3	20.2		0.3	0.05	
15-lb asphalt felt	14	1.0	5.6		1.0	0.18	
Single-kraft, double	3.2	31	42		0.032	0.024	
Liquid-applied coating materials	Thickness (in.)						
Commercial latex paints (dry film thickness)							
Vapor retarder paint	0.0031			0.45			2.22
Primer-sealer	0.0012			6.28			0.16
Vinyl acetate/acrylic primer	0.002			7.42			0.13
Vinyl-acrylic primer	0.0016			8.62			0.12
Semi-gloss vinyl-acrylic enamel	0.0024			6.61			0.15
Exterior acrylic house and trim	0.0017			5.47			0.18

Note: Exact values for permeance or permeability should be obtained from the manufacturer or from laboratory tests. The values shown indicate variations among mean values for materials that are similar but of different density, orientation, lot, or source. Permeance, resistance, permeability, and resistance per unit thickness values are given in the following units:

Permeance (Perm) = gr/h • sq ft • in. Hg

Resistance (Rep) = in. Hg • sq ft • h/gr

Permeability (Perm-in.) = gr/h • sq ft • (in. Hg/in.)

Resistance/unit thickness (Rep/in.) = (in. Hg • sq ft • h/gr)/in.

*Resistance and resistance/in. values have been calculated as the reciprocal of the permeance and permeability values.

†Depending on construction and direction of vapor flow.

‡Usually installed as vapor retarders, although sometimes used as exterior finish and elsewhere near cold side, where special considerations are then required for warm side barrier effectiveness.

§Low permeance sheets used as vapor retarders. High permeance used elsewhere in construction.

(Continued)

Temp. (°F)	Vapor pressure (in. Hg)	Temp. (°F)	Vapor pressure (in. Hg)
−20	0.01259	41	0.25748
−19	0.01333	42	0.26763
−18	0.01411	43	0.27813
−17	0.01493	44	0.28899
−16	0.01579	45	0.30023
−15	0.01671	46	0.31185
−14	0.01767	47	0.32386
−13	0.01868	48	0.33629
−12	0.01974	49	0.34913
−11	0.02086	50	0.36240
−10	0.02203	51	0.37611
−9	0.02327	52	0.39028
−8	0.02457	53	0.40492
−7	0.02594	54	0.42004
−6	0.02738	55	0.43565
−5	0.02889	56	0.45176
−4	0.03047	57	0.46840
−3	0.03214	58	0.48558
−2	0.03388	59	0.50330
−1	0.03572	60	0.52159
0	0.03764	61	0.54047
1	0.03967	62	0.55994
2	0.04178	63	0.58002
3	0.04401	64	0.60073
4	0.04634	65	0.62209
5	0.04878	66	0.64411
6	0.05134	67	0.66681
7	0.05402	68	0.69019
8	0.05683	69	0.71430
9	0.05978	70	0.73915
10	0.06286	71	0.76475
11	0.06608	72	0.79112
12	0.06946	73	0.81828
13	0.07300	74	0.84624
14	0.07670	75	0.87504
15	0.08056	76	0.90470
16	0.08461	77	0.93523
17	0.08884	78	0.96665
18	0.09327	79	0.99899
19	0.09789	80	1.0323
20	0.10272	81	1.0665
21	0.10777	82	1.1017
22	0.11305	83	1.1379
23	0.11856	84	1.1752
24	0.12431	85	1.2135
25	0.13032	86	1.2529
26	0.13659	87	1.2934
27	0.14313	88	1.3351
28	0.14966	89	1.3779
29	0.15709	90	1.4219
30	0.16452	91	1.4671
31	0.17227	92	1.5135
32	0.18035	93	1.5612
33	0.18778	94	1.6102
34	0.19456	95	1.6606
35	0.20342	96	1.7123
36	0.21166	97	1.7654
37	0.22020	98	1.8199
38	0.22904	99	1.8759
39	0.23819	100	1.9333
40	0.24767		

Standard vapor pressure at various temperatures. (*From Brick Institute of America*, Technical Note 7D, *BIA, Reston, Va.*)

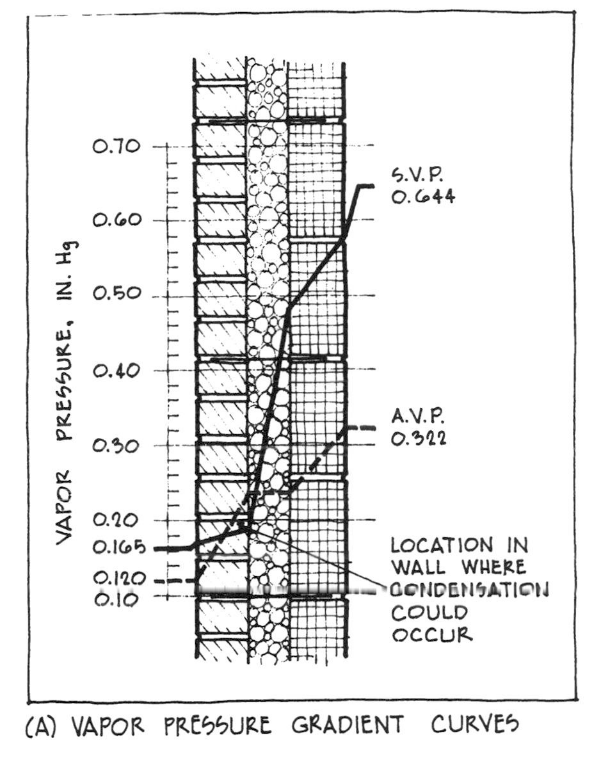

(A) VAPOR PRESSURE GRADIENT CURVES

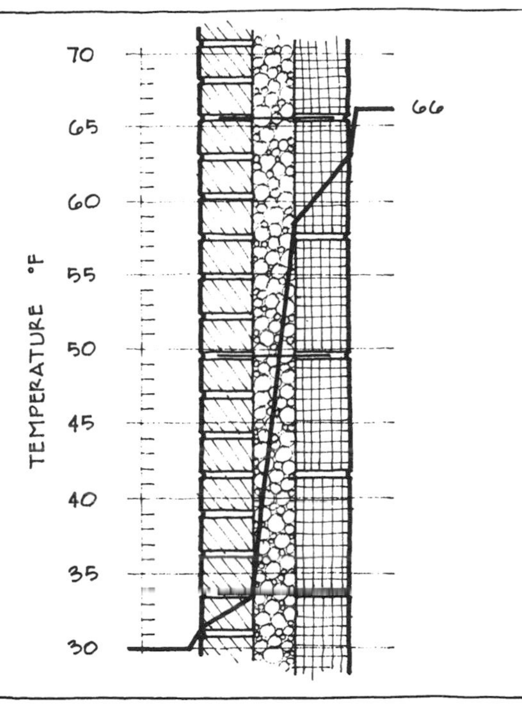

(B) TEMPERATURE GRADIENT CURVE

Material	Thermal analysis				Vapor pressure analysis				
	T.R.	%	T.D.	°F	S.V.P.	V.R.	%	V.P.D.	A.V.P.
Inside air				66	0.644				0.322
Inside film of still air	0.68	9	3	63	0.580				0.322
4 in. unglazed structural clay facing tile	1.11	14	5	58	0.486	0.678	40.5	0.082	0.240
$2\frac{1}{2}$ in. vermiculite insulation	5.50	70	25	33	0.188	0.016	1.0	0.002	0.238
4 in. high-density facing brick	0.44	5	2	31	0.172	0.977	58.5	0.118	0.120
Outside air at 15 mph	0.17	2	1	30	0.165				0.120
Total	7.90	100	36			1.671	100	0.202	

Note: T.R., thermal resistance; T.D., temperature difference in °F; S.V.P., saturated vapor pressure in inches mercury; V.R., vapor resistance in inches of mercury; V.P.D., vapor pressure difference in inches of mercury; A.V.P., actual vapor pressure in inches of mercury.

Condensation analysis for the design example. (*From Brick Institute of America,* **Technical Note 7D,** ***BIA, Reston, Va.***)

lower saturation temperature farther out in the wall (*see Fig. 9-37*). Conversely, the path of hot, humid air moving toward air-conditioned interior spaces is blocked at the cavity and condenses within the drainage space. Each design condition must be analyzed individually to determine the need and location for a vapor retarder within the wall assembly.

The infiltration and exfiltration of air through cracks and openings in a wall can also move substantial quantities of moisture vapor. In high-rise buildings, air leakage rates are increased by the stack effect—the inward

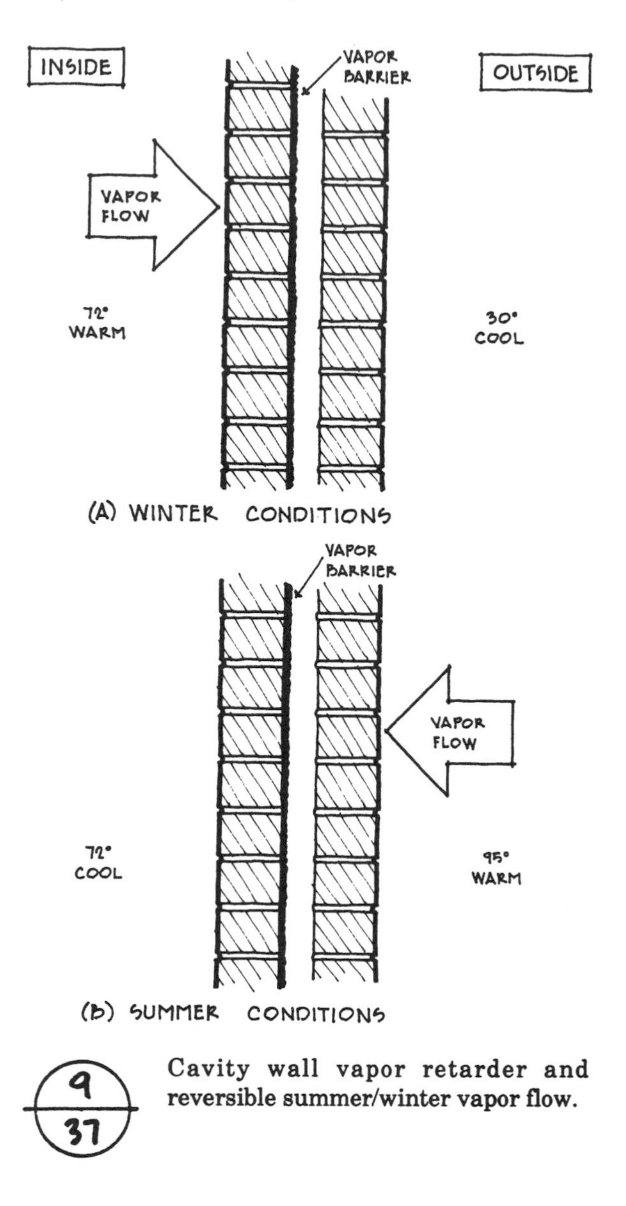

Cavity wall vapor retarder and reversible summer/winter vapor flow.

movement of air at lower stories and outward movement at upper stories due to pressure differentials. Seasonal patterns of efflorescence or dampness near the tops of buildings can be evidence of significant air leakage. Cracks between the backing wall and columns, slabs, or cross walls can provide paths for moisture-laden air to move back and forth between the wall cavity and the building interior. Perimeter sealant joints can provide air barriers at such locations.

9.4.6 Coatings

In walls of solid masonry or single-wythe construction, greater care must be taken to avoid trapping moisture. ASTM specifications for concrete brick and block do not specify water permeability, and protective coatings are often required to prevent water penetration. Some paint films and various other coatings are impervious to vapor flow and, if placed on the wrong side of the wall, can trap moisture inside the unit. Local climatic conditions must be evaluated in determining the direction of vapor flow.

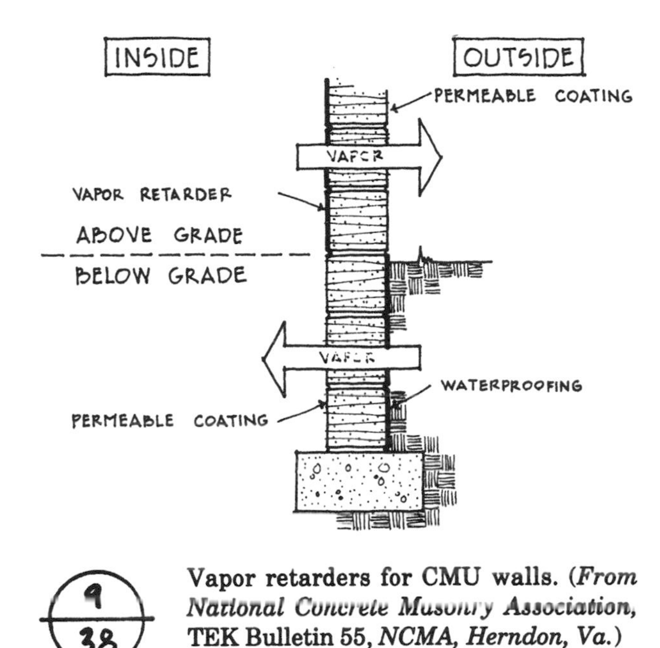

Vapor retarders for CMU walls. ***(From National Concrete Masonry Association,*** **TEK Bulletin 55,** ***NCMA, Herndon, Va.)***

The vapor pressure-temperature differentials discussed above also apply to concrete block walls.

NCMA recommendations for placement of vapor retarders are illustrated in *Fig. 9-38.* The retarder is placed on the moist side of the wall (the side with the highest vapor pressure). If the conditions shown for above-grade walls are revised and the highest vapor pressure is on the exterior, moisture will flow toward the interior spaces, so the retarder should be placed on the outside. In areas of warm summer temperatures and high humidity, such as those along the Gulf Coast, an unprotected concrete block wall with gypsum board and an impervious vinyl wall covering on the inside could mildew behind the vinyl. The moisture is able to enter the wall on the warm side, condenses on the cool side, but cannot escape and so is trapped behind the vinyl. The barrier must be on the moist side to prevent such damage.

In areas where summer and winter conditions cause a reverse in the direction of moisture flow, insulation must be used to prevent vapor condensation, or a cavity wall should be used in lieu of solid or single-wythe construction. Placing a vapor retarder and an insulation layer on the outside of the wall moves the saturated vapor pressure temperature enough to avoid coincidence with the actual vapor pressure, and stop the moisture flow on the warm side of the wall (*see Fig. 9-39*). This condition will hold true for both summer and winter conditions and for inward and outward flow of vapor. For various climatic differences, the thickness and type of insulation should be checked to ensure adequate thermal and vapor resistance.

The temperature differentials and temperature gradient curves shown are based only on ASHRAE steady-state calculation principles, and do not take into account the time lag created by the thermal mass of the masonry (refer to Chapter 8). The ability of masonry to store heat over relatively long periods of time can be advantageous in climates where the indoor and outdoor temperature and humidity conditions would normally reverse vapor flow one or more times a day. Dense wall materials such as masonry will keep the wall temperature at a more constant level and

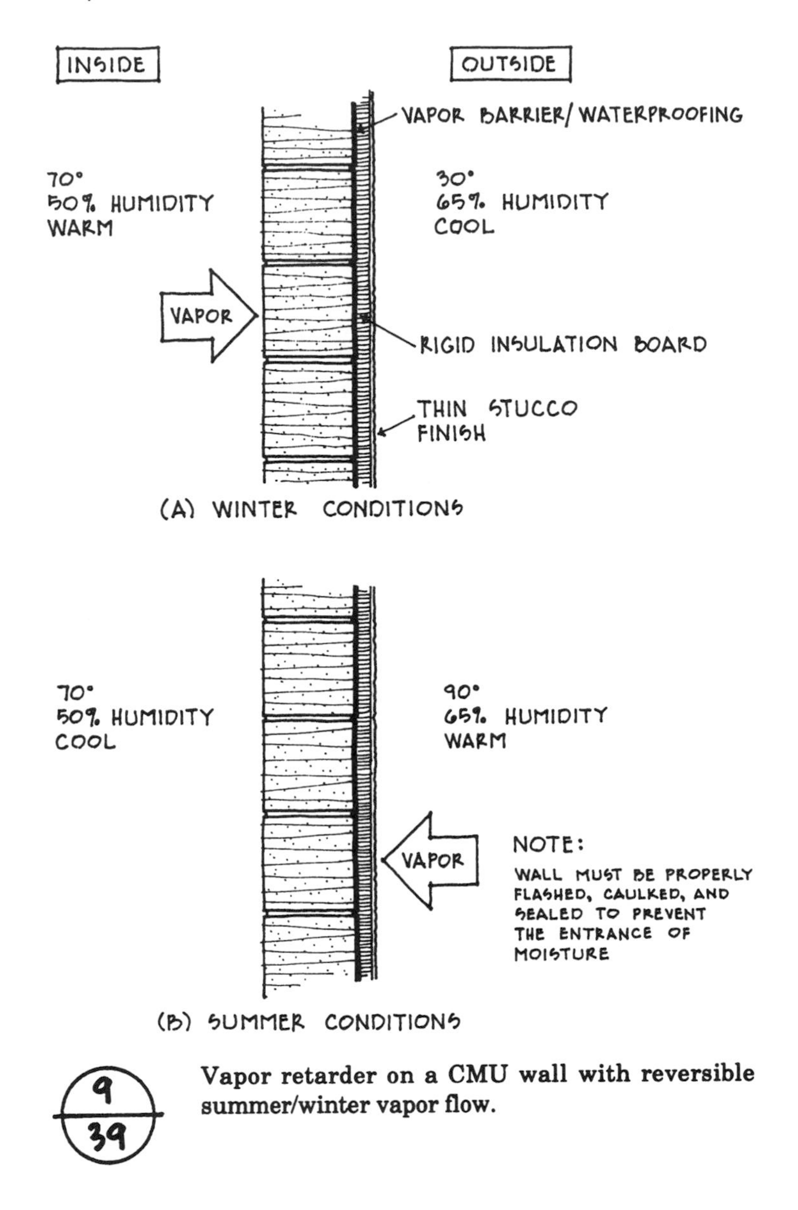

Vapor retarder on a CMU wall with reversible summer/winter vapor flow.

reduce the temperature fluctuation which occurs in lightweight thin-skinned construction.

This thermal inertia protects the wall from constantly shifting vapor movement and condensation points. Applied coatings still must be carefully selected on the basis of their permeability. Inadvertent use of an impermeable surface finish on the cold side of a wall can create problems that are difficult and expensive to correct. A nonporous paint film will prevent some rainwater from entering a wall but, more significantly, it will impede the escape of moisture entering the wall from other sources. Water may enter through pores in materials, partially filled mortar joints, improperly flashed copings, sills, parapet walls, and so on, through capillary contact with the ground, or any number of other means. Moisture escapes from a wall in only two ways: (1) through continuous cavities with adequate flashing and weepholes and (2) by evaporation at the wall face (breathing). In single-wythe construction, where there are no cavities or weepholes, the exterior wall face must breathe in order to dry out.

There are numerous types of paint suitable for masonry walls, including cement-based paints, water-thinned emulsions, fill coats, solvent-thinned paints, and high-build coatings. In selecting a paint finish system, there are several things to consider. Paint products that are based on drying oils may be attacked by free alkali from the units or mortar. Alkaline-resistant paints and primers are recommended to prevent this, or the masonry must fully cure before painting. Surface conditions must also be considered, and preparations suitable to the selected finish made. Efflorescence must be removed from a masonry surface and observed for recurrence prior to painting. New masonry must not be washed with acid cleaning solutions if paint is to be applied. If low-alkali portland cement is not used in the mortar, it may be necessary to neutralize the wall with a 2- to 3½-lb/gal solution of zinc chloride or zinc sulfate and water. Existing masonry must be cleaned of mold, moss, and mildew. Walls must be wetted before any cleaning solution is applied, and thoroughly rinsed afterward to prevent unfavorable paint reactions or chemical contribution to efflorescence.

Previously painted masonry surfaces normally require extensive preparation prior to repainting. All loose material, oil, and dirt must be removed. Penetrating primers may be required to improve adhesion to old paint. Emulsion paints and primers, which are nonpenetrating, generally require cleaner surfaces than do solvent-based paints. Existing cement-based paints may not be repainted with other types of coatings (although fresh cement-base paints may be used as primers for another paint when coverage will be within a relatively short time).

New brick masonry is seldom painted, as it is difficult to justify the initial cost and future maintenance expense. Once painted, an exterior brick wall will require costly, periodic repainting every 3 to 5 years. The benefits of the desired aesthetic effect must be carefully weighed against maintenance costs before a decision is made to paint over new brick.

Stucco can be used to reduce the permeability of old and new masonry walls, and is a popular finish on concrete block. A two-coat application of portland cement stucco may prove to be the most economical and satisfactory method for treatment of leaky walls where repointing and exterior wall treatment costs would be excessive. Good bond between the stucco and masonry depends on mechanical key and suction, and the texture of concrete masonry provides an excellent substrate. Lime may be added to the cement-sand mixture for plasticity, but should not exceed 10% by weight or 25% by volume of the cement. Total thickness of the stucco application should be a minimum of ⅝ in. Walls that are not reinforced against shrinkage and movement cracks can transmit excessive tensile stresses to monolithic stucco coatings and cause cracking of the finish surface as well.

Clear water repellents are usually advertised as a cure-all for masonry moisture problems, and they are often incorrectly referred to as "sealers" or "waterproof" coatings—which they are not. Water repellents generally change the capillary angle of pores in the face of the masonry to repel rather than absorb water, but they will not bridge hairline cracks or separations at the mortar-to-unit interface. Clear water repellents can reduce absorption through the face of the masonry and prevent soiling on light-colored units while still permitting the wall to breathe.

There are three types of clear water repellents: stearates, acrylics, and silicones. No single type is equally suitable or effective on all masonry substrates, because the physical and chemical properties of clay brick, concrete masonry, and stone vary so widely. Compatibility of substrate and surface treatment should always be evaluated on an individual basis.

Stearates, acrylics, and some silicone resins form a protective film on the masonry surface through resin deposition. The percent of solids content varies and should be selected on the basis of the porosity of the substrate to which it will be applied. A dense material treated with a high-solids compound will have a greatly reduced moisture vapor transmission (MVT) rate and will not breathe properly. Conversely, a porous material treated with a low-solids compound will not repel moisture effectively. Stearates and modified stearates generally have about 5% solids and are used for dense clay brick and stone surfaces. Acrylics range from 7.5 to 25% solids and are more suitable for concrete masonry. All acrylics will darken the masonry and change the natural matte finish. When solids exceed 10%, acrylics will leave a noticeable glossy sheen on the surface.

The most widely used water repellents are silane and siloxane compounds which impregnate the masonry surface and react chemically with water to form silicone resins. Although the extremely small molecular structure permits penetration to a depth of about $^3/_8$ in., the substrate pores are not completely blocked, so moisture vapor transmission remains high. Silanes and siloxanes also rely on a chemical bond to silicate minerals in the masonry, so they are not appropriate for application to limestone and marble. Silanes require the catalytic action of substrate alkalis to form the active silicone resin, so they are not appropriate on clay brick or natural stone. Siloxanes, on the other hand, are polymerized compounds which react independent of the substrate composition. Some proprietary water repellents are blends of silane, siloxane, stearates, and other compounds.

No water repellent, regardless of its chemical composition, will solve the problems of poorly designed or constructed masonry walls. Water repellents and other coatings are not a substitute for flashing and weep-hole systems in cavity wall or single-wythe designs. If a surface treatment is determined to be desirable on repair or renovation projects, defects such as leaky copings and roof flashing, defective sealant joints, and hairline cracks at the mortar-to-unit interface must be corrected before the treatment can be applied. Efflorescence must also be removed and the source of water which caused it found and repaired before applying either surface coating or penetrating treatments. Even though the MVT rate of the wall may be relatively unaffected by water repellents, if the masonry is efflorescing and the source of moisture has not been addressed, concealed interstitial salt crystals may be formed within the masonry. The partially obscured capillary pores allow moisture to evaporate through the surface, but block the natural escape of the salts which are deposited behind the treated area where they re-crystallize. The continuing action of this "subflorescence" gradually increases the salt concentration, and the expansive pressure of crystallization can be sufficient to spall the face of the masonry. Although a clear water repellent may initially appear to stop efflorescence, it may only be burying the problem below the surface.

Since water repellents cannot bridge even very small cracks, and since the primary path of moisture through the face of a masonry wall is through cracks, it is misleading to say that they "protect" the wall from moisture infiltration as many manufacturers claim. Depending on individual substrates and conditions, clear water repellents can help shed water from the face of masonry walls, decrease the absorption of porous units, and protect the materials from staining and from excessive absorption of acid rain. They will not, however, "waterproof," nor will they "seal" the surface. Water repellents, when used, should serve only as an adjunct to total system design, and not as the first and only line of defense.

10

NON-LOADBEARING WALLS AND VENEERS

The term "non-loadbearing" as it is used in masonry design means that the wall or element referred to does not carry the vertical compressive load of the structure. It may, however, be self-supporting and carry other applied loads from wind and seismic forces. Such non-loadbearing elements include partition walls, retaining walls, shaft enclosures, fire walls, and veneers.

Code requirements for non-loadbearing walls are based on standards and recommended procedures originally developed by the National Institute of Standards and Technology (NIST), the American National Standards Institute (ANSI), the Brick Institute of America (BIA), and the National Concrete Masonry Association (NCMA). Although there are some variations in detail, the design of unreinforced non-loadbearing masonry walls and partitions is generally governed by empirical lateral support requirements expressed as length- or height-to-thickness (h/t) ratios, or by the flexural tensile stress between the mortar and unit. Methods of providing lateral support are illustrated in *Fig. 10-1.* The *Uniform Building Code* (UBC) and the Masonry Standards Joint Committee (MSJC) *Building Code Requirements for Masonry Structures,* ACI 530/ASCE 6/TMS 402, prescribe a maximum h/t of 36 for interior non-bearing walls and a maximum h/t of 18 for exterior non-bearing walls. The *Standard Building Code* bases lateral support requirements on design wind pressure as indicated in the table in *Fig. 10-1.* Minimum required wall thickness varies among the codes.

Lateral support can be provided by cross walls, columns, pilasters, or buttresses where the limiting span is measured horizontally, or by floors, roofs, or spandrel beams where the limiting span is measured vertically. Anchorage between walls and supports must be able to resist wind loads and other lateral forces acting either inward or outward. All lateral support members must have sufficient strength and stability to transfer these lateral forces to adjacent structural members or to the foundation. All of the codes contain provisions stating that specific limitations may be waived if engineering analysis is provided to justify additional height or width, or

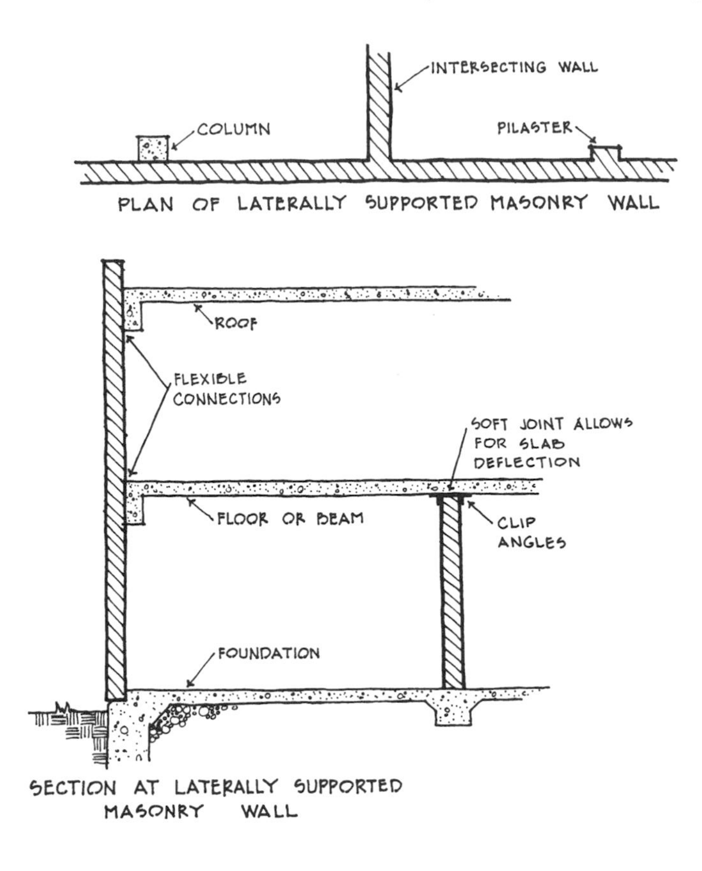

STANDARD BUILDING CODE TABLE 2105.1
LATERAL SUPPORT (*h/t*)* RATIOS
FOR EXTERIOR BEARING AND NON-BEARING WALLS†

	Design wind pressure (psf)§			
Wall construction‡	*15*	*20*	*25*	*30*
Grouted, solid, or filled-cell masonry	26	22	20	18
Hollow masonry or masonry-bonded hollow walls	23	20	18	16
Cavity walls¶	20	18	16	15

* *h* = clear height or length between lateral supports, *t* = nominal wall thickness.

† *h/t* ratios required for wind pressures greater than 30 psf must be determined by an engineering analysis in accordance with the Masonry Standards Joint Committee (MSJC) *Building Code Requirements for Masonry Structures,* ACI 530/ASCE 6/TMS 402, as prescribed in Standard Building Code section 2103.6.

‡ All masonry units shall be laid in Type M, S, or N mortar unless otherwise required (see *Standard Building Code* Table 2102.9. Where Type N mortar is used and the wall spans in the vertical direction, the ratios shall be reduced by 10%.

§ These wind pressures include shape factors from *Standard Building Code* Section 1205.

¶ In computing the *h/t* ratio for cavity walls, *t* shall be the sum of the nominal thicknesses of the inner and outer wythes.

Lateral support of masonry walls. (*Table taken from* Standard Building Code, *published by Southern Building Code Congress, Birmingham, Ala.*)

reduced thickness. The addition of reinforcing steel to relieve tensile stresses will also permit greater flexibility in design (see Chapters 12 and 13).

10.1 INTERIOR PARTITIONS

Partitions are interior, non-loadbearing walls one story or less in height, which support no vertical load other than their own weight. They may be separating elements between spaces, as well as fire, smoke, or sound barriers.

Based on a height/thickness ratio of 36 as prescribed in the UBC and the MSJC codes, a single-wythe 4-in. *brick* partition without reinforcing steel is limited to a 12-ft span, while a 6-in. brick partition can span 18 ft between supports, and an 8-in. hollow brick partition 24 ft. If the partition is securely anchored against lateral movement at the floor and ceiling, and if the height does not exceed these dimensions, there is no requirement for intermediate walls, piers, or pilasters along the length of the partition. If additional height is required, an 8-in. hollow brick can be used, or pilasters can be added at 12- or 18-ft intervals for the 4- and 6-in. walls, respectively. Lateral support is required in only one direction and can be either floor and ceiling anchorage or cross walls, piers, or pilasters, but need not be both.

Structural clay tile is often used for partitioning in schools, hospitals, food processing plants, sports facilities, airports, correctional facilities, and so on, where the imperviousness of a ceramic glazed surface, high durability, and low maintenance are required. Several different types of wall construction may be used depending on the aesthetic requirements for the facing. For the standard 4-, 6-, and 8-in. thicknesses, single units glazed one or both sides are available. Double wythes can be used to provide different colors or finishes on each side of the partition (*see Fig. 10-2*). The 6- and 8-in. walls are capable of supporting superimposed structural loads, but the 4-in. partitions are limited to non-loadbearing applications. Lateral support spacing is governed by the same length- or height-to-thickness ratio of 36, giving the same height limitations of 12, 18, and 24 ft without pilasters or cross walls.

Concrete block partitions are widely used as interior fire, smoke, and sound barriers. Decorative units can be left exposed, but standard utility block is usually painted, textured, plastered, or covered with gypsum board. Wood or metal furring strips can be attached by mechanical means as described in Chapter 7, or sheet materials may sometimes be laminated directly to the block surface. Code requirements for lateral support are the same as for brick and clay tile.

Single-wythe hollow brick, concrete masonry unit (CMU), and vertical cell tile partitions can all be internally reinforced to provide the required lateral support in lieu of cross walls or projecting pilasters (*see Fig. 10-3*). A continuous vertical core at the required interval is reinforced with deformed steel bars and then grouted solid to form an internal pilaster. Double-wythe cavity walls may be similarly reinforced without thickening the wall section.

Cavity walls also facilitate the placement of electrical conduit and piping for distribution of utilities within a building. The continuous collar joint easily accommodates horizontal runs. Although requirements vary from code to code, the thickness of cavity walls for computing lateral support requirements is normally taken as the net thickness of the two wythes without the width of the cavity.

10.2 SCREEN WALLS AND GARDEN WALLS

Perforated masonry *screen walls* may be built with specially designed block or clay tile units, with standard cored blocks laid on their sides, with lintel blocks laid upside down, or with solid brick units laid in an open pattern. As sun screens, the walls are often built along the outside face of a building to provide shading for windows. Screen walls are also used to provide privacy without blocking air flow, and to form interior and exterior area separations. The function of the wall influences finished appearance, from strong and heavy to light and delicate. Dark colors absorb more heat and reflect less light into interior spaces. Relatively solid wall patterns block more wind, and open patterns allow more ventilation.

Screen walls can generally be anchored only at the floor line or at vertical structural projections such as steel or masonry piers or pilasters (*see Fig. 10-4*). Screen walls are governed by the same h/t ratio lateral support requirements as interior partitions, but walls with interrupted bed joints should be designed more conservatively because of reduced flexural strength and lateral load resistance.

Concrete masonry screen wall units should meet the minimum requirements of ASTM C90. Brick should be ASTM C216, Grade SW, and clay tile units should be ASTM C530, Grade NB. Mortar for exterior screen walls should be Type M or Type S.

Solid, uncored brick are used to build what some call "pierced" walls by omitting the mortar from head joints and separating the units to form voids. The walls may be laid up in single- or double-wythe construction. In double-

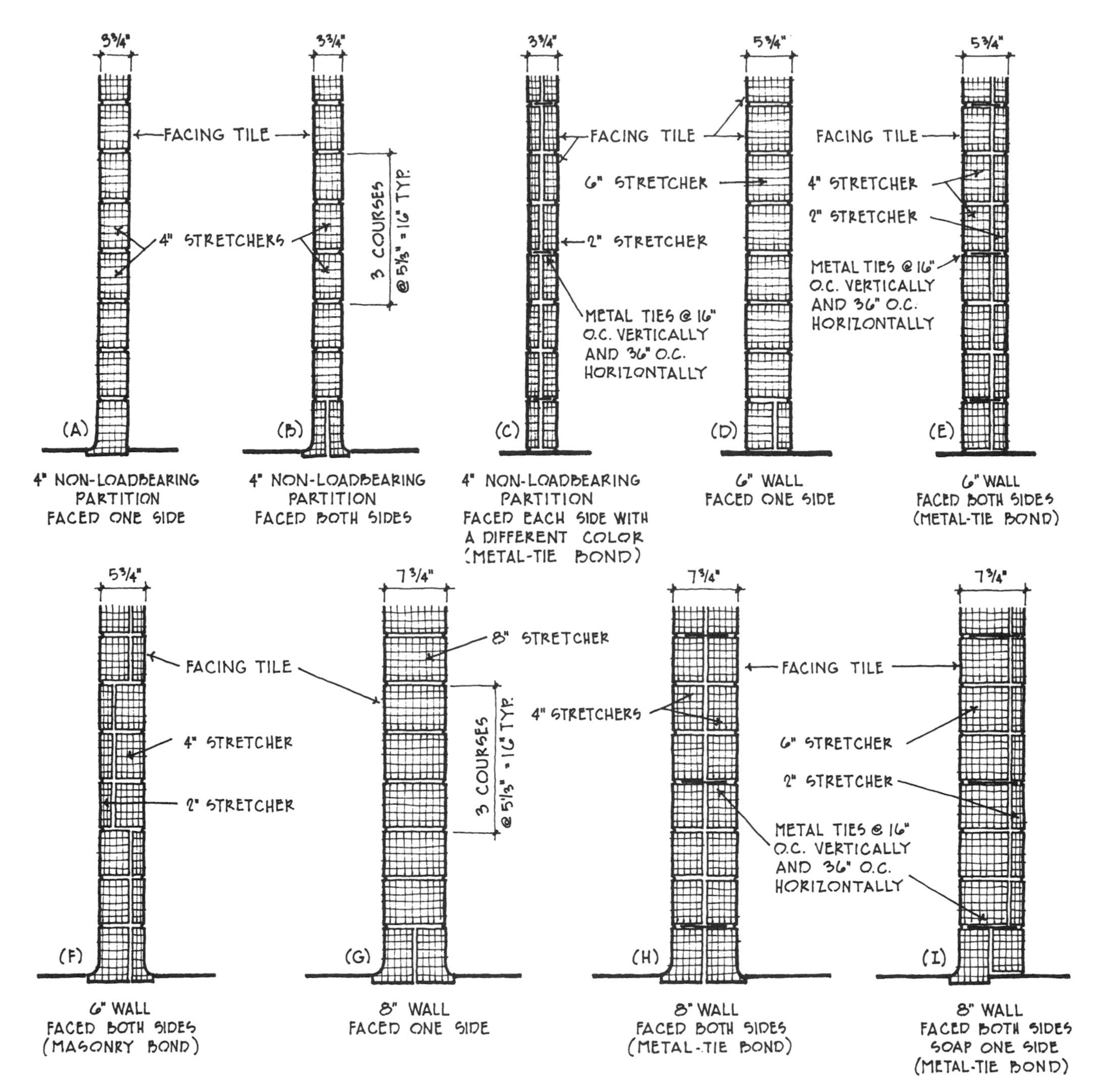

Non-loadbearing structural clay tile partitions. (*From Brick Institute of America,* **Technical Note 22,** ***BIA, Reston, Va.***)

wythe walls, separate header or rowlock courses alternate with stretcher courses to form different patterns. Double-wythe walls are more stable than single-wythe designs because of the increased weight, wider footprint, and through-wall bonding patterns. Piers may be either flush with the wall or projecting on one or both sides. The coursing of the screen panels must overlap the coursing of the piers to provide adequate structural connection.

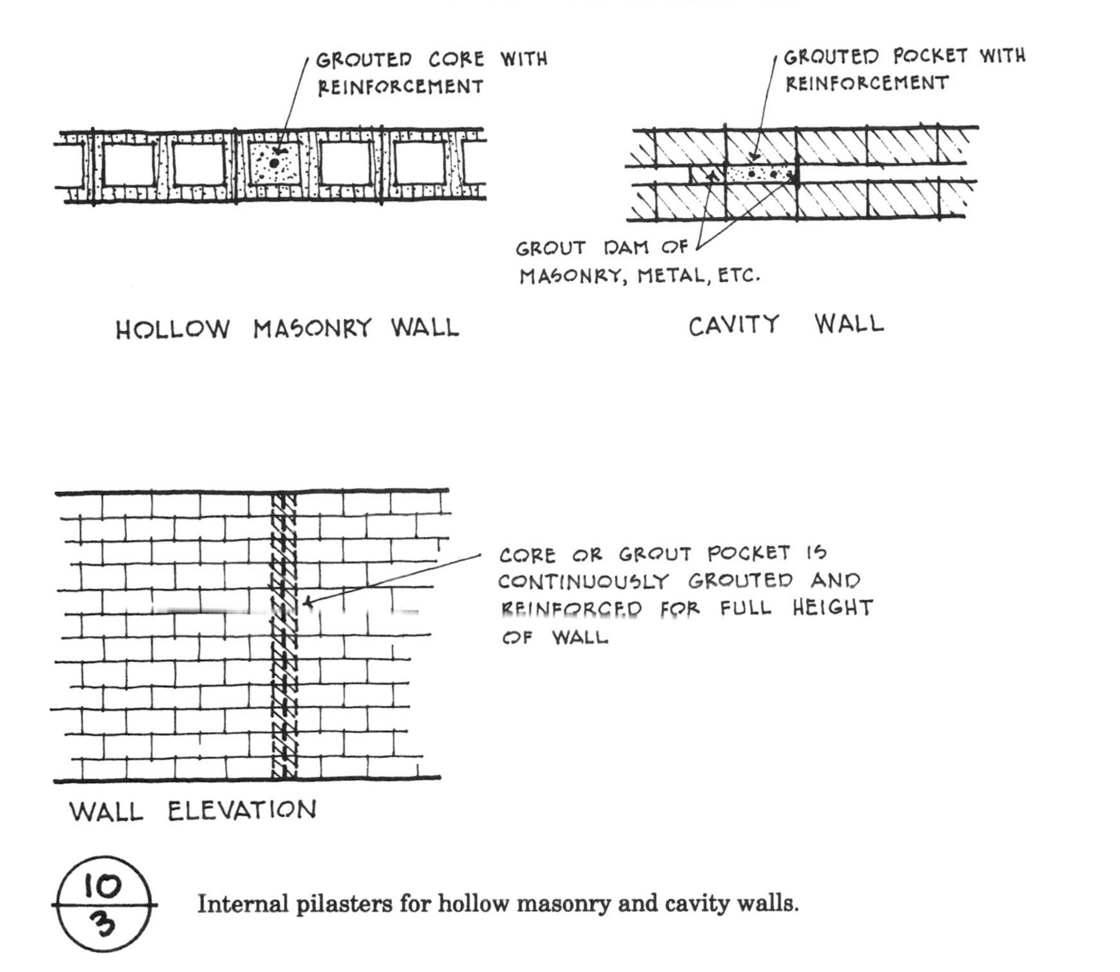

Internal pilasters for hollow masonry and cavity walls.

Regardless of exact design, however, the pattern of units in a pierced wall must provide continuous vertical paths for load transfer to the foundation, and the bearing width of these paths or "columns" should be at least 2 in.

Concrete screen block and clay screen tile are made with a decorative pattern of holes in the units, so it is not necessary to separate them with open head joints. Most unit types are designed to be laid with continuous vertical and horizontal mortar joints in stack bond patterns. The larger area of mortar bedding increases the lateral load resistance of the wall. The continuous bed joints accommodate the installation of horizontal joint reinforcement, and bond beam courses can be added at the top and bottom of the wall for even greater strength.

NCMA has done considerable research on concrete masonry screen walls, and as a result, more is known about this type of unit strength and wall performance than any other type of screen wall. Units should have a minimum compressive strength of 1000 psi (gross) when tested with the cores oriented vertically, and face shells and webs should be at least $^3/_4$ in. thick. Type S mortar is recommended, and truss-type joint reinforcement spaced 16 in. on center vertically.

The orientation of the cores in screen block affects the load carrying capacity of the units. NCMA tested a variety of screen block designs to determine their relative strengths when oriented in different directions. Compressive strength can range from 22 to 82% of that tested with the

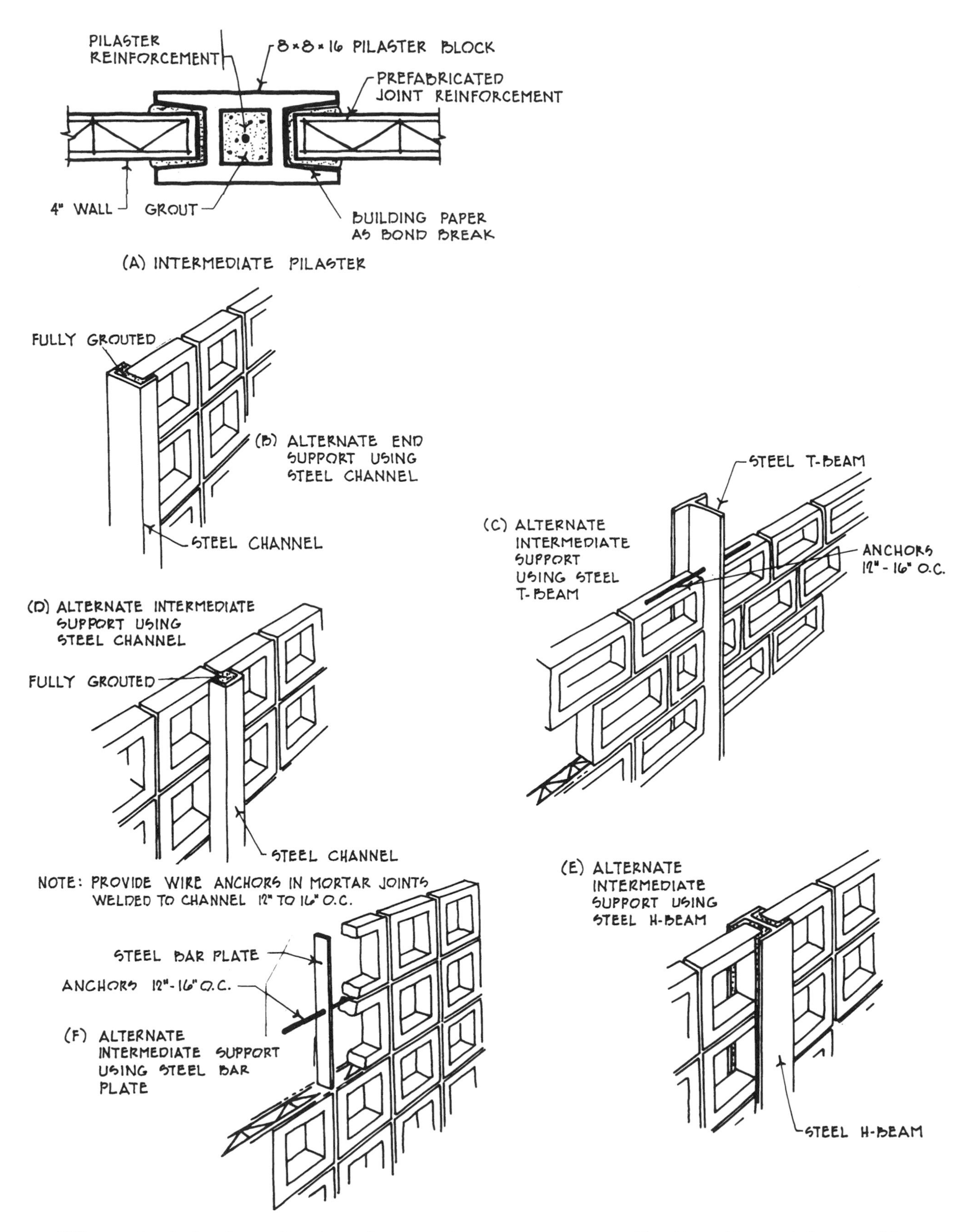

Methods of providing lateral support for masonry screen walls. (*From National Concrete Masonry Association,* TEK Bulletin 5, *NCMA, Herndon, Va.*)

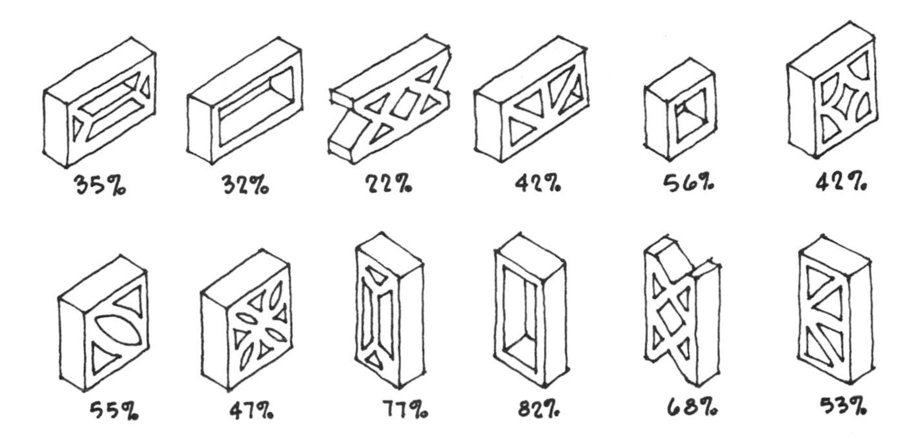

VALUES REPRESENT THE RELATIVE COMPRESSIVE STRENGTH OF UNITS LOADED AS SHOWN, EXPRESSED AS A PERCENTAGE OF STRENGTH WHEN LOADED WITH CELLS VERTICALLY ORIENTED.

Concrete masonry screen block.

Construction	Minimum nominal thickness, t (in.)	Maximum distance between lateral supports (height or length, not both)			
		Nominal thickness of wall (in.)			
		4	6	8	Other
Non-loadbearing					
Exterior	4	6′–0″	9′–0″	12′–0″	18*t*
Interior	4	12′–0″	18′–0″	24′–0″	36*t*
Loadbearing					
Reinforced	6	Not recom-mended	9′–0″	12′–0″	18*t*
Unreinforced	6		9′–0″	12′–0″	18*t*

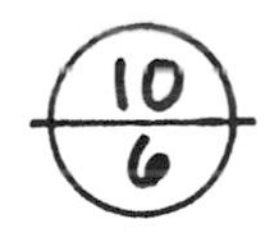

Maximum span for various concrete masonry screen walls, based on height- or length-to-thickness ratios.

cores vertical (*see Fig. 10-5*). Lower strength units should be limited to non-loadbearing applications. Loadbearing screen walls should not carry superimposed loads greater than 10% of the unit strength, or 50 psi on the gross area. If units are separated by open head joints, allowable stresses should be reduced in proportion to the reduction in mortar bedded area.

The table in *Fig. 10-6* shows NCMA recommendations for maximum span between lateral supports. Depending on local code requirements, design wind loads may be based on gross area (total wall surface area including voids), or net wall area (total surface area minus voids). When net wall surface area is used, a shape factor of 1.3 is recommended to provide a factor of safety and to allow for the resistance caused by the surface of the voids.

For most screen wall designs, water which penetrates the system rapidly drains or evaporates because of the openness of the design. Cored brick, however, should not be used because they can trap water in concealed locations. Frogged brick should be placed with the frog side down to avoid the same problem. Water penetration can be limited by through-wall flashing at the top of the wall and copings that are sloped or curved to shed water and provide a drip at both wall faces.

Lateral support for *concrete masonry garden walls and fences* is usually provided by reinforced pilasters or by internal vertical reinforcement. Pilaster spacing and reinforcement requirements are listed in *Fig. 10-7.* Foundations should be placed in undisturbed soil below the frost line. For stable soil conditions where frost heave is not a problem, a shallow continuous footing or pad footing provides adequate stability. Where it is necessary to go deeper to find solid bearing material, where location in relation to property lines restricts footing widths, or where the ground is steeply sloping, a deep pier foundation provides better support. In each instance, the supporting pilaster is tied to the foundation by continuous reinforcing steel. A vertical control joint should be provided on one side of each pilaster support. Reinforcing steel in the panel sections may run vertically or horizontally, depending on the type of units used and the bed joint design of the wall. The designs shown are based on wind loading conditions, but are not intended to resist lateral earth pressure as retaining walls.

Brick garden walls may take a number of different forms. A straight wall without pilasters must be designed with sufficient thickness to provide lateral stability against wind and impact loads. Rule of thumb is that for a 10-lb/sq ft wind load, the height above grade should not exceed three-fourths of the square of the all thickness ($h \leq \frac{3}{4}t^2$). If lateral loads exceed 10 lb/sq ft, the wall should be designed with reinforcing steel.

Brick "pier-and-panel" walls are composed of a series of thin panels (nominal 4 in.) braced intermittently by reinforced masonry piers (*see Fig. 10-8*). Reinforcing steel and foundation requirements are given in the tables in *Fig. 10-9.* Foundation diameter and embedment are based on a minimum soil bearing pressure of 3000 lb/sq ft. Reinforcing steel requirements vary with wind load, wall height, and span. Horizontal steel may be individual bars or wires, or may be prefabricated joint reinforcement, but must be continuous through the length of the wall with splices lapped 16 in.

Since the panel section is not supported on a continuous footing, it actually spans the clear distance between foundation supports, functioning as a deep wall beam (see Chapter 12). Masons build the sections on temporary 2×4 wood footings that can be removed after the wall has cured for at least 7 days.

Serpentine walls and "folded plate" designs are laterally stable because of their shape. This permits the use of very thin sections without the need for reinforcing steel or other lateral support. For non-loadbearing walls of relatively low height, rule-of-thumb design based on empirically derived geometric relationships is used.

Since the wall depends on its shape for lateral strength, it is important that the degree of curvature be sufficient. Recommendations for brick and CMU walls are illustrated in *Fig. 10-10.* The brick wall is based on a radius of curvature not exceeding twice the height of the wall above finished grade, and a depth of curvature from front to back no less than one-half of the height. A maximum height of 15 times the thickness is recommended for the

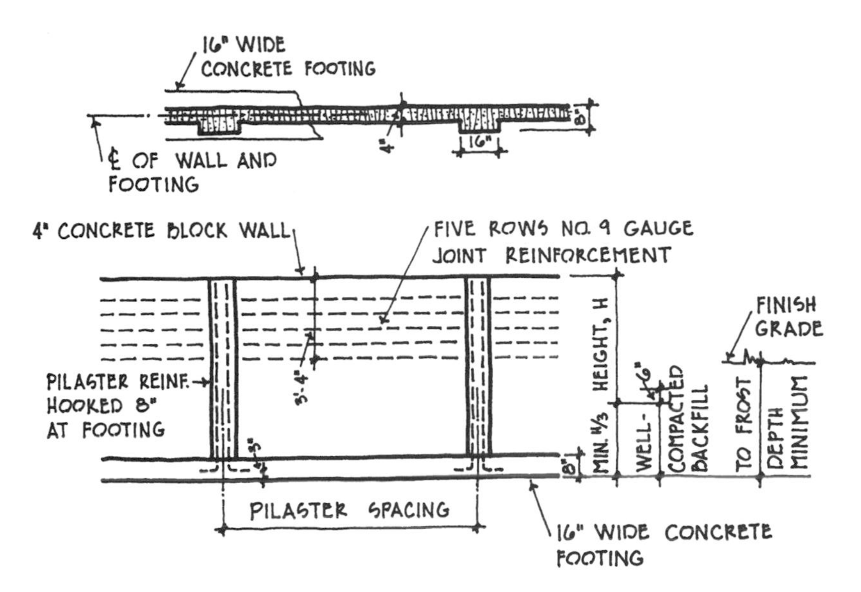

(A) Fence with pilasters

Pilaster spacing for wind pressure				H	Reinforcement for wind pressure			
5 psf	10 psf	15 psf	20 psf		5 psf	10 psf	15 psf	20 psf
19′-4″	14′-0″	11′-4″	10′-0″	4′-0″	1—No. 3	1—No. 4	1—No. 5	2—No. 4
18′-0″	12′-8″	10′-8″	9′-4″	5′-0″	1—No. 3	1—No. 5	2—No. 4	2—No. 5
15′-4″	10′-8″	8′-8″	8′-0″	6′-0″	1—No. 4	1—No. 5	2—No. 5	2—No. 5

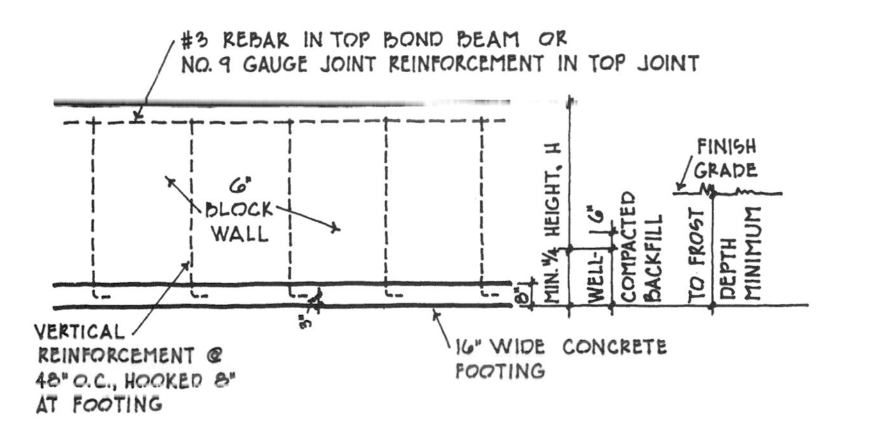

(B) Fence without pilasters

H	Reinforcement for wind pressure			
	5 psf	10 psf	15 psf	20 psf
4′-0″	1—No. 3	1—No. 3	1—No. 4	1—No. 4
5′-0″	1—No. 3	1—No. 4	1—No. 5	1—No. 5
6′-0″	1—No. 3	1—No. 4	1—No. 5	2—No. 4

Pilaster spacing and reinforcing requirements for concrete block fences. (*From Frank A. Randall and William C. Panarese,* Concrete Masonry Handbook for Architects, Engineers, and Builders, *Portland Cement Association, Skokie, Ill., 1991.*)

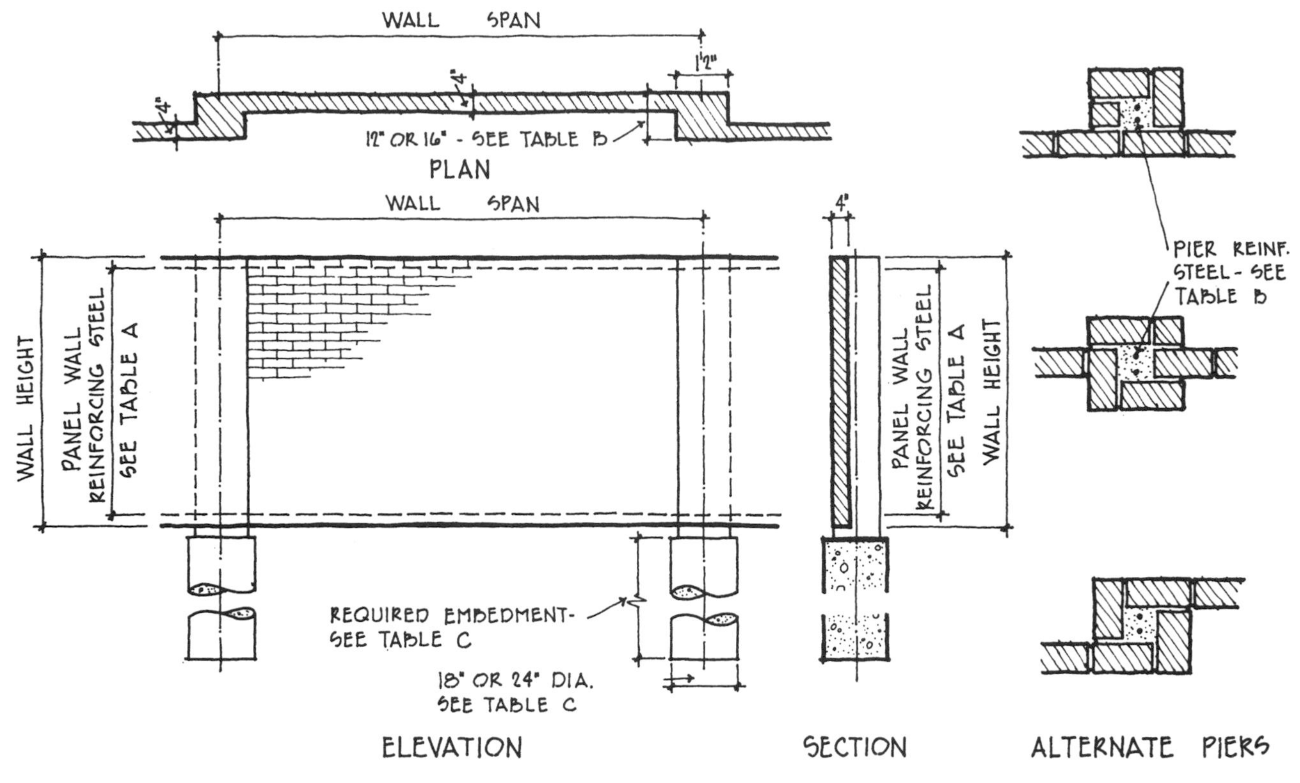

Brick pier-and-panel garden walls. (*From Brick Institute of America,* Technical Note 29A, *BIA, Reston, Va.*)

CMU wall, and depth-to-curvature ratios are slightly different. Free ends of a serpentine wall should be supported by a pilaster or short-radius return for added stability. Thicker sections and taller walls may be built if proper design principles are applied to resist lateral wind loads.

All freestanding masonry walls, regardless of thickness, must be properly capped to prevent excessive moisture infiltration from the top. The appearance and character of a wall is substantially affected by the type of cap or coping selected, including natural stone, cast stone, terra cotta, metal, brick, or concrete masonry (*see Fig. 10-11*). The thermal and moisture expansion characteristics of the wall and coping materials should be similar. Control and expansion joint locations should be calculated (refer to Chapter 9), and joints should be tooled concave to compress the mortar against the face of the units, and to decrease porosity at the surface. Copings should be sloped to shed water and should project beyond the face of the wall a minimum of ½ in. on both sides to provide a positive drip and prevent water from flowing down the face of the wall. Through-wall flashing should be installed immediately below the coping to prevent excessive water absorption, and the coping then secured with metal anchors. Grouting of hollow units, cavities, and hollow sections also increases durability and strength by eliminating voids where water can accumulate and cause freeze/thaw damage or efflorescence. The combined use of masonry and metal sections in constructing fences should allow for differential thermal expansion and contraction between the two materials.

Natural stone is used to build freestanding dry-stack and mortared

TABLE A PANEL WALL REINFORCING STEEL									
	Vertical Spacing (in.)*								
Wall span (ft)	*Wind load 10 psf*			*Wind load 15 psf*			*Wind load 20 psf*		
	A	*B*	*C*	*A*	*B*	*C*	*A*	*B*	*C*
8	45	30	19	30	20	12	23	15	9.5
10	29	19	12	19	13	8.0	14	10	6.0
12	20	13	8.5	13	9.0	5.5	10	7.0	4.0
14	15	10	6.5	10	6.5	4.0	7.5	5.0	3.0
16	11	7.5	5.0	7.5	5.0	3.0	6.0	4.0	2.5

*A, two - No. 2 bars; B, two - $\frac{3}{16}$-in. diam wires; C, two - 9 gauge wires.

TABLE B PIER REINFORCING STEEL*									
Wall span (ft)	*Wind load 10 psf*			*Wind load 15 psf*			*Wind load 20 psf*		
	Wall height (ft)			*Wall height (ft)*			Wall height (*ft*)		
	4	*6*	*8*	*4*	*6*	*8*	*4*	*6*	*8*
8	2#3	2#4	2#5	2#3	2#5	2#6	2#4	2#5	2#5
10	2#3	2#4	2#5	2#4	2#5	2#7	2#4	2#6	2#6
12	2#3	2#5	2#6	2#4	2#6	2#6	2#4	2#6	2#7
14	2#3	2#5	2#6	2#4	2#6	2#6	2#5	2#5	2#7
16	2#4	2#5	2#7	2#4	2#6	2#7	2#5	2#6	2#7

*Within heavy lines 12 by 16-in. pier required. All other values obtained with 12 by 12-in. pier.

TABLE C REQUIRED EMBEDMENT FOR PIER FOUNDATION*									
Wall span (ft)	*Wind load 10 psf*			*Wind load 15 psf*			*Wind load 20 psf*		
	Wall height (ft)			*Wall height (ft)*			*Wall height (ft)*		
	4	*6*	*8*	*4*	*6*	*8*	*4*	*6*	*8*
8	2′-0″	2′-3″	2′-9″	2′-3″	2′-6″	3′-0″	2′-3″	2′-9″	3′-0″
10	2′-0″	2′-6″	2′-9″	2′-3″	2′-9″	3′-3″	2′-6″	3′-0″	3′-3″
12	2′-3″	2′-6″	3′-0″	2′-3″	3′-0″	3′-3″	2′-6″	3′-3″	3′-6″
14	2′-3″	2′-9″	3′-0″	2′-6″	3′-0″	3′-3″	2′-9″	3′-3″	3′-9″
16	2′-3″	2′-9″	3′-0″	2′-6″	3′-3″	3′-6″	2′-9″	3′-3″	4′-0″

*Within heavy lines 24-in. diam. foundation required. All other values obtained with 18-in. diam. foundation.

Design tables for brick pier-and-panel garden walls. (*From Brick Institute of America*, Technical Note 29A, *BIA*, *Reston, Va.*)

walls. *Dry-stack walls* laid without mortar are generally 18 to 24 in. wide and depend only on gravity for their stability. Trenches are dug to below the frost line, and if the ground slopes, may take the form of a series of flat terraces. A concrete footing may be poured in the trench, but walls are often laid directly on undisturbed soil. Two rows of large stones laid with their top planes slightly canted toward the center will provide a firm base. All stones placed below grade should be well packed with earth in all the crevices. Stones should be well fitting, requiring a minimum number of

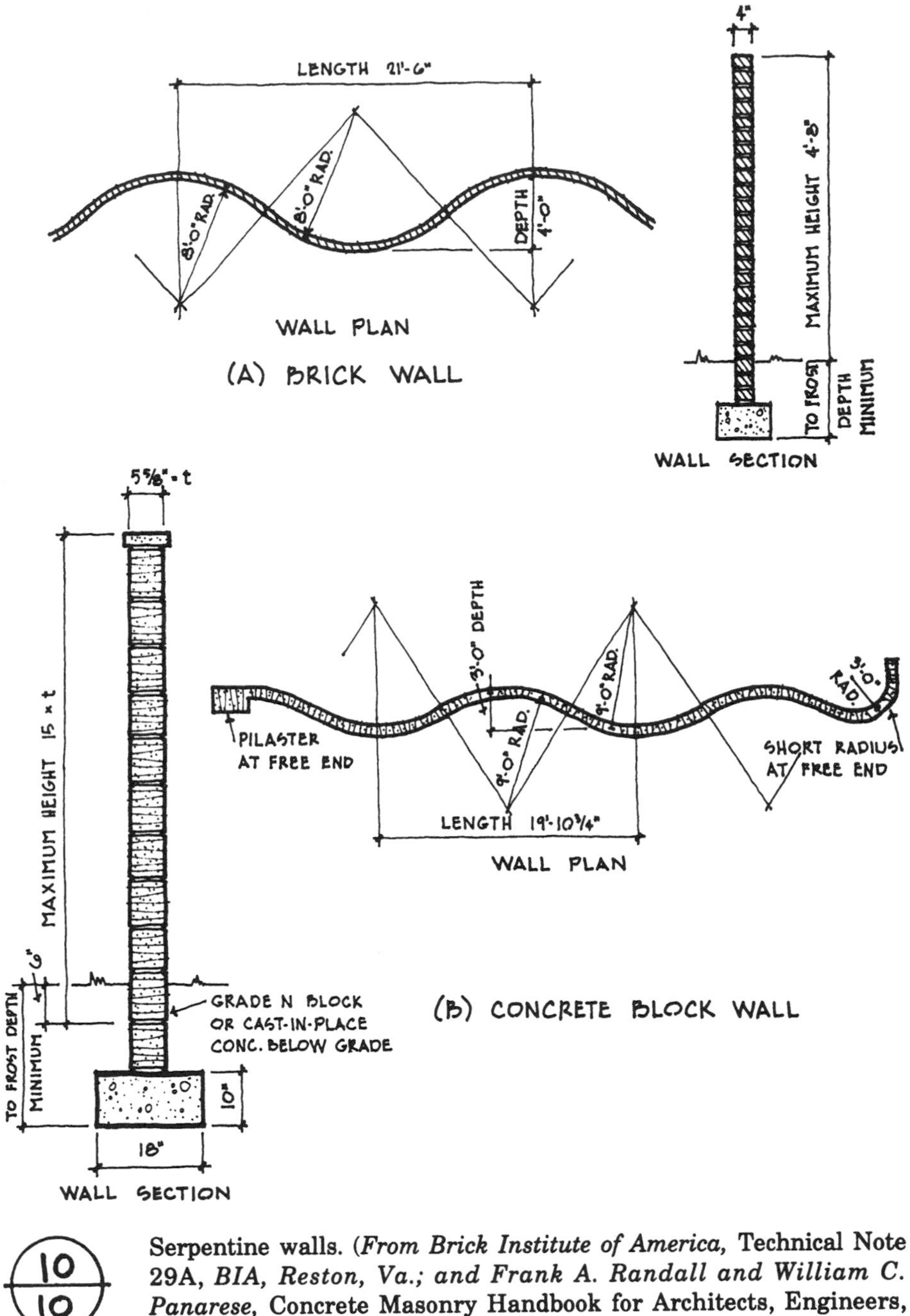

10/10 Serpentine walls. (*From Brick Institute of America,* Technical Note 29A, *BIA, Reston, Va.; and Frank A. Randall and William C. Panarese,* Concrete Masonry Handbook for Architects, Engineers, and Builders, *Portland Cement Association, Skokie, Ill., 1991.*)

shims. A bond stone equal to the full wall width should be placed every 3 or 4 ft in each course to tie the inner and outer wythes together. All of the stones should be slightly inclined toward the center of the wall so that the weight leans in on itself (*see Fig. 10-12*). Greater wall heights require more incline from base to cap. Wall ends and corners are subject to the highest stress and should be built with stones tightly interlocked for stability. Relatively flat slabs of roughly rectangular shape work best for cap stones. The top course should be as level as possible for the full length of the wall.

Mortared stone walls are laid on concrete footings poured below the

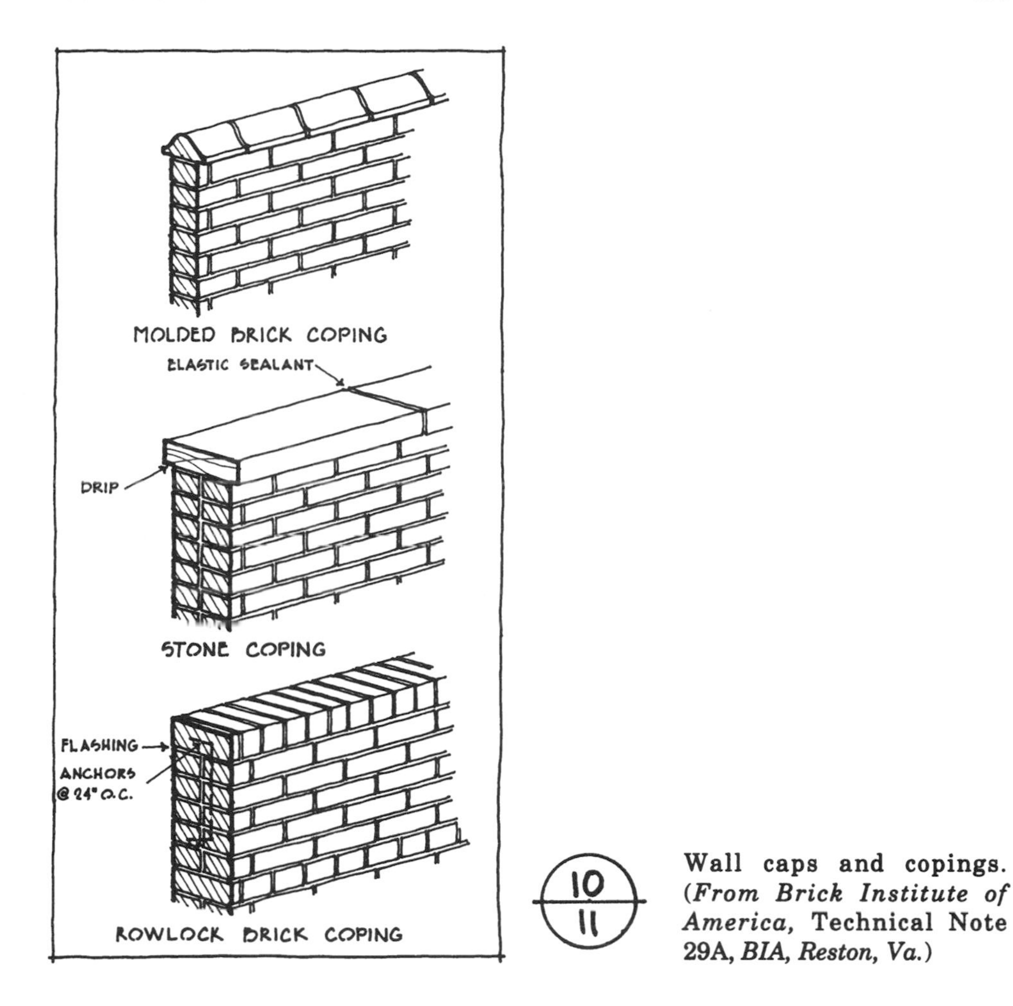

Wall caps and copings. *(From Brick Institute of America,* **Technical Note 29A,** ***BIA, Reston, Va.)***

frost line. Rubble stone or fieldstone walls are laid up in much the same way as dry-stack walls except that the voids and cavities are filled with mortar. Type S, M, or N mortar should be used, and each course should be laid in a full mortar bed for maximum bond and strength. In ashlar construction, building codes generally require that bond stones be uniformly distributed and account for no less than 10% of the exposed face area. Mortared rubble stone walls less than 24 in. thick must have bond stones at a maximum of 3 ft on center vertically and horizontally. For thicknesses greater than 24 in., provide one bond stone for each 6 sq ft of wall surface. For non-loadbearing conditions, the minimum thickness of the wall must be sufficient to withstand all horizontal forces and the vertical dead load of the stone itself. For relatively low structures, a mortared wall thickness of as little as 8 in. has proven satisfactory.

10.3 PANEL WALLS

A masonry panel wall is an exterior, non-loadbearing wall wholly supported at each floor by a concrete slab or beam, or by steel shelf angles. Masonry veneer constructed between relieving angles on multi-story buildings, and masonry infill between the floors and columns of structural frames are panel wall sections. Panel walls must be designed to resist lateral forces and transfer them to adjacent structural members. Design is governed by empirical length- or height-to-thickness ratios in most codes. Top and bottom anchorage must secure the panel against lateral loads acting in

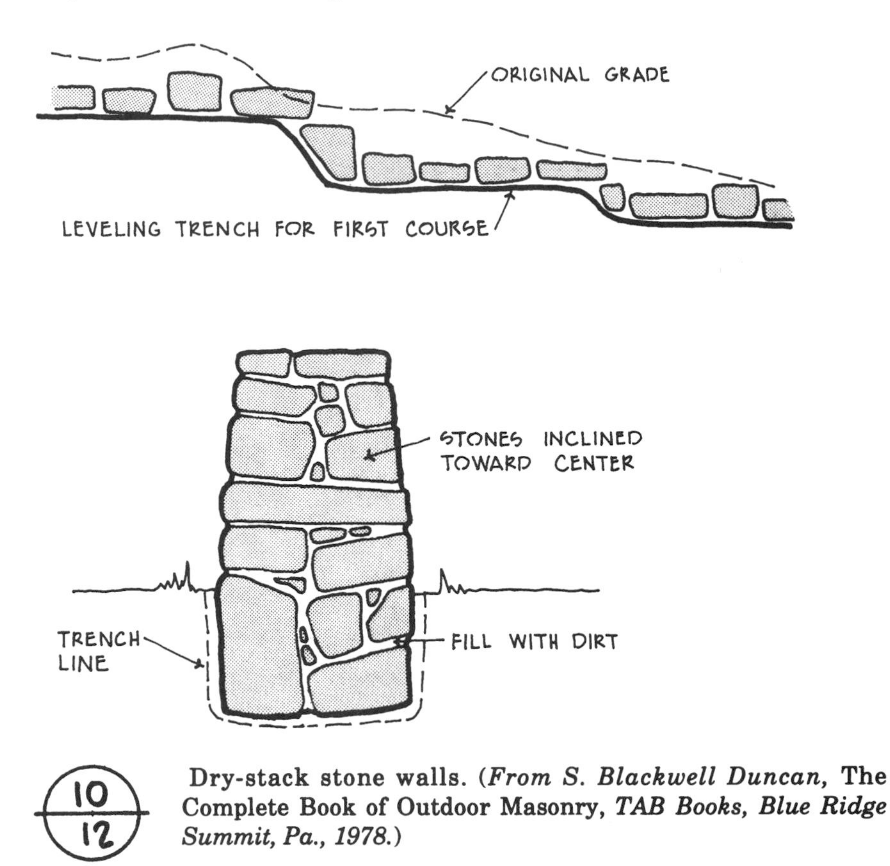

Dry-stack stone walls. (*From S. Blackwell Duncan,* **The Complete Book of Outdoor Masonry,** *TAB Books, Blue Ridge Summit, Pa., 1978.*)

either direction, but must also permit differential movement between the masonry and the steel or concrete framing. To accommodate movement, connections should be flexible, and allowance should be made for expansion and contraction (refer to Chapter 9). Four-inch veneer panels also require intermediate anchorage to a backing wall for lateral stability and load transfer.

Differential movement can be critical in panel wall construction. Steel and concrete are both subject to volume changes as well as permanent deflections caused by plastic flow. Without proper movement joints below a support, the masonry panel is actually squeezed between the frame members and can fail in bending under an axial load that it is not reinforced to withstand. Clay masonry is particularly vulnerable in these circumstances since it is subject to irreversible moisture expansion. If restrained by the concrete or steel structure, the brick panel can easily buckle or spall. Horizontal expansion joints will absorb the movement and protect the masonry. Calculations of minimum clearance should be based on expected environmental conditions, material characteristics, and span (refer to Chapter 9). The joint should be filled with a soft, compressible material and then caulked with an elastomeric sealant.

Shelf angles that are bolted to framing members are better able to move under stress than are angles welded rigidly in place. Bolted connections also permit adjustments for proper alignment in both the horizontal and vertical directions. Flashing should be installed immediately above all shelf angles or supports, and weepholes must be located just above the flashing to drain accumulated moisture.

10.3.1 Glass Block Panels

Glass block is used in non-loadbearing interior and exterior applications, and is most often installed as single-wythe, stack bond panel walls. The compressive strength of the units is sufficient to carry the dead load of the material weight for a moderate height. Intermediate supports at floor and roof slabs require care in detailing to allow expansion and contraction of dissimilar materials (*see Fig. 10-13*). Deflection of supporting members above or below glass block panels should be limited to $L/600$. Movement joints at the perimeter of the panels should be at least $\frac{3}{8}$ to $\frac{1}{2}$ in. Glass blocks are normally laid in Type S or Type N cement-lime mortar, and bed joints are reinforced with ladder-type horizontal joint reinforcement spaced a maximum of 16 in. on center vertically. Since the bond between mortar and glass block is relatively weak, head and jamb recesses or channel-type supports are usually required to increase the lateral resistance of the panel section. If jamb recesses or channels are not provided in the adjacent wall, jamb anchors are required at a maximum spacing of 16 in. on center (*see Fig. 10-14*).

Size and area limitations for glass block wall panels vary among the major building codes. The 1995 edition of the MSJC Code includes for the first time a chapter on glass unit masonry. The requirements prescribed by the various codes for standard $3\frac{7}{8}$ in. (nominal 4 in.) block are shown in *Fig. 10-15*. Requirements for $3\frac{1}{8}$ in. (nominal 3 in.) panels are more restrictive. Whenever panels exceed code requirements for area limitations, they must be subdivided by metal stiffeners and/or supports (*see Fig. 10-16*). Vertical stiffeners should also be installed at the intersection of curved and straight sections, and at every change in direction in multi-curved panels. All metal accessories, including joint reinforcement, jamb anchors, and stiffeners should be hot-dip galvanized after fabrication in accordance with ASTM A153, *Standard Specification for Zinc Coating (Hot-Dip) on Iron or Steel Hardware*. Panels constructed of solar reflective block must be protected from run-off of rainwater from concrete, masonry or metal materials located above the panel. Harmful substances may stain or etch the reflective block surface, so panels should be recessed a minimum of 4 in. and a drip provided at the edge of the wall surface above.

Using wedge-shaped head joints, panels can be curved at various radii depending on the size of the units. *Figure 10-17* shows the smallest achievable radius for each of four different block lengths. Ninety degree corners may be laid to a corner post of wood or steel, or may incorporate special shaped bullnose or hexagonal units (*see Fig. 10-18*).

10.3.2 Infill Panels

Masonry panel walls constructed between columns and between foundation or floor slab and the spandrel beam or floor slab above are called infill panels. These panels may form the exterior cladding, or serve as backup in a cavity wall.

Masonry infill panels are non-loadbearing, and are most often designed using empirical h/t ratios. However, analytical methods of design may be used to maximize efficiency and economy by reducing wall thickness and weight required by other parameters such as fire, sound, or thermal resistance. Flexible connection for lateral support at the top of the wall may be provided by clip angles, channels, or other mechanical anchorage (*see Fig. 10-19*). An open joint or compressible soft joint must also be provided at

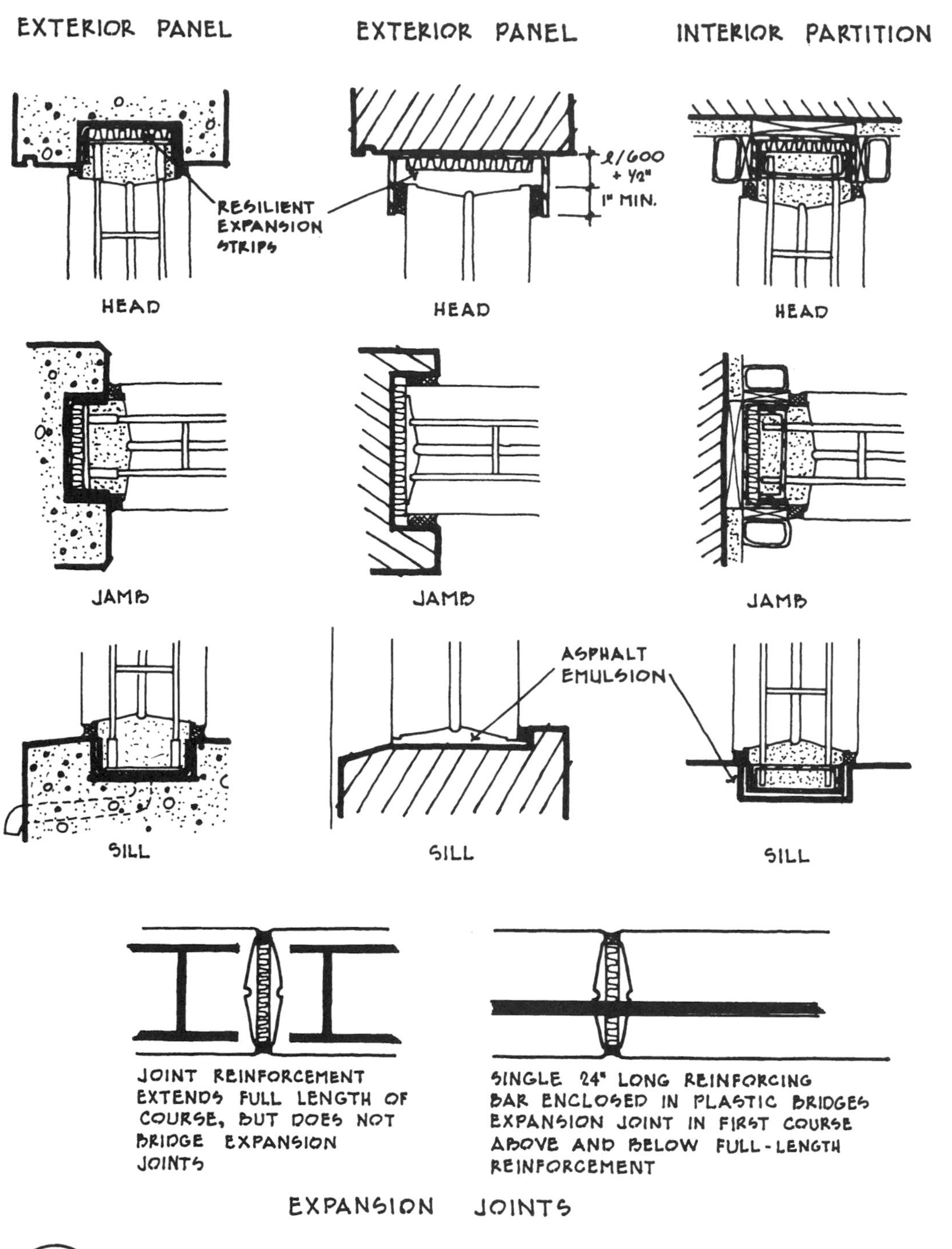

Glass block wall sections.

the top of the wall to prevent the weight of the structure from bearing on non-loadbearing walls after deflection and frame shortening have occurred. Non-loadbearing infill panels used as shear walls must be rigidly connected to the structural frame for load transfer. A computer program called *CavWal* is available from NCMA to assist in the design and detailing of both the infill and veneer portions of non-loadbearing masonry cavity

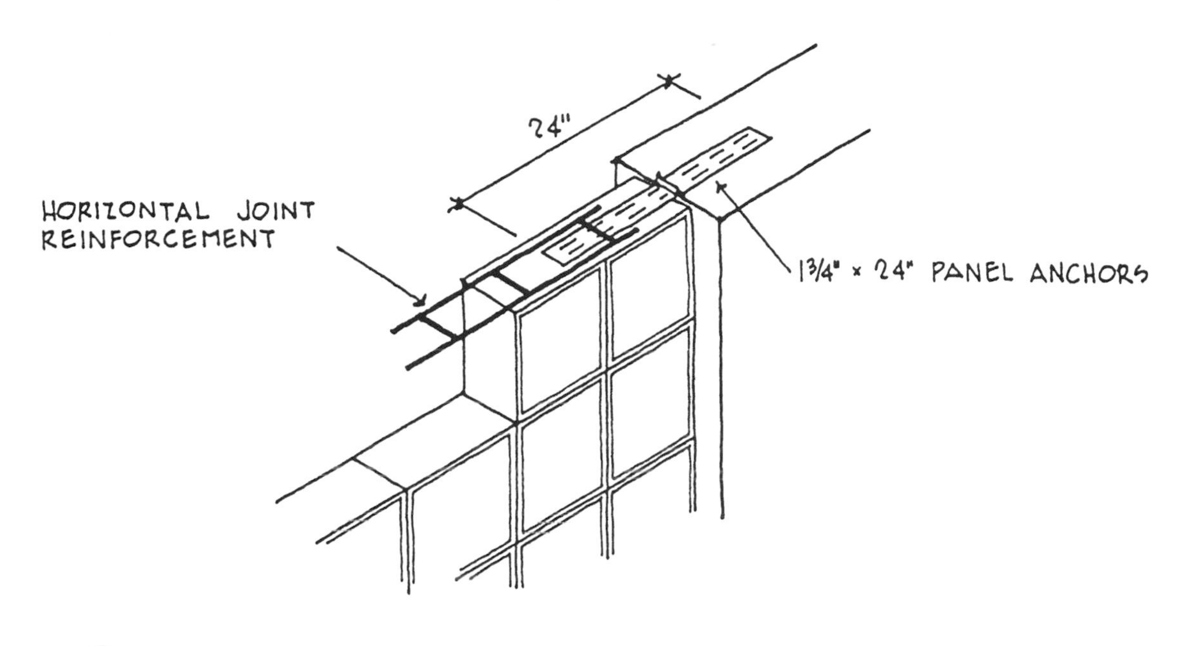

Glass block panel anchor.

walls. In addition to lateral seismic and wind load capacity, the program gives related information on water permeance, fire ratings, heat transfer, and crack control for each design solution.

10.4 MASONRY CURTAIN WALLS

Masonry curtain walls are designed to span horizontally or vertically between lateral connections without intermediate support. Horizontal curtain walls span across the face of columns or cross walls where they are connected for transfer of wind loads to the structure. Multi-story curtain walls are wholly supported at the foundation without intermediate shelf angles, and are connected only at the floors and roof for lateral load transfer. Masonry curtain walls can be designed by empirical methods or by engineering analysis. Empirical methods are governed by h/t ratios, but analytical design is not limited by such restrictions, so walls can be built to span multiple structural bays. Curtain walls may be single- or multi-wythe design, and may incorporate reinforcing steel to increase lateral load resistance or distance between lateral supports.

10.4.1 Empirical Design

Using empirical h/t ratios, an unreinforced 8-in. single-wythe hollow masonry curtain wall can be built as high as 15 ft 4 in. between lateral supports under *Standard Building Code* requirements, and 12 ft 0 in. under UBC or MSJC requirements. This type of wall is often used for single-story retail and warehouse construction over lightweight steel structural frames. Both hollow concrete block and hollow brick are used, depending on the aesthetic requirements of the project. Lateral support connections at the roof line must be flexible to permit deflection of the structural frame

MAXIMUM GLASS BLOCK PANEL SIZES						
	Exterior walls			*Interior walls*		
Building code	*Area (sq ft)*	*Height (ft)*	*Length (ft)*	*Area (sq ft)*	*Height (ft)*	*Length (ft)*
Uniform Building Code	144	15	15	250	25	25
BOCA National Code	250*	20	25	—	—	—
Standard Building Code	144	20	25	250	25	25
MSJC Code	See graph below			250	20	25

*Panels larger than 144 sq ft must have supplementary stiffeners.

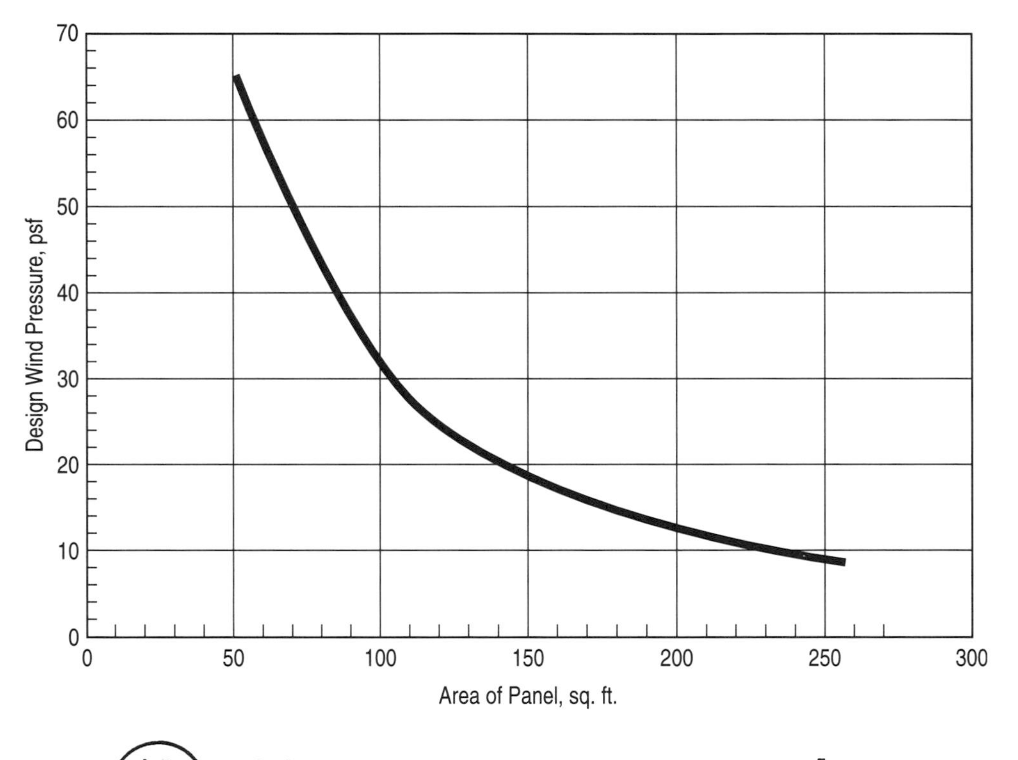

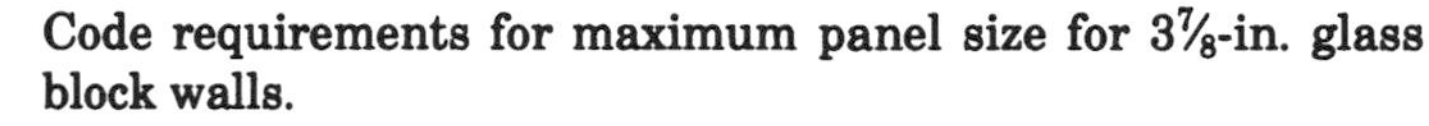

10/15 **Code requirements for maximum panel size for 3⅞-in. glass block walls.**

independent of the masonry, as well as differential thermal and moisture movement between the various materials (*see Fig. 10-20*). Six-inch units may also be used in single-wythe curtain wall construction, but h/t ratios generally limit spans to less than 12 ft between supports. Four-inch units must be reinforced to achieve practical spans.

Curtain walls of both hollow and solid units require the moisture protection of flashing at the top and bottom of the wall, as well as above and below any openings in the wall. General moisture protection should follow the recommendations given in Chapter 9.

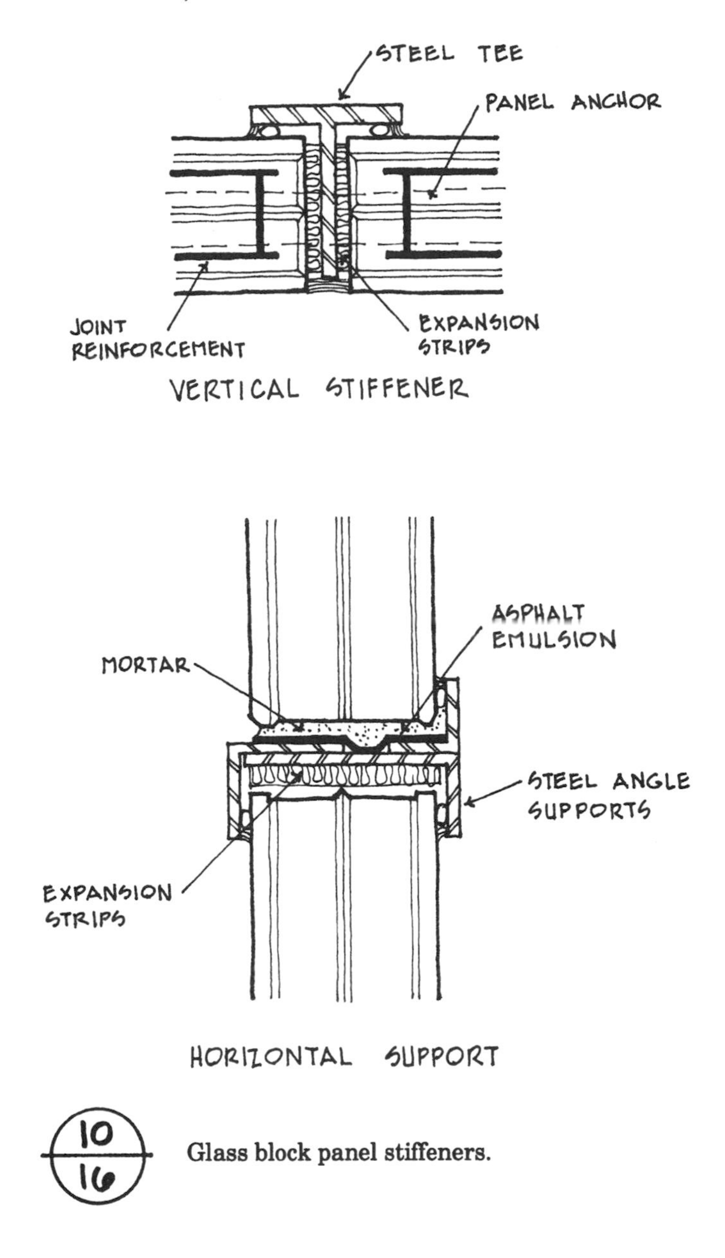

Glass block panel stiffeners.

10.4.2 Analytical Design

Masonry curtain walls which exceed allowable h/t ratios must be analytically designed, and may require reinforcing steel to resist flexural tensile stresses from either positive or negative wind pressures. The maximum moment in the wall (*see Fig. 10-21*) is used to calculate the amount of reinforcement needed to resist loads.

Long walls are usually designed to span horizontally between columns or cross walls, and must therefore resist bending and flexure in this direction. For horizontal spans, only horizontal steel is generally required to provide adequate resistance. Walls that span vertically from floor to roof or multi-story walls that span several floors generally require vertical steel to resist bending and flexure in the vertical direction. For thin walls, additional load distribution is provided by anchorage to a back-

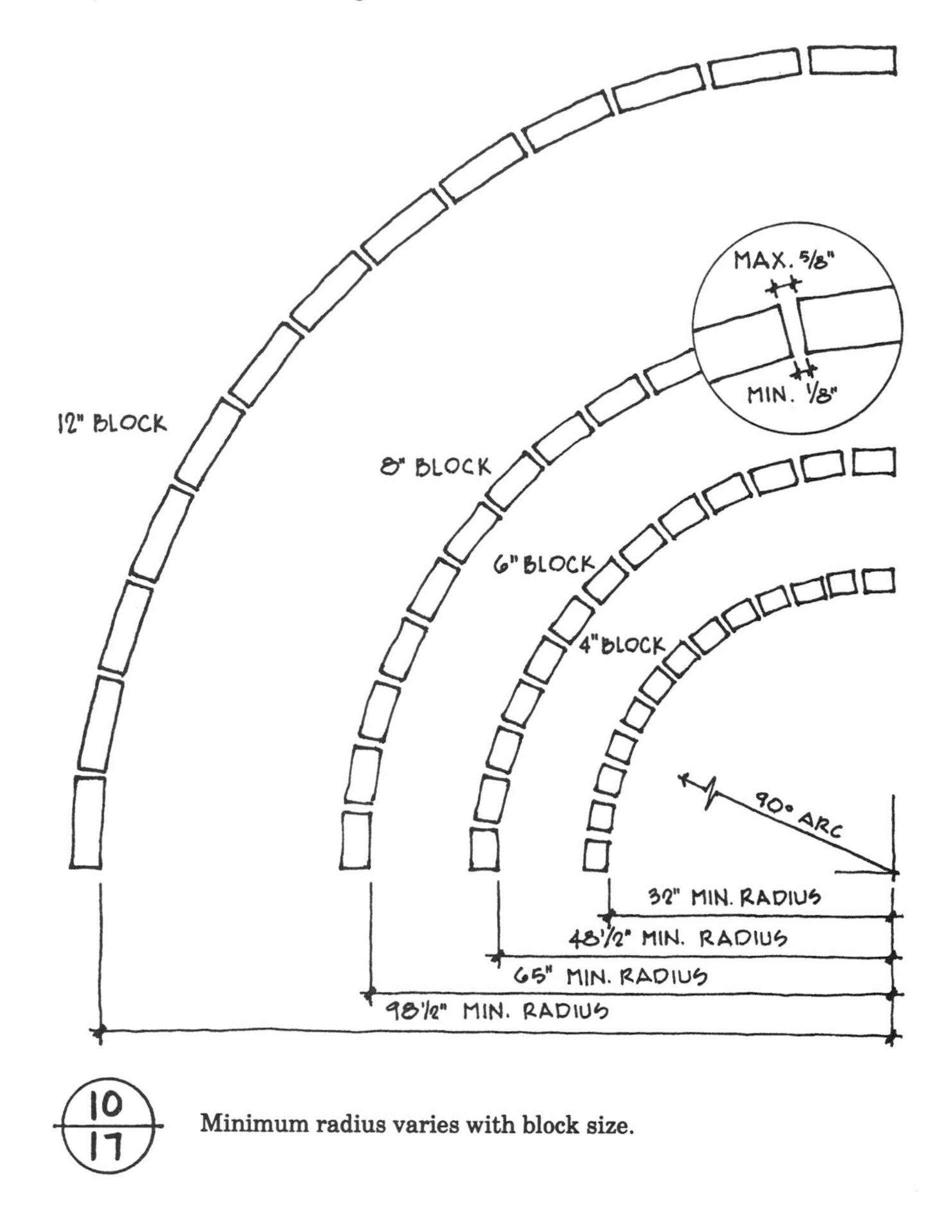

Minimum radius varies with block size.

ing wall of concrete, masonry, or stud construction. If the walls were loadbearing, the compressive load itself would increase flexural resistance, and therefore reduce the amount of steel required.

Analytical design should be in accordance with the code requirements outlined in Chapter 12. For nominal 4-in. brick curtain walls, BIA engineers have prepared the design tables in *Fig. 10-22,* based on a maximum deflection of *L*/200 for simply supported solid walls without openings. The area of reinforcing steel required is for each face of the wall, for various support conditions, span length, and wind loads. After the required area is determined from Table *A,* the bar size and spacing should be selected from Table *B.* Prefabricated joint reinforcement is best because the wires cannot easily be displaced or misaligned during construction. Special conditions at openings must be investigated on an individual basis since they reduce the section of the wall and its stiffness, and thus affect the area of reinforcement, strength, and distribution of lateral load.

All masonry curtain wall design calculations must be verifiable to

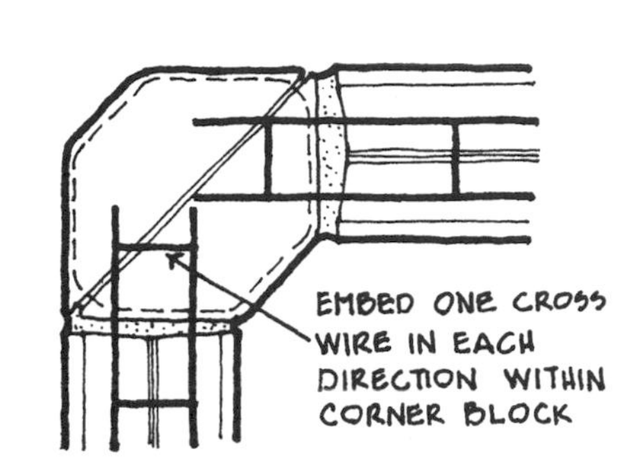

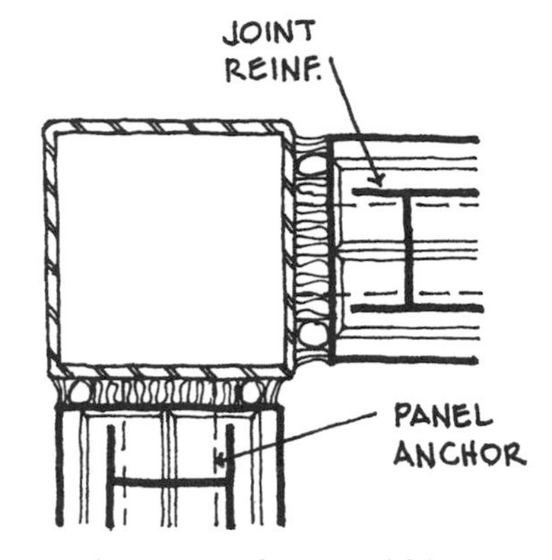

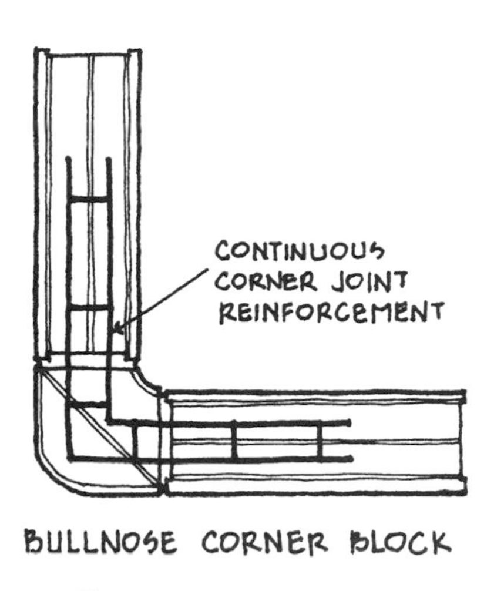

Glass block corner details.

code authorities for waiver of the h/t ratios. Flashing and weepholes for control of moisture which penetrates the wall should be detailed in accordance with the principles outlined in Chapter 9.

10.5 MASONRY VENEER AND CAVITY WALLS

The most typical non-loadbearing application of masonry is the construction of veneers over structural frames of concrete, steel or wood, or in masonry cavity wall systems. A veneer is defined as "a nonstructural facing attached to a backing for the purpose of ornamentation, protection, or insulation, but not bonded to the backing so as to exert a common reaction under load." A non-loadbearing masonry veneer mechanically anchored to a loadbearing or non-loadbearing masonry backing wall is commonly referred to as a cavity wall even though the outer wythe of masonry technically functions as a veneer. Connectors used to attach masonry veneers to backing walls of non-masonry materials are called anchors. Connectors used to attach masonry veneers to masonry backing walls are called ties. The terminology is often misapplied and can sometimes be confusing. Within the masonry industry, the term *veneer* is most often used to describe masonry cladding over non-

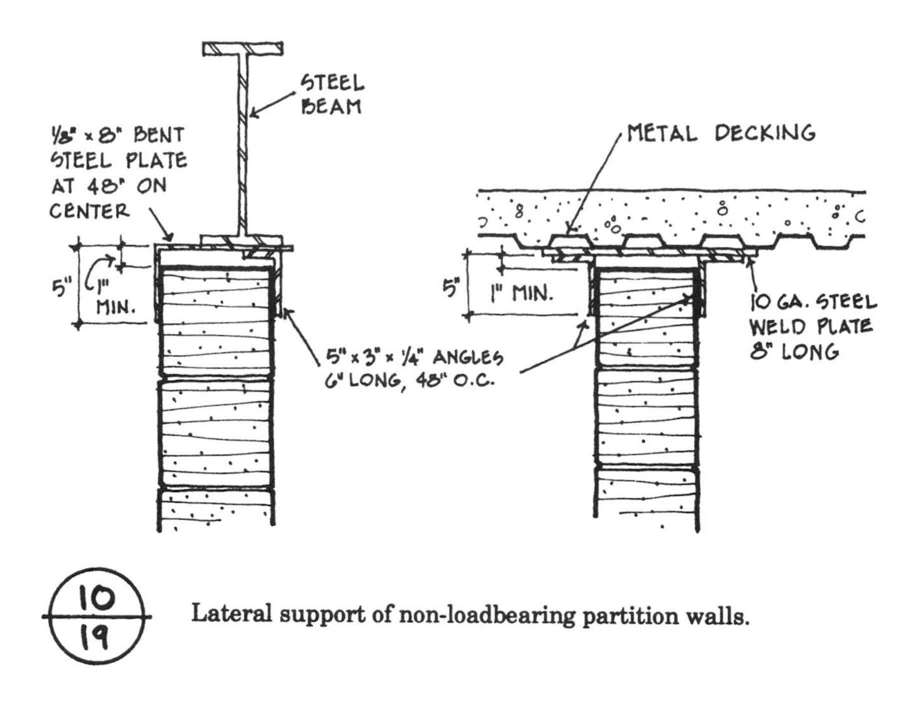

Lateral support of non-loadbearing partition walls.

masonry backing walls. The term *cavity wall* refers to multi-wythe masonry walls in which the wythes are separated by an air space and connected by metal ties, regardless of the bearing or non-bearing function of either wythe.

There are two basic methods of attaching masonry veneer. *Adhered veneer* is secured by adhesion with a bonding material to a solid backing. Adhered veneer does not support its own weight. *Anchored veneer* is secured by metal anchors attached to either a solid backing or structural frame. The weight of an anchored veneer is typically supported by the structure or by shelf angles attached to the structure, at every floor. An anchored veneer may also be fully supported at the foundation without intermediate shelf angles, and requires only lateral anchoring throughout its height to backing walls capable of transferring lateral loads to the structure.

10.5.1 Adhered Veneer

Adhesion attachment is normally limited to thin sections of terra cotta or stone facing. Codes generally limit the weight of the veneer to 15 lb/sq ft. That weighing more than 3 lb/sq ft may not exceed 36 in. in the greatest dimension or 720 sq in. in area. The bond of an adhered veneer to the supporting element must be designed to withstand a shearing stress of 50 psi. Differential thermal and moisture movement characteristics should be considered in selecting backing and facing materials. An expanding clay masonry facing and a contracting concrete backup are not compatible when relying exclusively on an adhesive bond.

For fully adhered applications, a paste of portland cement and water is brushed on the moistened surfaces of the backing and the veneer unit. Type S mortar is then applied to the backing and to the unit, resulting in a mortar thickness of not less than ½ in. or more than 1¼ in. If the surfaces are clean and properly moistened, the neat cement paste provides good bond to both surfaces, but a mechanical key formed by ribs or flanges on the back of the masonry helps support heavier units. Adhesion attachment

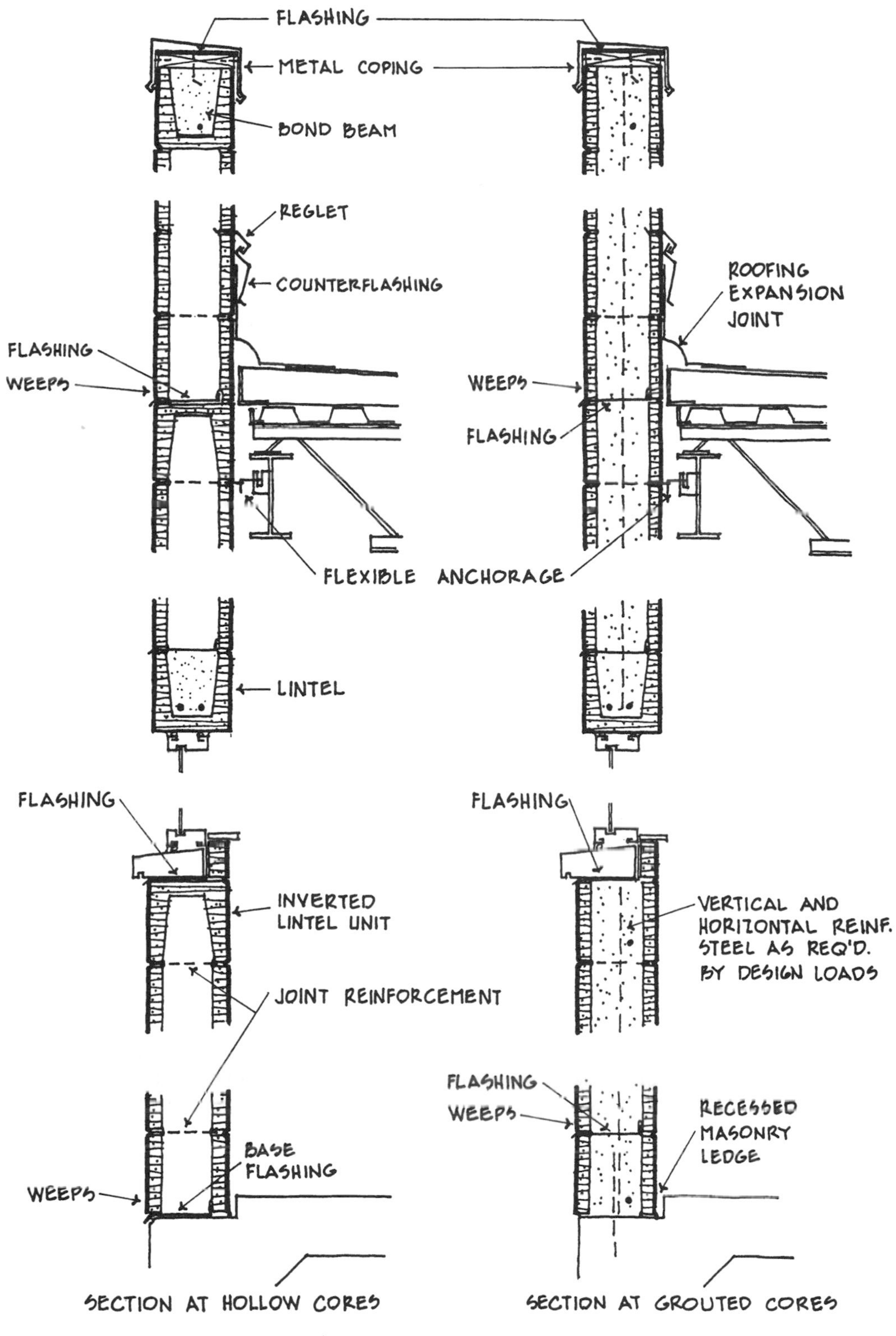

Single-wythe masonry curtainwall requires flexible connection at lateral support.

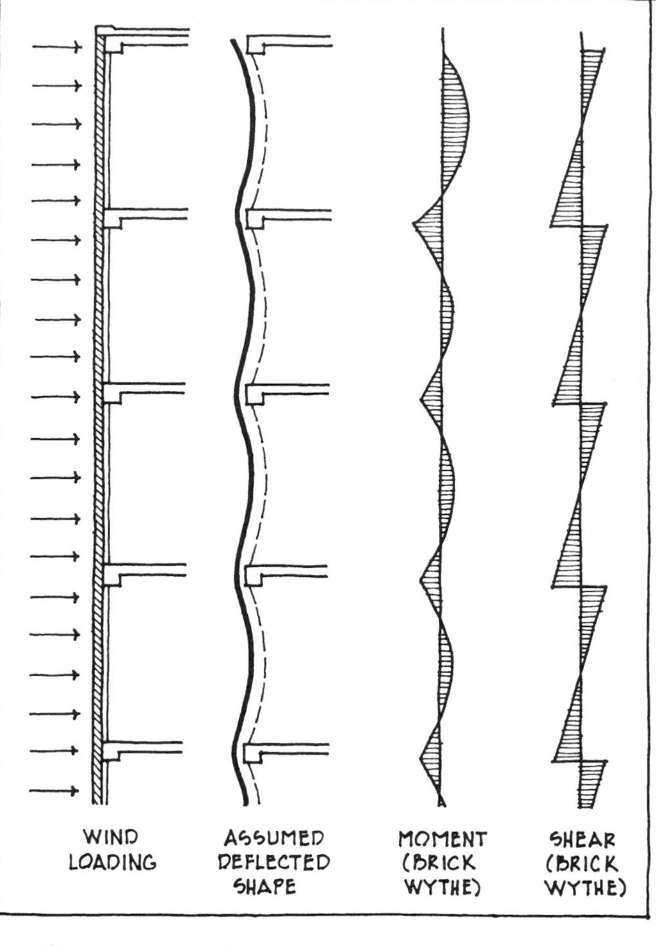

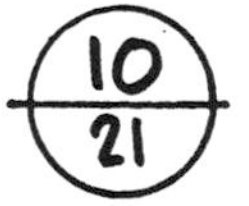

Lateral wind loading. (*From Brick Institute of America,* **Technical Note 28B,** ***BIA, Reston, Va.***)

is not common on wood or metal stud framing, but can be accomplished by first applying a scratch coat of cement plaster on metal lath over the studs.

Code requirements do not limit the length or height of adhered veneer except as necessary to control expansion and contraction. Any movement joints that occur in the backing or the frame must be carried through the bedding mortar and the veneer as well.

10.5.2 Anchored Veneer

Building codes regulate the design of anchored veneers by prescriptive requirements based on empirical data. The veneer chapter of the 1995 *MSJC Code* limits use of the prescriptive design to walls subject to design wind pressures of 25 psf or less. Higher wind pressures require analytical design. Noncombustible, noncorrosive lintels of masonry, concrete, stone, or steel must be provided over openings, with deflections limited to $L/600$ of the span. Although some codes require only a 1 in. clear cavity, the minimum recommended width of the open cavity between wythes of a cavity wall or between a veneer and its backing wall is 2 in. A narrower cavity is

TABLE A REQUIRED AREA OF REINFORCEMENT PER VERTICAL FOOT FOR *EACH FACE OF WALL* (SQ IN./FT)

Span (ft)	*Support condition*							
	Simple span or two continuous spans				*Three or more continuous spans*			
	Wind pressure (psf)				*Wind pressure (psf)*			
	15	*20*	*25*	*30*	*15*	*20*	*25*	*30*
10	0.035	0.047	0.059	0.070	0.028	0.038	0.047	0.056
12	0.051	0.068	0.084	0.101	0.041	0.054	0.068	0.081
14	0.069	0.092	0.115	0.138	0.055	0.074	0.092	0.110
16	0.090	0.120	0.150	0.180	0.072	0.096	0.120	0.144
18	0.114	0.152	0.190		0.091	0.122	0.152	0.182
20	0.141	0.188			0.113	0.150	0.188	
22	0.170				0.136	0.182		
24	0.203				0.162	0.216		
26					0.190			

TABLE B AREA OF REINFORCEMENT *PER FACE* OF WALL FOR VARIOUS BAR SIZES AND SPACING (SQ IN./VERTICAL FT) (BASED ON 3 COURSES PER 8-IN. HEIGHT)

Reinforcement per face	*Spacing of reinforcement*							
	Every course	*Every 2nd course*	*Every 3rd course*	*Every 4th course*	*Every 5th course*	*Every 6th course*	*Every 7th course*	*Every 8th course*
one #2 bar	0.225	0.112	0.075	0.056	0.045	0.037	0.032	0.028
one $\frac{3}{16}$-in. wire	0.124	0.062	0.041	0.031				
one #9 gauge wire	0.077	0.039	0.026					

TABLE C MAXIMUM *VERTICAL* SPACING OF WALL ANCHORS (IN.)

Wind pressure (psf)	*Horizontal wall span (ft)*								
	10	*12*	*14*	*16*	*18*	*20*	*22*	*24*	*26*
15	24	24	24	24	24	24	22	20	18
20	24	24	24	22	20	18	16	15	
25	24	24	20	18	16	14			
30	24	20	17	15	13				

Note: No. 6 gauge galvanized wire anchors. Load distribution assumed to be 600 lb per anchor. Maximum vertical spacing limited to 24 in. Spacing is for an interior support of a continuous wall. Spacing at an end support is 24 in.

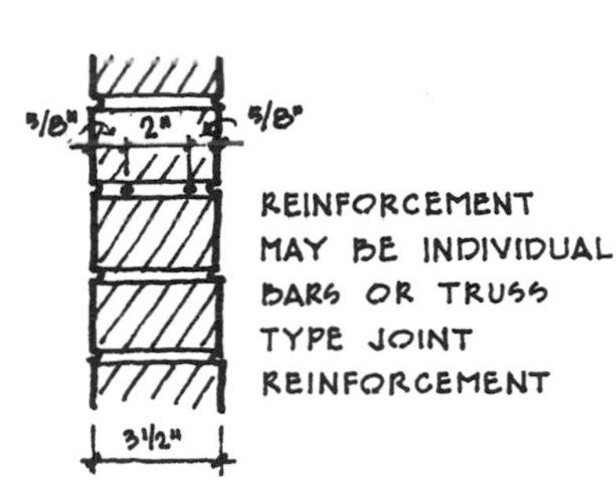

Reinforcing requirements for 4-in. brick curtainwalls spanning horizontally. (*From Brick Institute of America,* Technical Note 17L, *BIA, Reston, Va.*)

difficult for a mason to keep clear of mortar protrusions during construction (refer to Chapter 15). When rigid board insulation is to be installed in the cavity, the clear distance between the face of the insulation and the back of the facing wythe should be 2 in. Most codes limit the maximum distance between backing and facing to 4½ in. This limitation is based on the stiffness and load transfer capability of wire ties and anchors, and permits a maximum insulation thickness of 2½ in.

Empirical requirements limit the height permitted for anchored veneer. The *Uniform Building Code* and *Standard Building Code* set maximum heights at 30 ft, and the *BOCA National Code* at 25 ft. The UBC requires additional support at 12-ft intervals above 30 ft, the *Standard Building Code* at every story height. The *MSJC Code* limits anchored veneer backed by wood framing to a height of 30 ft, and veneer backed by cold-formed steel framing must be supported at every floor above 30 ft.

The BIA recommends, and some municipal building codes permit, anchored masonry veneer over concrete or masonry backing walls to be designed without shelf angles for heights of up to 100 ft. Flexible anchorage to the backing walls permits differential movement and transfers wind loads to the structure throughout the height of the building. Unlike the curtain walls described above, intermittent anchorage of the veneer eliminates the need for steel joint reinforcement. Proper detailing at parapets and other building elements is required to allow differential movement between the veneer and frame.

10.5.3 Cavity Walls

Cavity walls are among the strongest and most durable of exterior cladding systems. Although they may consist of brick backing and facing wythes or concrete masonry backing and facing wythes, cavity walls are most often constructed with concrete block as the backing wall and clay brick as the facing. The open cavity between the two wythes of masonry facilitates drainage of penetrated moisture. The wire ties used to connect the wythes are less prone to transferring moisture from the outer to inner surfaces than multi-wythe walls connected with masonry headers. Wire ties also create less thermal bridging than masonry headers.

Two-piece adjustable ties (see Chapter 7) permit differential thermal and moisture movements between the backing and facing wythes of a cavity wall. When constructed of dissimilar materials such as concrete and clay masonry, this differential movement can be significant. A concrete masonry backing wall experiences permanent moisture shrinkage as the latent moisture from the manufacturing process evaporates, and a clay masonry facing experiences permanent moisture expansion as the brick reabsorb atmospheric moisture after they are fired. These opposing movements can be accentuated when cavity insulation increases the temperature differential between inner and outer wythes.

The *MSJC Code* and UBC have the most stringent requirements for cavity wall ties (*see Fig. 10-23*). The two-piece adjustable ties, which permit differential movements to take place, are permitted a maximum vertical offset of 1¼ in., and maximum play of $\frac{1}{16}$ in. Adjustable ties must be spaced at a maximum of 16 in. on center vertically and horizontally, providing one tie for every 1.77 sq ft of wall area. Rigid rectangular or Z-ties of $\frac{3}{16}$-in. diameter (W2.8) wire are permitted a maximum spacing of 24 in. on center vertically by 36 in. on center horizontally, with one tie for every 4½ sq ft of wall area because the ties are stiffer than adjustable ties. Rectangular ties are for use

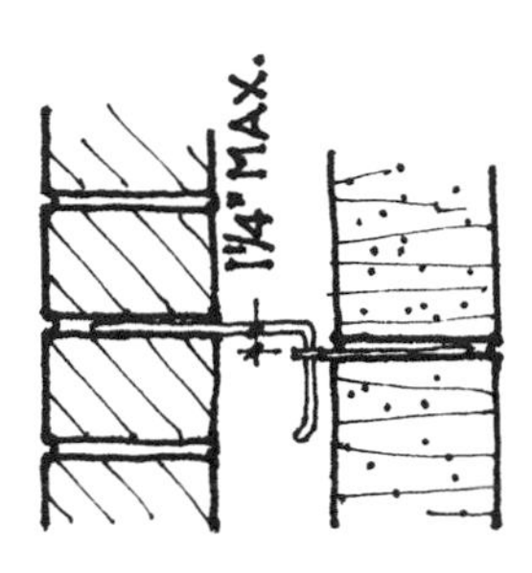

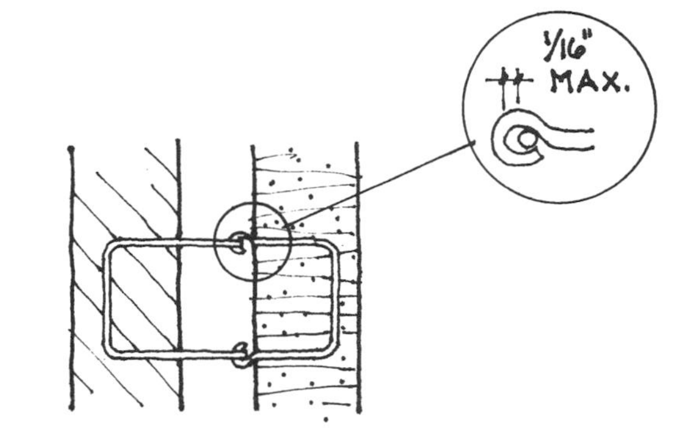

CAVITY WALL TIE SPACING		
Tie type and diameter	*Maximum spacing, horizontal × vertical (in.)*	*Wall area per tie, (sq ft)*
Rigid wire ties, minimum W2.8 gauge (3/16 in. dia.)	36×24	4.50
Three-wire joint reinforcement, or two-wire joint reinforcement with fixed tab ties, minimum W1.7 gauge (9 gauge)	36×24	2.67
Two-piece adustable wire ties, minimum W1.7 gauge	16×16	1.77
Two-wire joint reinforcement with adjustable ties, minimum W1.7 gauge	16×16	1.77

Cavity wall ties.

with hollow masonry units, and Z-ties for use with solid masonry units. Rigid ties should be limited to use in

- Cavity walls where the backing and facing wythes
  - Are of the same type of masonry with similar expansion and contraction characteristics
  - Are laid up at the same time
  - Are not separated by cavity insulation
  - Are constructed of units which course out at the same heights
- Multi-wythe walls grouted and reinforced so that backing and facing wythes react together under applied loads

When concrete masonry is used as the backing wythe in a cavity wall, the joint reinforcement required to control shrinkage cracking can be outfitted with adjustable ties to connect to a facing wythe of clay or concrete masonry. Three-wire joint reinforcement and joint reinforcement with fixed tab ties can also be used to connect the wythes of some types of cavity walls (*see Fig. 7-6*), but they do not provide as much flexibility for differential movement as adjustable ties. The spacing of joint reinforcement ties is less than that permitted for the heavier wire ties (*see Fig. 10-23*).

In most areas of the United States, the exterior brick wythes of cavity walls should be constructed of Grade SW units because the facing is isolated from the rest of the wall and therefore exposed to temperature extremes. Type N mortar is suitable for most cavity wall construction (refer to *Fig. 6-8* for mortar recommendations). Cavity walls should be protected against moisture penetration in accordance with the principles outlined in Chapter 9, relying primarily on a system of flashing and weepholes to collect and expel rain which enters through the facing wythe or moisture which condenses within the cavity.

10.5.4 Brick Veneer

Brick veneer is most commonly used over wood and metal stud backing in a steel or concrete structural frame. Flexible metal anchors permit horizontal and vertical movement parallel to the plane of the wall but resist tension and compression perpendicular to it. The veneer must transfer lateral wind loads to the backing, and these metal anchors and their mechanical fasteners are the weakest component of the system. Code requirements for spacing of veneer anchors are shown in *Fig. 10-24*. Additional anchors should be located within 12 in. of openings larger than 16 in. in either dimension. The spacing required for these additional anchors varies among the codes, ranging anywhere from 12 in. to 36 in. on center.

Wire anchors are used to attach veneer to structural steel. For concrete, wire or flat bar dovetail anchors are recommended. Wire anchors should be at least W2.8 gauge ($\frac{3}{16}$-in. diameter), with the wire looped and closed (see Chapter 7). Flat-bar dovetail anchors should be 6 gauge, $\frac{7}{8}$ in. wide, and fabricated so that the end embedded in the masonry is turned up $\frac{1}{4}$ in.

For securing brick veneer to residential wood frame construction, corrugated sheet metal anchors are often used. These should be 22-gauge galvanized steel, at least $\frac{7}{8}$ in. wide × 6 in. long. Corrosion-resistant nails should penetrate the stud a minimum of $1\frac{1}{2}$ in. exclusive of sheathing. The

Building code	*Maximum spacing, horizontal×vertical (in.)*	*Wall area per anchor (sq ft)*
Uniform	24×12	2
Standard	32×16	3
National	32×16	3
MSJC	32×18	Adjustable two-piece anchors of W1.7 (9 gauge) wire, and 22 gauge corrugated sheet metal anchors: 2.67 All other anchors: 3.5

Masonry veneer anchor spacing.

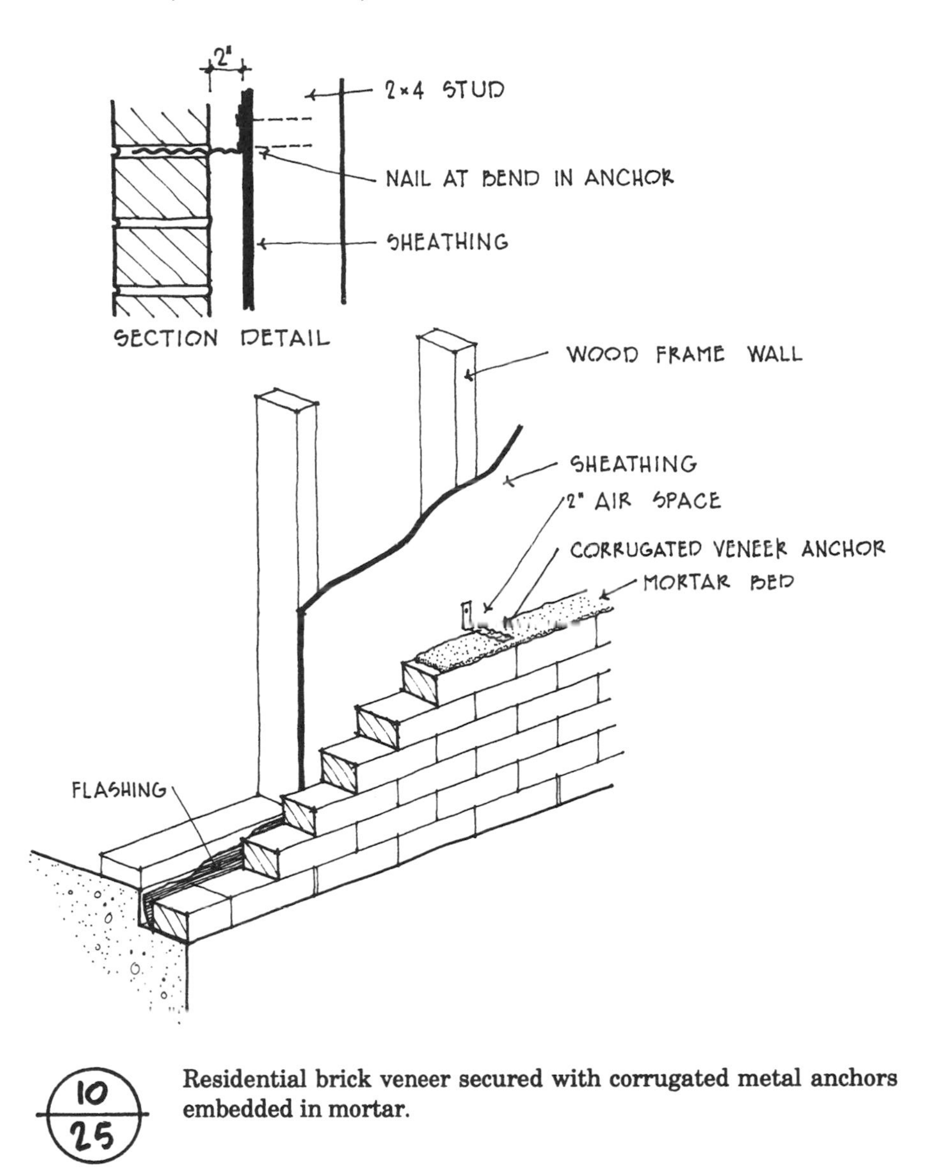

10/25 **Residential brick veneer secured with corrugated metal anchors embedded in mortar.**

free end of the anchor should be placed in the mortar rather than on top of the brick, and should extend at least 2 in. into the joint (*see Fig. 10-25*). Corrugated anchors are weak in compression, and provide load transfer only if the horizontal leg is properly aligned in plane with the mortar bed joint and the nail is positioned precisely at the 90° bend. Anchors randomly attached to the backing wall and bent out of plane to align with bed joints serve no useful purpose, and cracking failures are frequent. Corrugated anchors should be used only in low-rise construction, and only if wind loading is relatively light.

Brick veneer is anchored to metal stud frames with 9-gauge corrosion-resistant wire hooked through a slotted connector or looped eye for flexibility. Anchors are attached in one of two ways: (1) through the sheathing into the studs with corrosion-resistant, self-tapping screws (*see Fig. 10-26*) or (2) hooked to the studs through the joints in the sheathing. With the screw-attached anchors, additional moisture protection is provided by applying a layer of 15-lb asphalt saturated felt over the sheathing.

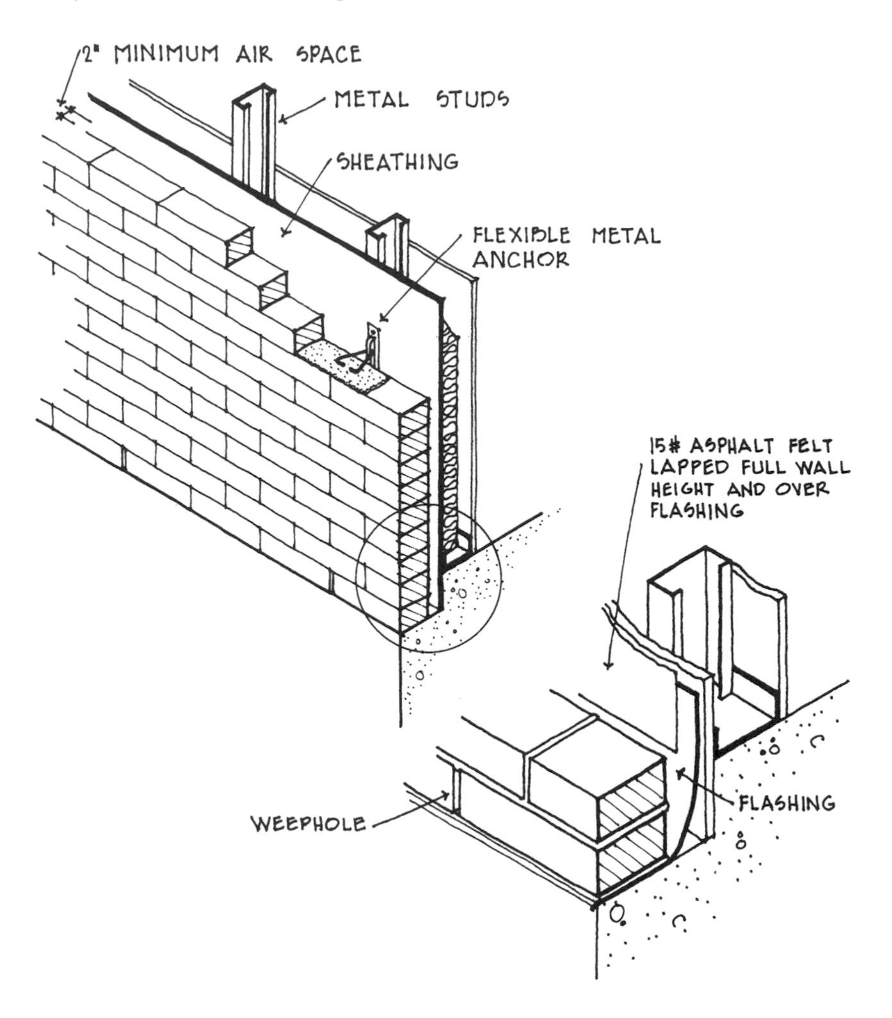

Typical brick veneer installation over a metal stud wall. (*From Brick Institute of America,* **Technical Note 28B,** ***BIA, Reston, Va.***)

The felt should be lapped shingle style in horizontal layers, and cover the top of the masonry flashing. With the clip anchors, the sheathing joints and penetrations must be sealed with tape and/or mastic, and the face of the sheathing must be water-repellent.

The use of brick veneer over metal stud backing is relatively recent in the long history of masonry construction. The system was first introduced as an economical substitute for CMU backup, but it was a false economy based on inadequate size and spacing of studs. Problems with cracking in the brick veneer raised questions about the relative rigidity of masonry veneer versus the flexibility of the metal frame in resisting lateral loads. BIA now recommends deflection limits of $L/600$ to $L/720$ to provide adequate stiffness in the studs. Lateral bracing or stiffeners in the stud wall may also be required for adequate rigidity to prevent veneer cracking and subsequent moisture intrusion. For one- and two-story buildings with limited floor-to-floor heights, deflection limits might be achieved with 18-gauge studs, depending on wind load factors. Increased floor-to-floor heights, higher wind loads, and taller buildings will generally require studs that are a minimum of 16 gauge. Stud spacing should not exceed 16 in., and galvanizing should be by hot-dip process, in accordance with ASTM A525, G60, or G90 coating.

Masonry veneers are designed as drainage wall systems because moisture will always be present, even with good design, good detailing, and good workmanship. Moist environments promote the corrosion of metals, so studs, tracks, and other components must be protected. When galvanized, self-tapping sheet metal screws are driven through galvanized metal studs, both contact surfaces are abraded of their coating, leaving the underlying steel unprotected from moisture corrosion from the outset of its service life. Since these fastener penetrations are often the initial site of corrosion, most metal stud manufacturers now recommend ASTM C954 cadmium-plated screws for attaching masonry veneer anchors. Although this provides a longer service life for the screw itself, questions still remain concerning the pullout strength from the stud if any corrosion is present. The best defense against such corrosion problems is adequate design for differential movement, proper detailing to limit moisture penetration, and good drainage through a system of flashing and weepholes.

Another method of attachment recognized by some building codes and by HUD "Minimum Property Standards" uses galvanized 16-gauge 2×2-in. paper-backed welded-wire mesh attached to metal studs with galvanized wire ties, or to wood studs with galvanized nails. Wire anchors are then hooked through the mesh, and the 1-in. space between veneer and backing is grouted solid. This procedure can eliminate the need for sheathing and, in some cases, increase the allowable veneer height because of greater stiffness. It is, however, generally limited to masonry units 5 in. or less in thickness (*see Fig. 10-27*).

Grade SW brick is recommended for exterior veneers in most areas of the United States because the facing is isolated from the rest of the wall and therefore exposed to temperature extremes. Type N mortar is suitable for most veneer construction (refer to *Fig. 6-8* for mortar recommendations).

Since the overall thickness of a brick veneer wall is approximately 10 in., a foundation wall of at least the same thickness is required for adequate support (Fig. 10-28). Many codes permit a nominal 8-in. masonry foundation provided that the top of the wall is corbeled as shown in Chapter 12. The total projection of the corbel cannot exceed 2 in., with individual corbels projecting not more than one-third the height of the unit. Brick veneer should

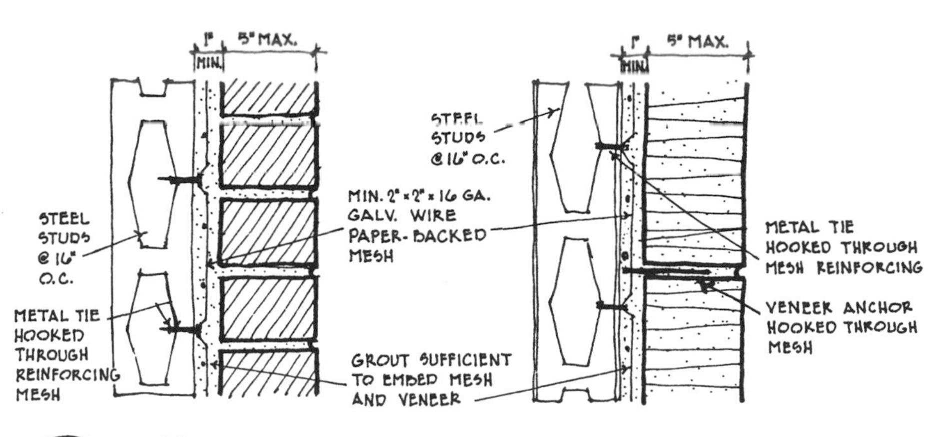

Alternative methods of attaching veneer to metal studs. (*From National Concrete Masonry Association,* TEK Bulletin 79, *NCMA, Herndon, Va.*)

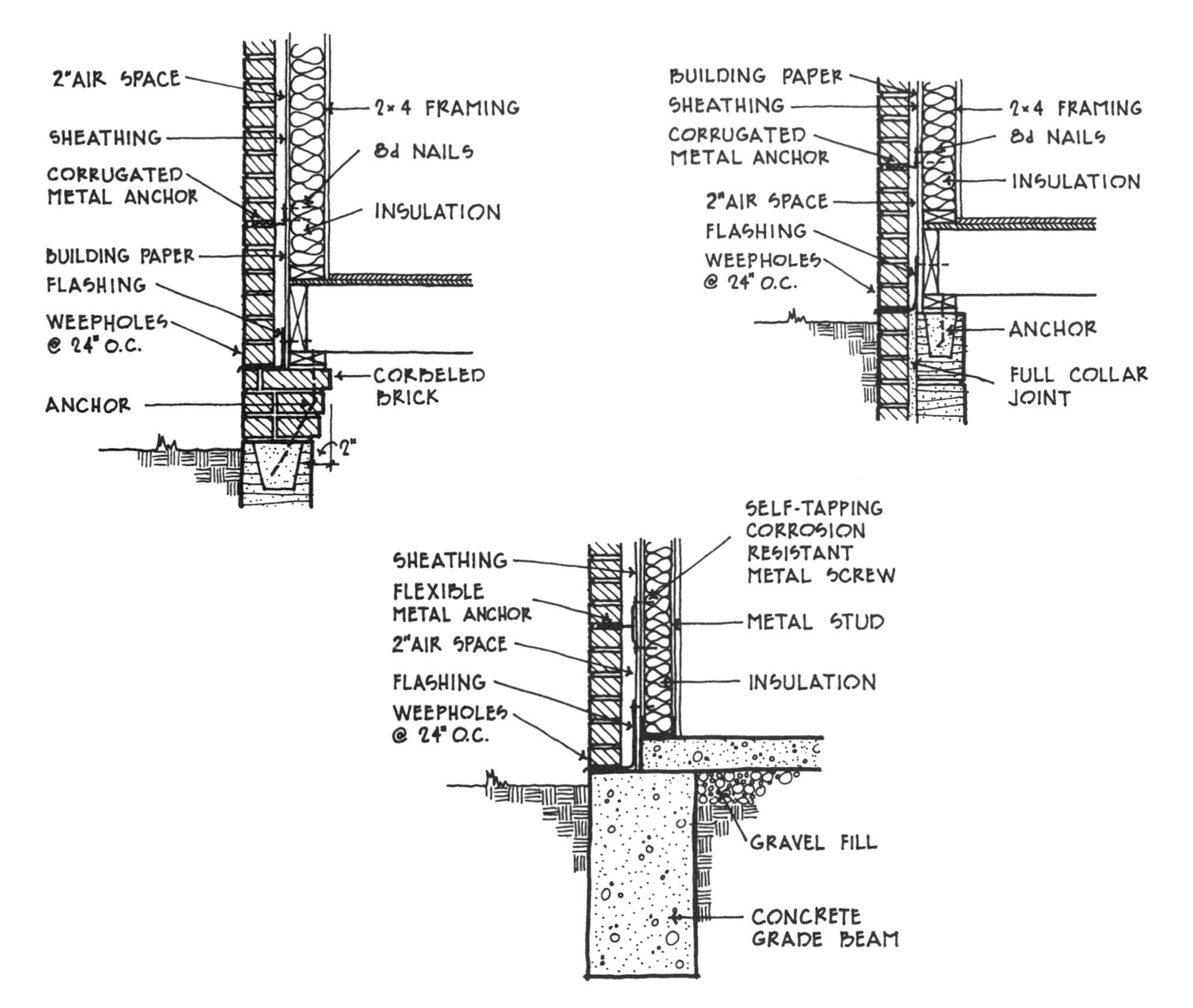

Foundation details for masonry veneer walls. (*From Brick Institute of America,* **Technical Note 28,** *BIA, Reston, Va.*)

start on a brick ledge below the finish floor line. Moisture entering the wall must be drained to the outside by flashing and weepholes located above grade at the bottom of the wall. Flashing should also be installed at the heads and sills of all openings (*see Fig. 10-29*). The flashing material should be of high quality, because replacement in the event of failure is very costly. Weep holes must be located in the masonry course immediately above all flashing, spaced n321 321o more than 24 in. on center horizontally (refer to Chapter 9 for additional flashing and weephole details).

In lieu of steel lintels over openings, brick veneer can be reinforced with ¼-in.-diameter deformed steel bars or joint reinforcement placed horizontally in the first bed joint above the opening. Where spans and loading permit, this method offers a more efficient use of materials (see Chapter 11 for design of masonry lintels).

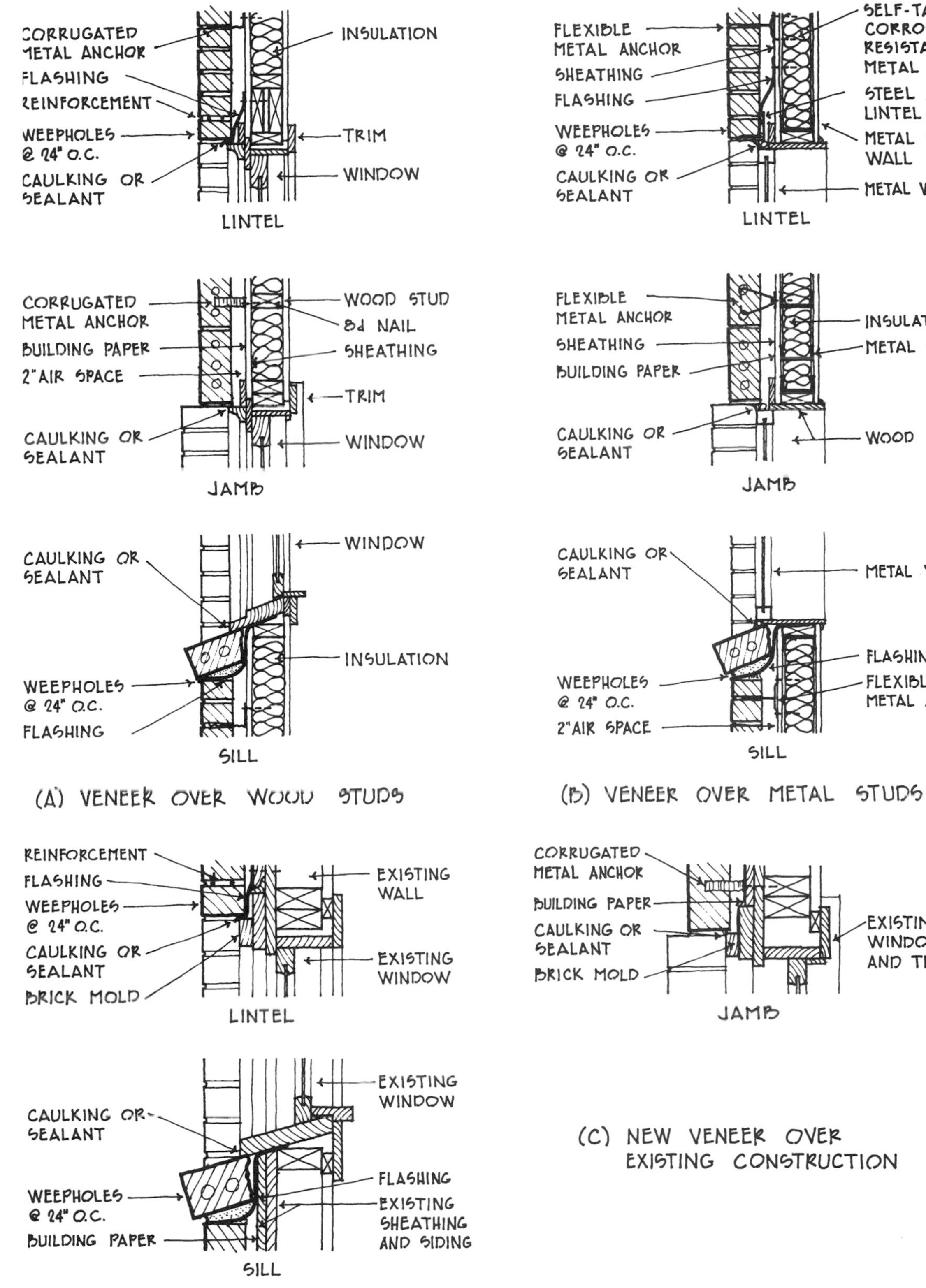

Typical veneer details. (*From Brick Institute of America,* Technical Note 28, *BIA, Reston, Va.*)

10.5.5 Concrete Brick and Concrete Block Veneer

This type of construction has increased in use with the variety of colors, textures, and patterns of decorative concrete masonry units now available. Thin veneers of less than 5 in. thickness may be attached in the same manner as clay brick, or in several other ways as shown in *Fig. 10-30.* "Spot-bedding" at anchors effectively increases their resistance to compressive

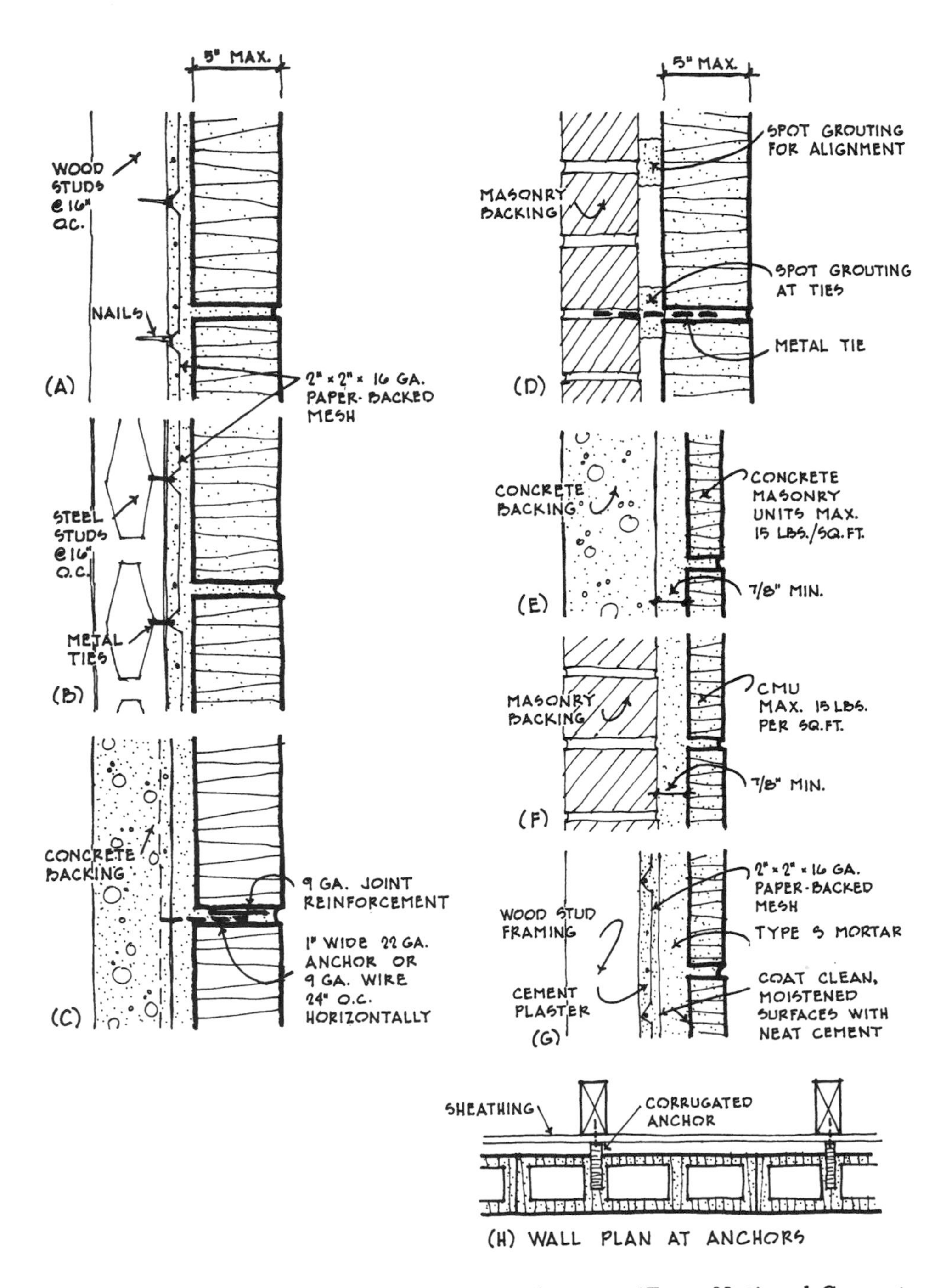

Concrete masonry veneer attachments. (*From National Concrete Masonry Association,* TEK Bulletin 79, *NCMA, Herndon, Va.*)

lateral loads and strengthens the wall at its weakest point. Hollow CMUs must be anchored in the intermittent mortar beds over the web flanges, or continuous joint reinforcement with wire tabs can be used. Solid concrete brick attaches to stud backing in the same manner as clay brick. Anchor spacing is the same as for clay masonry veneers (*see Fig. 10-24*).

Spans across window and door openings are easily accommodated with reinforced hollow concrete masonry. Steel lintels are not required with these units. U-shaped lintel blocks have a deep channel for placement of horizontal reinforcing bars and grout. Depending on the size of the unit and the amount of reinforcing, a masonry lintel can span relatively large openings (see Chapter 11).

Joint reinforcement is used in concrete masonry to control shrinkage cracks. As the stress increases, it is transferred to and redistributed by the steel. The effectiveness of joint reinforcement depends on the type of mortar used and the bond it creates with the wire. Greater bond strength means greater efficiency in crack control. Only Types M, S, and N mortars are recommended. Minimum mortar cover to the outside face of the block should be ⅝ in. for the exterior and ½ in. for the interior wall face. Prefabricated corner and T-type reinforcement is recommended for corners and intersecting walls. Splices should be lapped 6 in.

Joint reinforcement is normally located in the first bed joint of a wall, and at 8-, 16-, or 24-in. centers above that. In addition to normal placement, joint reinforcement should be located: (1) in the first and second bed joints immediately above and below wall openings, extending at least 24 in. past either side of the opening or to the end of the panel, whichever is less, and (2) in the first two or three bed joints above floor level, below roof level, and near the top of a wall. Reinforcement need not be located closer than 24 in. to a bond beam.

10.5.6 Terra Cotta Veneer

Architectural terra cotta is installed as both adhered veneer and anchored veneer. The backs of units which will be adhered must be scored or ribbed to form a mechanical key with the mortar bed. The units may have a maximum thickness of 1¼ in. including ribs, and may be attached over masonry, concrete, or stud backing.

Since the backing for adhered veneer must provide continuous support, studs must first be covered with metal lath and plaster. The height of the studs should not exceed 10 ft, and the backing should provide a degree of rigidity equal to that of 2×4 wood studs spaced 16 in. on center and covered with ⅜-in. ribbed lath and a ½-in. portland cement scratch coat (*see Fig. 10-31*). For moisture protection, the lath should be paper-backed, or applied over asphalt saturated felts laid up shingle style to shed water. Concrete and masonry backing walls should be clean and sound. Concrete surfaces should have a texture comparable to that obtained by using rough form boards. Any surface that is smooth and glossy, painted, effloresced, greasy, or oily or has loose surface material should be roughened by abrasive grinding or blasting, chipping, or scarifying. Since concrete masonry expands significantly when wetted and shrinks again as it dries, adhered terra cotta cannot be applied directly to the surface. Concrete block walls must first receive a waterproof membrane of asphalt saturated felts, reinforced asphalt paper, or polyethylene sheets, then expanded metal lath or self-furring wire fabric 2-in. or 1½ in. square and a ⅜-in. scratch coat of portland cement plaster.

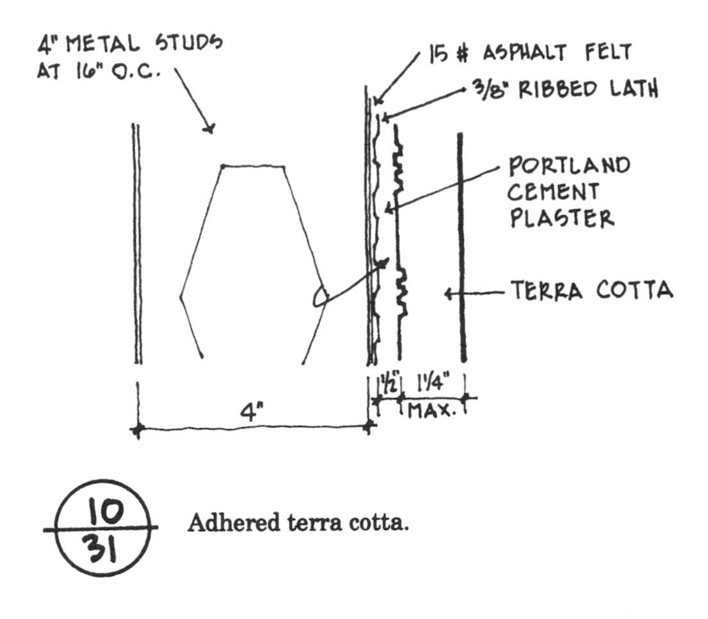

10/31 **Adhered terra cotta.**

Just prior to setting adhered veneer, units and backing surfaces other than plaster should be brushed with a uniform coat of neat cement. A setting mortar of 1 part portland cement (ASTM C150, Type I), ½ part hydrated lime (ASTM C207, Type S) and 4 to 4½ parts sand (ASTM C144) is then applied to both the wall and the back of the unit. The mortar should be screeded level on each surface to a minimum thickness of ½ in. Units are tapped into place with a rubber mallet to eliminate voids, producing a final mortar bed thickness of ½ to 1¼ in., with sufficient excess mortar squeezed out to fill the joints between units (*see Fig. 10-32*).

Anchored terra cotta must be tied to the backing with anchors of minimum 8-gauge galvanized wire spaced 12 to 18 in. on center in horizontal bed joints and vertical head joints. These wires are secured to ¼-in. galvanized pencil rods placed in vertically aligned loop anchors set into the backing wall (*see Fig. 10-33*). The anchors should have sufficient strength to support the full weight of the veneer in tension. A minimum 2-in. space between the wall and the unit ribs is then filled solidly with grout consisting of 1 part portland cement, 3 parts sand, and 2 parts graded pea gravel passing a ⅜-in. sieve.

At the beginning of each day, walls that are to be faced should be soaked with clean water by hose and spray nozzle, and soaked again within

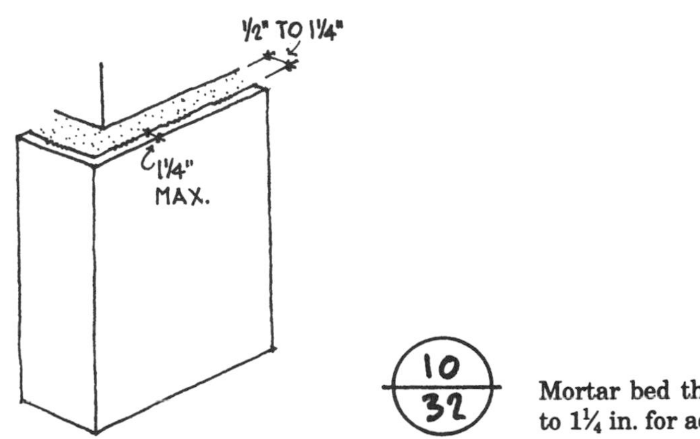

10/32 **Mortar bed thickness ranges from ½ in. to 1¼ in. for adhered terra cotta.**

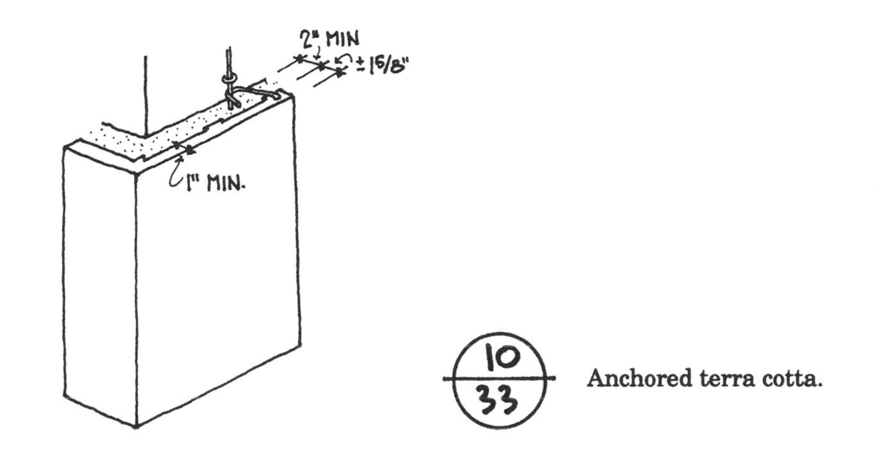

Anchored terra cotta.

an hour of actual setting or grouting. Just prior to installation, all architectural terra cotta must be soaked in clean water for at least one hour. After soaking, the units should be drained to eliminate surface water, but remain noticeably damp at the time of setting or grouting. Saturating both the wall and the units promotes maximum bond. Dry, windy conditions may require soaking closer to the time of setting. Shop drawings provided by the manufacturer should include setting diagrams that indicate all mortar joint sizes and locations, all anchors, hangers, and other supports, as well as the size and location of expansion joints.

Mortar joints may be tooled as the work progresses, or they may be raked to a depth of ½ in. and pointed later. Sealants should not be used to fill joints between glazed units, because the units' low permeability lets the wall breathe only through the mortar. This low permeability may also necessitate placement of a vapor retarder on the warm side of walls enclosing high-humidity areas. Venting tubes placed in the mortar joints will allow excess moisture vapor that may be forced through the wall by pressure differentials to escape easily.

It is also recommended that provision be made for differential expansion and contraction between the veneer and the backing. Expansion joints in terra cotta veneers should extend through the mortar setting bed to the backup. The required width and spacing of expansion joints should be calculated, not estimated.

Durable flashing materials should be installed beneath copings, sills, projected and recessed elements, and at horizontal supports. Wherever flashing is located, weepholes must be installed as well. All projecting features, including copings, should have drips to prevent water from running down the face of the wall where it can easily penetrate hairline cracks or cause stains.

As with any masonry system, the primary enemy of terra cotta veneers is trapped moisture. Because the unit glaze itself is impervious to moisture, terra cotta wall systems were once thought to be impervious as well. However, many historic structures have suffered severe damage because no provisions were thought necessary for flashing, drainage or venting. Leaks occurring at roof level, at the coping, at deteriorated window caulking, and other adjacent locations can allow copious amounts of moisture into the wall. Condensation of water vapor within the wall can also cause continuous saturation. Once trapped behind the ceramic glaze, water can cause freeze-thaw damage, spalling, and corrosion of metal anchors and structural framing. Once joints are deteriorated, or the glaze starts to pop off units, more water enters the wall and compounds the problem. Because the ceramic finish can hide damp conditions, internal deterioration is often not discovered until it is in an advanced stage.

10.5.7 Stone Veneer

There are two basic types of stone veneer: (1) rubble or cut stone laid in mortar beds, and (2) thin stone slabs mechanically or adhesively attached. Mortar bed construction is generally used in low-rise residential and commercial buildings (*see Fig. 10-34*). The stone may be laid up against a backing of concrete, wood or metal studs, or unit masonry. When the stones do not exceed 5 in. in thickness they may be installed without anchors in much the same way as thin concrete masonry (*see Fig. 10-30*). The bond between the stone and the mortar or grout must be sufficient to withstand a shearing stress of 50 psi. For larger stones up to 10 in. thick, wire or corrugated sheet metal anchors are required. The connections must be flexible enough to compensate for the irregularities of mortar bed height (*see Fig. 10-35*). Anchors should be spaced a maximum of 24 in. on center, and support no more than 2 sq ft of wall area. Metal anchors must receive full mortar coverage at the outside face of the wall to prevent rusting and corrosion. All joints and the space between the stone and the backing should be filled solidly with mortar.

Stone slab veneers are most commonly used on commercial buildings of low-, medium-, and high-rise construction. For adhered attachment without mechanical anchors, the individual stone units must not exceed 36 in. in the largest dimension, 720 sq in. in area, or 15 lb/sq ft in weight. They may be adhered to a concrete or masonry backing. The surfaces of the backing and the veneer must be cleaned, moistened, and brushed with neat cement paste. Type S mortar is then applied to both surfaces, and the units are tapped into place forming a collar joint ½ to 1¼ in. wide.

There are many ways to attach stone slab veneer with metal anchors. All anchoring systems must be designed to resist a horizontal force equal to twice the weight of the veneer. Some types of stone are drilled around the perimeter for insertion of corrosion-resistant metal dowels. Dowels are normally spaced 18 to 24 in. on center, with a minimum of four for each stone unit. The total area of the stone slab is dependent on the type of stone and its thickness. The UBC requires a minimum thickness of 2 in. and a maximum slab size of 20 sq ft. Each dowel is secured to the backing with wire or sheet metal anchors (*see Fig. 10-36*). In lieu of dowels, wire anchors may be bent into the holes and set with a cementitious mix. Dowel anchoring may not be used on stone slabs less than 1¼ in. thick. The space between the veneer and the backing surface may be solidly grouted or spot-bedded at anchor locations and for alignment. Stone slabs may also be sawed or kerfed at the edges to receive bent metal strap anchors (*see Fig. 10-37*). Carelessly cut kerfs can propagate cracking, and unless filled with a compatible elastomeric sealant, may also retain water. Joints are usually also filled with an elastomeric sealant rather than with mortar which might be subject to shrinkage cracking and subsequent moisture penetration. The sealant provides a weather-resistant joint and also permits slight movement of the units to relieve stress.

Relieving supports, or shelf angles, must be provided at least every 12 ft vertically. Stone fabricators generally prepare engineered shop drawings using anchorage details that are best suited for their particular type of stone.

Ultra-thin stone slabs (1½ in. or less) require much greater care in

Stone veneer mortar bed installation.

design and detailing because they are very brittle, and much more permeable than conventional thicknesses (*see Fig. 10-38*). Despite the additional risk, however, the same economic forces which led to the production of thin stone have also led to recommendations for reduced factors of safety in design.

Some experts have called safety factors "factors of ignorance" because they are traditionally larger when loads and stresses are uncertain, when

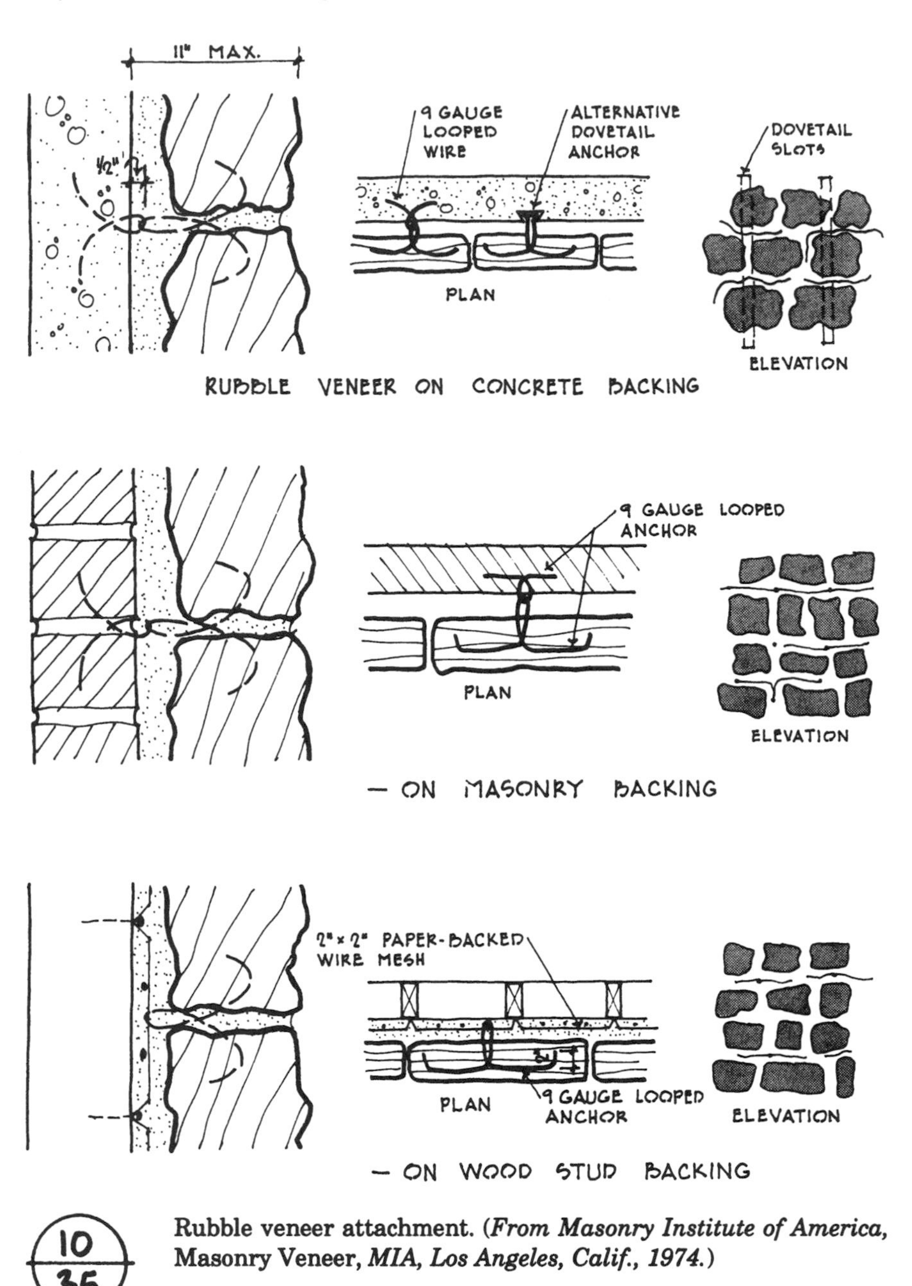

10/35 **Rubble veneer attachment. (*From Masonry Institute of America,* Masonry Veneer, *MIA, Los Angeles, Calif., 1974.*)**

the material strength is highly variable, and when the material is not very forgiving. Safety factors for stone have always been very conservative compared to those for ductile materials like steel. One reason for the conservatism is that stone is a natural material rather than a closely controlled, manufactured product, so physical properties can vary widely, even for the same type of stone from the same quarry (refer to *Fig. 5-1*). Some stones also lose strength after repeated heating-cooling and freeze-thaw cycles, and others gain or lose strength with wet-dry cycles.

The safety factors recommended by the various stone industry associations (*see Fig. 10-39*) are empirical. They are not derived from tests or from engineering analysis, but rather, have evolved as industry rules of thumb. Encouraged by intense competition in the curtain wall market,

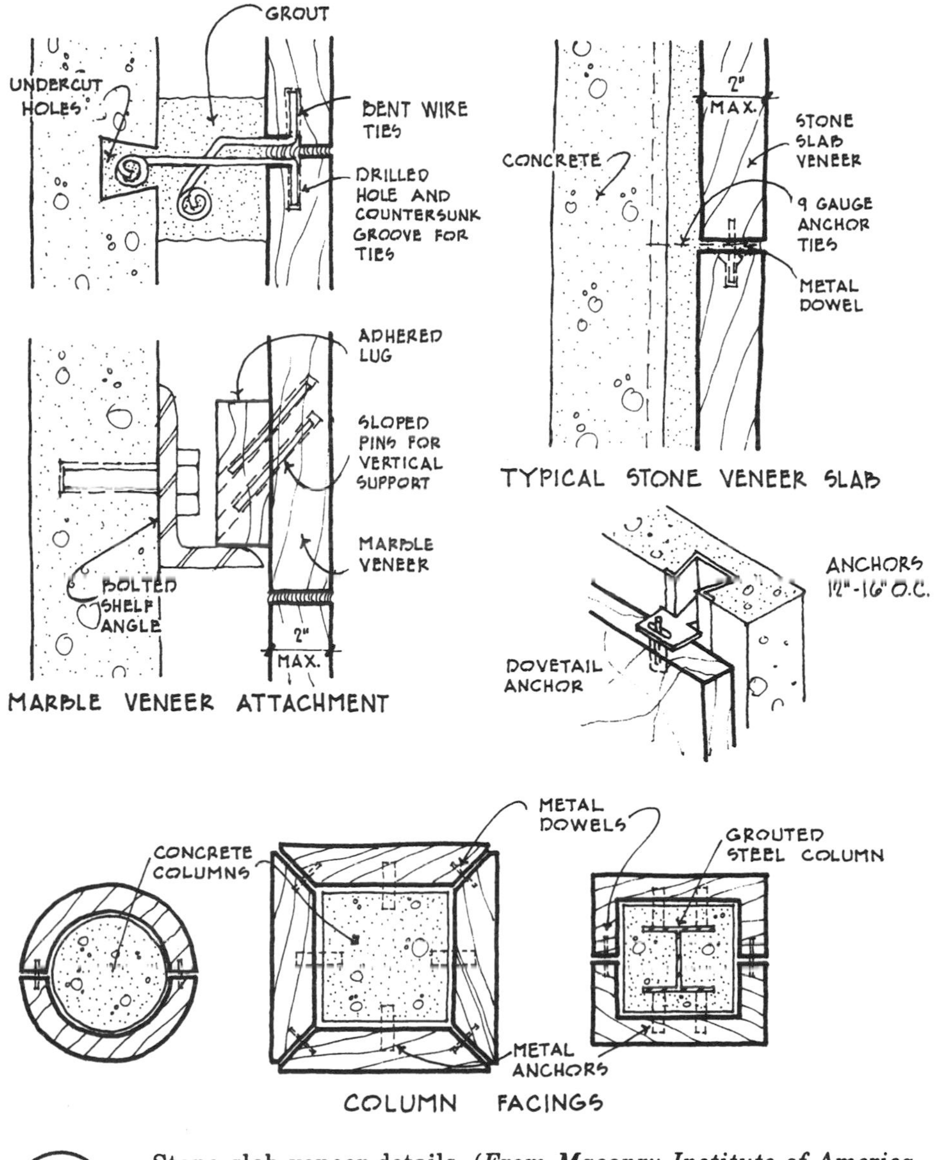

Stone slab veneer details. (*From Masonry Institute of America,* **Masonry Veneer,** *MIA, Los Angeles, Calif., 1974.*)

several experts have developed newer, rationally determined design criteria intended to make stone more efficient and more economical to use. Most of the proposed methods, however, are controversial and none has yet gained industry-wide acceptance. Based primarily on modulus of rupture and flexural strength in resisting wind loads, new recommendations for safety factors vary with the coefficient of variation of tested samples. The most thorough investigations to date have also included a weathering factor and statistical consideration of the size of sample tested. The numbers, however, vary significantly when such factors are taken into consideration, and the principle investigators ultimately recommend that traditional safety factors be used unless the designer has test data from a large sample and sufficient experience with a particular material to warrant prudent use of probabilistic design.

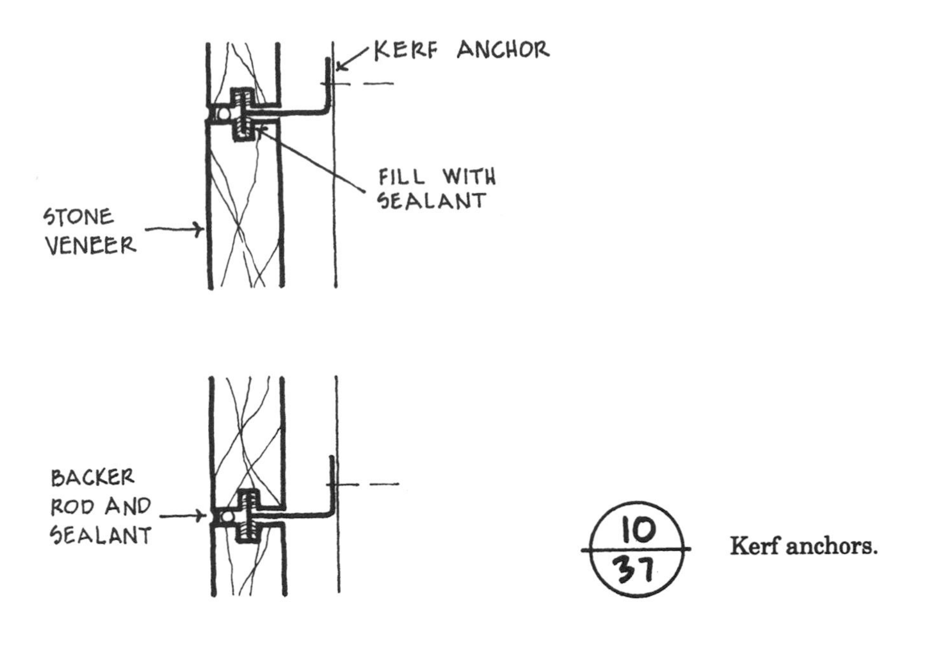

Kerf anchors.

PERMEABILITY (CU IN/SQ FT/HR FOR ½ IN. THICKNESS)			
	Pressure (psi)		
Type of stone	*1.1*	*50*	*100*
Granite	0.6–0.8	0.11	0.28
Limestone	0.36–2.24	4.2–44.80	0.9–109.0
Marble	0.06–0.35	1.3–16.8	0.9–28.0
Sandstone	4.2–174.0	51.2	221.0
Slate	0.006–0.008	0.08–0.11	0.1

Permeability of various commercial building stones. *(From Chin, Stecich, and Erlin, "The Design of Thin Stone Veneers on Buildings,"* **Building Stone Magazine,** *May–June 1986.)*

Because of the brittleness and permeability of thin stone, new methods of application based on prefabricated assemblies have been developed. Backing systems may be of steel frame, metal stud, precast concrete, or reinforced masonry. Prefabrication allows the assembly to be constructed with better environmental controls and working conditions, and the larger panels can be erected more quickly and economically than individually hand-set pieces.

Steel frames and metal stud frames may rely on mechanical anchorage to the framing members, or on a combination of anchors and structural silicone adhesive attachment to a surface diaphragm. Precast concrete and glass-fiber-reinforced concrete (GFRC) backing panels are often used with stone facings, but differential shrinkage and expansion must be accommodated by placing a slip sheet between the stone and the concrete surface.

Type of stone	*Recommended safety factor*
Granite	3*
Anchorage components in granite	4*
Marble veneers	5†
Limestone veneers	8‡

*National Building Granite Quarriers Association.
†Marble Institute of America.
‡Indiana Limestone Institute.

Safety factors recommended by stone industry associations.

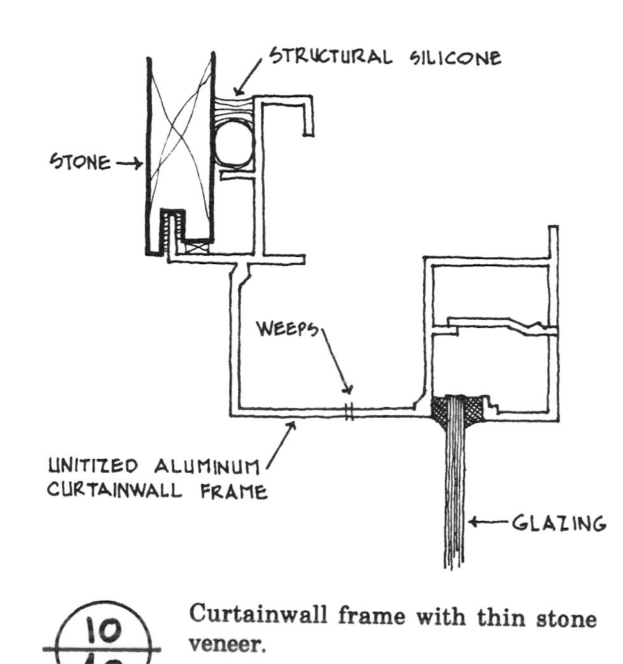

10
40

Curtainwall frame with thin stone veneer.

Stainless steel anchors are embedded in the concrete and fitted into predrilled holes in the back of the stone. The stone may be laid face down in the formwork with stainless steel anchors projecting from the back side, and the concrete cast on top of it. Stone is also being incorporated into conventional aluminum curtain wall systems of both "stick-built" and unitized design (*see Fig. 10-40*).

10.5.8 Parapets

For years, BIA has discouraged the use of masonry parapets because they are so often the source of leaks. In fact, most of the water that gets into masonry walls enters at either the parapet or the windows. A roof overhang is the best protection for the tops of masonry walls, but not all architectural styles lend themselves to such design.

Three problems are common with parapets—exposure, movement, and the roofing system interface. Exposure to the elements is more severe in parapets than in the walls beneath them. They are exposed on both sides, and so experience greater extremes of temperature in both winter and summer. In winter, snow collects on top of the parapet and drifts against the inside wall surface, keeping the masonry saturated throughout months of alternating freeze-thaw cycles. In blowing rain, parapets get wetter than the lower walls because of wind patterns at the tops and corners of buildings (refer to Chapter 9). Some parapet movement is directly related to this exposure. Greater temperature variations cause greater thermal expansion and contraction, and higher moisture contents contribute to greater moisture movement. The winds that drive the rain also induce greater lateral stress at the top of the wall.

Copings for masonry parapets should be selected on the basis of performance as well as compatibility of materials. Metal copings provide the best protection. They are impervious to moisture and can be installed in lengths requiring a minimum number of joints. Since every joint that occurs on the horizontal surface of a coping is an opportunity for a leak, the fewer joints there are, the greater the probability of keeping the wall dry. Metal copings should be designed with lap joints or with slip joints and cover or gutter plates to accommodate the differential movement between the masonry and the metal. The size and spacing of the joints will be affected by the movement characteristics of the masonry materials. That is, joints in the metal cap must be able to open to accommodate permanent expansion in brick walls, and close to accommodate permanent shrinkage in concrete masonry walls. The vertical legs of metal copings should extend at least 2 in. below the top course of the masonry, turn out to form a drip, and be caulked with a high-performance elastomeric sealant (*see Fig. 10-41*). Through-wall flashing should be installed below the metal cap, particularly over walls with open cores or cavities. If metal flashing is used, the material must be com-

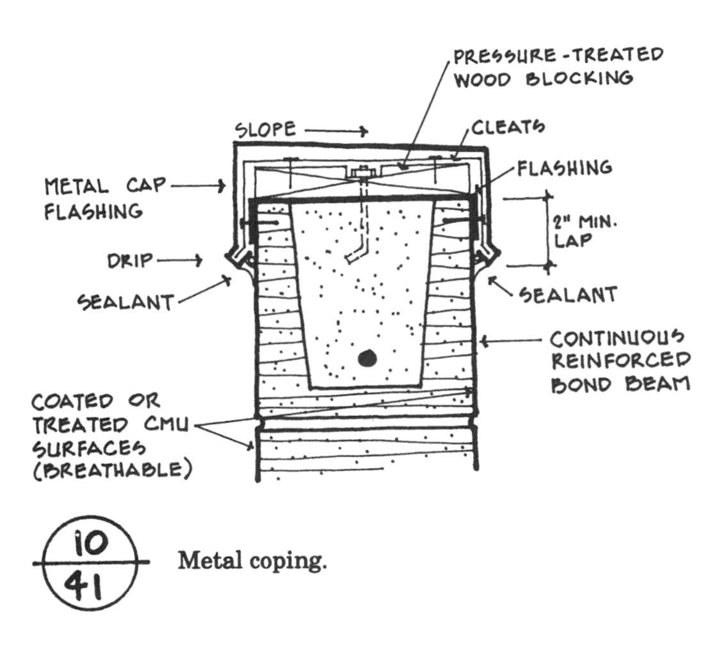

Metal coping.

patible with the metal of the coping itself. All penetrations through this flashing must be sealed with mastic.

In addition to the differential movement between the coping and the wall itself, differential movement between the back and the front of multi-wythe walls must also be considered. Popular details often include a brick veneer, a stone coping and a utilitarian concrete block backup, but this combination can spell disaster. The brick veneer is increasing in both height and length with permanent moisture expansion at the same time that the concrete masonry is experiencing permanent shrinkage in both directions. Lateral stresses will be highest at the corners, where a brick facing can literally slide off the edge of the building (*see Fig. 10-42*). Unless the mortar is very soft and flexible, the stone is ripped from its mortar bed and twisted out of place, opening joints along both the top and the face of the coping (*see Fig. 10-43*).

To minimize problems, the backing and facing wythes of parapet walls should be constructed of the same material. Tall parapets, or parapets which will be subjected to lateral loads from swing staging or window washing equipment, must be structurally reinforced and anchored to the roof slab (*see Fig. 10-44*). A fully reinforced parapet is more restrained

Differential brick expansion between parapet and building wall.

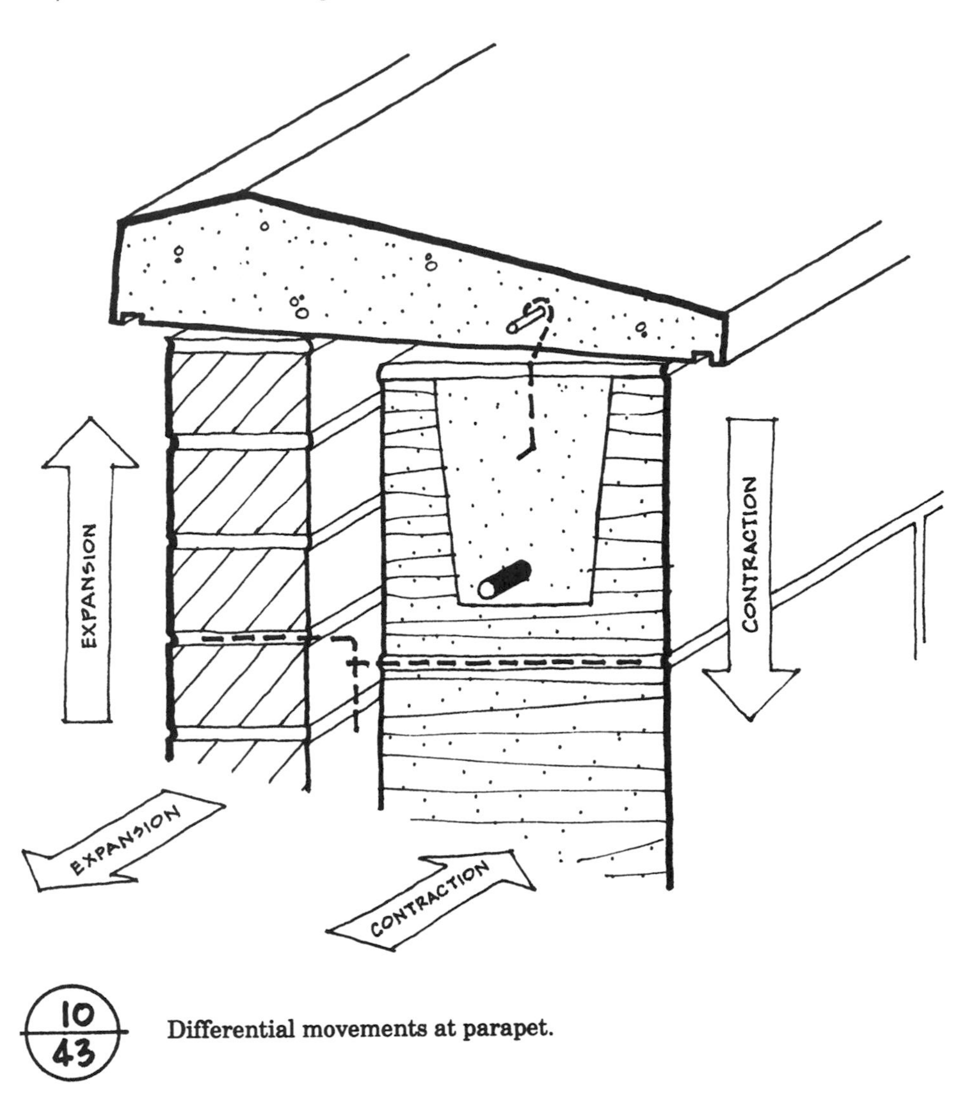

10/43 **Differential movements at parapet.**

against thermal and moisture expansion, but still requires accommodation of such movement. In unreinforced parapets, expansion and contraction can be limited by reinforced bond beams at the top of the wall. This grouted barrier will also protect against direct moisture penetration into hollow masonry and cavity walls. The backing and facing of multi-wythe brick parapets should be connected with rigid metal ties, and the wythes of concrete masonry parapets with continuous joint reinforcement.

To accommodate the greater lateral movement that is experienced in parapet walls compared to the building walls below, BIA has always recommended doubling the number of expansion joints in the parapet. For best performance in both brick and concrete masonry parapets, calculations of potential movement should always be based on expected service conditions, and control and expansion joints located accordingly (refer to Chapter 9).

When masonry copings are specified instead of metal cap flashing systems, select materials that have expansion and contraction characteristics similar to those of the wall materials. Precast concrete copings work best over concrete masonry parapets, natural or cast stone copings over stone walls, and terra cotta over brick or terra cotta walls. Avoid using brick

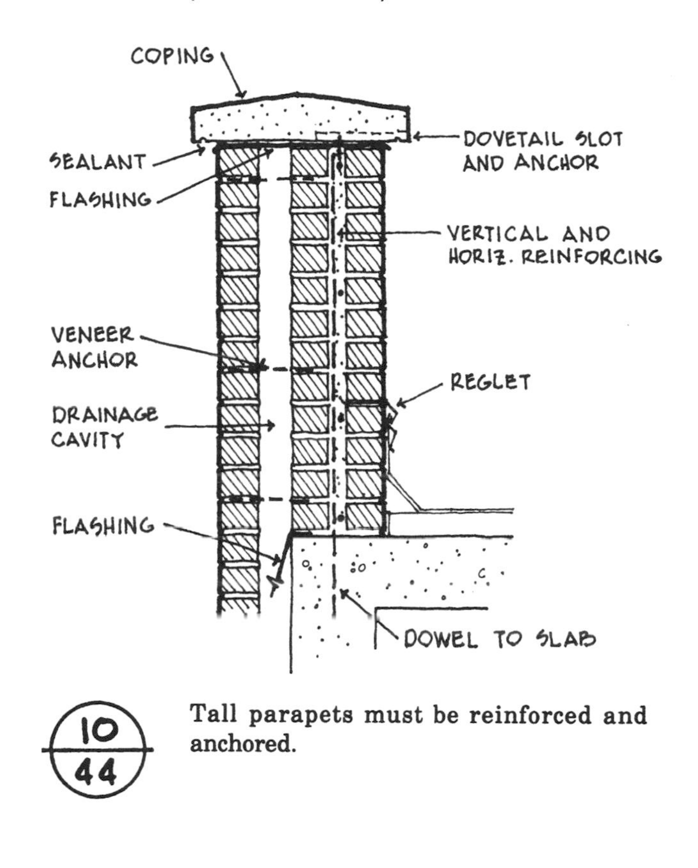

Tall parapets must be reinforced and anchored.

rowlock copings because the number of joints on the horizontal surface increases the probability of leaks. Precast, stone, and terra cotta copings can be formed in longer sections to minimize joint exposure. Masonry copings should always overhang both sides of the wall and have integral drips (*see Fig. 10-45*). Mechanical anchoring is required only in seismic areas.

Masonry coping joints are not impervious to water, so through-wall flashing must be installed underneath. Expansion joints should be located in the last joint in each run of coping or in the joints adjacent to each corner piece, as well as at calculated intervals along the length of the wall. All remaining joints in the coping, including front and back vertical surfaces should be raked out while the mortar is plastic and caulked with elastomeric sealant, because even hairline cracks or separations at the top of the wall act as funnels directing water to the interior.

It is not coincidence that roofs frequently leak at the intersection with masonry parapets, and masonry parapets often leak at the intersection with roofing. Where the work of two trades must interface to form a weather-resistant barrier, the blame for failure can often go either way. In the case of masonry parapets and roofing, it is not so much a matter of poor workmanship on the part of either trade, but the materials and systems used to form the interface.

Roof flashing must be turned up onto the face of the parapet wall and terminated above the level of the roof deck. Where it terminates, through-wall masonry flashing or smaller counterflashing sections are often specified to cap the membrane and to protect the masonry. When the flashing is later turned up to allow for installation of the roofing membrane, the process of bending breaks the mortar bond at the joint. Once bent, the flashing usually will not return to its original configuration, so a small lip

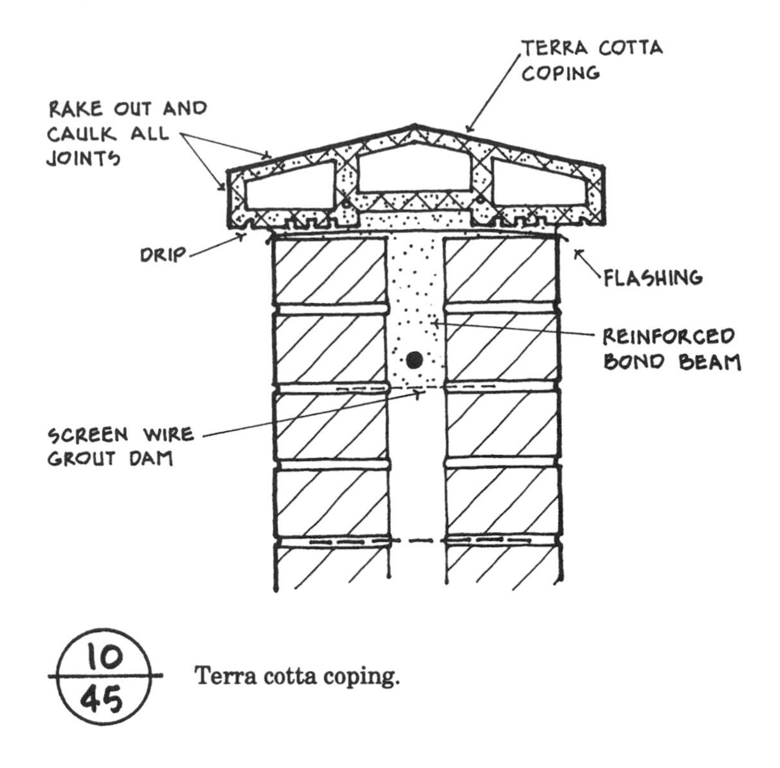

10/45 **Terra cotta coping.**

is formed which catches and holds water, funneling it into the fractured mortar joint. If the flashing inside the wall does not provide a continuous barrier, the water eventually finds its way to the interior.

Reglets with snap-in counterflashing legs provide a better separation between work of the two trades. Reglets designed to be placed in the mortar joint are installed by the mason. The roofing contractor removes and replaces the snap-in counterflashing leg, so the mortar joint is not disturbed by the roofing operation (*see Fig. 10-46*). Where the masonry and the roof slab must move independently, a flexible flashing connection or roof-edge expansion joint should be provided.

Where a thin veneer forms the outer face of the parapet, support is usually provided at roof level by a steel shelf angle. Flashing should bridge the cavity at this location to provide a final course of protection above the lower wall, and if the cavity is ungrouted, weepholes should be placed in the course immediately above the flashing (*see Fig. 10-47*). The flashing provides the slip plane needed to allow differential movement between the parapet and the building walls below.

10.5.9 Shelf Angles

Shelf angles may occur in masonry veneer walls, not only at the roof, but at every floor level as well. Shelf angles must be installed with a "soft joint" between the bottom of the angle and the top course of masonry below (*see Fig. 10-48*). This permits differential expansion and contraction of the veneer and structure to occur, as well as deflection and frame shortening, without the angle bearing on the veneer. At each location, flashing and weepholes must be installed to collect moisture and direct it to the outside. BIA recommends that the flashing be brought beyond the face of the wall

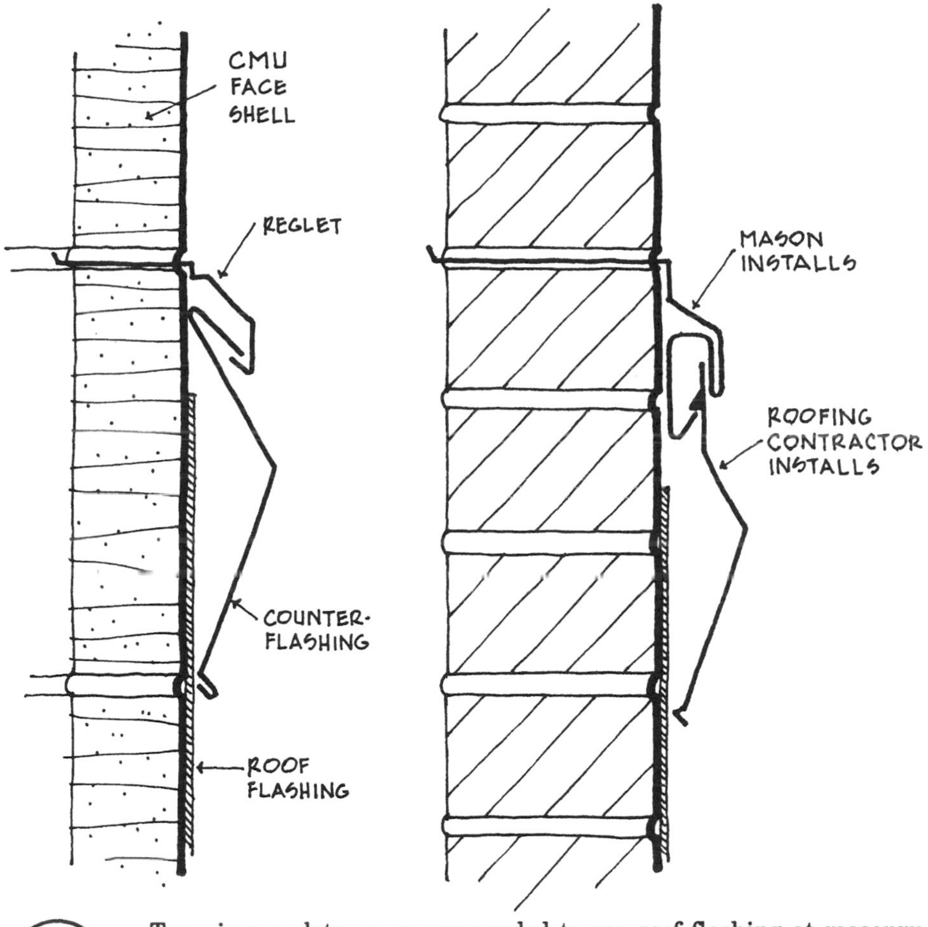

Two-piece reglets are recommended to cap roof flashing at masonry parapets.

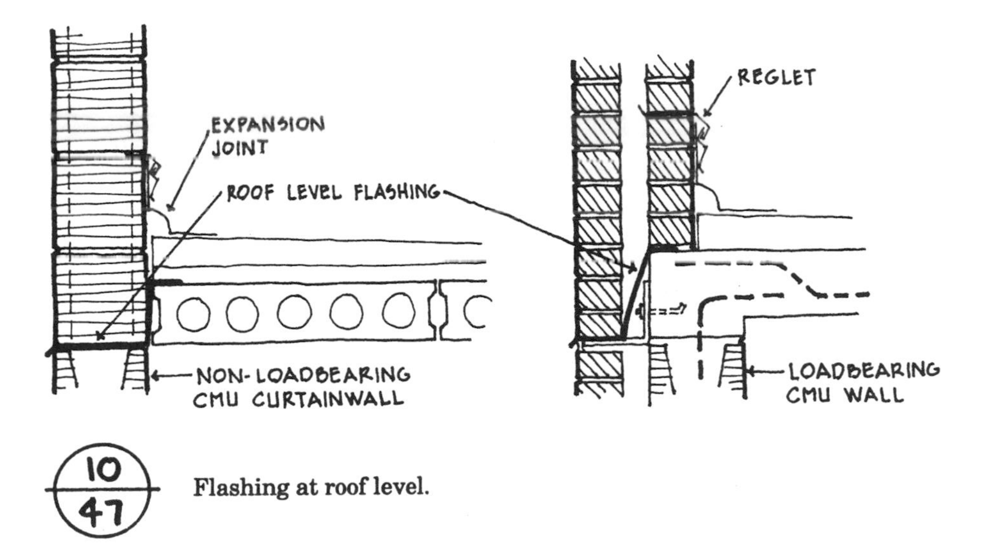

Flashing at roof level.

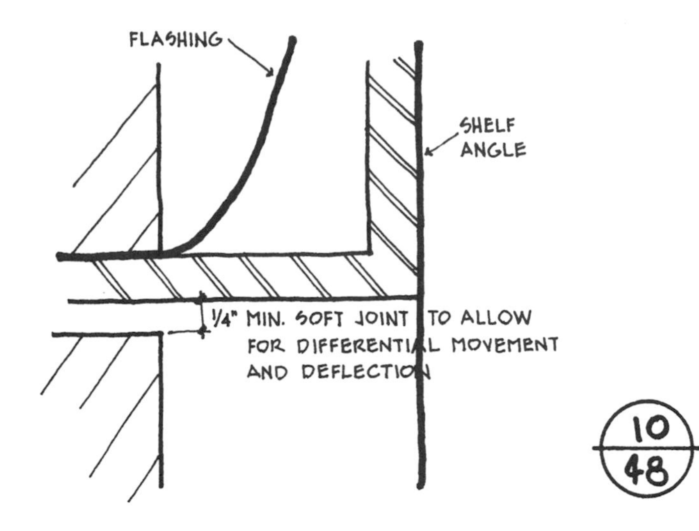

A soft joint is required below shelf angles.

and turned down to form a drip. A sealant joint below the flashing is required to prevent water from re-entering the joint and penetrating the wall below. The drip detail, however, makes it difficult to install the sealant (*see Fig. 10-49A*) Placing the sealant above the flashing forces water to drain only through the weepholes and eliminates additional drainage underneath the brick (*see Fig. 10-49B*). In any case, flashing should never be stopped short of the face of the wall.

A shelf angle detail developed by Smith, Hinchman and Grylls Architects in Detroit uses a two-piece metal flashing in a slightly different configuration to solve this problem (*see Fig. 10-50*). The outer leg of the flashing turns back and down to form a backing for the sealant joint, and uses a bond-breaker tape rather than a conventional backer rod. The two-piece design assures proper fit regardless of the effect of construction tolerances on the relative locations of the shelf angle and the masonry. Rubber flashings cannot be formed into a drip, but they should at least be extended beyond the face of the masonry and later trimmed flush, with a sealant joint installed below rubber flashing and bituminous flashings that cannot be exposed to sunlight may also be installed with a separate metal lip (refer to Chapter 9). Flashing that is stopped short of the face permits moisture to flow around and underneath where it can then pool in the cores of the brick or block, or enter the cavity of the wall below.

Since the soft joint and the thickness of the shelf angle make horizontal expansion joints quite wide, special detailing is sometimes used to disguise their visual impact. Some manufacturers make special shaped "lintel brick" with a lip designed to fit down over the end of the lintel and reduce the joint width. Using the lipped unit on top of the shelf angle creates an offset that is difficult for the flashing membrane to conform to, so many architects prefer to use the lipped unit in the course below the angle so that flashing installation is easier (*see Fig. 10-51*). Lipped units should be purchased as a special shaped brick. They should not be field cut because of a tendency with time for the lip to shear off. Horizontal joints can also be articulated using special shape units such as water table brick (*see Fig. 10-52*). This creates a strong shadow line in which the joint and flashing are hidden. The appearance of horizontal movement joints can also be minimized by changing the unit pattern or the unit color for a few courses above the shelf angle to create a strong horizontal band. The visual impact

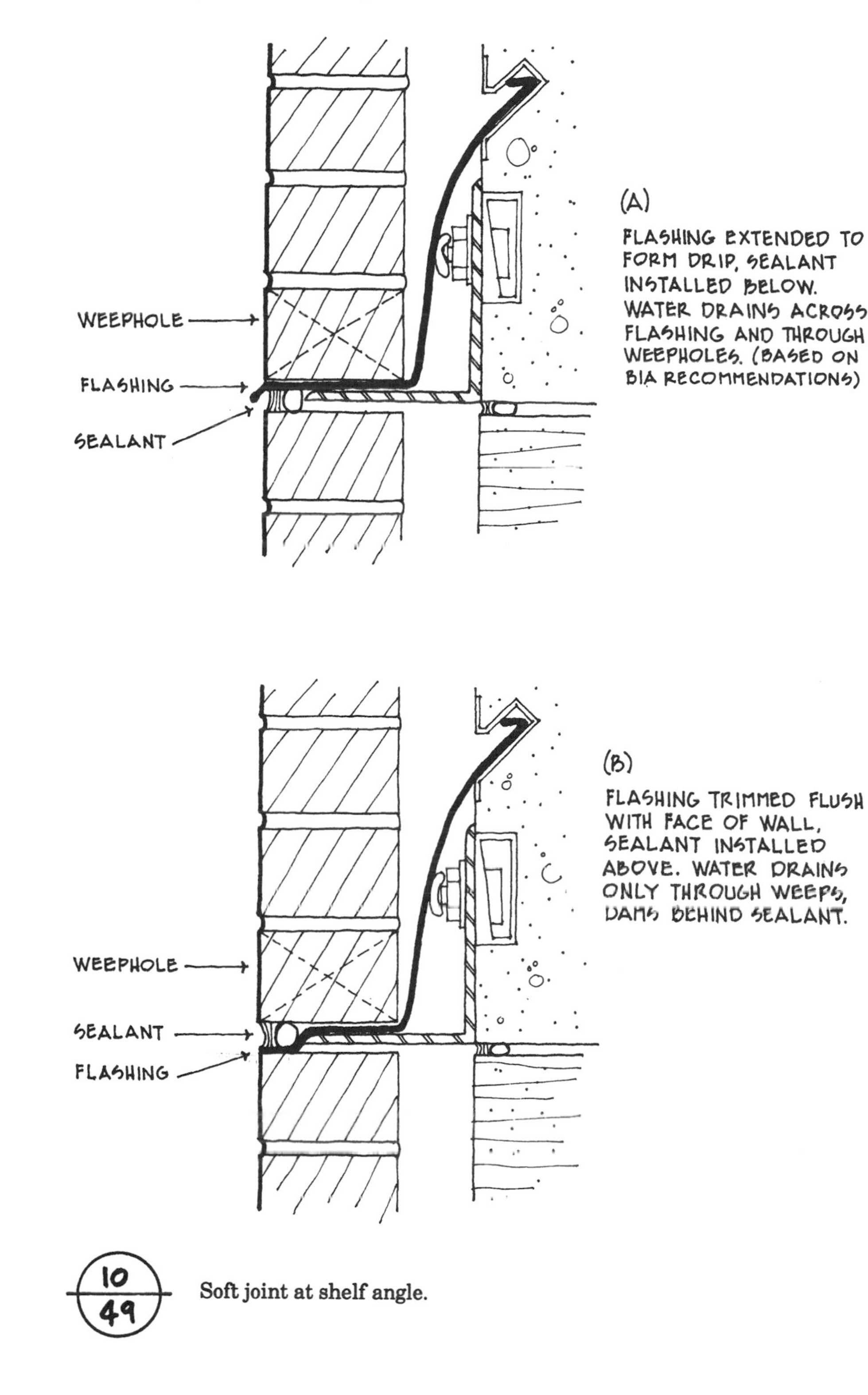

Soft joint at shelf angle.

of the decorative band distracts the eye from the soft joints, flashing and weeps above the shelf angle (*see Fig. 10-53*).

For architects who strongly object to the appearance of horizontal soft joints in a brick masonry facade, the best alternative is to design the veneer without any shelf angles at all. The veneer rests on the slab and is

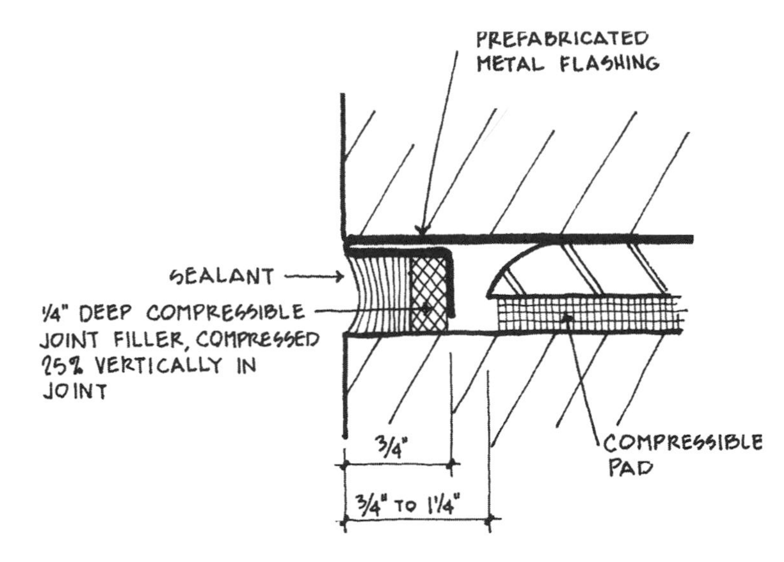

Shelf angle detail developed by Smith, Hinchman and Grylls Architects, Detroit.

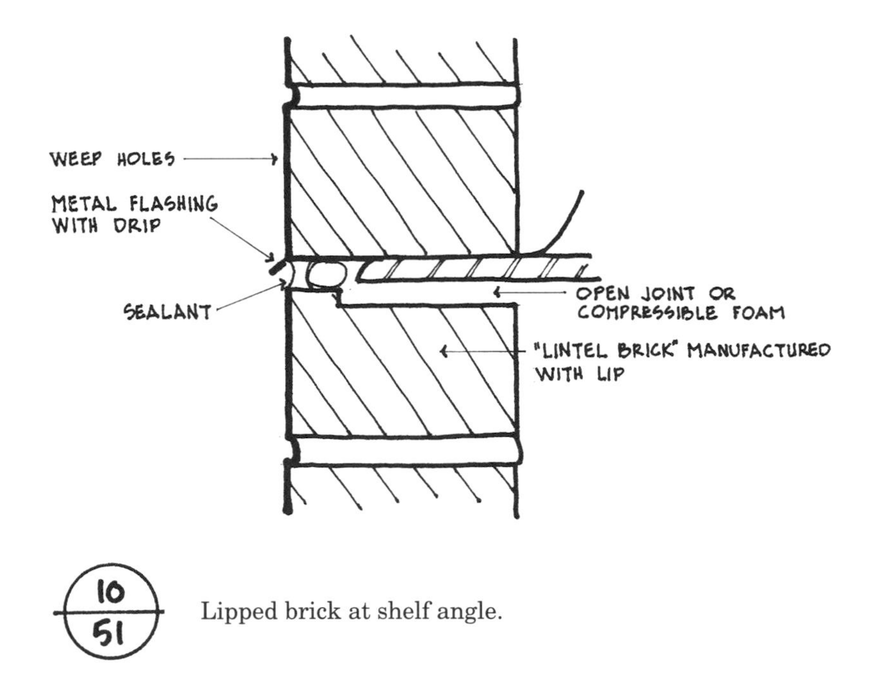

Lipped brick at shelf angle.

anchored to the backing wall in the usual way. Most building codes permit this type of construction up to a height of at least 100 ft. The compressive strength of the units is more than adequate to support the dead load of the masonry above. The parapet cap and any terminations at balconies or other protruding or recessed elements must be carefully detailed to allow for vertical expansion of the brick.

Flashing and weephole courses must also be installed above and below openings, and at the base of the wall (*see Figs. 10-20 and 10-29*).

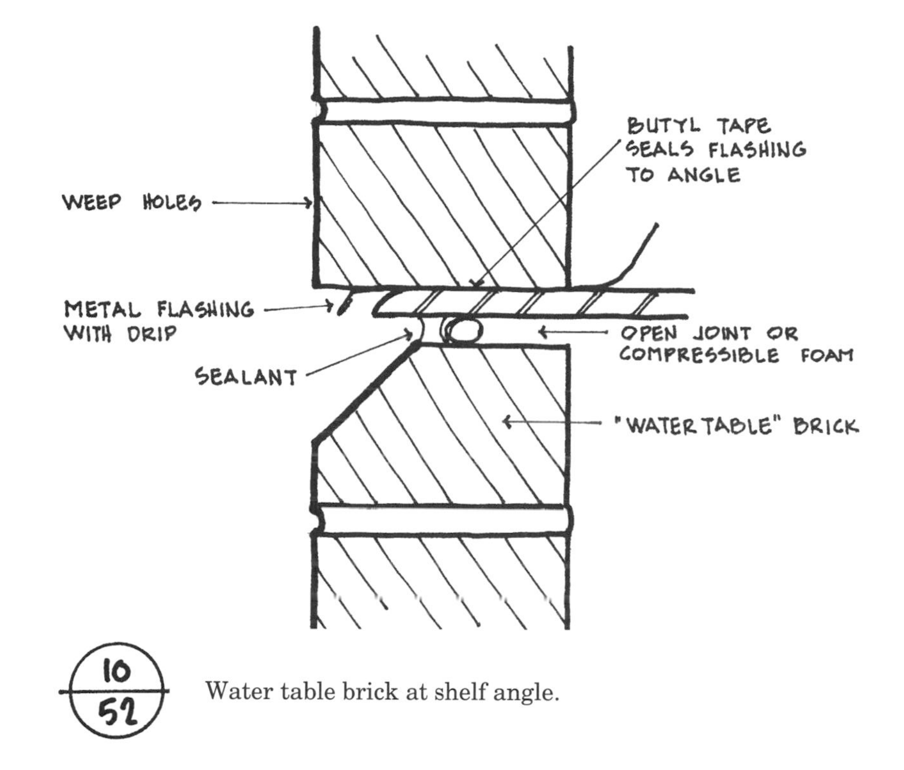

Water table brick at shelf angle.

Strong horizontal lines camouflage wide soft joints.

Windows and doors that are recessed even slightly from the face of the wall will be better protected against leaks than those installed flush with the masonry, particularly when construction tolerances cause the window locations to fall outside the plane of the wall (*see Fig. 10-54*). The base of the wall should always sit on a ledge below the level of the slab so that water draining through the cavity cannot enter at the floor level (*see Figs. 10-20, 10-26, and 10-29*).

Protruding window invites leakage at head.

11
LINTELS AND ARCHES

There are two ways to span openings in masonry walls. *Beams and lintels* are horizontal elements which carry loads as flexural members. Masonry *arches* may be flat or curved, but carry loads in compression because of the shape or orientation of the individual units.

Large wood or stone lintels were used in ancient Egypt and the Middle East to provide small window and door openings in massive loadbearing masonry walls. The strength of individual stones or timbers, however, limited the size of such openings. Early corbeled arches were constructed by progressively projecting the masonry units themselves across the top of an opening until they met at the apex, carrying the load essentially by cantilever action. True compressive arches were developed as early as 1400 B.C. in Babylonia and later perfected by the Romans, along with barrel vaults and domes. In more recent history, brick arches have been used for long spans with heavy loading as in the railway bridge at Maidenhead, England, built in 1835, which spans 128 ft with a rise of 24.3 ft. A railway bridge constructed in Baltimore in 1895 spans 130 ft with a rise of 26 ft.

This chapter discusses the design of steel, concrete, and masonry lintels and masonry arches. Structural masonry beams for large openings or heavy loads are discussed in Chapter 12.

11.1 LINTELS

Lintels of steel, reinforced masonry, stone, concrete, precast concrete, cast stone, and wood are still used today to span small openings in masonry walls. Lintels must be analyzed to determine the loads which must be carried and the resulting stresses which will be created in the member. Many of the cracks that appear over door and window openings result from excessive deflection of lintels which have been improperly or inadequately designed.

11.1.1 Load Determination

Regardless of the material used to form or fabricate a lintel, one of the most important aspects of design is the determination of applied loads.

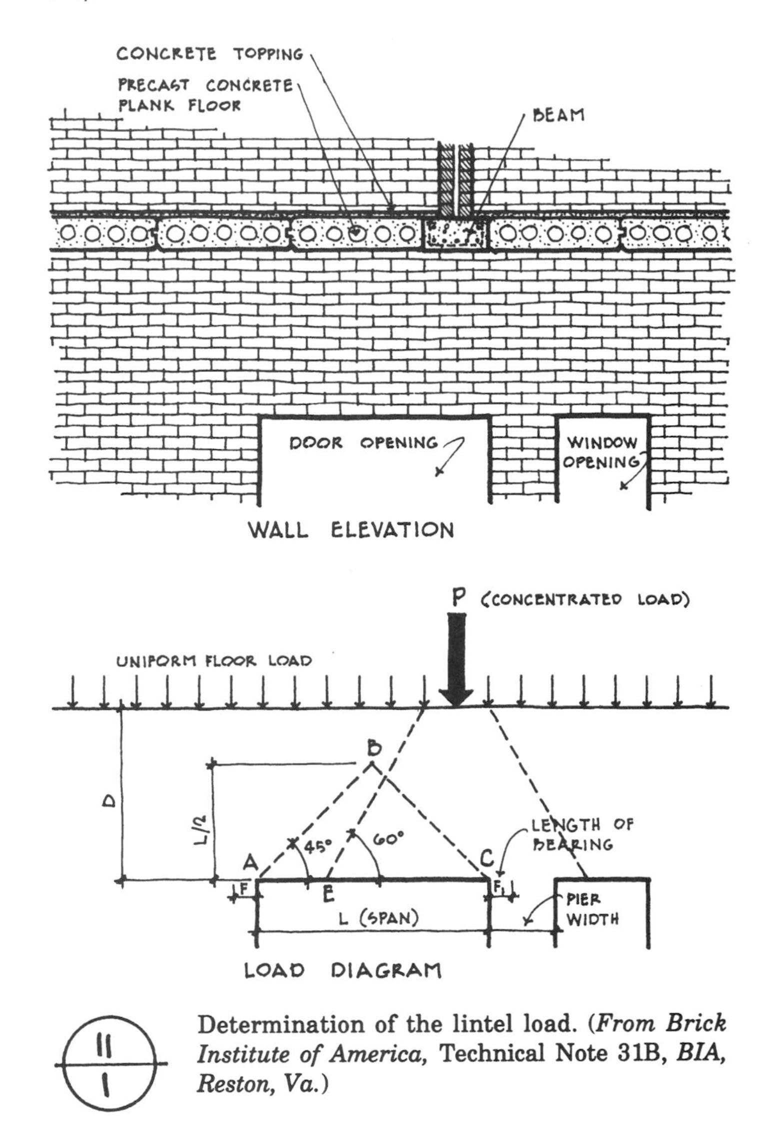

Determination of the lintel load. (*From Brick Institute of America,* **Technical Note 31B,** *BIA, Reston, Va.*)

Figure 11-1 shows an elevation of an opening with a concrete plank floor and concrete beam bearing on the wall, and a graphic illustration of the distribution of these loads. The triangular area (*ABC*) immediately above the opening has sides at 45° angles to the base and represents the area of wall weight actually carried by the lintel. "Arching action" of the masonry will carry other loads outside the triangle provided that the height of the wall above the apex is sufficient to resist arching thrusts. Arching action may be assumed only when the masonry is laid in running bond, or when sufficient bond beams distribute the loads in stack bond. For most lintels of ordinary thickness, load, and span, a depth of 8 to 16 in. above the apex is generally sufficient.

In addition to the dead load of wall contained within the triangular area, the lintel will also carry any uniform floor loads occurring above the opening and below the apex of the triangle. In *Fig. 11-1,* the distance D is greater than $L/2$, so the floor load may be ignored. If arching action does occur as described above, loads outside the triangle may be neglected.

Consideration must also be given to concentrated loads from beams, girders, or trusses which frame into the wall above the opening. These loads are distributed over a length of wall equal to the base of a trapezoid whose summit is at the point of load application and whose sides make an angle of 60° with the horizontal. In *Fig. 11-1,* the portion of concentrated load carried by the lintel is distributed over the length *EC* and is considered as a uniform load partially distributed. The sum of all loads is used to calculate the size of lintel required to span the opening.

11.1.2 Steel Lintels

Structural steel shapes are commonly used to span masonry openings. Steel angles are the simplest shapes and are suitable for openings of moderate width where superimposed loads are not excessive. For wider openings or heavy loads, steel beams with suspended plates may be required (*see Fig. 11-2*). The horizontal leg of a steel angle should be at least $3\frac{1}{2}$ in. wide to adequately support a nominal 4-in. wythe of brick. Generally, angles should be a minimum of $\frac{1}{4}$ in. thick to satisfy code requirements for exterior steel members.

The method of design for steel lintels is basically the same as for steel beams. Although the computations are relatively simple, proper steel lintel design should take into consideration loading, bending moment, reactions,

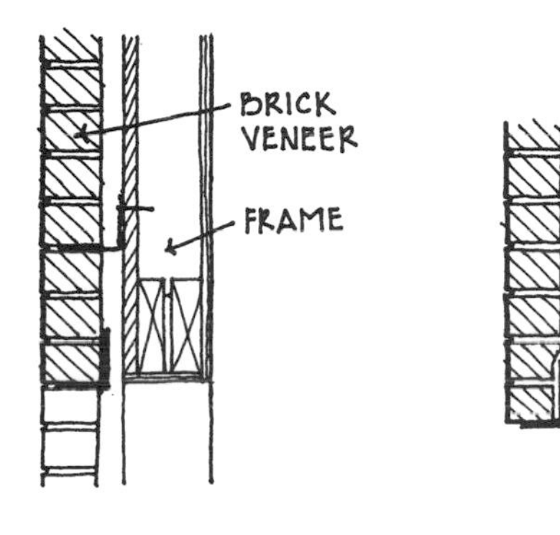

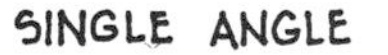

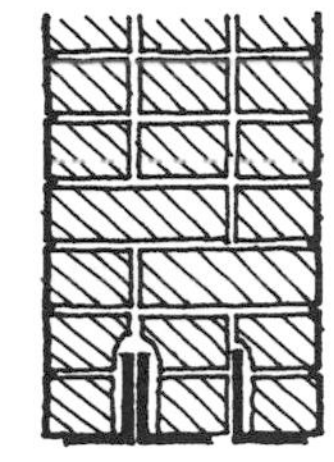

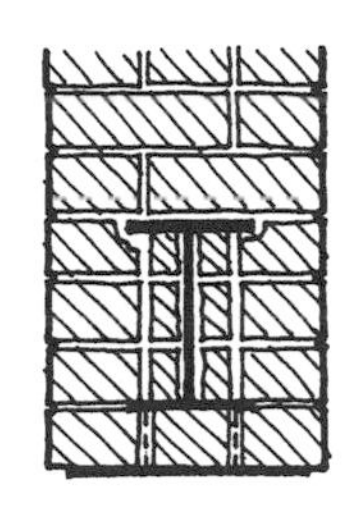

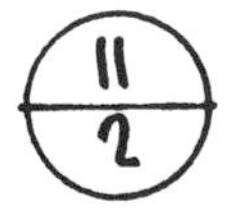

Steel lintels. (***From Brick Institute of America,* Technical Note 31B, *BIA, Reston, Va.***)

required section modulus, selection of section, bearing area, deflection, and shear. If steel members are subject to torsional loads, a special analysis should be made.

The *maximum bending moment* caused by uniformly distributed loads above an opening is determined by the formula

$$M = \frac{wL^2}{8} \tag{11.1}$$

where M = bending moment, ft-lb
w = total uniform load, lb/lin foot
L = span of lintel, center to center of end bearing, ft

To this should be added the bending moment caused by concentrated loads. Where such loads are located far enough above the lintel to be distributed as shown in *Fig. 11-1,* the bending moment formula for a uniform load partially distributed may be used. Otherwise, concentrated load bending moments are used [refer to the *Manual of Steel Construction,* published by the American Institute of Steel Construction (AISC)].

A steel lintel should be selected by first determining the required *section modulus* using the formula

$$S = \frac{12M}{F_s} \tag{11.2}$$

where S = section modulus, cu in.
F_s = allowable stress in bending of the steel, psi
M = bending moment, ft-lb

The allowable stress (F_s) for ASTM A-36 structural steel is 22,000 psi for members laterally supported. Under most conditions, masonry walls provide sufficient lateral stiffness to permit use of the full 22,000 psi, particularly for certain wall-floor framing conditions.

Using the tables in the AISC manual, an angle (or other steel shape) is selected which has an elastic section modulus equal to, or slightly greater, than that required. Within the limitations of minimum steel thickness and length of horizontal leg required, the lightest-weight member having the required section modulus should be chosen.

To determine overall lintel length, required *bearing area* must be calculated. Compressive bearing stresses in the supporting masonry should not exceed the allowable unit stress for the type of masonry used (see Chapter 12 for allowable stresses). The reaction at each end of the lintel is one-half the total uniform load plus a portion of any concentrated load or partially distributed uniform load. The required area may be found by the formula

$$A_b = \frac{R}{f_m} \tag{11.3}$$

where A_b = required bearing area, sq in.
R = reaction, lb
f_m = allowable compressive stress in masonry, psi

The width of the selected steel section divided into the required bearing area A_b will determine the length of bearing required (F and F_1 in *Fig. 11-1*). In no instance should this length be less than 4 in. The table in *Fig. 11-3* lists length of bearing required for angles with 3½- and 4-in. horizontal legs for various end reactions.

f_m (psi)	$3\frac{1}{2}$-in. leg horizontal, Length of bearing, in.					f_m (psi)	4-in. leg horizontal, Length of bearing, in.				
	4	5	6	7	8		4	5	6	7	8
400	5,600	7,000	8,400	9,800	11,200	400	6,400	8,000	9,600	11,200	12,800
350	4,900	6,125	7,350	8,575	9,800	350	5,600	7,000	8,400	9,800	11,200
300	4,200	5,250	6,300	7,350	8,400	300	4,800	6,000	7,200	8,400	9,600
275	3,850	4,813	5,775	6,738	7,700	275	4,400	5,500	6,600	7,700	8,800
250	3,500	4,375	5,250	6,125	7,000	250	4,000	5,000	6,000	7,000	8,000
225	3,150	3,938	4,725	5,513	6,300	225	3,600	4,500	5,400	6,300	7,200
215	3,010	3,763	4,515	5,268	6,020	215	3,440	4,300	5,160	6,020	6,880
200	2,800	3,500	4,200	4,900	5,600	200	3,200	4,000	4,800	5,600	6,400
175	2,450	3,063	3,675	4,288	4,900	175	2,800	3,500	4,200	4,900	5,600
160	2,240	2,800	3,360	3,920	4,480	160	2,560	3,200	3,840	4,480	5,120
155	2,170	2,713	3,255	3,798	4,340	155	2,480	3,100	3,720	4,340	4,960
150	2,100	2,625	3,150	3,675	4,200	150	2,400	3,000	3,600	4,200	4,800
140	1,960	2,450	2,940	3,430	3,920	140	2,240	2,800	3,360	3,920	4,480
125	1,750	2,188	2,625	3,063	3,500	125	2,000	2,500	3,000	3,500	4,000
115	1,610	2,013	2,415	2,818	3,220	115	1,840	2,300	2,760	3,220	3,680
110	1,540	1,925	2,310	2,695	3,080	110	1,760	2,200	2,640	3,080	3,520
100	1,400	1,750	2,100	2,450	2,800	100	1,600	2,000	2,400	2,800	3,200
85	1,190	1,488	1,785	2,083	2,380	85	1,360	1,700	2,040	2,380	2,720
75	1,050	1,313	1,575	1,838	2,100	75	1,200	1,500	1,800	2,100	2,400
70	980	1,225	1,470	1,715	1,960	70	1,120	1,400	1,680	1,960	2,240

11-3 **End reaction (in pounds) and required length of bearing (in inches) for steel angle lintels.** (*From Brick Institute of America,* Technical Note 31B, *BIA, Reston, Va.*)

Before a selected steel lintel section is incorporated into the final design, it should be checked for *deflection.* The allowable loads in *Fig. 11-4* are based on a maximum deflection of $\frac{1}{600}$ of the span. The table, which was developed by BIA engineers, contains tabulated load values to assist in the selection of proper steel lintel sizes.

Using steel lintels to span openings in masonry walls requires careful attention to flashing details, and to provisions for differential movement of the steel and masonry. Code requirements for fireproofing of steel members should also be thoroughly investigated. If fireproofing is required, it may be simpler to design the lintel as a reinforced masonry section. Steel lintels should be galvanized to prevent corrosion.

11.1.3 Concrete Masonry Lintels

Concrete masonry walls do not lend themselves to the use of steel angle lintels because of the size and weight of the units. Openings are more commonly spanned with U-shaped lintel blocks grouted and reinforced with deformed steel bars. Reinforced concrete masonry lintels not only cost less than structural steel lintels, but they eliminate the danger of steel corrosion and subsequent masonry cracking as well as the painting and maintenance of exposed steel.

In some instances, cast-in-place or precast concrete sections can also be used. *Cast-in-place lintels* are subject to drying shrinkage and have surface

Horizontal leg	Angle size	Weight per foot (lb)	Span in feet (center to center of required bearing)										Resisting moment (ft-lb)	Elastic section modulus
			3	4	5	6	7	8	9	10	11	12		
$3\frac{1}{2}$	$3 \times 3\frac{1}{2} \times \frac{1}{4}$	5.4	956	517	262	149	91	59					1082	0.59
	$\times \frac{5}{16}$	6.6	1166	637	323	184	113	73					1320	0.72
	$\times \frac{3}{8}$	7.9	1377	756	384	218	134	87	59				1558	0.85
$3\frac{1}{2}$	$3\frac{1}{2} \times 3\frac{1}{2} \times \frac{1}{4}$	5.8	1281	718	406	232	144	94	65				1448	0.79
	$\times \frac{5}{16}$	7.2	1589	891	507	290	179	118	80				1797	0.98
	$\times \frac{3}{8}$	8.5	1947	1091	589	336	208	137	93	66			2200	1.20
$3\frac{1}{2}$	$4 \times 3\frac{1}{2} \times \frac{1}{4}$	6.2	1622	910	580	338	210	139	95	68			1833	1.00
	$\times \frac{5}{16}$	7.7	2110	1184	734	421	262	173	119	85	62		2383	1.30
	$\times \frac{3}{8}$	9.1	2434	1365	855	490	305	201	138	98	71		2750	1.50
	$\times \frac{7}{16}$	10.6	2760	1548	978	561	349	230	158	113	82	60	3117	1.70
4	$4 \times 4 \times \frac{7}{16}$	11.3	2920	1638	1018	584	363	239	164	116	85	62	3299	1.80
	$\times \frac{1}{2}$	12.8	3246	1820	1141	654	407	268	185	131	95	70	3666	2.00
$3\frac{1}{2}$	$5 \times 3\frac{1}{2} \times \frac{1}{4}$	7.0	2600	1460	932	636	398	264	184	132	97	73	2933	1.60
	$\times \frac{5}{16}$	8.7	3087	1733	1106	765	486	323	224	161	119	89	3483	1.90
	$\times \frac{7}{16}$	12.0	4224	2371	1513	1047	655	435	302	217	160	120	4766	2.60
	$\times \frac{1}{2}$	13.6	4875	2736	1746	1177	736	488	339	244	179	134	5500	3.00
$3\frac{1}{2}$	$6 \times 3\frac{1}{2} \times \frac{1}{4}$	7.9	3577	2009	1283	888	650	439	306	221	164	124	4033	2.20
	$\times \frac{5}{16}$	9.8	4390	2465	1574	1090	798	538	375	271	201	151	4950	2.70
	$\times \frac{3}{8}$	11.7	5200	2922	1865	1291	945	636	443	320	237	179	5867	3.20
	$\times \frac{1}{2}$	15.3	6828	3834	2448	1695	1228	818	570	412	305	230	7700	4.20
4	$6 \times 4 \times \frac{1}{4}$	8.3	3739	2099	1340	928	679	458	319	231	171	129	4216	2.30
	$\times \frac{5}{16}$	10.3	4552	2556	1632	1129	827	562	391	283	209	158	5133	2.80
	$\times \frac{3}{8}$	12.3	5365	3012	1923	1331	974	665	463	335	248	187	6050	3.30
	$\times \frac{7}{16}$	14.3	6178	3469	2214	1533	1122	764	532	384	284	215	6967	3.80
	$\times \frac{1}{2}$	16.2	6990	3925	2506	1734	1270	857	597	431	319	241	7883	4.30

Note: Allowable loads to the left of the heavy line are governed by moment, and to the right by deflection.

Allowable uniform superimposed load (psf) for steel angle lintels (F_s = 22,000 psi). (*From Brick Institute of America,* Technical Note 31B, *BIA, Reston, Va.*)

textures that are not always compatible with the adjoining masonry. *Precast concrete lintels* and *cast stone lintels* are better in some respects since they are delivered to the job site ready for use, do not require temporary shoring, and can carry superimposed loads as soon as they are in place. These sections can be produced with surface textures closely matching that of the masonry, and can be scored vertically to simulate mortar joints. Precast lintels may be one-piece, or may be split into two thinner sections. Split lintels are relatively lightweight and easily handled. Split lintels, however, are not recommended to support combined wall and floor loads because of the difficulty involved in designing the heavily loaded inner section to match the deflection of the outer section, which may carry only wall loads. Differential deflection could cause critical stress concentrations in the wall. Mortar for bedding precast lintels should be the same quality as that used in laying the wall, and at least equal to ASTM C270, Type N.

Reinforced concrete masonry lintels are constructed with special shaped lintel units, bond beam units, or standard units with depressed, cut-out, or grooved webs to accommodate the steel bars (*see Fig. 11-5*). Individual units are laid end to end to form a channel in which continuous reinforcement and grout are placed. Among the major advantages of concrete masonry unit (CMU) lintels over steel are low maintenance and the elimination of differential movement between dissimilar materials. Con-

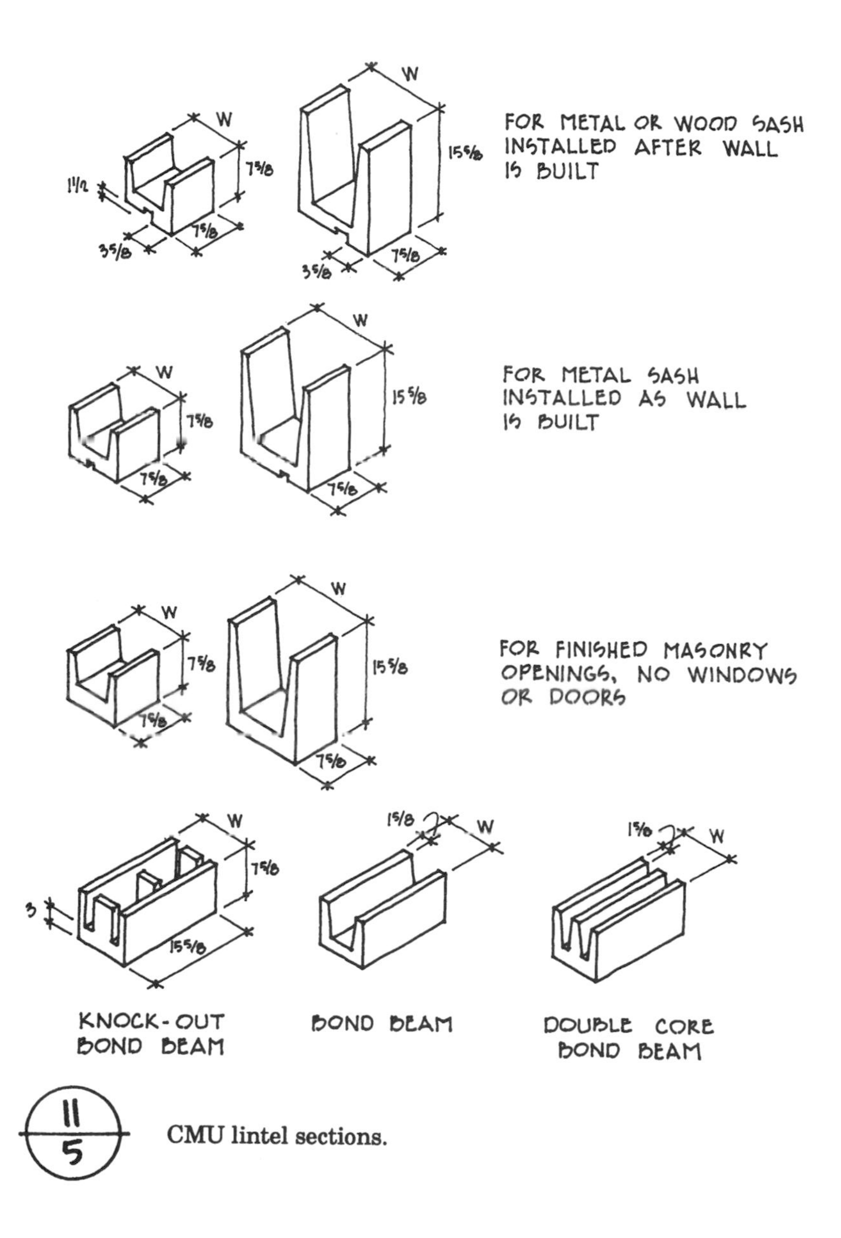

11-5 CMU lintel sections.

crete masonry lintels are often designed as part of a continuous bond beam course, which helps to further distribute shrinkage and temperature stresses in the masonry above openings. This type of installation is more satisfactory in areas subject to seismic activity.

Units used for lintel construction should comply with the requirements of ASTM C90, *Standard Specification for Loadbearing Concrete Masonry Units,* and should have a minimum compressive strength adequate to provide the masonry compressive strength ($f'm$) used in the design. Mortar should be equal to that used in constructing the wall and should meet the minimum requirements of ASTM C270, Type N. Grout for embedment of reinforcing steel should comply with ASTM C476, and maximum aggregate size should be dependent upon the size of the grout space (see Chapter 6). The first course of masonry above the lintel should be laid with full mortar bedding so that the cross webs as well as the face shells of the units bear on the lintel and reduce the shear stress between the grout-filled core and the face shells.

The magnitude of imposed loads is determined by the same method outlined above for steel lintels, with loads which occur outside the 45° triangle neglected because of natural arching action in the masonry. For ordinary wall thicknesses, loads, and spans, 8- to 16-in. of masonry above the apex of the triangle is required to resist arching thrusts. Arching action also requires adequate mass at each end of the lintel to resist lateral thrusts. When the end of a long-span lintel supporting relatively heavy loads is located near a wall corner or opening, it may be necessary to neglect arching action and design the lintel for all superimposed loads applied above the opening. If walls are laid in stack bond, no loads are distributed beyond a vertical joint unless structural bond beams are incorporated in the wall to effect such distribution. Generally, the loads from floor joists or rafters in residential and light commercial buildings may be considered as uniformly distributed when the height of the masonry between the lintel and the bearing plane is more than one-third of the joist or rafter spacing. If members bear directly on the lintel or are relatively heavy, they should be treated as concentrated loads.

A minimum end bearing of 8 in. is recommended for reinforced CMU lintels with relatively modest spans. For longer spans or heavy loads, bearing stresses should be calculated to ensure that the allowable compressive stress of the masonry is not exceeded. High stress concentrations may require the use of solid units or solidly grouted hollow units for one or more courses under the lintel bearing so that loads are distributed over a larger area.

The NCMA design table in *Fig. 11-6* is based on typical equivalent uniform loads of 200 to 300 lb/lin ft for wall loads, and 700 to 1000 lb/lin ft for combined floor and roof loads. The table can be used to determine required lintel size and reinforcing for various spans subject to this type of loading. Where concentrated loads occur, calculation of total shear and moment is required, and lintels may be sized on the basis of the table in *Fig. 11-7,* which lists resisting moments and shears for various unit sizes and prism strengths. A design example is given in *Fig. 11-8.*

11.1.4 Reinforced Brick Lintels

Standard brick masonry units are also adaptable to reinforced lintel design even though they do not have continuous channels for horizontal steel (*see Fig. 11-9*). Reinforcing may be located in bed joints or in a

Type of loading	Lintel section nominal size (in.)	Required reinforcing							
		Clear span							
		3′-4″	4′-0″	4′-8″	5′-4″	6′-0″	6′-8″	7′-4″	8′-0″
Wall loads	6 × 8	1-No. 3	1-No. 4	1-No. 4	2-No. 4	2-No. 5			
(200–300 lb/lin ft)	6 × 16					1-No. 4	1-No. 4	1-No. 4	1-No. 4
Floor and roof loads (700–1000 lb/lin ft)	6 × 16	1-No. 4	1-No. 4	2-No. 3	1-No. 5	2-No. 4	2-No. 4	2-No. 5	2-No. 5
Wall loads	8 × 8	1-No. 3	2-No. 3	2-No. 3	2-No. 4	2-No. 4	2-No. 5	2-No. 6	
(200–300 lb/lin ft)	8 × 16							2-No. 5	2-No. 5
Floor and roof loads	8 × 8	2-No. 4							
(700–1000 lb/lin ft)	8 × 16	2-No. 3	2-No. 3	2-No. 3	2-No. 4	2-No. 4	2-No. 4	2-No. 5	2-No. 5

Required reinforcement for simply supported CMU lintels. (*From National Concrete Masonry Association,* **TEK Bulletin 25,** *NCMA, Herndon, Va.*)

widened collar joint created by using half units. Horizontal cell clay tile are ideally suited for lintels since, like concrete blocks, they contain natural channels for the steel. Manufacturers of 8-in. hollow brick also produce lintel units similar to those of concrete masonry.

The physical dimensions of clay masonry lintels are partially determined by the type of wall that they support and the type of unit used in construction. The design width of brick lintels should be the same as the wall thickness. Design depth is, to a degree, determined by the course height of the units. Depth can be limited by the height of masonry above the opening, in which case compression steel may be required to provide adequate resistance to bending. *Figure 11-10B* shows a reinforced brick lintel capable of carrying the same loads as the three steel angles in *Fig. 11-10A.* The reinforced brick lintel is more economical and a more efficient use of structural materials.

The BIA has developed a tabulated set of resisting moments and shears as a design aid for clay masonry lintels. The tables in *Figs. 11-11 and 11-12* list resisting moments and shears for various reinforced brick and structural clay tile lintel with alternative amounts of reinforcing steel. Resisting moments are determined by either the masonry or the reinforcement, and the value listed is that which governs design. Where moment is governed by steel, the value is shown in italics. The two values of shear are diagonal tension in the masonry (V_m), and bond on the tensile steel (V_o). For lintels in which no steel stirrups are provided, the smaller resisting shear governs. V_m is based on an allowable working stress in diagonal tension of 50 psi. When web reinforcement is provided, the working stress increases to 150 psi, so resisting shears are tripled if stirrups are provided. However, before such high shearing stresses are reached, bond shear may govern the design. In Tables *A* and *B* in *Fig. 11-11,* bond governs the shear values and V_m is therefore not listed.

Where the resisting moment of the masonry governs design, it is often possible to increase this value by using units with higher compressive strengths than that assumed (see Chapter 12).

Tables *A* through *F* in *Fig. 11-11* for brick lintels are based on modu-

Steel reinf.	$f'_m = 1500$ psi						$f'_m = 2000$ psi					
	Bottom cover $1\frac{1}{2}$ in.			Bottom cover 3 in.			Bottom cover $1\frac{1}{2}$ in.			Bottom cover 3 in.		
	V_v (kip)	V_u (kip)	M (ft-kip)	V_v (kip)	V_u (kip)	M (ft-kip)	V_v (kip)	V_u (kip)	M (ft-kip)	V_v (kip)	V_u (kip)	M (ft-kip)
TABLE A 6-IN. LINTELS (8 IN. HIGH)												
1-#3	1.422	1.009	0.983*	1.063	0.743	0.685	1.642	1.021	0.995*	1.227	0.754	0.734†
1-#4	1.407	1.291	1.335	1.048	0.945	0.813	1.625	1.310	1.615	1.210	0.960	0.989
1-#5	1.392	1.554	1.501	1.033	1.132	0.899	1.608	1.582	1.834	1.193	1.152	1.106
1-#6	1.377	1.802	1.624	1.018	1.305	0.957	1.590	1.836	2.001	1.175	1.330	1.188
2-#5	1.392	2.965	1.802	1.033	2.154	1.059	1.608	3.024	2.238	1.193	2.201	1.326
2-#6	1.377	3.434	1.906	1.018	2.484	1.102	1.590	3.503	2.391	1.175	2.535	1.393
TABLE B 8-IN. LINTELS (8 IN. HIGH)												
1-#4	1.908	1.311	1.633	1.421	0.961	1.001	2.203	1.329	1.766†	1.640	0.976	1.210
1-#5	1.887	1.581	1.855	1.400	1.152	1.119	2.179	1.607	2.252	1.617	1.173	1.366
1-#6	1.867	1.837	2.025	1.380	1.331	1.202	2.156	1.870	2.477	1.594	1.356	1.481
1-#7	1.847	2.076	2.163	1.360	1.496	1.265	2.132	2.115	2.666	1.570	1.526	1.571
2-#6	1.867	3.505	2.422	1.380	2.536	1.411	2.156	3.576	3.013	1.594	2.588	1.770
2-#7	1.847	3.954	2.537	1.360	2.848	1.455	2.132	4.036	3.183	1.570	2.907	1.840
TABLE C 8-IN. LINTELS (16 IN. HIGH)												
1-#5	4.484	3.926	6.472†	3.997	3.482	5.739†	5.178	3.974	6.549†	4.616	3.526	5.811†
1-#6	4.464	4.614	8.755	3.977	4.087	7.228	5.155	4.677	9.177†	4.592	4.145	8.079†
2-#5	4.484	7.588	9.911	3.997	6.720	8.172	5.178	7.704	11.988	4.616	6.827	9.097
2-#6	4.464	8.881	10.981	3.977	7.856	9.019	5.155	9.032	13.379	4.592	7.993	11.016
2-#8	4.423	11.292	12.719	3.937	9.967	10.368	5.108	11.512	15.714	4.545	10.166	12.847
TABLE D 12-IN. LINTELS (8 IN. HIGH)												
1-#5	2.878	1.618	2.461	2.135	1.182	1.497	3.323	1.641	2.705†	2.466	1.200	1.812
1-#6	2.847	1.884	2.714	2.104	1.368	1.629	3.287	1.914	3.290	2.430	1.391	1.987
2-#5	2.878	3.111	3.073	2.135	2.265	1.841	3.323	3.165	3.751	2.466	2.307	2.263
2-#6	2.847	3.609	3.327	2.104	2.613	1.962	3.287	3.677	4.096	2.430	2.665	2.433
2-#8	2.785	4.513	3.679	2.042	3.237	2.103	3.216	4.606	4.599	2.358	3.304	2.651
TABLE E 12-IN. LINTELS (16 IN. HIGH)												
1-#5	6.838	3.994	4.937†	6.096	3.545	4.382†	7.896	4.035	4.987†	7.038	3.583	4.429†
1-#6	6.807	4.705	9.170†	6.065	4.171	8.130†	7.860	4.760	9.278†	7.003	4.222	8.229†
2-#5	6.838	7.755	12.782†	6.096	6.874	10.814	7.896	7.858	12.952†	7.038	6.969	11.486†
2-#6	6.807	9.099	14.643	6.065	8.055	12.078	7.860	9.235	17.688	7.003	8.180	14.622
2-#8	6.745	11.611	17.299	6.003	10.256	14.164	7.789	11.817	21.159	6.931	10.443	17.370

*Moment reduced by 25% since $p = A_s/bd$ is less than $100/f_y$, where f_y is yield strength of reinforcement.

†Moment governed by allowable tensile stress in steel (f_s = 20,000 psi).

Allowable shear and resisting moments for CMU lintels. (V_v, allowable shear based on shear stress; V_u, allowable shear based on bond stress.) (*From National Concrete Masonry Association,* TEK Bulletin 81, *NCMA, Herndon, Va.*)

Determine steel reinforcement required for the simply supported reinforced concrete masonry lintel shown.

Assume:

Effective span, $L = 6$ ft
Nominal wall thickness = 8 in.
Wall weight = 34 lb/sq. ft
Lintel weight (8×8 in.) = 54 lb/ft
Uniform floor load = 300 lb/ft

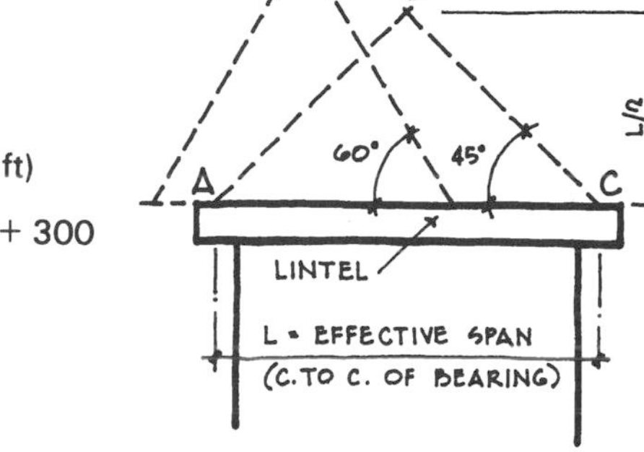

(A) Assuming no arching action ($D = 2$ ft)

Total uniform load, $w = 2(34) + 54 + 300$
$= 422$ lb/ft

$$\text{Moment} = \frac{wl^2}{8} = \frac{(422)(6)^2}{8}$$

$M = 1899$ ft-lb $= 1.899$ ft-kip
Assuming 8-in. bearing length, shear (at face of support),

$$V = \frac{wl}{2} - \frac{w}{3} = \frac{(422)(6)}{2} - \frac{422}{3}$$

$V = 1125$ lb $= 1.125$ kip

From Fig. 11-7 ($f'_m = 1500$ psi, cover = 1.5 in):
For 1 – No. 6 bar, $V_u = 1.837$ kips; $M = 2.025$ ft-kip
From Fig. 12-10 ($f'_m = 2000$ psi, cover = 1.5 in.):
For 1 – No. 5 bar, $V_u = 1.607$ kips; $M = 2.252$ ft-kip

(B) Assuming arching action ($D = 4$ ft)
Moment (due to uniform load of lintel)

$$M = \frac{(54)(6)^2}{8} = 243 \text{ ft-lb} = 0.243 \text{ ft-kip}$$

Moment (due to wall weight—triangular area)

$$M = \frac{(34)(6)^3}{24} = 306 \text{ ft-lb} = 0.306 \text{ ft-kip}$$

Total moment = 0.549 ft-kip

Shear (at face of support):

$$V = \frac{(54)(6)}{2} + \frac{(3)(3)(34)}{2} - \frac{54}{3} - \frac{(0.3)(0.15)(34)}{2}$$

$$= 162 + 153 - 18 - 0.8 = 296 \text{ lb} = 0.296 \text{ kip}$$

From Fig. 11-7 for 8×8 in. lintel with 1-No. 4 bar, allowable shear and moment for $f'_m = 1500$ psi and $f'_m = 2000$ psi are greater than actual calculated stresses. Therefore, 1-No. 4 bar will satisfy design requirements.

Design example. (*From National Concrete Masonry Association,* TEK Bulletin 81, *NCMA, Herndon, Va.*)

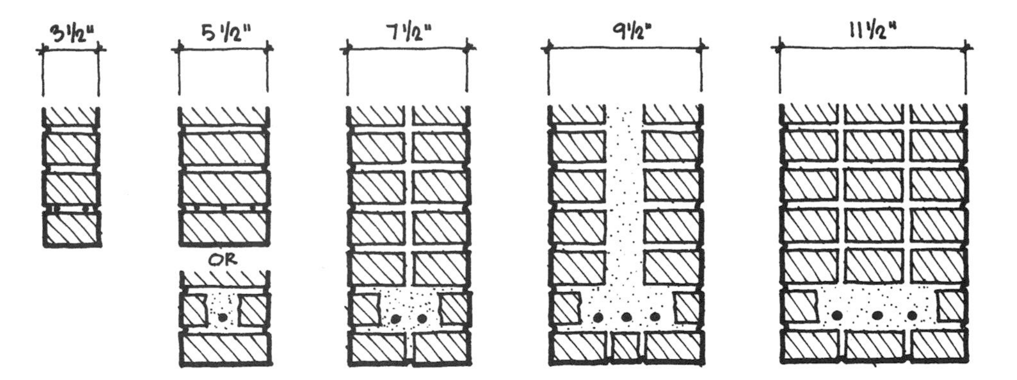

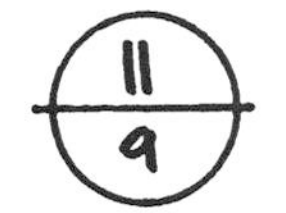

Reinforced brick lintels. (*From Brick Institute of America,* Technical Note 17H, *BIA, Reston, Va.*)

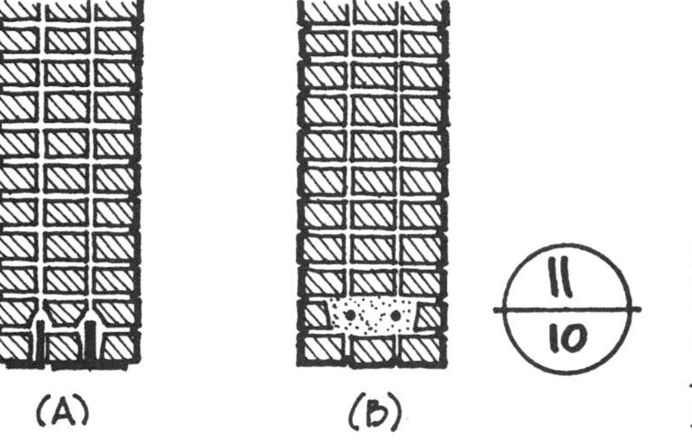

(A) Steel lintel and (B) reinforced brick lintel with the same load-carrying capacity. (*From Brick Institute of America,* Technical Note 17H, *BIA, Reston, Va.*)

lar units with ½-in. mortar joints where the steel is located in the bed joint. The effective depths in Tables *A* and *B* assume ¼-in. reinforcing bars centered within the bed joint. In Tables *C* through *F*, effective depth assumes reinforcement placed in grout with ½ in. clear between the steel and the top of the unit immediately below. True effective depth will therefore vary slightly with the bar diameter, but for the range of sizes shown here, the difference is not significant. Tables *G* through *I* in *Fig. 11-12* list resisting moments and shears for various cross sections of tile, and are also based on modular unit sizes with ½-in. mortar joints. The configurations shown serve as examples only, and the tables apply to all cross-sectional arrangements of grouted tile lintels of the same overall dimensions. The effective depths for the tile lintels assume that the reinforcement is completely embedded in grout with ¾ in. clear between the steel and the tile units.

11.1.5 Precasting

Reinforced brick lintels are normally built in place by using temporary shoring to support the wall weight until the section has cured sufficiently to carry superimposed loads. The soffit brick may be standard units or special shapes, and are laid with mortar in the head and collar joints only. Reinforced brick lintels may also be precast, however. This eliminates the need for shoring and allows work to proceed more rapidly.

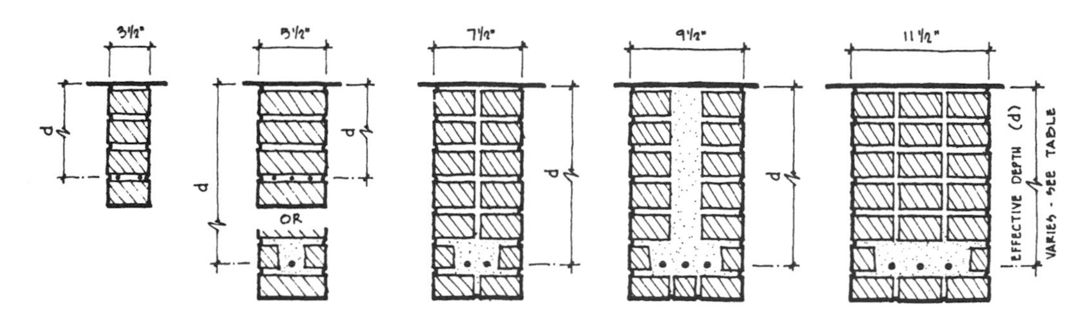

TABLE A NOMINAL 4-IN. LINTELS (REINFORCING IN BED JOINT)

Depth	d = 5.1" (2 courses)		d = 7.8" (3 courses)		d = 10.5" (4 courses)	
Reinf.	M	V_o	M	V_o	M	V_o
1-#2	*390*	220	605	455	820	620
2-#2	750	430	1180	890	1600	1210

TABLE B NOMINAL 6-IN. LINTELS (REINFORCING IN BED JOINT)

Depth	d = 5.1" (2 courses)		d = 7.8" (3 courses)		d = 10.5" (4 courses)		d = 13.1" (5 courses)		d = 15.8" (6 courses)	
Reinf.	M	V_o	M	V_o	M	V_o	M	V_o	M	V_o
2-#2	780	440	*1200*	905	*1630*	1230	*2060*	1550	*2490*	1400
3-#2	1140	645	1770	1340	*2410*	1820	*3040*	2290	*3670*	2080

TABLE C NOMINAL 6-IN. LINTELS (REINFORCING IN GROUT SPACE)

Depth	d = 4.6" (2 courses)			d = 7.3" (3 courses)			d = 10.0" (4 courses)			d = 12.6" (5 courses)			d = 15.3" (6 courses)			d = 18.0" (7 courses)		
Reinf.	M	V_m	V_o	M	V_m	V_o	M	V_m	V_o	M	V_m	V_o	M	V_m	V_o	M	V_m	V_o
1-#3	770	—	790	*1230*	—	1260	*1700*	—	1730	*2170*	—	2230	*2640*	—	2720	*3120*	—	3200
1-#4	1060	—	1010	*2160*	—	1630	*3000*	—	2260	*3840*	—	2890	*4700*	—	3550	*5530*	—	4180
1-#5	1190	1060	1210	2600	1730	1970	4400	2410	2750	*5810*	3090	3540	*7120*	3800	4330	*8410*	4480	5120

TABLE D NOMINAL 8-IN. LINTELS

Depth	d = 4.6" (2 courses)			d = 7.3" (3 courses)			d = 10.0" (4 courses)			d = 12.6" (5 courses)			d = 15.3" (6 courses)			d = 18.0" (7 courses)		
Reinf.	M	V_m	V_o	M	V_m	V_o	M	V_m	V_o	M	V_m	V_o	M	V_m	V_o	M	V_m	V_o
2-#3	1380	1540	1550	*2420*	2480	2500	*3350*	3420	3440	*4280*	4380	4400	*5220*	5340	5360	*6160*	6300	6340
3-#3	1570	1500	2260	3360	2430	3660	*4930*	3360	5070	*6310*	4310	6490	*7690*	5250	7920	*9100*	6210	9360
2-#4	1630	1470	1970	3530	2380	3200	*5890*	3310	4450	*7560*	4250	5700	*9230*	5190	6960	*10900*	6140	8240
2-#5	1800	1410	2350	4000	2300	3860	6820	3210	5380	10170	4140	6930	*13970*	5070	8500	*16600*	6010	10100
2-#6	1910	1350	2720	4350	2230	4480	7510	3120	6290	11300	4040	8110	15600	4950	9950	20500	5880	11800

TABLE E NOMINAL 10-IN. LINTELS

Depth	d = 7.3" (3 courses)			d = 10.0" (4 courses)			d = 12.6" (5 courses)			d = 15.3" (6 courses)			d = 18.0" (7 courses)			d = 20.6" (8 courses)		
Reinf.	M	V_m	V_o	M	V_m	V_o	M	V_m	V_o	M	V_m	V_o	M	V_m	V_o	M	V_m	V_o
2-#3	*2450*	—	2510	*3380*	—	3470	*4310*	—	4430	*5260*	—	5400	*6200*	—	6370	*7150*	—	7350
3-#3	*3600*	3110	3710	*4980*	4300	5120	*6380*	5510	6560	*7780*	6720	8020	*9190*	7930	9440	*10600*	9150	10900
2-#4	4130	3060	3230	*5950*	4240	4480	*7630*	5440	5760	*9320*	6640	7020	*11000*	7850	8300	*12700*	9060	9580
3-#4	4630	2990	4750	7960	4160	6610	*11200*	5340	8480	*13700*	6530	10400	*16300*	7730	12300	*18800*	8930	14200
2-#5	4720	2950	3900	7980	4120	5460	*11500*	5300	7010	*14100*	6490	8590	*16700*	7680	10200	*19300*	8890	11800
2-#6	5170	2870	4540	8850	4020	6380	13300	5180	8220	18300	6350	10100	*23300*	7540	12000	*26900*	8730	13900
3-#5	5320	2880	5710	9110	4030	8000	13700	5190	10300	18800	6360	12600	*24600*	7540	15000	*28500*	8720	17300
2-#7	5530	2780	5150	9600	3910	7250	14500	5060	9380	20200	6220	11500	26500	7400	13700	33400	8560	15900
3-#6	5750	2790	6660	9980	3910	9310	15100	5060	12000	21000	6210	14800	27500	7380	17600	34800	8550	20400

TABLE F NOMINAL 12-IN. LINTELS

Depth	d = 10.0" (4 courses)			d = 12.6" (5 courses)			d = 15.3" (6 courses)			d = 18.0" (7 courses)			d = 20.6" (8 courses)			d = 23.3" (9 courses)		
Reinf.	M	V_m	V_o	M	V_m	V_o	M	V_m	V_o	M	V_m	V_o	M	V_m	V_o	M	V_m	V_o
2-#4	*5990*	—	4530	*7680*	—	5810	*9380*	—	7070	*11100*	—	8370	*12800*	—	9650	*14500*	—	10900
3-#4	*8840*	5080	6670	*11300*	6520	8540	*13800*	7970	10400	*16400*	9420	12400	*18900*	10900	14300	*21500*	12400	16200
2-#5	*9060*	5040	5500	*11600*	6480	7070	*14200*	7920	8660	*16800*	9370	10200	*19500*	10800	11800	*22100*	12300	13400
4-#4	9800	5010	8750	14800	6430	11200	*18200*	7870	13800	*21600*	9310	16300	*25000*	10800	18800	*28300*	12200	21400
2-#6	10100	4920	6450	15000	6330	8300	*19800*	7770	10200	*23500*	9210	12100	*27100*	10600	14000	*30800*	12100	15900
3-#5	10400	4940	8080	15500	6350	10400	*21000*	7780	12800	*24900*	9220	15100	*28700*	10700	17500	*32600*	12100	19800
2-#7	11000	4800	7340	16500	6190	9470	23000	7610	11600	30100	9030	13800	*36400*	10500	16000	*41400*	11900	18200
4-#5	11300	4850	10600	17000	6250	13600	23500	7670	16800	30700	9090	19900	*37800*	10500	23000	*43000*	12000	26100
3-#6	11400	4800	9440	17200	6200	12200	23800	7600	14900	31200	9020	17700	39400	10400	20500	*45500*	11900	23400
4-#6	12400	4710	12300	18800	6090	16000	26100	7480	19600	34300	8880	23300	43400	10300	27000	53200	11700	30700
3-#7	12400	4670	10700	18800	6040	13900	26200	7430	17100	34500	8840	20300	43500	10300	23500	53300	11700	26800
4-#7	13300	4590	14000	20300	5940	18200	28400	7310	22400	37500	8690	26600	47600	10100	30900	58600	11500	35100

Note: Resisting moments in italics are controlled by the steel, others are controlled by the masonry. Resisting moments are given in foot-pounds. Resisting shears are given in pounds.

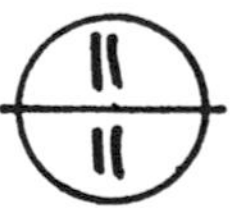

Design tables for reinforced brick lintels. (*From Brick Institute of America,* **Technical Note 17H,** ***BIA, Reston, Va.***)

TABLE G NOMINAL 4 IN. LINTELS

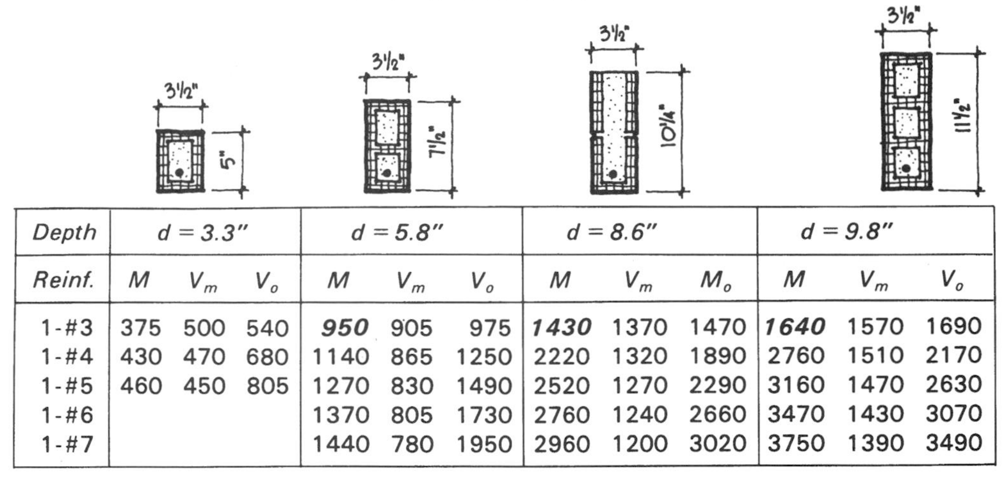

Depth	$d = 3.3''$			$d = 5.8''$			$d = 8.6''$			$d = 9.8''$		
Reinf.	M	V_m	V_o	M	V_m	V_o	M	V_m	M_o	M	V_m	V_o
1-#3	375	500	540	***950***	905	975	***1430***	1370	1470	***1640***	1570	1690
1-#4	430	470	680	1140	865	1250	2220	1320	1890	2760	1510	2170
1-#5	460	450	805	1270	830	1490	2520	1270	2290	3160	1470	2630
1-#6				1370	805	1730	2760	1240	2660	3470	1430	3070
1-#7				1440	780	1950	2960	1200	3020	3750	1390	3490

TABLE H NOMINAL 6-IN. LINTELS

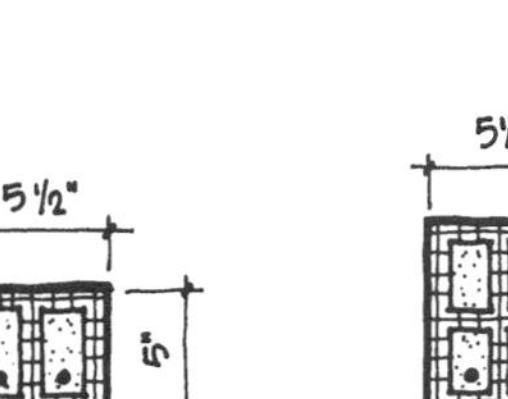
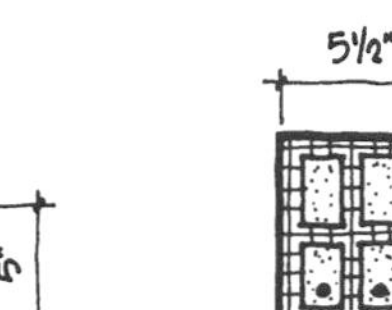
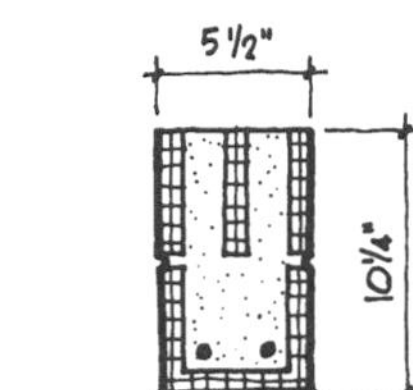
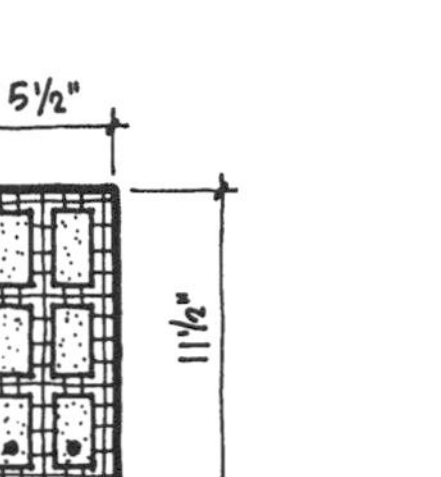
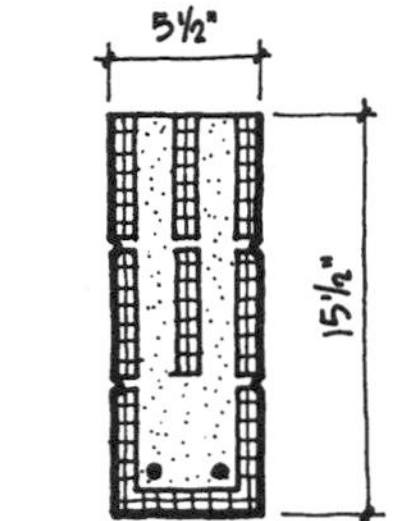

Depth	$d = 3.3''$			$d = 5.8''$			$d = 8.6''$			$d = 9.8''$			$d = 13.9''$		
Reinf.	M	V_m	V_o	M	V_m	V_o	M	V_m	V_o	M	V_m	V_o	M	V_m	V_o
2-#3	635	780	1060	1630	1410	1930	***2830***	2120	2910	***3250***	2430	3340	***4680***	3510	4810
2-#4	720	730	1340	1930	1340	2450	3770	2040	3730	4710	2350	4290	***8260***	3410	6220
2-#5	760	695	1580	2130	1290	2950	4260	1970	4500	5330	2270	5190	9760	3310	7560
2-#6				2270	1240	3410	4620	1910	5240	5840	2210	6050	10800	3230	8840

TABLE I NOMINAL 8-IN. LINTELS

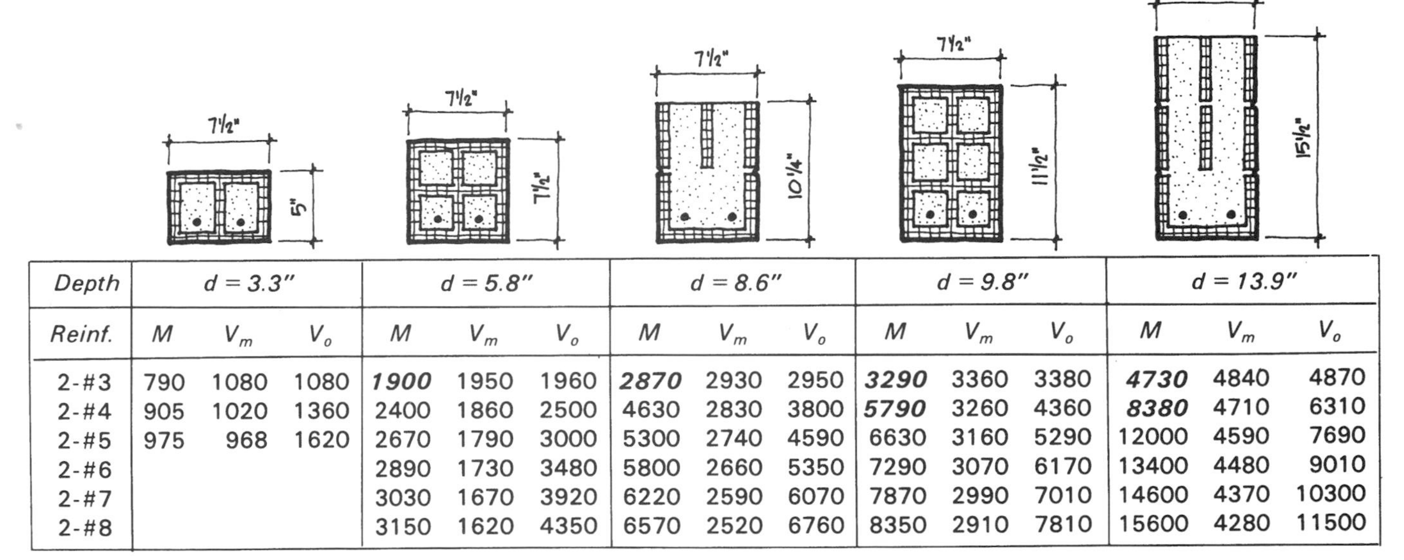

Depth	$d = 3.3''$			$d = 5.8''$			$d = 8.6''$			$d = 9.8''$			$d = 13.9''$		
Reinf.	M	V_m	V_o	M	V_m	V_o	M	V_m	V_o	M	V_m	V_o	M	V_m	V_o
2-#3	790	1080	1080	***1900***	1950	1960	***2870***	2930	2950	***3290***	3360	3380	***4730***	4840	4870
2-#4	905	1020	1360	2400	1860	2500	4630	2830	3800	***5790***	3260	4360	***8380***	4710	6310
2-#5	975	968	1620	2670	1790	3000	5300	2740	4590	6630	3160	5290	12000	4590	7690
2-#6				2890	1730	3480	5800	2660	5350	7290	3070	6170	13400	4480	9010
2-#7				3030	1670	3920	6220	2590	6070	7870	2990	7010	14600	4370	10300
2-#8				3150	1620	4350	6570	2520	6760	8350	2910	7810	15600	4280	11500

Note: Resisting moments in italics are controlled by the steel, others are controlled by the masonry.

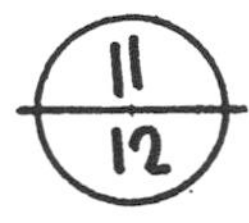

Design tables for reinforced clay tile lintels. (*From Brick Institute of America,* Technical Note 17H, *BIA, Reston, Va.*)

11.2 ARCHES

Arches may be constructed in various forms such as *segmental, elliptical, Tudor, Gothic, semicircular,* and *parabolic* to *flat* or *jack arches* (*see Fig. 11-13*). The primary advantage of an arch is that under uniform loading conditions, the induced stress is principally compression rather than tension. For this reason, an arch will frequently provide the most efficient structural span. Since masonry's resistance to compression is greater than to other stresses, it is an ideal material for the construction of arches.

Arches are divided structurally into two categories. *Minor arches* are those whose spans do not exceed 6 ft with a maximum rise/span ratio of 0.15, with equivalent uniform loads on the order of 1000 lb/ft. These are most often used in building walls over door and window openings. *Major arches* are those whose spans or loadings exceed the maximum for minor arches. With larger spans and uniformly distributed loads, the parabolic arch is often the most structurally efficient form.

11.2.1 Minor Arch Design

In a fixed masonry arch, three conditions must be maintained to ensure the integrity of the arch action: (1) the length of span must remain constant, (2) the elevation of the ends must remain unchanged, and (3) the inclination of the skewback must be fixed. If any of these conditions are altered by sliding, settlement, or rotation of the abutments, critical stresses can develop and may result in structural failure. Adequate foundations and high-quality mortar and workmanship are essential to proper arch construction. The compressive and bond strength of the mortar must be high, and only Types M, S, or N are recommended. It is also particularly important that mortar joints be completely filled to assure maximum bond and even distribution of stresses.

Arches are designed by assuming a shape and cross section based on architectural considerations or empirical methods, and then analyzing the shape to determine its adequacy to carry the superimposed loads. The following discussion of arch design is taken from BIA *Technical Notes* Series 31. Minor arch loading may consist of live and dead loads from floors, roofs, walls, and other structural members. These may be applied as concentrated loads or as uniform loads fully or partially distributed. The dead load on an arch is the weight of the wall area contained within a triangle immediately above the opening. The sides of the triangle are at 45° angles to the base, and its height is therefore one-half of the span. Such triangular loading is equivalent to a uniformly distributed load of 1.33 times the triangular load. Superimposed uniform loads above this triangle are carried beyond the span of the opening by arching action of the masonry wall itself when running bond patterns are used. Uniform live and dead loads below the apex of the triangle are applied directly on the arch for design purposes. Minor concentrated loads bearing directly or nearly directly on the arch may safely be assumed as equivalent to a uniformly distributed load twice the magnitude of the concentrated load. Heavy concentrated loads should not be allowed to bear directly on minor arches (especially jack arches).

There are two basic theories for verification of the stability of an assumed arch section. The *elastic theory* considers the arch as a curved beam subject to moment and shear, whose stability depends on internal stresses. For arches subject to nonsymmetrical loading which can cause tensile stress development, the elastic theory provides the most accurate method of analysis. There are many methods of elastic analysis for arch design, but in most instances, their application is complicated and time-

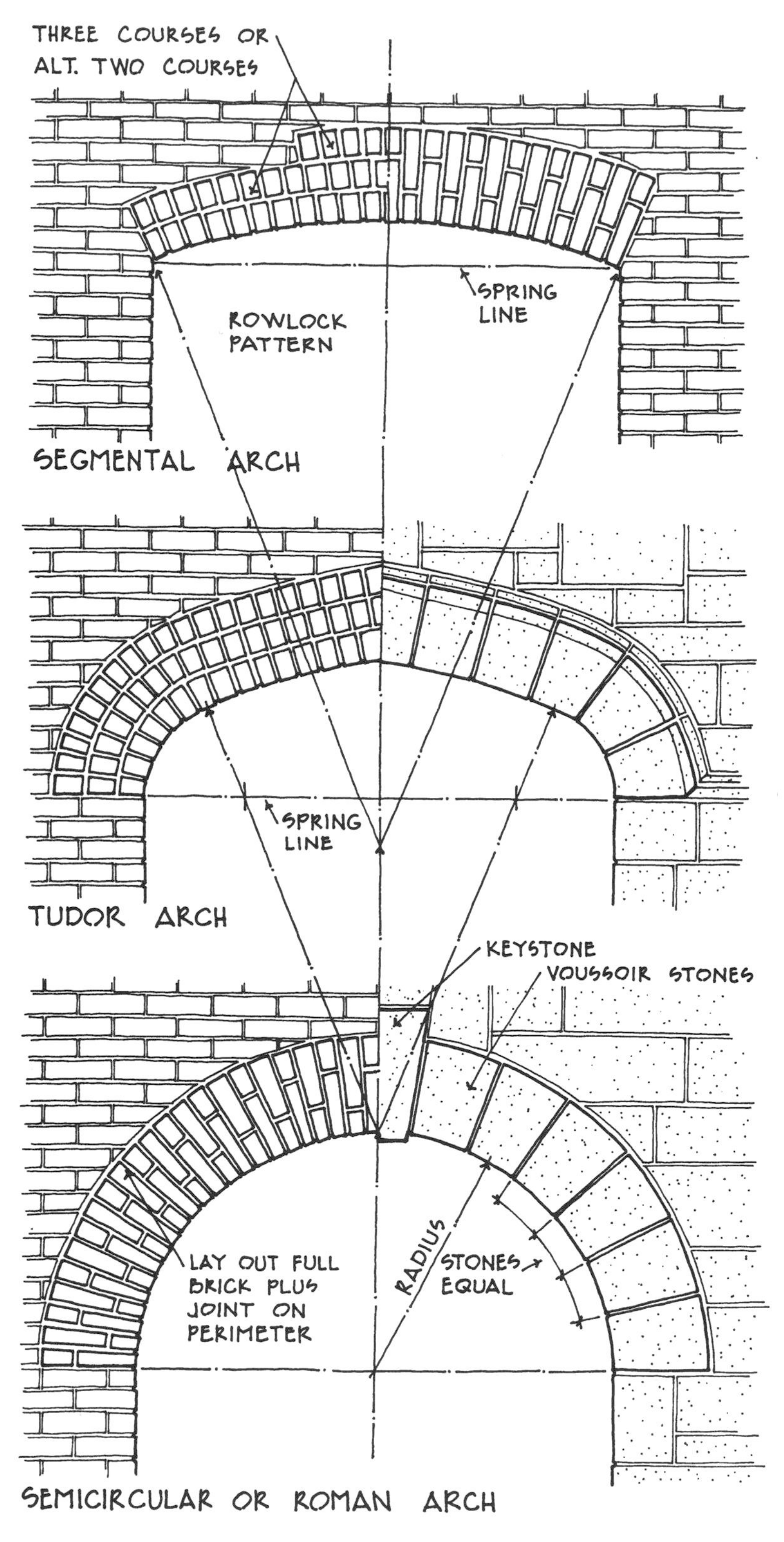

Masonry arch forms. (*From Brick Institute of America,* Technical Note 31, *BIA, Reston, Va.*)

consuming. Such detailed engineering discussions are beyond the scope of this book, and the reader is referred to Valerian Leontovich's *Frames and Arches* for further information.

A second theory of analysis is the *line-of-thrust method,* which considers the stability of the arch ring to be dependent on friction and the reactions between the several arch sections or voussoirs. In general, the line-of-thrust method is most applicable to symmetrical arches loaded uniformly over the entire span or subject to symmetrically placed concentrated loads. For such arches, the line of resistance (which is the line connecting the points of application of the resultant forces transmitted to each voussoir) is required to fall within the middle third of the arch section so that neither the intrados nor extrados of the arch will be in tension (*see Fig. 11-14* for arch terminology).

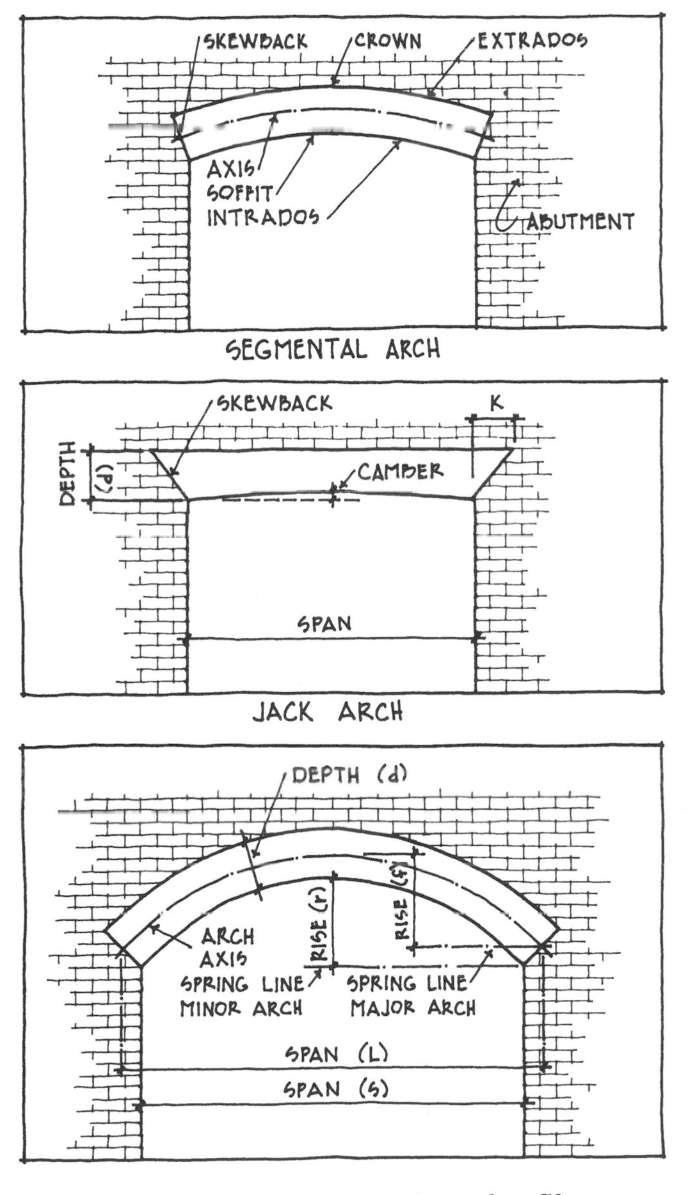

Arch terminology (see the Glossary, Appendix A). (*From Brick Institute of America,* Technical Note 31A, *BIA, Reston, Va.*)

11.2.2 Graphic Analysis

The simplest and most widely used line-of-thrust method is based on the hypothesis of "least crown thrust," which assumes that the true line of resistance of an arch is that for which the thrust at the crown is the least possible consistent with equilibrium. This principle can be applied by static methods if the external forces acting on the arch are known and the point of application and direction of crown thrust are assumed. Normally, the direction of the crown thrust is assumed as horizontal and its point of application as the upper extremity of the middle one-third of the section (i.e., two-thirds the arch depth from the intrados). This assumption has been proven reasonable for symmetrical arches symmetrically loaded, but is not applicable to unsymmetrical or partially distributed uniform loads.

With these assumptions, the forces acting on each section of an arch may be determined by analytical or graphic methods. The first step in the procedure is to determine the joint of rupture. This is the joint for which the tendency of the arch to open at the extrados is the greatest and which therefore requires the greatest crown thrust applied to prevent the joint from opening. At this joint, the line of resistance of the arch will fall on the lower extremity of the middle third of the section. For minor segmental arches, the joint of rupture is ordinarily assumed to be the skewback of the arch. (For major arches with higher rise/span ratios, this will not be true.) Based on the joint of rupture at the skewback and the hypothesis of least crown thrust, the magnitude and direction of the reaction at the skewback may be determined graphically (*see Fig. 11-15*).

In this analysis, only one-half of the arch is considered since it is symmetrical and uniformly loaded over the entire span. *Figure 11-15A* shows

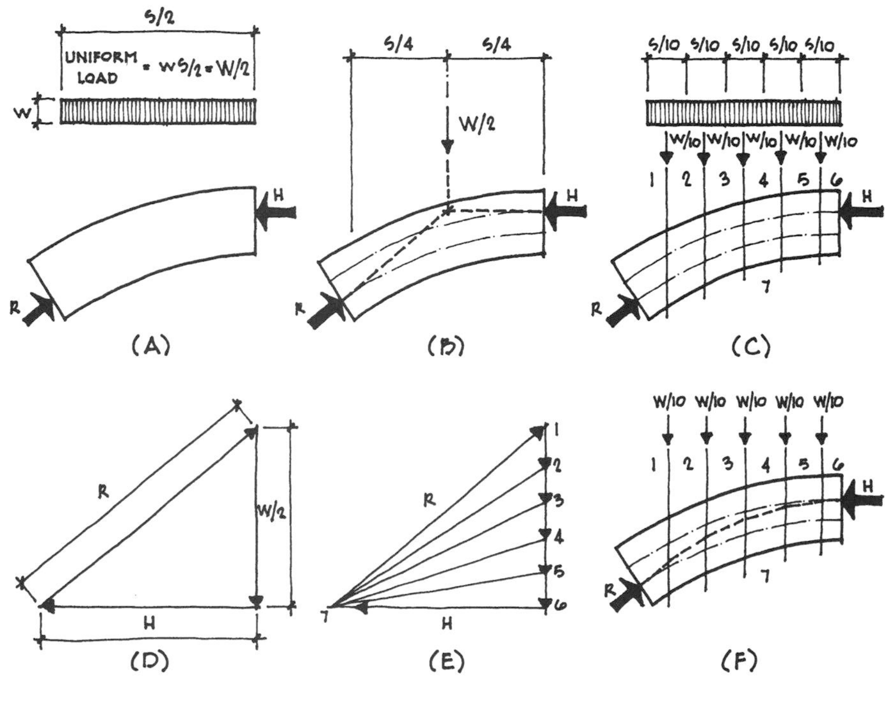

Graphic arch analysis. (*From Brick Institute of America,* Technical Note 31, *BIA, Reston, Va.*)

the external forces acting on the arch section. For equilibrium, the lines of action of these three forces ($W/2$, H, and R) must intersect at one point as shown in *Fig. 11-15B*. Since the crown thrust (H) is assumed to act horizontally, this determines the direction of the resisting force (R). The magnitude of the resistance may be determined by constructing a force diagram as indicated in *Fig. 11-15D*. The arch is divided into voussoirs and the uniform load transformed into equivalent concentrated loads acting on each section (*see Fig. 11-15C*). Starting at any convenient point (in this example between the reaction and the first load segment past the skewback) numbers are placed between each pair of forces, so that each force can subsequently be identified by a number (i.e., 1–2, 5–6, 7–1, and so on). The side of the force diagram which represents $W/2$ (*Fig. 11-15D*) is divided into the same number of equivalent loads, and the same numbers previously used for identification are placed as shown in *Fig. 11-15E* to identify the forces in the new force diagram. Thus, the line 7–1 is the skewback reaction, 6–7 the horizontal thrust, and so on. From the intersection of H and R (7–1 and 6–7) a line is drawn to each intermediate point on the leg representing $W/2$.

The equilibrium polygon may now be drawn. First extend the line of reaction until it intersects the line of action of 1–2 (*see Fig. 11-15F*). Through this point, draw a line parallel to the line 7–2 until it intersects the line of action of 2–3. Through this point, a draw a line parallel to the line 7–3, and so on, and complete the polygon in this manner. If the polygon lies completely within the middle third of the arch section, the arch is stable. For a uniformly distributed load, the equilibrium polygon, which coincides with the line of resistance, will normally fall within this section, but for other loading conditions it may not.

The eccentricity of the voussoir reactions will produce stresses which differ from the axial stress H/A, where A is the cross-sectional area of the arch ($A = bd$). These stresses are determined by the formula

$$f_m = \frac{H}{bd} \pm \frac{6He}{bd} \tag{11.4}$$

where f_m = maximum compressive stress in the arch, psi
H = horizontal thrust, lb
b = thickness of the arch, in.
d = depth of the arch, in.
e = the perpendicular distance between the arch axis and the line of action of the horizontal thrust

Maximum allowable compressive stresses in an arch are determined on the basis of the compressive strength of the units and mortar, and are governed by the same code requirements as other masonry construction (see Chapter 12).

A number of mathematical formulas have been developed for the design of minor arches. Among the structural considerations are three methods of failure of unreinforced masonry arches: (1) by rotation of one section of the arch about the edge of a joint, (2) by the sliding of one section of the arch on another or on the skewback, and (3) by crushing the masonry.

11.2.3 Rotation

The assumption that the equilibrium polygon lies entirely within the middle third of the arch section precludes the rotation of one section of the arch about the edge of a joint or the development of tensile stresses in either the intrados or extrados. For conditions other than evenly distributed uniform loads where the polygon may fall outside the middle third, however, this method of failure should be considered.

11.2.4 Sliding

The coefficient of friction between the units of a masonry arch is at least 0.60 without considering the additional resistance to sliding resulting from the bond between the mortar and the masonry units. This corresponds to an angle of friction of approximately 31°. If the angle between the line of resistance and a line perpendicular to the joint between sections is less than the angle of friction, the arch is stable against sliding. This angle may be determined graphically as shown above. For minor segmental arches, the angle between the line of resistance and a line perpendicular to the joint is greatest at the skewback. This is also true for jack arches if the joints are radial about a center at the intersection of the planes of the skewbacks. However, if the joints are not radial about this center, each joint should be investigated separately for resistance to sliding. This can be most easily accomplished by constructing an equilibrium polygon.

11.2.5 Crushing

A segmental arch is one whose curve is circular but is less than a semicircle. The minimum recommended rise for a segmental arch is 1 in. per foot of span. The horizontal thrust developed depends on the span, depth, and rise of the arch.

The graph in *Fig. 11-16* identifies thrust coefficients (H/W) for seg-

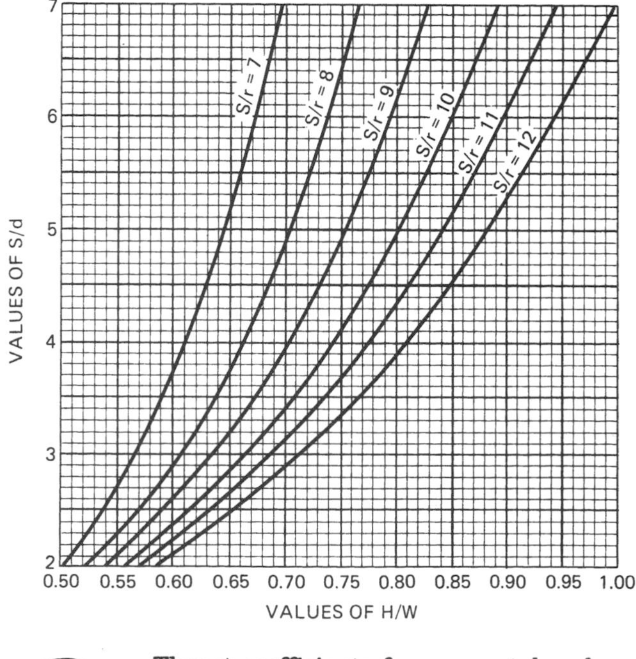

Thrust coefficients for segmental arches. (*From Brick Institute of America,* **Technical Note 31A,** *BIA, Reston, Va.*)

mental arches subject to uniform loads over the entire span. Once the thrust coefficient is determined for a particular arch, the horizontal thrust (H) may be determined as the product of the thrust coefficient and the total load (W). To determine the proper thrust coefficient, first determine the characteristics S/r and S/d of the arch, where S is the clear span in feet, r is the rise of the soffit in feet, and d is the depth of the arch in feet. If the applied load is triangular or concentrated, the same method may be used, but the coefficient H/W is increased by one-third for triangular loading and doubled for concentrated loads.

Once the horizontal thrust has been determined, the *maximum compressive stress* in the masonry is determined by the formula

$$f_m = \frac{2H}{bd} \tag{11.5}$$

This value is twice the axial compressive stress on the arch due to the load H because the horizontal thrust is located at the third point of the arch depth.

The common rule for *jack arches* is to provide a skewback (K measured horizontally; see *Fig. 11-14*) of ½ in. per foot of span for each 4 in. of arch depth. Jack arches are commonly constructed in depths of 8 and 12 in. with a camber of in. per foot of span. To determine the *horizontal thrust* at the spring line for jack arches, the following formulas may be used:

For uniform loading over full span:

$$H = \frac{3WS}{8d} \tag{11.6}$$

For triangular loading over full span:

$$H = \frac{WS}{2d} \tag{11.7}$$

Maximum compressive stress may be determined by the formulas

$$f_m = \frac{2H}{6d} \tag{11.8}$$

For uniform loading over full span:

$$f_m = \frac{3WS}{4bd} \tag{11.9}$$

For triangular loading:

$$f_m = \frac{WS}{6d} \tag{11.10}$$

11.2.6 Thrust Resistance

The horizontal thrust resistance developed by an arch is provided by the adjacent mass of the masonry wall. Where the area of the adjacent wall is substantial, thrust is not generally a problem. However, at corners and openings where the resisting mass is limited, it may be necessary to check the resistance of the wall to this horizontal force. The diagram in *Fig. 11-17* shows how the resistance is calculated. It is assumed that the arch thrust attempts to move a volume of masonry enclosed by the boundary lines *ABCD*. For calculating purposes, the area *CDEF* is equivalent in resis-

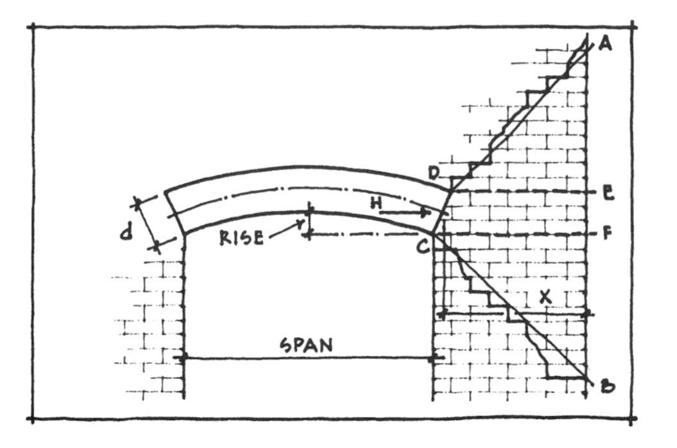

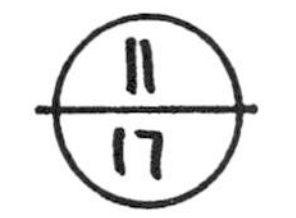

Horizontal thrust resistance. (*From Brick Institute of America,* **Technical** Note 31A, *BIA, Reston, Va.*)

tance. The thrust is thus acting against two planes of resistance, *CF* and *DE*. Resistance is determined by the formula

$$H_1 = v_m nxt \tag{11.11}$$

where H_1 = resisting thrust, lb
v_m = allowable shearing stress in the masonry wall, psi
n = number of resisting shear planes
x = distance from the center of the skewback to the end of the wall, in.
t = wall thickness, in.

By using this principle, the minimum distance from a corner or opening at which an arch may be located is easily determined. To do so, write the formula to solve for *x,* substituting actual arch thrust for resisting thrust:

$$x = \frac{H}{v_m nt} \tag{11.12}$$

11.2.7 Design Tables for Semicircular Arches

The semicircular and segmental are perhaps the most popular and widely used arch forms in contemporary design and construction. The following design tables have been developed by the BIA to simplify the analysis of these arches for normal loading conditions.

Since masonry arches are usually integral with the wall rather than free-standing, basic assumptions can be made which assist in design analysis. For instance, the spring line is assumed to be located on a horizontal line one-fourth of the span length above the horizontal axis. The arches are assumed to be fully restrained at the spring line and that portion of a semicircular arch which is above this line is analyzed in a manner similar to that for parabolic arches. In the determination of the capacity for uniform loads, the limiting factor is the compressive strength of the masonry. In determining the capacity for concentrated loads, the limiting factors are bending at the centerline of span, shear at the spring line, and maximum compressive stress. Tensile stresses are not permitted to develop at midspan. Since axial forces develop in the arch ring from the concentrated and uniform loads, interaction formulas were developed for each loading condition. These formulas combine the axial stresses with the bending stresses.

In all of the formulas given here for use with the BIA design tables, the arch depth and thickness are measured in inches, the span in feet, and the following loading conditions were considered in analyzing a semicircular arch.

For *uniform loads,* the tables in *Figs. 11-18 through 11-21* give the allowable uniform loads occurring over the entire span length for a 1-in.-thick arch ring. *Figure 11-22* shows design requirements and limitations.

The table in *Fig. 11-23* gives allowable *concentrated loads* occurring at the center line of span for a 1-in.-thick arch ring, and *Fig. 11-24* shows design requirements and limitations.

	d = 3.5 in.		d = 7.5 in.		d = 11.5 in.		d = 15.5 in.	
L	W	H	W	H	W	H	W	H
2	810	697	1520	1496	2071	2295	2509	3094
4	424	686	857	1489	1230	2289	1556	3089
6	277	659	591	1474	870	2277	1124	3078
8	193	611	444	1447	669	2257	875	3061
10	134	533	349	1406	538	2227	713	3036
12	86	420	280	1347	445	2185	597	3001
14	—	—	226	1268	374	2128	510	2956
16	—	—	180	1164	317	2055	440	2898
18	—	—	141	1034	269	1964	383	2825
20	—	—	105	873	228	1852	335	2737

Allowable uniform loads on semicircular arches: f_m=300 psi, t=1 in. (*From Brick Institute of America,* Technical Note 31C, *BIA, Reston, Va.*)

	d = 3.5 in.		d = 7.5 in.		d = 11.5 in.		d = 15.5 in.	
L	W	H	W	H	W	H	W	H
2	1082	930	2028	1996	2762	3061	3347	4127
4	569	919	1145	1989	1642	3055	2077	4121
6	376	892	792	1973	1164	3043	1501	4111
8	268	844	599	1947	897	3023	1172	4094
10	195	766	474	1906	724	2993	957	4068
12	137	653	385	1847	602	2951	804	4034
14	87	498	317	1767	510	2894	690	3988
16	—	—	261	1664	437	2821	600	3930
18	—	—	212	1533	377	2730	526	3858
20	—	—	170	1373	325	2618	464	3770

Allowable uniform loads on semicircular arches: f_m=400 psi, t=1 in. (*From Brick Institute of America,* Technical Note 31C, *BIA, Reston, Va.*)

	d = 3.5 in.		*d* = 7.5 in.		*d* = 11.5 in.		*d* = 15.5 in.	
L	*W*	*H*	*W*	*H*	*W*	*H*	*W*	*H*
2	1353	1163	2536	2495	3453	3827	4185	5159
4	714	1152	1434	2488	2055	3821	2597	5154
6	475	1125	993	2473	1458	3809	1879	5143
8	343	1077	753	2446	1125	3789	1468	5126
10	255	1000	600	2405	911	3759	1201	5101
12	188	886	491	2347	760	3717	1012	5066
14	131	731	408	2267	647	3660	870	5021
16	80	527	341	2164	558	3587	759	4962
18	—	—	284	2033	485	3496	669	4890
20	—	—	234	1872	423	3384	594	4802

Allowable uniform loads on semicircular arches: f_m = 500 psi, t = 1 in. (*From Brick Institute of America,* Technical Note 31C, *BIA, Reston, Va.*)

	d = 3.5 in.		*d* = 7.5 in.		*d* = 11.5 in.		*d* = 15.5 in.	
L	*W*	*H*	*W*	*H*	*W*	*H*	*W*	*H*
2	1624	1396	3044	2995	4145	4593	5023	6192
4	859	1385	1722	2988	2467	4587	3118	6186
6	573	1359	1194	2973	1752	4575	2257	6175
8	418	1310	908	2946	1353	4555	1765	6158
10	315	1233	725	2905	1097	4525	1445	6133
12	238	1120	596	2846	918	4483	1219	6099
14	174	964	499	2767	783	4426	1050	6053
16	118	760	421	2663	678	4353	918	5995
18	—	—	356	2532	592	4262	812	5922
20	—	—	299	2372	520	4150	723	5834

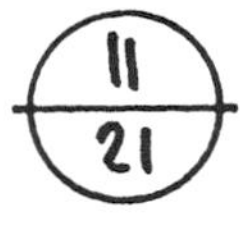

Allowable uniform loads on semicircular arches: f_m = 600 psi, t = 1 in. (*From Brick Institute of America,* Technical Note 31C, *BIA, Reston, Va.*)

When uniform loads are combined with concentrated loads, the concentrated load capacity of the arch ring increases. This additional capacity is due to the compressive stress from the uniform load equalizing the tensile bending stress at midspan from the concentrated load ($M/S = P/A$). The additional capacity may be expressed by the formula

$$P^* = \frac{H_{DL}d}{1.34L} \tag{11.13}$$

where H_{DL} is the horizontal thrust caused by a uniform dead load. The values of P' and H' in *Fig. 11-25* are the allowable capacities governed by compression or shear. They should be used only as a check for combined loadings. In all cases, the actual load must be less than P', and less than the allowable load P plus the additional capacity, P^*. The total horizontal

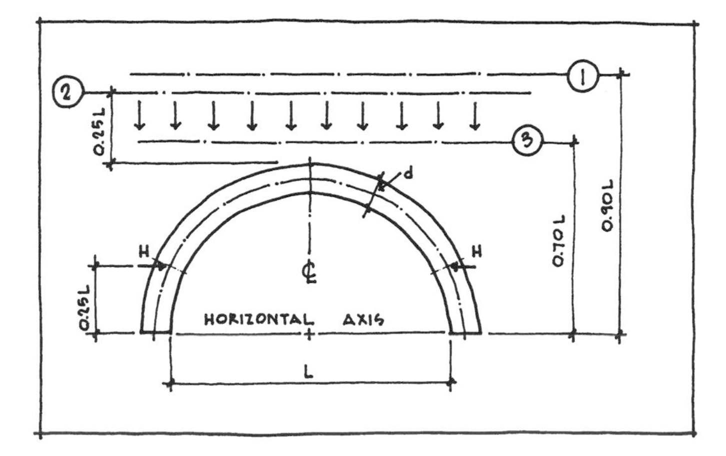

Uniform load on a semicircular arch. (*From Brick Institute of America,* Technical Note 31C, *BIA, Reston, Va.*)

	f_m = 300 to 600 psi							
	d = 3.5 in.		d = 7.5 in.		d = 11.5 in.		d = 15.5 in.	
L	P	H	P	H	P	H	P	H
2	15	12	703	547	1078	841	1451	1131
4	9	9	126	101	1075	841	1449	1131
6	13	16	38	35	418	332	1445	1132
8	20	25	40	41	97	86	967	764
10	28	36	48	52	88	84	199	172
12	38	49	58	67	93	95	162	150
14	48	65	70	83	104	111	160	156
16	61	83	84	102	117	130	168	171
18	75	103	100	124	133	152	181	191
20	90	125	117	148	151	177	198	215

Note: Values may be linearly interpolated except where horizontal lines occur. At these lines, the allowable load is

$$\frac{0.241(L + 0.083d)^3 + 0.134(L + 0.083d)^2 d}{1.34(L + 0.083d) - 0.778d}$$

or the value above the line, whichever is smaller. The horizontal thrust is

$$0.778P + 0.134(L + 0.083d)$$

or the value above the line, whichever is smaller.

Allowable concentrated loads on semicircular arches (t = 1 in.). (*From Brick Institute of America,* Technical Note 31C, *BIA, Reston, Va.*)

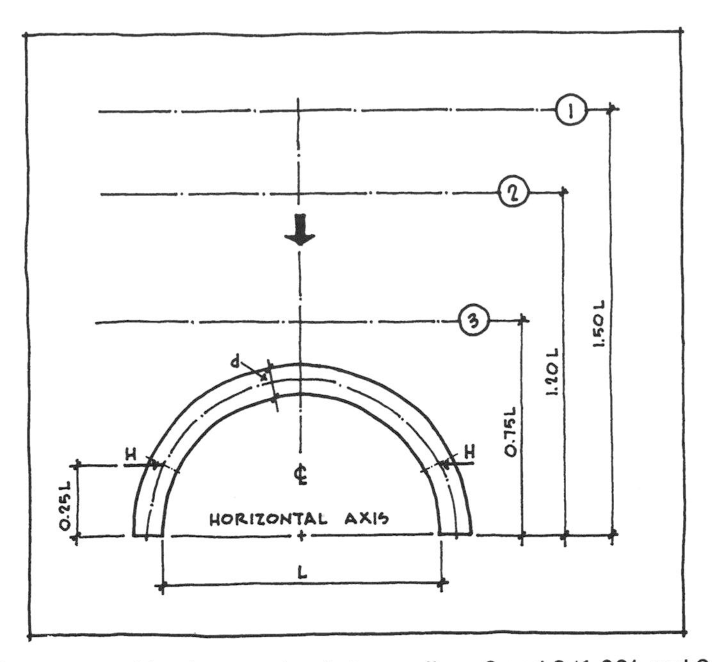

(1) Concentrated loads occurring between lines 2 and 3 (1.20*L* and 0.75*L*) are those provided for in the table.

(2) Concentrated loads occurring between lines 1 and 2 may be divided by the span length (*L*) and considered as equivalent uniform loads.

(3) Concentrated loads occurring above line 1 (1.50*L*) may be ignored (at the discretion of the designer) provided arching action occurs in the masonry above the arch ring.

(4) In all cases, condition (4) for uniform loads (*see Fig. 11-22*) must be used, with the resulting thrusts added to those of the concentrated loads.

(5) There must be a minimum height of masonry (line 3) equal to 0.75*L* above the horizontal axis. No superimposed loads are permitted below this line.

(6) In all cases, the horizontal thrust (*H*) must be checked. For a given arch, the horizontal thrust is directly proportional to the concentrated load.

(7) The proportion of wall which resists the horizontal thrust is assumed to be nonyielding to any lateral movement.

Concentrated load on a semicircular arch. (*From Brick Institute of America,* Technical Note 31C, *BIA, Reston, Va.*)

thrust must be checked and should be less than the maximum allowable for a uniform load.

Any segmental arch where f/L' is greater than 0.29 but less than 0.50 is considered as an equivalent semicircular arch, as shown in *Fig. 11-26.* Twice the radius is the equivalent L for use with the tables.

The method of analysis presented here for the design of semicircular and segmental arches is a simplified but conservative approach to a complex structural problem. An analysis for all possible assumptions and loading conditions is beyond the scope of this book, but most loading conditions encountered will be similar to those in *Figs. 11-18 and 11-24.* To load an arch asymmetrically defeats the benefits of its natural load-carrying structure and induces bending stresses which may cause failure. If openings in

	f_m = 300 psi				f_m = 400 psi				f_m = 500 psi				f_m = 600 psi				f_m = 300 to 600 psi			
	d = 3.5		d = 7.5		d = 3.5		d = 7.5		d = 3.5		d = 7.5		d = 3.5		d = 7.5		d = 11.5 in.		d = 15.5 in.	
L	P'	H'	P'	H'	P'	H'	P'	H'	P'	H'	P'	H'	P'	H'	P'	H'	P'	H'	P'	H'
2	328	256	703	547	328	256	703	547	328	256	703	547	328	256	703	547	1078	841	1451	1131
4	326	256	701	547	326	256	701	547	326	256	701	547	326	256	701	547	1075	841	1449	1131
6	324	256	698	548	324	256	698	548	324	256	698	548	324	256	698	548	1072	842	1445	1132
8	273	203	695	549	321	257	695	549	321	257	695	549	321	257	695	549	1065	842	1442	1133
10	235	170	691	550	310	227	691	550	316	257	691	550	316	257	691	550	1060	842	1440	1134
12	211	143	685	551	275	194	685	551	310	258	685	551	310	258	685	551	1055	843	1436	1135
14	195	125	685	504	252	169	676	552	304	258	676	552	304	258	676	552	1050	843	1431	1136
16	187	110	630	453	237	149	670	553	287	188	670	553	297	258	670	553	1045	843	1422	1137
18	183	98	587	410	228	133	661	554	273	169	661	554	289	259	661	554	1040	842	1411	1137
20	184	88	555	374	225	120	723	505	266	151	652	555	280	259	652	555	1025	842	1400	1138

Maximum concentrated load on semicircular arches under combined loading conditions (t = 1 in.). (*From Brick Institute of America,* Technical Note 31C, *BIA, Reston, Va.*)

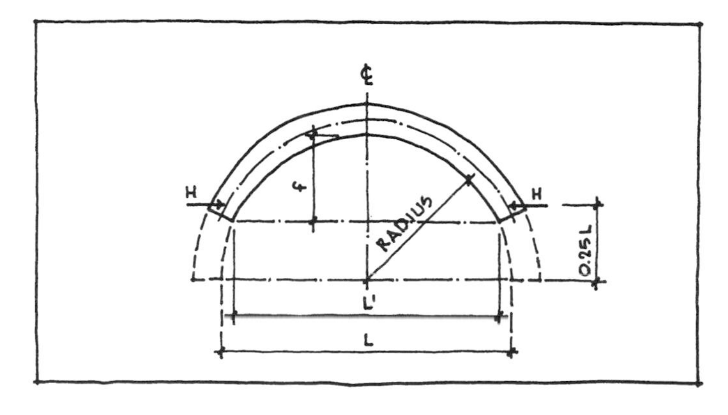

Segmental arch equivalent. (*From Brick Institute of America,* Technical Note 31C, *BIA, Reston, Va.*)

a masonry wall will be loaded asymmetrically, or if design assumptions and limitations do not comply with those given above, consideration should be given to other structural forms such as lintels or beams. If conditions for arched openings are other than those covered here, special engineering analysis should be made.

11.2.8 Major Arch Design

Major arches are those with spans greater than 6 ft or rise-to-span ratios of more than 0.15 (*see Figs. 11-27 and 11-28*). The design of these elements is a much more complicated structural problem than minor arches because of increased loading and span conditions. Leontovich's book, *Frames and Arches,* gives formulas for arches with rise-to-span ratios (f/L) ranging from 0.0 to 0.6. These are straightforward equations by which redundant

First National Bank of Fayetteville, Arkansas. Polk Stanley Gray, architects. (*Photo courtesy BIA.*)

moments and forces in arched members may be determined. The equations are based on a horizontal and vertical grid coordinate system originating at the intersection of the arch axis and the left skewback. Each set of equations depends on the conditions of loading. Moments, shears, and axial thrusts are determined at various increments of the span. No tensile stresses should be permitted in unreinforced masonry arches under static loading conditions. For a detailed analysis of major structural arch design, see Leontovich, *Frames and Arches.* For relatively high-rise, constant-section arches, his Method A of Section 22 applies.

United Bank Tower in Austin, Texas. Zapalac and Associates, architects. (*Photo courtesy BIA.*)

12 STRUCTURAL MASONRY

Extensive structural engineering design is beyond the intended scope of this text. This chapter discusses only the general concepts of masonry bearing wall design. For detailed methods of analysis, design formulas, and sample calculations, the reader should consult Schneider and Dickey's *Reinforced Masonry Design*; Amrhein's *Masonry Design Manual*; Hamid, Drysdale, and Baker's *Masonry Structures Behavior and Design*; or the *Masonry Designer's Guide* based on the Masonry Standards Joint Committee Code and Specifications published by the American Concrete Institute (ACI).

12.1 MASONRY STRUCTURAL SYSTEMS

The general concept of a masonry bearing wall structure is combined action of the floor, roof, and walls in resisting applied loads. The bearing walls can be considered as continuous vertical members supported laterally by the floor and roof systems. Vertical live loads and dead loads are transferred to the walls by the floor and roof systems acting as horizontal flexural members. The floor and roof systems also act as diaphragms to transfer lateral loads to the walls. Vertical and lateral loads applied from only one side of a wall will induce bending moment. The total moment is a result of the combined loading. Since compressive loads counteract some of the tension from this bending moment, the primary stresses that control loadbearing systems are compression and shear. *Figures 12-1 and 12-2* illustrate the typical forces acting on masonry bearing walls.

12.1.1 Axial Load Distribution

Normal axial load distribution in masonry is based on the units being laid in running bond pattern with a minimum overlap between units of one-fourth the unit length. Units laid in stack bond must be reinforced with bond beams or joint reinforcement to achieve the same distribution of axial loads.

When a superimposed axial load is applied to a masonry wall laid in running bond, it is assumed to be distributed uniformly through a triangular section of the wall (*see Figs. 12-3 and 12-4*). Bearing pads or plates are used to distribute concentrated load stresses and to permit any slight differential lateral movement which might occur. When a joist, beam, or gird-

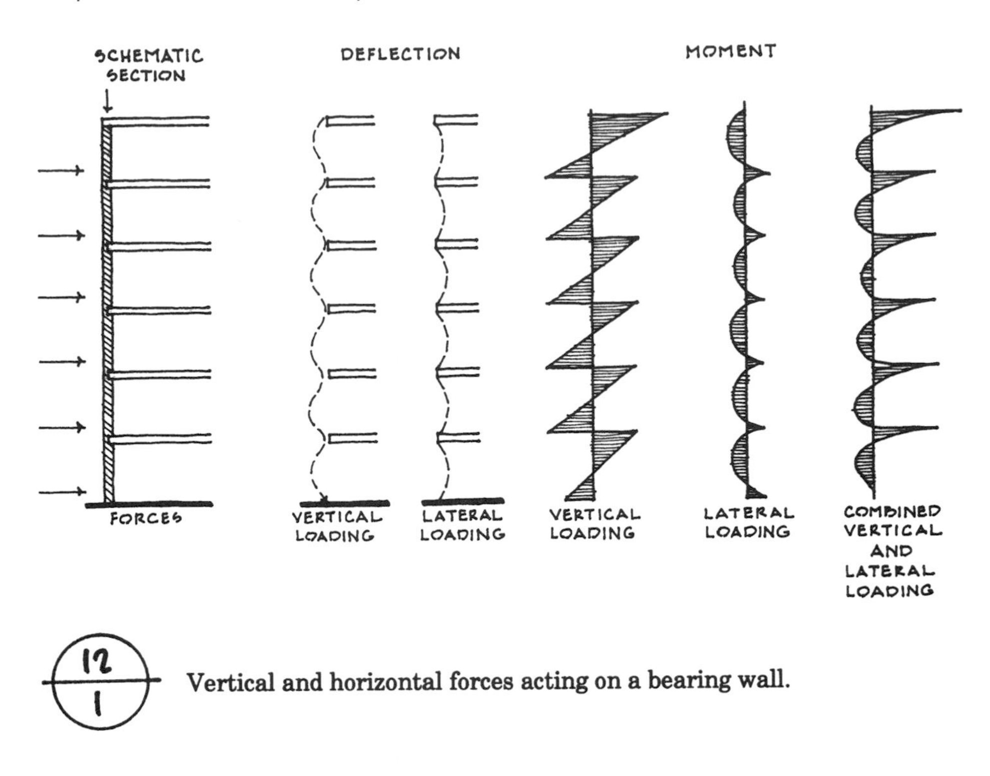

Vertical and horizontal forces acting on a bearing wall.

er bears directly on a masonry wall, the reaction will not generally occur in the center of the bearing area because deflection of the bearing member moves the reaction toward the inner face of the support. If significant eccentricity develops, the addition of reinforcing steel may be required to resist the tensile bending stresses which result.

12.1.2 Lateral Load Distribution

Skeleton frame systems transfer lateral loads from wind and seismic forces through rigid connections at column and beam intersections, and through diagonal bracing. This results in a concentration of stresses at joints in the frame and at the foundation. In a bearing wall system, or box frame, the structural floors and walls constitute a series of intersecting planes with the resulting forces acting along continuous lines rather than at intermittent points. The use of masonry in multistory loadbearing applications is dependent on the cohesive action of the structure as a whole. Floor and roof framing systems must be sufficiently rigid to function as horizontal diaphragms transferring lateral loads to shear walls without excessive in-plane deflection.

Seismic forces are caused by a stress buildup within the earth's crust. An earthquake is the sudden relief of this stress and consequent shifting of the earth mass along an existing fault plane. Primary vibration waves create a push-pull effect on the ground surface. Secondary waves traveling at about half the speed set up transverse movements at right angles to the first shock. Structures may experience severe lateral dynamic loading under such conditions, and must be capable of absorbing this energy and withstanding the critical temporary loading of seismic ground motion.

Framing systems designed to withstand seismic and high wind loads may be either *flexible structures* with low damping characteristics such as

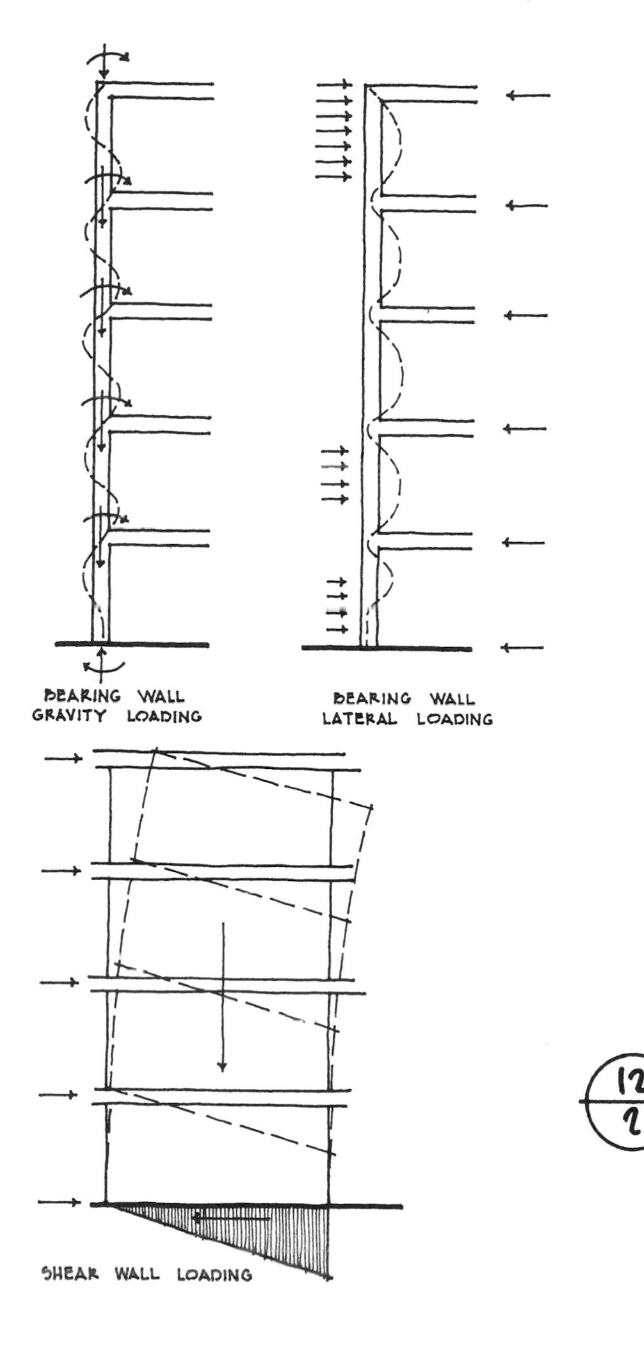

12/2 Loading on masonry bearing walls and shear walls. (*From Brick Institute of America,* **Recommended Practice for Engineered Brick Masonry,** *BIA, Reston, Va., 1969.*)

concrete and steel frame buildings, or *rigid structures* with high damping such as masonry buildings.

Damping is the ability of a structure to diminish its amplitude of vibration with time through dissipation of energy by internal frictional resistance. It is generally recognized that structural response to earthquake motion is influenced by the building's fundamental period of vibration (the time it takes to complete its longest cycle of vibration). Low-rise masonry buildings typically have fundamental periods of vibration of 0.3 seconds or less, compared to 0.6 seconds for low-rise flexible frame buildings. The damping effect of masonry construction and its resulting low period of vibration accounted for good performance of many low-rise masonry buildings in the 1985 Mexico City earthquake which registered 8.1 on the Richter Scale. The soft clay soil under the city caused long-last-

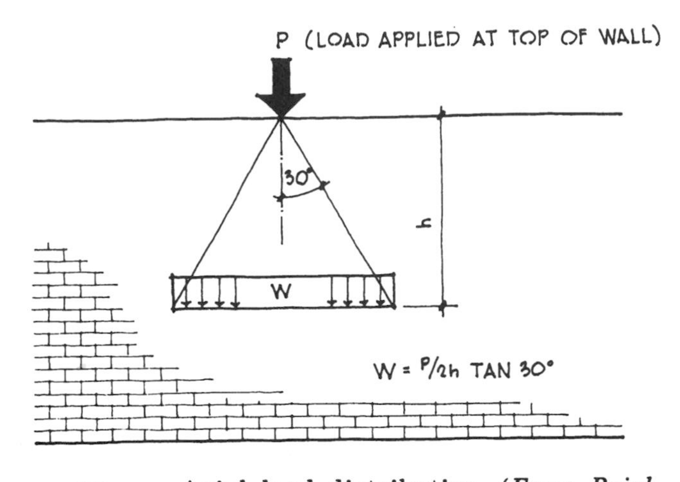

Axial load distribution. (*From Brick Institute of America,* **Recommended Practice for Engineered Brick Masonry,** *BIA, Reston, Va., 1969.*)

ing ground motions with long periods of vibration. Such movements collapsed many buildings of 5 to 20 stories in height that had long periods of vibration in the same range as those of the ground motion. Base motions were greatly amplified in the upper stories of these buildings because the similar periods of vibration set up a condition of resonance. Rigid, low-rise masonry structures, however, suffered little damage, including many unreinforced historic structures.

Different soil and rock conditions produce different periods of ground motion, but the capacity of masonry structures to absorb seismic energy through damping is such that unit stresses remain extremely low, factors of safety very high, and damage negligible.

Steel and concrete skeleton frames are generally classified as moment-resisting space frames in which the joints resist forces primarily by flexure. This flexibility, although effective in dissipating the energy of the seismic loads, can cause substantial secondary, nonstructural damage to windows, partitions, piping, and mechanical equipment. Structural frame buildings "ride" through an earthquake because they can deflect in response to loads, but such deflection breaks glass and damages plaster, drywall partitions, stairs, mechanical piping, and other costly and potentially dangerous elements.

Loadbearing masonry buildings are classified as rigid box frames in which lateral forces are resisted by shear walls. A box frame structural system must provide a continuous and complete path for all of the assumed loads to follow from the roof to the foundation, and vice versa. This is achieved through the interaction of floors and walls securely connected along their planes of intersection. Lateral forces are carried by the floor diaphragms to vertical shear walls parallel to the direction of the load. The shear walls act as vertical cantilevered masonry beams subject to concentrated horizontal forces at floor level, and transfer these lateral forces to the foundation by shear and flexural resistance. The amount of horizontal load carried by a shear wall is proportional to its relative rigidity or stiffness. The rigidity of a shear wall is inversely proportional to its deflection under unit horizontal load, and resistance is a function of wall length. The load transfer induces shear stresses in the wall.

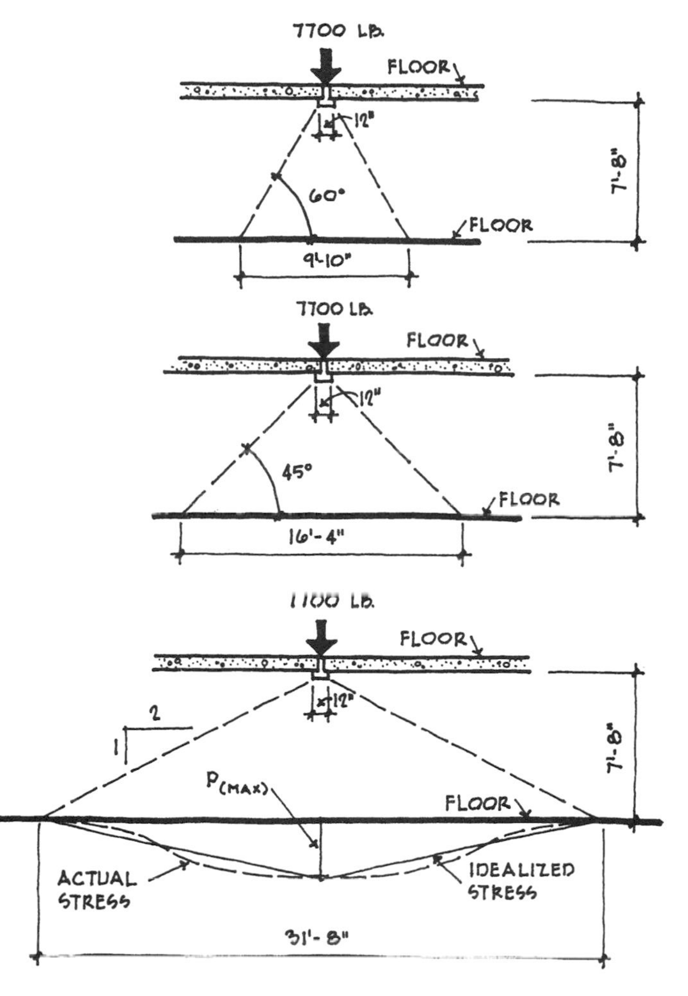

Axial load distribution in concrete masonry. *(From Theodore Lebra, Jr.,* **Design Manual—The Application of Non-reinforced Concrete Masonry Load-Bearing Walls in Multistory Structures,** *National Concrete Masonry Association, Herndon, Va.)*

12.1.3 Diaphragms

A roof or floor diaphragm must have limited deflection in its own plane in order to transmit lateral forces without inducing excessive tensile stress or bending in the walls perpendicular to the direction of the force. The stiffness of the diaphragm also affects the distribution of lateral forces to the shear walls parallel to the direction of the force (*see Fig. 12-5*). No diaphragm is infinitely rigid, and no diaphragm capable of carrying loads is infinitely flexible. For the purposes of analysis, diaphragms are classified as *rigid, semirigid or semiflexible,* or *flexible.* Cast-in-place concrete and flat precast concrete slabs are considered rigid. Steel joists with structural concrete deck are considered semirigid or semiflexible, and steel or wood joists with wood decking are considered flexible. Rigid and semirigid diaphragms do not experience excessive deflection under load. They distribute lateral loads to the shear walls in proportion to their relative rigidity compared to that of the walls.

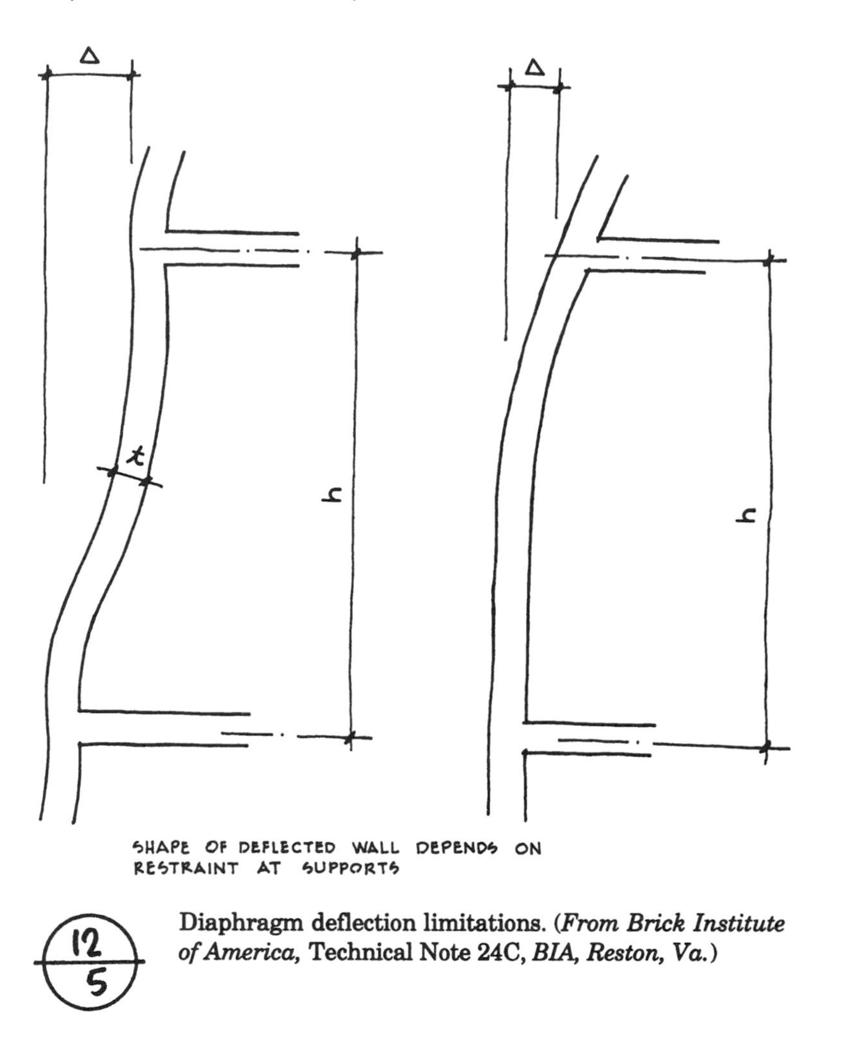

Diaphragm deflection limitations. (*From Brick Institute of America,* Technical Note 24C, *BIA, Reston, Va.*)

Diaphragms may be constructed of concrete, wood, metal, or combinations of materials. Design criteria for materials such as steel and reinforced concrete are well known and easily applied once the loading and reaction conditions are known. Where a diaphragm is made up of distinct units such as plywood panels, precast concrete planks, or steel deck sections, its characteristics are largely dependent on methods of attachment to one another and to supporting members. Such attachments must resist shearing stresses and provide proper anchorage to the supporting shear walls.

12.1.4 Masonry Shear Walls

The lateral load absorbed by a floor or roof diaphragm is transferred to shear walls. Shear walls are designed to resist lateral forces applied parallel to the plane of the wall (*see Fig. 12-6*). Shear walls perform best when they are also loadbearing, because the added loading offers greater resistance to overturning moments. The orientation of the bearing walls in a building can minimize lateral load stresses and take advantage of the natural compressive and shear resistance of the masonry. If all of the bearing walls in a building are oriented in one direction, they will resist lateral loads only in that direction, and additional non-bearing shear walls may be needed in other orientations. Two directional and multidirectional wall configurations

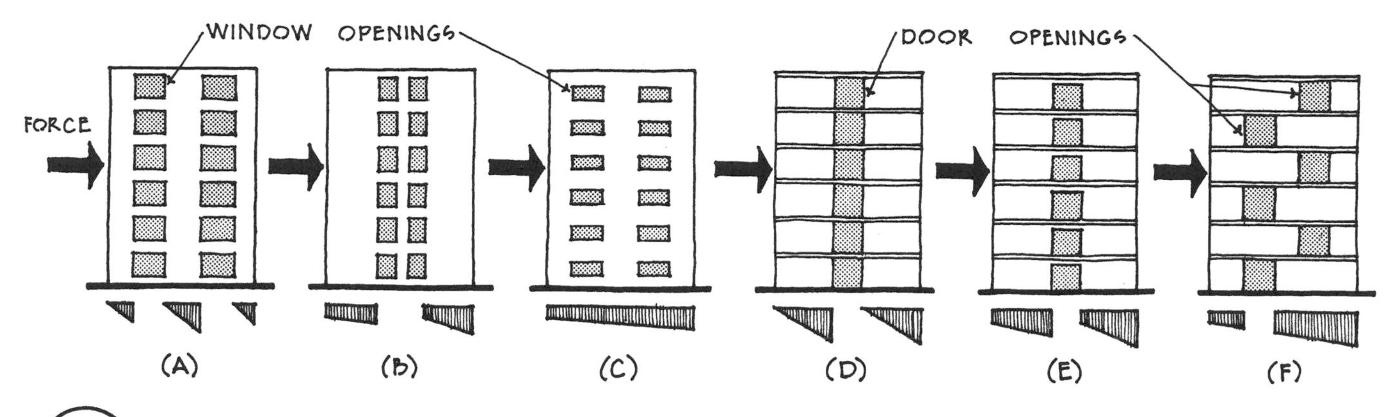

Coupled shear walls. (*From Brick Institute of America,* Technical Note 24C, *BIA, Reston, Va.*)

can resist lateral loads and shear from more than one direction. Both loadbearing and non-loadbearing shear walls can be stiffened by adding flanges that have a positive connection to the intersecting shear walls.

If analysis indicates that tension will be developed in unreinforced masonry shear walls, the size, shape, or number of walls must be revised, or the walls must be designed as reinforced masonry.

The lateral load resistance of masonry structures is basically a function of the orientation, area, and strength of the shear walls. If the cumulative area of shear wall is sufficient, the building can elastically resist even strong earthquakes without reaching the yield point of the steel reinforcement.

Some of the most important aspects of shear wall and seismic design are qualitative elements regarding symmetry and location of resisting members, relative deflections, anchorage, and discontinuities. Shear walls resist horizontal forces acting parallel to the plane of the wall through resistance to overturning and shearing resistance. The location of shear walls in relation to the direction of the applied force is critical. Since ground motion may occur in perpendicular directions, the location of resisting elements must coincide with these forces. Shear walls may be either loadbearing or non-loadbearing elements. It is best to combine the functions of such members and, whenever possible, design the building with both transverse and longitudinal loadbearing shear walls. Designing all the loadbearing walls to resist lateral forces improves overall performance because the increased number of shear walls distributes the load and lowers unit stresses. Shear walls that are also loadbearing have greater resistance to seismic forces because of the stability provided by increased axial loads.

If loadbearing walls also function as shear walls, then general building layout becomes a very important aspect of seismic design. The building shear wall layout should be symmetrical to eliminate torsional action. Several compartmented floor plans are shown in *Figs. 12-7 and 12-8.* The regular bay spacing lends itself to apartment, hotel, hospital, condominium, and nursing home occupancies where large building areas are subdivided into smaller areas. By changing the span direction of the floor, shear walls in two or more directions can become loadbearing. The radial walls of the round building can actually absorb seismic shocks from any direction and dissipate the earthquake energy with very low levels of stress. In

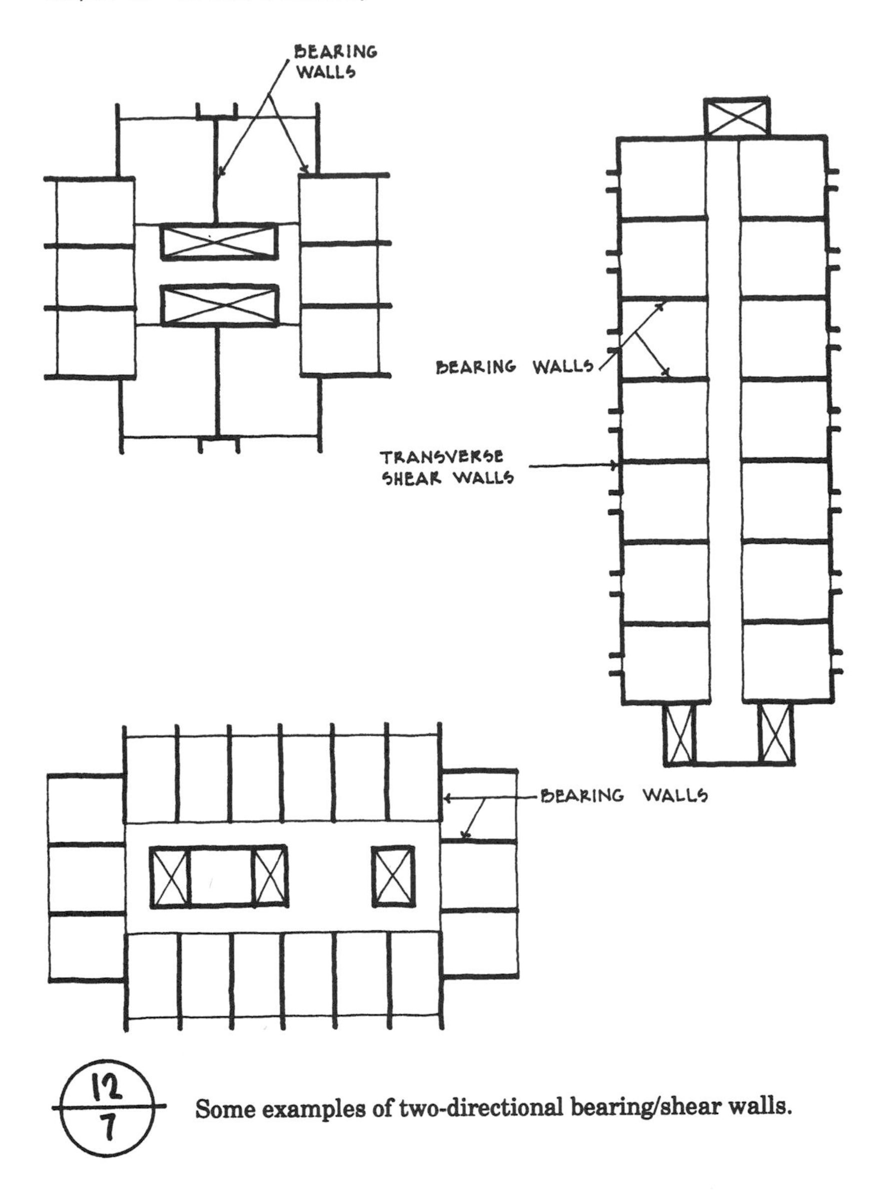

12/7 Some examples of two-directional bearing/shear walls.

skeleton frame buildings, elevator and stair cores of concrete or masonry are often used as shear walls even though they may not have axial load-bearing capacity.

Providing a good balance between the amount of shear wall along each of the principal building axes will provide greatest economy and best performance. It is also best to design wall lengths that are uniform between openings, and to provide wall returns or flanges wherever possible so that variations in relative rigidity are minimized and maximum shear is decreased.

One method of increasing the stiffness of shear walls as well as their resistance to bending is the use of intersecting walls or flanges (*see Fig. 12-9*). Although the effective length of flanges is limited, walls with *L, Z, I,* or *C* shapes have better flexural resistance for loads applied perpendicular to their flange surface. Shear stresses at the intersection of the walls are dependent on the type of bonding used to tie the two elements together.

Another method that may be used to increase the stiffness of a bearing wall building is the coupling of co-planar shear walls. The illustrations in

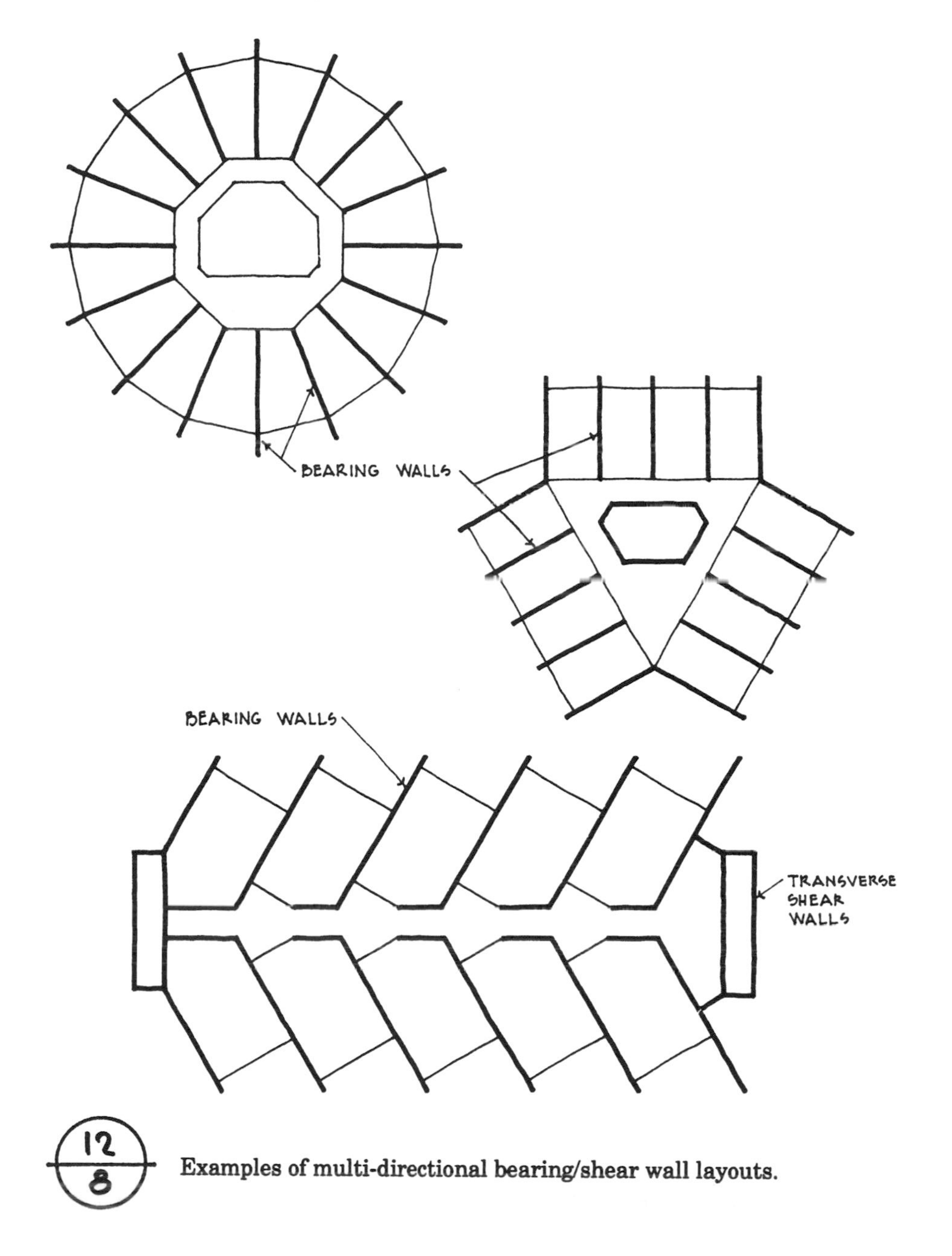

12/8 **Examples of multi-directional bearing/shear wall layouts.**

Fig. 12-6 indicate the effect of coupling on stress distribution from forces parallel to the wall. In parts *A* and *D,* a flexible connection between the walls is assumed so that they act as independent vertical cantilevers in resisting the lateral loads. Walls *B* and *E* assume the elements to be connected with a more rigid member capable of shear and moment transfer so that a frame-type action results. This connection may be made with a steel, reinforced concrete, or reinforced masonry section. The plate action in parts *C* and *F* assumes an extremely rigid connection between walls such as full-story-height sections or deep rigid spandrels.

The type of floor-to-wall connection also influences the transfer of stress, any many seismic failures in masonry buildings occur because of inadequate connections. When the various elements of a structure are rigidly connected to one another and one element is deflected by a force,

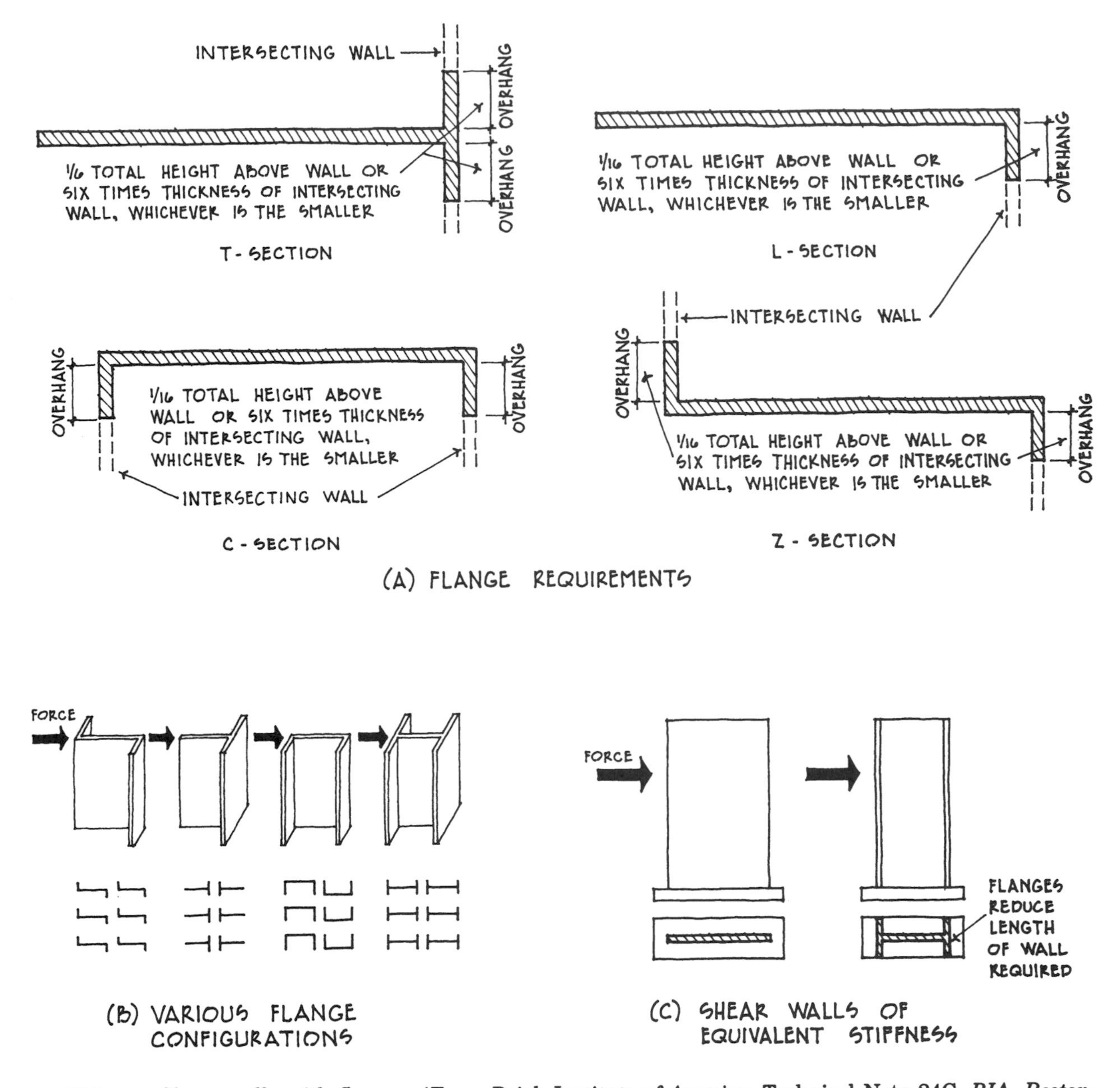

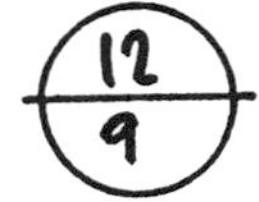

Shear walls with flanges. (*From Brick Institute of America,* **Technical Note 24C,** *BIA, Reston, Va.; and Theodore Leba, Jr.,* **Design Manual—The Application of Non-reinforced Concrete Masonry Load-Bearing Walls in Multistory Structures,** *National Concrete Masonry Association, Herndon, Va.*)

the other elements must move equally at the point of connection or failure will occur somewhere in the system.

Analysis of the damage to masonry walls caused by earthquake stress shows that cracks in shear walls typically follow a diagonal path. The plane of failure extends from near the top corner where the maximum load is applied, diagonally downward toward the bottom support (*see Fig. 12-10*). This is the same mode of failure produced by diagonal tension or racking tests, in which 4×4-ft masonry panels are subjected to diagonal loading at opposing corners. Shear strength at the joints is independent of unit properties such as initial rate of absorption, but it is affected by mortar type, com-

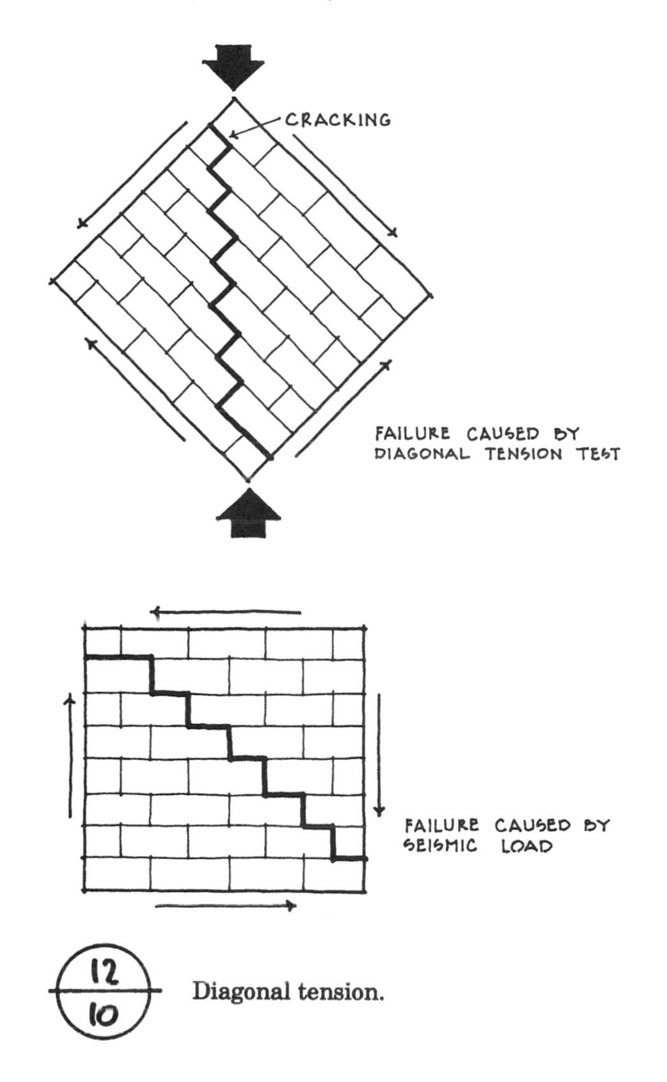

Diagonal tension.

pressive strength, and workmanship. Since the failure is in tension, masonry with weak mortar bond also has low diagonal tensile or shear strength. Failure, however, occurs without explosive popping or spalling of unit faces. The steel reinforcement holds the wall together after a shear failure to prevent the panel from separating after joint cracking occurs. This stability under maximum stress prevents catastrophic structural failure and increases the factor of safety in the aftermath of seismic disturbances.

Where masonry is used as non-loadbearing infill panels in steel or concrete frame structures, this diagonal shear strength increases the building's lateral stiffness (*see Fig. 12-11*). In the 1985 Mexico City earthquake, medium- and high-rise reinforced concrete frame structures with masonry infill performed much better than similar buildings without masonry infill. Even when the infill panels served only as a backup for veneer systems and were not designed as structural elements, were not reinforced, and were not correctly anchored to the surrounding frames, they absorbed large amounts of seismic energy and, in many cases, apparently prevented structural collapse. When infilling was omitted in lower stories to provide access to retail businesses, the buildings proved more susceptible to damage at the lower levels. Corner buildings in which the masonry infill was omitted on two sides for retail access suffered severe damage.

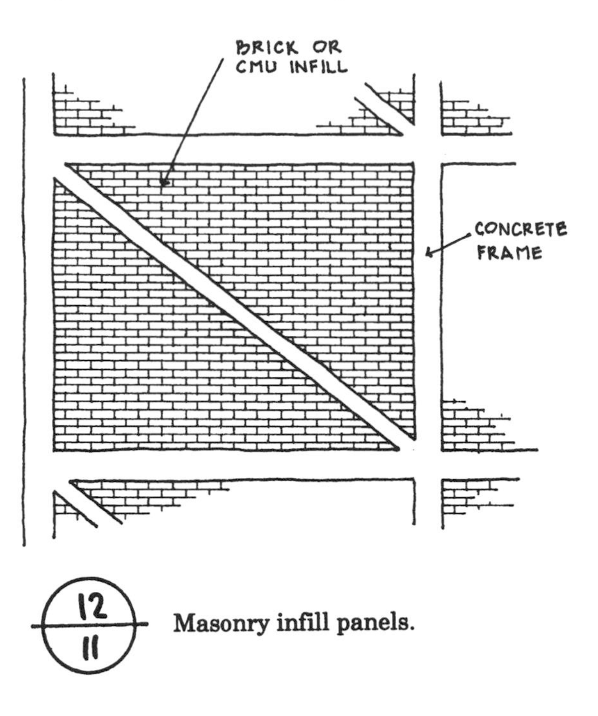

12/11 **Masonry infill panels.**

12.1.5 Beams and Girders

The use of reinforcing steel in masonry construction permits the design of flexural masonry members such as lintels, beams, and girders to span horizontal openings (*see Fig. 12-12*). This gives a continuity of materials, finishes, and fire ratings by eliminating the introduction of other materials solely for flexural spans.

The design of reinforced masonry beams and girders is based on the straight line theory of stress distribution. The required steel is determined by actual calculated stress on the member. The reinforcement needed to resist this stress is then provided in the necessary amounts and locations. The member must be designed to resist at all sections the maximum bending moment and shears produced by dead load, live load, and other forces determined by the principles of continuity and relative rigidity. Building codes place deflection limits on all flexural members which support unreinforced masonry.

The concept of deep masonry wall beams is based on a wall spanning between columns or footings instead of having continuous line support at the bottom as in conventional loadbearing construction (*see Fig. 12-13*). If soil bearing capacities permit this type of concentrated load, the wall may be designed as a flexural member and must resist forces in bending rather than in direct compression.

Deep wall beams may also be used to open up the ground floor of a loadbearing structure. The bearing wall on the floor above can be supported on columns to act as a deep wall beam and transfer its load to the supports. This alternative permits the design of larger rooms and open spaces that might not be possible with regularly spaced bearing walls. Bearing walls, non-bearing walls, and shear walls may all use this principle to advantage in some circumstances.

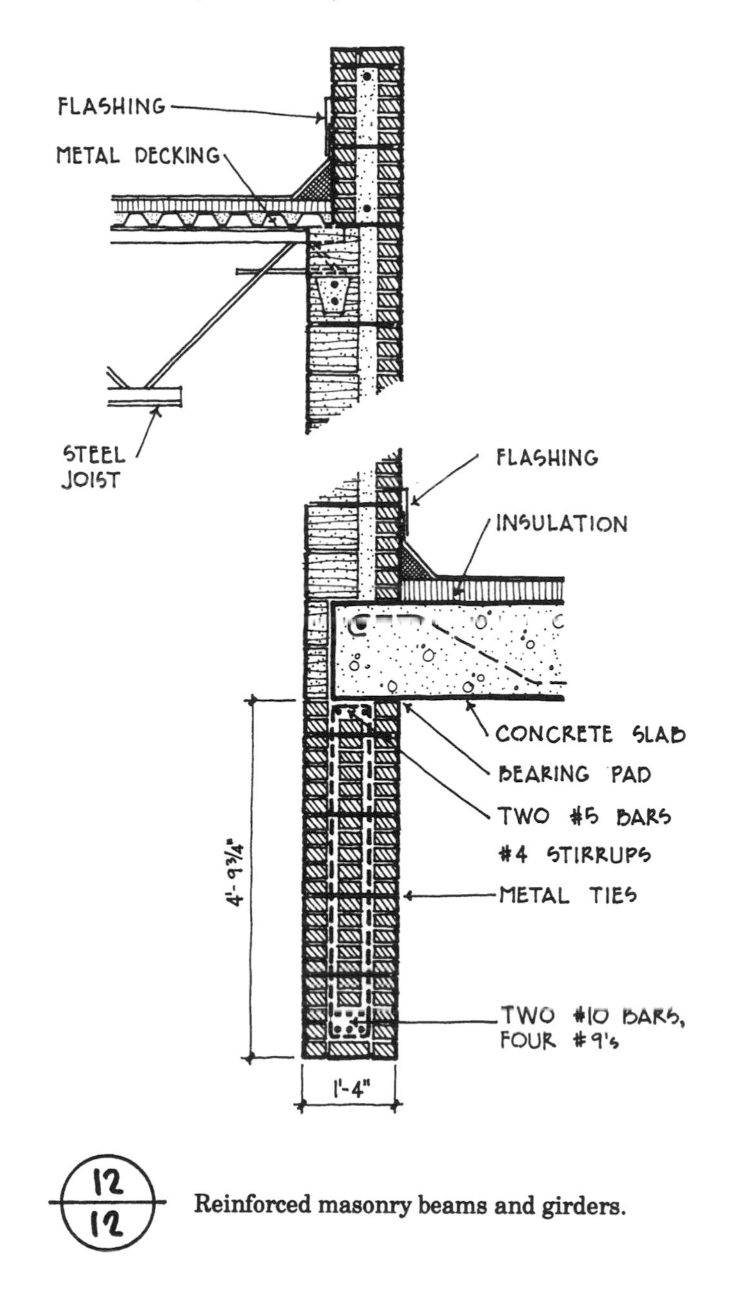

Reinforced masonry beams and girders.

12.1.6 Connections

The box frame system of lateral load transfer requires proper connection of shear walls and diaphragms. Connections may be required to transmit axial loads, shear stresses, and bending moments acting separately or in combination with one another. Connections can be made with anchor bolts, reinforcing dowels, mechanical devices, or welding, and may be either fixed or hinged. Although neither complete restraint nor a completely hinged condition actually exists, these assumptions may be made for purposes of calculation. Each individual condition will dictate the type of connection needed, and a variety of solutions can usually be designed for a given problem. The design and detailing of structural connections are covered in depth in the engineer-

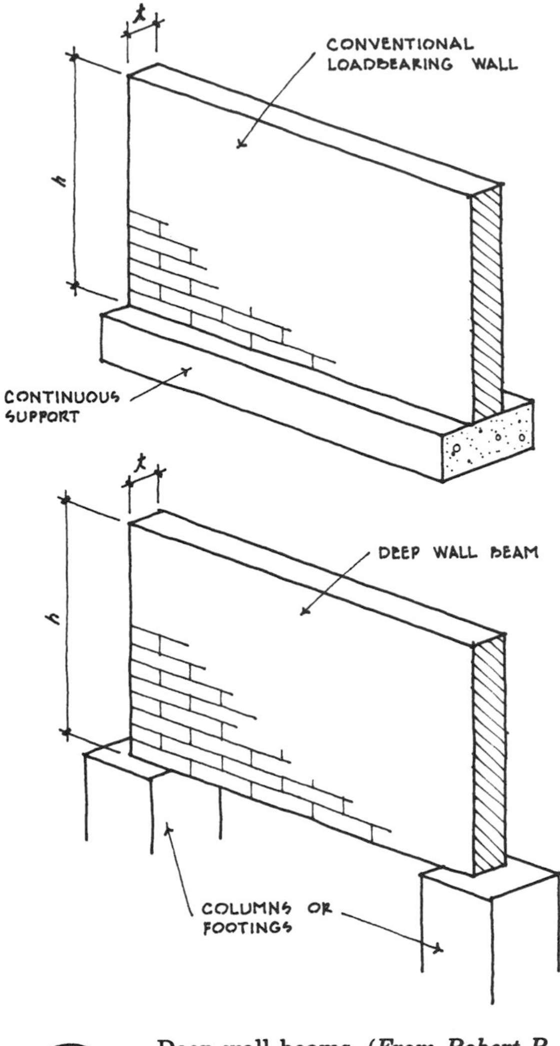

Deep wall beams. (*From Robert R. Schneider and Walter L. Dickey,* **Reinforced Masonry Design,** *2d edition, © 1987, p. 305. Reprinted by permission of Prentice-Hall, Inc., Englewood Cliffs, N.J.*)

ing texts listed at the beginning of this chapter. *Figures 12-14 through 12-18* show some examples of floor and roof system connections.

12.1.7 Foundations

Although the weight of a loadbearing structure is greater than that of a similar frame building, the required soil bearing capacity is often less because the bearing walls distribute the weight more evenly. Bearing wall structures are compatible with all of the common types of foundations, including grade beams, spread footings, piles, caissons, and mats. Foundation walls below grade may be of either concrete or masonry, but must be doweled to the footing to assure combined action of the wall and the foundation.

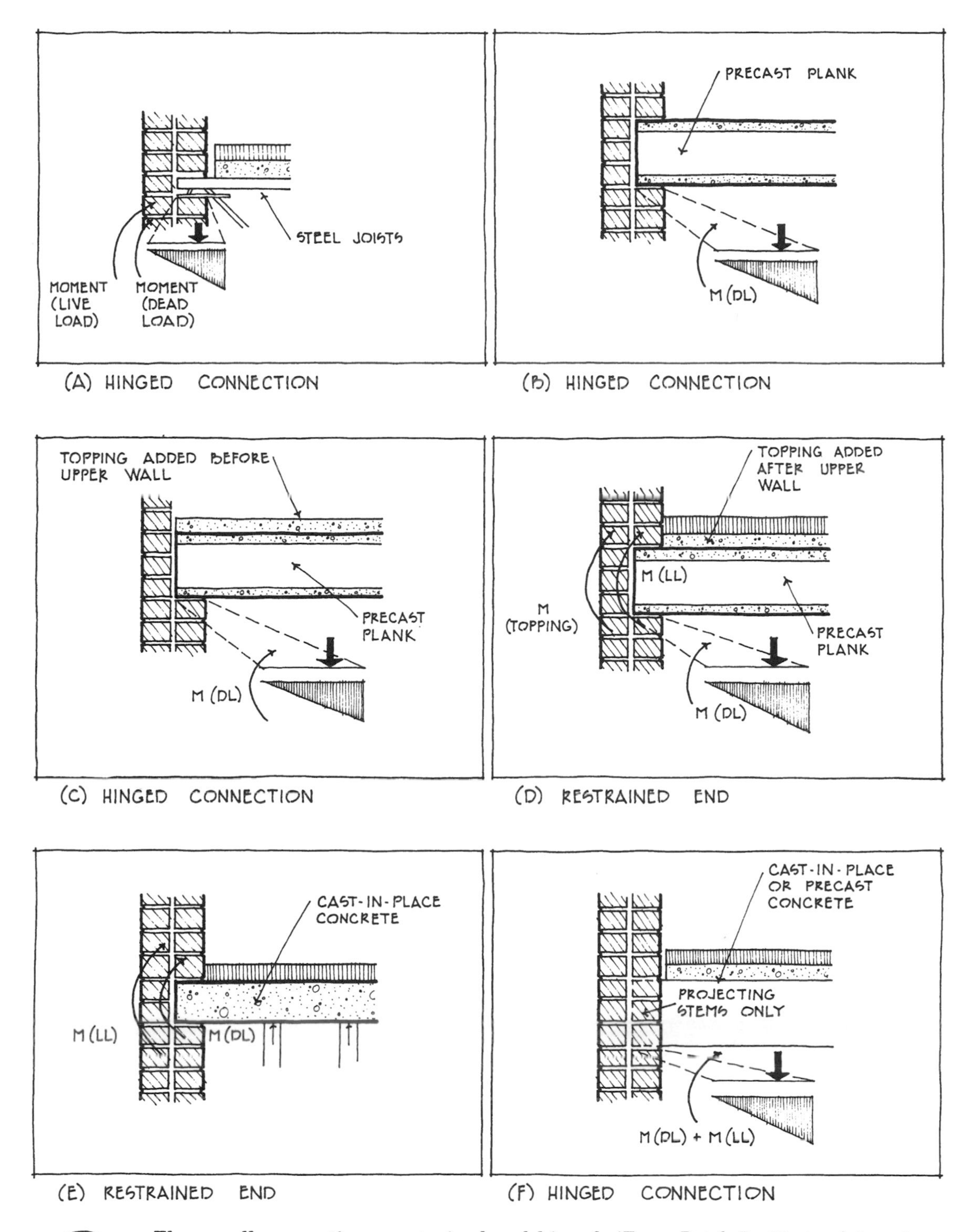

Floor–wall connections, restrained and hinged. (*From Brick Institute of America,* Technical Note 24B, *BIA, Reston, Va.*)

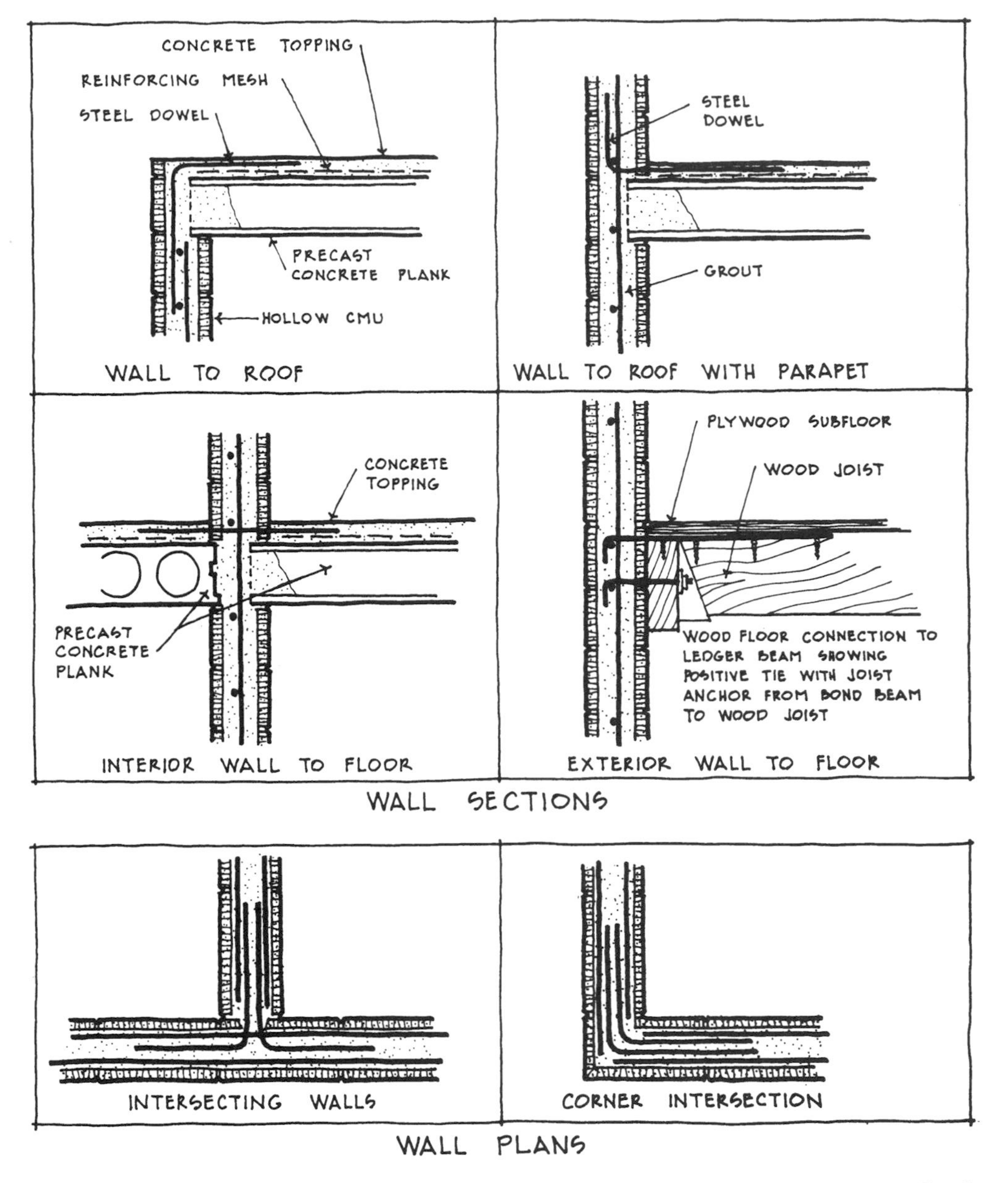

Connection details. (*From Masonry Institute of America,* **Reinforcing Steel in Masonry,** *MIA, Los Angeles, Calif.*)

12.1.8 Reinforced and Unreinforced Masonry

Plain masonry contains no reinforcing steel that is designed to resist applied loads. It is very strong in compression, but weak in tension and shear. Small lateral loads and overturning moments are resisted by the mass of the wall, but if lateral loads are higher, resistance to shear and flexural stresses is limited by the bond between mortar and units and the precompression effects of vertical loading. Shearing stresses in bearing wall buildings, however, seldom control the wall type and thickness. Although flexural stresses may sometimes control the design of non-bearing and shear walls, compressive stresses generally govern in loadbearing structures.

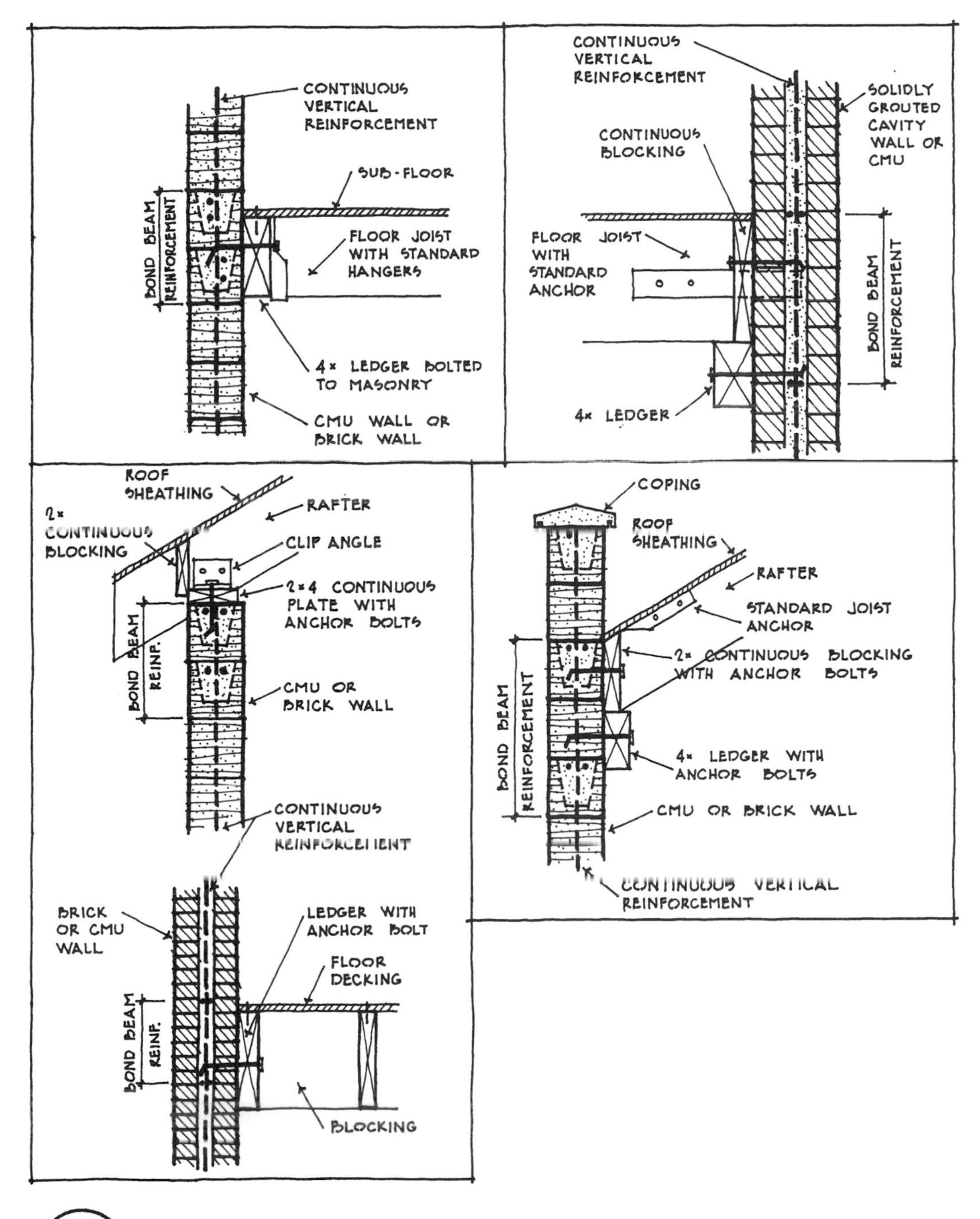

12/16 **Connecting masonry to a wood frame.**

Where lateral loads are a significant factor in the design of structural masonry, flexural strength can be increased by placing steel reinforcement in mortar bed joints, bond beams, grouted cells, or cavities. The hardened mortar and grout bind the masonry units and steel together so that they act as a single element in resisting applied forces. Reinforcement may be added to resist isolated stresses wherever design analysis indicates that excessive flexural tension is developed. The reinforcing steel is then designed

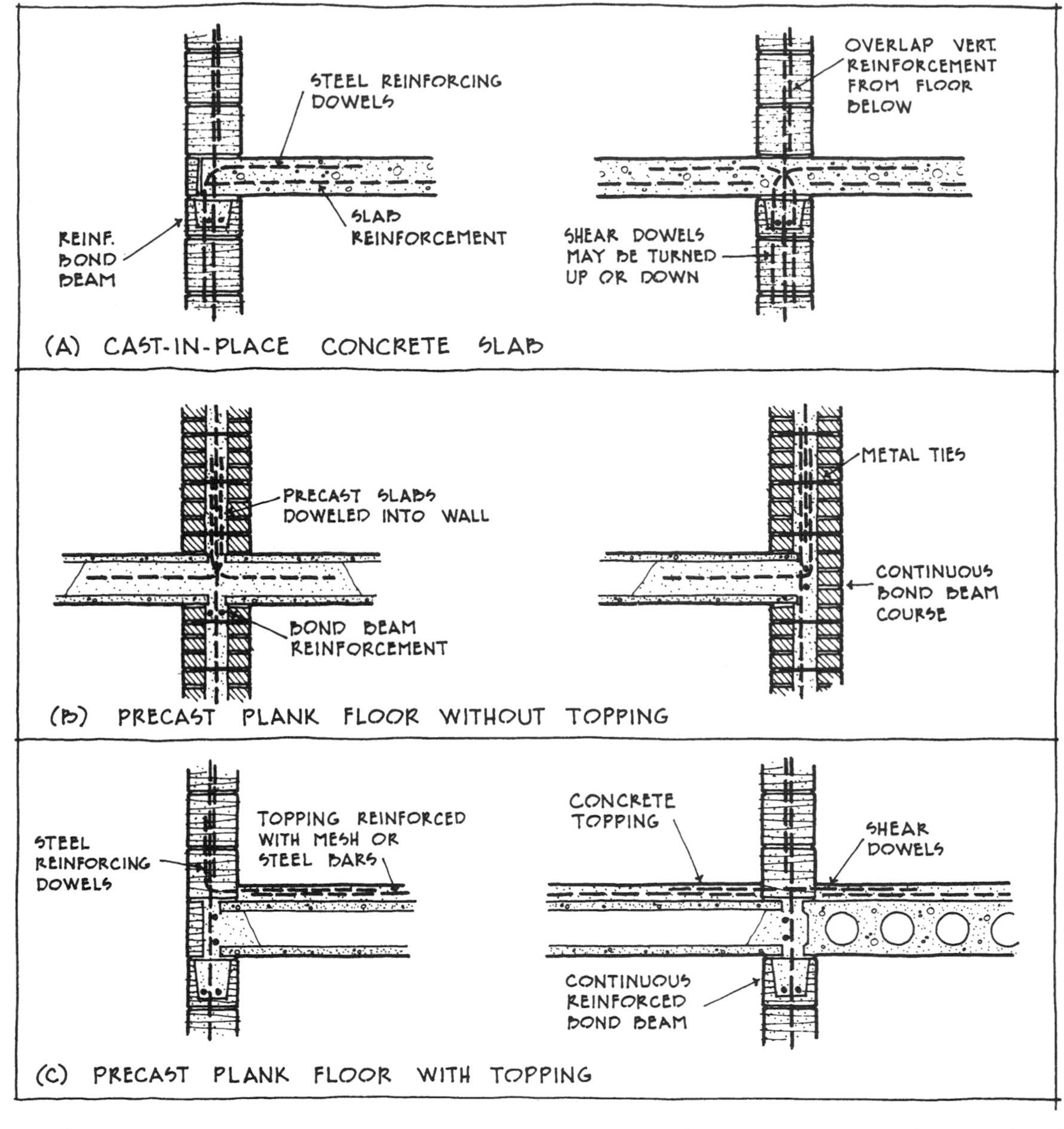

Wall-to-floor connections. *(From Robert R. Schneider and Walter L. Dickey,* **Reinforced Masonry Design,** *2d edition, © 1987, pp. 441, 443. Reprinted by permission of Prentice-Hall, Inc., Englewood Cliffs, N.J.)*

to resist all of the tensile stresses and the flexural strength of the masonry is neglected entirely.

12.1.9 Empirical and Analytical Design

The analytical design of loadbearing masonry buildings by the allowable stress method is based on a general analysis of the structure to determine the magnitude, line of action, and direction of all forces acting on the various members. All dead loads, live loads, lateral loads, and other forces, such as those resulting from temperature changes, impact, and unequal

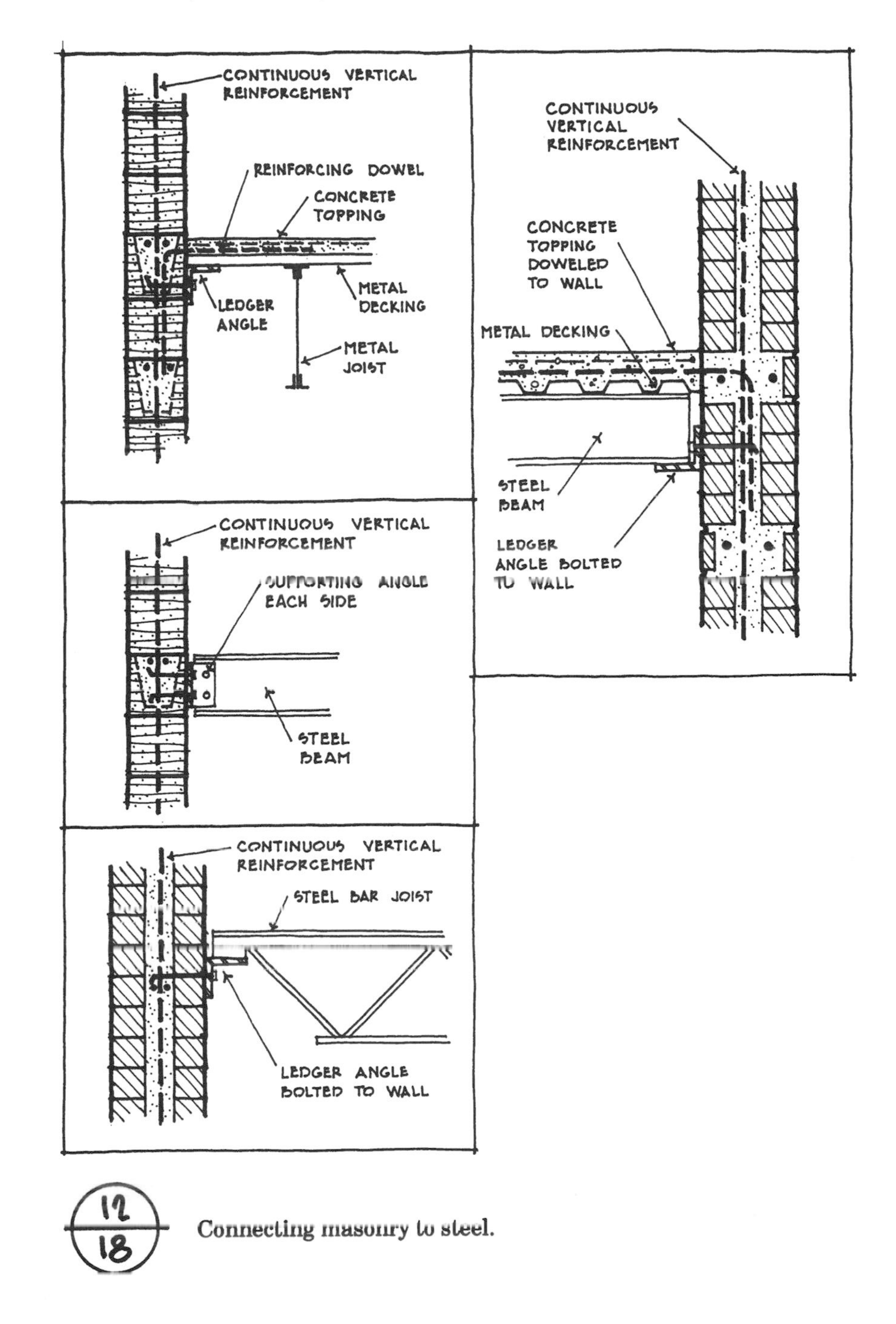

12/18 Connecting masonry to steel.

settlement, are considered. The combinations of loads that produce the greatest stresses are used to size the members. The performance of load-bearing masonry walls, pilasters, and columns can be predicted with reasonable accuracy using allowable stress methods. The complexity of the analysis will depend on the complexity of the building with respect to height, shape, wall location, and openings in the wall.

While empirical design is restricted by arbitrary limits on the ratio of wall height to thickness, analytical design determines the actual thickness required to resist service loads on walls of any desired height. Empirical design codes arbitrarily dictate the spacing of lateral supports, but analytical design calculates loads and provides lateral support as needed to resist specific forces and provide stability. Analytical design establishes allowable compressive stresses based on the characteristics of the units and mortar selected, the slenderness of the wall and the eccentricity of applied loads. Analytical design also takes into account shear and flexural stresses, which are not considered in empirical design.

Empirical design methods may be used for low-rise buildings, but analytical design will produce more efficient and economical results for both unreinforced and reinforced masonry. Empirical requirements are very simplistic in their application. Height or length-to-thickness ratios are used in conjunction with minimum wall thicknesses to determine the required section of a given wall. Analytical design by the allowable stress method is based on the properties of the component materials in resisting calculated stresses rather than on arbitrary empirical limits, and may be used for high-rise or low-rise construction.

12.1.10 Code Requirements

Design requirements for both the empirical and analytical methods are governed by two codes. The *Uniform Building Code* [International Conference of Building Officials (ICBO)] includes detailed requirements for structural masonry, including material standards, quality control procedures, general design requirements, and seismic requirements. The *Standard Building Code* [Southern Building Code Congress (SBCC)] and *National Building Code* [Building Officials and Code Administrators (BOCA)] both require use of the Masonry Standards Joint Committee (MSJC) *Building Code Requirements for Masonry Structures* (*ACI 530/ASCE 5/TMS 402*) and *Specifications for Masonry Structures* (*ACI 530.1/ASCE 6/TMS 602*), which are jointly written by the American Concrete Institute (ACI), the American Society of Civil Engineers (ASCE), and The Masonry Society (TMS). Sometimes referred to simply as *ACI 530,* the MSJC Code replaced six standards previously referenced by SBCC and BOCA which were published separately by the American National Standards Institute (ANSI), Brick Institute of America (BIA), National Concrete Masonry Association (NCMA), and ACI.

The *Uniform Building Code* (UBC) and MSJC Code impose restrictions and quality standards on materials, design, and construction. They establish the type of units; control reinforcement, connectors, mortar and grout types; require inspection; and impose limits on construction tolerances.

12.2 EMPIRICAL DESIGN

Masonry buildings built before the twentieth century, including all of the historic masonry buildings throughout the world, are unreinforced, empirically designed structures. These traditional loadbearing designs used massive walls and buttresses to resist lateral loads, including those induced by roof thrusts, arches, and large domes. Empirical masonry today is limited in its application because it is now economical only for small structures. However, in buildings of limited height where wind loads are low and seismic loading is not a consideration, empirically designed masonry is still sometimes used.

Empirical design is based on historical precedent and rules of thumb rather than detailed analysis of loads and stresses, and calculated structural response. Empirically designed buildings do not incorporate reinforcing steel for load resistance, but may include joint reinforcement for control of shrinkage cracking and thermal movement. Elements that do not contribute to the primary lateral-force-resisting system in masonry structures, and masonry elements in steel or concrete frame buildings may be designed empirically.

The basic restrictions on empirically designed masonry differ only slightly between the UBC and MSJC code:

- *Uniform Building Code.* Empirically designed masonry prohibited in Seismic Zones 2, 3, and 4 and in areas where design wind loads exceed 80 mph.
- *MSJC Code.* Empirically designed buildings prohibited in Seismic Performance Category D and E and in areas where design wind loads exceed 25 psf; in Seismic Performance Categories B and C, empirically designed masonry may not be part of the lateral force-resisting system.

Both codes limit the height of empirically designed buildings which rely on masonry walls for lateral load resistance to 35 ft above the foundation or supporting element.

12.2.1 Allowable Compressive Stresses

Both codes list allowable compressive stresses which vary with unit and mortar type. Service loads must be limited so that the maximum average compressive stress in the wall does not exceed the allowable values shown in *Fig. 12-19.* To determine compressive stresses in the masonry, the combined effects of vertical dead loads plus live loads (exclusive of wind and seismic forces) must be considered.

12.2.2 Lateral Support Requirements

In lieu of analytical design, prescriptive requirements are given for the ratio of the unsupported height or length to the nominal thickness of masonry bearing walls and non-bearing partitions. Lateral support must be provided in *either* the horizontal or the vertical direction within the limits shown in *Fig. 12-20.* Cross walls, pilasters, buttresses, columns, beams, floors, and roofs may all be used to provide the required support. Typical configurations for unreinforced masonry pilasters are shown in *Fig. 12-21.* Typical reinforced pilasters are shown in *Figs. 10-3 and 15-9.* In computing the ratio for metal-tied cavity walls, the value for thickness is taken as the sum of the nominal thicknesses of the inner and outer wythes only, excluding the width of the cavity itself. In composite walls which combine different types of units or construction, lateral support should be provided as required for the weakest of the combinations.

Lateral support gives the wall sufficient strength to resist wind loads and other horizontal forces acting either inward or outward. Members providing lateral support must be adequately bonded or anchored to the masonry, and must be capable of transferring forces to adjacent structural members or directly to the ground. Pilasters may be either bonded into the wall or mechanically connected to it across a continuous movement joint (*see Fig. 12-21*).

Where lateral support is provided by the walls of the masonry structure itself, they must provide shear resistance in the direction parallel to the assumed direction(s) of the lateral load. Shear walls may be loadbearing or non-loadbearing.

Construction: compressive strength of unit, gross area	Allowable compressive stresses* gross cross-sectional area	
	Type M or S mortar	Type N mortar
Solid masonry of brick and other solid units of clay or shale; sand-lime or concrete brick:		
8000 plus psi	350	300
4500 psi	225	200
2500 psi	160	140
1500 psi	115	100
Grouted masonry, of clay or shale; sand-lime or concrete:	(MSJC)(UBC)	
4500 plus psi	225 275	200
2500 psi	160 215	140
1500 psi	115 175	100
Solid masonry of solid concrete masonry units:		
3000 plus psi	225	200
2000 psi	160	140
1200 psi	115	100
Masonry of hollow loadbearing units:		
2000 plus psi	140	120
1500 psi	115	100
1000 psi	75	70
700 psi	60	55
Hollow walls (cavity or masonry bonded)† solid units:		
2500 plus psi	160	140
1500 psi	115	100
Hollow units	75	70
Stone ashlar masonry:		
Granite	720	640
Limestone or marble	450	400
Sandstone or cast stone	360	320
Rubble stone masonry		
Coarse, rough, or random	120	100
Unburned clay masonry	30	—

*Linear interpolation may be used for determining allowable stresses for masonry units having compressive strengths which are intermediate between those given in the table.

†Where floor and roof loads are carried upon one wythe, the gross cross-sectional area is that of the wythe under load. If both wythes are loaded, the gross cross-sectional area is that of the wall minus the area of the cavity between the wythes.

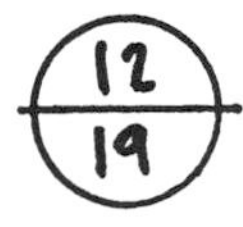

UBC and MSJC Code allowable compressive stresses for empirically designed masonry.

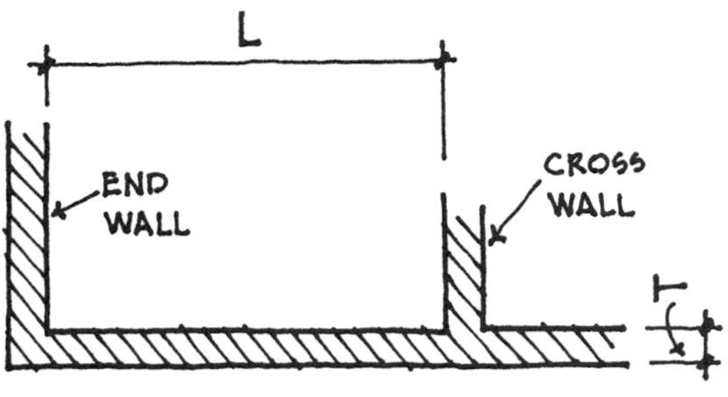

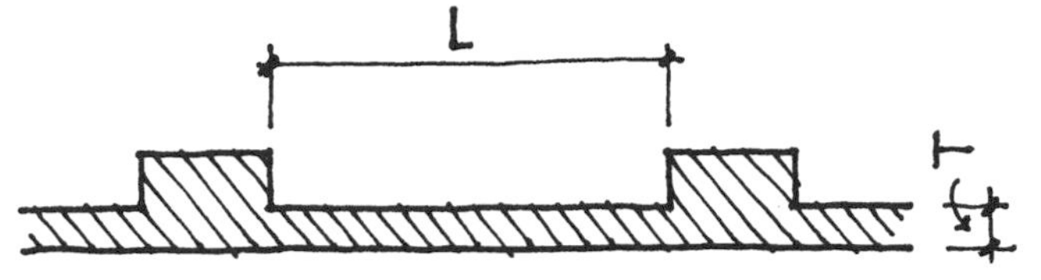

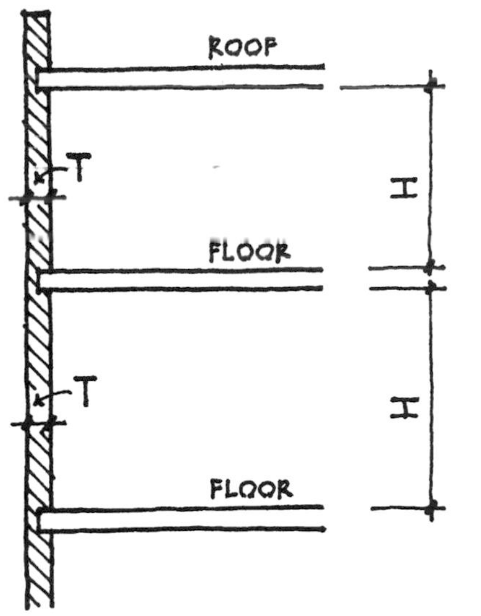

MAXIMUM RATIO OF UNSUPPORTED HEIGHT OR LENGTH TO NOMINAL THICKNESS	
Construction	*Maximum l/t or h/t*
Bearing walls	
Solid or solidly grouted	20
All others	18
Non-bearing walls	
Exterior	18
Interior	36

Lateral support, height-to-thickness ratios, and minimum thickness requirements for empirically designed masonry walls.

Shear walls must have a minimum nominal thickness of 8 in., and their minimum cumulative length in each required direction, exclusive of openings, must be 40% of the long dimension of the building (*see Fig. 12-22*). BIA also recommends that only walls whose length is greater than or equal to one-half the story height be used in calculating cumulative length. The required spacing of shear walls is based on the type of floor and roof provided, because diaphragm rigidity varies with each system. Stiffer elements permit wider spacing (*see Fig. 12-23*).

12.2.3 Wall Thickness

Bearing walls of one-story buildings must have a nominal thickness of 6 in. UBC requires that these walls be constructed of solid masonry units or fully grouted hollow units. Bearing walls of buildings more than one story in height must have a nominal thickness of 8 in. Parapet walls must also be 8 in. thick, and their height is limited to 3 times the nominal thickness (taller parapets must be designed analytically). Foundation walls must

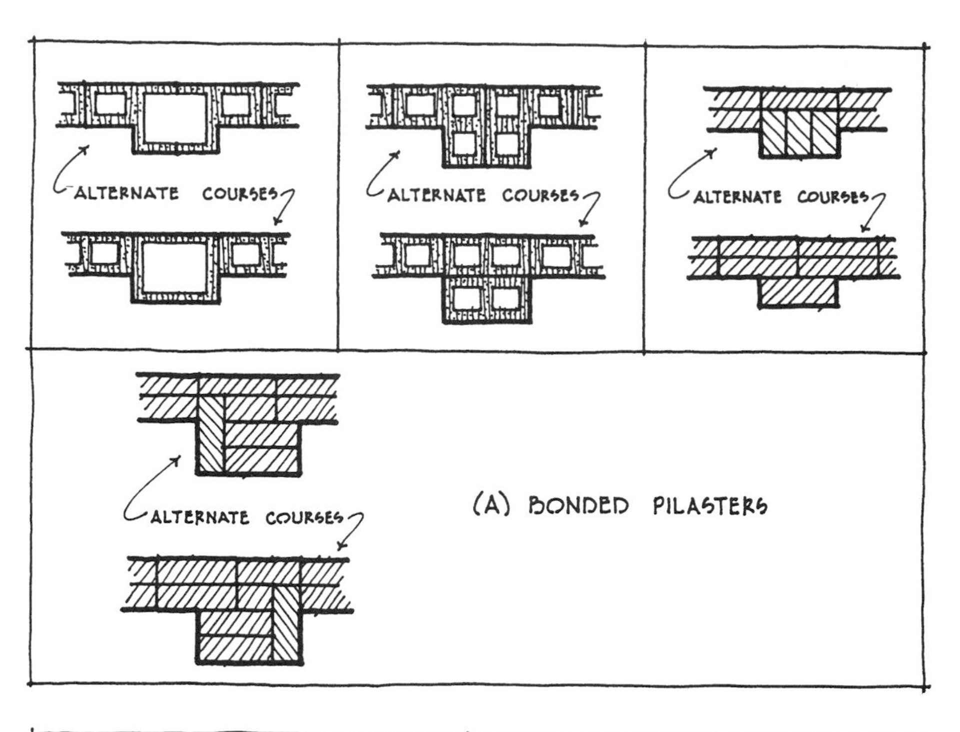

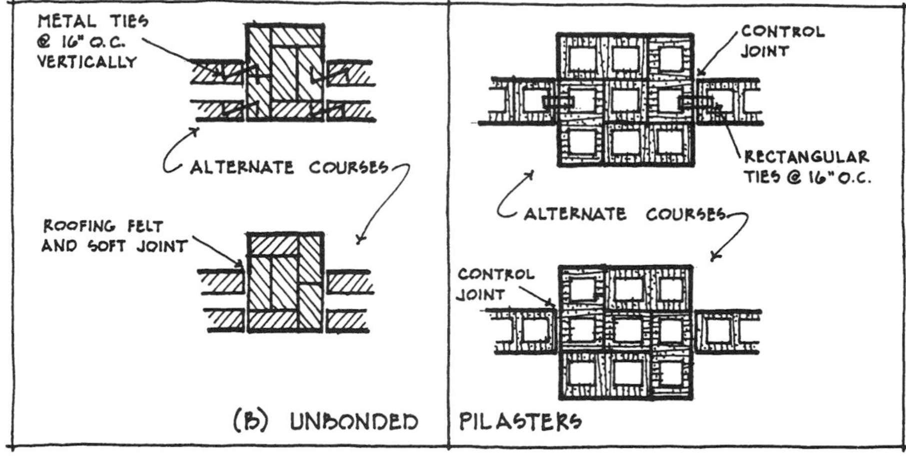

Masonry pilasters. (*From Harry C. Plummer,* **Brick and Tile Engineering,** *Brick Institute of America, Reston, Va., 1962.*)

meet the thickness requirements shown in *Fig. 12-24,* and must be constructed with Type M or S mortar. If wall height, lateral support, or unbalanced fill conditions exceed code limits, foundation walls must be analytically rather than empirically designed.

12.2.4 Bonding

Multi-wythe walls may be bonded with masonry headers (*see Fig. 12-25*), metal ties, or prefabricated joint reinforcement. There must be one metal tie for each $4\frac{1}{2}$ sq ft of wall area, with a maximum vertical spacing of 24 in. and a maximum horizontal spacing of 36 in. Metal ties in alternate courses

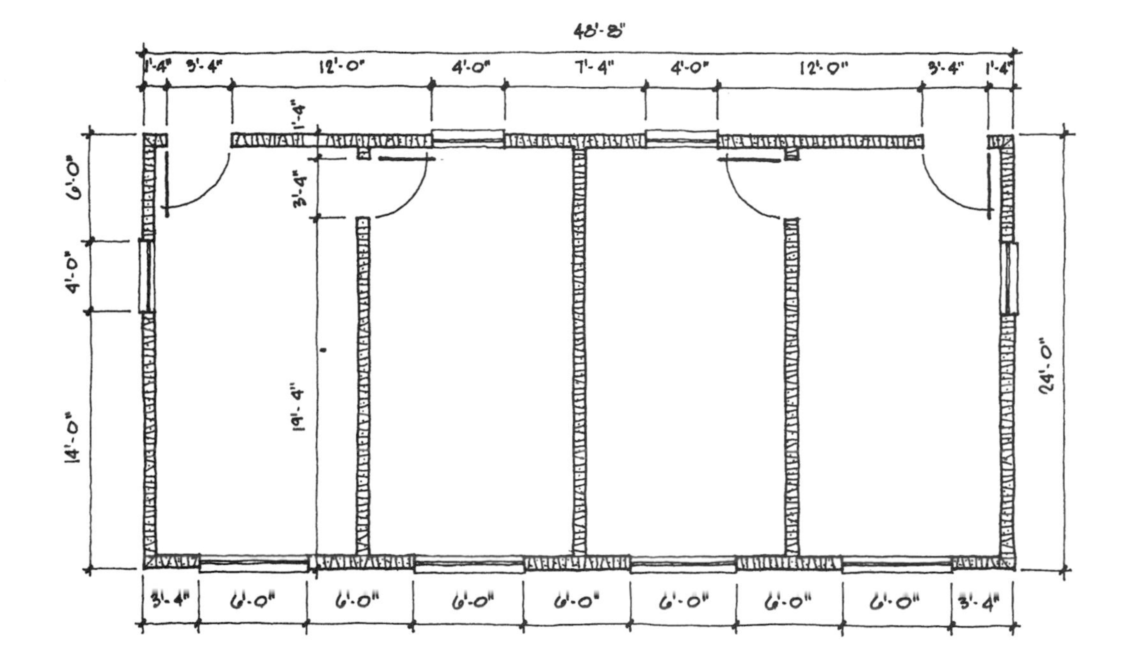

MINIMUM CUMULATIVE SHEAR WALL LENGTH = 0.4 × LONG DIMENSION

MINIMUM L = 0.4 × 48'-8" = 19'-4"
X DIRECTION: L = 12'-0" + 7'-4" + 12'-0" + 3'-4" + 6'-0" + 6'-0" + 6'-0" + 3'-4" = 56'-0" > 19'-4" OK
Y DIRECTION: L = 14'-0" + 6'-0" + 19'-4" + 24'-0" + 19'-4" + 14'-0" + 6'-0" = 102'-8" > 19'-4" OK

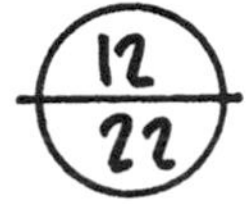

MSJC Code minimum shear wall requirements.

Floor or roof construction	*Maximum ratio, shear wall spacing/ shear wall length*
Cast-in-place concrete	5:1
Precast concrete	4:1
Metal deck with concrete fill	3:1
Metal deck with no fill	2:1
Wood diaphragm	2:1

UBC requirements for shear wall spacing in empirically designed masonry. (*Reproduced from the 1994 edition of the* Uniform Building Code, *copyright 1994, with permission of the publisher, the International Conference of Building Officials.*)

Foundation wall construction	Nominal thickness (in.)	Maximum depth of unbalanced fill (ft)	
Ungrouted hollow masonry units	8	4	
	10	5	
	12	6	
Solid masonry units	8	*UBC*	*MSJC*
	10	5	5
	12	6	7
		7	7
Hollow or solid masonry units fully grouted	8	7	
	10	8	
	12	8	
Hollow masonry units reinforced vertically with No. 4 bars and grout at 24 in. on center. Bars located not less than $4\frac{1}{2}$ in. from pressure side of wall.	8 (UBC only)	7 (UBC only)	

UBC and MSJC Code requirements for foundation wall thickness for empirically designed masonry.

must be staggered horizontally (*see Fig. 12-26*). Additional ties must be provided at all openings, spaced not more than 3 ft apart around the perimeter, and within 12 in. of the opening itself. Prefabricated wire joint reinforcement used to provide bond between multiple wythes must have cross wires of 9-gauge steel, be spaced a maximum of 16 in. on center vertically, and provide one cross wire for every $2\frac{2}{3}$ sq ft of wall area.

12.2.5 Anchorage

Intersecting walls that depend on one another for lateral support must have at least 50% of the units at the intersection laid in an overlapping masonry bond (*see Fig. 12-27*). If the walls are built up separately, the first must be regularly toothed or blocked with 8-in. maximum offsets. Metal anchors 2 ft long with bent legs and a minimum section of $\frac{1}{4} \times 1\frac{1}{2}$ in. are then used to tie the joints at 4-ft vertical intervals. Intersecting walls may also be bonded with L or T sections of prefabricated joint reinforcement spaced 8 in. on center vertically. Anchorage for floors and roofs, as well as anchorage to structural framing, must also comply with specific code requirements.

12.2.6 Corbeling

Only solid masonry units may be used for corbeling. The maximum corbeled projection beyond the face of the wall is limited to one-half the wall thickness for solid walls, or one-half the wythe thickness for cavity walls. The maximum projection of any individual unit may not exceed half the unit height or one-third its thickness (*see Fig. 12-28*).

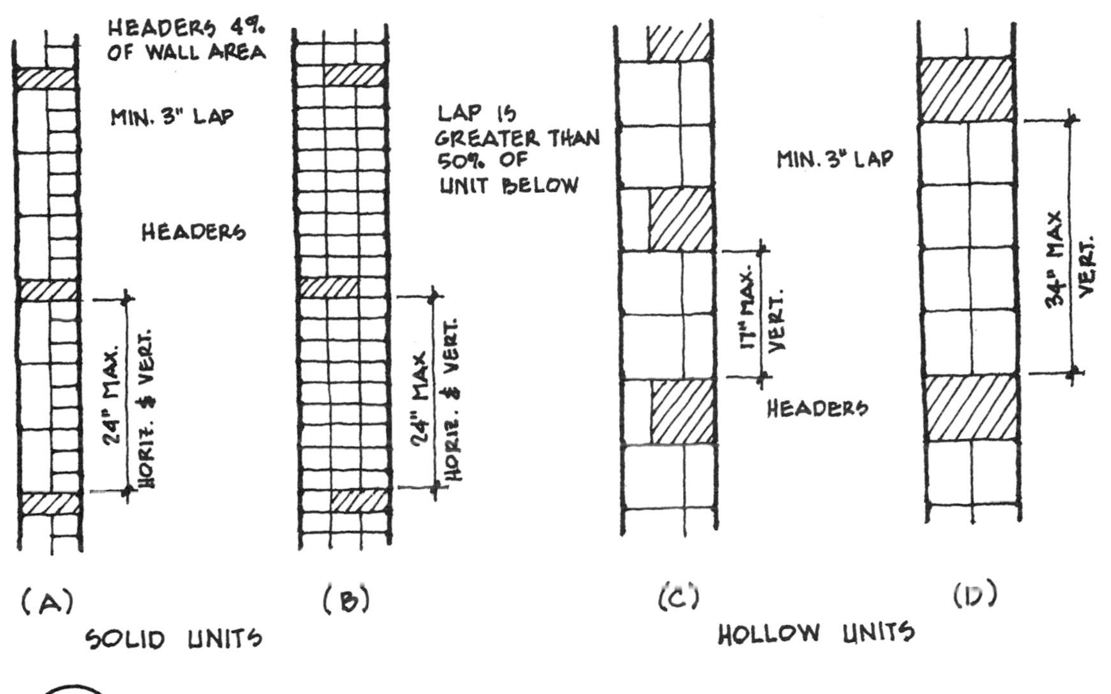

12/25 MSJC Code requirements for masonry bonding.

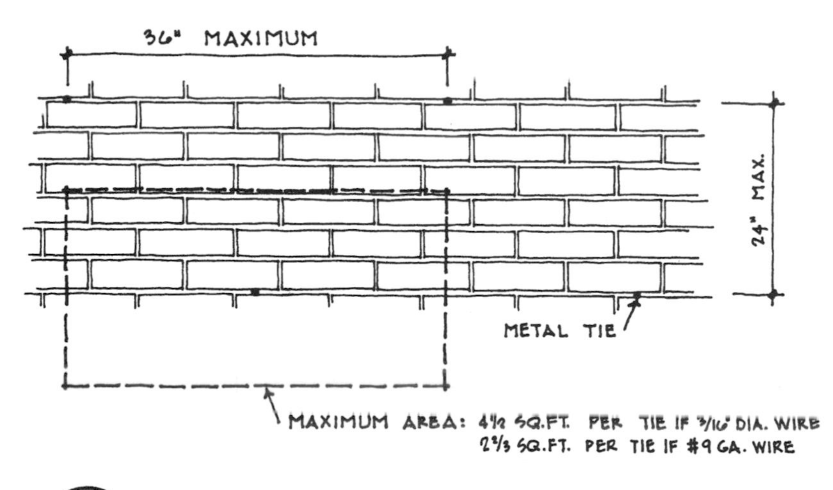

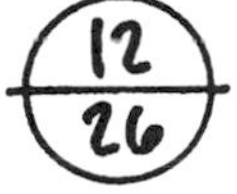

Requirements for metal tie spacing.

12.3 ALLOWABLE STRESS DESIGN

Analytical design in both the UBC and MSJC are based on the allowable stress method in which the computed stresses resulting from service loads may not exceed allowable stresses dictated by the codes. The allowable stresses used are quite conservative, generally resulting in safety factors which range from 3 to 5. The two codes are very similar in their requirements except that MSJC *requires* inspection and UBC treats inspection as optional, with allowable stresses reduced by ½ when special inspection is not provided.

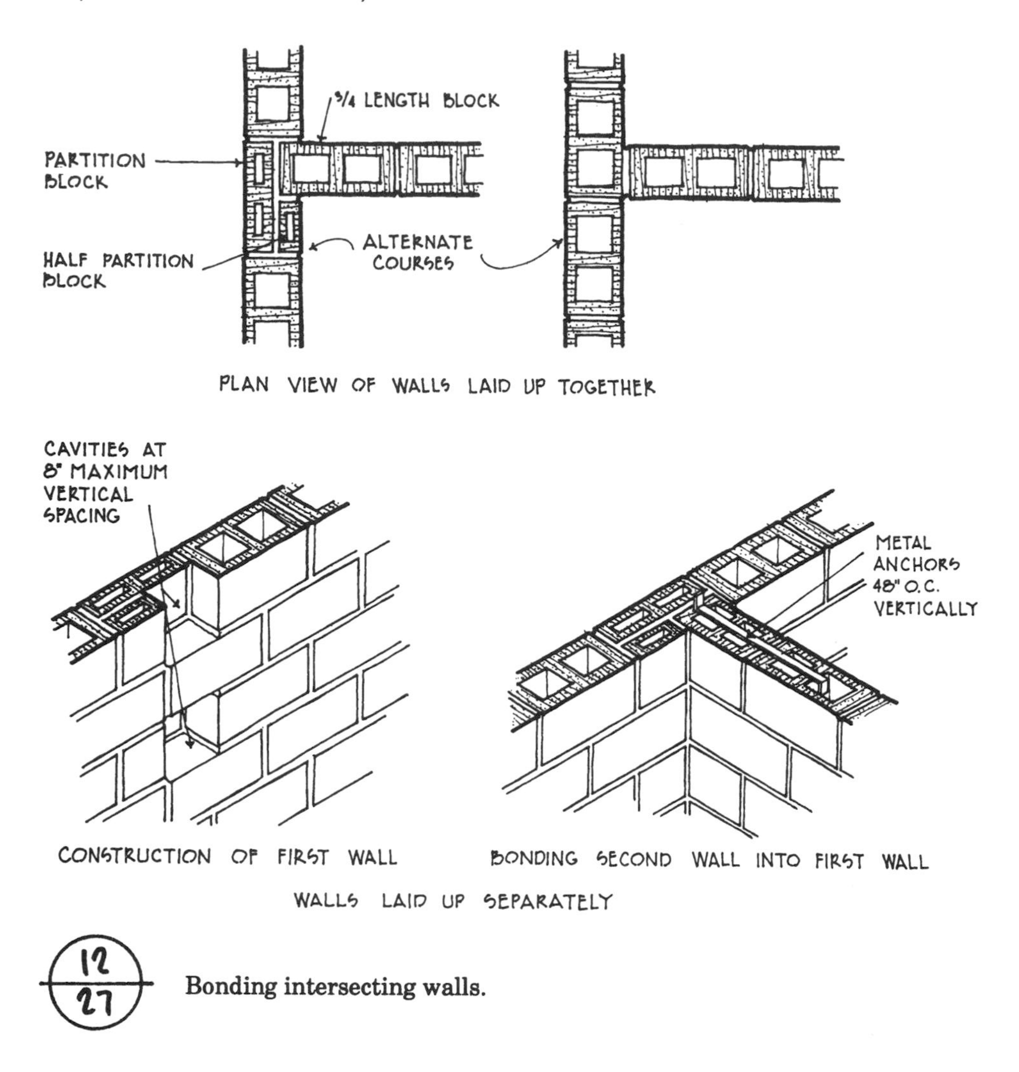

Bonding intersecting walls.

The *Uniform Building Code* gives the most comprehensive guideline in the industry for an inspection program on structural masonry projects. The UBC permits higher allowable stresses when the owner employs a certified "special inspector" to (1) observe the preparation of masonry wall prisms, sampling and placing of all masonry units, and placement of reinforcement; (2) inspect grout spaces immediately prior to closing cleanouts; and (3) observe all grouting operations. UBC requires *continuous* inspection during preparation of masonry wall prisms, sampling and placing of all masonry units, and placement of reinforcement, and immediately prior to and during all grouting operations. *Periodic* inspection for placing of units is permitted for:

- Closed-end hollow unit masonry where the specified compressive strength of the masonry does not exceed 1500 psi for concrete masonry units or 2600 psi for clay masonry units, and cleanouts are provided at the bottom course of every grout pour at each vertical bar
- Open-end hollow unit masonry where cleanouts are provided at the bottom course of every grout pour at each vertical bar

The intent is to provide qualified inspection during construction in addition to inspections by the building official.

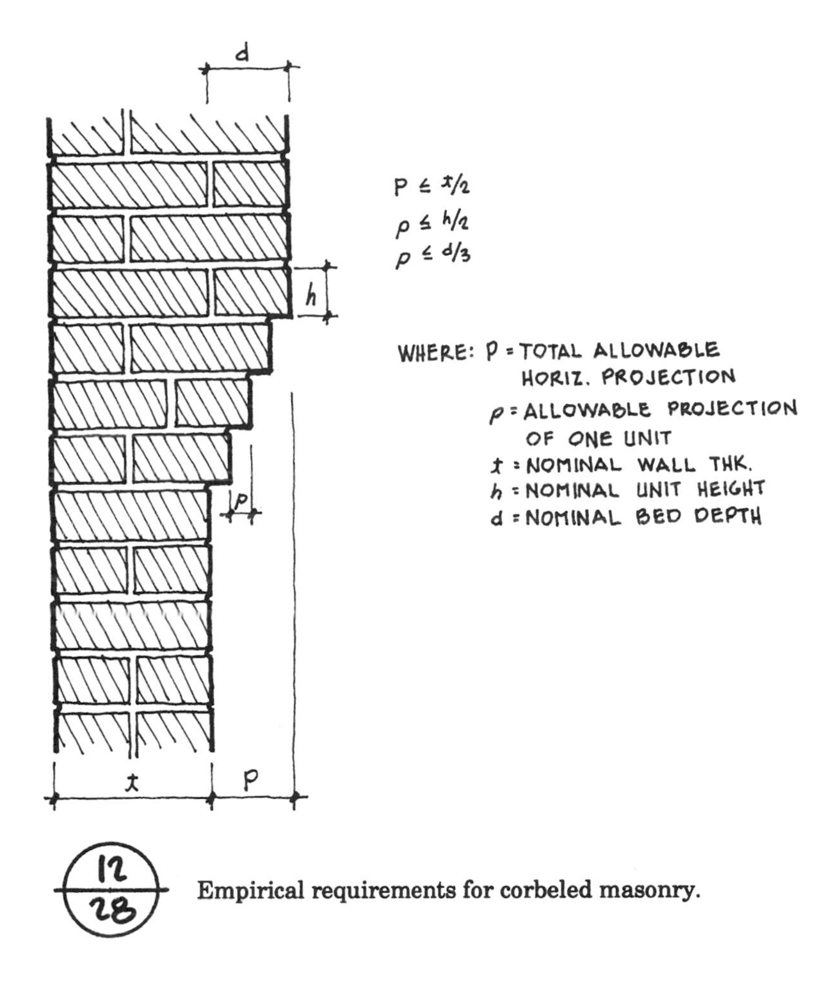

12/28 **Empirical requirements for corbeled masonry.**

The MSJC Code, however, bases its allowable stresses on required inspection to evaluate quality and acceptability of materials, equipment, and procedures. There is no option for uninspected work. The Commentary to the MSJC Specifications goes into some detail about the need for and purpose of quality assurance and quality control programs, including testing, inspection, documentation, and resolution of noncomplying conditions. The necessary level of inspection, though, is not defined in the MSJC Specification, nor indeed has it been well established within the industry in regions outside the western United States. Some feel that the level of inspection should vary with the size and complexity of the structure, and others favor requiring the same level of inspection regardless of project size. Proposed changes to the MSJC Code may provide for different levels of inspection depending on the design procedure used (empirical or analytical) and the type of structure being built (essential or nonessential facility).

All allowable stress design requirements are based on units laid in running bond with a minimum overlap between units of one-fourth the unit length. Units laid in stack bond must be reinforced with bond beams or prefabricated joint reinforcement spaced 48 in. on center vertically, and the area of steel must equal 0.0003 times the vertical cross-sectional area of the wall.

In determining stresses, the effects of all dead and live loads must be taken into account, and stresses must be based on actual rather than nominal dimensions. Consideration must be given to the effects of lateral load, eccentricity of vertical load, nonuniform foundation pressure, deflection, and thermal and moisture movements. All critical loading conditions must be calculated. Fixity, or end restraint, must also be considered, as it affects resistance to applied loads.

Flexural, shear, and axial stresses resulting from wind or earthquake forces must be added to the stress of dead and live loads, and connections must be designed to resist such forces acting either inward or outward. For combined stresses due to these temporary or transient lateral forces plus the dead and live loads, allowable stresses may be increased 33⅓% provided that the strength of the resulting section is not less than that required for dead loads and vertical live loads alone. For purposes of stress calculation, wind and earthquake forces may be assumed never to occur simultaneously.

Materials. Brick units must comply with ASTM C62, ASTM C216, or ASTM C652 for building brick, face brick, or hollow brick. Concrete masonry must meet the requirements of ASTM C90 for loadbearing units, or ASTM C55 for concrete brick. Mortar must comply with ASTM C270, and grout with ASTM C476. Metal anchors, ties, and joint reinforcement must be of corrosion-resistant finish. When different kinds or grades of units or mortars are used in composite walls, stresses may not exceed the allowable for the weakest of the combinations of units and mortars unless a higher strength is substantiated by laboratory test.

Masonry Compressive Strength. An engineer bases the design of masonry structural systems on a certain "specified compressive strength of masonry" (f'_m), on which the allowable axial, flexural, compressive, shear, and bearing stresses are based. Under UBC and MSJC Code requirements, the contractor must verify to the engineer by one of two methods that the proposed materials and construction will meet or exceed this strength. The contractor may elect to use the *unit strength method* based on the combined strength of the masonry units and mortar as determined by tables in the code, or the *prism test method.*

Projects that are not large enough to justify the cost of prism testing will generally use the unit strength method. Through submittals, the contractor certifies that the proprietary masonry units specified in the contract documents or the generic masonry units selected to comply with specified ASTM standards, are of sufficient strength to produce the "specified compressive strength" when combined with ASTM C476 grout and either the specified ASTM C270 mortar Type (M, S, or N), or the ASTM C270 mortar Type (M, S, or N) selected to produce the "specified compressive strength" (*see Figs. 12-29 and 12-30*). The proportion specification of ASTM C270 (the default) would govern, as well as the proportion specification of ASTM C476. Both the proportion specifications of ASTM C270 and C476 and the unit strength method of determining "specified compressive strength" are very conservative and usually produce mortar and masonry of greater strength than the minimum required by the design. This is a built-in high safety factor, but does not produce the most cost-efficient design.

For large projects where efficiency in design can produce a cost savings by optimizing the use of materials in mortar and grout designs, the contractor may elect to use the prism test method to verify "specified compressive strength" (*see Fig. 12-31*). Using this method, the contractor hires the testing laboratory to produce mortar and grout mix designs in accordance with the minimum property specification of ASTM C476 and ASTM C270 respectively which, when combined with the specified or selected masonry units, will produce the "specified compressive strength" when

Compressive strength of clay masonry units† (psi)	Specified compressive strength of masonry (f'_m)*	
	Type M or S mortar‡ (psi)	Type N Mortar‡ (psi)
14,000 or more	5300	4400
12,000	4700	3800
10,000	4000	3300
8,000	3350	2700
6,000	2700	2200
4,000	2000	1600
Compressive strength of concrete masonry units§, (psi)	**Specified compressive strength of masonry (f'_m**	
	Type M or S mortar‡ (psi)	**Type N mortar‡ (psi)**
4800 or more	3000	2800
3750	2500	2350
2800	2000	1850
1900	1500	1350
1250	1000	950

*Compressive strength of solid clay masonry units is based on gross area. Compressive strength of hollow clay masonry units is based on minimum net area. Values may be interpolated. When hollow clay masonry units are grouted, the grout shall conform to the proportions in [UBC] Table 21-B.

†Assumed assemblage. The specified compressive strength of masonry f'_m based on gross area strength when using solid units or solid grouted masonry and net area strength when using ungrouted hollow units.

‡Mortar for unit masonry, proportion specification, as specified in [UBC] Table 21-A. These values apply to portland cement lime mortars without added air entraining materials.

§Values may be interpolated. In grouted concrete masonry, the compressive strength of grout shall be equal to or greater than the compressive strength of the concrete masonry units.

UBC specified compressive strength of masonry f'_m (psi) based on specifying the compressive strength of masonry units. (*Reproduced from the 1994 edition of the* Uniform Building Code, *copyright 1994, with permission of the publisher, the International Conference of Building Officials.*)

prisms are laboratory-tested in accordance with ASTM E447 or ASTM C1314. Method A of ASTM E447 is used for preconstruction testing to select mix designs and to compare various combinations of units, mortar, and grout. Method B of ASTM E447 is used to test field-constructed prisms to verify that they provide the "specified compressive strength." The specifications should stipulate that the contractor construct the Method B prisms at the job site in order to factor in workmanship. The ASTM C1314 test method is similar to that in ASTM E447 except that ASTM C1314 does not require the collection of any extraneous information other than that required for verification of the specified compressive strength.

The process of verifying the compressive strength of masonry is similar to specifying concrete for small projects by stipulating a certain number of sacks of cement per cubic yard and a certain water:cement ratio (comparable to the unit strength method), and to specifying concrete for

TABLE A COMPRESSIVE STRENGTH OF MASONRY BASED ON THE COMPRESSIVE STRENGTH OF CLAY MASONRY UNITS AND TYPE OF MORTAR		
Net area compressive strength of clay masonry units (psi)		*Net area compressive strength of masonry (psi)*
Type M or S mortar	*Type N mortar*	
2,400	3,000	1000
4,400	5,500	1500
6,400	8,000	2000
8,400	10,500	2500
10,400	13,000	3000
12,400	—	3500
14,400	—	4000

TABLE B COMPRESSIVE STRENGTH OF MASONRY BASED ON THE COMPRESSIVE STRENGTH OF CONCRETE MASONRY UNITS AND TYPE OF MORTAR		
Net area compressive strength of concrete masonry units (psi)		*Net area compressive strength of masonry* (psi)*
Type M or S mortar	*Type N mortar*	
1250	1300	1000
1900	2150	1500
2800	3050	2000
3750	4050	2500
4800	5250	3000

*For units of less than 4 in. height, 85% of the values listed.

MSJC Code requirements for compressive strength of masonry based on compressive strength of clay and concrete masonry units and type of mortar used in construction. [*From the Masonry Standards Joint Committee (MSJC)* Specifications for Masonry Structures (ACI 530.1/ASCE 6/TMS 602), *1995 edition, ACI.*]

large projects by requiring a minimum compressive strength for which the contractor proposes a laboratory mix design (comparable to preconstruction prism testing) that is verified by cylinder tests of field-sampled concrete (comparable to field-constructed prism testing).

If the prism test method is used, ASTM C780 is used for preconstruc-

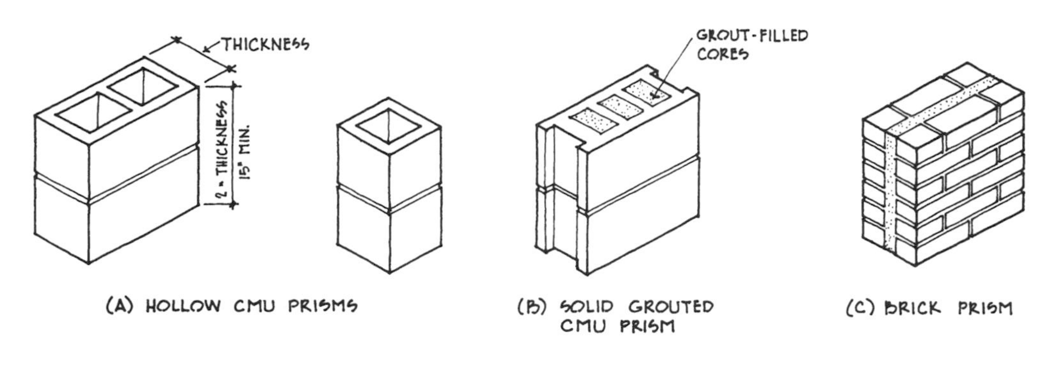

Masonry test prisms.

tion and construction evaluation of mortar mixes, ASTM C1019 is used for preconstruction and construction evaluation of grout mixes, and ASTM E447 or C1314 is used for prism tests. If the unit strength method is used, no testing is required.

Differential Movement. Buildings are dynamic structures whose successful performance depends on allowing the differential movement of adjoining or connecting elements to occur without excessive stress or its resulting damage. All building materials experience volumetric changes from temperature fluctuations, and some experience moisture-related movements as well. Structural movements include column shortening, elastic deformation, creep, and wind sway. The differential rates and directions of movements must be accommodated by flexible connections and movement joints (refer to Chapter 9).

Structural connections may be required to permit movement in some directions and transfer applied loads in others. For example, at the joint between interior bearing walls and exterior non-bearing wall shear flanges, the exterior wall will undergo more thermal movement and less elastic and creep induced movement than the interior wall. The connection between these elements must accommodate such movement or transfer shear stress. Each building must be analyzed for differential movement characteristics and provisions made to relieve the resulting stresses. MSJC is the first masonry code to include design coefficients for thermal expansion, moisture expansion, shrinkage, and creep.

12.3.1 Unreinforced Masonry

Although the compressive strength of unreinforced masonry is high, its flexural strength depends on three things: (1) the type and design of the masonry unit, (2) the type of mortar and its materials, and (3) the quality of workmanship. Higher flexural strengths are developed with solid masonry units because the wider mortar bed provides more bonding surface. Failure in lateral loading usually results from bond failure at a ruptured bed joint, so factors that affect mortar bond also affect the flexural strength of the wall. Full, unfurrowed joints, good mortar flow and consistency, proper unit absorption, and moist curing all improve bond (see Chapter 15).

When lateral loads are applied perpendicular to a wall, they are transmitted to vertical and horizontal edge connections. The proportion of the load transmitted either vertically or horizontally will depend on the flexural resistance and rigidity of the wall in each direction, the degree of fixity or restraint developed at the edges, the horizontal-to-vertical span ratio, and the distribution of the loads as they are applied to the wall. The lateral stability of loadbearing masonry walls is greater than that of non-bearing or lightly loaded walls. Applied vertical loads produce compressive stresses which must be overcome by the tensile stress of the lateral load before failure can occur. In the lower stories of loadbearing buildings, compressive stresses generally counteract tension, but in the upper stories of tall buildings where wind loads are higher and axial loads smaller, the allowable flexural tensile stresses may be exceeded, thus requiring steel reinforcement to be added at those locations.

Allowable axial compression and flexure, flexural tension, and shear in unreinforced masonry are dictated by code requirements based on analytical design formulas and slenderness reduction factors for height-to-thickness (h/t) ratios or radius of gyration. Joint reinforcement incorporated in the wall for control of shrinkage cracking or thermal movements is not considered to increase flexural tensile strength. Isolated columns in unreinforced structures must be designed as reinforced masonry.

12.3.2 Reinforced Masonry

Reinforced masonry is used where compressive, flexural, and shearing stresses are higher than those allowed for unreinforced masonry. Designs that incorporate reinforcing steel neglect the flexural strength contribution of the masonry altogether, and rely on the steel to resist 100% of the tensile loads.

Reinforced masonry walls may be of double-wythe construction with a grouted cavity to accommodate the steel reinforcing, or of single-wythe hollow units with grouted cores. Steel wire reinforcement may be laid in horizontal mortar joints as long as code requirements for wire size and minimum mortar cover are met. Vertical reinforcement must be held in position at the top and bottom of the bar as well as at regular vertical intervals.

Vertical reinforcing steel contributes to the load-carrying capacity of masonry columns and pilasters because lateral ties are required to prevent buckling and confine the masonry within. The vertical steel in walls does not take any of the axial load unless it is also restrained against buckling. Reinforcing steel also helps masonry resist volume changes due to temperature and moisture variations, and its effect should be considered in the calculation of differential movement and the location and spacing of movement joints.

The size and placement of steel reinforcement is determined by design analysis of service load conditions. Different combinations of bar sizes and spacing can give the same ratio of steel area to masonry area, and some consideration must be given to economically achieving the analytical requirements. Steel spaced too closely will slow construction, can inhibit grout placement, and may be unnecessarily expensive. For instance, No. 3 bars at 8 in. on center give the same area of steel per foot of wall as No. 7 bars at 48-in. centers, but the closer spacing requires more labor expense than is necessary to produce the same result. Bar size and spacing, however, must also take into consideration the size of the grout space and the code requirements for minimum protective coverage of reinforcement.

12.3.3 Wind and Seismic Loads

Buildings must be capable of resisting all lateral seismic forces, assumed to act nonconcurrently in the direction of each of the main axes of the structure. In addition to the calculation and resolution of total seismic forces, consideration must also be given to individual structural and nonstructural elements of the building. Parts or portions of structures, nonstructural components, and their anchorage to the main structural system must be designed for lateral seismic forces.

Distribution of horizontal shear must be in proportion to the relative rigidities of the resisting elements and must consider the rigidity of the diaphragm. Overturning moments caused by seismic loads must be distributed to the resisting elements in the same proportion as the distribution of shears.

The UBC includes special provisions for designing masonry structures in seismic zones 2, 3, and 4 (*see Fig. 12-32*) and the MSJC Code for structures in Seismic Performance Categories C, D, and E. To increase the elastic response of masonry structures in seismic risk areas, both codes prescribe minimum requirements for size, area, and placement of reinforcing steel (*see Figs. 12-33 and 12-34* for UBC requirements). Anchorage requirements are also spelled out, and limitations prohibit the use of some lower strength mortars.

The amount of steel required is based on test results and empirical

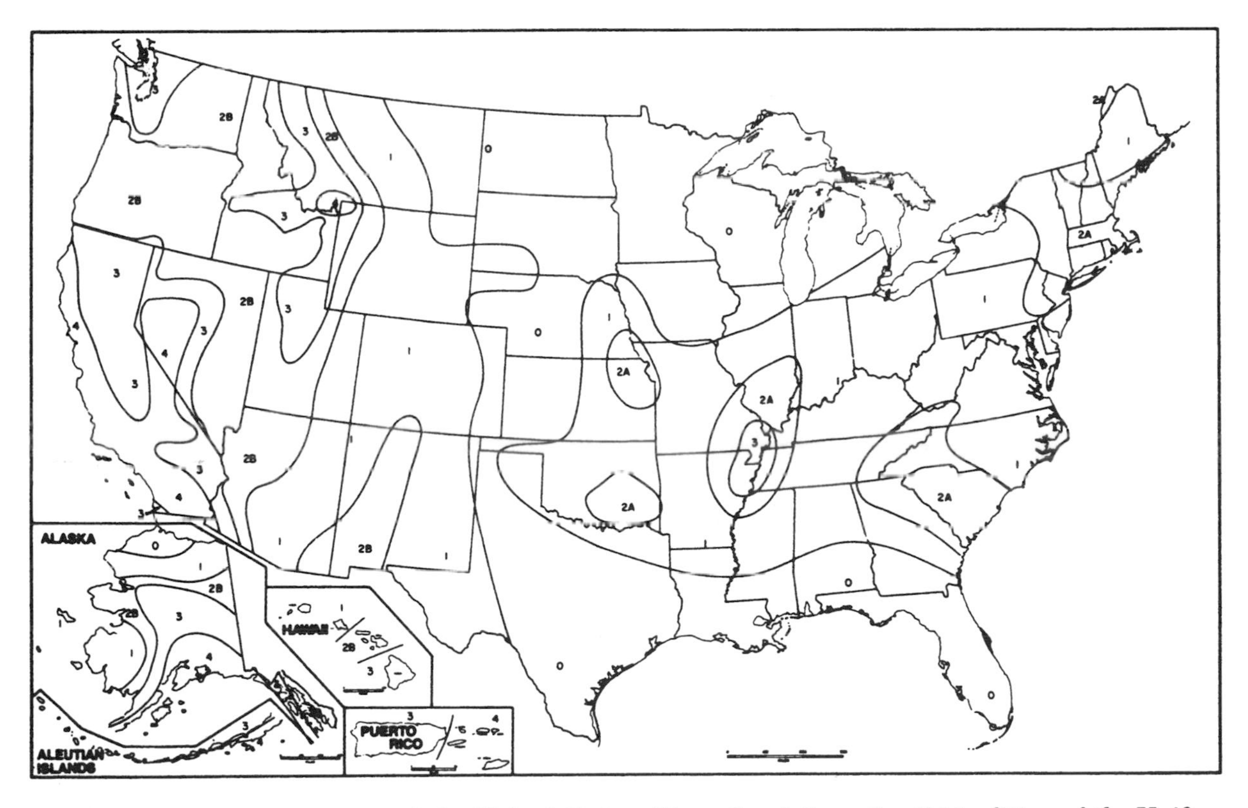

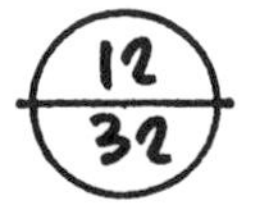

Seismic zone map of the United States. (*Reproduced from the 1994 edition of the* Uniform Building Code, *copyright 1994, with permission of the publisher, the International Conference of Building Officials.*)

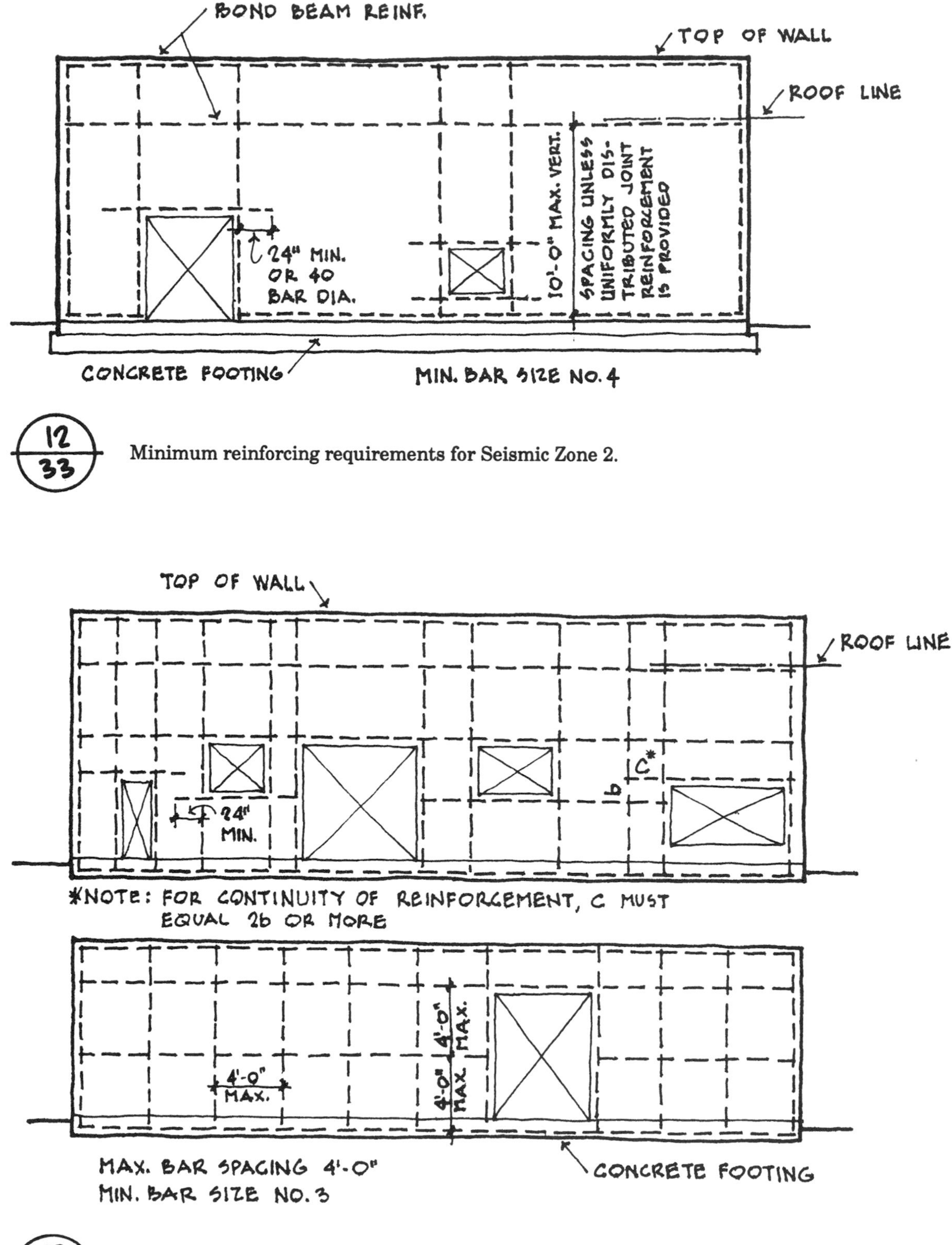

Minimum reinforcing requirements for Seismic Zone 2.

Minimum reinforcing requirements for Seismic Zones 3 and 4.

judgment rather than engineering analysis of stress or performance. The prescriptive ratio of steel to wall area is provided to increase the ductility of the masonry structure in seismic events. Where analytical design indicates that more steel is required, the prescriptive minimum may be included as part of the total. If analysis indicates that less steel is required, the prescriptive minimum seismic reinforcing must still be provided. Joint reinforcement cannot be used for seismic resistance. All reinforcing steel designed to resist seismic loads must be fully embedded in grouted cores, cavities, or bond beams.

Reinforced masonry structures designed in compliance with modern building code requirements have successfully withstood substantial seismic forces, and the rigidity inherent in the masonry systems often reduces or eliminates secondary damage. Reinforced masonry buildings as tall as 10 stories survived near the epicenter of the 7.1 Loma Prieta, California, earthquake in 1989 and the 6.7 Northridge, California, earthquake in 1994 without structural damage, glass breakage, pipe separations, or even cracking in the drywall or door jambs. Such secondary safety is critical in the construction of essential facilities such as hospitals, fire stations, communications centers, and other facilities required for emergency response. The only masonry buildings to sustain significant damage were older unreinforced masonry structures built before modern building code requirements and unretrofitted to meet stricter performance criteria. Older buildings which had been retrofitted in accordance with the City of Los Angeles Division 88 ordinance fared much better.

12.4 LIMIT STATES DESIGN

Allowable stress design methods are very conservative. Even though they are primarily based on engineering analysis, they also include arbitrary limits on length or height-to-thickness (h/t) ratios of walls, and prescriptive requirements for seismic reinforcement. Although allowable stress design codes arbitrarily limit the height of masonry walls, tests have shown that tall, slender masonry walls are much stronger and more ductile than such limits would indicate. Test panels of reinforced brick and reinforced concrete masonry walls experienced static load deflections of as much as 19 in. without failure, while carrying axial loads of up to 860 plf and lateral loads of 100 psf and higher (*see Fig. 12-33*). The panels, constructed 24 ft 8 in. high, represented h/t ratios ranging from 30 to 57 for the various 4-, 6-, 8-, and 10-in. wall thickness used (compared to a maximum h/t of 20 under allowable stress design code requirements). Limit states design (or strength design) permits the use of such tall, slender walls by more accurately predicting behavior than the allowable stress method.

Limit states design methods for masonry are beginning to appear, and some strength design procedures for masonry have recently been adopted by UBC. Limit states design normally includes minimum and maximum reinforcement ratios, maximum allowable deflections under load, and ultimate moment capacity requirements.

Even more important than the economy and efficiency which limit states design provides by using masonry to its full structural capabilities is the fact that the performance of masonry is more accurately predicted than with the allowable stress method. Under severe earthquake loading where wall response may be inelastic because of flexural yielding of the reinforcement under combined axial, bending, and shear loads, limit states design provides greater safety because of its accuracy.

In California, under UBC Seismic Zone 4, 8-in. grouted and reinforced block walls designed by the limit states method have been built to 28 ft in height before lateral support bracing is provided at the roof connection. In Utah under Seismic Zone 2 requirements, 6-in. grouted and reinforced hollow brick walls have been built to 25 ft and in Illinois in tornado alley, 12-in. reinforced masonry cavity walls to more than 42 ft before lateral support is required.

Research is still ongoing to document the ultimate behavior of masonry structures under static and dynamic loading, and to develop the basis for a full limit states design code.

Wall thickness (in.)	*h/t ratio*	*Deflection at yield (in.)*	*Deflection at failure (in.)*
Single-wythe CMU			
5.63	51.2	9.0	17.7
7.63	38.0	5.8	14.8
7.63	38.0	6.5	11.2
9.63	30.0	5.5	17.1
9.63	30.7	6.3	19.0
Single-wythe hollow brick			
3.50	57.0	10.0	17.0
5.50	52.4	8.0	19.3
Two-wythe brick			
7.50	30.3	10.5	14.8
9.60	30.3	9.3	16.8

Note: Compressive strength of masonry (f'_m) ranged from 2460 psi to 6243 psi. Lateral loads ranged from 56 psf to 162 psf.

Results of tests on tall, slender masonry walls. (*Photo courtesy James E. Amrhein, Masonry Institute of America, Los Angeles, Calif.*)

13
BASEMENT AND RETAINING WALLS

Masonry walls may be used to retain earth in landscape applications, below-grade building structures, and even in swimming pools.

13.1 GENERAL CONSIDERATIONS

Basement and retaining wall design must be concerned with lateral soil pressures, surcharge loads occurring during construction and in service, overturning moments, and sliding.

13.1.1 Lateral Earth Pressure

The magnitude and direction of soil pressure on the wall is dependent on the height and shape of the surface, and on the nature and physical properties of the backfill. The simplest way of determining lateral earth pressure is the *equivalent fluid method.* This method assumes that the retained earth will act as a fluid, and the wall is designed to withstand the pressure of a liquid assumed to exert the same pressure as the actual backfill material. Assumed equivalent-fluid unit weights vary with the nature of the soil in the backfill. Most building codes specify fluid pressures for various types of soil.

13.1.2 Surcharge

Additional loads are created by operating automobiles, trucks, or construction equipment on the soil surface behind a retaining wall or basement wall. If activities of this nature are anticipated, the design must make allowance for the increased lateral pressures that will be imposed on the wall.

13.1.3 Overturning and Sliding

Retaining walls must safely resist overturning and sliding forces induced by the retained earth. Unless otherwise required by code, the factor of safety against overturning should not be less than 2.0, and against sliding 1.5. In addition, the bearing pressure under the footing should not exceed the allowable soil bearing pressure. In the absence of controlled tests substantiating the actual bearing capacity of the soil, each of the model building codes lists allowable pressures for different types of soil. Local requirements may vary slightly and should be checked to assure design conformance.

13.1.4 Expansion and Control Joints

The size and location of expansion or control joints should be calculated on the basis of expected movement. Joints should always be provided at wall offsets and at abrupt changes in height or thickness. Joints should be designed with a shear key for lateral stability, but still allow for longitudinal movement (refer to Chapter 9). Weep hole openings should be protected at the back to prevent clogging with backfill material.

13.1.5 Materials

Brick masonry for earth retaining structures should be ASTM C62, ASTM C216, or ASTM C652, Grade SW, with a minimum strength of 5000 psi. Hollow concrete units should be ASTM C90 normal weight, with an oven-dry density of 125 lb/cu ft or more. Mortar should comply with the requirements of ASTM C270, Type M, and grout with ASTM C476. Mortar and grout should be moist-cured for 7 days before backfilling. Concrete for footings should have a minimum compressive strength of 2000 psi, or as required by structural analysis.

13.2 BASEMENT WALLS

For both residential and commercial buildings, it is often economical to use masonry walls to enclose basement space or underground parking areas. Footings are set below the frost line in undisturbed soil. The masonry walls provide low thermal conductivity and may easily be waterproofed against moisture infiltration and dampness. Below-grade masonry walls also offer excellent enclosures for underground or earth-sheltered buildings. analytically designed reinforced masonry permits the construction of deep basement walls, walls supporting heavy vertical loads, and walls where unsupported height or length exceeds lateral support requirements for empirically designed, unreinforced masonry.

13.2.1 Design and Construction

Concrete masonry is much more widely used in below-grade construction than brick, and much research has been done to test its capability and performance. General design considerations must include (1) maximum lateral load from soil pressure, (2) vertical loads from building superstructure, (3) minimum wall thickness required by code, and (4) length or height of wall between lateral supports. Basement walls supporting bearing wall construction must usually support relatively heavy compressive loads in addition to earth pressure or other lateral loads. In skeleton frame construction, columns may extend down to separate footings and carry most of the dead and live loads of the superstructure. In such cases, the basement walls may be subject to appreciable lateral load, but little vertical load. If the columns are closely spaced, or if pilasters are added, the wall may be designed to transmit these lateral loads horizontally and vertically as two-way slabs. If the vertical supports are widely spaced, and the first-floor construction cannot be considered as providing lateral support, a design cantilever action will be required (i.e., retaining wall design).

It is normally assumed that the stresses created in basement walls by soil pressure against their exterior face are resisted by bending of the walls in the vertical span. This means that the wall behaves like a simple beam supported at top and bottom. Support at the top is provided by the first-floor construction, and bottom support by the footing and basement floor slab. If the first floor is to contribute lateral support, backfilling should be delayed until this construction is in place.

A portion of the lateral earth load is carried by the wall acting as a beam in the horizontal span. The distribution of the total lateral load horizontally and vertically will depend on wall height and length as well as stiffness in both directions. If the length of the wall between supports is no greater than its height, the load is generally divided equally between vertical and horizontal spans.

The overall stability of a below-grade wall may be enhanced by increasing the stiffness in either direction, or by reducing the length of the horizontal span. *Horizontal stiffness* can be increased by incorporating bond beams into the design, or by placing prefabricated joint reinforcement in the mortar joints at vertical intervals of not more than 16 in. Bond beams are most advantageously located at or near the top of the wall, and built to extend continuously around the perimeter of the building. When used in this manner, they will also serve to distribute concentrated vertical loads. The increase in flexural strength achieved with horizontal joint reinforcement is limited by the practical amount of steel that can be embedded in the joints, and by the amount of bond strength developed between mortar, reinforcement, and masonry units.

Vertical stiffness may be increased in one of two ways: (1) steel reinforcement may be grouted into hollow cells or (2) pilasters may be added (*see Fig. 13-1*). Pilasters should project from the wall a distance equal to approximately one-twelfth of the wall height. Pilaster width should be equal to approximately one-tenth of the horizontal span between supports. The distance between pilasters or between end walls or cross walls and pilasters should not exceed 18 ft for unreinforced walls 10 in. thick, or 15 ft for walls 8 in. thick.

In relying on floors and footings for lateral bracing, proper anchorage of members must be provided to assure transfer of loads. Steel dowels should connect walls securely to the footing. Pilasters, cross walls, and end walls must be bonded with interlocking masonry units or with metal ties. Sill plates should be anchored to the wall at 6-ft maximum intervals (*see Fig. 13-2*). Intersecting walls should be anchored with $24\times\frac{1}{4}\times1\frac{1}{2}$-in. metal straps spaced not more than 32 in. on center vertically. If a partition does not provide lateral support, strips of metal lath, galvanized hardware cloth, or joint reinforcement may be substituted for the heavier straps. Mortar joints at the intersection of cross walls with exterior below-grade walls

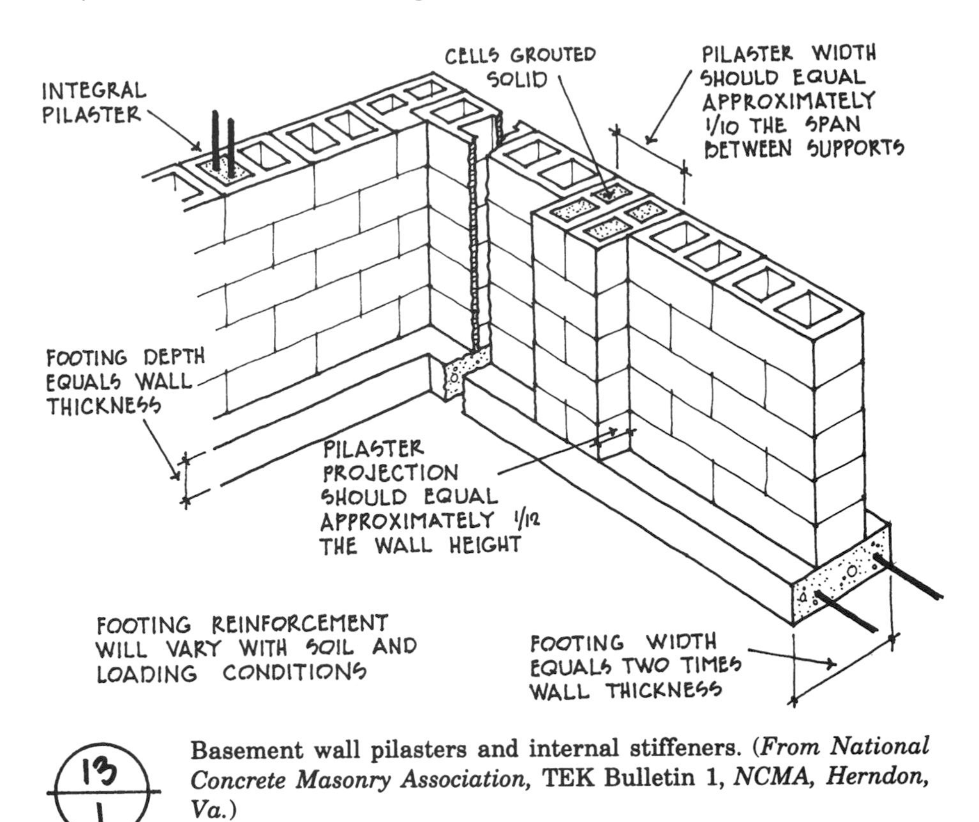

Basement wall pilasters and internal stiffeners. (*From National Concrete Masonry Association,* **TEK Bulletin 1,** *NCMA, Herndon, Va.*)

should be raked out and caulked to form a control joint. Sill plates should be connected with ½-in.-diameter bolts extended at least 15 in. into the filled cells of the masonry, and spaced to within 12 in. of the end of the plate.

Where beams bear on a basement wall, at least two block cores in the top course below the end of the member should be filled with mortar or grout, or a bearing plate installed to distribute loads. Pilasters may be bonded to the wall at beam locations to provide additional support, and should be grouted solid in the top course. Ends of floor joists should be anchored at 6-ft intervals (normally every fourth joist). At least the first three joists running parallel to a wall should also be anchored to it at intervals not exceeding 8 ft.

The axial load on basement walls is usually transmitted eccentrically at some point between the centerline of the wall and the inner surface, thus inducing a bending moment. Additional moment is induced at any point where flexural members are restrained by their connection to the wall. These moments tend to counteract the bending moments from lateral earth pressures at the exterior face. In other words, vertical compressive loads are effective in reducing the tensile stresses developed in resisting lateral loads. In this regard, it is important to remember that only dead loads may be safely considered as opposing lateral bending stresses, since live loads may be intermittent. Precautions must also be taken in construction scheduling to ensure that the amount of dead load calculated for such resistance is actually present before the lateral load is applied. If early backfill is unavoidable, temporary bracing must be provided to prevent actual stresses from exceeding those assumed in the design.

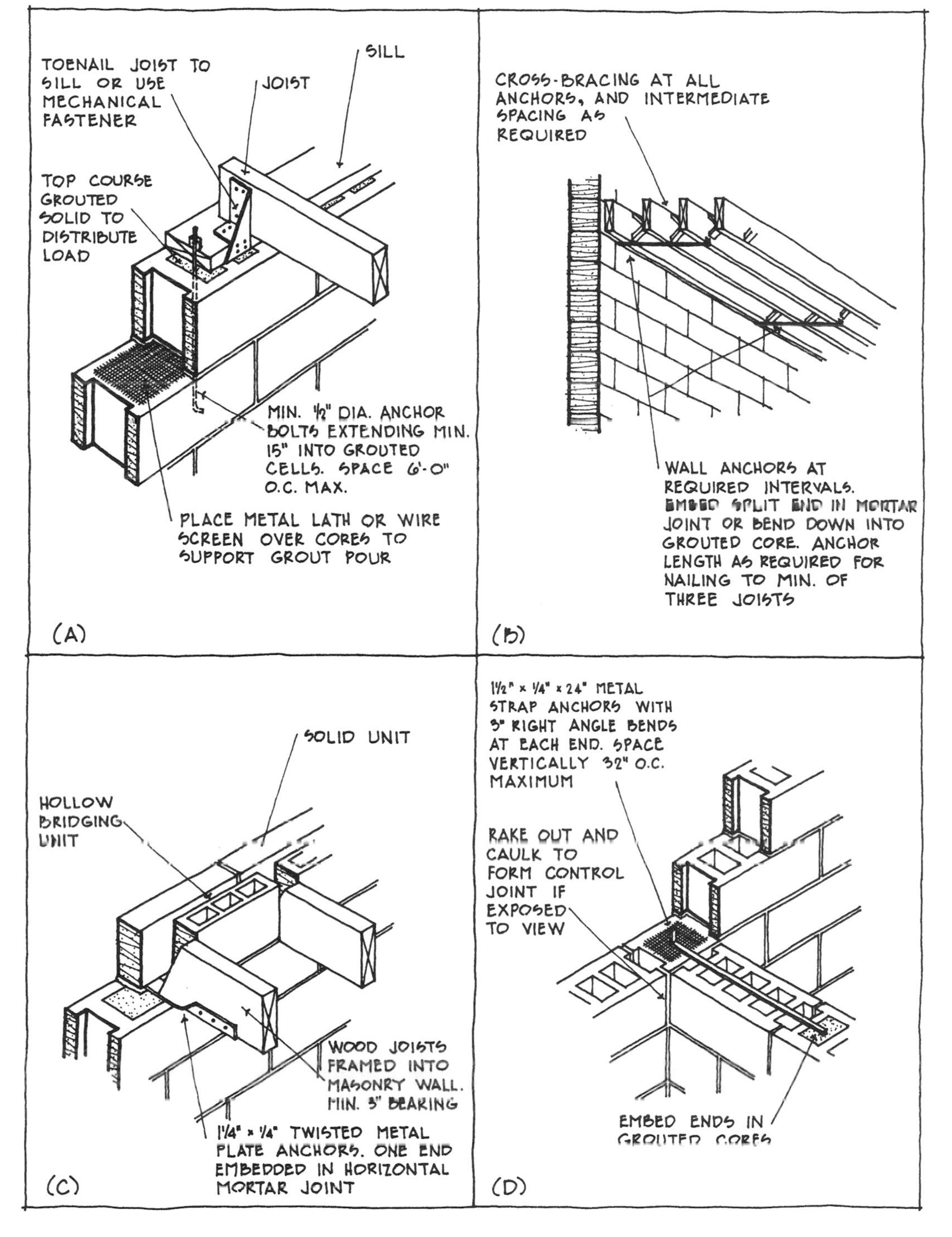

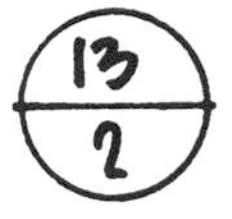

Basement wall anchorage details. (*From National Concrete Masonry Association,* TEK Bulletin 1, *NCMA, Herndon, Va.*)

Other loads applied to below-grade walls may be variable, transient, or moving, such as surcharge, wind, snow, earthquake, or impact forces. The pressures from wind that ordinarily affect basement and foundation walls are those transmitted indirectly through the superstructure: compressive, uplift, shearing, or racking loads. Stresses developed in resisting overturning are not often critical except for lightweight structures subject to high wind loads, or for structures having a high ratio of exposed area to depth in the direction of wind flow. Analytical design procedures may be used to analyze and calculate such forces so that the masonry structure will adequately resist all applied loads and induced stresses.

13.2.2 Unreinforced Walls

Unreinforced masonry basement walls may be designed by the empirical or analytical procedures described in Chapter 12. Using analytical design, exact load determinations will dictate spacing of pilasters or placement of reinforcing at critical locations. Compressive and flexural strength, slenderness coefficients, and eccentricity ratios are all considered in the analysis, and the wall designed for safety and efficiency.

13.2.3 Reinforced Walls

If unreinforced masonry walls are not adequate to resist the anticipated service loads, reinforced walls can be designed to accommodate higher compressive and flexural stresses. Reinforced walls are analytically designed so that the steel reinforcement resists flexural stresses higher than those permitted for unreinforced walls. A detailed structural analysis will determine exact requirements and criteria for placement, but for preliminary planning and estimating, tables developed by the National Concrete Masonry Association (NCMA) can be used to approximate final conditions. The information in *Fig. 13-3* covers 8-in, 10-in, and 12-in. walls with compressive strengths (f'_m) of 1200 or 1600 psi. Walls may be as much as 12 ft below grade, but only one soil pressure equivalent is included.

In the preparation of this table, it was assumed that the walls would act as simple beams supported at top and bottom in resisting lateral earth pressures, and that any vertical compressive loads could be considered as axially applied and uniformly distributed over the entire wall length at a maximum of 5000 plf. Design assumptions for materials and for allowable stresses are as listed. If the wall under consideration meets the given criteria, the table may be used to determine moment and shear and to select appropriate reinforcement.

The table in *Fig. 13-4* was developed using the same design assumptions given above, except that the combinations of steel bar sizes and spacing given will satisfy code requirements for the minimum area of steel for Seismic Zones 3 and 4. This table may be used as a design aid whenever the wall being considered meets the basic criteria for size, strength, materials, and so on.

13.2.4 Footings

Some general rules of thumb may be applied to the preliminary design of footings for below-grade concrete masonry walls: (1) cast-in-place concrete

Design Assumptions

A. Materials: 1. *Concrete Masonry Units*—Grade N hollow load-bearing units conforming to ASTM C90 except that compressive strength of units should not be less than that required for applicable f'_m; see design standard noted in B below.

2. *Mortar*—Type S conforming to ASTM C270.

3. *Grout*—Fine or coarse grout (ASTM C476) with an ultimate compressive strength (28 days) of at least 2500 psi.

4. *Reinforcement*—Deformed Billet-Steel Bars (ASTM A615); Rail-Steel Deformed Bars (A616); or Axle-Steel Deformed Bars (A617).

B. Allowable Stresses—In accordance with "Specification for the Design and Construction of Load-Bearing Concrete Masonry," NCMA.

Soil equiv.-fluid wt. = 25 pcf

Recommended Vertical Reinforcement with Axial Compressive Load (P) Equal to or Less Than 5000 lb/lin ft

8-in. Walls

Wall Height H (ft)	Wall Depth below Grade, h (ft)	Earth Pressure: Moment (in. lb/ft)	Earth Pressure: Shear (lb/ft)	f'_m = 1200 psi, Spacing of Reinforcement (in.): No. 3	No. 4	No. 5	f'_m = 1600 psi, Spacing of Reinforcement (in.): No. 3	No. 4	No. 5
10	4	2240	175	24	40	40	24	40	56
	5	3690	260		24	32	16	24	40
	6	6250	360			16		16	24
12	4	2380	180	16	24	40	16	24	40

10-in. Walls

Wall Height H (ft)	Wall Depth below Grade, h (ft)	Moment (in. lb/ft)	Shear (lb/ft)	f'_m = 1200 psi: No. 4	No. 5	No. 6	No. 7	f'_m = 1600 psi: No. 4	No. 5	No. 6	No. 7
10	6	6250	360	32	40	48	48	40	56	64	72
	7	9000	470	24	32	32	40	24	40	48	56
	8	12200	585	16	24	24	32	16	24	32	40
12	5	4300	270	40	48	56	56	40	56	72	88
	6	6900	375	24	32	40	40	24	40	48	64
	7	10100	495	16	24	32	32	16	24	32	40
14	5	4500	275	32	40	48	56	32	48	56	72
	6	7400	390	16	24	32	40	16	32	40	48

12-in. Walls

Wall Height H (ft)	Wall Depth below Grade, h (ft)	Moment (in. lb/ft)	Shear (lb/ft)	f'_m = 1200 psi: No. 4	No. 5	No. 6	No. 7	f'_m = 1600 psi: No. 4	No. 5	No. 6	No. 7
10	7	9000	470	40	48	56	56	40	56	80	80
	8	12200	585	32	40	40	40	32	48	56	64
	9	15700	710	24	32	32	32	24	40	48	48
	10	19200	835	16	24	24	32	16	32	40	40
12	6	6900	375	40	56	64	64	40	56	80	96
	7	10100	495	32	40	48	48	32	48	56	72
	8	13800	625	24	32	32	40	24	32	40	56
	9	18100	760	16	24	32	32	16	24	32	40
14	6	7400	390	32	48	56	56	32	48	64	80
	7	11000	510	24	32	40	40	24	32	48	56
	8	15200	650	16	24	32	32	16	24	32	40
16	6	7700	395	24	40	48	56	24	40	56	64
	7	11550	525	16	24	32	40	16	32	40	48
	8	16200	670		16	24	32		24	24	32

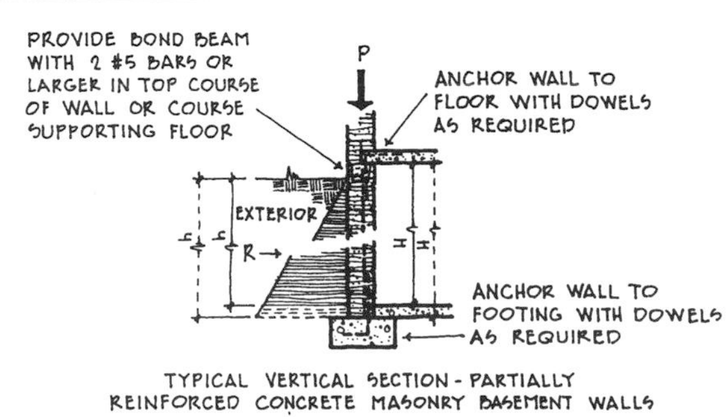

TYPICAL VERTICAL SECTION - PARTIALLY REINFORCED CONCRETE MASONRY BASEMENT WALLS

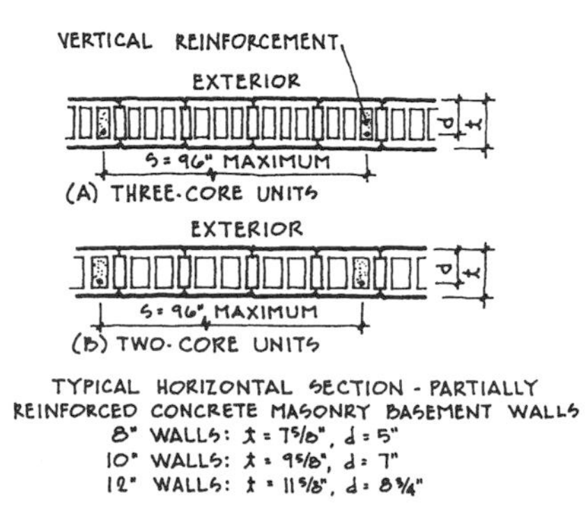

TYPICAL HORIZONTAL SECTION - PARTIALLY REINFORCED CONCRETE MASONRY BASEMENT WALLS

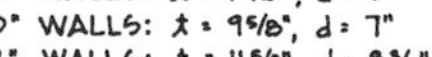

8" WALLS: t = 7 5/8", d = 5"
10" WALLS: t = 9 5/8", d = 7"
12" WALLS: t = 11 5/8", d = 8 3/4"

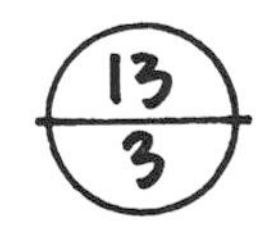

Design table for reinforced CMU basement walls in Seismic Zones 0, 1, and 2. (*From National Concrete Masonry Association*, TEK Bulletin 56, *NCMA, Herndon, Va.*)

Design Assumptions

A. Materials: 1. *Concrete Masonry Units*—Grade N hollow load-bearing units conforming to ASTM C90 except that compressive strength of units should be not less than that required for applicable f'_m; see design standard noted in B below.

2. *Mortar*—Type S conforming to ASTM C270.

3. *Grout*—Fine or coarse grout (ASTM C476) with an ultimate compressive strength (28 days) of at least 2500 psi.

4. *Reinforcement*—Deformed Billet-Steel Bars (ASTM A615); Rail-Steel Deformed Bars (A616); or Axle-Steel Deformed Bars (A617).

B. Allowable Stresses—In accordance with "Specification for the Design and Construction of Load-Bearing Concrete Masonry," NCMA.

Soil equiv.-fluid wt. = 25 pcf. Recommended Vertical Reinforcement with Axial Compressive Load (P) Equal to or Less Than 5000 lb/lin ft

8-in. Walls

Wall Height H (ft)	Wall Depth below Grade, h (ft)	Earth Pressure Moment (in. lb/ft)	Earth Pressure Shear (lb/ft)	$f'_m = 1200$ psi, Spacing of Reinforcement (in.) No. 4	No. 5	No. 6	No. 7	$f'_m = 1600$ psi, Spacing of Reinforcement (in.) No. 4	No. 5	No. 6	No. 7
10	5	3690	260	16	24	32	40	16	24	40	48
	6	6250	360		16	24	24	16	24	32	40
	7	9000	470			16	16		16	24	24
	8	12200	585								16
12	5	4300	270	16	24	32	32	16	24	32	40
	6	6900	375			16	24		16	24	24

10-in. Walls

Wall Height H (ft)	Wall Depth below Grade, h (ft)	Earth Pressure Moment (in. lb/ft)	Earth Pressure Shear (lb/ft)	$f'_m = 1200$ psi, Spacing of Reinforcement (in.) No. 5	No. 6	No. 7	$f'_m = 1600$ psi, Spacing of Reinforcement (in.) No. 5	No. 6	No. 7
10	7	9000	470	24	32	40	24	32	40
	8	12200	585	24	24	32	24	32	40
	9	15700	710	16	24	24	24	24	32
	10	19200	835		16	16	16	24	24
12	7	10100	495	24	32	32	24	32	40
	8	13800	625	16	24	24	16	24	32
	9	18100	760		16	16		16	24
14	6	7400	390	24	32	40	24	32	40
	7	11000	510	16	24	32	16	24	32
	8	15200	650		16	24		16	24

12-in. Walls

Wall Height H (ft)	Wall Depth below Grade, h (ft)	Earth Pressure Moment (in. lb/ft)	Earth Pressure Shear (lb/ft)	$f'_m = 1200$ psi, Spacing of Reinforcement (in.) No. 5	No. 6	No. 7	No. 8	$f'_m = 1600$ psi, Spacing of Reinforcement (in.) No. 5	No. 6	No. 7	No. 8
10	8	12200	585	16	24	32	48	16	24	32	48
	9	15700	710	16	24	32	40	16	24	32	48
	10	19200	835	16	24	32	32	16	24	32	48
12	8	13800	625	16	24	32	40	16	24	32	48
	9	18100	760	16	24	32	32	16	24	32	48
	10	22800	905	16	24	24	24	16	24	32	40
	11	28000	1050		16	16	24	16	24	24	32
	12	33100	1200						16	24	24
14	8	15200	650	16	24	32	32	16	24	32	48
	9	20300	800	16	24	24	24	16	24	32	40
	10	26000	955		16	24	24		16	24	32
	11	31800	1115				16			16	24
16	8	16200	670	16	24	32	32	16	24	32	40
	9	21800	825		16	24	24		16	24	32
	10	28200	990			16	16			16	24

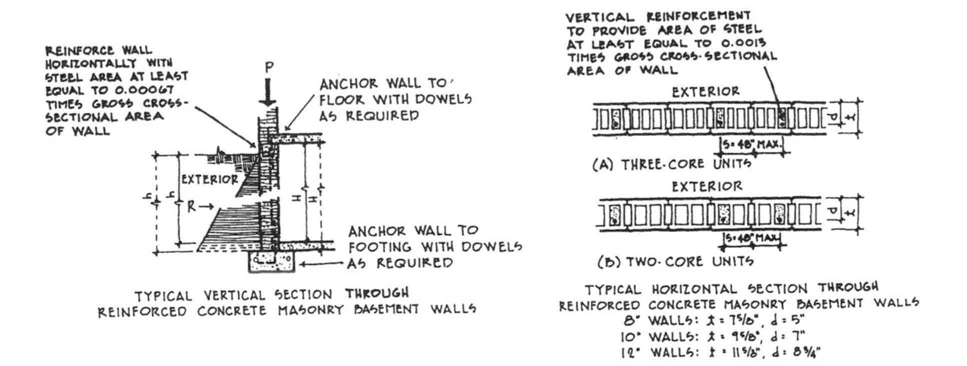

Design table for reinforced CMU basement walls in Seismic Zones 3 and 4. (*From National Concrete Masonry Association,* TEK Bulletin 56, *NCMA, Herndon, Va.*)

should have a minimum compressive strength of 2000 psi, (2) footing depth should be equal to the wall thickness and footing width equal to twice the wall thickness (*see Fig. 13-1*), and (3) the bottom of the footing should be placed in undisturbed soil below the frost line. Allowable soil bearing pressure, of course, must be checked against actual loads in the final design.

13.2.5 Drainage and Waterproofing

Proper drainage and waterproofing of below-grade walls is essential, not only to prevent buildup of hydrostatic pressure, but also to maintain dry conditions and eliminate dampness in interior spaces.

A 4- to 6-in. bed of crushed stone or gravel should be laid inside the footing covering most of the slab area, with the thickness increased at the edges (*see Fig. 13-5*). Drain tiles should be installed in the thicker perimeter areas as shown. These drains should lead to a lower surface outlet or to a storm sewer. Weepholes should be installed every 32 in. to drain water into the crushed stone over the drain tile. Floor slabs should be cast at a level above the weepholes.

The exterior of the walls should be waterproofed. Alternative procedures include the use of a heavy troweled-on coat of fiber-reinforced asphalt mastic, built-up membranes of fabric and mastic, liquid-applied polymer membranes, bentonite clay waterproofing, or EPDM membranes. No pinholes, cracks, or open joints should be permitted. The finish grade adjacent to a below-grade wall should slope away from the building to prohibit the accumulation of rainwater or other moisture and prevent its seepage down along the surface of the wall.

Damp or wet conditions in basements are often mistakenly attributed to seepage of moisture through the walls. The trouble is more often due to condensation of moisture from the air inside the space. Condensation is most frequent in warm, humid areas, but can occur wherever the temperature and relative humidity of the inside air are maintained at high levels, either artificially or due to atmospheric conditions. Water vapor will form anytime the surface temperature of the wall is below the dew-point temperature of the interior air.

Condensation can be controlled by regulating ventilation or by the proper location and installation of wall insulation (see Chapters 8 and 14).

13.3 RETAINING WALLS

Because retaining walls are often used in landscape applications and may not enclose habitable space, attention to design and detailing is often cursory. However, the walls are exposed to extremes of weather, are in contact with earth, often saturated with moisture, and must resist significant lateral forces. Such severity of use and exposure demands careful attention to design and details.

13.3.1 Traditional Retaining Wall Types

There are four basic types of traditional masonry retaining walls (*see Fig. 13-6*). Reinforced cantilever walls *B* offer the most economical design, and are most commonly used. The vertical stem is reinforced to resist tensile stresses. The concrete footing anchors the stem and resists overturning and sliding due to both vertical and lateral forces.

Proprietary systems of interlocking concrete masonry units also offer

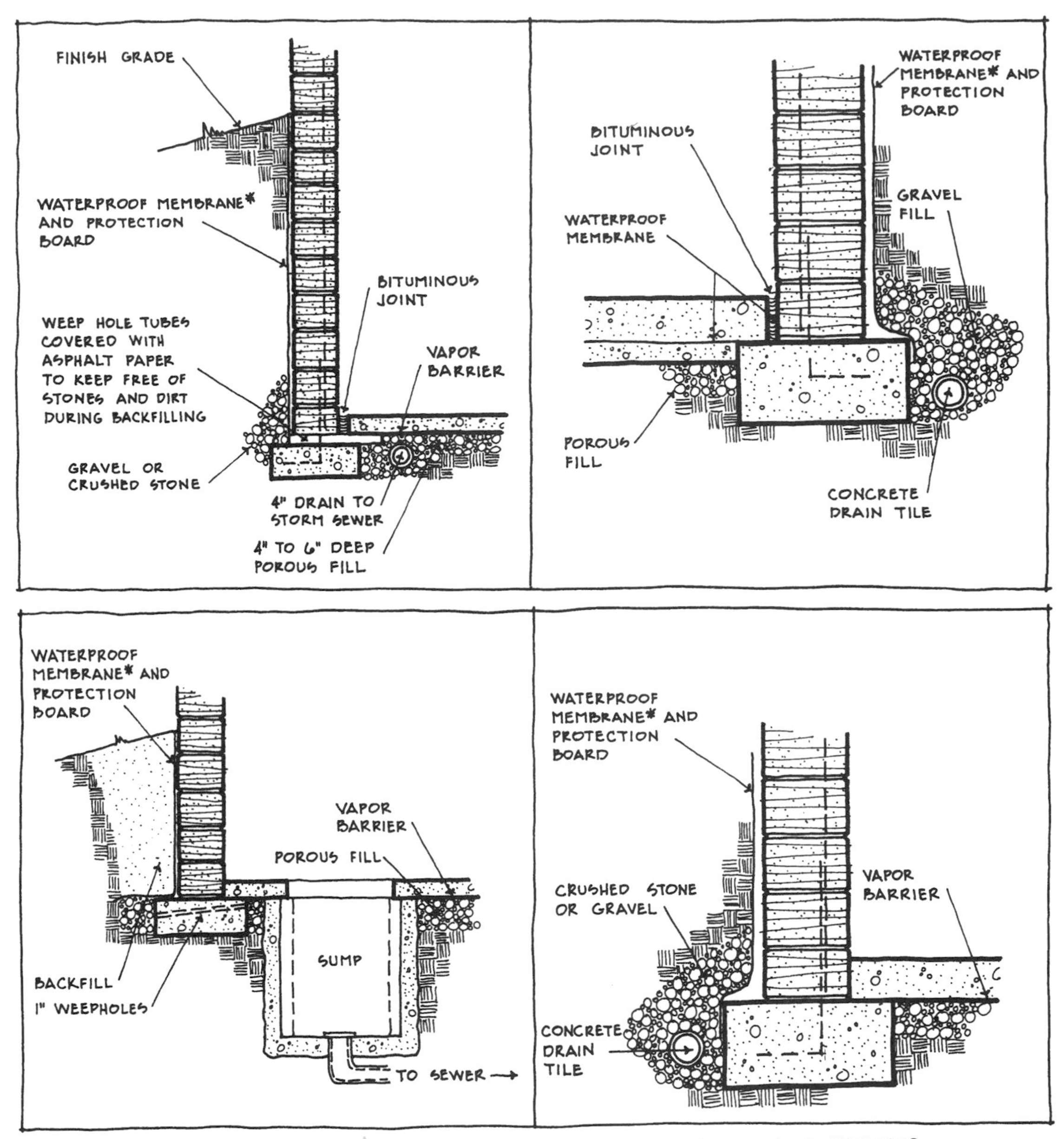

* WATERPROOF MEMBRANE MAY BE PARGE OR PLASTER COAT WITH BITUMINOUS MEMBRANE SURFACE; FIBER-REINFORCED ASPHALT MASTIC; LIQUID-APPLIED POLYMER MEMBRANE; BENTONITE CLAY; EPDM MEMBRANE; OR OTHER SUITABLE MATERIAL. USE PROTECTION BOARD WHERE REQUIRED TO PREVENT TEARS OR PUNCTURES DURING BACKFILLING OPERATIONS.

Basement wall drainage. (*From National Concrete Masonry Association,* TEK Bulletins 1, 43, *NCMA, Herndon, Va.*)

economical and attractive solutions for unreinforced retaining wall applications of moderate height.

Some of the primary considerations in retaining wall design should be (1) a proper cap or coping to prevent water collecting or standing on top of the wall, (2) a waterproof coating on the back of the wall to prevent saturating the masonry, (3) permeable backfill behind the wall to collect water and prevent soil saturation and increased hydrostatic pressure, (4) weep-

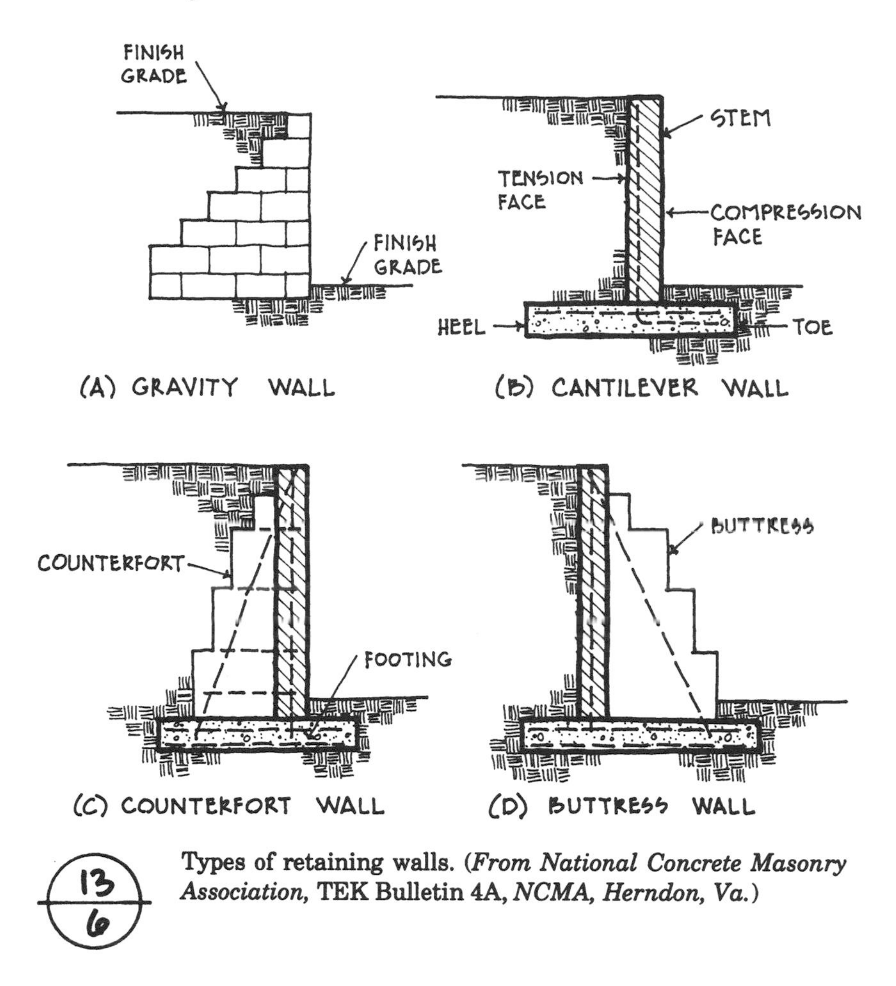

Types of retaining walls. (*From National Concrete Masonry Association,* TEK Bulletin 4A, *NCMA, Herndon, Va.*)

holes or drain lines to drain moisture, and (5) expansion or control joints to permit longitudinal thermal and moisture movement.

13.3.2 Cantilever Walls

Reinforced concrete masonry unit (CMU) cantilever retaining walls are designed to resist overturning and sliding with resultant forces that fall within the middle third of the footing. Many design tables have been developed to simplify selection of wall dimensions and steel reinforcing in CMU retaining walls, including those by the NCMA and the Masonry Institute of America. Graphic examples of design parameters for 4- to 10-ft-high walls are given in Newman's *Standard Structural Details for Building Construction.* The tables and drawings in *Fig. 13-7* represent the most commonly recommended solutions.

Figure 13-8 shows three different methods of locating vertical reinforcement in a brick masonry cantilever retaining wall. Each offers certain advantages depending on wall thickness, bar spacing, and unit type. Where 8-in. hollow units are available, as shown in part (*B*), they can often be less expensive than a double-wythe wall. If only standard units are available, grout pockets may be used (part *A*). A double-wythe grouted cavity wall, however, offers greatest flexibility because bar spacing is not limited by the fixed dimensions of the units (part *C*).

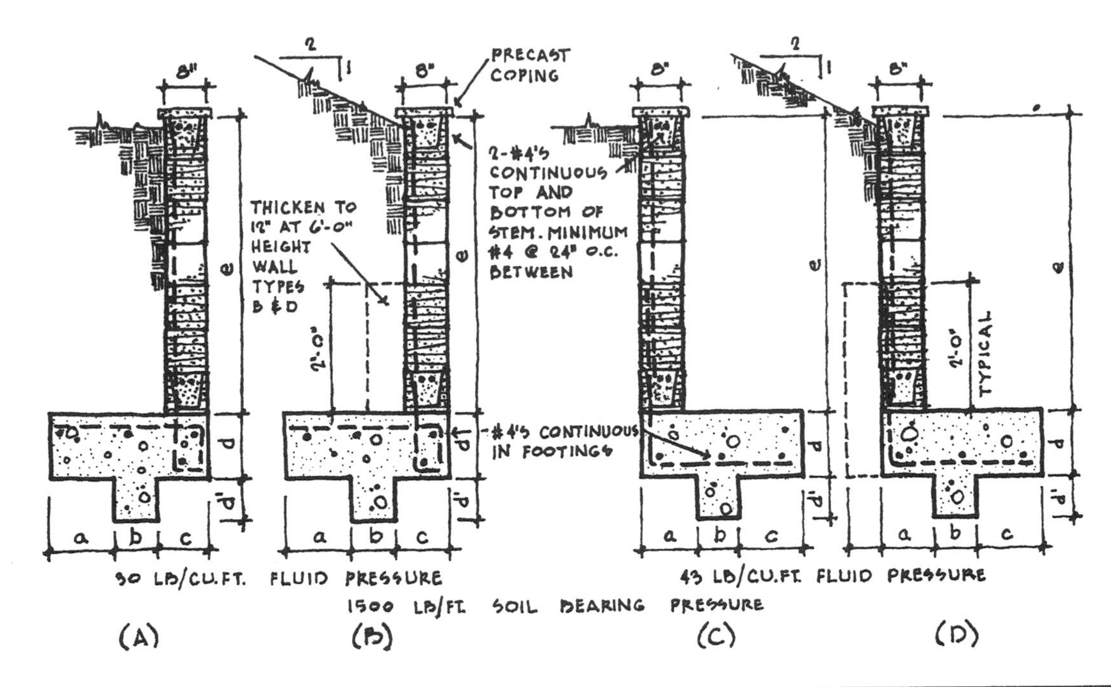

	Wall type			
	A	B	C	D
Maximum stem height, *e*, 4'-0"				
Footing dimensions				
a	1'-4"	1'-2"	1'-6"	1'-0"
b	8"	1'-0"	8"	1'-0"
c	6"	1'-0"	6"	1'-2"
d/d'	12"/12"	12"/12"	12"/8"	12"/12"
Vertical stem reinforcing	No. 4's at 24"	No. 4's at 16"	No. 4's at 16"	No. 4's at 16"
dowels	No. 4's at 24"	No 4's at 16"	No. 4's at 16"	No. 4's at 16"
Maximum stem height, *e*, 6'0"				
Footing dimensions				
a	1'-10"	3'-8"	2'-0"	2'-0"
b	12"	12"	8"	1'-0"
c	1'-0"	6"	10"	1'-0"
d/d'	12"/12"	1'-4"/1'-0"	12"/8"	12"/12"
Vertical stem reinforcing	No. 5's at 24"	No. 5's at 16"	No. 4's at 16"	No. 5's at 16"
dowels	No. 5's at 16"	No. 5's at 8"	No. 4's at 16"	No. 5's at 16"

Design of CMU cantilever retaining walls.

The Brick Institute of America (BIA) recommends horizontal truss or ladder-type reinforcement at a maximum of 16 in. on center vertically. In grouted cavity walls, one No. 3 bar at 20 in. on center vertically may be substituted.

The table in *Fig. 13-9* was developed for preliminary design of brick walls with a maximum height of 6 ft. The following assumptions are used, and materials must meet these requirements: (1) brick strength must be in excess of 6000 psi in compression; (2) steel design tensile strength must be

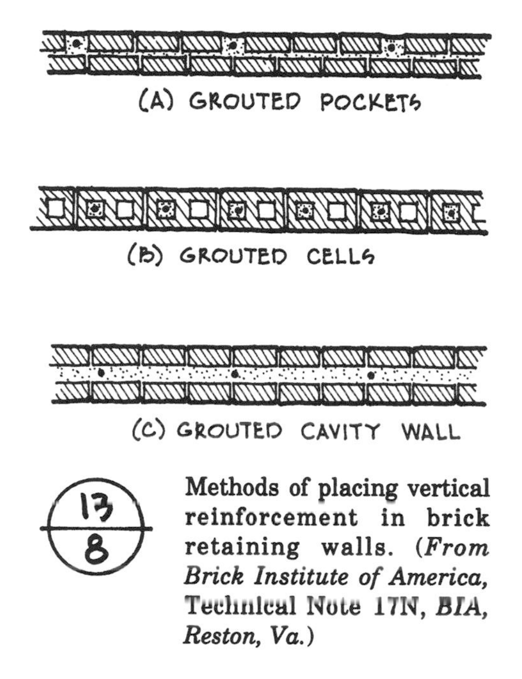

Methods of placing vertical reinforcement in brick retaining walls. *(From Brick Institute of America,* **Technical Note 17N,** *BIA, Reston, Va.)*

20,000 psi; (3) no surcharge is permitted; and (4) wall thickness is 10 in. ($9\frac{1}{2}$ in. actual thickness) with reinforcing steel located at the center line of the wall.

For significantly taller walls (6 to 20 ft), the details and tables in *Figs. 13-10* and *13-11* are used. Grout pockets for these walls should generally be spaced 4 ft on center except when adjacent to expansion joints. Expansion joints (at wall offsets and every 30 ft maximum) should be centered between pockets spaced no more than 3 ft apart.

The bottom of the footing must be below the frost line and at an elevation where the soil is of sufficient strength to withstand the toe pressure shown in the tables. If the footing elevation must vary, steps must be located adjacent to a grout pocket with the pocket extending to the top of the lower footing. Once the elevation for the bottom of the footing has been established, the $(H + D)$ dimension to finished grade on the high side is known, and the H dimension for entering the design tables may be determined.

All brick retaining walls should be laid in running bond pattern with masonry headers or metal ties every sixth course.

13.3.3 Footings

Concrete footings for retaining walls should be placed on firm undisturbed soil. In areas subject to freezing, they should also be placed below the frost line to avoid heave and possible damage to the wall. If the soil under the footing consists of soft or silty clay, it may be necessary to place compacted fill before pouring the concrete.

13.3.4 Drainage and Waterproofing

Failure to drain the backfill area behind retaining walls causes a buildup of hydrostatic pressure which can quickly become critical if rainfall is

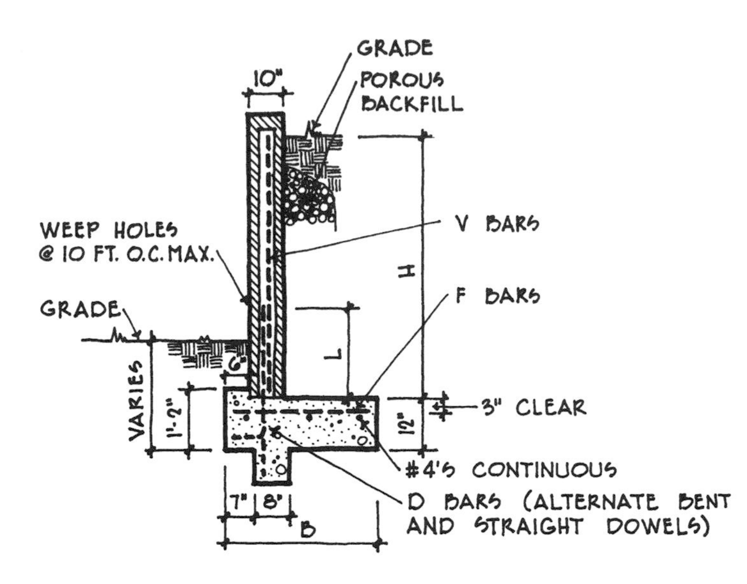

H	B	L	D-bars	V-bars	F-bars
2′-0″	1′-9″	1′-10″	No. 3 at 40″	—	No. 3 at 40″
2′-6″	1′-9″	2′-4″	No. 3 at 40″	—	No. 3 at 40″
3′-0″	2′-0″	2′-10″	No. 3 at 40″	—	No. 3 at 40″
3′-6″	2′-0″	3′-4″	No. 3 at 40″	—	No. 3 at 40″
4′-0″	2′-4″	1′-4″	No. 3 at 27″	No. 3 at 27″	No. 3 at 27″
			No. 4 at 40″	No. 3 at 40″	No. 3 at 40″
4′-6″	2′-8″	1′-6″	No. 3 at 19″	No. 3 at 38″	No. 3 at 19″
			No. 4 at 35″	No. 3 at 35″	No. 3 at 35″
5′-0″	3′-0″	1′-8″	No. 3 at 14″	No. 3 at 28″	No. 3 at 14″
			No. 4 at 25″	No. 3 at 25″	No. 3 at 25″
			No. 5 at 40″	No. 4 at 40″	No. 4 at 40″
5′-6″	3′-3″	1′-10″	No. 3 at 11″	No. 3 at 22″	No. 3 at 11″
			No. 4 at 20″	No. 4 at 40″	No. 3 at 20″
			No. 5 at 31″	No. 4 at 31″	No. 4 at 31″
6′-0″	3′-6″	2′-0″	No. 3 at 8″	No. 3 at 16″	No. 3 at 8″
			No. 4 at 14″	No. 4 at 28″	No. 3 at 14″
			No. 5 at 20″	No. 5 at 40″	No. 4 at 20″

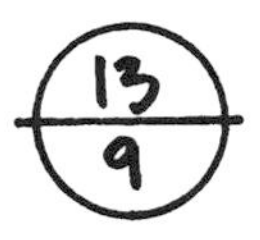

Simplified design of brick cantilever retaining walls and footings. (*From Brick Institute of America,* Technical Note 17N, *BIA, Reston, Va.*)

heavy. In mild climates, weepholes at the base of the wall should be provided at 4- to 8-ft intervals. In areas where precipitation is heavy or where poor drainage conditions exist, prolonged seepage through weepholes can cause the soil in front of the wall and under the toe of the footing to become saturated and lose some of its bearing capacity. In these instances, a continuous longitudinal drain of perforated pipe should be placed near the base with discharge areas located beyond the ends of the wall (*see Fig. 13-12*).

BIA recommends that backfill against brick retaining walls from the top of the footing to within 12 in. of finished grade should be coarse gravel, 2 ft wide, and run the entire length of the wall. To prevent the infiltration of fine fill material or top soil, a layer of 50-lb roofing felt is laid along the

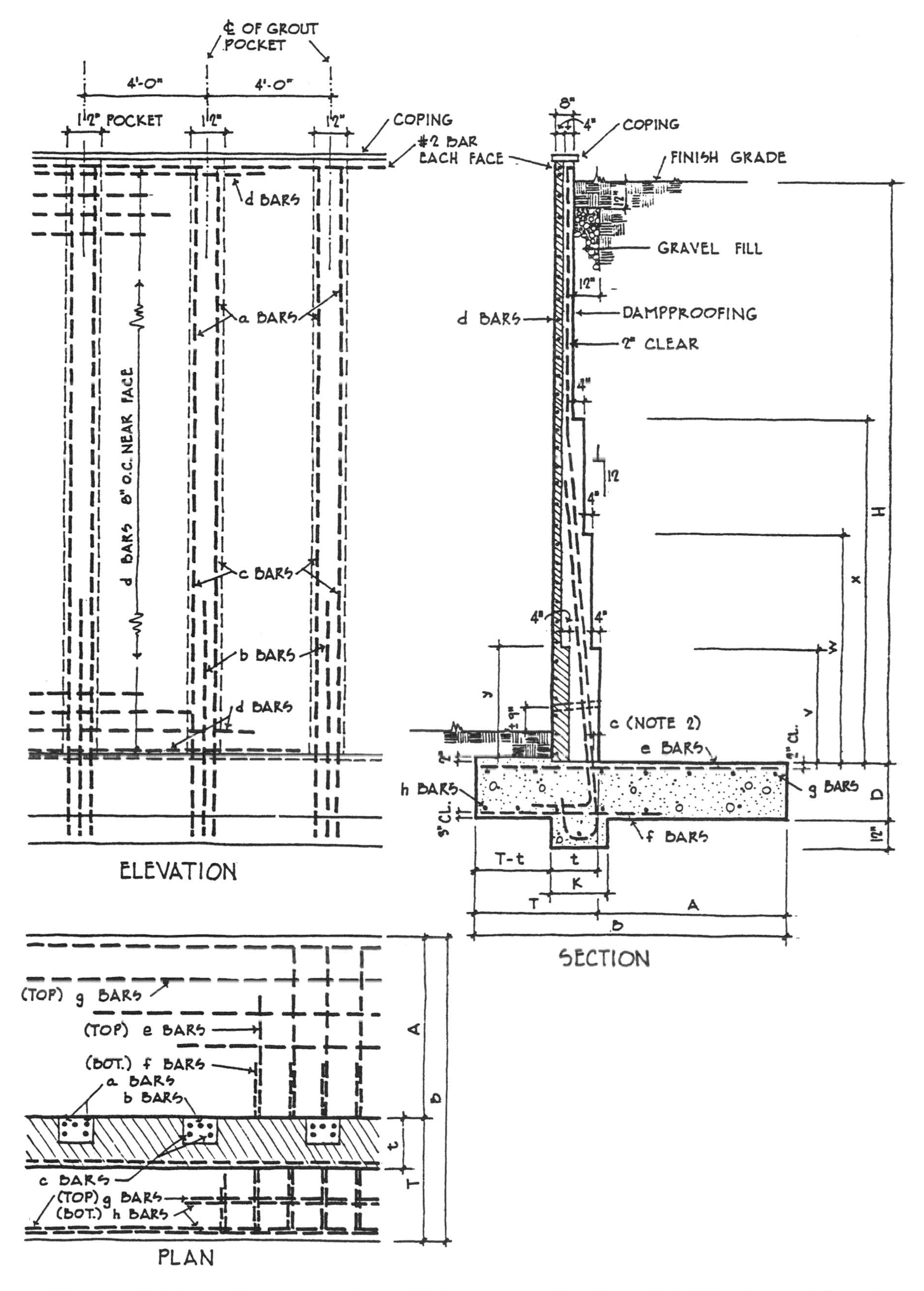

Brick cantilever retaining wall design example. (*From Brick Institute of America,* Technical Note 17G, *BIA, Reston, Va.*)

TABLE A STEM									TABLE B FOOTING							
Dimensions							Loads		Dimensions					Loads		
H (ft)	Nom. t (in.)	c (in.)	v (ft)	w (ft)	x (ft)	y (ft)	V (kips per 4 ft)	M (ft-kips per 4 ft)	B	T	A	D	K	P (kips per ft)	M_{OT} (ft-kips per ft)	Toe pressure (kips/sq ft)
6	8	2	—	—	—	—	2.06	4.13	3′-6″	1′-3″	2′-3″	1′-0″	—	0.70	1.63	1.33
7	8	2	—	—	—	—	2.80	6.55	4′-0″	1′-3″	2′-9″	1′-0″	—	0.92	2.45	1.54
8	8	2	—	—	—	—	3.66	9.78	4′-6″	1′-3″	3′-3″	1′-0″	—	1.16	3.48	1.74
9	12	5	—	—	1	—	4.63	13.92	5′-0″	1′-6″	3′-6″	1′-0″	—	1.43	4.76	2.00
10	12	4	—	—	2	—	5.72	19.10	5′-6″	1′-9″	3′-9″	1′-0″	—	1.73	6.35	2.11
11	12	3	—	—	3	—	6.92	25.42	6′-0″	2′-0″	4′-0″	1′-3″	1′-3″	2.14	8.75	2.31
12	12	2	—	—	4	—	8.24	33.00	6′-6″	2′-3″	4′-3″	1′-3″	1′-3″	2.52	11.10	2.37
13	16	5	—	1	5	—	9.67	41.96	7′-0″	2′-6″	4′-6″	1′-3″	1′-3″	2.90	13.80	2.62
14	16	4	—	2	6	—	11.21	52.41	7′-6″	2′-9″	4′-9″	1′-6″	1′-6″	3.43	17.70	2.82
15	16	3	—	3	7	—	12.87	64.46	8′-0″	3′-0″	5′-0″	1′-6″	1′-6″	3.89	21.40	2.87
16	16	2	—	4	8	—	14.64	78.23	8′-6″	3′-3″	5′-3″	1′-6″	1′-6″	4.51	26.70	3.04
17	20	5	1	5	9	1	16.53	93.84	9′-0″	3′-6″	5′-6″	1′-9″	1′-9″	5.03	31.40	3.32
18	20	4	2	6	10	2	18.53	111.39	9′-6″	3′-9″	5′-9″	1′-9″	1′-9″	5.72	38.20	3.49
19	20	3	3	7	11	3	20.65	131.01	10′-0″	4′-0″	6′-0″	1′-9″	1′-9″	6.15	42.50	3.52
20	20	2	4	8	12	4	22.88	152.80	10′-6″	4′-3″	6′-3″	2′-0″	2′-0″	6.92	50.80	3.65

TABLE C REINFORCEMENT																
Bars	a (2 per pocket)		b (1 per pocket)		c (2 per pocket)		d		e		f		g		h	
H (ft)	Bar size	Extension above top of footing	Bar size	Extension above top of footing	Bar size	Extension above top of footing	Bar size	No. of bars	Bar size	Length of bar	Bar size	Length of bar	Bar size	No. of bars	Bar size	No. of bars
6	#5	Full height of pocket	—	—	—	—	#2	10	#3	3′-2″	#3	2′-3″	#3	4	#3	3
7	#6		—	—	—	—	#2	11	#4	3′-8″	#3	2′-3″	#3	4	#3	3
8	#6		#6	3′-0″	—	—	#2	13	#4	4′-2″	#3	2′-3″	#3	4	#3	3
9	#7		#6	3′-0″	—	—	#2	14	#5	4′-8″	#3	2′-6″	#3	4	#3	3
10	#7		#7	4′-0″	—	—	#2	16	#5	5′-2″	#3	2′-9″	#3	5	#3	4
11	#8		#7	4′-0″	—	—	#2	17	#6	5′-8″	#3	3′-0″	#3	5	#3	5
12	#8		#8	5′-0″	—	—	#2	19	#6	6′-2″	#3	3′-3″	#3	5	#3	5
13	#8		#8	6′-0″	—	—	#2	20	#7	6′-8″	#4	3′-9″	#4	5	#4	5
14	#9		#10	6′-0″	—	—	#2	22	#7	7′-2″	#4	4′-0″	#4	6	#4	5
15	#10		#9	5′-0″	—	—	#2	23	#7	7′-8″	#4	4′-3″	#4	6	#4	5
16	#10		#11	6′-6″	—	—	#2	25	#8	8′-2″	#5	4′-9″	#4	6	#4	5
17	#10		#8	4′-0″	#8	7′-0″	#2	26	#8	8′-8″	#5	5′-0″	#4	6	#4	5
18	#10		#9	4′-0″	#9	8′-0″	#2	28	#8	9′-2″	#5	5′-3″	#4	7	#4	6
19	#10		#9	4′-0″	#10	9′-0″	#2	29	#9	9′-8″	#6	6′-0″	#4	7	#4	6
20	#10		#11	5′-6″	#10	10′-0″	#2	31	#9	10′-2″	#6	6′-3″	#4	7	#4	6

Design table for grouted, reinforced brick masonry cantilever retaining walls. (*From Brick Institute of America,* Technical Note 17G, *BIA, Reston, Va.*)

top of this course. Weepholes or drain lines at the bottom of the wall to relieve moisture buildup in the gravel fill should extend the full length of the wall.

Waterproofing requirements for the back face of a retaining wall will depend on the climate, soil conditions, and type of masonry units used.

Seepage through a brick wall can cause efflorescence if soluble salts are present, but a waterproof membrane will prevent this water movement. Walls of porous concrete units should always receive waterproof backing because of the excessive expansion and contraction that accompanies variable moisture content. In climates subject to freezing, a waterproof membrane can prevent the potentially destructive action of freeze-thaw cycles when moisture is present in the units.

13.3.5 Segmental Retaining Walls

One of the newest developments in the concrete masonry industry is the dry-stacked, interlocking concrete block retaining wall, sometimes referred to as the segmental retaining wall (SRW). A variety of proprietary units and systems are available (refer to *Fig. 4-18*). The units are stepped back slightly in each course, or battered, toward the embankment. Some units interlock simply by their shape, while others use pins or dowels to connect successive courses. Because they are dry-stacked without mortar, interlocking retaining wall systems are simple and fast to install. The open joints allow free drainage of soil moisture, and the stepped-back designs reduce overturning stresses.

The National Concrete Masonry Association has published the *Design Manual for Segmental Retaining Walls,* which presents a thorough engineering methodology, guide specifications, test methods, and design tables for two types of retaining walls. *Conventional SRWs* are structures that resist external destabilizing forces from the retained soils solely through self-weight and batter of the SRW units (*see Fig. 13-13*). Conventional SRWs may be either single or multiple unit depths. *Soil-reinforced SRWs* are composite systems consisting of SRW units in combination with a mass of retained soil stabilized by horizontal layers of geosynthetic reinforcement materials (*see Fig. 13-14*).

Because they are dry-stacked, segmental retaining walls are relatively flexible, and can tolerate movement and settlement without distress. Typically supported on flexible aggregate leveling-bed foundations, SRWs also permit water to drain directly through the face of the wall, preventing any buildup of hydrostatic pressure. Supplemental drainage is provided by gravel backfill and collection pipes at the base of the wall. The maximum height that can be constructed using a single-unit-depth conventional SRW is directly proportional to its weight, width, and vertical batter for any given soil type and site geometry conditions. The height can be increased by using multiple unit depths. Soil-reinforced SRWs use geosynthetic reinforcement to enlarge the effective width and weight of the gravity mass. The reinforcement (either geogrids or geotextiles) extends through the interface between the SRW units and into the soil to create a composite gravity mass structure. This composite structure offers increased resistance for taller walls, surcharged structures, or more difficult soil conditions.

Both conventional and soil-reinforced SRWs function as gravity retaining walls. To be stable, a gravity retaining wall must have sufficient weight (mass) and width to resist both sliding at the base and overturning of the mass about the toe of the structure. Stability calculations that involve forces acting on the boundary of the gravity structure are called external stability calculations. For soil-reinforced SRWs and multiple-depth conventional SRWs, a set of internal stability calculations are also

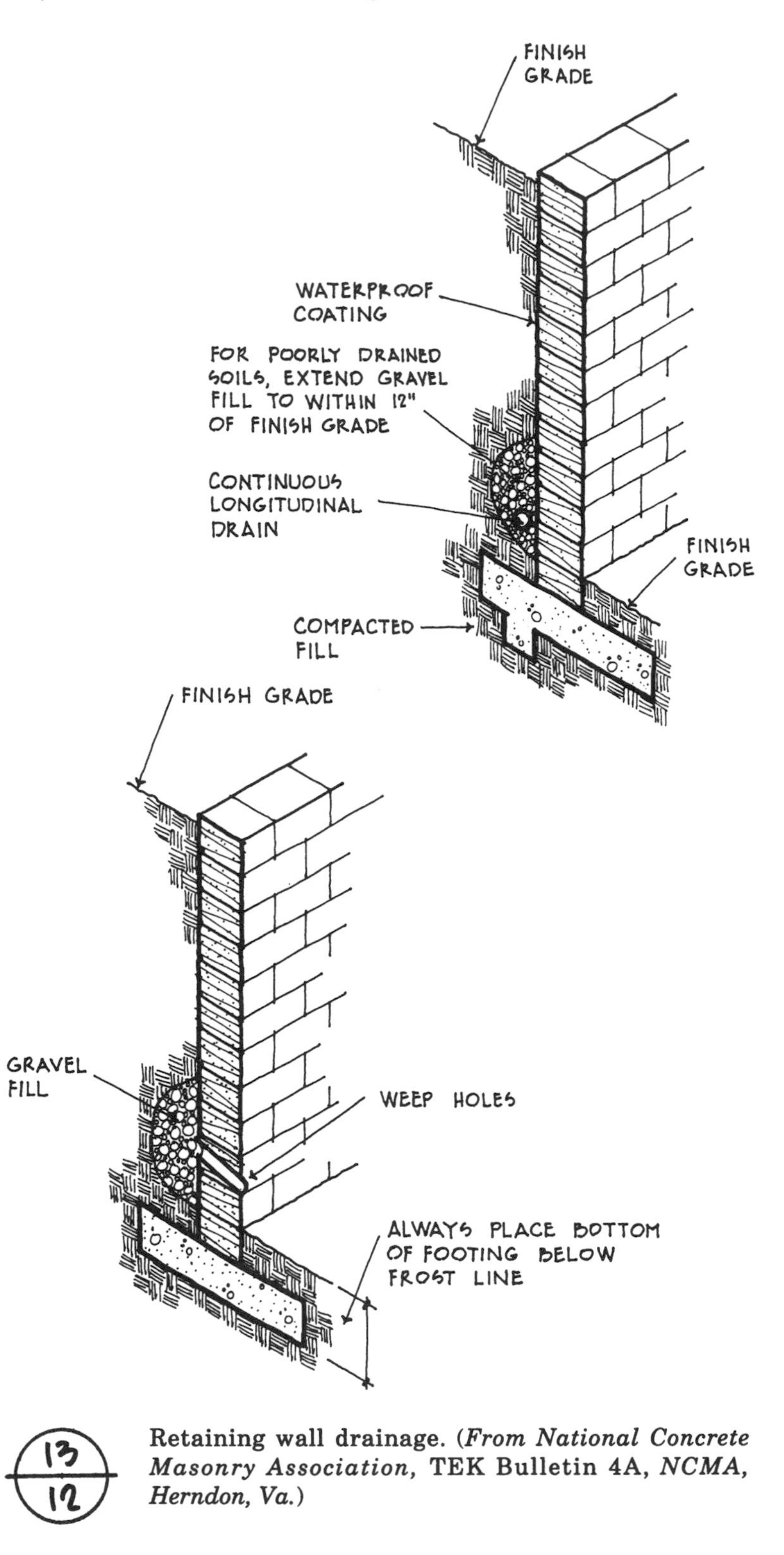

Retaining wall drainage. (*From National Concrete Masonry Association,* **TEK Bulletin 4A,** *NCMA, Herndon, Va.*)

required to ensure that there is adequate strength and width to create the stable monolithic gravity mass. The local stability of the dry-stacked column of units must also be analyzed. The NCMA *Design Manual for Segmental Retaining Walls* provides complete engineering calculations, soil information, test methods, guide specifications, and design tables.

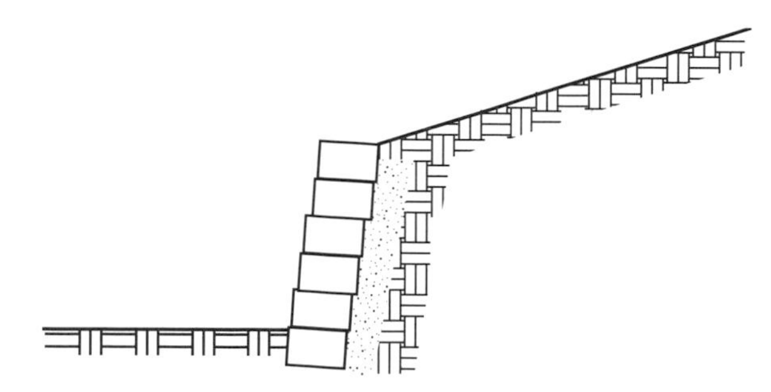

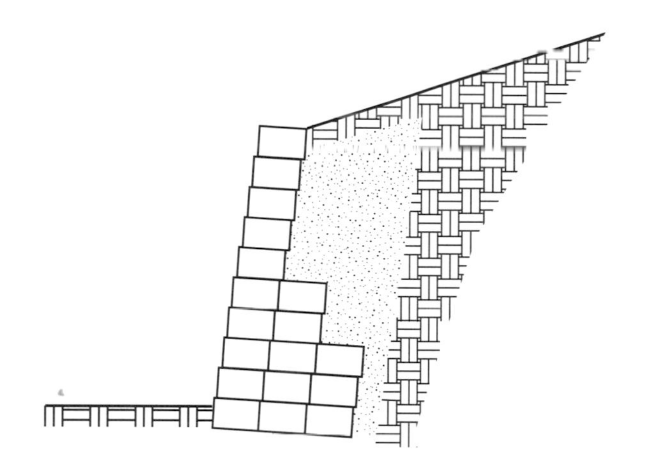

Conventional segmental retaining wall. (*From National Concrete Masonry Association,* Design for Segmental Retaining Walls, *NCMA, Herndon, Va., 1993*)

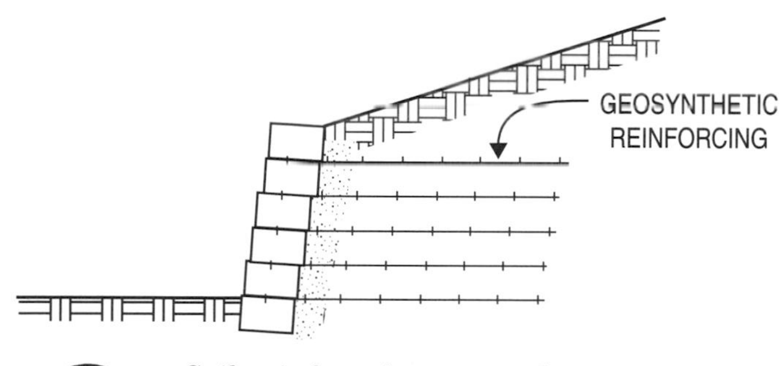

Soil-reinforced segmental retaining wall. (*From National Concrete Masonry Association,* Design for Segmental Retaining Walls, *NCMA, Herndon, Va., 1993*)

14 MASONRY PAVING AND FIREPLACES

Two of the most common nonstructural masonry applications are paving and fireplaces. Masonry's noncombustibility and heat storage capacity have always made it the material of choice for functional fireplaces, and its aesthetic warmth is symbolic even in purely decorative nonfunctional hearths.

Roads made of stone paving blocks were built by the Romans over 2000 years ago, some of which are still in use today. Many cities in the United States and Europe also have brick streets which continue in service after many years of heavy use, and have proven both durable and easy to maintain. After World War II, economical methods of manufacturing high-strength concrete paving units were developed. Since their introduction, concrete pavers have been used extensively in Europe and increasingly in the United States and Canada.

14.1 MASONRY PAVING

Clay, concrete, and stone paving units may all be used for interior or exterior applications, and may be installed over different sub-bases suitable for residential and commercial buildings, walkways, patios, streets, and parking areas. Paving assemblies are classified in accordance with the type of base used and the rigidity or flexibility of the paving itself. *Rigid paving* is defined as units laid in a bed of mortar and with mortar joints between the units. *Flexible paving* contains no mortar below or between the units.

Base supports may be rigid, semirigid, flexible, or suspended. A *rigid base diaphragm* is a reinforced concrete slab on grade, and can accommodate either rigid or flexible paving. A *semirigid continuous base* usually consists of asphalt or other bituminous road pavement, and is suitable for flexible paving. A *flexible base* is compacted gravel or a damp, loose, sand-cement mixture which is tamped in place. Only flexible paving should be laid over this type of base. *Suspended diaphragm* bases are structural floor or roof deck assemblies, the composition of which will vary depending on the type of structural system used. Either flexible or rigid paving may be used on suspended decks.

Selection of the type of paving system to be used will depend to a large extent on the desired aesthetic effect and the intended use. There are a number of design considerations which must be taken into account, particularly for outdoor paving. Heavy vehicular traffic generally requires rigid concrete diaphragms or semirigid asphalt bases. Lighter vehicular and pedestrian traffic may be supported on flexible bases and flexible paving. Traffic patterns, which dictate the size and shape of a paved area, may also influence the choice of base and cushion material. Successful installations always depend on proper subgrade preparation and removal of all vegetation and organic materials from the area to be paved. Soft spots of poor soil should be removed and filled with suitable material, then properly compacted.

Site preparation and system selection should also take into consideration the location of underground utilities and storm drainage systems. With rigid concrete bases and rigid masonry paving, access must be provided by means of manholes, cleanout plugs, and so on. If semirigid or flexible bases are used with flexible paving, however, the user may gain unlimited access to underground pipes and cables without incurring the expense of extensive surface repairs. This fact is generally cited as one of the major advantages of flexible masonry paving, which allows utility repairs and alterations by simply removing, stockpiling, and then replacing the paving units and base material. No air hammers or concrete pours are required to complete the work, and there is reduced danger of damage to utilities by the elimination of such equipment.

14.1.1 Outdoor Paving

Drainage is very important in the consideration of all outdoor paving systems, and excessive runoff is a legitimate environmental concern. In addition to mortarless paving systems which permit a degree of water percolation through the joints, masonry paving units have been developed which lessen the impact of storm drainage even further. These concrete masonry grid pavers (refer to *Fig. 4-15*) contain open spaces designed for growth of indigenous grasses to maintain soil permeability while providing a stable base for vehicular traffic. Grid units have also been used in a variety of applications for soil stabilization, erosion control, and aesthetic treatment of drainage and access. Installations include shoulder slopes along highways and under bridges, the lining of canals, construction of mobile home parks, boat launch ramps, fire lanes adjacent to apartments and hotels, and erosion control of steep embankments (*see Fig. 14-1*).

14.1.2 Bases

The successful performance of masonry paving systems depends to a great extent on proper base preparation for the type of pedestrian or vehicular traffic anticipated (*see Fig. 14-2*). *Gravel bases* provide maximum drainage efficiency and prevent the upward flow of moisture by capillary action. Clean, washed gravel should be specified. Bases of unwashed gravel mixed with fine clay and stone dust are popular low-cost systems, but they will cause a loss of porosity and effective drainage due to hardening with the absorption of moisture. Masonry units in direct contact with such contami-

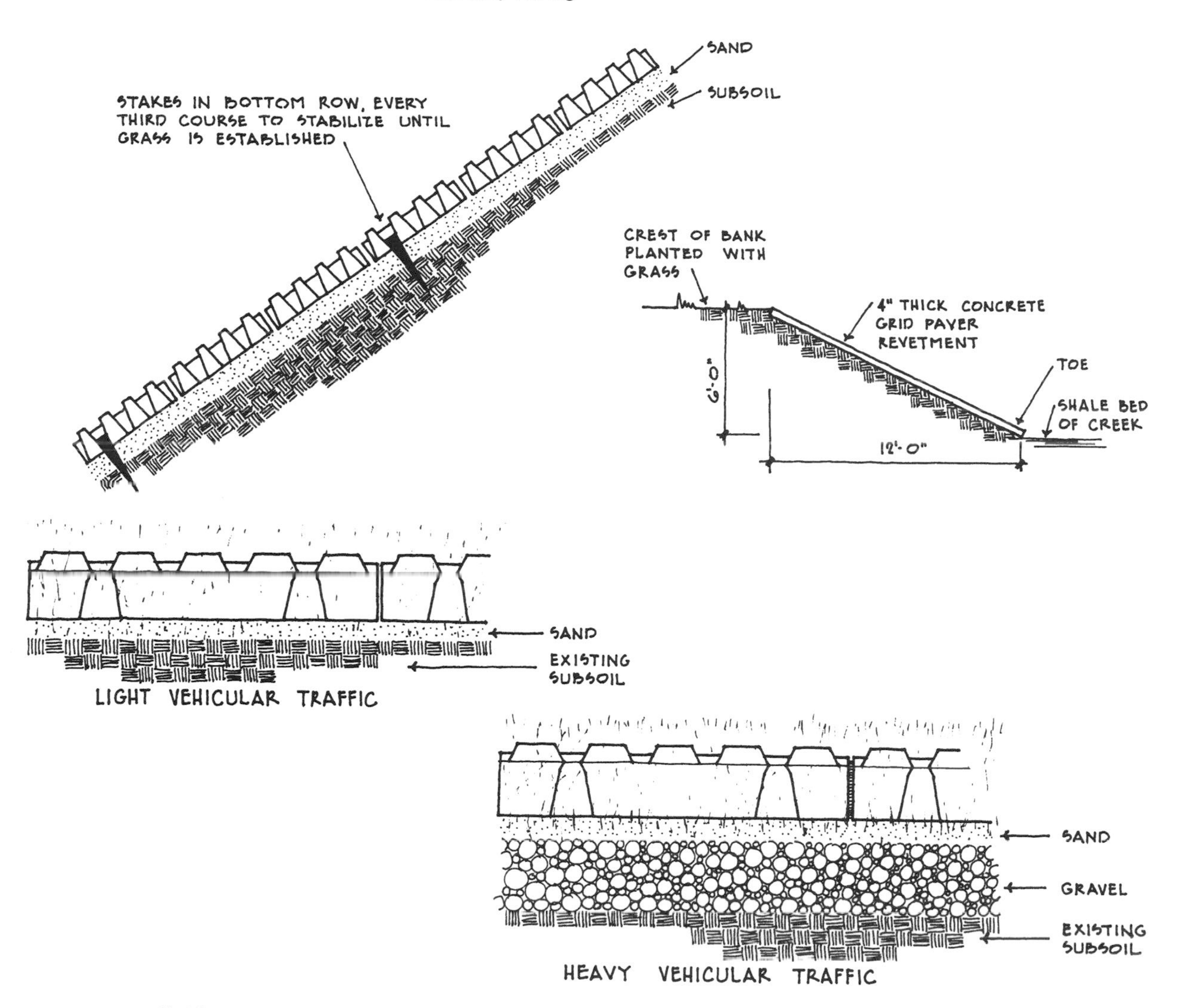

Grid paver installation examples. (*From National Concrete Masonry Association,* **TEK Bulletin 91,** *NCMA, Herndon, Va.*)

nated materials may also be susceptible to efflorescence as a result of soluble salts leached to the surface (see Chapter 16). Gravel bases are suitable only for flexible paving, but can accommodate both pedestrian and light vehicular traffic. Ungraded gravel bases have better interlocking qualities, where graded gravel has a tendency to roll underfoot. Stone screenings compact better than pea gravel, but provide poorer subsurface drainage. Large gravel generally requires the use of heavy road construction equipment for proper preparation, whereas stone screenings of fine gradation lend themselves to compaction with hand tools. If the paving units are turned on edge for greater compressive depth, thin bases of fine screenings can accommodate the same light vehicular loads as thicker beds of coarse gravel. Gravel bases range in thickness from 3 to 12 in., depending on the expected load and the paver thickness.

Concrete bases with either flexible or rigid paving can support heavy vehicular traffic. Existing concrete surfaces may be used, but major cracks

Application	*Thickness of sub-base (in.)*		Thickness of concrete masonry paving units (in.)
	Well-drained dry areas	*Low wet areas*	
Light duty Residential Driveways Patios Pool decks Walkways Parking Bicycle path Erosion control Temporary paving	0–3	4–8	$2\frac{1}{2}$–3
Medium duty Sidewalks Shopping malls Residential streets Public parking Bus stops Service roads Cross walks Parking lots Camping areas Mobile home parks Canal lining Safety zones Maintenance areas Farm equipment storage	4–6	10	3–4
Heavy duty City streets Intersections Gas stations Loading docks Loading ramps Industrial floors Stables	8	12	4–6

Note: The sand bed between the sub-base and the concrete paving units is always made 2 in. thick.

Recommendations for base and paver thickness for concrete masonry pavers. (*From National Concrete Masonry Association,* TEK Bulletin 75, *NCMA, Herndon, Va.*)

must be properly filled and stabilized. If a mortar leveling bed is used over new or existing slabs, the surface should be raked or floated to facilitate good bond. If a non-cementitious leveling bed or cushion is used, the surface of new concrete need only be screeded.

Asphalt bases (new or used) can support flexible paving systems for heavy vehicular traffic. Mortar leveling beds can be used, but there will be little or no bond between the mortar and the asphalt. Mortar leveling beds should not be placed on hot or warm asphalt or flash setting of the mortar will occur. Any major defects in an existing asphalt pavement should be repaired before installing masonry pavers.

14.1.3 Setting Beds

Mortar setting beds may be used for rigid paving over concrete bases. A Type M portland cement–lime mortar is generally recommended for outdoor use. Thickness of the bed may vary from ½ to 1 in. Bituminous setting beds composed of aggregate and asphaltic cement may be used over concrete or bituminous bases for flexible paving installations. The mix is generally designed and prepared at an asphalt plant and delivered to the job site ready for application.

Cushion material is generally placed between mortarless pavers and the base as a leveling layer which compensates for minor irregularities of the surface or the units. Sand for this purpose should be specified in accordance with ASTM C144. Under extremely wet conditions, however, sand cushions will provide poor drainage. Sand cushions over gravel bases require a membrane to prevent settlement. Dry mixtures of 1 part portland cement and 3 to 6 parts damp, loose sand may also be used. The higher sand ratio mixtures will provide little or no bond between paver and cushion. Roofing felt (15- to 30-lb weight) provides some compensation for minor irregularities, can be installed rapidly, and adds a degree of resilience for pavers installed over concrete bases. Several examples of masonry paving installations are shown in *Figs. 14-3 through 14-6.*

To prevent horizontal movement of mortarless paving, a method of containment must be provided around the perimeter of the paved areas. A soldier course set in mortar or concrete, new or existing retaining walls, building walls, or concrete curbs will all provide the required stability. Any new edging that must be installed should be placed prior to the

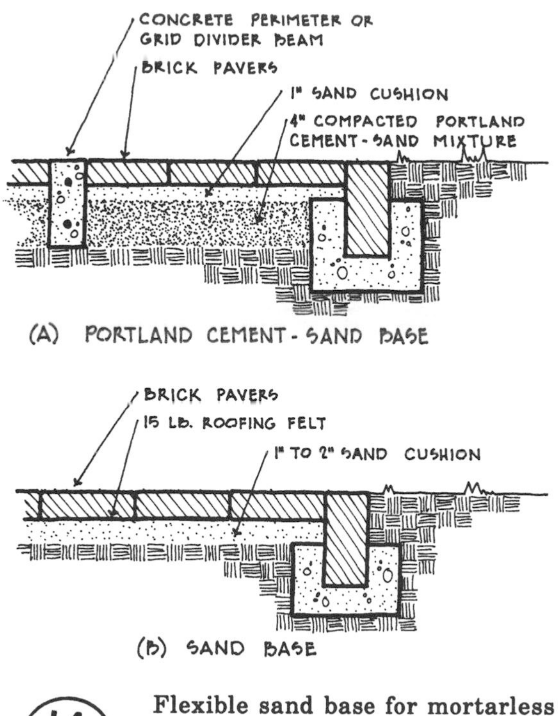

Flexible sand base for mortarless paving. (*From Brick Institute of America,* **Technical Note 14A,** ***BIA, Reston, Va.***)

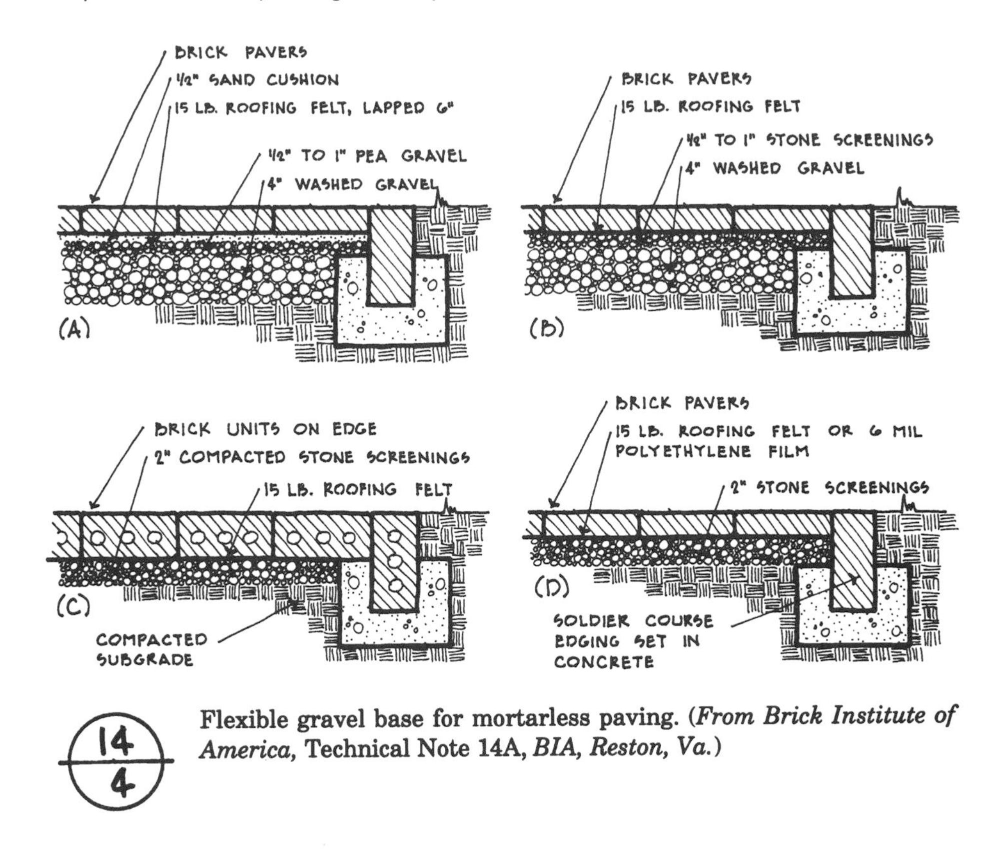

14/4 **Flexible gravel base for mortarless paving.** (*From Brick Institute of America,* Technical Note 14A, *BIA, Reston, Va.*)

paving units, and the pavers worked toward the established perimeters. Modular planning in the location of perimeter edging can eliminate or reduce the amount of cutting required to fit the units.

14.1.4 Paving Joints

Installing masonry paving with mortar joints may be done in one of three ways.

- Using a conventional mason's trowel, the pavers may be buttered and shoved into a leveling bed of mortar.
- The units may be placed on a mortar leveling bed with ⅜- to ½-in. open joints into which a grout mixture is then poured. Grout proportions are normally the same as for the mortar, except that the hydrated lime may be omitted. Special care must be taken in pouring this mixture, to protect the unit surfaces from spills and stains that would require special cleaning.
- Masonry pavers may also be laid on a cushion of 1 part portland cement and 3 to 6 parts damp, loose sand, and the open joints broomed full of the same mixture. After excess material has been removed from the surface, the paving is sprayed with a fine water mist until the joints are saturated. The installation should then be maintained in a damp condition for 2 or 3 days to facilitate proper curing.

Mortarless masonry paving may be swept with plain dry sand to fill between units, or with a portland cement–sand mixture equivalent to the proportions for Type M mortar. Pavers are generally butted together with

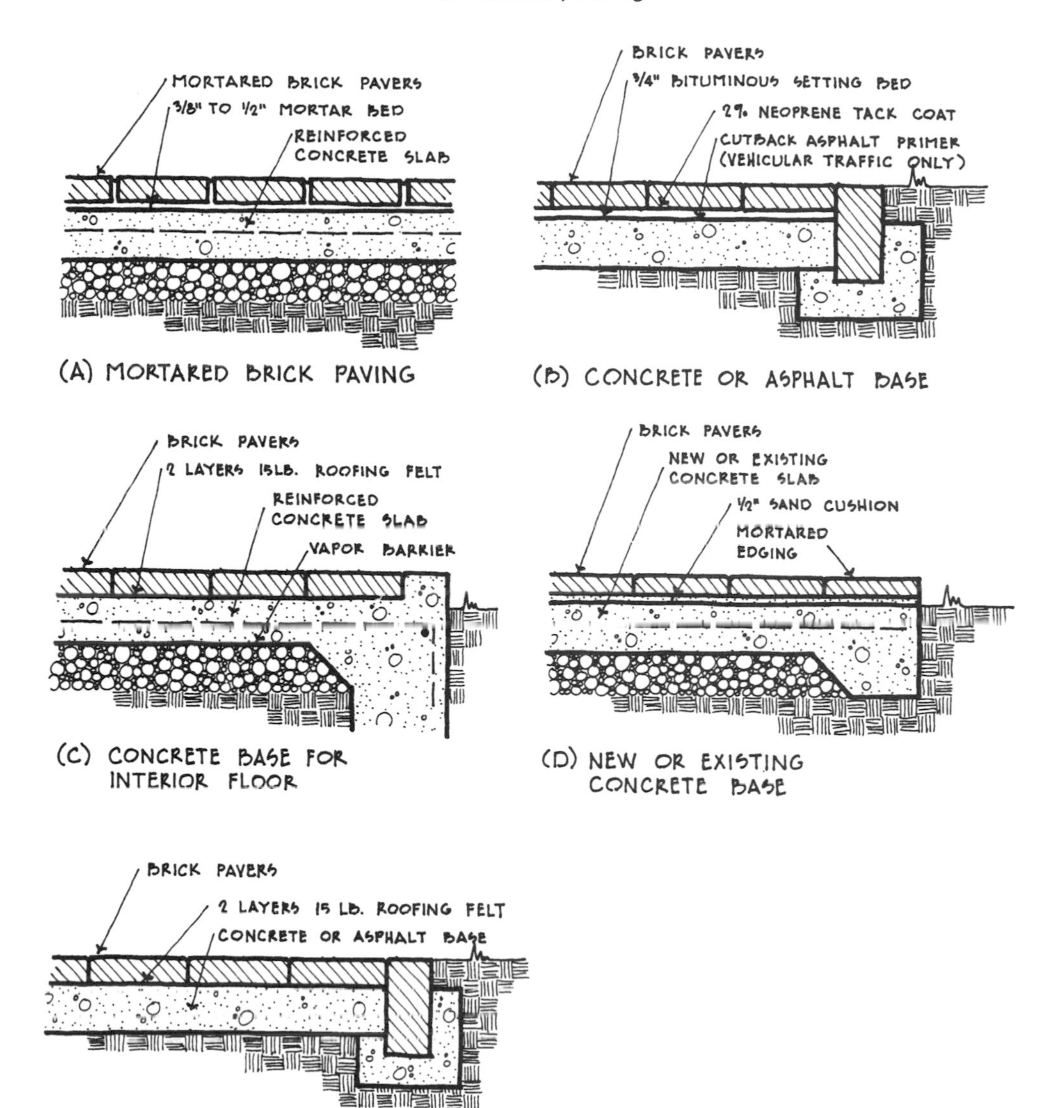

14-5 **Rigid base for mortared or mortarless paving.** (*From Brick Institute of America*, Technical Note 14A, *BIA, Reston, Va.*)

only the minimal spacing between adjacent units caused by irregularities of size and shape.

Expansion joints must be provided in rigid masonry paving to accommodate thermal and moisture movements. Joints should generally be located parallel to curbs and edges, at 90° turns and angles, and around interruptions in the paving surface. Fillers for these joints must be compressible, and of materials not subject to rot or vermin attack. Solid or preformed materials of polyvinyl chloride, butyl rubber, neoprene, and other elastic compounds are suitable (*see Fig. 14-7*). Even though mortarless masonry paving is flexible and has the ability to move slightly to accom-

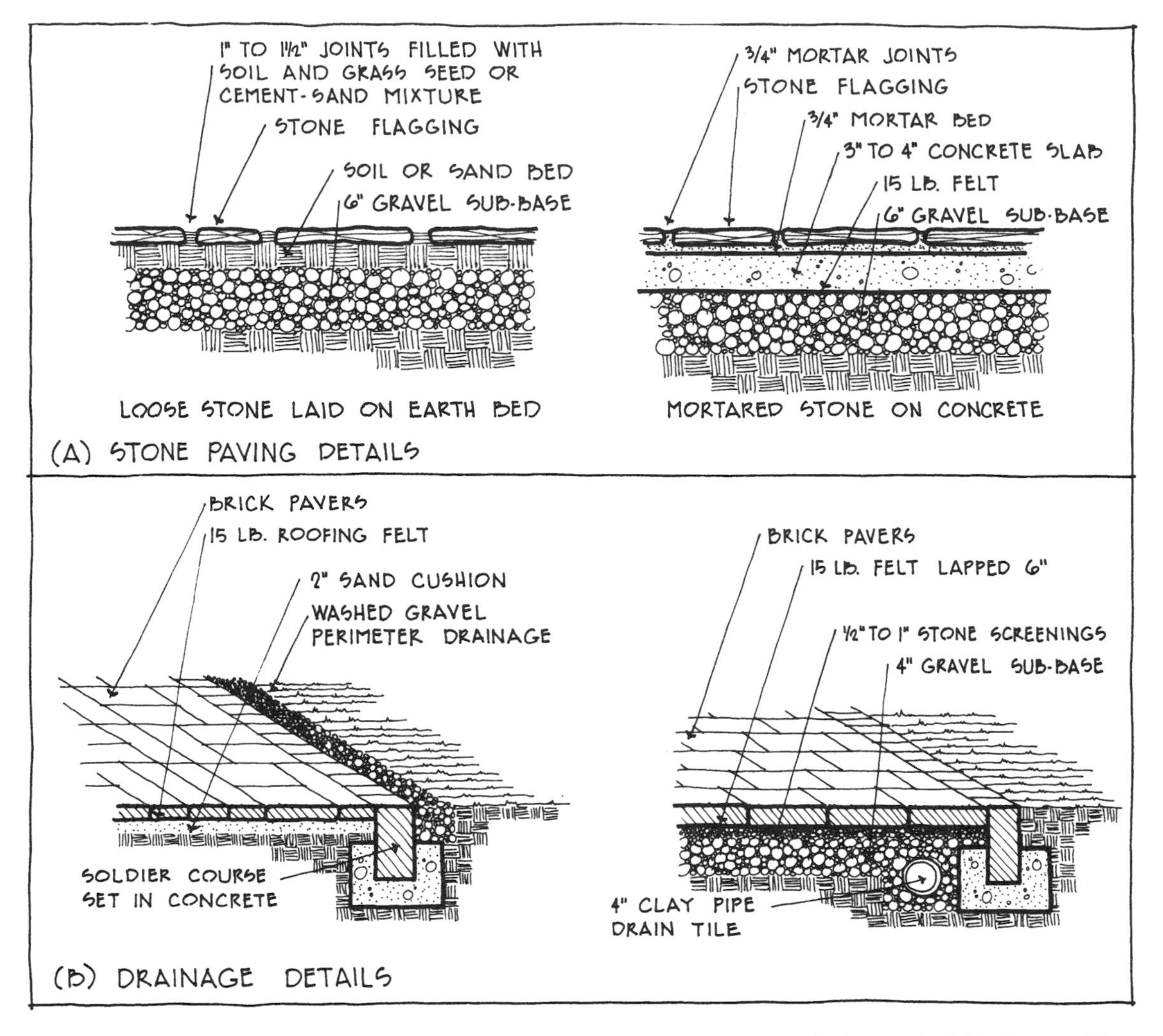

Paving details. ***(From Brick Institute of America,*** **Technical Note 14A,** ***BIA, Reston, Va.; and Charles G. Ramsey and Harold S. Sleeper,*** **Architectural Graphic Standards,** ***6th ed., ed. Joseph N. Boaz. Copyright © 1970 by John Wiley & Sons, Inc. Reprinted by permission of John Wiley & Sons, Inc.)***

modate expansion and contraction, it is recommended that expansion joints be placed adjacent to fixed objects such as curbs and walls.

14.1.5 Membrane Materials

Membranes of sheet or liquid materials are installed in some paving applications to reduce or control the passage of moisture, to discourage weed growth, or to form a separating layer or bond break to accommodate differential movement. Roofing felt, polyethylene film, vinyl, neoprene, rubber, asphaltic liquids, modified urethane, or polyurethane bitumens are all suitable since they are moisture- and rot-resistant. Liquid types, if installed properly, have some advantages over sheet materials because they are seamless and will conform to irregular surfaces. Precautions should be taken during construction to avoid membrane damage, particularly for roof deck installations, where resistance to moisture penetration is critical.

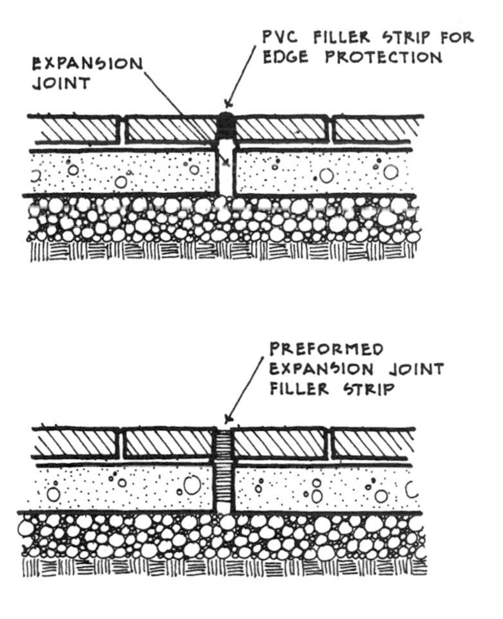

Expansion joints for masonry paving. (*From Brick Institute of America,* Technical Note 14A, *BIA, Reston, Va.*)

14.1.6 Masonry Units

The materials available for masonry paving systems have a wide range of structural and aesthetic capabilities. Solid concrete pavers are manufactured in a variety of shapes and sizes (refer to *Fig. 4-15*). ASTM C936, *Standard Specification for Solid Concrete Interlocking Paving Units* requires that minimum compressive strength be 8000 psi. Lightweight and normal weight aggregates may be used. Dense compaction assures a minimum of voids, and absorption must be less than 5%. The interlocking patterns transfer loads and stresses laterally by an arching or bridging action. Distribution of loads over a larger area in this manner reduces point loads, allowing heavier traffic over sub-bases that would normally require greater strength. Some pavers are manufactured with chamfered edges at the top to reduce stress concentrations and chippage. Chamfers should not, however, reduce the area of the bearing surface by more than 30%.

Solid units are made in thicknesses ranging from 2½ to 6 in. The thicker units are used for heavy service loads, and the thinner ones for light-duty residential areas. The National Concrete Masonry Association (NCMA) has developed recommendations for concrete pavers for various uses and base preparations (*see Fig. 14-2*). Individual solid paving units are small in size to facilitate manual installation. The irregular shapes prevent accurate description by linear measure, but units typically will range in size from 15 sq in. to a maximum of about 64 sq in. Some manufacturers are now casting from 6 to 12 units together in clusters which are designed to be laid at the job site by machine or by a team of two workers. After the units are in place, the individual pavers break apart along preformed cleavage lines. This method increases production and can result in reduced labor costs.

The concrete paver shapes shown in *Fig. 4-15* are full-size units for field areas. Most designs also include edge units and half-length units to reduce the amount of job site cutting required. Units are generally produced in a number of colors, but both color and shape will depend on local availability. Before planning the size and layout of a paved area, check with local manufacturers to verify design related information.

Concrete grid pavers are popular for applications where soil stabilization is required but a natural grass covering is also desirable. ASTM C1319, *Standard Specification for Concrete Grid Paving Units,* requires an average compressive strength of 5000 psi and a maximum average water absorption of 10%. Physically, the units must are permitted a maximum 50% void area, and must have a minimum 1¼ in. web thickness. Durability standards are based solely on proven field performance in maintaining the required compressive strength and water absorption characteristics after 3 years in service. Minimum nominal thickness is 3⅛ in., and maximum dimensions are 24×24 in. Features such as weight classification, surface texture, color, and finish are not covered in the standard and must be specified separately. Availability of specific unit patterns, shapes, and colors will vary locally.

Clay brick paving units (refer to *Fig. 3-10*) also come in a number of shapes, sizes, thicknesses, colors, and textures. Coarser-textured, slip-resistant units are recommended for outdoor installations exposed to rain, snow, and ice. This type of exposure also calls for units that are highly resistant to damage from freezing in the presence of moisture. For residential and light commercial applications such as patios, walkways, floors, plazas, and driveways, brick pavers should meet or exceed the requirements of ASTM C902, *Standard Specification for Pedestrian and Light Traffic Paving Brick.* Three weathering classifications are given, Classes SX, MX, and NX, roughly corresponding to the three grades of face brick for severe weathering, moderate weathering, and no weathering exposure. Traffic Type I, II, or III may be specified for exposure to extensive abrasion, intermediate traffic, or low traffic, respectively.

For paving in areas with a high volume of heavy vehicular traffic such as streets, commercial driveways, and aircraft taxiways, brick pavers should meet or exceed the requirements of ASTM C1272, *Standard Specification for Heavy Vehicular Paving Brick.* The standard covers two types of pavers. Type R (rigid paving) is intended to be set in a mortar setting bed supported by an adequate concrete base or an asphalt setting bed supported by an adequate asphalt or concrete base. Type F (flexible paving) is intended to be set in a sand bed with sand joints. Three different applications are also covered, PS, PX, and PA, roughly corresponding to appearance types FBS, FBX, and FBA for clay facing brick. Because they are intended to be installed over a flexible base, Type F pavers must have the highest compressive strength at 10,000 psi. Type R pavers which will be supported on an asphalt or concrete base must have a minimum average compressive strength of only 8000 psi. Heavy vehicular paving brick are obviously intended to be more rugged and durable in commercial, municipal, and industrial applications than pedestrian and light traffic pavers.

The Brick Institute of America (BIA) does not recommend the use of salvaged or used brick in paving installations. Older manufacturing processes did not assure uniformity in the quality of materials or performance, and units may spall, flake, pit, and crack when exposed to outdoor freeze-thaw cycles. Although used brick may be adequate for small residential jobs, and may provide a pleasing rustic effect, materials of unknown origin and composition should not be used for larger installations unless performance criteria can be tested and verified.

14.1.7 Paving Patterns

Many different effects can be achieved with standard rectangular pavers by varying the bond pattern in which the units are laid (*see Fig. 14-8*). It is important to specify the proper size of unit required for a particular pattern. Any of the patterns shown can be achieved with 4×8-in. *actual dimension* units laid dry and tight, or with nominal 4×8-in. units ($3\frac{5}{8}\times7\frac{5}{8}$ in. actual size) laid with $\frac{3}{8}$-in. mortar or sand joints. Patterns that require the width of the unit to be exactly one-half the length may not be laid dry and tight using nominal dimension units designed for mortar joints, and vice versa. The interlocking and herringbone patterns provide greater stability and lateral stress transfer. Designs that result in continuous joints (especially longitudinal joints in the direction of traffic flow) are more subject to shoving, displacement, and the formation of ruts.

14.1.8 Brick Floors

Interior brick floors and masonry roof deck plazas can be installed over diaphragm bases in both commercial and residential buildings. Structural design of suspended wood joist, steel frame, or reinforced concrete bases must take into consideration the weight of the masonry and the maximum allowable deflection. The dead weight of mortared or mortarless brick pavers is approximately 10 lb/sq ft per inch of thickness. For the $1\frac{5}{8}$-in.- and $2\frac{1}{4}$-in.-thick units, the weight would therefore be 16.25 and 22.5 lb/sq ft, respectively. Diaphragm action is important in maintaining the integrity of mortar joints, and deflections should be limited to $\frac{1}{600}$ of the span for mortared paving. Mortarless brick paving may be installed over bases designed for deflections of $\frac{1}{360}$ of the span (*see Fig. 14-9*). Metal decking supported on steel framing can serve as an economical means of constructing brick floors over open spans. The decking serves as combined form and reinforcement. The bricks are placed on a bed of mortar and the vertical joints filled with mortar or grout. For continuous spans, negative steel is placed in the joints (*see Fig. 14-9*).

Masonry paving is often used as a decorative wear surface on pedestrian roof deck plazas. For mortarless brick paving, sloping membranes in conjunction with porous base layers will permit rapid drainage and prevent possible damage from alternate freezing and thawing of trapped water. Consideration should be given to differential horizontal movement between supporting base and waterproof membrane. Bituminous membranes are nonelastic, but seamless liquid waterproofing and rubber sheet membranes are usually elastic and capable of adjusting to differential horizontal movement. Since the masonry pavers will also change dimensionally with temperature variations, a slip plane is recommended between pavers and membrane. Porous gravel cushions and asphalt-impregnated protection boards can withstand both horizontal abrasive movement and vertical traffic loadings (*see Fig. 14-10*).

Reinforced brick masonry can be used to span open spaces as floor assemblies or to span across fill material that may tend toward uneven

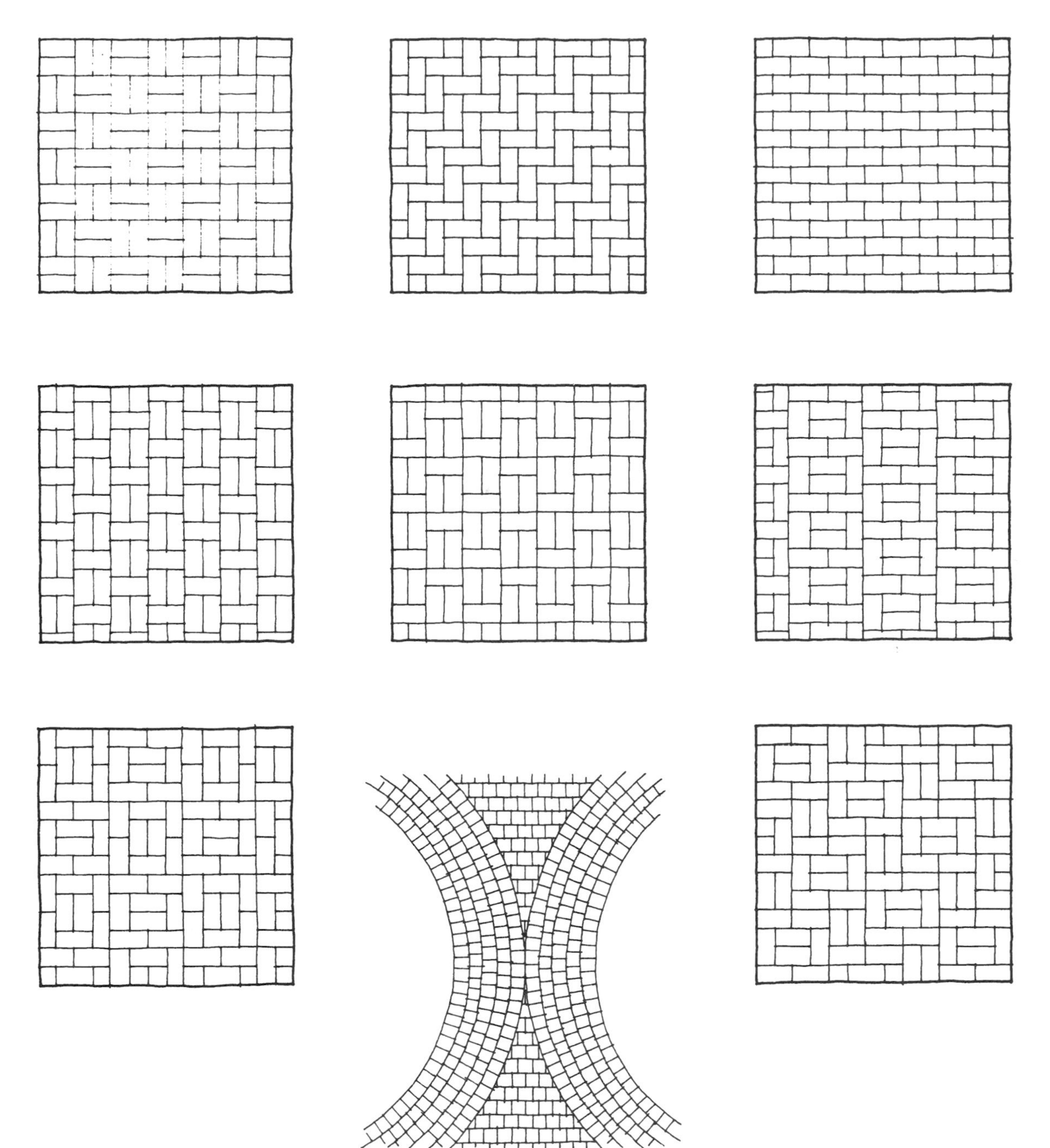

Bond patterns for rectangular paving units. (*From Brick Institute of America,* **Technical Note 14 Rev.**, *BIA, Reston, Va.*)

settlement. Reinforcement within the masonry section eliminates the need for a separate reinforced concrete slab or other base. These masonry slabs are very practical, especially over short spans, and different designs can satisfy design loadings for both pedestrian and vehicular traffic (*see Fig. 14-11*). Brick floors and roofs are still widely used throughout Mexico and the Middle East, and can be constructed without reinforcing steel if designed as a flat arch (refer to Chapter 11).

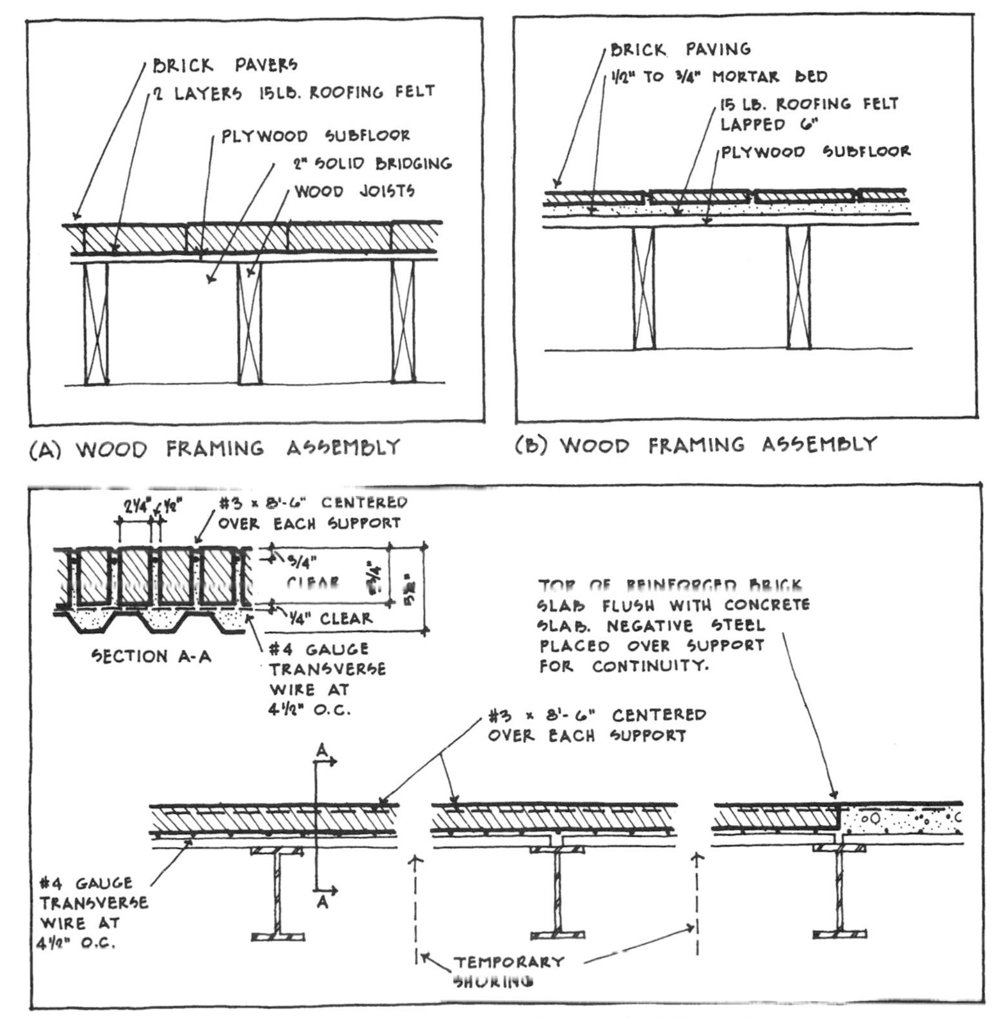

(A) WOOD FRAMING ASSEMBLY

(B) WOOD FRAMING ASSEMBLY

(C) STEEL DECKING - REINFORCED BRICK FLOOR ASSEMBLY

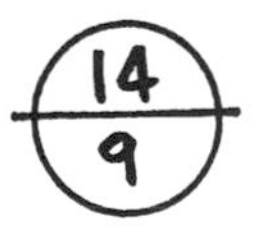

Mortared, mortarless, and reinforced brick flooring. (*From Brick Institute of America,* Technical Note 14B, *BIA, Reston, Va.*)

14.2 FIREPLACES

Residential fireplace design has evolved over the centuries toward standardization of the functional elements that assure successful operation. A fireplace must have proper fuel combustion, good chimney draw, and maximum heat radiation. Design should also provide simplicity of construction and fire safety, particularly when adjacent building elements are of combustible materials.

The proper functioning of a fireplace is related to the shape and relative dimensions of the combustion chamber or fire box, the proper location of the fireplace throat in relation to the smoke shelf, and the ratio of the flue area to the area of the fireplace opening (*see Figs. 14-12 through 14-14*). The shape of the combustion chamber influences both the draft and the amount of heat radiated into the room. The dimensions recommended in the tables may be varied slightly to correspond with brick coursing for modular and non-modular unit sizes, but it is inadvisable to make significant changes. A multifaced fireplace can be a highly effective unit, but presents certain prob-

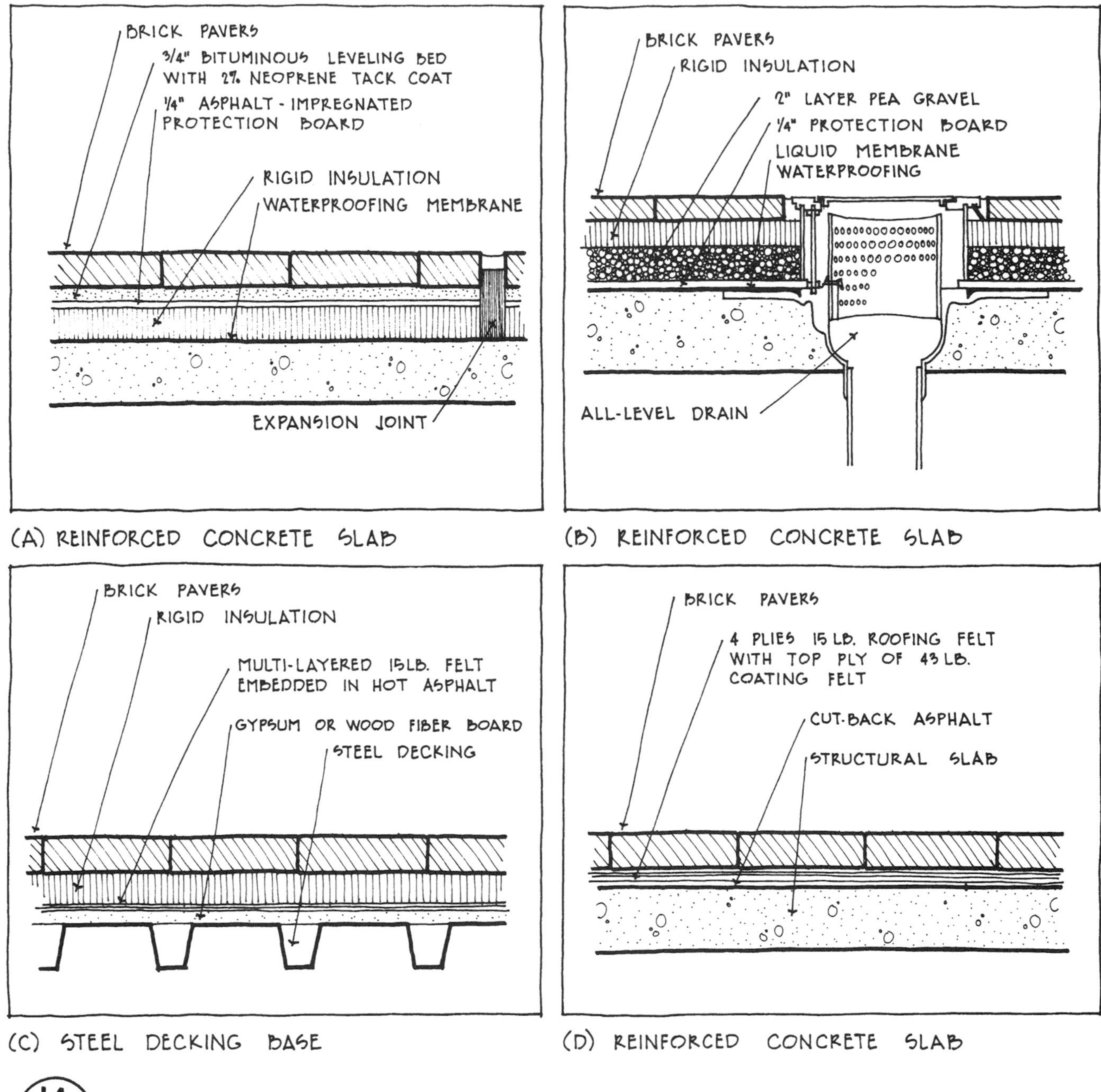

Masonry paver roof decks. (*From Brick Institute of America,* Technical Note 14B, *BIA, Reston, Va.*)

lems of draft and opening size that must sometimes be solved on an individual basis. The single-face fireplace is the most common and the oldest design, and the majority of the standard detail information is based on this type. ASTM C315, *Standard Specification for Clay Flue Linings,* covers minimum material requirements, and ASTM C1283, *Standard Practice for Installing Clay Flue Lining,* covers minimum installation requirements for residential masonry chimneys not exceeding 40 ft. in height.

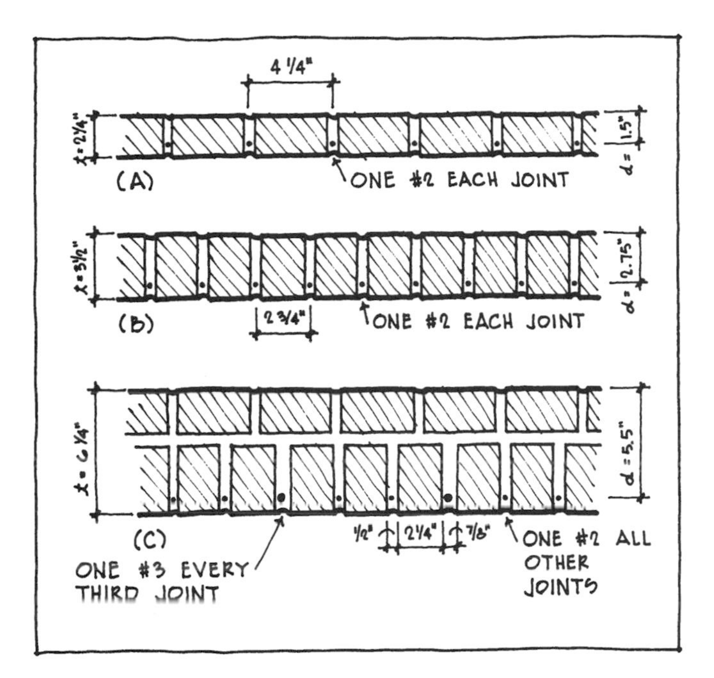

Live load (psf)	Maximum clear span		
	$t = 2\frac{1}{4}$ in. 1 No. 2 each joint [Fig. 14.11(A)]	$t = 3\frac{1}{2}$ in. 1 No. 2 each joint [Fig. 14.11(B)]	$t = 6\frac{1}{4}$ in. 1 No. 3 every 3rd joint 1 No. 2 remaining joints [Fig. 14.11(C)]
30	6′-10″	10′-5″	14′-5″
40	6′-3″	9′-9″	13′-8″
50	5′-10″	9′-2″	13′-1″
100	4′-6″	7′-3″	10′-11″
250	1′-10″	5′-0″	7′-10″

Note: Design parameters are as follows: The brick compressive strength average is 8000 psi. The mortar is type M (1 : $\frac{1}{4}$: 3), portland cement, lime and sand. Reinforcement steel is ASTM A82-66, f_s = 20,000 psi. A simple span loading condition was assumed:

$$M = \frac{wl^2}{8}$$

All mortar joints are $\frac{1}{2}$ in. thick for the slabs shown, except as noted.

Reinforced brick floor slabs. (*From Brick Institute of America,* Technical Note 14B, *BIA, Reston, Va.*)

The structural design of residential wood-burning fireplaces and chimneys is a fairly simple and well-known empirical procedure. For commercial and industrial chimneys, however, the sizes are sufficiently larger that a number of different forces are in effect. The principal forces acting on a large masonry chimney are (1) wind pressure; (2) the weight of the chimney, including its lining and foundation; (3) the foundation reaction; and (4) the expansion and contraction of the chimney walls due to the dif-

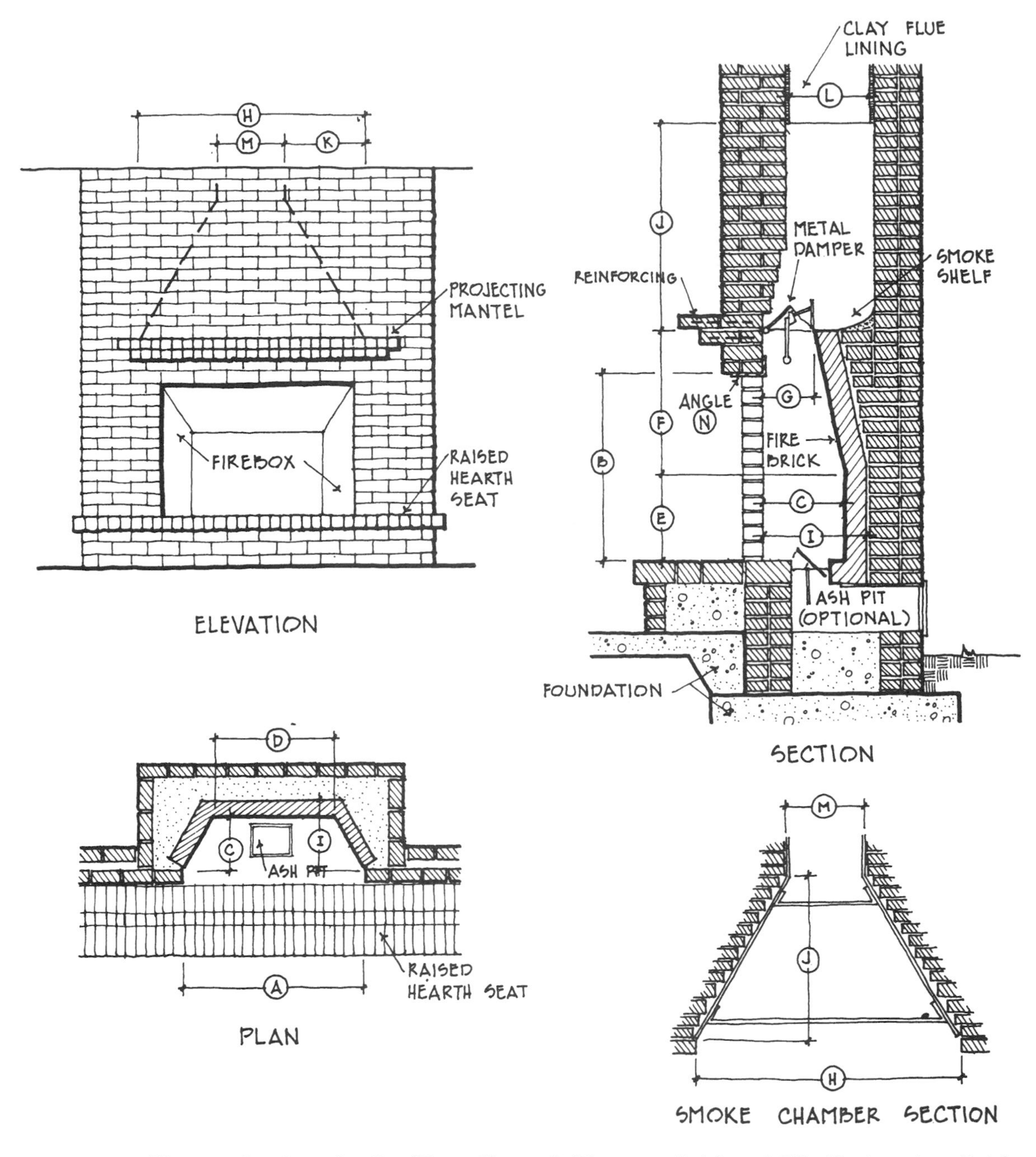

Masonry fireplace details. (*From Harry C. Plummer,* **Brick and Tile Engineering,** *Brick Institute of America, Reston, Va., 1962.*)

ference in temperature between the combustion gases inside and the surrounding air.

Design wind loads are listed in most building codes by geographic area and height zone. Since industrial chimneys can often be as tall as 100 ft, the potential lateral loads are significant. Chimneys constructed in areas subject to earthquakes must be designed as fully reinforced masonry. A sample design of a reinforced brick chimney 80 ft tall is shown in *Fig. 14-15.* The chimney shell measures 15 in. at the base and 9 in. in the upper section,

Finished fireplace opening							Rough brick work and flue sizes								Steel angles†
										New sizes*			Old sizes		
A	B	C	D	E	F	G	H	I	J	K	L M	R‡	K	L M	N
24	24	16	11	14	18	$8\frac{3}{4}$	32	20	19	10	8×12	8	$11\frac{3}{4}$	$8\frac{1}{2}\times 8\frac{1}{2}$	A-36
26	24	16	13	14	18	$8\frac{3}{4}$	34	20	21	11	8×12	8	$12\frac{3}{4}$	$8\frac{1}{2}\times 8\frac{1}{2}$	A-36
28	24	16	15	14	18	$8\frac{3}{4}$	36	20	21	12	8×12	10	$11\frac{1}{2}$	$8\frac{1}{2}\times 13$	A-36
30	29	16	17	14	23	$8\frac{3}{4}$	38	20	24	13	12×12	10	$12\frac{1}{2}$	$8\frac{1}{2}\times 13$	A-36
32	29	16	19	14	23	$8\frac{3}{4}$	40	20	24	14	12×12	10	$13\frac{1}{2}$	$8\frac{1}{2}\times 13$	A-42
36	29	16	23	14	23	$8\frac{3}{4}$	44	20	27	16	12×12	12	$15\frac{1}{2}$	13 ×13	A-42
40	29	16	27	14	23	$8\frac{3}{4}$	48	20	29	16	12×16	12	$17\frac{1}{2}$	13 ×13	A-48
42	32	16	29	14	20	$8\frac{3}{4}$	50	20	32	17	16×16	12	$18\frac{1}{2}$	13 ×13	A-48
48	32	18	33	14	26	$8\frac{3}{4}$	56	22	37	20	16×16	15	$21\frac{1}{2}$	13 ×13	B-54
54	37	20	37	16	29	13	68	24	45	26	16×16	15	25	13 ×18	B-60
60	37	22	42	16	29	13	72	27	45	26	16×20	15	27	13 ×18	B-66
60	40	22	42	16	31	13	72	27	45	26	16×20	18	27	18 ×18	B-66
72	40	22	54	16	31	13	84	27	56	32	20×20	18	33	18 ×18	C-84
84	40	24	64	20	28	13	96	29	61	36	20×24	20	36	20 ×20	C-96
96	40	24	76	20	28	13	108	29	75	42	20×24	22	42	24 ×24	C-108

*New flue sizes conform to modular dimensional system. Sizes shown are nominal. Actual size is $\frac{1}{2}$ in. less each dimension.
†Angle sizes: A, 3 by 3 by $\frac{3}{16}$ in.; B, $3\frac{1}{2}$ by 3 by $\frac{1}{4}$ in.; C, 5 by $3\frac{1}{2}$ by $\frac{5}{16}$ in.
‡Round flues.

Standard fireplace dimensions *(refer to Fig. 14-12). (From Harry C. Plummer,* **Brick and Tile Engineering,** *Brick Institute of America, Reston, Va., 1962.)*

Fireplace width	Rectangular liners				Round Liners	
	Standard		Modular			
	Outside dimensions	Effective area	Nominal outside dimensions	Effective area	Inside diameter	Effective area
24	$8\frac{1}{2}\times 8\frac{1}{2}$	52.56	8 × 12	57	8	50.26
			8 × 16	74		
30–34	$8\frac{1}{2}\times 13$	80.50			10	78.54
			12 × 12	87		
	$8\frac{1}{2}\times 18$	109.69	12 × 16	120	12	113.00
36–44	13 × 13	126.56	16 × 16	162	15	176.70
46–56	13 × 18	182.84	16 × 20	208		
58–68	18 × 18	248.06	20 × 20	262	18	254.40
70–84			20 × 24	320	20	314.10
			24 × 24	385	22	380.13
					24	452.30

Recommended flue sizes. *(From Harry C. Plummer,* **Brick and Tile Engineering,** *Brick Institute of America, Reston, Va., 1962.)*

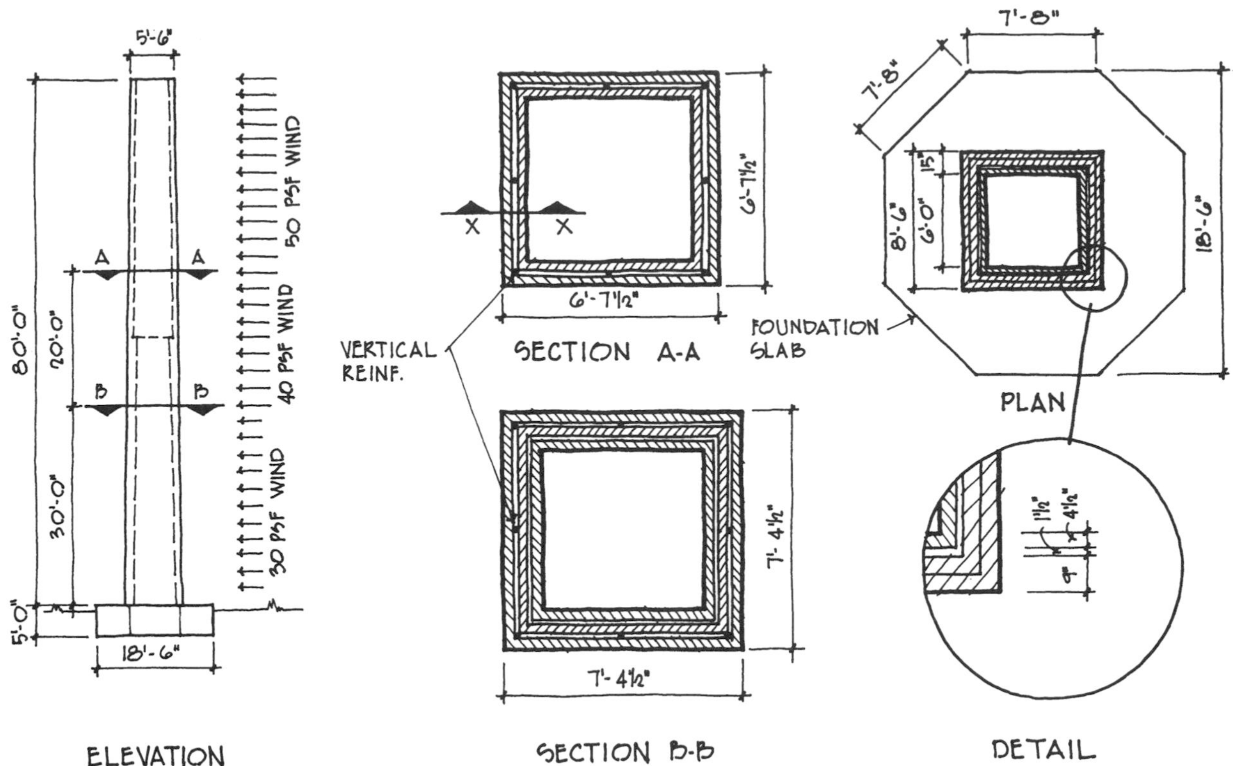

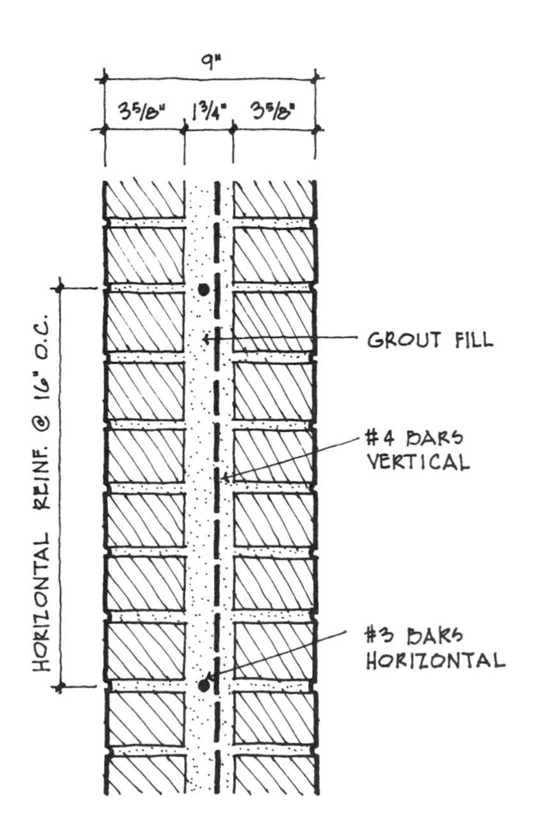

Reinforced brick masonry industrial chimney. (*From Brick Institute of America,* Technical Note 19B, *BIA, Reston, Va.*)

with the profile tapering toward the top. Design for an identical chimney of unreinforced construction is shown in *Fig. 14-16.* Shell thickness at the base of this chimney is almost 2 ft exclusive of the flue lining. Many unreinforced chimneys perform satisfactorily without horizontal reinforcement to resist thermal stresses. However, as an added safety factor for high flue gas temperatures, two No. 2 bars placed in bed joints 2 in. from the interior and exterior faces of the wall and spaced 16 in. on center vertically may be used.

Unreinforced masonry chimneys can be built to withstand wind pressures exceeding 30 lb/sq ft, but in areas where considerable lateral forces may be anticipated, the addition of steel reinforcement usually offers a more economical solution. In areas not subject to seismic activity, and where wind pressures are less severe, unreinforced chimneys may be safely and economically built to a height approaching 100 ft.

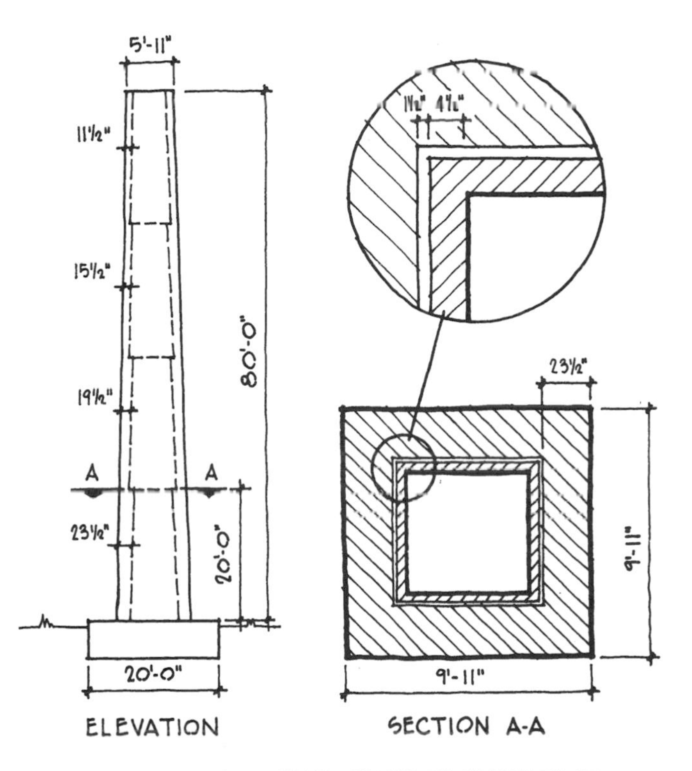

Unreinforced brick masonry industrial chimney. (*From Brick Institute of America,* Technical Note 19B, *BIA, Reston, Va.*)

CONSTRUCTION PRACTICE

15

INSTALLATION AND WORKMANSHIP

Masons and bricklayers belong to one of the oldest crafts in history. The rich architectural heritage of many civilizations attests to the skill and workmanship of the trade, and the advent of modern technological methods and sophisticated engineering practices has not diminished the importance of this aspect of masonry construction. The best intentions of the architect or engineer will not produce a masterpiece unless the workmanship is of the highest order and the field practices are as exacting and competent as the detailing.

15.1 MOISTURE RESISTANCE

Workmanship has a greater effect on the moisture resistance of masonry than any other single factor. Key elements in the quality of workmanship include:

- Proper storage and protection of materials
- Consistent proportioning and mixing of mortar ingredients
- Full mortar joints
- Complete mortar-to-unit bond
- Continuity of flashing
- Unobstructed weepholes
- Tooled joint surfaces
- Protection of uncompleted walls

Among these elements, mortar placement ranks high in limiting the amount of moisture which penetrates through the wall face. Such leakage can usually be traced to either capillary passages at the mortar-to-unit interface, partially filled mortar joints, or cracks caused by unaccommodated building movements. Virtually all masonry walls suffer some moisture penetration because of joint defects and other design, construction, or workmanship errors. It is for this reason that the installation of flashing and weepholes is critical in collecting and discharging any water which does enter the wall. This backup drainage system provides redundancy in moisture control and allows the construction to be somewhat forgiving of defects. Since it *is* the backup system, though, the flashing installation itself cannot tolerate defects or discontinuities without providing avenues for moisture penetration directly to the interior of the building.

15.2 PREPARATION OF MATERIALS

Field quality control begins with the proper storage and protection of materials. Preparations necessary prior to construction will vary according to the specific materials and conditions involved.

15.2.1 Material Storage and Protection

Proper storage and protection of masonry materials at the project site are critical to the performance and appearance of the finished construction. Materials properly stored and covered will remain in good condition, unaffected by weather. Improper procedures, however, can result in damage to units and contamination or degradation of mortar and grout ingredients.

Brick and block units should be stored off the ground to prevent staining from contact with the soil, and absorption of moisture, soluble salts, or other contaminants which might cause efflorescence in the finished work. Stored units should be covered for protection against the weather. Cut stone usually requires stacking on wooden pallets or frames with spacers between panels to allow air circulation. Treated wood may contain chemicals which stain light-colored stone. Handling of all masonry should avoid chipping or breakage of units.

Mortar and grout aggregates should also be covered to protect against contamination from rain, snow, ice, and blowing dust and debris. Different aggregates should be stockpiled separately. Packaged mortar and grout ingredients should be received in their original containers with labels intact and legible for easy identification. Broken packages, open containers, and materials with missing or illegible identification should be rejected. All packaged materials should be stored off the ground and covered to prevent moisture penetration, deterioration, and contamination.

15.2.2 Mortar and Grout

The mortar mix required in the project specifications must be carefully controlled at the job site to maintain consistency in performance and appearance. Consistent measurement of mortar and grout ingredients should ensure uniformity of proportions, yields, strengths, workability, and mortar color from batch to batch. Volumetric rather than weight proportioning is most often called for, and most often miscalculated because of variations in the moisture content of the sand. Common field practice is to use a shovel as the standard measuring tool for dry ingredients. However, moisture in the sand causes a "bulking" effect, and the same weight of wet sand occupies more volume than dry sand. Such variables often cause over- or undersanding of the mix, which affects both the strength and bonding characteristics of the mortar. Oversanded mortar is harsh and unworkable, provides a weak bond with the masonry units, and performs poorly during freeze-thaw cycles.

Simple field quality control measures can be instituted through the use of 1-cu-ft. batching boxes. The mixer may then determine the exact number of shovels full which equal 1 cu ft. Since the moisture content of the sand will vary constantly because of temperature, humidity, and evaporation, it is good practice to check the volume measurement at least twice a day and make adjustments as necessary. For even greater consistency, a site-constructed or proprietary batching box can be set to discharge as much as 9 cu ft of sand directly into the mixer.

The other dry ingredients in masonry mortar are normally packaged and labeled only by weight. Regardless of weight, however, these cementitious materials are usually charged into the mixer in whole- or half-bag measures. Each bag of portland cement or masonry cement equals 1 cu ft regardless of its labeled weight, and each bag of hydrated mason's lime equals 1¼ cu ft regardless of its weight. In some regions, additional convenience is provided by preblended and bagged portland cement–lime mixes.

The amount of mixing water required is not stated as part of the project specifications. Unlike concrete, however, masonry mortar and grout require the maximum amount of water consistent with characteristics of good flow and workability. Excess water is rapidly absorbed by the masonry units, reducing the water-cement ratio to normal levels and providing a moist environment for curing. Optimum water content is best determined by the mason's feel of the mortar on the trowel. A mortar with good workability is mixed with the proper amount of water.

Mortar with good workability should spread easily, cling to vertical unit surfaces, extrude easily from joints without dropping or smearing, and permit easy positioning of the unit to line, level, and plumb. Dry mixes do not spread easily, produce poor bond, and may suffer incomplete cement hydration. Mixes that are too wet are also difficult to trowel, and allow units to settle after placement. Thus mixing water additions are self-regulating. The water proportion will vary for different conditions of temperature, humidity, unit moisture content, unit weight, and so on.

The necessary water content for grout is significantly higher than that for mortar, because grout must flow readily into the cores and cavities and around reinforcement and accessories. Grout consistency should be such that, at the time of placement, the grout has a slump of 8 to 11 in.

Recent innovations in masonry technology include ready-mixed mortars and pre-batching of dry mortar ingredients to eliminate the field variables that often affect the quality and consistency of job-mixed mortar. This moves the mixing operation to a controlled batching plant where ingredients can be accurately weighed and mixed, then delivered to the job site. Ready-mixed mortars are delivered trowel ready in trucks or sealed containers, without the need for additional materials or mixing. Extended-life set retarders, which keep the mix plastic and workable for up to 72 hours, must be absorbed by the masonry units before cement hydration can begin, so unit suction can affect set time and construction speed. Pre-batched dry ingredients are delivered to the site in weathertight silos ready for automatic mixing (*see Fig. 15-1*). Both methods improve uniformity and offer greater convenience and efficiency, but sand bulking can still be a problem with dry-batched mixes unless the sand is oven-dried. Ready-mixed mortars are governed by ASTM C1142, *Standard Specification for Ready-Mixed Mortar for Unit Masonry*.

There are two traditional methods of mixing mortar on the job site. For very small installations, *hand mixing* may be most economical. It is accomplished using a mason's hoe and a mortar box. Sand, cement, and lime are spread in the box in proper proportions and mixed together until an even

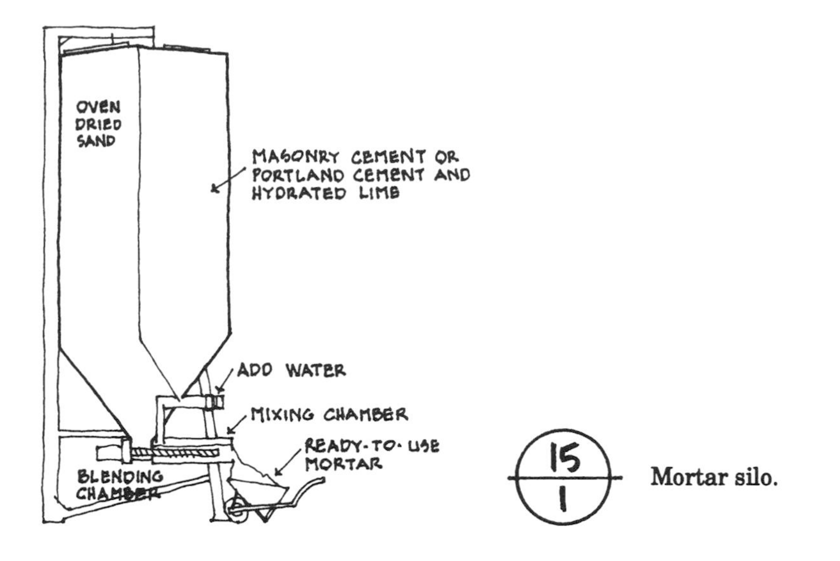

Mortar silo.

color is obtained. Water is then added, and mixing continues until the consistency and workability are judged to be satisfactory.

More commonly, *machine mixing* is used to combine mortar ingredients. The mechanical drum or paddle-blade mixers used are similar to but of lighter duty than concrete mixers. Normal capacity ranges from 4 to 7 cu ft. About three-fourths of the mixing water, half the sand, and all of the cementitious ingredients are first added and briefly mixed together. The balance of the sand is then added, together with the remaining water. After all the materials and water have been combined, grout should be mixed a minimum of 5 minutes, and mortar a minimum of 3 and a maximum of 5 minutes. Less mixing time may result in nonuniformity, poor workability, low water retention, and less than optimum air content. Overmixing causes segregation of materials and entrapment of excessive air, which may reduce bond strength. Specified admixtures and pigments should be added in the approved quantities only after all other ingredients are mixed. Pigments should always be pre-batched for consistency in color.

To avoid excessive drying and stiffening, mortar batches should be sized according to the rate of use. Loss of water by absorption and evaporation can be minimized on hot days by wetting the mortar board and covering the mix in the mortar box. Within the first 1½ to 2½ hours of initial mixing, the mason may add water to replace evaporated moisture (refer to Chapter 6). *Retempering* is accomplished by adding water to the mortar batch and thoroughly remixing. Sprinkling of the mortar is not satisfactory. Mortars containing added color pigment should not be retempered, as the increased water will lighten the color and thus cause variation from batch to batch.

15.2.3 Masonry Units

Concrete masonry units are cured and dried at the manufacturing plant, and when delivered to the job site, their moisture content should be within the specified limits. Concrete units should never be moistened before or during placement because they will shrink as they dry out. If this shrinkage is restrained, as it normally is in a finished wall, stresses can develop that will cause the wall to crack. Proper storage and protection of the units

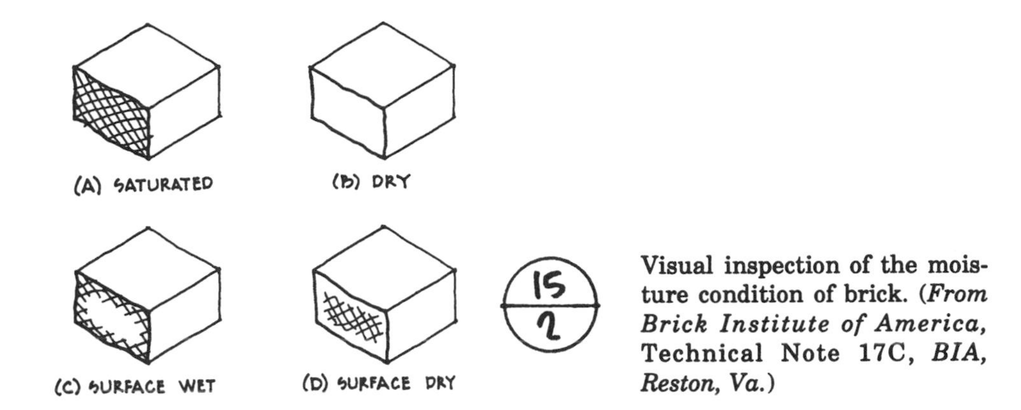

Visual inspection of the moisture condition of brick. (*From Brick Institute of America,* Technical Note 17C, *BIA, Reston, Va.*)

is essential to prevent rain or groundwater absorption from affecting the controlled moisture content.

There may be some instances in which concrete units should ideally be dried beyond the specified limits. Where walls will be exposed to relatively high temperature and low humidity, as in the interior of heated buildings, the units should be dried to the approximate average air-dry condition to which the finished construction will be exposed.

The moisture content of clay brick is not controlled in the manufacturing process, and will vary for different units and conditions. The initial rate of absorption (see Chapter 3) will affect the amount of capillary action that takes place between the brick and the mortar. A high-suction brick, if laid dry, will absorb excessive water from the mortar, thus impairing the hydration process, weakening the bond of the finished masonry and resulting in water-permeable joints. The amount of water that will enter a brick by capillary action varies greatly. Optimum bond is produced with units having initial rates of absorption between 5 and 25 g/minute/30 sq in. (refer to Chapter 3). If the initial rate of absorption (IRA) is greater than 30 g/minute/30 sq in., the units should be thoroughly wetted by spray, dip, or soaker hose, and then allowed to surface-dry before they are laid. A surface film of water will cause the brick to float on the mortar bed and will prevent proper bonding. Visual inspection of a broken unit will indicate whether or not moisture is evenly distributed throughout the unit (*see Fig. 15-2*).

Brick and architectural concrete masonry units must also be properly blended for color to avoid uneven visual effects. Units from four different cubes or pallets should be used at the same time, and brick manufacturers often provide unstacking instructions for even color distribution. For single-color units, this takes advantage of the subtle shade variations produced in the manufacturing process, and on a blend of colors, will prevent stripes or patchy areas in the finished wall (*see Fig. 15-3*).

All masonry units should be clean and free of contaminants such as dirt, oil, or sand which might inhibit bond.

15.2.4 Accessories

Steel reinforcement, anchors, ties, and other accessories should be cleaned to remove oil, dirt, ice, and other contaminants which could prevent good bond with the mortar or grout. Careful storage and protection will minimize cleaning requirements. Flashing materials should be protected from

15
3

Masonry units must be blended for even color distribution.

damage or deterioration prior to placement, and insulation materials protected from wetting.

15.2.5 Layout and Coursing

The design of masonry buildings should take into consideration the size of the units involved. The length and height of walls as well as the location of openings and intersections will greatly affect both the speed of construction and the appearance of the finished work. The use of a common module in determining dimensions can reduce the amount of field cutting required to fit the building elements together.

A number of the common brick sizes available are adaptable to a 4- or 6-in. module, and dimensions based on these standards will generally result in the use of only full- or half-size units. Similarly, a standard 16-in. concrete block layout may be based on an 8-in. module with the same reduction in field cutting (*see Fig. 15-4*). In composite construction of brick and concrete block, unit selection should be coordinated to facilitate the anchorage of backing and facing wall, as well as the joining and intersecting of the two systems. The tables shown in *Fig. 15-5* indicate the height, width, length, and coursing of various brick sizes, and the relationship between the brick units and nominal 8×8×16-in. concrete blocks. For instance, three courses of Standard Modular brick equal the height of one concrete block course, five courses of Engineer brick equal two courses of concrete block, and so on. As shown in *Fig. 15-6,* the brick and block units work together in both plan and section, thus increasing the speed with which the mason can lay up a wall and improving the general quality, workmanship, and appearance of the job.

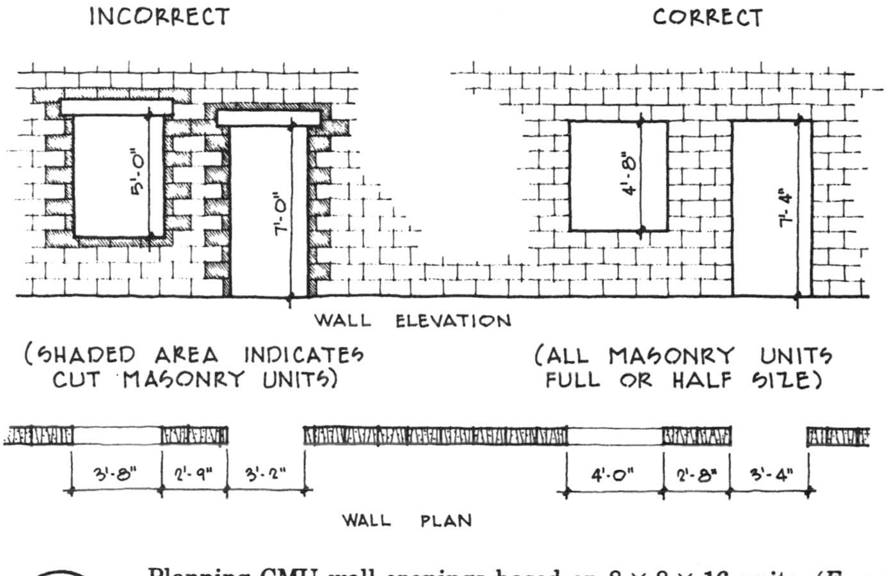

Planning CMU wall openings based on 8 × 8 × 16 units. (*From National Concrete Masonry Association,* **TEK Bulletin 14,** *NCMA, Herndon, Va.*)

Corners and intersections in masonry walls can be critical both structurally and aesthetically, and proper planning can facilitate construction of these elements. When masonry shear walls are used to transfer wind loads and seismic forces, they must be securely anchored to the transverse elements with steel reinforcing, and the coursing and layout of the units affects the ease with which the steel can be placed. Corner intersections are often points of high stress, and must also be aesthetically pleasing from the exterior if the masonry units are to be left exposed. The use of masonry pilasters as integral stiffening elements in a wall must also be carefully considered in the layout, and dimensions properly set so that the pilaster fits in with the regular coursing. The examples in *Figs. 15-7 through 15-12* illustrate various common methods of detailing.

The laying up of masonry walls is a very ordered and controlled process. Units must remain in both vertical and horizontal alignment throughout the height and length of the structure in order for the coursing to work out with opening locations, slab connections, anchorage to other structural elements, and so on. Laying out of the first course is critical, since mistakes at this point would be difficult, if not impossible, to correct later. The first course must provide a solid base on which the remainder of the walls can rest.

After locating the corners of the structure, it is a good idea to check dimensions by either measuring or actually laying out a dry course of units. Chalk lines are used to establish initial alignment on the foundation, and string lines are used once the walls are up in the air. Base courses must always be laid in a full bed of mortar even if face shell bedding [for hollow brick or concrete masonry units (CMU's)] is to be used in the rest of the wall.

Corner units are laid first and walls worked from outside corners and openings toward the center. The corners are usually built four or five courses higher than the center of the wall, and as each course is laid, it is checked

TABLE A NOMINAL MODULAR SIZES OF BRICK

Unit designation	Thickness (in.)	Face dimensions: Height (in.)	Face dimensions: Length (in.)	Number of courses in 16 in.
Standard	4	2 2/3	8	6
Engineer	4	3 1/5	8	5
Economy 8 or Jumbo Closure	4	4	8	4
Double	4	5 1/3	8	3
Roman	4	2	12	8
Norman	4	2 2/3	12	6
Norwegian	4	3 1/5	12	5
Economy 12 or Jumbo Utility	4	4	12	4
Triple	4	5 1/3	12	3
SCR brick	6	2 2/3	12	6
6-in. Norwegian	6	3 1/5	12	5
6-in. Jumbo	6	4	12	4
8-in. Jumbo	8	4	12	4

TABLE B LENGTH OF CMU WALLS WITH STRETCHER UNITS

Number of stretchers	Wall length*
1	1'4"
1 1/2	2'0"
2	2'8"
2 1/2	3'4"
3	4'0"
3 1/2	4'8"
4	5'4"
4 1/2	6'0"
5	6'8"
5 1/2	7'4"
6	8'0"
6 1/2	8'8"
7	9'4"
7 1/2	10'0"
8	10'8"
8 1/2	11'4"
9	12'0"
9 1/2	12'8"
10	13'4"
10 1/2	14'0"
11	14'8"
11 1/2	15'4"
12	16'0"
12 1/2	16'8"
13	17'4"
13 1/2	18'0"
14	18'8"
14 1/2	19'4"
15	20'0"
20	26'8"

*Based on units 15 5/8 in. long, and half units 7 5/8 in. long, with 3/8 in.-thick head joints.

TABLE C HEIGHT OF CMU WALLS BY COURSES

Number of courses	Wall height: 3/8-in. bed joint, 8-in. block	3/8-in. bed joint, 4-in. block	7/16-in. bed joint, 8-in. block	7/16-in. bed joint, 4-in. block	1/2-in. bed joint, 8-in. block	1/2-in. bed joint, 4-in. block
1	8"	4"	8 1/16"	4 1/16"	8 1/8"	4 1/8"
2	1'4"	8"	1'4 1/8"	8 1/8"	1'4 1/4"	8 1/4"
3	2'0"	1'0"	2' 3/16"	1' 3/16"	2' 3/8"	1' 3/8"
4	2'8"	1'4"	2'8 1/4"	1'4 1/4"	2'8 1/2"	1'4 1/2"
5	3'4"	1'8"	3'4 5/16"	1'8 5/16"	3'4 5/8"	1'8 5/8"
6	4'0"	2'0"	4' 3/8"	2' 3/8"	4' 3/4"	2' 3/4"
7	4'8"	2'4"	4'8 7/16"	2'4 7/16"	4'8 7/8"	2'4 7/8"
8	5'4"	2'8"	5'4 1/2"	2'8 1/2"	5'5"	2'9"
9	6'0"	3'0"	6' 9/16"	3' 9/16"	6'1 1/8"	3'1 1/8"
10	6'8"	3'4"	6'8 5/8"	3'4 5/8"	6'9 1/4"	3'5 1/4"
15	10'0"	5'0"	10' 15/16"	5' 15/16"	10'1 7/8"	5'1 7/8"
20	13'4"	6'8"	13'5 1/4"	6'9 1/4"	13'6 1/2"	6'10 1/2"
25	16'8"	8'4"	16'9 9/16"	8'5 9/16"	16'11 1/8"	8'7 1/8"
30	20'0"	10'0"	20'1 7/8"	10'1 7/8"	20'3 3/4"	10'3 3/4"
35	23'4"	11'8"	23'6 3/16"	11'10 3/16"	23'8 3/8"	12' 3/8"
40	26'8"	13'4"	26'10 1/2"	13'6 1/2"	27'1"	13'9"
45	30'0"	15'0"	30'2 13/16"	15'2 13/16"	30'5 5/8"	15'5 5/8"
50	33'4"	16'8"	33'7 1/8"	16'11 1/8"	33'10 1/4"	17'2 1/4"

Modular coursing of masonry units. (*From Brick Institute of America,* **Principles of Brick Masonry**, *BIA, Reston, Va., 1989; and Frank A. Randall and William C. Panarese,* **Concrete Masonry Handbook for Architects, Engineers, and Builders**, *Portland Cement Association, Skokie, Ill., 1991.*)

for level, plumb, and alignment. For filling in between the corners of a wall, a string line is stretched from end to end and the top outside edge of each unit is laid to this line. Use of the mason's level between corners is then limited to checking the face of the units to assure that they are in the same plane. This speeds construction time and assures greater accuracy.

The story pole or corner pole is used to simplify the accurate location of course heights. They are generally of metal with adjustable coursing

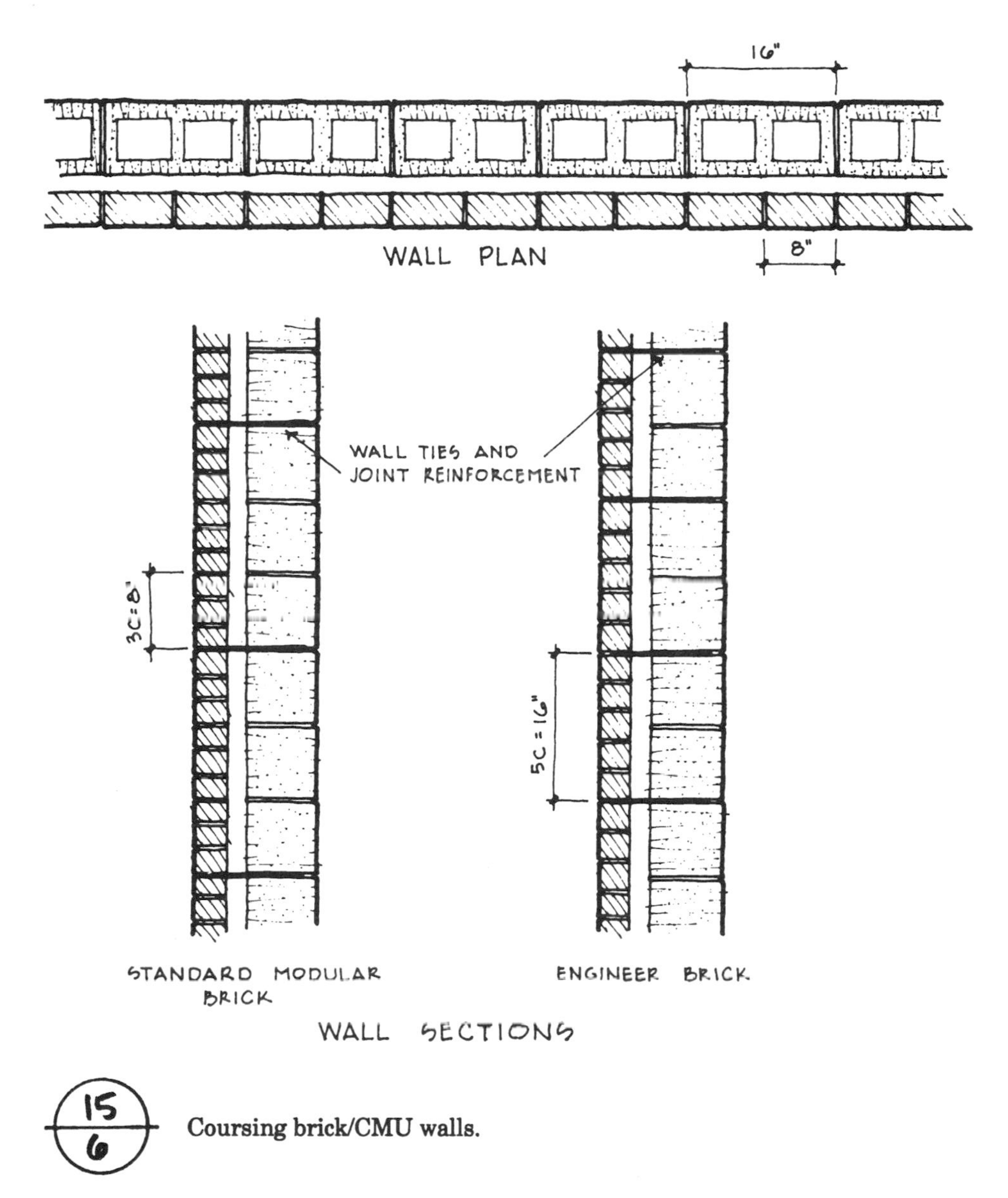

Coursing brick/CMU walls.

scales attached. The poles must be rigid enough to resist bending when a string line is pulled from one side, easily attachable to the masonry walls at the corners, and easily plumbed and maintained for the height of the wall. Using corner poles assures that the brick coursing is uniform and level, that the wall is plumb, and that joint thicknesses are more uniform than attainable by normal methods.

15.3 INSTALLATION

Masonry construction includes the placement of mortar, units, anchors, ties, reinforcement, grout, and accessories. Each element of the construction performs a specific function, and should be installed in accordance with recommended practice.

15.3.1 Mortar and Unit Placement

Mortar is the cementitious material that bonds units, connectors, and reinforcement together for strength and weather resistance. Although it con-

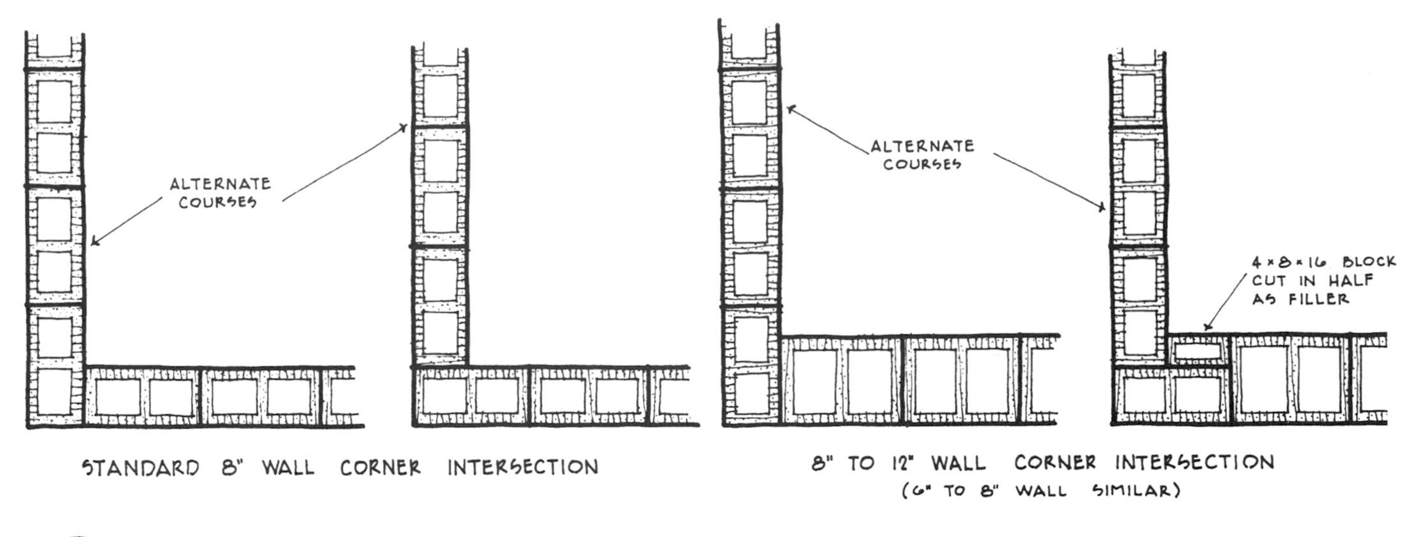

Modular corner layouts.

tributes to the compressive strength of the assemblage, its primary functions are in providing flexural and tensile bond and in sealing the joints between units against the passage of air and water. To perform these functions, it must be properly mixed and placed to achieve intimate contact with the unit surface and form both a mechanical and chemical interlock (refer to Chapter 6).

Tests conducted by the National Institute of Standards and Technology have proven that masonry walls with full head and bed joints are stronger and less likely to leak than walls with furrowed bed joints and lightly buttered head joints. Partially filled mortar joints reduce the flexural strength of masonry by as much as 50 to 60%, offer only minimal resistance to moisture penetration, and may contribute to spalling and cracking if freezing occurs when the units are saturated. Bed joints should be laid full and unfurrowed, only slightly beveled away from the cavity to avoid mortar droppings (*see Fig. 15-13*). The ends of the units should be fully buttered with mortar so that when they are shoved into place, mortar is extruded from the joint (*see Fig. 15-14*). Concrete block should always be laid with the thicker end of the face shell up to provide a larger mortar bedding area. For face shell bedding of hollow CMUs, only the end flanges of the face shells are buttered with mortar. Because of their weight and difficulty in handling, masons often stand several units on end and apply mortar to the flanges of three or four units at one time. Each block is then individually placed in its final position, tapped down into the mortar bed, and shoved against the previously laid block, thus producing well-filled vertical head joints at both faces of the masonry. When installing the last closure unit in a course, all edges of the opening and all vertical edges of the unit should be buttered and the unit carefully lowered into place. If any of the mortar falls out, leaving a void in the joint, the closure unit should be removed and the operation repeated.

Bed joint mortar should be spread only a few units at a time so that the mortar will not dry excessively before the next course of units is placed. For both brick and block, a long mason's level is used as a straightedge to assure correct horizontal alignment. Units are brought to level and

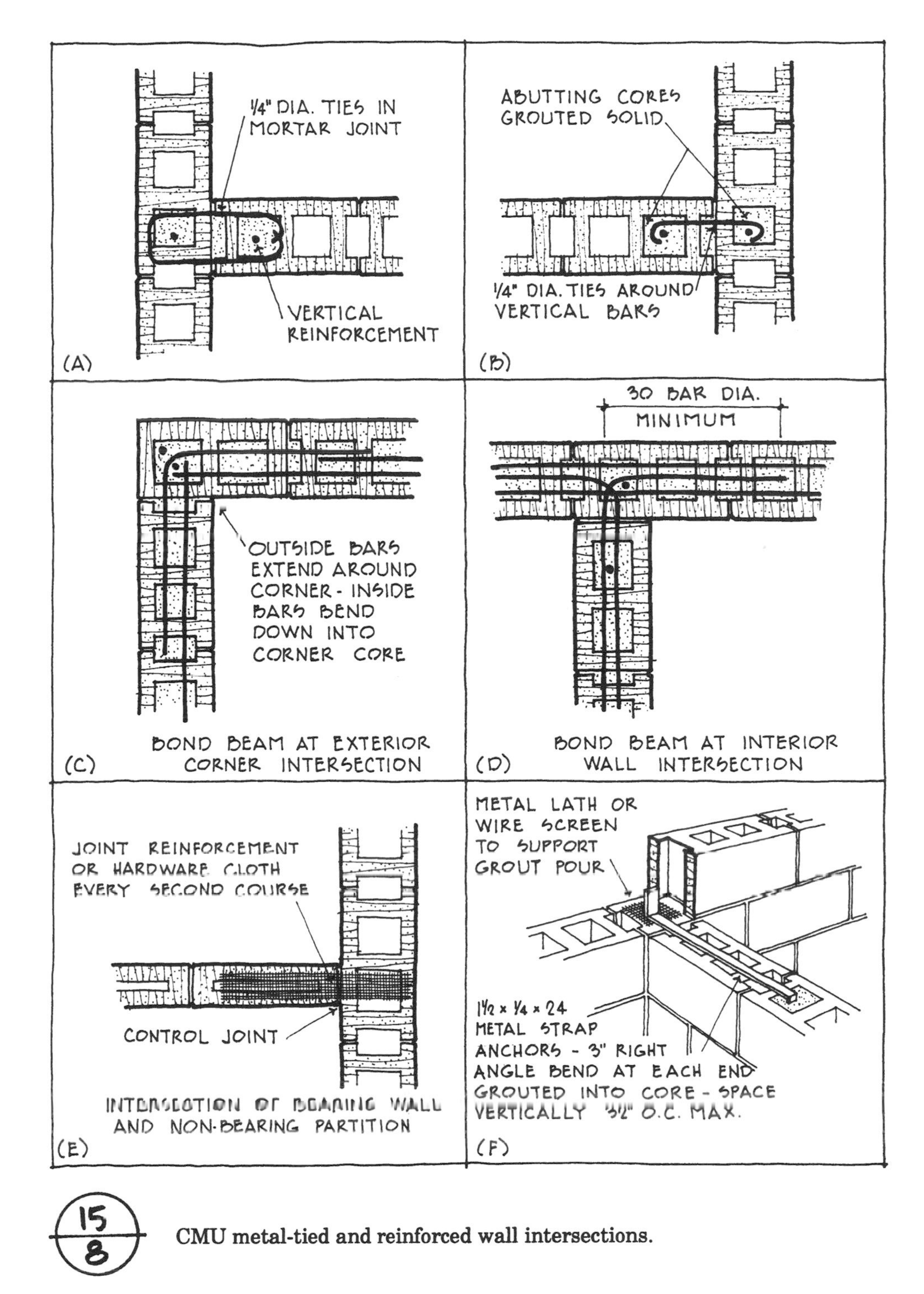

CMU metal-tied and reinforced wall intersections.

made plumb by light tapping with the trowel handle. This tapping, plus the weight of the unit and those above helps form a good bond at the bed joint. Once the units have been laid, they cannot be adjusted or realigned by tapping without breaking the bond. If it is necessary to reposition the masonry, all the old mortar must be removed and replaced with fresh.

In cavity wall and veneer wall construction, it is extremely important that the cavity between the outer wythe and the backing wall be kept

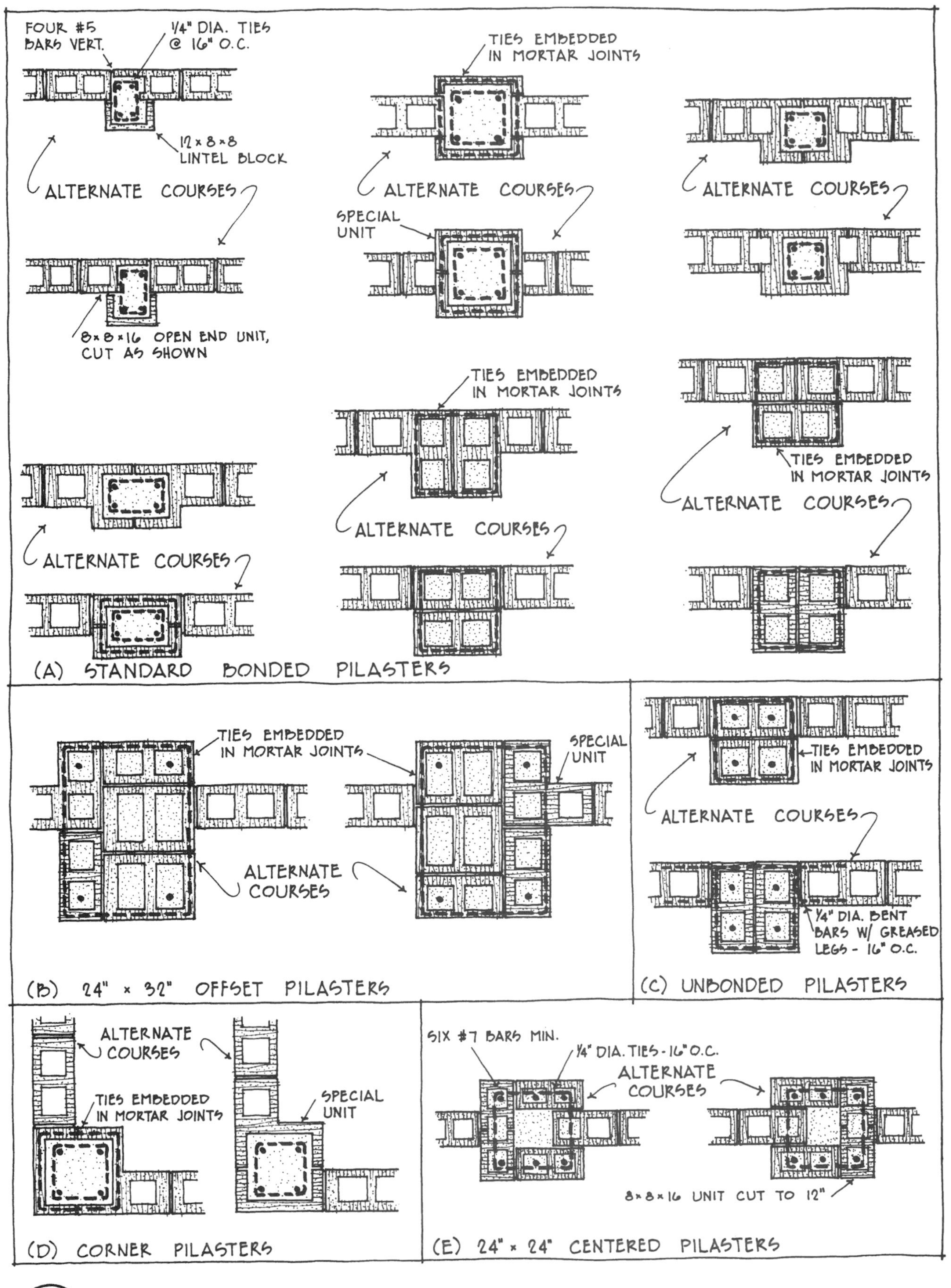

CMU pilaster coursing.

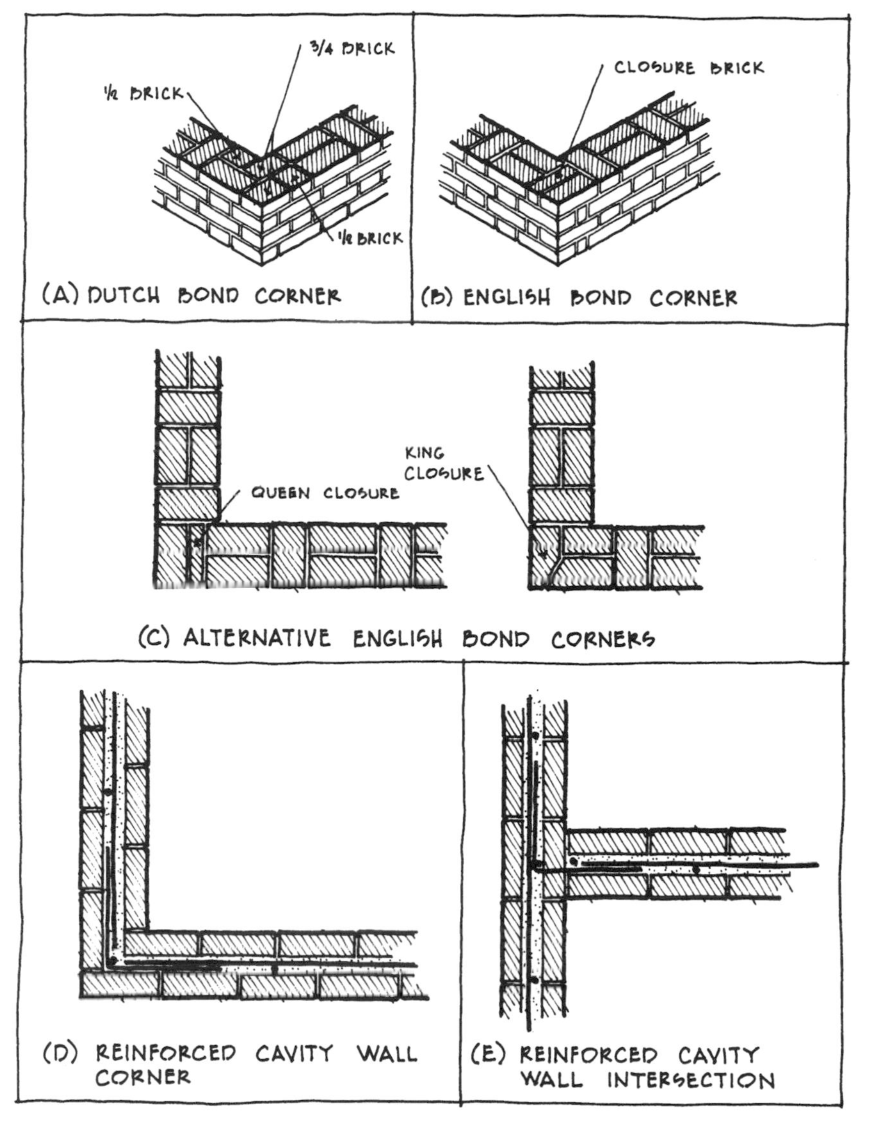

Brick corners and intersecting walls (A, B, and C seldom used today except in restoration work). (*From Harry C. Plummer,* **Brick and Tile Engineering,** *Brick Institute of America, Reston, Va., 1962.*)

clean to assure proper moisture drainage. If mortar clogs the cavity, it can form bridges for moisture passage, or it may block weepholes. Some masons use a removable wooden strip to temporarily block the cavity as the wall is laid up and prevent mortar droppings. However, beveling the mortar bed as shown in *Fig. 15-13* allows little mortar to extrude toward the cavity. Any mortar fins that may protrude into the cavity should be cut off or flattened to prevent interference with the placement of reinforcing steel, grout, or insulation.

Use of the various types of insulation covered in Chapter 8 will affect the manner in which the masonry walls are laid up. In veneer construction over wood frame, the board or batt insulation and the corrugated metal ties are placed against the frame before the masonry work is begun. If

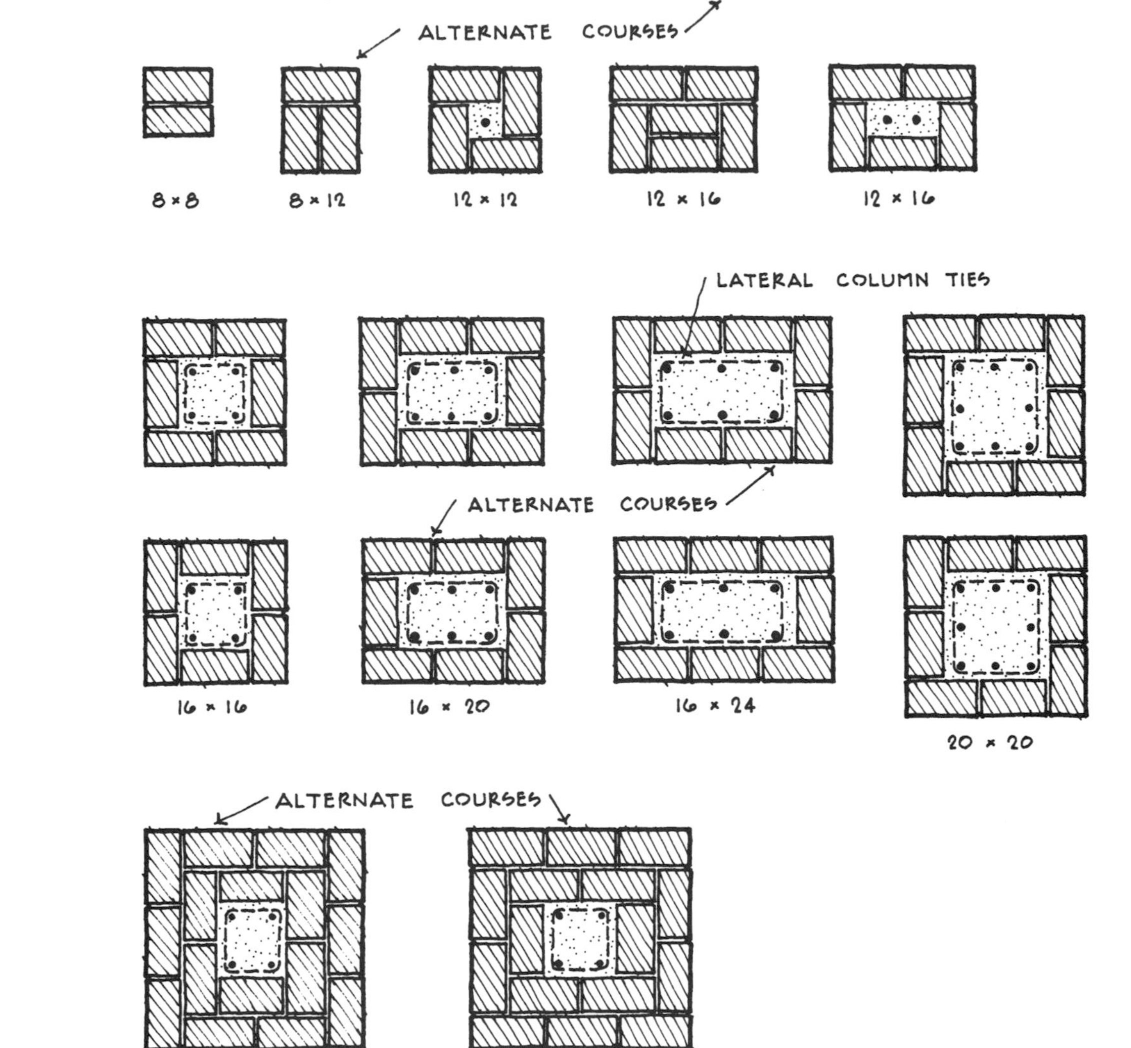

Brick column coursing. (*From Harry C. Plummer,* **Brick and Tile Engineering,** ***Brick Institute of America, Reston, Va., 1962; and Brick Institute of America,*** **Principles of Brick Masonry,** ***BIA, Reston, Va., 1989.***)

rigid board insulation is used in insulated masonry cavity walls, the backing wall must be laid up higher than the facing wall so that the boards may be attached to it before the facing wythe covers it. If the masons are working overhand from inside the building (as they often do on multistory construction), this makes the insulating process more awkward, and therefore less economical. In these cases, the masons would work better from scaffolding on the outside of the building, but the cost of the installation would increase.

Loose fill insulation does not require that the two wythes of masonry

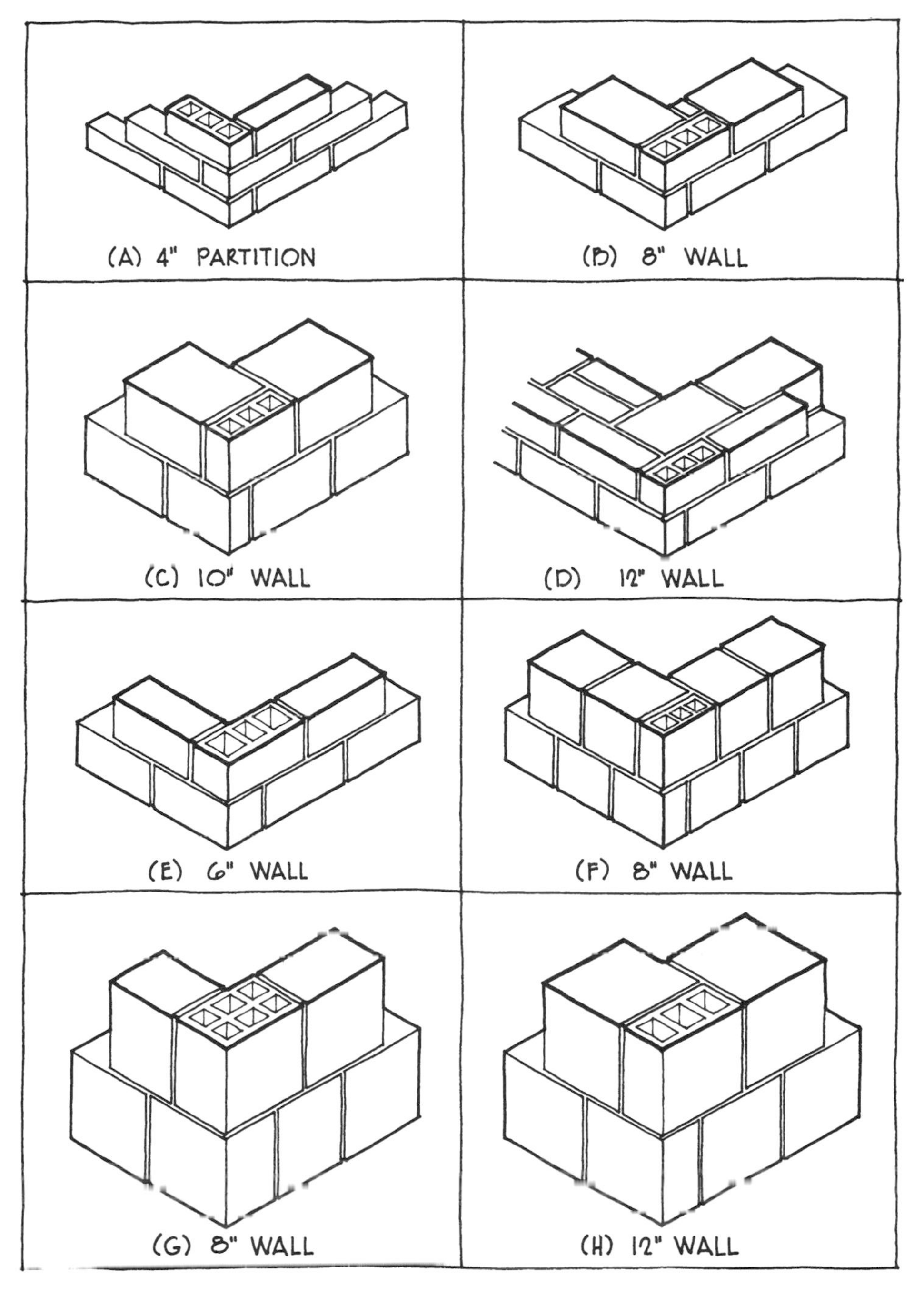

Structural clay tile corner details. (*From Brick Institute of America,* **Technical Note 22,** *BIA, Reston, Va.*)

be laid up separately. Both the inner and outer wythes can be laid up simultaneously, and the insulation poured or pumped into place at designated vertical intervals.

To add visual interest to masonry walls, units may be laid in different positions as shown in *Fig. 15-15,* and arranged in a variety of patterns (*see Figs. 15-16 and 15-17*). The patterns were originally conceived in connection

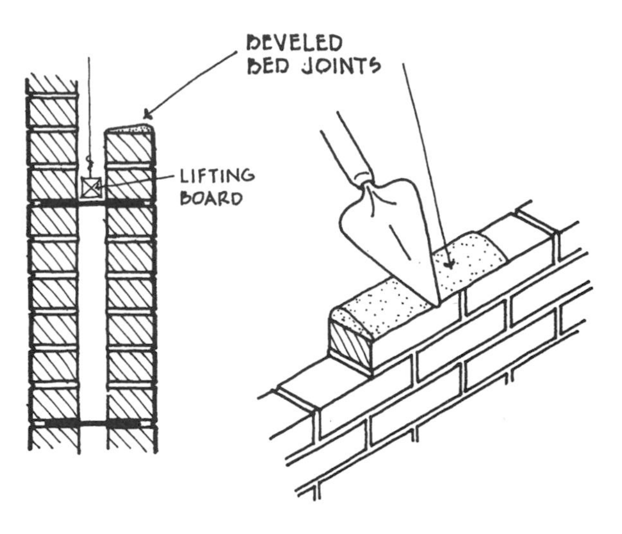

Beveled bed joints help prevent mortar droppings from filling the cavity between wythes. (*From Brick Institute of America,* Technical Note 21C, *BIA, Reston, Va.*)

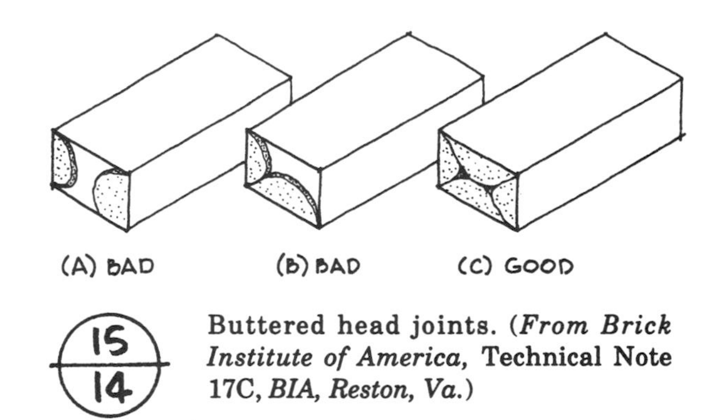

15
14

Buttered head joints. (*From Brick Institute of America,* Technical Note 17C, *BIA, Reston, Va.*)

with masonry wall bonding techniques that are not widely used today. In older work constructed without metal ties or reinforcement, rowlock and header courses were used to structurally bond the wythes of a wall together. Most contemporary buildings use the one-third or one-half running bond, or stack bond with very little decorative pattern work. In cavity wall construction, half rowlocks and half headers may be used for aesthetic effect on the exterior without the unit actually penetrating the full thickness of the wall (*see Fig. 15-18*).

Brick soldier and sailor courses should be installed carefully to prevent mortar from slumping in the tall head joints, leaving voids which might be easily penetrated by moisture. Units used for sailor or shiner courses must be solid and uncored. Vertical coursing between backing and facing wythes must also be coordinated to accommodate ties and anchors.

Masonry arches may be built of special brick or stone shapes to obtain mortar joints of constant thickness, or of standard brick units with joint thickness varied to obtain the required curvature. The method select-

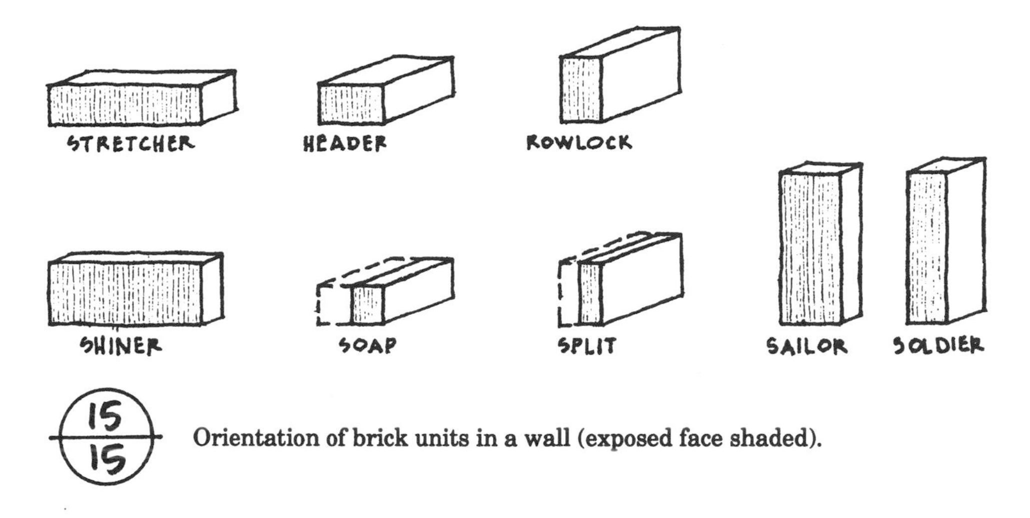

Orientation of brick units in a wall (exposed face shaded).

ed should be determined by the arch dimensions and by the desired appearance. It is especially important in a structural member such as an arch that all mortar joints be completely filled. Brick arches are usually built so that units at the crown will be laid in soldier bond or rowlock header bond (refer to *Fig. 11-13*). Under many circumstances, it is difficult to lay units in soldier bond and still obtain full joints. This is especially true where the curvature of the arch is of short radius with mortar joints of varying thickness. In such cases, the use of two or more rings of rowlock headers is recommended. In addition to facilitating better jointing, rowlock headers provide a bond through the wall to strengthen the arch.

Mortar color and joint type can be just as important in determining the appearance of a wall as the selection of a unit type or color, and should be carefully considered in the design of the building. Sample panels at the job site can help in evaluating workmanship and appearance of the finished work, and should always be specified on jobs of any mentionable size to assure that the desired effect can be achieved.

Variations in aesthetic effect can be achieved by using different types of mortar joints. Two walls with the same brick and the same mortar color can have a completely different appearance depending on the joint treatment used. There are several types of joints common today (refer to *Fig. 9-31*).

Concave or V tooled joints are recommended for use in areas subject to heavy rains and high winds. *Rough-cut* or *flush* joints are used when other finish materials, such as stucco, gypsum board, or textured coatings, are to be applied over the masonry. *Weathered* joints are more difficult to form since they must be formed from below, but some compaction does occur, and the joint sheds water naturally. *Struck* joints are easily cut with a trowel point, but the small ledge created collects water, which may then penetrate the wall. *Raked* joints are made by scraping out a portion of the mortar while it is still soft, using a square-edged tool. Even though the mortar is slightly compacted by this action, it is difficult to make the joint weather-resistant, and it is not recommended where rain, high winds, or freezing are likely to occur. The cut of the joint does form a shadow, and tends to give the wall a darker appearance. *Weeping* joints leave excess mortar protruding from the joint to give a rustic appearance, but again are not weather-resistant. Other, more specialized effects can be achieved with tools to bead or groove the joint.

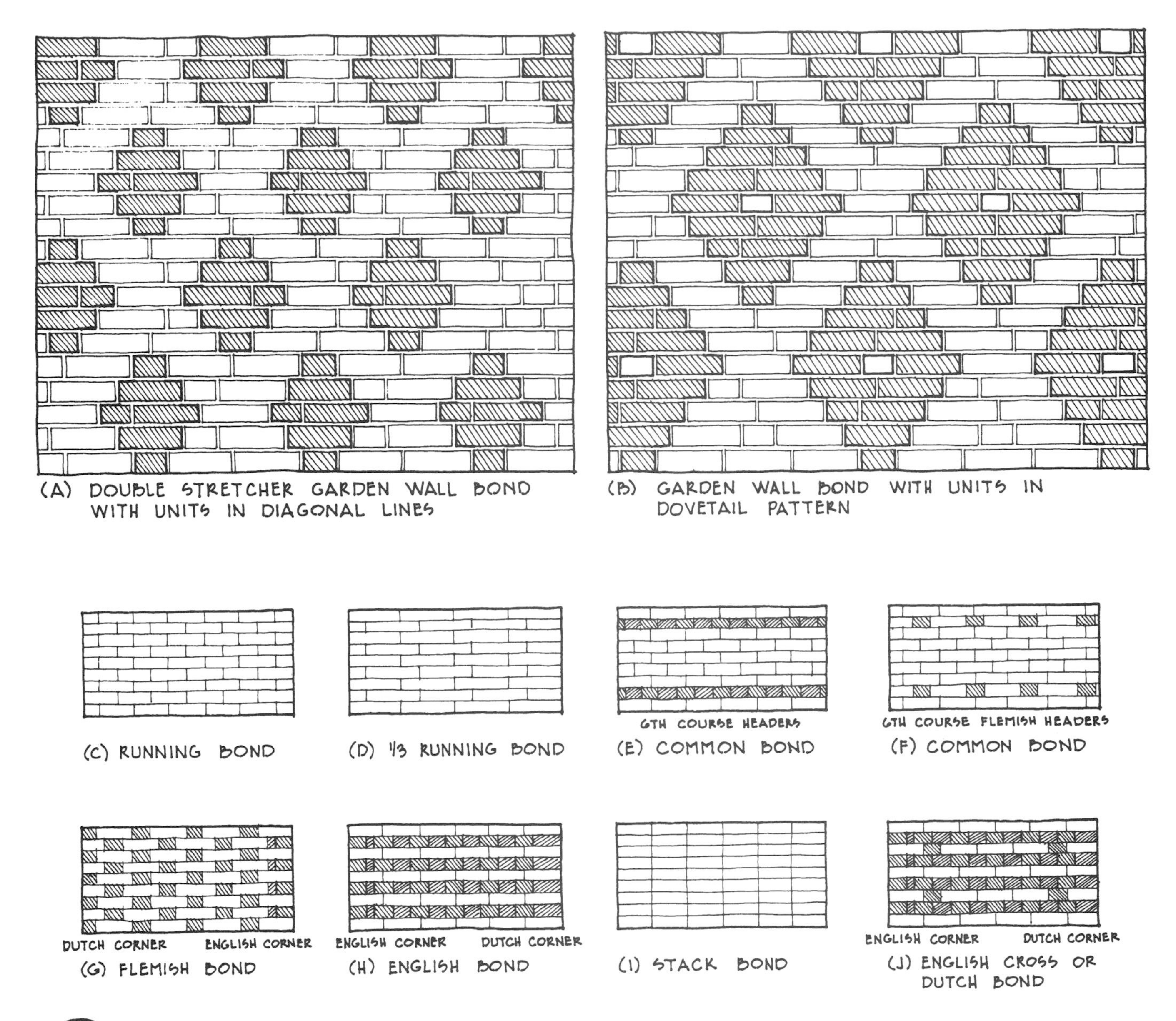

Brick bond patterns. (*From Brick Institute of America,* Technical Note 30, *BIA, Reston, Va.*)

The most effective and moisture resistant joints are the *concave* and *V-shaped* tooled joints. Mortar squeezes out of the joints as the masonry units are set in place, and the excess is struck off with a trowel. After the mortar has become "thumbprint" hard (i.e., when a clear thumbprint can be impressed and the cement paste does not stick to the thumb), joints are finished with a jointing tool slightly wider than the joint itself. As the mortar hardens, it has a tendency to shrink slightly and separate from the edge of the masonry unit. Proper tooling compresses the mortar against the unit and compacts the surface making it more dense and more resistant to moisture penetration. However, full head and bed joints and good mortar bond are more critical to moisture resistance than tooling. Other joint treat-

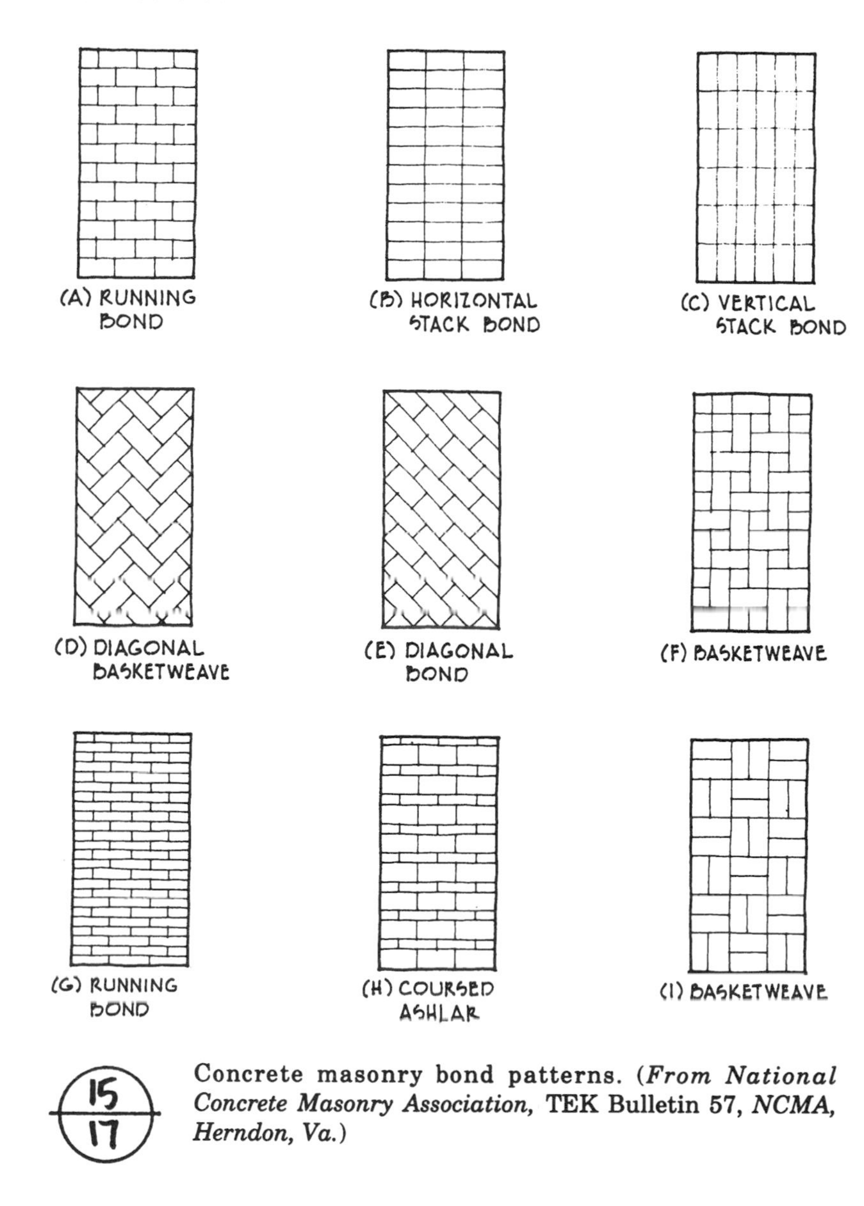

15/17 **Concrete masonry bond patterns. (*From National Concrete Masonry Association,* TEK Bulletin 57, *NCMA, Herndon, Va.*)**

ments may be used in mild to moderate exposures (refer to Chapter 9) if the workmanship is good, the bond between units and mortar is complete and intimate, and the flashing and weephole system is properly designed and installed.

Horizontal joints should be tooled before vertical joints, using a jointer that is at least 22 in. long and upturned on one end to prevent gouging. Jointers for vertical tooling are small and S-shaped. Although the material most commonly used for these tools is steel, plastic jointers are used to avoid darkening or staining white or light-colored mortars. After the joints have been tooled, mortar burrs or ridges should be trimmed off flush with the face of the unit with a trowel edge, or by rubbing with a burlap bag, a brush, or a piece of carpet.

It is important that the moisture content of the mortar be consistent at the time of tooling, or color variations may create a blotchy appearance in the wall (*see Fig. 15-19*). Drier mortar tools darker than that with a higher moisture content. Along with time and weather conditions, unit

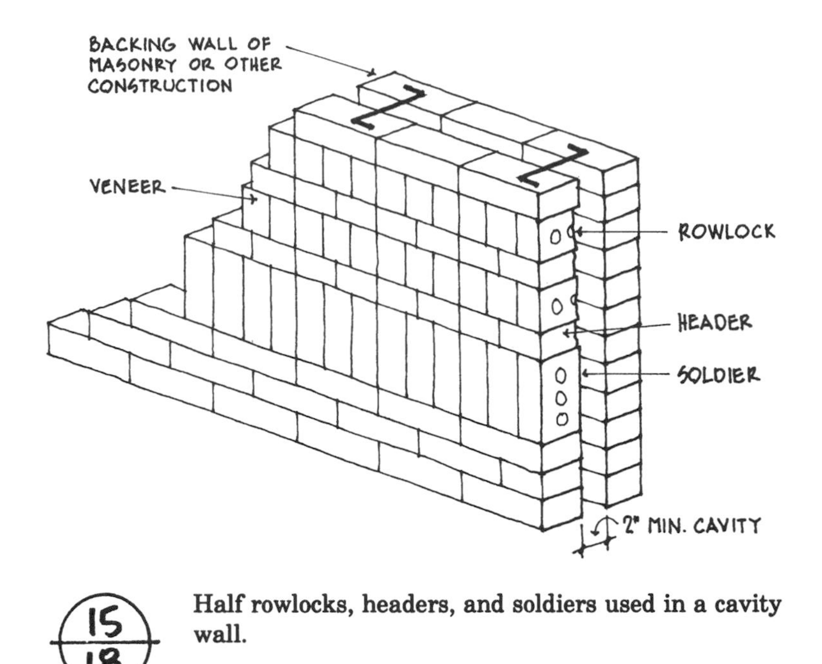

15-18 **Half rowlocks, headers, and soldiers used in a cavity wall.**

suction affects the rate at which the mortar loses its mixing water. Units that have not been protected from accidental wetting at the construction site will have inconsistent suction, as those at the top and sides of pallets absorb rain or melting snow. When placed in the wall, units with varying suction will cause inconsistent mortar drying rates and inconsistent color in the finished joints. Keeping the units covered will help prevent variations in the color of the tooled mortar joints.

Even with high quality workmanship, some patching or repair of mortar joints may be expected. In addition, any holes left by nails or line pins must be filled with fresh mortar before tooling. The troweling of mortar into joints after the units are laid is known as *pointing*. It is preferable that pointing and patching be done while the mortar is still fresh and plastic, and before final tooling of the joints is performed. If however, the repairs must be made after the mortar has hardened, the joint must be raked or chiseled out to a depth of about ½ in., thoroughly wetted, and repointed with fresh mortar.

15.3.2 Flashing and Weepholes

Flashing must be installed in continuous runs with all seams and joints lapped 4 to 6 in. and sealed with a nonhardening mastic or caulking material. Unsealed lap joints will allow water to flow around the end of the flashing and penetrate the wall. Inside and outside corners can be fabricated of metal, or preformed rubber corner boots can be used, even with metal flashing systems (*see Fig. 15-20*). At lateral terminations where the flashing abuts other construction elements, and at terminations on each side of door and window lintels and window sills, flashing must be turned up to form an end dam. Metal flashing can be cut, folded, and soldered or sealed with mastic to form a watertight pan, and flexible flashing can be folded into place. Without end dams, water that collects on the flashing is free to run off the ends and down into the wall (*see Fig. 15-21*).

Drier mortar tools to a darker color than mortar that is wetter when tooled.

Flashing should never be stopped short of the face of the wall, or water may flow around the edge and back into the wall. Metal flashing should be brought out beyond the wall face and turned down to form a drip. A hemmed edge will give the best appearance. Flexible flashing cannot be formed in the same way, but should be extended beyond the face of the wall and later trimmed flush with the joint. Some designs may call for flexible or bituminous flashing to be lapped and sealed over the top of a separate metal drip edge (*see Fig. 15-22*). Two-piece flashing can also be used, even with all metal systems, to accommodate construction tolerances in the necessary length of the horizontal leg. The vertical leg of the flashing should be turned up several inches to form a back dam and be placed in a mortar joint in the backing wythe, in a reglet, or behind the sheathing of the backing wall (see *Fig. 15-23*).

Weepholes are required in masonry construction at the base course and at all other flashing levels (such as shelf angles, sills, and lintels) so that water which is collected on the flashing may be drained from the wall as quickly and effectively as possible. Weepholes should be spaced 16 to 24 in. on center depending on the method used:

- Oiled rods, rope, or pins placed 24 in. on center in the head joints and removed before final set of the mortar (*see Fig. 15-24*)
- Cotton sash cord or other suitable wicking material placed 16 in. on center in the head joint (*see Fig. 15-25*)
- Metal or plastic tubes placed 16 in. on center in the head joints (*see Fig. 15-26*)
- Mortar omitted from head joints at 24 in. on center to leave a large drainage void (*see Fig. 15-27*).

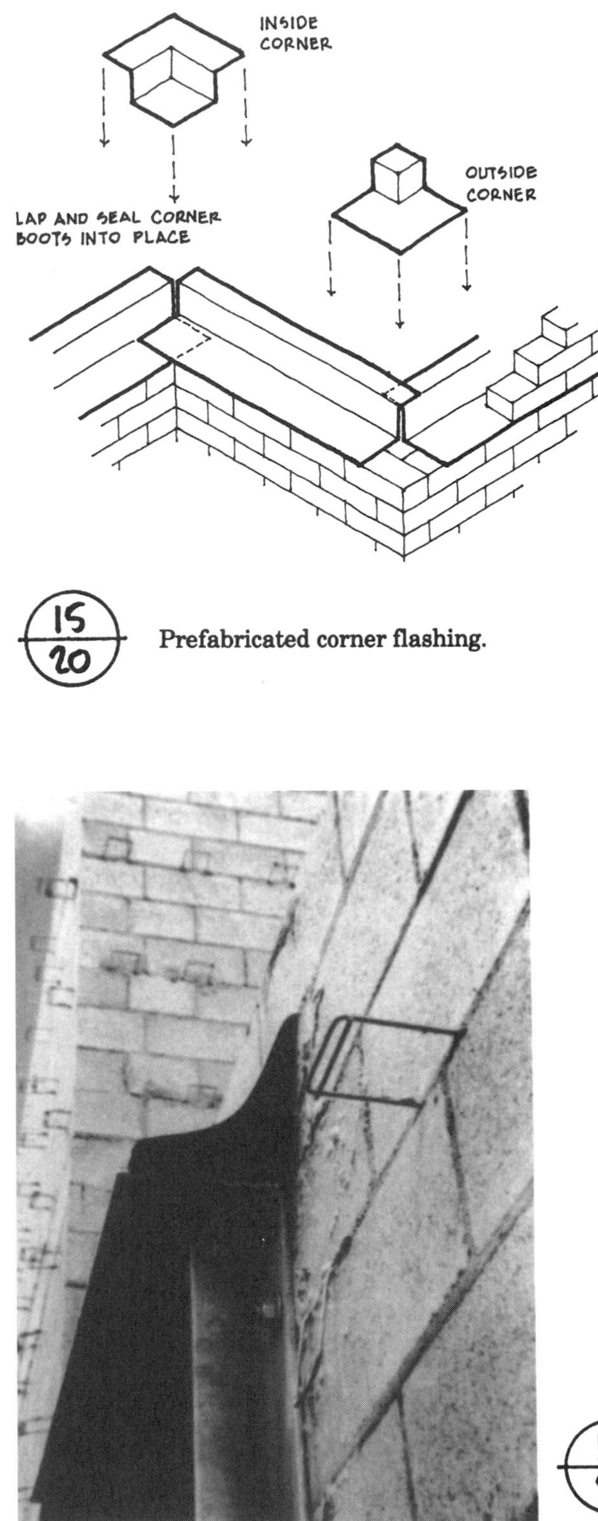

15/20 Prefabricated corner flashing.

15/21 Water runs off the ends of flashing that is not dammed at terminations.

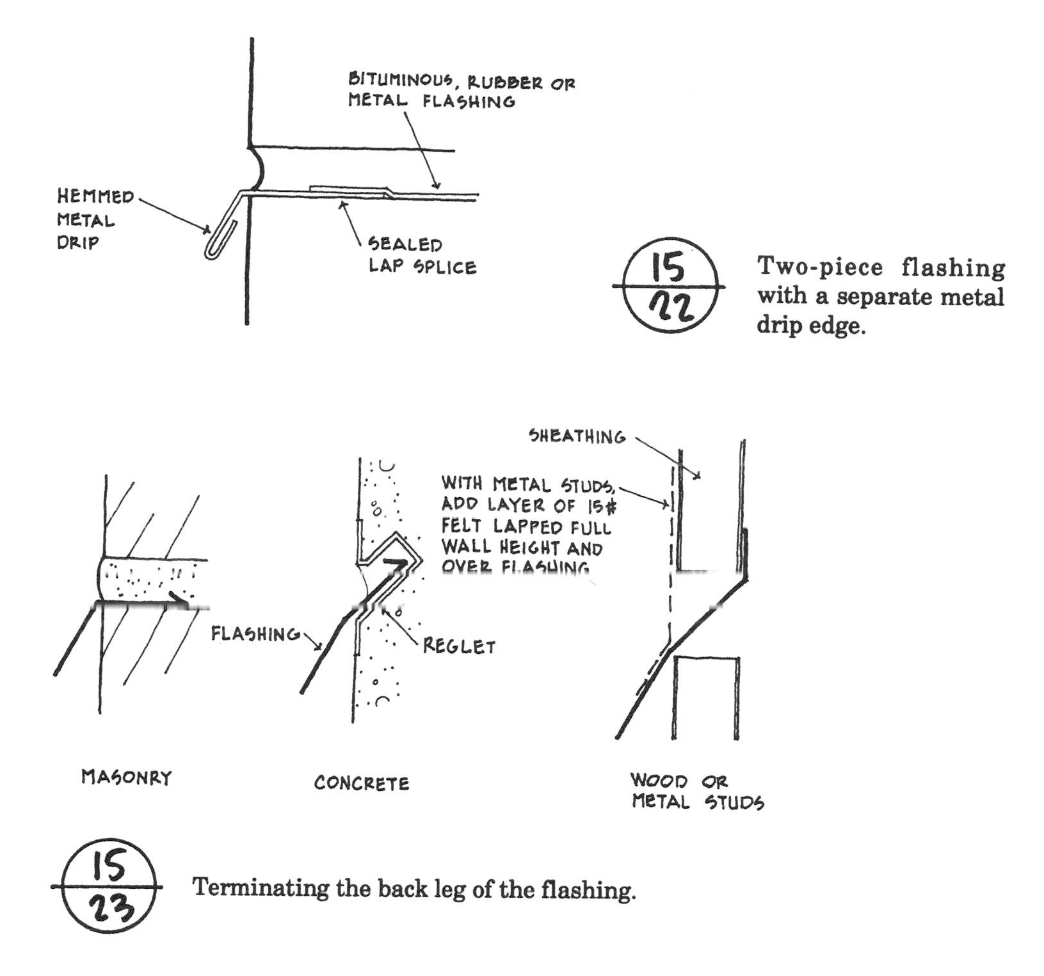

Two-piece flashing with a separate metal drip edge.

Terminating the back leg of the flashing.

To function properly, weepholes must be unobstructed by mortar droppings or other debris. Blocked or missing weepholes can cause saturation of the masonry just above the flashing as moisture is dammed in the wall for longer periods of slow evaporation. Efflorescence, staining, corrosion of steel lintels or studs, and freeze-thaw damage can result (*see Fig. 15-28*). Weephole tubes are most vulnerable to blockage, even when gravel drainage beds are used. Chapter 7 discusses the advantages and disadvantages of the different weephole drainage methods and identifies various accessory items used for functional or aesthetic purposes.

15.3.3 Control and Expansion Joints

Control joints and expansion joints are used to relieve stresses caused by differential movement between materials, and by thermal and moisture movement in the masonry itself (refer to Chapter 9). *The terms control joint and expansion joint are not interchangeable.* The two types of joints are different in both function and configuration.

Control joints are continuous vertical head joints constructed with or without mortar, to accommodate the permanent moisture *shrinkage* which all *concrete masonry* units experience. When shrinkage stresses are sufficient to cause cracks, the cracking will occur at these weakened joints rather than at random locations. Shear keys are used to provide lateral stability against wind loads (refer to *Fig. 9-12*), and elastomeric sealants are used to provide a

Oiled rope weeps are removed after the mortar has set. (*Photo courtesy BIA.*)

Cotton sash cord or rope wicks water from the cavity. (*Photo courtesy BIA.*)

water tight seal (*see Fig. 15-29*). Mortared control joints must be raked out to a depth which will allow placement of a backer rod or bond-breaker tape to prevent three-sided adhesion, and a sealant joint of the proper depth. Concrete masonry shrinkage always exceeds expansion because of the initial moisture loss after manufacture. So even though control joints contain hardened mortar, they can accommodate reversible thermal expansion and contraction because it occurs after the initial curing shrinkage.

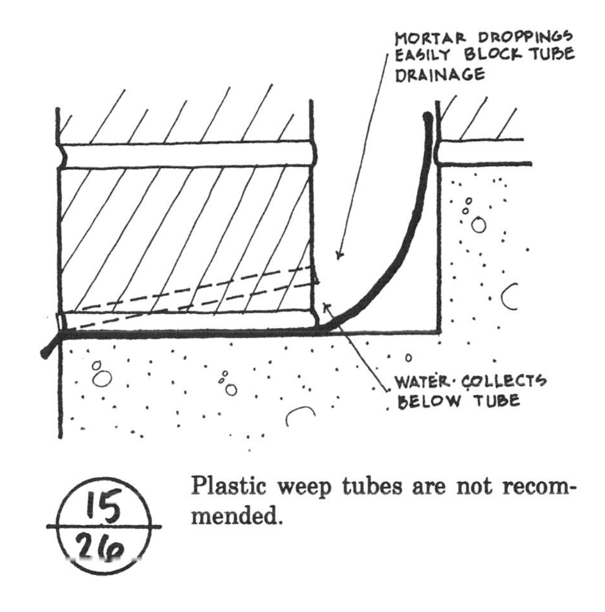

15/26 Plastic weep tubes are not recommended.

15/27 Open head joint weepholes drain moisture quickly.

Expansion joints are used in *brick, terra cotta,* and *structural clay tile* construction to accommodate the permanent moisture *expansion* which all clay masonry products experience as they reabsorb atmospheric moisture after firing. Expansion joints are also used in stone masonry to accommodate thermal movement. Clay masonry moisture expansion always exceeds reversible thermal expansion and contraction, so expansion joints cannot contain mortar or other hard materials (*see Fig. 15-30*). Lateral support is provided in veneer construction by placing an anchor on either side of the joint. During construction, plywood strips can be used to prevent mortar

Moisture is trapped in the wall when weepholes are obstructed by mortar droppings, and may cause efflorescence, corrosion of metal components, or leakage.

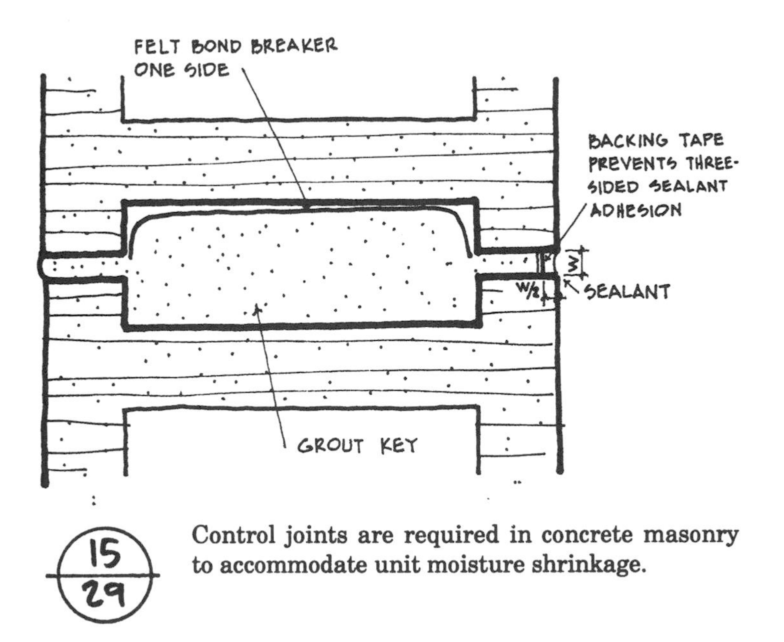

15
29

Control joints are required in concrete masonry to accommodate unit moisture shrinkage.

from bridging the expansion joint and restricting subsequent movement, but such rigid materials *must* be removed when the masonry construction is complete. Soft joint filler materials such as neoprene rubber sponge may also be used to keep mortar out of the joint during construction, but may be left in place only if they are sufficiently compressible to allow calculated movement to occur. Joint fillers that are left in place must be set deep enough in the joint to allow room for a backer rod and a sealant joint of the proper depth (refer to Chapter 9). Filler materials should not be used as

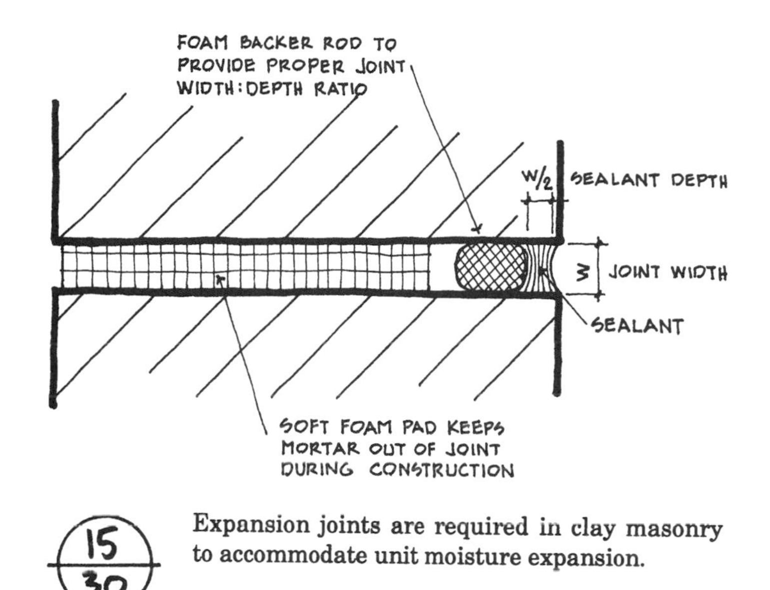

Expansion joints are required in clay masonry to accommodate unit moisture expansion.

backing for the sealant because of potential problems with compatibility, adhesion, and consistent joint width to depth ratios.

In cavity wall construction of brick with block backup, control joints and expansion joints in the backing and facing wythes, respectively, should occur at approximately the same locations. Joint reinforcement should not continue across movement joints.

When control or expansion joints are located adjacent to openings spanned by loose steel lintels, movement of the lintel itself must be taken into consideration. The joint may be offset to the end of the lintel, or the lintel may be isolated between two sheets of flashing to permit slippage of the steel angle independent of the masonry (*see Fig. 15-31*). Mortar is omitted from the horizontal joint at the face of the lintel, and replaced with sealant and backer rod. This allows the movement joint to be aligned with the edge of the opening without causing the mortar to spall and expose the steel to weathering and corrosion. The lintel depth from the face of the wall must be sufficient to permit placement of a backer rod, or the sealant may suffer cohesive failure because of three-sided adhesion.

15.3.4 Accessories and Reinforcement

Metal ties, anchors, horizontal joint reinforcement, and steel reinforcing bars are all placed by the mason as the work progresses. Anchors, ties, and joint reinforcement must be corrosion-resistant, properly spaced, and placed in the mortar to assure complete encapsulation and good bond. Since mortar is spread only a limited distance along bed joints to avoid excessive evaporation, long sections of joint reinforcement are usually laid directly on the units and lifted slightly after the mortar is placed. All metals should be protected by a minimum ⅝-in. mortar cover at exterior joint faces (*see Fig. 15-32*).

Vertical reinforcement in a cavity wall is easily placed, and the masonry built up around it. Spacers are required at periodic intervals to hold the reinforcing bars in vertical alignment. If horizontal steel is required in the cavity, it is tied to the vertical members or may rest on the spacers at the proper intervals (*see Fig. 15-33*).

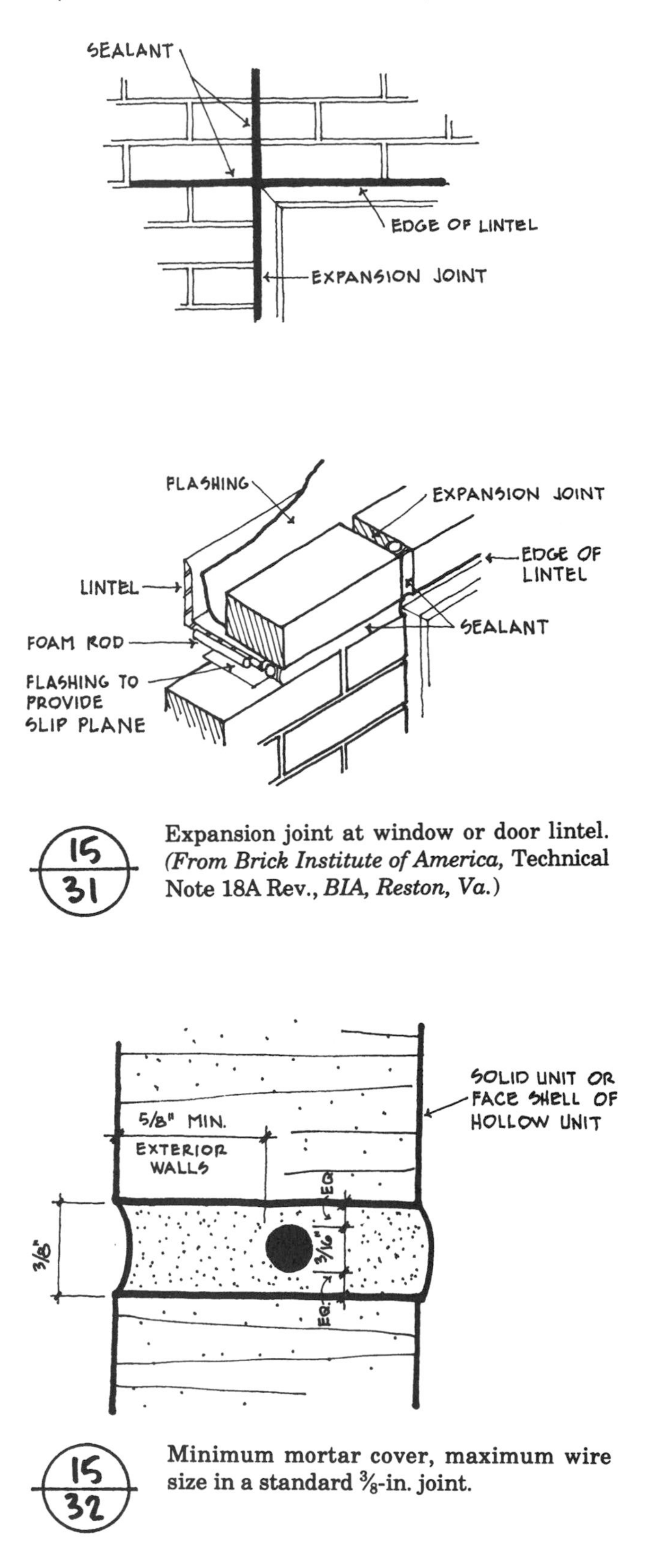

Expansion joint at window or door lintel. *(From Brick Institute of America,* Technical Note 18A Rev., *BIA, Reston, Va.)*

Minimum mortar cover, maximum wire size in a standard ⅜-in. joint.

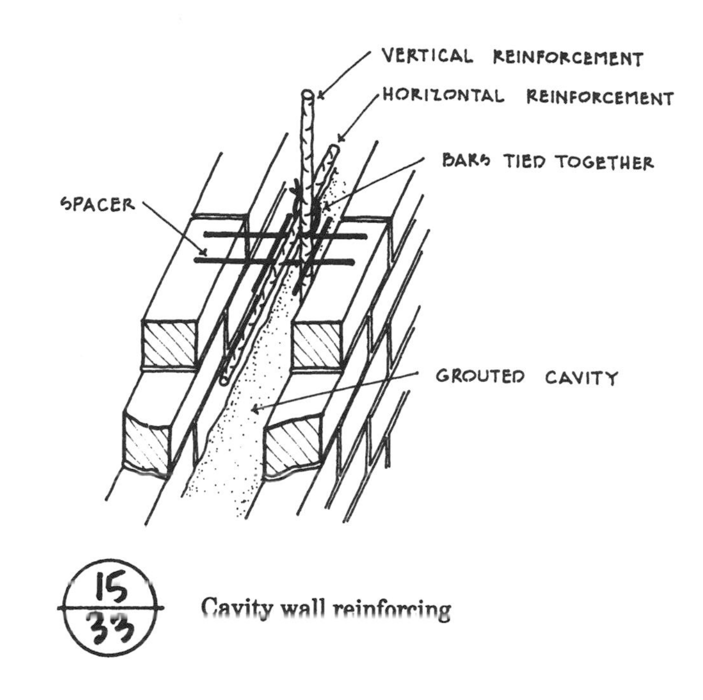

15/33 Cavity wall reinforcing

For single-wythe reinforced CMU walls, special open-end units are made so that the block may be placed around the vertical steel rather than threaded over the top of the bar (*see Fig. 15-34*). Some specially designed blocks have been produced which can accommodate both vertical and horizontal reinforcing without the need for spacers. The proprietary block shown in *Fig. 15-34* not only has open ends, but also incorporates notches in the webs for placement of horizontal bars. This type of unit is very economical for grouted, reinforced CMU walls, particularly when the design utilizes wall beams and bond beams requiring large quantities of horizontal steel.

During the course of construction, the mason also places anchorages and cutouts required to fit the work of other trades. These items are furnished and located by others, but incorporated into the wall by the mason. Steel or precast lintels for small openings are also placed by the mason if reinforced masonry lintels are not used in the design. Metal shims used for alignment of steel lintels and shelf angles should be the full height of the vertical leg to prevent rotation.

15.3.5 Grouting

In reinforced masonry construction, the open collar joint of a double-wythe wall, or the vertical cells of hollow units, must be pumped with grout to secure the reinforcing steel and bond it to the masonry.

For hollow masonry construction, the cells that are to be grouted must be fully bedded in mortar, including the webs and face shell flanges. In both brick and CMU work, the importance of keeping the cavity clean has been stressed before, but should be reemphasized here. Protrusions or fins of mortar which project into the cavity will interfere with proper flow and distribution of the grout, and could prevent complete bonding. The spacers used to maintain alignment of vertical reinforcing will assure complete coverage of the steel and full embedment in the grout for proper structural performance. If bond beams or isolated in-wall columns are to be poured in a double-wythe

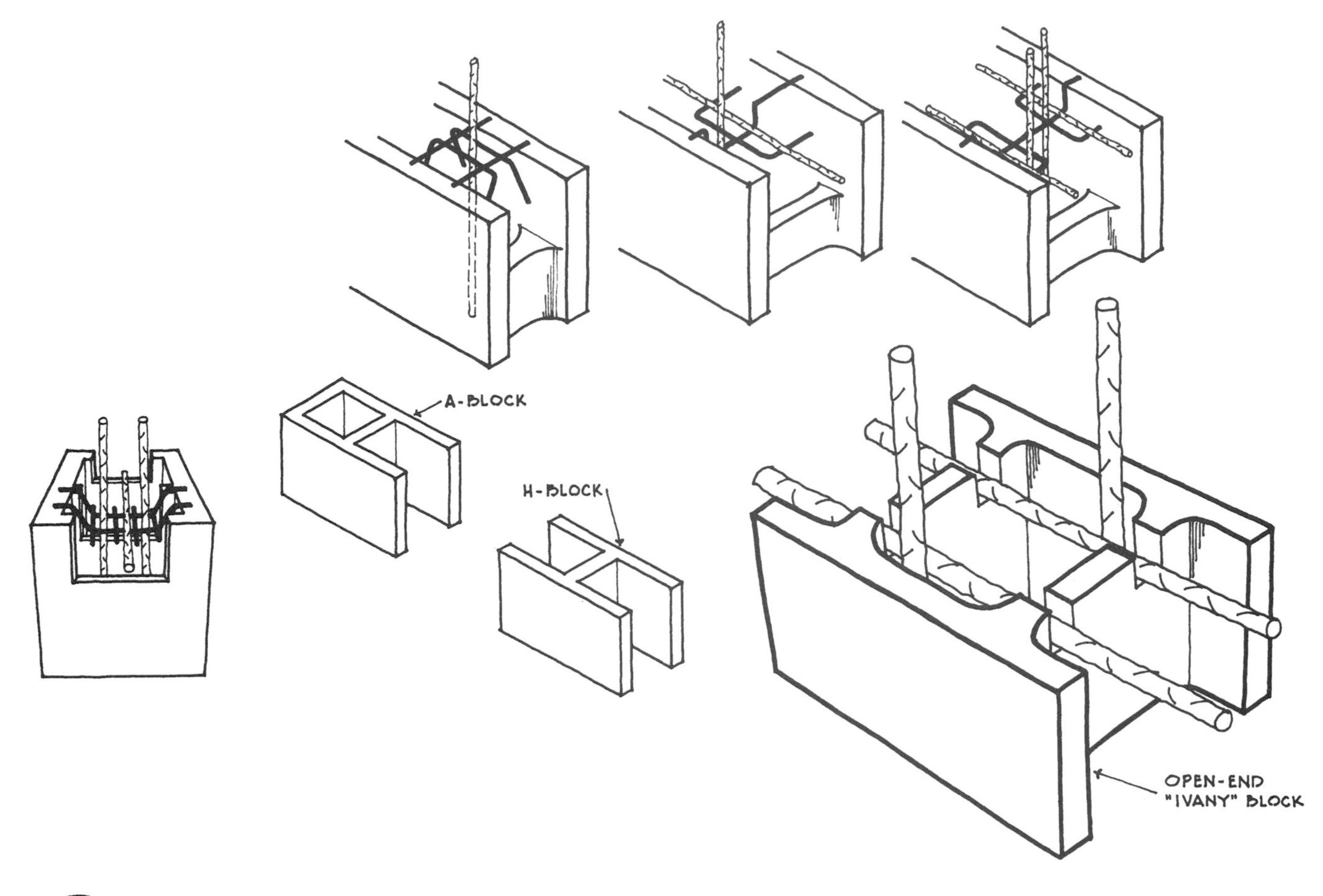

Concrete block reinforcing.

wall, material must be placed below and/or to either side of the area to prevent the grout from flowing beyond its intended location. For example, if a bond beam is to be poured in a double-wythe brick wall, expanded metal lath or metal screen should be placed in the bed joint below to contain the pour (refer to *Fig. 9-18*). Grouting of concrete masonry should be performed as soon as possible after the units are placed so that shrinkage cracking at the joints is minimized, and so that the grout bonds properly with the mortar.

The *low-lift method* of grouting a wall is done in 8-in. lifts as the wall is laid up. For double-wythe wall construction, the first wythe is laid up, followed by the second wythe, which is generally left 8 to 12 in. lower. Grout should be well mixed to avoid segregation of materials, and carefully poured to avoid splashing on the top of the brick, since dried grout will prevent proper mortar bond at the succeeding bed joint. At least 15 minutes should elapse between pours to allow the grout to achieve some degree of stiffness before the next layer is added. If grout is poured too quickly, and the mortar joints are fresh, hydrostatic pressure can cause the wall to bulge out of plumb. A displacement of as little as ⅛ in. will destroy the bed joint bond, and the work must be torn down and rebuilt. The joint rupture will cause a permanent plane of weakness and cannot be repaired by simply realigning the wall.

Bed joints can also be broken by rotation of the brick from uneven suction. To avoid this, the grout level should be kept at or below the center of the top course during construction. If operations are to be suspended for more than 1 hour, however, it is best to build both wythes to the same level, and pour the grout to within ¾ in. of the top of the units to form a key with the next pour. Grout that is in contact with the masonry hardens more rapidly than that in the center of the grout space. It is therefore important that agitation or puddling of the grout take place immediately after the pour and before this hardening begins.

In single-wythe hollow-unit construction, walls are built to a maximum 4 ft height before grout is pumped or poured into the cores. These vertical cores must have the minimum area and dimensions required by the governing building code. Grout is placed in the cores, and then agitated to ensure complete filling and solid embedment of steel.

High-lift grouting operations are not performed until the wall is laid up to full story height. In multi-wythe walls, one wythe is built up not more than 16 in. above the other, and vertical grout barriers of solid masonry are placed a maximum of 25 ft apart. The cross webs of hollow units in single-wythe walls are fully embedded in mortar to contain the grout within the designated area.

Cleanouts must be provided at the base of the wall by leaving out every other unit in the bottom course of the section being poured. In single-wythe hollow unit walls, cleanout openings at least 3×4 in. are located at the bottom of every core containing dowels or vertical reinforcement, and in at least every second core that will be grouted, but has no steel. In solidly grouted, unreinforced single-wythe walls, every other unit in the bottom course should be left out. Building codes generally specify exact cleanout requirements, and should be consulted prior to construction.

A high-pressure air blower is used to remove any debris which may have fallen into the core or cavity. The cleanout plugs are filled in after inspection of the cavity, but before the grouting begins. The mortar joints in a wall should be allowed to cure for at least 3 days to gain strength before grouting by this method. In cold, damp weather, or during periods of heavy rain, curing should be extended to 5 days.

Grout should be placed in a continuous operation with no intermediate horizontal construction joints within a story height. Four-foot maximum lifts are recommended, with 30 to 60 minutes between pours to allow for settlement, shrinkage, and absorption of excess water by the units. In each lift, the top 12 to 18 in. are reconsolidated before or during placement of the next lift.

It is critical that the grout consistency be fluid, and that it be mechanically vibrated into place. When the grout is stiff, it hangs up on the side walls and the reinforcing bars, leaving voids in which the steel is not properly bonded.

15.3.6 Protections

High-lift grouting requires that walls be temporarily braced until the mortar and grout has fully set. Partially completed walls should also be braced during construction against lateral loads from wind or other forces applied before full design strength is attained or before permanent supporting construction is completed (*see Fig. 15-35*). Partially completed structures may be subject to loads which exceed their structural capabilities. Wind pressure, for instance, can create 4 times as much bending stress in a new, free-standing wall as in the wall of a completed building. Fresh masonry with

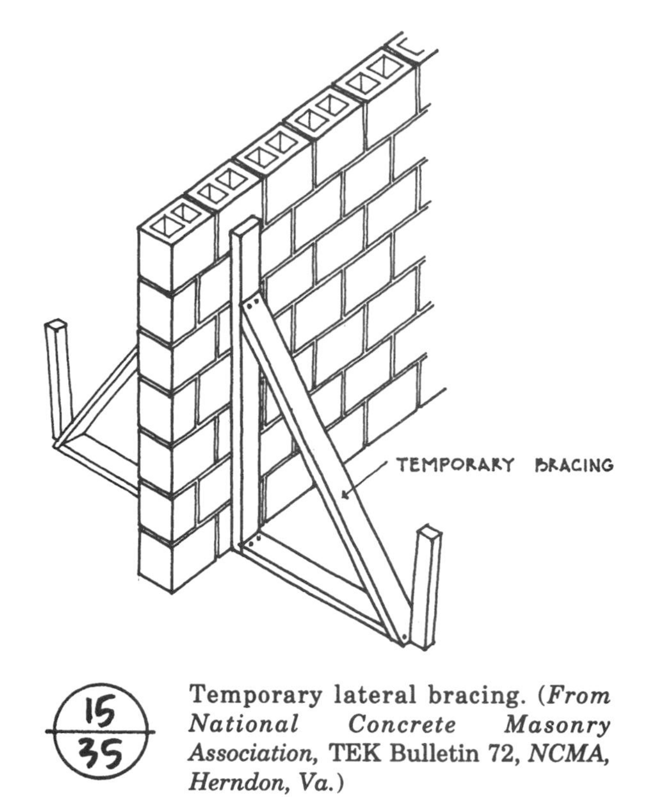

15-35 **Temporary lateral bracing.** (*From National Concrete Masonry Association,* **TEK Bulletin 72,** *NCMA, Herndon, Va.*)

uncured mortar has no tensile strength to resist such lateral forces. Most codes require that new, uncured, unanchored walls be braced against wind pressure. Bracing should be provided until the mortar has cured and the wall has been integrally tied to the structural frame of the building. Bracing should be designed on the basis of wall height and expected wind pressures.

Arches are constructed with temporary shoring or centering to carry the dead load of the material and other applied loads until the arch itself is completed and has gained sufficient strength (*see Fig. 15-36*). Temporary bracing should never be removed until it is certain that the masonry is capable of carrying all imposed loads. For unreinforced masonry arches, it is generally recommended that centering remain in place for 7 days after the completion of the arch. Where loads are relatively light, or where the majority of the wall load will not be applied until some later date, it may be possible to remove the centering earlier.

Masonry walls should be covered at the end of each day and when work is not in progress. Excess moisture entering the wall during construction can cause saturation of units which may take weeks or months to dry out. Such prolonged wetting will take even slightly soluble salts into solution and may result in efflorescence. Prolonged wetting will also prolong cement hydration, producing large amounts of calcium hydroxide which may also be taken into solution and leached to the surface to cause calcium carbonate stains.

Covers such as water-repellent tarps or heavy plastic sheets should extend a minimum of 2 ft down each side of the wall and be held securely in place. During construction, scaffold planks should also be turned on edge at the end of each day so that rain will not splash mortar droppings or dirt onto the face of the masonry.

15/36 Arches require temporary support.

15.4 SPECIAL CONSTRUCTION

Nontraditional forms of masonry include alternative bonding techniques without mortar, and prefabrication of units into panelized sections. These specialized systems differ significantly from typical masonry construction.

15.4.1 Surface-Bonded CMU

In 1967, the U.S. Department of Agriculture introduced surface bonding of concrete block walls as an economical technique for construction of low-cost housing. The system calls for dry-stacking of the units without mortar, then coating both sides of the wall with a thin layer of fiberglass-reinforced cement plaster. Mortar is used in bed joints only at the base of the wall and at bond beam courses. The $\frac{1}{16}$- to $\frac{1}{8}$-in.-thick surface coating binds the units together in a strong composite construction, adds a protective weathering surface, and imparts great tensile strength against lateral loads.

Less time and skill are required for building surface-bonded walls, and productivity may be increased by as much as 70%. The waterproofing characteristics of the coating material are excellent, and with the addition of colored pigments, the walls may be completely finished in one operation without the added labor involved in painting.

Strength in bending and flexure is equal to and sometimes much greater than conventionally built walls. Normal units provide slightly less resistance to compressive loads than conventional masonry, but if the bearing surfaces of the blocks are ground, the surface-bonded wall is as strong in compression as the conventional wall. The compressive strength of any CMU wall is directly related to the strength of the units used in its construction. In conventional construction with mortar joints, the wall strength will equal about 50% of the unit strength, compared to 30% for unground surface-bonded units. The lower strength developed in surface-bonded walls is due to the natural roughness of the block, which prevents solid bearing contact between courses. The mortar bed in conventional construction compensates for this roughness and provides uniform bearing, as does precision surface grinding of the dry-stacked units. The table in *Fig. 15-37* reflects these design variables in the allowable stresses for ground and unground units in surface-bonded construction. The maximum compressive stress on the gross area of the wall is 45 psi for unground units but increases to 85 psi for ground units, the same as permitted by many codes for conventional construction. Unground units, however, are more than adequate to support the superimposed loads normally encountered in a two- or three-story building.

The method of constructing a surface-bonded wall is simple. The bottom course is set in mortar as a leveling course to compensate for any unevenness in the footings or slab, and to obtain a good, sound base for the remainder of the wall. The base course should not include mortared head joints, however, as this would upset the modular alignment of the succeeding courses. Dry-stacking of the rest of the units begins at the corners, and unit ends should be butted tightly together. Shims may occasionally be required with unground units to maintain the wall level and plumb. Control joints and bond beams are used to control temperature and moisture movements since there are no mortar joints in which to embed horizontal reinforcement (*see Fig. 15-38* for construction details).

Job-site mixing and application of proprietary bonding mortars should be in accordance with the manufacturer's recommendations. The units are lightly sprayed with water just prior to application of the surface bonding mortar. Premixed mortars are smooth textured and easy to apply with a hand trowel or power sprayer. Sprayed-on coatings have slightly less tensile

Compressive:	45 psi based on gross area with unground masonry bearing surfaces 85 psi based on gross area with ground masonry bearing surfaces
Shear:	10 psi based on gross area
Flexural:	*Horizontal Span:* 30 psi based on gross area when units are drystacked in interlocking (running bond) pattern 18 psi based on gross area when units are drystacked in noninterlocking (stack bond) pattern *Vertical Span:* 18 psi based on gross area

Allowable stresses for surface-bonded masonry. (*From National Concrete Masonry Association,* TEK Bulletin 74, *NCMA, Herndon, Va.*)

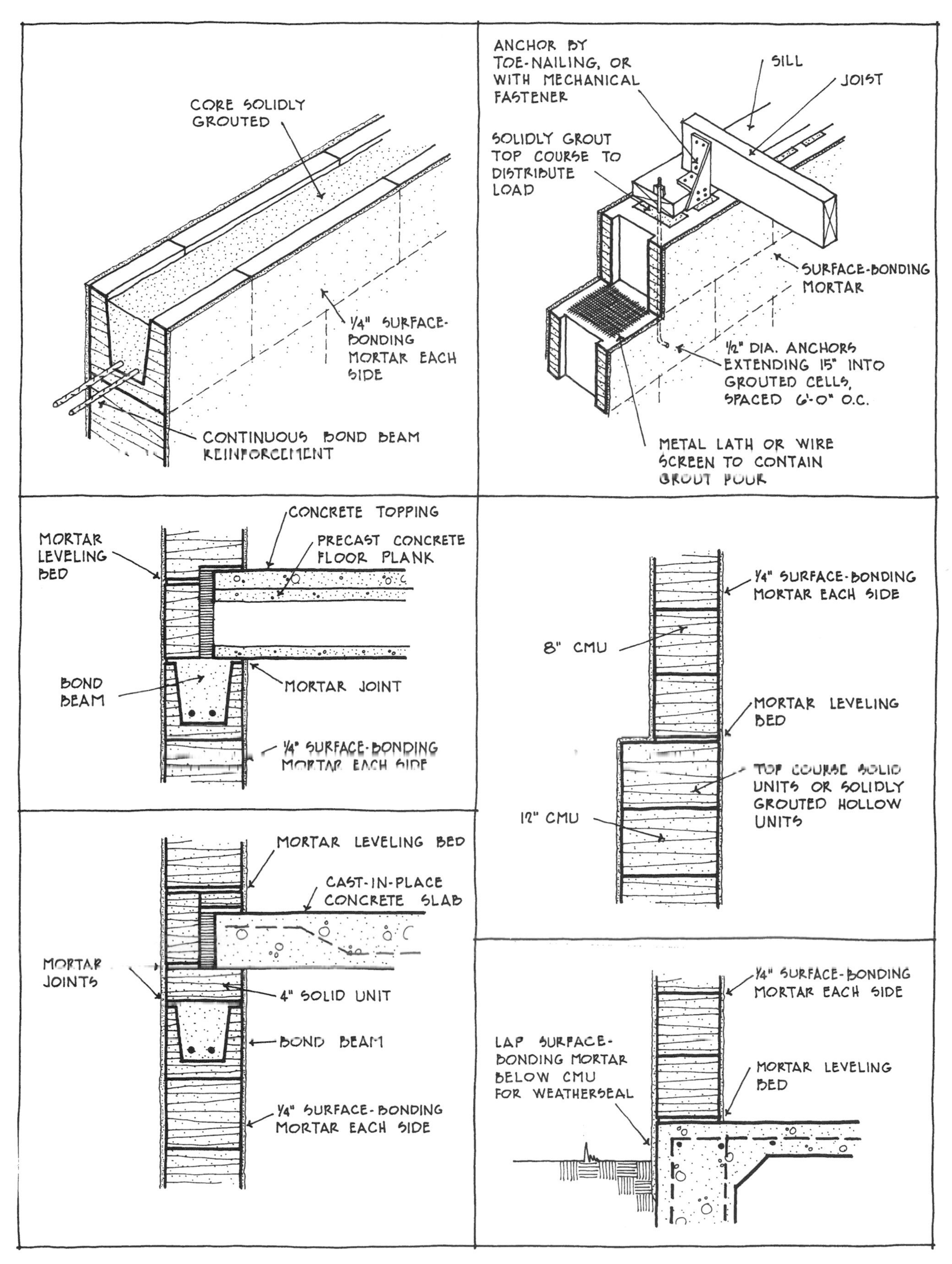

Details for surface-bonded masonry. (*From National Concrete Masonry Association,* TEK Bulletin 88, *NCMA, Herndon, Va.*)

strength than trowel coatings. Troweling tends to orient the glass fibers in the same plane as the block surface, thus giving greater strength. The surface coating must be a minimum of $\frac{1}{16}$ in. thick, but preferably should average $\frac{1}{8}$ in. In some cases, two-coat spray applications have been made with supplemental troweling of the first coat for strength, and the second coat left as a stipple finish. The finishing techniques obviously lend themselves to regional architectural styles of the Southwest, where stucco is a commonly used material.

Since the basic cementitious material in surface bonding mortars is portland cement, the coating should be damp-cured for 24 hours after its application to prevent premature evaporation of moisture before the hydration process is complete.

15.4.2 Prefabricated Masonry

Techniques for prefabricating brick masonry were developed in France, Switzerland, and Denmark during the 1950s, and adopted in the United States in the early 1960s. Reducing on-site labor led the construction industry to the use of prefabricated building components, but the masonry industry was a late entrant into the field. The evolution of analytical design methods for masonry, together with improved units and mortars, has made masonry prefabrication a feasible and economical masonry alternative to other systems such as precast concrete (*see Fig. 15-39*).

Prefabrication methods are used most successfully on brick and stone wall panels and spandrel sections backed by lightweight metal frames. A major requirement for the economic feasibility of preassembly is the repetition of design elements in the structure. Large numbers of identical sections may be mass-produced in environmentally favorable locations, then hoisted into place at the job site for field connections. The need for on-site scaffolding can be eliminated, and panelization allows the construction of complicated shapes without the need for expensive falsework and shoring. Quality control is more easily maintained under factory conditions by automating mortar batching systems and standardizing curing conditions. Prefabrication may also shorten the total construction time, allow earlier occupancy, and benefit the owner by increased income and lower interim financing costs.

Panel connections for facing materials generally combine the use of shelf angles and welded, bolted, or masonry tie anchors, depending on the type of structural frame used. Allowance must still be made for differential movement between masonry and concrete or steel.

15.5 CONSTRUCTION TOLERANCES

Historically, most construction was done on site and custom-fitted. Within generous limits, brick and stone could be laid to fit existing conditions; roof timbers were cut to fit whatever the masons built; hand-made doors and windows could be made to accommodate the peculiarities of any opening. Today we have metals that are fabricated at the mill, stone that is cut and dressed at the quarry, and concrete that is cast before erection. These prefabricated components are not easily customized on the job, and they must be fit to site-built frames. Suddenly construction tolerances become very important in ensuring that the puzzle pieces fit together with reasonable accuracy—puzzle pieces that may come from a dozen different manufacturers in a half-dozen competing industries.

Lifting a prefabricated masonry panel into place. (*Photo courtesy BIA.*)

Little is exact in the manufacturing, fabrication, and construction of buildings and building components. Tolerances allow for the realities of fit and misfit of the various parts as they come together in the field and ensure proper technical function such as structural safety, joint performance, secure anchorage, moisture resistance, and acceptable appearance. Webster defines tolerance as "the allowable deviation from a standard, especially the range of variation permitted in maintaining a specified dimension."

Each construction trade or industry develops its own standards for acceptable tolerances based on economic considerations of what is reasonable and cost-effective. Few, if any, construction tolerances are based on hard data or engineering analysis. There has also never been any coordination among various industry groups even though steel and concrete are used together, masonry is attached to or supports both, windows must fit into openings in all three, and sealants are expected to fill all the gaps left between adjacent components.

Different materials and systems, because of the nature of their physical properties and manufacturing methods, have greater or lesser relative allowances for the manufacture or fabrication of components and the field assembly of parts. Masonry includes a variety of materials and unit types, each of which has its own set of tolerances.

15.5.1 Masonry Size Tolerances

Allowances for the variation in sizes of clay brick are covered in ASTM C216 for face brick, ASTM C62 for building brick, and ASTM C126 for glazed clay facing tile and facing brick. Face brick allowances are divided into Type FBX and Type FBS. The quality of the units is the same, but Type FBX are required to have tighter size tolerances so that when a designer wishes to create a crisp, linear appearance such as the stack bond, the units and mortar joint variations are kept to a minimum. Type FBS brick are more popular for both commercial and residential masonry using running bond and other patterns. The size tolerances for ASTM C62 building brick are the same as for ASTM C216 face brick Type FBS.

Type FBA brick are not governed by size tolerances because they are supposed to vary significantly from one unit to the next so that they often look like rough, hand-molded brick. Type FBA brick are very popular for residential masonry and for projects in historic areas.

Glazed clay masonry units are the most precisely sized masonry products, and are often used to provide easily cleanable surfaces in hospitals and food handling or preparation areas. Since mortar absorbs stains, odors, and bacteria more easily than the glazed unit surface, joint widths must be kept to a minimum, which means that unit sizes must be closely controlled.

Concrete block dimensional variations are covered in ASTM C90 and C129 for loadbearing and non-loadbearing units, respectively. For both types, the standards permit a maximum size variation of $\pm \frac{1}{8}$ in. from the specified standard dimensions (defined as the "manufacturer's designated dimensions").

Although there are no industrywide standards for dimensional variation of cut and dressed limestone, the Indiana Limestone Institute publishes dimensional tolerances for several different types of finishes (refer to Chapter 5). For granite, the National Building Granite Quarries Association publishes recommended tolerances (refer to Chapter 5). For marble, the Marble Institute of America recommends fabrication tolerances based on the thickness of the panels (refer to Chapter 5). For thin stone curtain walls constructed with sealant joints instead of mortar, dimensional variations are critical to the proper performance of the sealant.

15.5.2 Mortar Joints

Unit masonry size tolerances are accommodated by varying the thickness of the mortar joints. Modular units are designed to be laid with standard $\frac{3}{8}$-in. joints. The Masonry Standards Joint Committee (MSJC) *Specifications for Masonry Structures* (*ACI 530.1-92/ASCE 6-92/TMS 602*) sets allowable variation in joint thickness based on the structural performance of the masonry (*Fig. 15-40*).

Aesthetic rather than structural requirements govern tolerances for non-loadbearing masonry veneers. These tolerances will vary according to the type of units specified and the dexterity and skill of the mason. For example, Type FBX brick can be laid with the most uniform joint thickness because the unit size tolerances are very tight. This characteristic lends itself to stack bond patterns where alignment of the head joints is critical to appearance. Usually, all of the units in a shipment are either over- or undersized, but not both, so the *range* of variation will be smaller and the joint width more consistent. Type FBA brick will require considerable variation in joint thickness because of the greater unit size variations, but this is part of the charm and the reason for the popularity of this brick type.

Joint	Allowable tolerance (in.)
Bed joint	$\pm\frac{1}{8}$
Head joint	$-\frac{1}{4}, +\frac{3}{8}$
Collar joint	$-\frac{1}{4}, +\frac{3}{8}$

Structural tolerances for mortar joints. [*From the Masonry Standards Joint Committee (MSJC)* Specifications for Masonry Structures ACI 530.1-95/ASCE 6-95/TMS 602-95, *ACI, ASCE, TMS, 1995.*]

15.5.3 Sealant Joints

The proper extension and compression of sealants and the performance of sealant joints in maintaining a weather seal are dependent on correct joint size and shape. Stone panel size tolerances are accommodated by variations in the width of the sealant joints, and expansion and control joints in unit masonry construction are affected by unit size tolerances. CSI Monograph 07M900, *Joint Sealers,* contains a complete set of tables for calculating sealant joint sizes. Once you have calculated combined thermal and moisture movement and added a factor of safety for construction tolerances, you can adjust joint size to change spacing, or adjust spacing to change joint size. The fewer joints you provide in a building facade, the wider they must be to equal or exceed the total calculated movement. The more joints you provide, the narrower they may be—up to a point.

Elastomeric sealants require a minimum joint width of ¼ in. for proper extension and compression. While ¼-in. joints are achievable in some types of stone panel construction, allowable tolerances in masonry unit and joint sizes make thin joints impractical to achieve in the field. A more realistic minimum for sealant joints in unit masonry construction is ⅜-in. to match the width of the mortar joints.

15.5.4 Connections

Structural frame tolerances are based on structural performance, accidental eccentricities, and member-to-member connection methods. Tolerances for cladding such as masonry veneers are based on stability, method of anchorage, and aesthetic perceptions. Allowable tolerances for concrete and steel structural frames are much greater than for the masonry panels, curtain walls, or veneers attached to them.

Allowable construction tolerances are much greater for concealed concrete and steel structural frames than for exposed cladding systems such as masonry veneer. *Figure 15-41* shows the differences in out-of-plumb tolerances for steel frame, concrete frame, and brick veneer. Where exposed veneer is permitted only ½ in. latitude in either direction, the frames to which it must be connected may vary 2 to 4 times that much. Although the cast-in-place concrete tolerances shown are much more restrictive than those for steel, the few field studies that have been done indicate that the steel tolerances are more realistic in their relationship to actual field con-

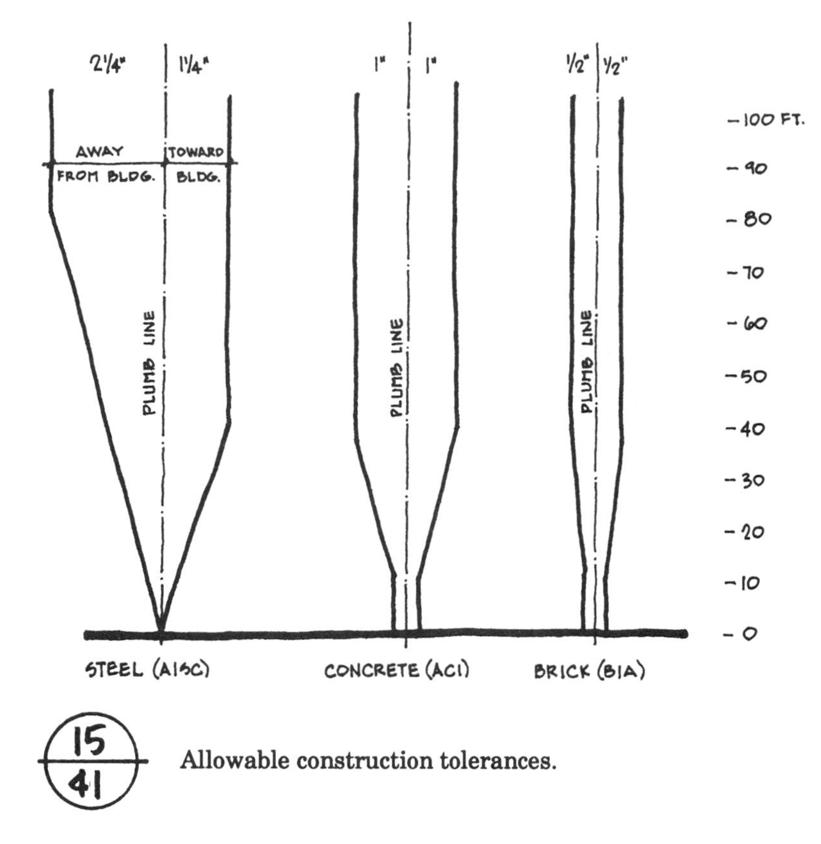

Allowable construction tolerances.

struction. Conflicts between structural frame and masonry veneer tolerances affect anchor embedment, support at shelf angles, and flashing details.

When a masonry veneer or curtain wall is attached to a structural frame that is alternately recessed or projected, the adjustments necessary to maintain a plumb line across the facade must be taken up in the anchorages. Different anchor lengths, however, create variable conditions of stiffness, deflection, and load transfer across a building elevation or throughout its height. Varying clearances between the edge of a slab or beam and the back of a curtain wall or veneer also affect the size and placement of thermal insulation, sprayed fireproofing, and fire-safing insulation.

A steel shelf angle with a 5-in. horizontal leg is usually adequate to span a 2-in. cavity width and support a single wythe of 4-in. modular brick veneer. If vertical and/or lateral displacement of the slab or beam to which the shelf angle is attached causes misalignment of the veneer surface, extreme measures are sometimes taken in an attempt to maintain the veneer alignment within its tolerances (*see Fig. 15-42*). Some contractors have been known to cut the brick or the shelf angle leg, or even to chip back the face of structural concrete members so far as to expose the reinforcing steel. Such drastic field alterations can sometimes threaten the safety of the building.

Narrowing or eliminating the open cavity behind the veneer jeopardizes proper wall drainage. When the cavity is wider than planned, longer anchors are required to achieve proper embedment in the mortar joints, and shelf angles may be too short to provide adequate support. Shelf angles that are too long may rotate, causing eccentric loading on the

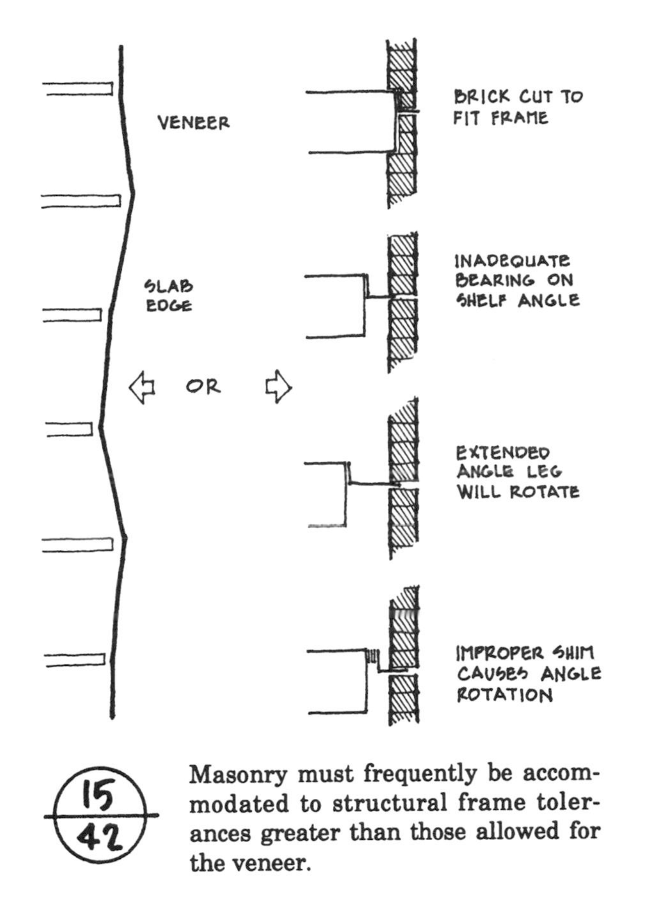

Masonry must frequently be accommodated to structural frame tolerances greater than those allowed for the veneer.

masonry courses below and spalling of the unit faces. Increasing the angle thickness to compensate for rotation will create differential stiffness and deflection conditions at random locations in the facade. To accommodate such problems, specify:

- Bolted rather than welded connections for steel shelf angles, with slotted holes for field adjustments and wedge inserts where attaching to concrete frames
- That the contractor provide a variety of anchor lengths as necessary to accommodate construction tolerances and provide minimum ⅝-in. mortar cover on outside wall face and minimum 1½-in. embedment in solid masonry units, or minimum ½-in. embedment into face shell of hollow units
- Two-piece flashing to accommodate varying cavity widths
- Horseshoe shims that are the full height of the vertical leg of the shelf angle and of a compatible metal, for shimming the angle up to a maximum of 1 in.

15.5.5 Grout and Reinforcement

For reinforced masonry, tolerances are allowed for the placement of the steel bars (*Fig. 15-43*) and the size of the grout spaces (*Fig. 15-44*). The most important thing is to assure complete embedment of the steel within the grout so that full strength is developed. To assure that the reinforcement is not displaced during the grouting operation, specify reinforcing bar spacers or special units that hold the steel in place.

Element	Distance from centerline of steel to opposite face of masonry		
	≤ 8 in.	> 8 in. but ≤ 24 in.	> 24 in.
Walls and flexural elements	±½ in.	±1 in.	±1¼ in.
Walls	For vertical bars, 2 in. from location along length of wall indicated on project drawing		

Reinforcement placement tolerances. [*From the Masonry Standards Joint Committee (MSJC) Specifications for Masonry Structures ACI 530.1-95/ASCE 6-95/TMS 602-95, ACI, ASCE, TMS, 1995.*]

WITH TOLERANCE OF +⅜ or −¼ in.			
Grout type*	Maximum grout pour height (ft.)	Minimum width of grout space (in.)†‡	Minimum grout space dimensions for grouting cells of hollow units (in.)‡§
Fine	1	¾	1½×2
	5	2	2×3
	12	2½	2½×3
	24	3	3×3
Coarse	1	1½	1½×3
	5	2	2½×3
	12	2½	3×3
	24	3	3×4

*Fine and coarse grouts are defined in ASTM C476. Grout shall attain a minimum compressive strength of 2000 psi at 28 days.

†For grouting between masonry wythes.

‡Grout space dimension is the clear dimension between any masonry protrusion and shall be increased by the diameters of the horizontal bars within the cross section of the grout space.

§Area of vertical reinforcement shall not exceed 6% of the area of the grout space.

Grout space requirements. [*From the Masonry Standards Joint Committee (MSJC) Specifications for Masonry Structures ACI 530.1-95/ASCE 6-95/TMS 602-95, ACI, ASCE, TMS, 1995.*]

15.6 COLD WEATHER CONSTRUCTION

Cold weather causes special problems in masonry construction. Even though sufficient water may be present, cement hydration and strength development in mortar and grout will stop at temperatures below 40°F. Construction may continue during cold weather, however, if the mortar and grout ingredients are heated and the masonry units and structure are protected during the initial hours after placement. As temperatures drop, additional protective measures are required.

Mortar and grout mixed using cold but unfrozen ingredients have different plastic properties from those mixed under normal conditions. For a given consistency, the mix will contain less water, will exhibit longer setting and hardening times, and have higher air content and lower early strength. Heating the ingredients prior to mixing, however, will produce mortar with performance characteristics identical to those in a more moderate ambient temperature range. Frozen mortar assumes the outward appearance of being hardened, but it is not actually cured and will not develop full design strength or complete bond until it is thawed and liquid water is again available for hydration. Frozen mortar is easily scratched from joints, has a "crows feet" pattern on the surface of tooled joints, and may flake at the surface.

Cement hydration will resume only when the temperature of the mortar or grout is raised above 40°F and its liquid moisture content exceeds 75%. When these conditions are maintained, ultimate strength development and bond will be the same as those attained under moderate conditions.

The rapidity with which masonry freezes is influenced by the severity of ambient temperature and wind, the temperature and absorption characteristics of the units, the temperature of reinforcing steel and metal accessories, and the temperature of the mix itself at the time of placement.

The water content of mortar and grout significantly affect their freezing characteristics. Wet mixes experience more expansion than drier ones, and expansion increases as the water content increases. During freezing weather, low moisture content mixes and high suction units are desirable, but mortar and grout consistency must maintain good workability and flow so that surface bond is maximized.

Cold *masonry units* exhibit all the performance characteristics of heated units except that volume is smaller and the potential for thermal expansion within the wall is greater. Wet, frozen units show decreased moisture absorption. Preheated units, on the other hand, will withdraw more water from the mortar because of the absorptive characteristics of a cooling body, but if they are too wet, may still have inadequate absorption. Highly absorptive units, by withdrawing water from the mortar, will increase bond and lower the moisture content, decreasing the potential disruptive expansion which might occur with initial freezing. Units that are dry, but excessively cold, will also withdraw heat from the mortar and increase the rate of freezing.

During cold-weather construction, it may be desirable to use a Type III, high-early-strength portland cement because of the greater protection it will provide the mortar. So-called "antifreeze" additives are not recommended. If used in quantities that will significantly lower the freezing point of the mortar, these additives will rapidly decrease compressive and bond strength. Accelerators that hasten the hydration process are more widely used, but may also have damaging side effects. Calcium chloride is the major ingredient in proprietary accelerators, and although it is effective, it has a highly corrosive effect on metal reinforcement and accessories. High salt contents of accelerating admixtures may also contribute to efflorescence and cause spalling of the units. In general, the use of set accelerators is not recommended, but when used, such admixtures should be limited to those containing non-chloride ingredients.

Masonry *materials should be stored and protected* at the job site to prevent damage from wet, cold, or freezing weather. Bagged materials and masonry units should be stored elevated to prevent moisture migration from the ground, and covered to protect the sides and tops. Consideration should be given to the method of stockpiling sand to permit heating of the materials if required.

As the temperature falls, the number of different materials requiring heat will increase. Mixing water is easily heated. If none of the other materials are frozen, mixing water may be the only ingredient requiring artificial heat. It should be warmed sufficiently to produce mortar and grout temperatures between 40 and 70°F at the time of placement. Water temperatures above 180°F can cause cement to flash-set, so sand and water should be mixed first to moderate high temperatures before the cement is added. Masonry sand, which contains a certain amount of moisture, should be thawed if frozen to remove ice. Sand should be warmed slowly to avoid scorching, and care should be taken to avoid contamination of the material from the fuel source. Dry masonry units should be heated if necessary to a temperature above 20°F at the time of use. Wet, frozen masonry units must be thawed without overheating.

The degree of protection against severe weather which is provided for the work area is an economic balance between mason productivity and cost of the production. Protective apparatus may range from a simple windbreak to an elaborate heated enclosure. Each job must be evaluated individually to determine needs and cost benefits, but some general rules do apply.

Characteristics such as strength, durability, flexibility, transparency, fire resistance, and ease of installation should be considered in selecting protective materials. Canvas, vinyl, and polyethylene coverings are often used. In most instances, a windbreak or unheated enclosure will reduce the chill factor sufficiently to provide the degree of protection required. Precautions must also be taken to safeguard workers against injury, and enclosures must be adequate to resist wind, snow, and uplift loads. The ACI 530 Code requires cold weather protection measures when the ambient temperature or the temperature of the units is below 40°F. The chart in *Fig. 15-45* summarizes heating and protection requirements for various work temperatures.

15.7 HOT WEATHER CONSTRUCTION

Hot weather conditions also pose special concerns for masonry construction. High temperatures, low humidity, and wind can adversely affect performance of the masonry. Rapid evaporation and the high suction of hot,

Workday temperature	*Construction requirement*	*Protection requirement*
Above 40°F	**Normal masonry procedures**	*Cover walls with plastic or canvas at end of workday to prevent water entering masonry*
40–32°F	**Heat mixing water to produce mortar temperatures between 40–120°F**	*Cover walls and materials to prevent wetting and freezing; covers should be plastic or canvas*
32–25°F	**Heat mixing water and sand to produce mortar temperatures between 40–120°F**	*With wind velocities over 15 mph provide windbreaks during the workday and cover walls and materials at the end of the workday to prevent wetting and freezing; maintain*
25–20°F	**Mortar on boards should be maintained above 40°F**	*masonry above freezing for 16 hours using auxiliary heat or insulated blankets*
20–0°F and below	**Heat mixing water and sand to produce mortar temperatures between 40 and 120°F**	*Provide enclosures and supply sufficient heat to maintain masonry enclosure above 32°F for 24 hours*

Cold-weather masonry construction requirements. (*From International Masonry Industry All-Weather Council,* **Recommended Practices and Guide Specifications for Cold Weather Masonry Construction,** *International Masonry Institute, Washington, D.C., 1977.*

dry units can quickly reduce the water content of mortar and grout mixes so that cement hydration actually stops.

When ambient temperatures are above 100°F, or above 90°F with wind velocities greater than 8 mph, the ACI 530 Code requires that protective measures be taken to assure continued hydration, strength development, and maximum bond. Whenever possible, materials should be stored in a shaded location, and aggregate stockpiles covered with black plastic sheets to retard moisture evaporation. High suction brick can be wetted to reduce initial absorption, and metal accessories such as reinforcing steel, anchors and ties, mixers, mortar boards and wheelbarrows can be kept cool by spraying with water.

Additional mixing water may be needed in mortar and grout, and additional lime will increase water retentivity (refer to Chapter 6). Increasing the cement content in the mix accelerates early strength gain and maximizes hydration before evaporative water loss. Adding ice to the mixing water can also lower the temperature of the mortar and grout and slow evaporation. Water that is too hot can cause the cement to flash set. Approved set-retarding or water-reducing admixtures may also be used. Retempering should be limited to the first 1½ hours after mixing. Mortar beds should not be spread more than 4 ft ahead of the masonry, and units should be set within 1 minute of spreading the mortar.

Sun shades and wind screens can modify the effects of hot, dry weather, but consideration should also be given to scheduling work during the cooler parts of the day.

15.8 MOIST CURING

Cement hydration cannot occur if the temperature of the mortar or grout is below 40°F or if the moisture content of the mix is less than 75%. Both hot and cold weather can produce conditions which cause hydration to stop before curing is complete. These *dryouts* occur most frequently in concrete masonry construction and under winter conditions, but may also occur in brick construction and in hot, dry weather. Dryouts are reactivated by higher temperatures and the subsequent introduction of natural rainwater, but pending these actions, construction is temporarily limited in compressive strength, bond, and weather resistance.

Moist curing methods similar to those used in concrete construction can help prevent masonry dryouts. Periodically wetting the finished masonry with a fine water spray for several days will usually assure that adequate moisture is available for curing, strength development, and good bond. Covering the walls with polyethylene sheets will also retard evaporation and create a greenhouse effect that aids in moist curing. Extreme winter conditions may also require the application of heat inside these enclosures to maintain minimum temperatures. Even concrete masonry can be moist-cured, because the restraining conditions of the joint reinforcement and surrounding construction minimize the effects of moisture shrinkage in the units.

16

MASONRY CLEANING AND RESTORATION

New masonry construction should be cleaned after completion to remove mortar smears and construction-related stains. Periodically throughout its life, the masonry may require additional cleaning if heavy industrial or urban pollutants discolor the surface. Cleaning may also become a diagnostic tool in the repair of structures whose surface defects may be obscured by soil or grime. But cleaning should always be evaluated for necessity and appropriateness, and any cleaning method selected should always be the gentlest possible.

16.1 CONSTRUCTION CLEANING

Cleaning new brick and concrete masonry is easiest if some simple protective measures are taken during construction. But even with protections in place, some mortar smears and splatters will have to be cleaned after the work is complete.

The finished appearance of masonry walls depends to a great extent on the attention given to the surfaces during construction and during the cleaning process. Care should always be taken to prevent mortar smears or splatters on the face of the wall, but if such stains do occur, daily cleaning can help prevent permanent discoloration. Excess mortar and dust can be brushed from the surface easily when the work is still fresh. For brick walls, a brush of medium-soft bristle is preferable. Any motions that rub or press mortar particles into the unit face should be avoided. On concrete block walls, mortar droppings are easier to remove after they have dried.

16.1.1 Protections

Other precautions that may be taken during construction include (1) protecting the base of the wall from rain-splashed mud or mortar droppings by using straw, sand, sawdust, or plastic sheeting spread out on the ground and up the wall surface; (2) turning scaffold boards on edge at the end of the day to prevent rain from splashing mortar or dirt directly onto the wall; (3) covering the tops of unfinished walls at the end of the day to prevent saturation or contamination from rain; and (4) protecting masonry units and packaged mortar ingredients from groundwater or rainwater contamination by storing off the ground, protected with waterproof coverings.

16.1.2 Cleaning Methods

The cleaning process itself can be a source of staining if chemical or detergent cleansing solutions are improperly used, or if windows, doors, and trim are not properly protected from possible corrosive effects. New masonry may be cleaned by bucket-and-brush hand scrubbing with water, detergent, muriatic acid solution, or proprietary cleaning compounds. Cleaning should be scheduled as late as possible in the construction, and the mortar must be thoroughly set and cured. However, long periods of time should not elapse between completion of the masonry and the actual cleaning, because mortar smears and splatters will cure on the wall and become very difficult to remove. Most surfaces should be thoroughly saturated with water before beginning (saturated masonry will not absorb dissolved mortar particles). Confine work to small areas that can be rinsed before they dry. Environmental conditions will affect the drying time and reaction rate of acid solutions, and ideally the cleaning crew should be just ahead of the sunshine to avoid rapid evaporation. Walls should be cleaned only on dry days.

Detergent solutions will remove mud, dirt, and soil accumulations. One-half cup dry measure of trisodium phosphate and ½ cup dry measure of laundry detergent dissolved in 1 gal of water is recommended. *Acid cleaners* must be carefully selected and controlled to avoid both injury and damage. Hydrochloric acid dissolves mortar particles, and should be used carefully in a diluted state. Muriatic acid should be mixed with at least 9 parts clean water in a nonmetallic container, and metal tools or brushes should not be used. Acid solutions can cause green vanadium or brown manganese stains on some clay masonry, and should not be used on light-colored, brown, black, or gray brick which contains manganese coloring agents. *Proprietary cleaning compounds* should be carefully selected for compatibility with the masonry material, and the manufacturer's recommended procedures and dilution instructions should be followed.

Some contractors use *pressurized water or steam* cleaning combined with detergents or cleaning compounds. If the wall is not thoroughly saturated before beginning, high-pressure application can drive the cleaning solutions into the masonry, where they may become the source of future staining problems. High-pressure washing can also damage soft brick and mortar and accelerate deterioration. *Abrasive sandblasting* should not be used to clean masonry.

All cleaning methods should be tested on a small, inconspicuous area to determine both the effect and the effectiveness of the process. For cleaning new masonry, the Brick Institute of America (BIA) has established guidelines for the selection of methods depending on the type of brick used (*see Fig. 16-1*). The American Society for Testing and Materials (ASTM) Committee E6 on Building Performance is in the process of developing a standard guide for the selection of cleaning techniques for masonry, concrete, and stucco surfaces which will address identification and characterization of substrates, identification of soiling and staining, selection criteria, cleaning techniques, testing, and evaluation.

Brick category	*Cleaning method*	*Remarks*
Red and red flashed	Bucket and brush hand cleaning High-pressure water Sandblasting	Hydrochloric acid solutions, proprietary compounds, and emulsifying agents may be used. *Smooth texture:* Mortar stains and smears are generally easier to remove; less surface area exposed; easier to presoak and rinse; unbroken surface, thus more likely to display poor rinsing, acid staining, poor removal of mortar smears. *Rough texture:* Mortar and dirt tend to penetrate deep into textures; additional area for water and acid absorption; essential to use pressurized water during rinsing.
Red, heavy sand finish	Bucket and brush hand cleaning High-pressure water	Clean with plain water and scrub brush, or *lightly* applied high pressure and plain water. Excessive mortar stains may require use of cleaning solutions. *Sandblasting is not recommended.*
Light-colored units, white, tan, buff, gray, specks, pink, brown, and black	Bucket and brush hand cleaning High-pressure water Sandblasting	*Do not use muriatic acid!!* Clean with plain water, detergents, emulsifying agents, or suitable proprietary compounds. Manganese-colored brick units tend to react to muriatic acid solutions and stain. Light-colored brick are more susceptible than darker units to "acid burn" and stains.
Same as light-colored units, etc., plus sand finish	Bucket and brush hand cleaning High-pressure water	Lightly apply either method. (See notes for light-colored units, etc.) *Sandblasting is not recommended.*
Glazed brick	Bucket and brush hand cleaning	Wipe glazed surface with soft cloth within a few minutes of laying units. Use soft sponge or brush plus ample water supply for final washing. Use detergents where necessary and acid solutions only for *very difficult* mortar stain. Do not use acid on salt glazed or metallic glazed brick. Do not use abrasive powders.
Colored mortars	Method is generally controlled by the brick unit	Many manufacturers of colored mortars do not recommend chemical cleaning solutions. Most acids tend to bleach colored mortars. Mild detergent solutions are generally recommended.

Cleaning guide for new brick masonry. (*From Brick Institute of America,* Technical Note 20 Rev., *BIA, Reston, Va.*)

16.1.3 Cleaning Fresh Mortar Smears

Although hydrochloric acid solutions are highly effective in removing mortar stains, they are not recommended for concrete masonry. Acid solutions remove the stain by dissolving the cement, but they also dissolve the cement matrix in the unit and etch the surface, leaving it porous and highly absorptive. As the cement is dissolved, more aggregate is exposed, changing both the color and the texture of the block.

Dry rubbing is usually sufficient for removing mortar stains from concrete masonry. To prevent smearing, mortar droppings and splatters should be almost dry before being removed. Large droppings can be pried off with a trowel point, putty knife, or chisel. The block surface can then be rubbed with another small piece of block, and finally with a stiff fiber-bristle or stainless steel brush (*see Fig. 16-2*).

On brick and other clay masonry units, the mortar must be thoroughly set and cured before it can be properly removed. Trying to clean uncured

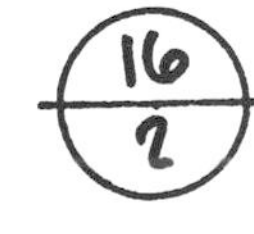

Removing fresh mortar smears from the face of concrete masonry. (*Photos courtesy Portland Cement Association, 5420 Old Orchard Rd., Skokie, Ill.*)

mortar from the surface presses the cement paste into the unit pores, making it harder to clean. Wooden paddles or nonmetallic scrapers should be used to remove large mortar droppings. For small splatters, stains, or the residue from larger pieces, a medium-soft fiber-bristle brush is usually sufficient.

Mortar that cures too long is harder and more expensive to remove than fresh splatters, and may require acid cleaning. Mild acid solutions easily dissolve thin layers of mortar. Large splatters should be scraped off first and, if necessary, the residue removed with acid. Hydrochloric acid (commonly known as muriatic acid) is suitable for cleaning clay masonry if it is diluted to a 5 or 10% solution (1 part acid to 20 parts water or 1 part acid to 10 parts water).

Mud, dirt, and soil can usually be washed away with a mild detergent solution consisting of ½ cup dry measure of trisodium phosphate (TSP) and ½ cup dry measure of laundry detergent to 1 gal clean water. Dried mud may require the use of pressurized water or a proprietary "restoration" type cleaner containing hydrofluoric acid and phosphoric acid. Hydrofluoric acid, however, etches polished surfaces such as glass, marble, and granite, so adjacent materials must be protected from accidental contact. Hydrofluoric acid is not suitable for cleaning mortar stains and splatters because it cannot dissolve portland cement products.

All cleaning solutions, even detergent, should be tested for adverse effects on a small, inconspicuous area of the wall. Some detergents contain soluble salts that can contribute to efflorescence. Muriatic acid can leave a white scum on the wall if the residue of dissolved cement is not thoroughly rinsed after a brief dwell time and light scrubbing. White scum can be removed only with special proprietary compounds, or it may have to simply wear off. Detergent and acid solutions usually are applied by bucket and brush, but large jobs may require low-pressure spray application. The masonry should be thoroughly saturated from the top down before cleaning to prevent absorption of the acid or the dissolved mortar particles. Failure to adequately pre-wet a wall, or using an acid solution that is too strong will cause acid burn—a chemical reaction that changes the color of the masonry. Nonmetallic buckets, brushes, and tools must always be used with acid cleaners because the metals react with acid, leaving marks on the wall that can oxidize and leave stains. Muriatic acid can also "bleach" colored mortars.

16.2 EFFLORESCENCE AND STAINS ON UNIT MASONRY

White, brown, and green stains can appear on unit masonry surfaces because of excessive moisture in the wall, or improper cleaning methods. Stains can also be caused by other materials such as paint or welding splatter. Each type of stain has an appropriate cleaning method.

16.2.1 Efflorescence and Calcium Carbonate Stains

Efflorescence and calcium carbonate stains are the two most common forms of surface stains on masonry. Both are white and both are activated by excessive moisture in the wall, but beyond that, there are no similarities. Efflorescence is a powdery salt residue, while calcium carbonate stains are hard, sometimes shiny, and much more difficult to remove.

Efflorescence occurs when soluble salts in the units or mortar are taken into solution by water entering through joint separation cracks, faulty copings, leaky window flashing, or other construction defects. As the

Efflorescence.

wall begins to dry, the salt solution migrates toward the surface through capillary pores. When the water evaporates, the salts are deposited on the face of the wall (*see Fig. 16-3*).

Hot summer months are not as conducive to efflorescence because the wetting and drying of the wall is generally quite rapid. In late fall, winter, and early spring, particularly after rainy periods, when evaporation is slower and temperatures cooler, efflorescence is more likely to appear.

Three simultaneous conditions must exist in order for efflorescence to occur: (1) soluble salts must be present within the masonry assembly; (2) there must be a source of water sufficiently in contact with the salts to form a solution; and (3) the wall construction must be such that paths exist for the migration of the salt solution to a surface where evaporation can take place (*see Fig. 16-4*). In conventional masonry construction exposed to weather, it is virtually impossible to ensure that no salts are present, no water penetrates the masonry, and no paths exist for migration. The most practical approach to the prevention and control of efflorescence is to reduce all of the contributing factors to a minimum.

Soluble salts may be present in either the masonry or the mortar, or may be absorbed into the wall through rain or groundwater. Since efflorescence usually appears on the face of the units, they are generally assumed to be at fault. This, however, is not usually the case. Virtually all clay brick contains at least some salts, but their efflorescing potential is small. The degree of probability may be easily determined by the wick test

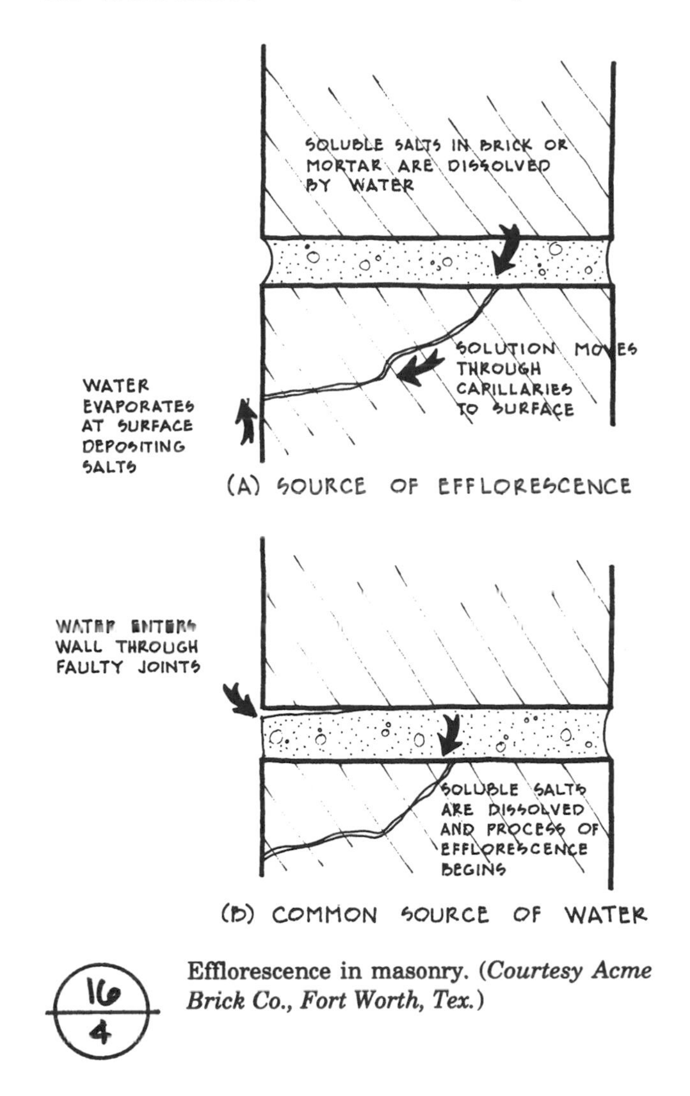

Efflorescence in masonry. (*Courtesy Acme Brick Co., Fort Worth, Tex.*)

included in ASTM C67, *Standard Methods for Sampling and Testing Brick and Structural Clay Tile.* Brick units relatively free from impurities are readily available throughout the United States. Dense to moderately absorptive units are least troublesome. Researchers differ in their opinions on concrete masonry, some saying that they have even less effloresing potential than clay products, and others recording 2 to 7 times as much soluble material.

Mortars also vary in the amounts of soluble salts they contain, depending on the type of cement used. Cements are generally the greatest source of soluble materials that contribute to efflorescence. Those with a high alkali content and limestone impurities are most likely to cause problems. Some companies have developed special "low alkali" and "nonstaining" cements for use in masonry mortars. Hydrated limes are relatively pure and generally have 4 to 10 times less efflorescing potential than cements. Therefore, lime is one of the lesser sources, along with well-washed sand and clean, potable water. Soluble salts from the soil may be absorbed into masonry in contact with the ground through the capillary action of groundwater migrating upward into the units. Sulfurous gases in the atmosphere in highly industrialized areas may also contaminate the masonry with soluble salts through soaking with "acid rain."

The source of moisture necessary to produce efflorescence may be either rainwater or the condensation of water vapor within the assembly. Water may also be present because unfinished walls were not properly protected from rain and snow during construction. "New building bloom" (efflorescence which occurs within the first year of the building's completion) is often traced to slow evaporation of such moisture.

The most common cause of efflorescence is faulty design and construction practices. Regardless of impurities in the materials, it is unlikely that efflorescence will occur if proper precautions and high-quality workmanship are employed. Some of the more common malpractices are (1) failure to store masonry units off the ground and protect with waterproof covers, (2) failure to cover and protect unfinished walls, (3) inadequately flashed copings and parapet walls, (4) absence of drips on cornices or projecting members, (5) poorly filled mortar joints, (6) absence of dampproof courses at ground level, (7) failure to repair or patch cracked or broken mortar joints, and (8) use of dense units and mortar which absorb moisture through unrepaired cracks and are then slow to dry out.

To minimize the possibility of efflorescence, the following measures are of greatest importance: (1) use only units of low to moderate absorption or specify that the brick be tested for efflorescing potential in accordance with ASTM C67 and rated as "not effloresced"; (2) use only low alkali, nonstaining cements in the mortar; (3) properly protect materials before and during construction; (4) install flashing and weepholes, caulking, and sealants at strategic locations to expedite the removal of moisture that has entered the wall; (5) achieve good bond with compatible units and mortar; and most important of all, (6) construct full mortar joints. These precautions are particularly important in regions with high annual rainfall. ASTM Committee C15 on Manufactured Masonry Units is in the process of developing a standard guide for the reduction of efflorescence potential in new masonry walls, which is proposed to address issues of moisture penetration, moisture drainage, and construction practices.

Efflorescence will often disappear with normal weathering if the source of moisture is located and stopped. Efflorescence can also be dry-brushed, washed away by a thorough flushing with clean water, or scrubbed away with a brush.

Clear water repellents are often recommended as a solution to efflorescing problems. However, if the water repellent is applied to a wall that still contains both moisture and salts, the resulting problems may be even more damaging than the stain. The water in the wall will still take the salts into solution, and as it migrates toward the outer face, most of it will stop at the inner depth of the water repellant. The water will then evaporate through the surface and deposit the salts inside the masonry unit. This interior crystalline buildup (sometimes called subflorescence) can exert tremendous pressure capable of spalling the unit face (*see Fig. 16-5*). Clear water repellent applications are generally not recommended as a treatment for efflorescence unless the chain of contributory conditions (moisture, salts, and migration paths) is also broken.

Calcium carbonate stains occur when calcium hydroxide from the mortar is leached to the surface where it reacts with atmospheric carbon dioxide to form calcium carbonate. The calcium hydroxide (lime) is present not only in portland cement–lime mortars, but in masonry cement mortars as well, because it is a natural by-product of the cement hydration process itself. As the cement cures, it produces 12 to 20% of its weight in calcium hydroxide. Extended saturation of the wall through construction defects

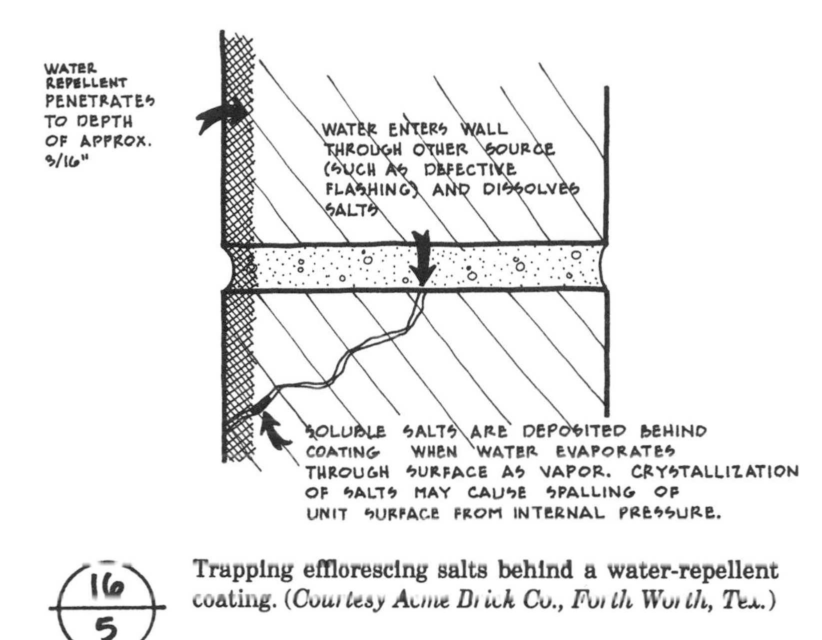

Trapping efflorescing salts behind a water-repellent coating. (*Courtesy Acme Brick Co., Fort Worth, Tex.*)

prolongs the curing process and maximizes the amount of lime produced. The excess moisture also carries the calcium hydroxide to the wall surface where it reacts with carbon dioxide in the air to form calcium carbonate (limestone). The stains usually occur as hard, encrusted streaks coming from the mortar joints, and are sometimes referred to as "lime deposits" or "lime run" (*see Fig. 16-6*).

Before calcium carbonate stains can be removed, the source of moisture must be located and stopped. Once that is done, the stain and surrounding area should be saturated with water, and a dilute solution of 1

Calcium carbonate stain.

part hydrochloric (muriatic) acid to 12 parts water applied. With a stiff fiber-bristle brush, the stain can be scrubbed away and the wall thoroughly rinsed with water to remove the acid and residue.

16.2.2 Vanadium and Manganese Stains

Two stains that are peculiar to clay products are green or yellow vanadium stains and brown manganese stains. *Vanadium* salts originate in the raw materials used to manufacture brick, and the stains occur on white or light-colored units. The chloride salts of vanadium require highly acidic leaching solutions, and the problem of green stain often does not occur unless the walls are washed down with a muriatic acid solution. To minimize the occurrence of green stain, do not use acid solutions to clean light-colored brick, and follow the recommendations of the brick manufacturer for the proper cleaning compounds and solutions. If green stains do appear as a result of acid washing, flush the wall with clean water and then wash or spray with a solution of 2 lb of potassium or sodium hydroxide to 1 gal of water to neutralize the acid. After the solution has been on the wall for 2 or 3 days, the white residue may be hosed off with clean water.

Manganese stain may occur on the surface of mortar or bricks containing manganese coloring agents. The stain may be tan, brown, or gray, is oily in appearance, and may streak down over the face of the wall. The manufacturing process chemically changes the manganese into compounds that are soluble in weak acid solutions. The staining may occur because of acid cleaning procedures, or even because of acid rain in some industrial areas. Muriatic acid solutions should never be used to clean tan, brown, black, or gray brick or mortar unless the wall is thoroughly saturated with water before application and the acids are washed away with a rinsing operation.

Brown manganese stains can be removed with a 1:1:6 solution of acetic acid (80% or stronger), hydrogen peroxide (30 to 35%), and water. Wet the wall thoroughly, and brush or spray on the solution, but do not scrub. The reaction is generally very quick and the stain rapidly disappears. After the reaction is complete, rinse the wall thoroughly with water. Although this solution is very effective, it is dangerous to mix and use, and proper precaution should be taken to protect workers and adjacent surfaces. Manganese stains often recur after they are first removed, and the process must be repeated. To avoid manganese stains, always request and follow the recommendations of the manufacturer in cleaning brick that contains manganese coloring agents.

16.2.3 Stains from External Sources

The method of removing externally caused stains will depend on the type of material that has been splattered on, or absorbed by the masonry. Many stains can be removed by scrubbing with ordinary kitchen cleansers. Others require the use of a *poultice* or paste made with a solvent or reagent and an inert material. The stain is dissolved, and the solution leached into the poultice. After drying, the powdery substance remaining is simply brushed off. Although repeated applications may be required, the poultice will prevent the stain from spreading during treatment, by actually pulling it from the pores of the masonry. Some of the more common stains and cleaning methods are listed below.

Paint stains on both brick and concrete masonry may be removed with a commercial paint remover, or a solution of 2 lb of trisodium phosphate in 1 gal of water. Apply the liquid with a brush, allow it to remain and soften the paint, then remove with a scraper and wire brush. Rinse the surface afterward with clear water.

Iron stains or welding splatter are removed from clay and from concrete masonry in different ways. On clay brick, spray or brush the area with a solution of 1 lb of oxalic acid crystals, 1 gal of water, and ½ lb of ammonium bifluoride to speed the reaction. This solution should be used with caution because it generates hydrofluoric acid, which will etch the brick surface. The etching will be more noticeable on smooth masonry. An alternative method, which may also be used on concrete masonry, uses 7 parts lime-free glycerine with a solution of 1 part sodium citrate in 6 parts lukewarm water mixed with whiting to form a poultice. Apply a thick paste and scrape off when dry. Repeat the process until the stain has disappeared, then rinse the area thoroughly with clear water.

Copper or bronze stains are removed from both clay and concrete masonry by a mixture in dry form of 1 part ammonium chloride and 4 parts powdered talc, with ammonia water added to make a thick paste. Apply the paste over the stain and remove when it is dry using a scraper or, on glazed masonry, a wooden paddle.

Smoke stains are difficult to remove. Scrubbing with a scouring powder that contains bleach, using a stiff-bristle brush, will generally work well. Small, stubborn stains are better dealt with using a poultice of trichloroethylene and talc, but the area should be well ventilated to avoid a buildup of harmful fumes. In some instances where large areas have been stained, alkali detergents and commercial emulsifying agents may be brush- or spray-applied or used in steam cleaners. If given sufficient time to work, this method will work well.

For information on identifying unknown stains and determining appropriate cleaning methods, consult Grimm's handbook, *Cleaning Masonry—A Review of the Literature.* All proposed cleaning methods should be tested on a small area before general application is made to a wall or surface.

16.3 CLEANING STONE MASONRY

Mortar is sometimes smeared on stone surfaces during construction. Mortar smears can usually be removed by scrubbing with stone dust and fiber brushes wetted with white vinegar. To avoid smearing mortar across the stone surface, allow the mortar to take its initial set, and then remove it with a trowel rather than wiping with a cloth. Mortar can also be placed into head joints with a grout bag to minimize the amount of wet mix coming in contact with the stone surface. Acids or chemical cleaners are not usually required to clean new stone. If stubborn dirt or other foreign substance has become embedded in the surface, mild abrasive cleaners will usually remove them. If more aggressive methods are required, consult the stone fabricator about the most appropriate cleaning chemicals and procedures. Cleaning methods for existing stone surfaces should achieve a balance between removal of dirt and stains and protection of the stone. Processes that are too abrasive can destroy the stone's natural protection and expose more surface area to the environment. Existing stone should be cleaned in accordance with the methods recommended under historic masonry below.

16.4 CLEANING HISTORIC MASONRY

There are more than just cosmetic reasons to clean and maintain historic masonry buildings. In fact, cosmetic reasons alone may not always be sufficient justification for a full-scale cleaning program. The weathered patina of masonry often becomes as much a part of a building's character as the materials themselves. The unnecessary cleaning of otherwise undamaged or lightly soiled walls may do more harm if harsh chemicals or abrasive action remove too much of the "protective crust" that has formed on the surface. As long as it does not contribute to or conceal deterioration, it should be preserved. The body of the brick or stone underneath may be too soft to withstand the attack of twentieth-century urban pollutants.

On the other hand, excessive soiling can disguise or even contribute to physical damage of the masonry. A heavy dirt buildup may easily conceal cracks and other signs of deterioration that warrant investigation and repair, and a thorough investigation may not be effectively accomplished without first cleaning the surface.

Dirt may also cause or aggravate deterioration of the masonry. Its presence significantly increases the amount of moisture that is attracted to and held on a wall surface, and impedes natural drying after a rain. Prolonged dampness tends to enhance the chemical reactivity of the masonry with common atmospheric pollutants. It also increases the risk of freeze-thaw damage in the winter and the growth of "micro-vegetation" in warmer conditions. Water that gets into the wall from other sources is also trapped because it cannot evaporate at the surface, so concealed metal components and structural supports are subject to accelerated corrosion and failure. And moisture damage, of course, can go beyond the masonry wall itself to interior finishes and other adjacent elements.

If cleaning has been determined as a necessary and desirable part of the restoration or preservation process, the first step in developing a cleaning program and specification must be one of testing and evaluation. Rudimentary field examinations and laboratory chemical analysis can determine the relative inertness or reactivity of the masonry and the nature and composition of the dirt or stains.

Dirt (or soiling) generally refers to particulate surface deposits, while *stains* are produced by foreign matter that has penetrated into or permeated the masonry. Dirt may include such solids as dust, sand, grit, carbon soot, and inorganic sulfates. Stains include those of metallic origin such as iron or copper; industrial stains of grease, oil, and tar; biological and plant stains caused by lichens, moss, algae, and fungal growth such as mildew; and internally activated stains such as efflorescence, calcium carbonate, vanadium, and manganese. Surface coatings such as paint, wax, or water repellents may also be present.

There is no such thing as typical urban dirt, nor is there typical masonry when dealing with historic buildings. An extraordinary variety of geological and man-made materials have been used in masonry construction, and often in combination with one another. A single facade may incorporate several textures and colors of brick, terra cotta copings or decorative elements, and two or more types of stone used as lintels, sills, cornices, or belt courses. Side and rear elevations that are less exposed to public view may also be of less expensive, softer materials. The degree of soiling also varies with geographic orientation, location relative to street and pedestrian traffic, height above ground level, and configuration of projecting elements. The cleaning program must be designed to preserve the integrity of the entire building fabric (including non-masonry materials such as wood, glass, and metal), as well as to protect adjacent buildings, the surrounding landscape, occupants, workers, and passersby. Each building presents a unique set of problems—some known and some unexpected—and each requires a unique solution. There *are* no standard specifications. The Construction Specifications Institute (CSI) and the Association for Preservation Technology International (APT) have jointly published a technical document entitled *Guide to Preparing Design and Construction Documents for Historic Projects* (CSI Document TD-2-8) which provides in-depth information on documenting existing conditions and preparing drawings and specifications for the restoration or rehabilitation of historic structures.

16.4.1 Testing

A cleaning program should be initiated with carefully planned, on-site testing of specific materials and cleaning methods, begun well in advance of necessary completion dates. An experienced preservation consultant or cleaning contractor should be hired to perform the testing separate and apart from the cleaning contract itself, even if the same contractor will be used for the actual cleaning.

Because of the number of unforeseeable factors and the uncertainty of the results, most test patches should be located in an inconspicuous area of the building. Paint removal testing, however, should be done near the front entrance to the building where the most layers of paint are likely to be. Test patches should also be representative of the different types of substrates involved, and the (often dissimilar) substances to be removed. To ensure the most accurate test results, remove as much of the dirt or stain as possible by hand scraping with wooden paddles or brushing with nonmetallic bristle brushes before test cleaning—and follow the same procedure when full-scale cleaning begins.

Start with what the Secretary of the Interior's Standards for historic rehabilitation call "the gentlest means possible." Carefully document each tested procedure as to number of applications, cleaning material and equipment, dwell time, and wash/rinse pressures. Even small buildings may require a variety or a combination of cleaning methods. The best approach is to find the gentlest technique that will remove the prevailing substance, and augment it with more aggressive localized cleaning in difficult areas. It is always better to underclean rather than overclean. If you are testing chemical cleaners, nonstaining pH papers should be held on the surface of the masonry before and after to determine if any acidic or alkaline residue remains.

Test patches serve as the standard by which full-scale cleaning is judged. But do not evaluate the test areas until they are dry and have weathered as long as possible. Ideally, exposure to a complete 1-year weathering cycle will give the most accurate and reliable information. When this is not feasible, a minimum of 1 month should be allowed, during which there are several wetting cycles and a number of temperature variations. Tests should also be conducted under weather conditions similar to those anticipated during actual cleaning, particularly when chemical compounds that are affected by weather are used. The dilution ratios and dwell times used successfully in one season may not be as effective in another. Remember, too, that tests are usually performed under optimum conditions. It is always easier to effectively clean small areas at ground level than to achieve the same results from a scaffolding or swing stage at higher wind elevations on a Friday afternoon when everyone is tired. Expectations should be realistically based on actual field conditions.

16.4.2 Cleaning Methods

There are several different levels of intervention which can be implemented using prudent combinations of water, hand scrubbing, detergents, and chemicals. *Do not use abrasives.* Grit blasting, wet or dry, whether it uses sand, crushed nut shells, rice hulls, egg shells, silica flour, ground corncobs, or any other medium, removes dirt and stains by tearing away the surface of the substrate itself. It accelerates deterioration of the brick or stone, disintegrates mortar joints, and irreversibly damages the masonry, shortening the remaining life of the building.

Grinding and power sanding can be equally destructive. Most historic brick is soft by today's standards. Any cleaning method that removes or abrades the durable outer layers formed in the firing kiln or the protective crust formed by weathering exposes the soft inner body to harsh environmental deterioration. The cost is prohibitive in terms of damage to historic building materials that are neither indestructible nor renewable.

Water washing methods include soaking, pressure washing, pressure washing supplemented with detergents or surfactants, and steam. Most masonry can be cleaned with simple water washing without the need for more aggressive measures. The amount of soiling will determine the level at which testing should begin.

For light to moderate soiling, particularly on rough-textured brick or stone, *water spray* at moderate pressure (200 to 600 psi) may be needed. Nonionic detergents applied by bucket and brush, or added to the power spray, can hasten the cleaning process and reduce the amount of water that must be applied to the wall, but they must be thoroughly rinsed to remove any film or residue left on the surface.

Water soaking is effective for carbon or sulfate encrustations which often build up in protected areas under cornices, eaves, and overhangs where rain cannot keep the wall clean. A fine mist sprayed on the wall for a prolonged period softens the crust by causing the dirt deposits to swell and loosen their grip on the masonry. The continuous application of water then rinses the deposits away, simulating the natural washing action of rain. A low- to moderate-pressure rinse may be needed as a final step. The volume of water required for cleaning can be enormous (9.8 million gal on Chicago's Field Museum). Precautions must be taken to prevent moisture damage to other parts of the building. Repair open mortar joints, replace deteriorated sealant joints, and check windows for glass that is loose before beginning work. In water soak applications, it is best to cover windows, doors, and lower courses of masonry to keep most of the water out.

Steam cleaning requires much less water, and is used almost exclusively for interior work. Steam is dangerous because it burns and because it obscures the visibility of the equipment operator. It can be very effective, though, on delicate, ornately carved stonework.

Chemical compounds are usually needed to remove heavy dirt buildup, wax coatings, water repellents, and paint. Acid-based cleaners are most effective for removing dirt, and alkaline cleaners for paint removal. The lime mortars used in historic masonry construction, however, are acid-sensitive, and acid can also damage brick. To prevent the chemicals from penetrating beyond the layers of surface dirt, the walls must be thoroughly presoaked with water before application, and thoroughly rinsed with clean water afterward. Test patches must be used to determine the exact chemical concentrations and dwell times required for specific surfaces and specific soiling conditions.

Alkaline paint strippers are very effective in removing layers of paint from masonry buildings. It is important, however, to first determine whether the paint should be removed at all. Painted brick buildings were popular during several historic periods. Many were painted immediately after construction, sometimes to protect soft, inexpensive brick. If the underlying substrate is soft, low-fired brick, paint removal may be more damaging than beautifying in the long run.

16.4.3 Precautions

Because water is used in all masonry cleaning procedures, provisions must be made for adequate site drainage. The high volume of water also precludes cleaning during cold weather when there is any danger of the masonry freezing before it dries out. Effluent control and handling must also be provided when contaminants such as lead-based paints are involved, and chemical cleaning should not be performed under windy conditions when overspray or drift could be a problem. Caution is essential to all phases of historic masonry cleaning, not only in the handling and treatment of the sometimes delicate building fabric itself.

16.5 REPAIR AND REPOINTING

The first priority in repairs to historic buildings should be identifying and treating the cause, rather than the effects, of deterioration and damage. Before stones cracked by settlement are repaired or replaced, the foundation itself must be stabilized, and roof leaks must be stopped before repairing moisture damaged walls. If the symptoms rather than the disease are treated, the problems will recur.

16.5.1 Repairs

Begin with major structural work before undertaking minor repairs, and provide permanent weather protection as soon as possible. Be wary of methods that have not been extensively tested. Historic buildings should not be the testing grounds for new materials and procedures. Avoid the tendency to rush work, because shortcuts and poor craftsmanship compare poorly with the original work and result in jobs requiring additional repair. The most cost-effective approach is not necessarily the least expensive, but the one that will last longest, is technically best, and requires the least change to the historic property.

No matter how soiled the masonry itself is, or how much moisture it retains on the surface, much larger quantities of water will enter the wall through cracked or broken mortar joints, defective copings, and leaky roofs and gutters. Joint integrity must be maintained, not only for aesthetic reasons, but for structural soundness and weather protection as well. But repointing will not be effective if there are other sources of moisture that have not been identified and repaired. If the roof leaks, fix it. If the coping is ineffective in stopping water infiltration, repair it. Then repoint the mortar as needed. To finish the job, look carefully at caulking and sealant joints. There is a limit to the effective service life of these materials, even under optimum conditions. Periodic maintenance and repair is always necessary and should be scheduled regularly to avoid more costly long-term damage from water.

16.5.2 Joint Preparation

Deteriorated mortar joints in existing masonry can be cleaned out and repointed with fresh mortar. Mortar joints deteriorate for many reasons:

- Water, wind, and pollution erode the mortar.
- Historic mortars made with little or no portland cement are soft, and often more susceptible to weathering.
- Uneven settlement can cause cracks to form in the mortar joints.
- Mortar joints are only partially filled during construction, or mortar inadequately bonds to the units, allowing excessive moisture to penetrate the wall.
- Walls saturated by moisture can freeze and thaw again, eventually spalling both the mortar and the masonry.

Raking refers to the process of removing or cutting out the old mortar. Mortar joints should be raked out to a depth of ½ in to ¾ in (*see Fig. 16-7*). If the mortar is still unsound, cut the joint deeper. Mortar can be removed with a hand-held grinder, a small mason's chisel, or with a special raking tool. If a grinder is used to rake vertical head joints, be careful not to cut into the brick courses above or below. Before repointing, brush all loose fragments and dust from the joint or flush them out with a stream of water or pressurized air. ASTM Committee E6 on Building Performance is in the process of developing a standard guide for repointing masonry which will address evaluation, mortar removal techniques, mortar selection, execution, and evaluation of the work.

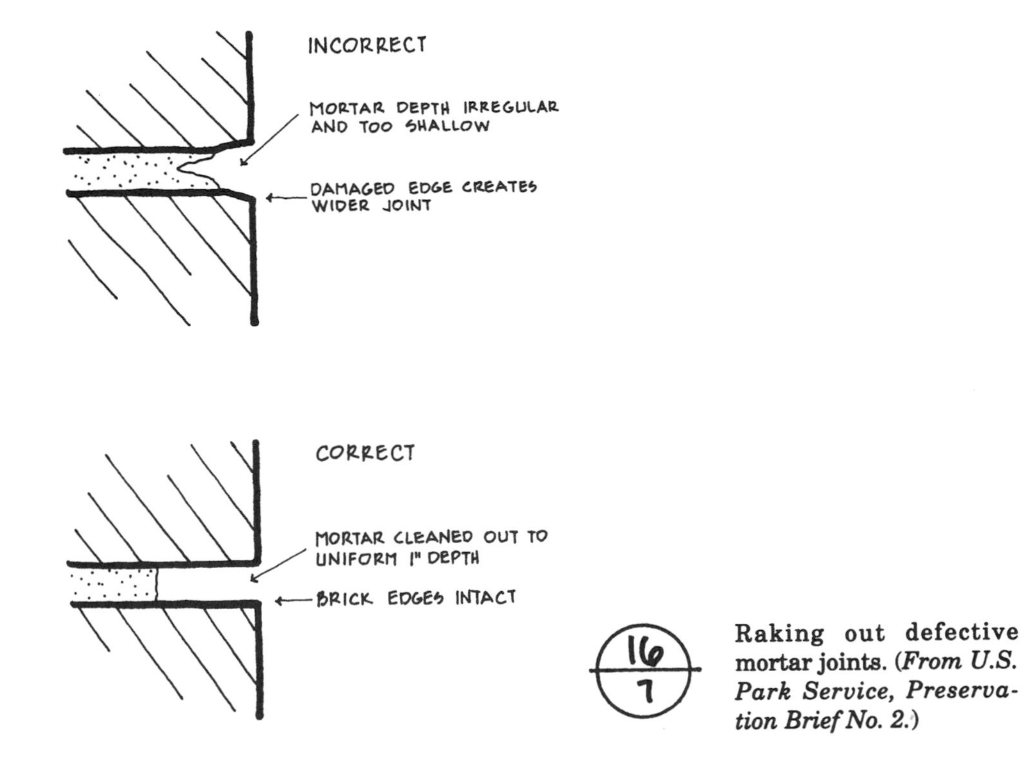

Raking out defective mortar joints. (*From U.S. Park Service, Preservation Brief No. 2.*)

16.5.3 Mortar

Repointing mortar should match the existing material as closely as possible in strength, hardness, color, and texture. Historic mortars were generally soft and may have been mixed from clay, gypsum, lime, natural cement, and some later ones with portland cement. Laboratory analysis can be used to determine the exact ingredients and proportions. Modern mortars containing portland cement are much harder than these older mixtures, and in some cases, harder than the brick or stone itself. Ideally, the new mortar should have the same density and absorbency as the stones or bricks in the wall. A hard mortar used with soft brick or stone can cause deterioration of the masonry because the two components do not respond to temperature and moisture changes at the same rate, or to the same degree. The softer material will absorb more movement stress and more moisture, and hard mortar can act as a wedge, breaking the edges off the units (*see Fig. 16-8*). Many buildings have been irreparably damaged in this manner. Strong portland cement mortars may also shrink, leaving minute cracks at the mortar-to-unit interface.

Recommendations for repointing mortar vary with almost every source consulted. The Preservation Assistance Division of the U.S. National Park Service recommends lime-sand mortars with the optional addition of

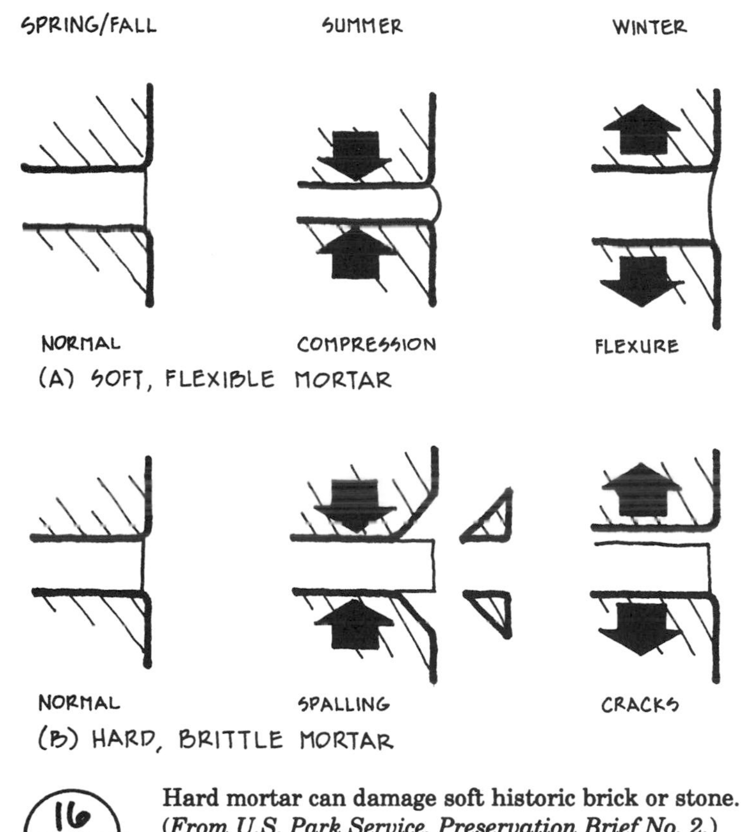

Hard mortar can damage soft historic brick or stone. (*From U.S. Park Service, Preservation Brief No. 2.*)

	Anticipated exposure		
Masonry material	***Sheltered***	***Moderate***	***Severe***
Highly durable granite or hard brick	C	B	A
Moderately durable stone or brick	D	C	B
Poorly durable soft brick or friable stone	E	D	C

Mortar designation	Mix proportion (cement:lime:sand)
A	1:½:4–4½
B	1:1:5–6
C	1:2:8–9
D	1:3:10–12
E	0:2:5

Recommendations for re-pointing mortar. (*Adapted from the Ontario Ministry of Citizenship and Culture,* **Annotated Master Specifications for the Cleaning and Repointing of Historic Masonry Structures,** *1985.*)

portland cement not to exceed 20% of the total volume of cement and lime. For brick masonry restoration in which the ingredients of the original mortar are unknown, BIA recommends an ASTM C270, Type N or Type O mortar mixed proportionally with 1 part portland cement, 1 part lime, and 6 parts sand. The appendix to ASTM C270 recommends Type N, O, or K depending on conditions of exposure. The recommendations in *Fig. 16-9* from the Ontario Ministry of Citizenship and Culture are based on both expected weather exposure and the type of masonry involved.

To compensate for shrinkage in mortars containing portland cement, prehydrate the mortar by first mixing the dry ingredients with only enough water to produce a damp, unworkable mix (one that will retain its form when pressed into a ball). Keep the mortar in this damp condition for 1 to 2 hours and then add enough water to bring it to a working consistency somewhat drier than conventional mortar. The drier mix is easier to place, and does not flow to the bottom of the joints as easily. To see if the color of the new mortar matches the color of the old, test a sample area in an inconspicuous location, using a garden hose to soak a portion of the wall. The color of the new mix should match the darker, wet color of the existing mortar. Minor adjustments can be made by adding or subtracting sand or cement, but records should be kept of exact proportions so that the selected color can be reproduced consistently throughout the job.

16.5.4 Pointing

To ensure good bond with brick and stone, dampen the cleaned joints with water just before beginning work. Mortar is placed by using a small mortar board, called a "hawk," and a tuckpointer's trowel which looks like a jointer, but has a flat blade. Vertical joints are filled first, then horizontal joints, with the mortar applied in thin ¼-in. layers that are allowed to become "thumbprint hard" before the next layer is placed. Joints should not be overfilled to the point where mortar hides the edges of the units.

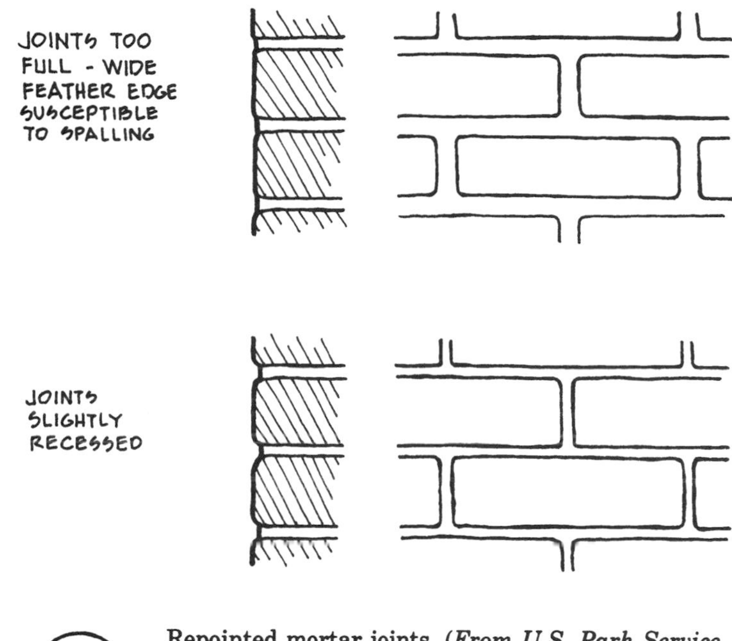

Repointed mortar joints. (*From U.S. Park Service, Preservation Brief No. 2.*)

This makes the joint appear too wide, and the edges break off too easily, leaving voids through which moisture can penetrate. For brick or stone that has weathered to a rounded profile, the new mortar joint should be slightly recessed from the unit surface and tooled concave to avoid "feathered" edges (*see Fig. 16.10*). Stippling joints with the bristles of a stiff, nonmetallic brush while the mortar is still soft will give it a worn appearance. Moist curing may be necessary in hot, dry weather to assure hydration of the cement and good mortar bond to the units.

16.6 COLOR MATCHING BRICK

Once the brick or stone in a wall begins to crumble, the deterioration will continue, often at an accelerated rate. The condition can be remedied only by replacement of the affected unit. Where damage or deterioration is extensive, replacement of entire sections of masonry may be required. The brick, stone, terra cotta, or clay tile used for such repairs should match the original material characteristics as closely as possible. Where damage has been caused by excessive moisture penetration from groundwater migration, the installation of a waterproof membrane as a dampproof course may be possible as sections of the original masonry are removed.

Whenever replacement of brick masonry is required, or when additions to existing buildings are necessary, consideration must be given to color matching. Matching the brick on an existing building involves more than just picking a unit color. To achieve an acceptable result, the mortar mix, the joint tooling, and the moisture content of the brick must also be controlled.

The first consideration is whether to match the brick as is (dirt and all) or to clean the masonry and then match the original color. The question may be at least partially answered by the age of the building itself. If constructed in the last 10 to 20 years, it is likely that the original brick is still available. Cleaning the existing walls at a cost of 15 to 25 cents/sq ft may prove to be less expensive than a custom clay blend to match the soiled brick. Older brick may be unavailable, documentation may not exist to identify the original units, or cleaning may be unwarranted or undesirable from a historic perspective. In such cases, some brick manufacturers will do custom color matching.

Contact the brick manufacturer 6 to 10 months ahead of construction, to give time for several test runs if necessary. Make two separate counts of the percentages of light and dark range units in a typical blend area close to the location of the repair or the planned addition. Request at least 48 loose sample brick in the selected colors. These samples should be numbered on the back, and the manufacturer should retain duplicates at the plant. Compare the samples to the existing walls and view them at distances of 20 ft and 60 ft. Do not be influenced by slight size, finish, or texture variations that are noticeable only at close range—color is the most important factor. When comparing the units, be sure that both the samples and the existing walls are surface-dry. If you are not satisfied with the samples, order more.

Deliberately choose a color range that has slightly *less* contrast than the existing wall. That is, make sure the darkest sample you choose is a little lighter than the darkest brick in the wall, and the lightest sample slightly darker than the lightest brick in the wall. When you are satisfied with the color range and blend of the units, return 10 approved samples to the plant, packaged so that they do not become separated in transit.

The mortar must also match the existing to achieve good results. If you can identify the original mortar, specify the same type, proportion, and materials (i.e., masonry cement or portland and lime mix). If a colored mortar is needed, the manufacturer can help in selecting or custom blending the pigments. Specify controls on job site mortar mixing to ensure consistent color. Each batch of mortar must be mixed in exactly the same way using exactly the same ingredients. The type and amount of each ingredient affects mortar color. Use only one brand of cement, one brand of lime, and one source and color of sand throughout the entire job. Mix only full batches of mortar, maintain a consistent water content, and mix each batch the same length of time. When you cannot exactly match the existing brick color, construct several sample panels with varying mortar colors to find the one that minimizes the difference.

To achieve consistency in color pattern on the wall, units with a pronounced range of colors, or mixes which contain more than one color must be properly blended. Blending problems are rare in historic buildings because careful hand blending at the job site was necessary after the shipment was dumped from the back of a wagon or cart. Manufacturers today routinely perform hand blending operations at the plant and ship the brick on pallets that hold the equivalent of 500 standard modular units. Since the units typically go to the scaffold in batches of only about 100 though, the masonry contractor must unstack the brick according to instructions on each pallet and distribute them to the bricklayers uniformly. Narrow color ranges present fewer potential problems than wider ranges or blends of more than one color. Always specify that units be laid in the wall in a blend that will result in even color without patchy areas.

Establish a standard of acceptable workmanship for color blending by specifying a sample panel for approval. This panel should be constructed in addition to any that were used in the selection process, because it must be built with brick from the actual production run used in the building.

The sample panel can also be used to set standards for joint tooling. The joint size and shape, of course, must match the existing, but tooling should also be consistent to avoid the patchy effect of light and dark joints. Soft mortar tools to a lighter color than drier mortar because more cement paste is drawn to the surface. Tooling therefore, must occur at the same mortar consistency throughout the job rather than after a set number of courses, brick, or elapsed time.

Brick moisture content at the time of laying affects mortar curing time. An *inconsistent* moisture content therefore affects the color of the finished joint. If an unprotected pallet of brick, for instance, becomes partially wet during an overnight rain, the wet units will cause patches of lighter colored joints because their higher moisture content keeps the mortar softer for a longer period of time than adjacent areas (refer to Chapter 15). Always specify weather protection, not only for unfinished walls, but also for units and mortar ingredients.

Visual separation elements can also help minimize the effect of any contrast between new and existing walls. An entry, a decorative panel with a different joint type, or even a small projecting pilaster will help isolate the different work instead of abutting them for critical comparison.

17
QUALITY ASSURANCE AND QUALITY CONTROL

Every building owner wants to be sure that he is getting what he pays for in terms of the quality of the building's construction. Quality assurance programs and quality control procedures are used toward that end.

Quality is defined by the International Organization for Standardization (ISO) as the "totality of features and characteristics of a product or service that bear on its ability to satisfy stated or implied needs." In most construction projects, the particular standard of quality which will apply in a given case is established by and measured in terms of the contract document requirements. The owner initially establishes a general standard of quality which is then developed by the architect/engineer into specific terms and incorporated into the contract documents. The standard of quality required on a given project will vary depending on the needs of the owner, the project type, and the established schedule and budget.

What is the difference between the terms *quality assurance* and *quality control?* The American Society for Testing and Materials (ASTM) defines quality assurance as "all planned and systematic actions necessary to provide adequate confidence that an item or a facility will perform satisfactorily in service" and quality control as "those quality assurance actions which provide a means to control and measure the characteristics of an item, process, or facility to established requirements." In specification language, to assure means "to give confidence to," and to ensure means "to make certain in a way that eliminates the possibility of error." Quality assurance in construction could therefore be considered as the architect/engineer's (A/E's) administrative process for *as*suring that the work will conform to the standard of quality established by the contract documents, and quality control as the procedures for testing, inspecting, checking and verifying to *en*sure that the work meets the required standard of quality established by the contract documents. The A/E can endeavor to "assure" quality, but the contractor alone has control over construction means, methods, techniques, sequences, and procedures, and therefore, only the contractor can "control" quality.

17.1 STANDARD OF QUALITY

The American Institute of Architects (AIA) Document A201, *General Conditions of the Contract for Construction* (1987 edition), contains very broad quality assurance and quality control provisions. The general requirements in Division 1 of a project specification expand on these provisions, but are still written broadly enough to apply to the work of all the specification sections. Each technical section of Divisions 2 through 16 of the specifications includes whatever specific quality assurance and quality control measures may be required for that particular section.

The AIA General Conditions establish very broadly that materials and equipment must be of "good quality and new," and that work must be free of defects and conform to requirements of the contract documents. More detailed requirements may be elaborated in supplementary conditions, and in the project specifications. The General Requirements contained in Division 1 of the project specifications may include several sections which help define project quality standards. The general requirements also establish administrative and procedural requirements which are just as important in achieving the required project quality as the technical standards in Divisions 2 through 16.

A standard of quality may be established in different ways, depending on the method of specifying. *Descriptive specifications* identify exact properties of materials and methods of installation without using proprietary names. *Proprietary specifications* list specific products, materials, or manufacturers by brand name, model number, and other proprietary information. *Reference standard specifications* stipulate minimum quality standards for products, materials, and processes based on established industry standards. *Performance specifications* establish a standard of quality by describing required results, the criteria by which performance will be judged, and the method(s) by which it can be verified.

17.2 QUALITY ASSURANCE PROGRAM

The MSJC Code states that "A quality assurance program shall be used to ensure that the constructed masonry is in conformance with the Contract Documents." Neither the Code nor the MSJC Specifications, however, elaborate specifically on what is required to be included in this program. The *Uniform Building Code* requires quality assurance "that materials, construction and workmanship are in compliance with the plans and specifications," including assurance that:

- Masonry units, reinforcement, cement, lime, aggregate, and all other materials meet the requirements of the applicable standards of quality and that they are properly stored and prepared for use
- Mortar and grout are properly mixed according to specified ingredients and proportions, and that the method of measuring materials are such that the proportions of the materials are controlled
- Construction details, procedures, and workmanship are in accordance with the plans and specifications
- Placement, splices, and bar diameters are in accordance with the plans and specifications

AIA Document A201 requires that the architect/engineer (A/E) "endeavor to guard the Owner against defects and deficiencies in the Work." The A/E does not have "control" of the work, and therefore cannot "control" quality. But the A/E can attempt to "assure" that the specified standard of quality is attained by developing and implementing a quality assurance program. A quality assurance program includes establishing

- Administrative procedures, rights, and responsibilities
- Required submittals, inspections, and tests
- Methods for resolving non-conforming conditions
- Required records

17.2.1 Quality Assurance Requirements in the General Conditions

The foundation of the quality assurance program is laid in the AIA General Conditions which stipulate that, when required by the A/E, the contractor must "furnish satisfactory evidence as to the kind and quality of materials and equipment" to be furnished under the contract. The A/E usually requires that such evidence be submitted in the form of shop drawings, product data, and samples. Through review of such submittals, the A/E has an opportunity before work begins to verify that the materials and equipment proposed are in conformance with the requirements of the contract documents. Division 1 sections establish procedural requirements for such submittals, but the informational content and the physical examples themselves help assure that the design intent and required standard of quality are understood and will be met.

17.2.2 Quality Assurance Requirements in the Specifications

Division 1 may include several project specification sections that establish quality assurance procedures, including project meetings, submittals, and product options and substitutions. Divisions 2 through 16 of the specifications are composed of individual technical sections, each addressing a distinct subject area. With each subject come specific issues and requirements concerning quality assurance and quality control. Each part of the CSI three-part Section Format™ addresses different types of requirements. Since quality assurance is an administrative process, quality assurance provisions are addressed in Part 1 General, including articles entitled "Submittals" and "Quality Assurance."

The "Submittals" article may include paragraphs on:

- Shop drawings, product data, and samples
- Quality assurance and quality control submittals (which may include design data, test reports, certificates, manufacturer's instructions, and manufacturer's field reports deemed necessary to assure that the quality of the work meets the requirements of the contract documents)
- Closeout submittals (which may include project record documents, operation and maintenance data, and warranty information)

Quality control submittals are different from shop drawings, product data, and samples. They are generally required to document the results of source quality control and field quality control procedures specified in Part 2 and Part 3. Quality control submittals are required as part of a quality assurance program to verify and document that the required quality control procedures have been performed.

The quality assurance article requirements include prerequisites, standards, limitations, and criteria which establish an overall level of quality for products and workmanship under a particular specification section. On any given project, requirements might include:

- Qualifications
- Regulatory requirements
- Certifications
- Field samples
- Mockups
- Pre-installation meetings

17.3 QUALITY CONTROL PROCEDURES

Quality control includes the systematic performance of inspection and testing to ensure conformance with the required standard of quality. Specific quality control procedures may be required by the contract documents, but other quality control measures are self-imposed by responsible contractors as a normal part of good business practice.

The contract documents establish a standard of quality for various materials, products, and procedures required for the work. To ensure that the specified standard of quality is achieved, the contractor, materials suppliers, manufacturers, fabricators, and installers execute required quality control procedures, and implement other quality control measures they may deem necessary. Quality control includes not only laboratory and field testing and third-party inspection, but simple checking and verifying to ensure that the materials, products, systems, and equipment supplied conform to the specified requirements.

The contractor's quality control begins with the General Conditions stipulation that field measurements and site conditions be correlated with the contract documents before beginning work. Errors, inconsistencies, or omissions discovered must be reported to the A/E for resolution. The General Conditions also establish the right of the owner to require independent testing and inspection, and the contractor's responsibility for securing jurisdictional testing, inspection, and approvals.

17.3.1 Quality Control Requirements in the General Conditions

The AIA General Conditions require the contractor to achieve the specified standard of quality and prevent defective work through control of construction means, methods, techniques, sequences and procedures. The General Conditions also assign the contractor the responsibility for coordinating, supervising, and directing the work. The contractor is not relieved of these obligations by the activities or duties of the A/E during the contract administration process.

17.3.2 Quality Control Requirements in the Specifications

The Division 1 sections on Quality Control include testing laboratory services, inspection services, field samples, mockups, contractor's quality control, and manufacturer's field services. These general requirements should cover only the administrative and procedural aspects of quality control that are applicable to all sections of the specifications. Requirements for specific tests, services, or field samples should be covered in the technical sections to which they apply.

The technical sections in Divisions 2 through 16 of the specifications will vary in the need for, requirements of, and applicability of specific quality control procedures. Part 1 should list specific administrative and procedural requirements that apply to a particular section. Part 2 should specify source quality control, and Part 3 should cover field quality control.

PART 1 GENERAL

- *Quality control submittals:* Should list the submittals required for this section of the work, including, as appropriate, design data, test reports, certificates, manufacturer's instructions, and manufacturer's field reports

PART 2 PRODUCTS

- *Source quality control:* Involves checking material or product quality prior to incorporation in the project. Material suppliers implement quality control procedures prior to shipping to manufacturers. Manufacturers incorporate quality control procedures in their manufacturing processes. Manufactured components may in turn be fabricated into larger units that may also be subject to quality control requirements. An example would be a system involving components from several specification sections which together must meet specific test criteria related to fire resistance ratings.
- *Fabrication tolerances:* Establish a dimensional or statistical range of acceptability for products which the manufacturer or fabricator must control to ensure proper fit and coordination.
- *Source testing:* May involve the manufacturer or fabricator periodically obtaining or performing tests for verification of conformance with quality standards. For example, sieve analysis of soil materials or aggregates, compressive strength tests for masonry units, acoustical rating analyses of doors, or thermal transmission characteristics of windows. May also include requirements for owner's independent agent to perform tests of materials sampled at the plant, shop, mill or factory.
- *Source inspection:* May require owner's independent agent (such as a testing agency or the A/E) to perform inspections at the plant, shop, mill or factory.
- *Verification of performance:* May require compliance with specified performance criteria before items leave the shop or plant.

PART 3 EXECUTION

- *Acceptable installers:* Quality can be enhanced by the contractor's use of subcontractors and craftsmen with qualifications and experience in a particular trade or with certain expertise for special products or systems types. This article is suggested for use in sections for historical projects, testing and balancing of mechanical systems, special finishes, and others which may require a high quality of workmanship.
- *Site tolerances:* Establish an acceptable range of deviation from specified dimensions. Contractor is required to control dimensional tolerances, and verification may be required if the deviation appears visually unacceptable

or interferes with performance requirements. Dimensional tolerances may involve such issues as surface flatness, levelness, plumbness, or alignment. Frequency of the deviation from tolerances is sometimes controversial. A tolerance which indicates that deviation shall not exceed ¼ in. in 10 ft may be questioned as to direction and whether the deviation is cumulative (e.g., ¾ in. in 30 ft), fragmentary (e.g., ¼ in. in 1 ft), or multiple (e.g., ¼ in. every 6 in. or so).

- *Field quality control:* Represents the last form of verification and may form the basis for decisions about defective work during or after installation.
  - *Site tests:* Usually involves quality control of variable conditions. Test methods, frequency interval, and location of testing are important issues. Field testing may involve soil compaction, load tests, compression tests, and various forms of nondestructive testing. Field testing may not always be performed at the site but may be performed on samples taken from the site.
  - *Inspection:* May involve simple visual observation for conformance with certain specified criteria, or may require third-party inspection of construction to verify conformance with contract requirements.
  - *Manufacturers' field service:* May require an authorized manufacturer's representative to visit the site to instruct or supervise installers in the installation/application of a product or system, or for the training, start-up or demonstration of specialized equipment. The manufacturers' field service may be required to determine or verify compliance with manufacturer's instructions. A manufacturer's field report as a quality control submittal should be required for these services.

17.4 FIELD OBSERVATION AND INSPECTION

The term *field observation* is often used as an architectural term denoting the type of periodic site visits associated with the services of a standard design services contract with a building owner. The term *inspection* more often implies special services that are more time intensive and typically associated with structural engineering services and construction quality control. The items and issues which should be covered under an architect's field observation services are outlined in Chapter 18. The following discussion covers engineering inspection of structural masonry. Refer also to Chapter 12 for more on structural masonry inspection.

The owner may engage independent testing laboratories and special inspectors to test, inspect, and verify the quality of work. Testing laboratories and inspectors are normally selected for their qualifications in a particular area of expertise. Testing laboratories for masonry construction should be accredited in accordance with ASTM C1093, *Standard Practice for Accreditation of Testing Agencies for Unit Masonry.* Quality control testing and inspection may also be required by governmental authorities having jurisdiction over the project. This may involve tests, inspections, and approvals of portions of the work as required by laws, ordinances, rules, regulations, or orders of public authorities. Any required certificates of testing, inspection, or approvals are secured by the contractor and delivered to the A/E as quality control submittals. The contractor may also obtain independent testing and inspection services as a part of its own quality control program.

Analytical design in both the UBC and MSJC Code are based on the allowable stress method in which the computed stresses resulting from service loads may not exceed allowable stresses dictated by the codes. The allowable stresses are based on inspected construction. The MSJC requires structural inspection, and UBC treats inspection as optional, with allowable stresses reduced by one-half when "special inspection" is not provided (refer to Chapter 12).

The language of the MSJC Code has been made more specific regarding the specification and implementation of a quality assurance program, but the Specification remains vague and undefined. MSJC does not yet define "inspection," but proposed changes to the code may provide for different levels of inspection depending on the design procedure used (empirical or analytical) and the type of structure being built (essential or nonessential facility). In the 1995 edition of the MSJC Code, the responsibility for deciding what tests should be required and how often inspections should be made still remains with the design professional. Since the design and complexity of the masonry will vary from project to project, this allows the engineer of a high-rise loadbearing masonry structure to require a higher level of quality assurance than the architect of a low-rise masonry veneer project would typically need.

17.5 INDUSTRY STANDARDS FOR MASONRY

Industry standards such as those developed and published by the American Society for Testing and Materials (ASTM) are an important part of quality assurance and quality control in construction. Some standards establish minimum requirements for products or systems, and others outline standardized testing procedures for verifying compliance with the requirements stated in the contract documents.

At last count, there were more than 80 ASTM standards on masonry and masonry-related products, with more in development. Most project specifications, however, require reference only to a core group of standards that apply to the most frequently used products and systems. Because there are so many different products and materials that fall under the umbrella of the term *masonry,* there are perhaps more standards than for other construction systems. Some standards, however, are embedded references within other standards and ordinarily do not require specific citation in project specifications. Others apply to specialty products such as sewer brick, chemical-resistant units and mortar, high-temperature refractory brick, and clay flue liners that are outside the scope of the typical design project. Still other standards are used primarily for research and product development rather than building construction.

Many ASTM standards cover more than one grade, type, or class of material or product from which the specifier must choose. Some also contain language designating which requirements govern by default if the project specifications fail to stipulate a preference. The following summary of standards should serve as a checklist in preparing project specs and developing a quality assurance and quality control testing program.

The *Uniform Building Code* does not use references to ASTM standards. Instead, they draft their own standards as needed, usually based on those already published by ASTM. For projects in UBC jurisdictions, if there is an equivalent UBC standard, it should be cited along with or instead of the ASTM document. The following list of standards commonly cited in project specifications includes the name and number of the equivalent UBC standard wherever appropriate.

17.5.1 Standards for Clay Masonry Units

- ASTM C216, *Standard Specification for Facing Brick (Solid Masonry Units Made of Clay or Shale)*
- UBC 21-1, *Building Brick, Facing Brick and Hollow Brick (Made From Clay or Shale)*

Face brick are solid clay units for exposed applications where the appearance of the brick is an important consideration in the design. "Solid units" are defined as those with a maximum cored area of 25%. ASTM C216 covers two grades and three types of face brick. Brick *type* designates size tolerance and allowable chippage and distortion based on desired appearance.

Type FBS (Standard) is the industry standard and the type of face brick used in most commercial construction. Type FBX (Select) has tighter size tolerances and less allowable chippage for use in applications where a crisp, linear appearance is desired such as stack bond masonry. Type FBA (Architectural) is nonuniform in size and texture, producing characteristic "architectural" effects such as that typical of, or required to simulate, hand-made brick. Type FBA is popular for residential masonry styles because of its softer profile and less commercial look. All three brick types must meet the same strength and physical property criteria, but brick type is not related to color or color range. If the project specifications do not identify a specific proprietary product or designate brick type, ASTM C216 states that Type FBS standards shall govern.

Brick *grade* classifies units according to their resistance to damage from freezing when they are wet. The property requirements for Grades SW and MW are given in a table that covers minimum compressive strength, maximum water absorption, and minimum saturation coefficient (see Chapter 3).

These properties are tested in accordance with ASTM C67, *Standard Test Methods of Sampling and Testing Brick and Structural Clay Tile.* Since ASTM C67 is referenced in ASTM C216, it is not necessary for the specifier to list ASTM C67 as a separate reference standard. If the brick is specified to meet the requirements of ASTM C216, that automatically requires that the units be tested for compliance in accordance with ASTM C67 methods and procedures.

In general, Grade SW (Severe Weathering) should be specified when a "high and uniform" resistance to damage from cyclic freezing is required and when the brick is likely to be frozen when it is saturated with water. Grade MW (Moderate Weathering) should be specified where only moderate resistance to damage from cyclic freezing is required and when the brick may be damp but not saturated when freezing occurs. ASTM C216 includes a table of grade recommendations for various types of exposure and a related map of geographic weathering regions (see Chapter 3). If the project specifications do not designate the required grade, Grade SW is the default standard, and Grade SW may be substituted by the supplier if Grade MW is specified.

Grade SW brick is required by ASTM C216 to have a minimum average *gross area* compressive strength of 3000 psi, and Grade MW brick 2500 psi. These strengths are more than adequate for most non-loadbearing applications, and the majority of brick produced in the United States and Canada is much stronger. If a specific unit strength requirement greater than the standard minimum is required, that compressive strength should be required by the project specifications.

ASTM C216 also requires that brick be tested for efflorescence in accordance with ASTM C67 and be rated "not effloresced." Color is not covered in this standard, so the specifier must designate the desired color, by specifying a proprietary product, with color and color range verified with a sample panel or mock-up panel.

- ASTM C62, *Standard Specification for Building Brick (Solid Masonry Units Made From Clay or Shale)*
- UBC 21-1, *Building Brick, Facing Brick and Hollow Brick (Made From Clay or Shale)*

Building brick (sometimes called common brick) is used primarily for utilitarian applications or as a backing for other finishes, where strength and durability are more important than appearance. ASTM C62 covers Grades SW and MW based on the same physical requirements for durability and resistance to freeze-thaw weathering as face brick. Building brick is also available in Grade NW (No Weathering) which is permitted only for interior work where there will be no weather exposure.

This standard lists permissible variations in size, but does not classify units by various types. The size tolerances listed apply to all ASTM C62 brick. Since these units are generally used in unexposed applications, there is no requirement for efflorescence testing. The discussion of compressive strength requirements under ASTM C216 above also applies to building brick.

- ASTM C652, *Standard Specification for Hollow Brick (Hollow Masonry Units Made From Clay or Shale)*
- UBC 21-1, *Building Brick, Facing Brick and Hollow Brick (Made From Clay or Shale)*

ASTM C652 covers hollow brick with core areas between 25 and 40% (Class H40V) and between 40 and 60% (Class H60V). The two grades listed correspond to the same requirements for durability as for face brick—Grade SW (Severe Weathering) and Grade MW (Moderate Weathering). Types HBX (Select), HBS (Standard), and HBA (Architectural) are comparable to face brick types FBX, FBS, and FBA respectively. Another type, HBB, is for general use where appearance is not a consideration and greater variation in size is permissible. Type HBB is the hollow brick equivalent of ASTM C62 building brick. When the project specification does not designate brick type, requirements for Type HBS govern. The default standard for brick grade is SW. This standard does include requirements for efflorescence testing the same as that for ASTM C216 face brick. The discussion of compressive strength requirements under ASTM C216 also applies to hollow brick.

Color is not covered in this standard, so the specifier must designate the desired color, by specifying a proprietary product, with color and color range verified with a sample panel or mockup panel.

- ASTM C126, *Standard Specification for Ceramic Glazed Structural Clay Facing Tile, Facing Brick, and Solid Masonry Units*

Glazed brick and structural clay tile are fired with clear or colored ceramic coatings to produce a matte or gloss finish. ASTM C126 covers properties of the ceramic finish including imperviousness, chemical resistance, crazing, and limitations on unit distortion and dimensional variation. Durability and weather resistance are not covered by this standard, so for exterior use, the body of glazed brick should be specified to conform to the requirements for ASTM C216 face brick, Grade SW, except with unit tolerances and surface glaze in accordance with ASTM C126. Glazed brick and tile may suffer severe freeze-thaw damage in cold climates if not adequately protected from moisture penetration, and are not recommended for copings or other horizontal surfaces in any climate. ASTM C126 covers Grade S (select) and Grade SS (select sized, or ground edge), where a high degree of mechanical perfection and minimum size variation are required. When unit grade is not specified, the requirements for Grade S govern by default. Units may be either Type I, single-faced, or Type II, double-faced (opposite or adjacent faces glazed). When unit type is not specified, the requirements for Type I govern. ASTM C126 includes tests for imperviousness, chemical resistance, crazing, and opacity of the finish, and references ASTM C67 for compressive strength testing.

17.5.2 Standards for Concrete Masonry Units

- ASTM C90, *Standard Specification for Load-Bearing Concrete Masonry Units*
- UBC 21-4, *Hollow and Solid Loadbearing Concrete Masonry Units*

Hollow and solid loadbearing concrete block are covered in this standard. Type I units have limits on linear shrinkage (0.03% or less to 0.065% maximum) and limits on moisture content at the time of delivery to the job site based on the mean relative humidity conditions at the project site (above 75%, 50 to 75%, or less than 50%—refer to Chapter 4). Type I units must be protected from wetting while stored at the manufacturer's plant, while in transit, and at the job site. Type II units have no moisture content limits, but shrinkage is limited to 0.065%. Recommended control joint spacing is based in part on unit Type (see Chapter 9) and amount of restraining joint reinforcement. Weight classifications are divided into Light Weight (less than 105 lb/cu ft oven dry weight of concrete), Medium Weight (105 to less than 125 lb/cu ft), and Normal Weight (125 lb/cu ft or more). Unit weight affects water absorption, sound absorption, sound transmission, and thermal and fire resistance. There are no default requirements in ASTM C90, so the specifier must designate unit type and weight classification if these properties are important to the design.

The minimum *net area* compressive strength required for Type I and Type II units and for all three weight classifications for ASTM C90 loadbearing units is 1900 psi. Compressive strength is largely a function of the characteristics of the aggregate used in the units, and may vary regionally according to the types of aggregates available. Aggregates in some areas may routinely produce units with much higher compressive strengths without a cost premium. If a specific unit strength requirement greater than the standard minimum is required, that compressive strength should be required by the project specifications. See side bar for a discussion of masonry compressive strength.

Compliance with the requirements of ASTM C90 is verified by testing in accordance with ASTM C140, *Standard Method of Sampling and Testing Concrete Masonry Units.* C140 is referenced in the C90 standard and need not be listed separately in the project specification. ASTM C90 also references ASTM C33, *Standard Specification for Aggregates for Concrete,* and ASTM C331, *Lightweight Aggregates for Concrete Masonry Units,* as well as standards for the cementitious materials that are permitted in these units.

It is not necessary for the specifier to list these referenced standards separately.

Size tolerances and limits on chippage and cracking are covered in the text of the standard. These requirements are more liberal than those for clay brick because of the nature of the material and the method of manufacture. For exposed architectural units such as split-faced, ribbed, or ground-faced units, these requirements may not be appropriate. Rough units may require greater tolerances and ground-faced units tighter tolerances. For such products, it may be more appropriate to consult local manufacturers for tolerance requirements. Color is not covered in this standard, so the specifier must designate the desired color, by specifying a proprietary product, with color and color range verified with a sample panel or mockup panel.

- ASTM C129, *Standard Specification for Non-Load-Bearing Concrete Masonry Units*
- UBC 21-5, *Non-Loadbearing Concrete Masonry Units*

The requirements of this standard are similar to those of C90 except that the units are designed for non-loadbearing applications. Unit types and weight classifications are the same, as are referenced standards for aggregates, cements, sampling, and testing. Since the units are designated as non-loadbearing, the minimum requirements for net area compressive strength are lower than for ASTM C90 units at an average of only 600 psi. For typical non-loadbearing applications, this strength is more than adequate, but stronger units may be commonly available without a cost premium in some areas. Color requirements are not covered in the specification, and should be specified in the same way as that recommended for ASTM C90 units.

- ASTM C55, *Standard Specification for Concrete Building Brick*
- UBC 21-3, *Concrete Building Brick*

Concrete brick can be loadbearing or non-loadbearing. Grading is based on strength and resistance to weathering. Grade N provides "high strength and resistance" to moisture penetration and severe frost action. Grade S has only "moderate strength and resistance" to frost action and moisture penetration. Each grade may be produced as Type I, moisture-controlled, or Type II, non-moisture-controlled, designated as N-I, N-II, S-I, or S-II. For moisture-controlled units, the standards limit the moisture content at the time of construction according to the relative humidity conditions typical of the project site. The moisture content of units affects the potential for shrinkage and therefore the spacing of control joints (see Chapter 9). Minimum gross area compressive strength for Grade N units is 3500 psi and for Grade S units, 2500 psi.

ASTM C55 does not include requirements for color, texture, weight classification, or other special features. These properties must be covered separately in the project specifications. Sampling and testing is referenced to ASTM C140, and standards for aggregates and cements are also referenced, so the specifier need not list these separately.

17.5.3 Standards for Masonry Mortar and Grout

- ASTM C270, *Standard Specification for Mortar for Unit Masonry*
- UBC 21-15, *Mortar for Unit Masonry and Reinforced Masonry Other Than Gypsum*

This standard covers four types of masonry mortar made from a variety of cementitious materials, including portland cement (ASTM C150 or UBC 19-1) and masonry cement (ASTM C91 or UBC 21-11), as well as blended hydraulic cement and slag cement (ASTM C595 or UBC 19-1), quicklime (ASTM C5 or UBC 21-12) and hydrated masonry lime (ASTM C207 or UBC 21-13). UBC 21-15 also lists mortar cement (UBC 21-14 or ASTM C1329). These material standards are referenced in ASTM C270 and UBC 21-15, so the specifier need not list them separately. If any materials are to be excluded for any reason, this should be noted in the project specifications. Requirements for mortar aggregates are referenced to ASTM C144.

Types M, S, N, and O mortar may be specified to meet either the proportion requirements or the property requirements of ASTM C270. If the project specifications do not designate which method the contractor must use, then the proportion method governs by default. The proportion method is the most conservative, and will usually produce mortars with higher compressive strengths than those required by the property method. Since bond strength decreases as compressive strength increases, it is generally not desirable to use mortar that is stronger in compression than the application requires. To optimize mix design, property-specified mortar requires preconstruction laboratory testing in accordance with the test methods included in ASTM C270. These test methods are not suitable for testing of field-sampled mortar during construction, and cannot be compared to the results of such tests. If field testing of mortar will be required, then *both* preconstruction and construction phase testing should be performed in accordance with ASTM C780 rather than ASTM C270. There is no test method for measuring the composition or physical properties of hardened mortar removed from a masonry structure.

Recommendations for appropriate use of the four basic mortar types are included in a nonmandatory Appendix X1 to ASTM C270 and summarized in Chapter 6.

- ASTM C476, *Standard Specification for Grout for Masonry*
- UBC 21-19, *Grout for Masonry*

This standard covers two types of masonry grout—fine and coarse. The same standards for cementitious materials are referenced as those in ASTM C270 and UBC 21-15, but aggregates must conform to ASTM C404. Fine grout is used for small grout spaces and coarse grout for economy in larger grout spaces (see Chapter 6). Masonry grout may be specified to meet the proportion requirements included in the standard, or it may be required to have a minimum compressive strength of 2000 psi when sampled and tested in accordance with ASTM 1019 or UBC 21-18. If higher compressive strength is required for structural masonry, the required strength should be indicated in the contract documents.

ASTM C476 permits the use of "pumping aid" admixtures in cases where the brand, quality, and quantity are approved in writing. Such admixtures are commonly used in high-lift grouting projects, as are certain other types of admixtures.

- ASTM C1142, *Standard Specification for Ready-Mixed Mortar for Unit Masonry*

This standard covers four types of ready-mixed masonry mortar—RM, RS, RN, and RO. These are the equivalent of the four basic mortar types covered in ASTM C270, except that they are mixed at an off-site batching plant and delivered to the project ready to use. This standard does not include specific recommendations for use of the ready-mixed mortar types, but the recommendations in the appendix of ASTM C270 apply to these ready-mixed mortars as well (see Chapter 6). Standards for cementitious materials and aggregates are referenced in ASTM C1142 and need not be cited separately by the specifier.

17.5.4 Standards for Masonry Accessories

- ASTM A36, *Standard Specification for Structural Steel*

This standard covers the type of steel used for angle lintels and shelf angles in masonry construction. It also applies to heavy bent bar or strap anchors that are often used to structurally connect intersecting masonry walls. Requirements for shop coating or for galvanized corrosion protection should be specified separately for this as well as any other metal accessory used in masonry.

- ASTM A82, *Standard Specification for Cold Drawn Steel Wire for Concrete Reinforcement*

This standard covers steel wire that is used in concrete masonry joint reinforcement and some types of masonry anchors and ties. It includes strength requirements and permissible variations in wire size, but does not include any options which the specifier must designate in the project documents.

- ASTM A951, *Joint Reinforcement for Masonry*
- UBC 21-10, *Joint Reinforcement for Masonry*

These standards cover material properties, fabrication, test methods, and tolerances for prefabricated wire joint reinforcement for masonry. The specifier must designate corrosion protection as Brite Basic, Mill Galvanized, Class I Mill Galvanized (minimum 0.40 oz zinc per sq ft of surface area), Class III Mill Galvanized (minimum 0.80 oz zinc per sq ft of surface area), or Hot-Dipped Galvanized (minimum 1.50 oz zinc per sq ft of surface area). The hot-dip galvanizing is the same as that required for joint reinforcement under ASTM A153 as listed below.

- ASTM A153, *Standard Specification for Zinc Coating (Hot-Dip) on Iron or Steel Hardware*

This standard covers hot-dip galvanized coatings that are required to provide corrosion resistance in exterior wall applications for masonry accessories such as steel joint reinforcement, anchors, and ties. Minimum coating weight is given in four classes based on the size and type of item being coated. Masonry accessories of various sizes are covered under Class B.

- ASTM A167, *Standard Specification for Stainless and Heat Resisting Chromium-Nickel Steel Plate, Sheet and Strip*

This standard covers stainless steel of the type that is used for masonry anchors, ties, and flashing. There are more than two dozen types of stainless steel included in the standard, varying according to chemical and mineral composition. Type 304 is the type most commonly used in masonry construction. Type 316 is also sometimes used in masonry.

- ASTM A 366, *Standard Specification for Steel, Carbon, Cold-Rolled Sheet, Commercial Quality*

This standard covers sheet steel of the type used for sheet metal anchors and ties used in masonry. It also covers the type of sheet metal used in masonry veneer anchors that are a combination of metal plates and wire rods.

- ASTM A615, *Standard Specification for Deformed and Plain Billet-Steel Bars for Concrete Reinforcement*

This standard covers deformed steel reinforcing bars of the type most commonly used in reinforced concrete and reinforced masonry construction. Although there are other types that are acceptable (such as rail steel and axle steel), ASTM A615 is most prevalent in the industry. Even in unreinforced masonry projects, there may be some isolated uses of reinforcing steel such as in lintels over window and door openings.

17.5.5 Standards for Laboratory and Field Testing

- ASTM C780, *Standard Test Method for Preconstruction and Construction Evaluation of Mortars for Plain and Reinforced Unit Masonry*

ASTM C780 covers methods for sampling and testing mortar for plastic and hardened properties either before or during construction. If construction phase testing of mortar will be required, there must be some basis for comparison of the results of such tests. The compressive strengths and other requirements listed under the property specification of ASTM C270 or C1142 cannot be used for comparison because the test methods are different. The laboratory test methods used in ASTM C270 mix mortar samples with a relatively low water-cement ratio. Field-mixed mortars, however, use much higher water-cement ratios in order to overcome the initial absorption of the masonry units. When compared with one another, the field-mixed mortars would appear to be much weaker than the C270 test results. To provide an "apples-to-apples" comparison, the preconstruction design mix must also be tested with a high water-cement ratio to simulate that which will actually be prepared during construction. Using ASTM C780 to obtain a preconstruction mortar mix design provides a basis for acceptance or rejection of field-sampled mortars during construction.

- ASTM C1019, *Standard Test Method for Sampling and Testing Grout*
- UBC 21-18, *Method of Sampling and Testing Grout*

This standard covers both field and laboratory sampling for compressive strength testing of masonry grout. This standard should be referenced in the project specifications for loadbearing masonry construction if the compressive strength of the masonry construction is to be verified by either the unit strength method or prism test method.

- ASTM E447, *Standard Test Methods for Compressive Strength of Masonry Prisms*
- UBC 21-17, *Test Method for Compressive Strength of Masonry Prisms*

This standard covers compressive strength testing of masonry prisms for loadbearing masonry construction. Two methods of tests are included.

Method A prescribes a single-wythe specimen laid in stack bond with $\frac{3}{8}$-in. mortar joints that are struck flush with the face of the masonry. Method B prescribes that test prisms be constructed, in so far as possible, in the same manner as that being used in the structure. Method A is appropriate for research and development in determining comparative data for masonry using different units or mortars. Method B is appropriate for preconstruction testing of the compressive strength of a particular set of materials to be used in construction. Project specs should stipulate Method B. This is the specific method of test required by the MSJC Code.

- ASTM C1314, *Standard Test Method for Constructing and Testing Masonry Prisms Used to Determine Compliance with Specified Compressive Strength of Masonry*

In structural masonry projects, the engineer must indicate on the drawings the required compressive strength of masonry (f'_m) on which the design is based. The contractor must verify to the engineer that the construction will achieve this minimum compressive strength. This verification may be provided in one of two ways—the unit strength method or the prism test method. The unit strength method is very conservative, and is based on the empirical assumption that the combination of certain mortar types with units of a certain compressive strength will produce masonry of a given strength. If the manufacturer submits certification of the unit compressive strength and the mortar is specified by the ASTM C270 proportion method, compressive strength verification can be provided by a table in the code without any preconstruction or construction testing of any kind. If the mortar was specified by the ASTM C270 property method, the mortar test discussed above along with the manufacturer's certification of unit strength are sufficient to verify compressive strength compliance. If f'_m must be verified by the prism test method, an assemblage of the selected units and mortar must be constructed and tested in accordance with ASTM C1314. This test may be used both for preconstruction and construction evaluation of the masonry. Although the ASTM C1314 test method is similar to that in ASTM E447 described immediately above, ASTM C1314 does not require any extraneous information other than that required for verification of the specified compressive strength.

17.6 MASONRY SUBMITTALS

Submittals are a time-consuming but important part of construction projects. Submittals are used to help assure that the work meets the requirements of the contract documents and that the contractor achieves the standard of quality established by the specifications. For each project, the architect or engineer must decide what submittals are needed for each portion of the work. Submittals require time and money to prepare and process (for both the A/E and the contractor), so it is important that only those submittals that are appropriate and necessary to the work be required.

The types of submittals that are appropriate or necessary will vary from project to project according to the nature of the construction, both aesthetic and structural. For structural masonry projects designed under the Masonry Standards Joint Committee (MSJC) *Building Code Requirements for Masonry Structures (ACI 530/ASCE 5/TMS 402)*, some submittals are mandatory and others are optional. Projects that are nonstructural, but aesthetically important may lean more toward submittal of unit and mortar samples than test reports. Each project is unique in its requirements.

17.6.1 Specifying Submittals

According to the CSI *Manual of Practice,* administrative and procedural requirements for submittals should be specified in Division 1—General Requirements, because they apply to all project submittals. CSI's Master Format™ designates Section 01300 as the proper location for these requirements which would include information such as the number of copies required, how much time should be allowed for review, and to whom reviewed submittals should be distributed.

Specific submittals required for a masonry project should be specified in the appropriate technical section in Division 4. Each of the technical sections should include in Part 1 a complete list of the submittals required for that portion of the work. Submittals may include shop drawings, product data, samples, and quality assurance/quality control submittals. Each type of submittal has a different function, and is applicable to different types of materials, products, or systems. *Figure 17-1* lists all of the types of submittals and submittal information that might be included in a masonry specification. The list will vary as appropriate to the project, the type of construction, and the wishes of the architect or engineer.

17.6.2 Submittal Procedures

Submittals must be reviewed and approved before construction can begin. Material and equipment cannot be ordered or fabricated until specified submittals are approved by both the contractor and the A/E. The general contractor is responsible for submitting required information to the A/E for review and approval. Many of the required submittals may actually be prepared by subcontractors, suppliers, fabricators, or manufacturers. The general contractor must check all submittals, stamp, and sign them, assemble them with transmittal forms, and submit them to the A/E for review. Submittals that are not approved must be resubmitted with the required changes, reviewed, and approved before construction can begin. Both the A/E and the contractor should maintain a submittal log to track the progress of all project submittals. A copy of all approved submittals should be kept with the record documents at the job site until the project is complete. Both the A/E and the contractor usually retain copies of approved submittals as part of their permanent project records.

In masonry construction, it is the responsibility of the masonry subcontractor to prepare or assemble the required masonry submittals and turn them over to the general contractor. Manufacturer's literature on masonry accessories, product certifications on masonry units, or metal flashing details may sometimes be prepared by the supplier, manufacturer, or fabricator, respectively, for submittal by the masonry subcontractor to the general contractor.

17.6.3 Shop Drawings

Shop drawings are prepared to illustrate some detail of the construction. They are typically prepared by a manufacturer or fabricator for use in producing items, and as an aid to the contractor in coordinating the work with adjacent construction.

For example, structural engineering drawings typically show reinforcing steel only diagrammatically in plans and sections. The shop draw-

Shop drawings

- ❑ Fabrication dimensions and placement locations for reinforcing steel and accessories
- ❑ Flashing details
- ❑ Temporary wall bracing

Product data

- ❑ Proprietary mortar ingredients
  - ❑ Portland cement
  - ❑ Masonry cement
  - ❑ Mortar cement
  - ❑ Lime
  - ❑ Admixtures
- ❑ Accessory items
- ❑ Joint reinforcement
- ❑ Shear keys
- ❑ Weephole ventilators
- ❑ Cleaning agents

Samples

- ❑ Units
- ❑ Mortar color
- ❑ Connectors
- ❑ Accessories

Quality assurance/quality control submittals

- ❑ Design data
  - ❑ Mortar mix designs
  - ❑ Grout mix designs
- ❑ Test reports
  - ❑ Preconstruction testing
  - ❑ Field testing
  - ❑ Source quality control testing
- ❑ Certifications
  - ❑ Compliance with specified requirements
  - ❑ Compliance with specified ASTM standards
  - ❑ Brick IRA
- ❑ Inspection reports
  - ❑ Materials
  - ❑ Protection measures
  - ❑ Construction procedures
  - ❑ Reinforcements
  - ❑ Grouting
- ❑ Manufacturers' instructions
  - ❑ Cleaning agents
  - ❑ Mortar coloring pigments
- ❑ Manufacturer's field reports
  - ❑ Cleaning operations
- ❑ Proposed hot and/or cold weather construction procedures

Masonry submittals checklist.

ings show each size, dimension, and type of re-bar, its configuration, and its splice details as well as a key to its plan location and the quantity required. These drawings are used then in the steel fabricator's shop to prepare the individual elements needed at the project site. The engineer reviews the shop drawings for conformance to design and contract document requirements, but does not generally check the quantities. Projects under the jurisdiction of the MSJC Code are required to have shop drawings for structural reinforcing steel.

The A/E may also wish to have shop drawings submitted to illustrate flashing details such as end dams, corners, lap seals, and abutments with other construction. These drawings can then be used to fabricate the required flashing sections in the sheet metal shop for installation at the project site by the masons. Requiring shop drawings for flashing can help assure that the contractor has anticipated and planned for all field installation conditions and has properly interpreted the drawing and specification requirements.

Loose steel angle lintels and prefabricated concrete lintels should require the submittal of shop drawings for verification of dimensions and coordination with masonry coursing. Projects with cut stone may have extensive shop drawings that identify each size and shape of stone, its anchorage conditions, and placement location. In grouted construction, the engineer may also require shop drawings showing the type of temporary construction that will be used to brace uncompleted walls.

17.6.4 Product Data

Fabricated products such as the accessories used in masonry construction typically require the submittal of manufacturers' product data rather than shop drawings. Many specifications list the products of several different manufacturers that are acceptable for use in the construction. Others specify products only by description or by reference standard without mentioning proprietary names. These methods of specifying make it necessary to require the submittal of proprietary product data to verify that the products which the contractor proposes to use meet the specified requirements. Masonry product data might include catalog sheets or brochures for anchors, ties, re-bar positioners, joint reinforcement, weephole ventilators, and shear keys. The masonry contractor or supplier who prepares the submittal should mark data sheets that include more than one item to clearly show which item or items are proposed for use. If there are various model numbers, materials, sizes, etc., these too should be marked to show the appropriate selection.

Manufactured products such as cement, admixtures, mortar coloring pigments, and cleaning agents may also be included in the A/E's list of required submittals. If more than one brand of proprietary masonry cement or mortar cement is approved for use on the project, the manufacturer's product literature should be submitted to indicate which particular products the contractor is proposing to use, and to verify their conformance to contract document requirements. Product data on approved types of admixtures should clearly indicate the chemical ingredients included to assure that they contain no calcium chloride or other harmful substances. Product data on mortar coloring pigments and proprietary cleaning agents other than hydrochloric or hydrofluoric acids should also be submitted for review and approval.

17.6.5 Samples

Samples may be required for masonry units, colored mortars, and some selected accessory items. Unit samples are most often reviewed for color selection purposes during earlier design phases, but if the masonry has been specified on a unit price basis, or only by ASTM reference standard, the A/E must approve samples submitted by the contractor. Cut stone, brick, and architectural CMU samples should indicate the full range of color, texture, shape, and size. Any project requirements for sample panels or mockups should be specified under the Quality Assurance article of Part 1 rather than under this article, which is reserved for individual unit or material samples.

17.6.6 Quality Assurance/Quality Control Submittals

Quality assurance and quality control submittals include test reports, manufacturer's or contractor's certifications, and other documentary data. These submittals are usually for information only. They are processed in the same manner as shop drawings and product data, but do not always require review and approval.

If mortar is specified by ASTM C270 property requirements for compressive strength rather than the default proportion specification, mix *design data* should be submitted for review, along with the results of *preconstruction tests* verifying compliance with the required compressive strength. Grout mixes that are required to attain a specified compressive strength should also require mix design and test result submittals. The results of preconstruction tests must be available for comparison with the results of any *field tests* that may be required, because they are the only valid criteria against which field test results can be compared. For structural masonry projects on which the contractor chooses to verify the strength of masonry by *prism tests,* these results should also be submitted.

The A/E may sometimes require that a manufacturer or fabricator perform testing of a specific product lot, run, shipment, etc. For example, masonry unit manufacturers might be required to submit test results verifying compliance with specified properties such as compressive strength or absorption. For structural masonry projects on which the contractor chooses to verify the strength of masonry by the unit strength method, these unit strength test results should be compared to minimum requirements listed in the code tables. This type of submittal is called a *source quality control submittal.*

Instead of laboratory test results, some products may be submitted with written *certification* from the manufacturer that the item complies with specified requirements. Certifications are usually in the form of a letter, and require the signature of an authorized company representative. The MSJC Specification requires that in addition to reinforcing steel shop drawings, certifications of compliance be submitted for each type and size of reinforcement, anchor, tie, and metal accessory to be used in the construction. Certification of unit, mortar, and grout materials may also be required instead of test results for some projects that do not involve structural masonry elements.

On projects where field inspection is provided by someone other than the project engineer, the specifications should require submittal of *inspection reports* on materials, protection measures, construction procedures, reinforcing steel placement, and grouting operations. If the project engineer is doing field inspections, the same type of information may be kept on file as field notes.

For some products such as cleaning agents or mortar coloring pigments, the A/E may require submittal of *manufacturer's instructions* for application, mixing, or handling of materials. Hazardous materials should require submittal of Material Safety Data Sheets (MSDS). *Manufacturer's field reports* are also sometimes required if the A/E wants to verify that a representative of the cleaning agent manufacturer, for instance, has visited the project site to inspect substrate conditions or to instruct the contractor in the application of certain materials or cleaning methods.

Finally, the A/E may require the submittal of proposed *hot and/or cold weather construction procedures.* The specifications usually call only for compliance with the International Masonry Industry All-Weather Council *Recommended Practices and Guide Specifications for Cold Weather Masonry Construction.* The contractor's submittal should describe the specific methods and procedures proposed to be used to meet these requirements.

17.6.7 Closeout Submittals

Closeout submittals include such things as record documents, extended warranty information, maintenance instructions, operating manuals, and spare parts. Masonry construction does not usually involve this type of documentation. If coatings or clear water repellents are used to reduce the surface absorption of the masonry, and if those materials carry a manufacturer's extended warranty, the information would be submitted by the applicator of the material rather than by the masonry contractor.

17.7 SAMPLE PANELS AND MOCKUPS

Sample panels and mockups are an important part of quality assurance programs. They can also be an effective tool of communication between the design office and the job site in setting both technical and more qualitative aesthetic standards. For aesthetic criteria, sample and mockup panels are the *only* practical and effective method of establishing a fair and equitable procedure for evaluating the completed work. For technical criteria, mockups provide a well-defined yardstick for measuring performance without dispute.

A sample panel is defined as a site-constructed panel of masonry to be used as a basis of judgment for *aesthetic* approval of the appearance of the materials and workmanship. Sample panels should not routinely be used to make design decisions on color, bond pattern, or joint type unless the work of constructing multiple sample panels has been contracted separately from the project construction based on a unit price per panel. Color matching masonry on renovation or rehabilitation projects may require numerous panels to make such decisions. Design panels should be constructed very early to allow time for procurement of the selected materials.

A mockup is more than just the units and mortar of the traditional masonry sample panel. Mockups also incorporate all of the basic elements of the project masonry including as appropriate, reinforcing steel, shelf angles or supports, ties or anchors, joint reinforcement, flashing, weepholes, and control or expansion joints. Design elements such as windows or parapets which may be considered critical aesthetically or from a performance standpoint can also be incorporated at the discretion of the architect or engineer.

Mockups should be used instead of sample panels whenever the acceptability of the masonry will be judged on more than just finished appearance, and construction observation or inspection will be provided to verify conformance. Mockups can be used not only to verify size, chippage, and warpage tolerance of units, but also to establish aesthetic criteria such as unit placement, joint tooling, joint color uniformity, and the even distribution or blending of different color units or units with noticeable color variations. Because they incorporate other elements, however, mockups are perhaps most valuable for establishing acceptable workmanship and procedural requirements for such items as placement of reinforcement, embedment of connectors, installation of flashing, and prevention of mortar droppings in wall cavities.

Since many of the items required in the mockup will be concealed, and since acceptance is based on procedure as well as appearance, the architect or engineer should be present during construction of the panel to observe the work and to answer questions about specified requirements. Documentation of concealed elements and procedural items may best be accomplished by photographing the work-in-progress. A cursory examination of a completed mockup panel will tell the observer nothing about what is inside the wall (or is not inside the wall). Acceptance on such a basis does not give adequate criteria on which to accept or reject the project masonry. The proper evaluation and comparison of the project masonry with the standards of the mockup requires on-site observation or inspection by the architect, engineer, or independent inspector. The person who will evaluate and accept or reject the work, if different from the design architect or engineer, should also be present for construction of the mockup. Depending on its size, construction of the mockup could be incorporated into a preconstruction conference. Both the meeting and the mockup can be instrumental in clarifying project requirements, understanding design intent, and resolving potential problems or conflicts.

The mockup should be constructed by a mason or masons whose work is typical of that which will be provided in the project, because it establishes the standard of workmanship by which the balance of the masonry will be judged. If the mock-up is constructed by the foreman or foremen, the record photographs can be used to communicate to the crew precisely what is expected of them and the standards by which their work will be judged.

18

SPECIFICATIONS AND FIELD OBSERVATION

The quality, durability, and cost-effectiveness of masonry systems are affected by decisions made throughout the design and documentation phases, and by field observation and inspection practices. Project specifications establish standards of quality which should be strictly enforced to ensure structural integrity, weather resistance, and long service.

18.1 ECONOMIC CONSIDERATIONS

Exterior envelope materials are usually selected on the basis of both cost and aesthetics. An architect or building owner may begin with a mental image of the project that is related to its context, its corporate identity, and its budget. Masonry is very cost-competitive as an envelope material, with stone cladding at the high end and brick veneer and architectural concrete masonry units (CMUs) at the lower end (*see Fig. 18-1*), but the decision to use masonry of one type or another is usually an aesthetic one. Material selections are made based on color, texture, and scale.

The relative cost of different types of brick or different types of architectural block is related primarily to unit size and labor production. Typical union production rates for several types of brick, block, and structural clay tile are listed in *Fig. 18-2*. Within a selected size, however, aesthetic preference should govern unit selections, because the cost of materials has only a small effect on the cost of the completed envelope. According to a 1989 study, doubling the cost of brick from $425/M to $850/M added less than $2.00/sq ft to the wall cost. There are a number of other design and specification decisions which affect masonry cost, and which can be used to minimize budget limitations.

18.1.1 Factors Affecting Cost

- Larger face-size units increase the area of wall completed each day, even though the mason may lay fewer units because of greater weight. This

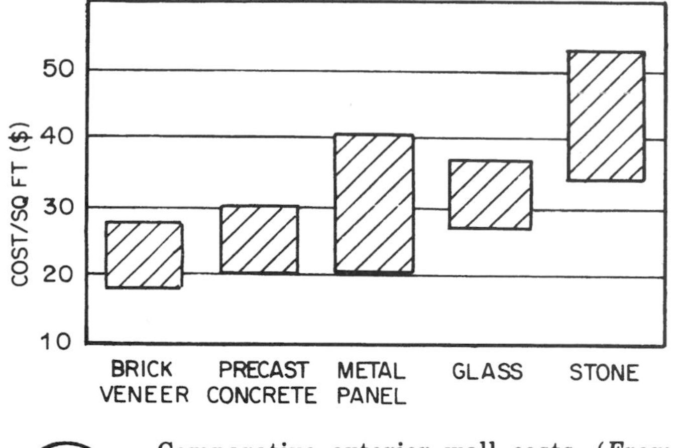

Comparative exterior wall costs. (*From* **Building Design and Construction Magazine,** *September 1988.*)

Unit			Wall		
Description (size and type)	*Weight (lb)*	*Surface area (sq ft)*	*Nominal thickness (in.)*	*Productivity (sq ft/8 hr)*	*Relative productivity*
4" × $2\frac{2}{3}$" × 8" Standard brick	—	0.148	4	160*	1.00
4" × $2\frac{2}{3}$" × 12" Norman brick	—	0.222	4	207	1.28
4" × 4" × 8" Economy brick	—	0.222	4	218	1.36
4" × 4" × 12" Economy brick	—	0.333	4	242	1.51
8" × 5" × 12" Hollow tile	13.7	0.444	8	257	1.61
4" × 5" × 12" Hollow tile	8.6	0.444	4	266	1.66
6" × 5" × 12" Hollow tile	12.4	0.444	6	287	1.79
4" × 8" × 12" Hollow tile	15.2	0.667	4	308	1.93
8" × 8" × 12" Hollow tile	25.3	0.667	8	360	2.25
12" × 8" × 16" Hollow block	31.9	0.889	12	376	2.35
8" × 8" × 16" Hollow block	26.2	0.889	8	467	2.92
4" × 8" × 16" Hollow block	16.4	0.889	4	494	3.09
6" × 8" × 16" Hollow block	21.1	0.889	6	526†	3.29
4" × 12" × 12" Hollow tile	18.9	1.000	4	413	2.58

*Lowest value.
†Highest value.

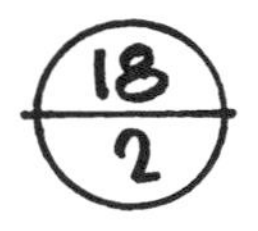

Mason productivity with various unit sizes. (*From National Concrete Masonry Association,* **TEK Bulletin 54,** *NCMA, Herndon, Va.*)

option is simple and cost-effective. The higher price of larger units can be offset by lower labor costs and by earlier completion of the work. For some designs, larger masonry units may actually give better proportional scale with the size of the building as well.

- Type FBS and FBA brick require "standard" workmanship and are less expensive to install than FBX units which require precision placement and jointing to look their best (refer to *Fig. 3-2*).

- Raked mortar joints are more expensive than concave or V-tooled joints, and are generally considered less weather resistant.
- Colored mortar costs more than ordinary gray mortar without added pigments, but white portland cement can be used for lighter shades.
- Field cutting units can be expensive if the number of cuts required is excessive because of poor modular or dimensional planning.
- Stack bond patterns are sometimes less expensive than one-half or one-third running bond patterns because of reduced cutting required at openings and corners.
- Special unit shapes can be less expensive than field cutting if sufficient quantities are required to justify production expenses.
- Rounded corners with a radius of as little as 2 ft can be turned using standard rowlock, header, or soldier instead of curved units.
- Door jambs covered by wood bucks or metal trim eliminate the need for saw-cut or bullnose units and reduce costs.
- Analytically designed brick or CMU curtainwalls can eliminate the need for shelf angles on buildings up to 100 ft or more in height.
- Mechanical and electrical lines and conduit are less expensive to place in double-wythe cavity walls than in single-wythe walls unless special concrete block units are used.
- Openings spanned with masonry arches or reinforced masonry lintels eliminate the need for steel angle lintels and the associated maintenance costs they include.

Careful detailing and thoughtful design can enhance the cost economy of any building system. Conscientious planning and material selection, attention to detail, thorough specifications, and on-site field observation and inspection can all contribute to lower masonry costs.

18.1.2 Value Engineering

In estimating the total cost of a building system or product, future as well as present costs must be considered. Value engineering and life-cycle costing methods evaluate expenses throughout the life of a building. For example, the fire resistance of masonry structures means lower insurance rates and lower repair costs if interior spaces do sustain damage from fire (refer to Chapter 8). Masonry thermal characteristics reduce energy consumption for heating and air conditioning, and the durability and finish of the surfaces also minimize maintenance costs.

Maximizing the structural and functional capabilities of the masonry will also reduce initial costs. Strength design of reinforced masonry permits construction of tall, slender walls with significant reductions in lateral support requirements. Using double-wythe walls for loadbearing applications multiplies their functional role to that of structure, acoustic and fire separation, mechanical and electrical chase, exterior envelope, and interior finish.

18.2 SPECIFICATION GUIDELINES

Specifications are an important part of quality assurance and quality control in masonry construction. To achieve quality workmanship and proper performance, the architect or engineer must carefully outline the products and standards of construction required. Reference standards should be used to govern the quality of specified products. ASTM standards cover all of the mortars, unit types, and varieties of stone (see Appendix B), and are widely accepted throughout the industry.

Lump-sum or unit-price allowances may be used for specifying masonry units, but the specifications should also include sufficient information (including unit size, grade, type, and texture) so that the contractor can accurately bid the labor required for installation. Most ASTM standards for masonry products cover two or more grades and types of units, so the project specifications must identify what is required. Omitting this information makes it impossible for bidders to accurately estimate cost.

The size of unit required should always be included in the specifications, preferably giving actual rather than nominal dimensions to avoid ambiguity. In some industries, "nominal" means approximate, but in modular masonry, it means the manufactured dimension plus the thickness of the mortar joint. A nominal 8-in. modular brick can be manufactured at $7\frac{1}{2}$ in. for use with $\frac{1}{2}$-in. joints, or $7\frac{5}{8}$ in. for use with $\frac{3}{8}$-in. joints. Dimensions should be listed with thickness first, followed by the face dimensions of height and then length.

Color and texture are not included in ASTM standards, so requirements must be established by the specifications. If an allowance method is used, the final selection may be made from samples submitted by the contractor or supplier. If trade names are used to identify a color range and finish, or if descriptions are not given in the project specifications, samples of acceptable materials should be available to the contractors for inspection prior to bidding.

The specification guidelines which follow may be used as a reminder list for the primary items requiring attention in the specifications. If more than one masonry system is used on a project, sections should be combined to include the mortar, units and accessories for each system under a separate heading (i.e., 04320 Veneer Masonry System, or 04340 Reinforced Unit Masonry System). This makes it clear to the contractor which anchors or ties go where, what mortar type or flashing to use, and so on.

18.2.1 Mortar and Grout

- Portland cement: ASTM C150 Type I, or Type III for cold weather, low alkali content, nonstaining
- Masonry cement: ASTM C91 or UBC 21-11 (list acceptable manufacturers)
- Mortar cement: ASTM C1329 or UBC 21-14 (list acceptable manufacturers)
- Hydrated lime: ASTM C207 or UBC 21-13, Type S
- Sand: ASTM C144, clean and washed
- Grout aggregates: ASTM C404
- Water: clean and potable
- Admixtures: no calcium chloride permitted (list others permitted or prohibited)
- Mortar type: ASTM C270 or UBC 21-15, Type (M, S, N, O, or K), proportion specification (default), or property specifications (minimum compressive strength for structural masonry)
- Grout type: ASTM C476 or UBC 21-19 (fine or coarse)

18.2.2 Masonry Accessories

- Metals: cold drawn steel wire, ASTM A82; welded steel wire fabric, ASTM A185 or A497; sheet metal, ASTM A366; plate, headed, and bent bar ties, ASTM A36; reinforcing steel (billet steel deformed bars, ASTM A615; rail steel deformed bars, ASTM A616; axle steel deformed bars, ASTM A617)
- Corrosion protection: stainless steel, ASTM A167, Type 304; hot-dip galvanized steel, ASTM A153, Class B
- Masonry ties: manufacturer, model number, type of metal
- Veneer anchors: manufacturer, model number, type of metal
- Fasteners: list appropriate types
- Joint reinforcement: ASTM A951 or UBC 21-10, wire gauge, type (ladder or truss), corrosion protection
- Accessories: through-wall flashing, weephole accessories, control joint shear keys, compressible expansion joint filler, cleaning agents

18.2.3 Masonry Units

- Facing brick: ASTM C216 or UBC 21-1, Grade (MW or SW), Type (FBX, FBS, or FBA), unit size, color and texture, manufacturer, minimum compressive strength
- Glazed brick: ASTM C216 body, Grade (MW or SW); ASTM C126 glaze, Grade (S or SS), Type (I or II), unit size, color and texture, manufacturer, minimum compressive strength
- Building brick: ASTM C62 or UBC 21-1, Grade (SW, MW, or NW), unit size, minimum compressive strength
- Hollow brick: ASTM C652 or UBC 21-1, Grade (SW or MW), Type (HBX, HBS, HBA, or HBB), unit size, color and texture, manufacturer, minimum compressive strength
- Hollow or solid loadbearing CMU: ASTM C90 or UBC 21-4, Type (I or II), weight (normal, medium, or light), unit size, color and texture (architectural block only), minimum compressive strength; UBC 21-4 also requires designation of Grade N (for general use in exterior walls above and below grade) or Grade S (for uses limited to above-grade installation in exterior walls with weather-protective coatings, or in walls not exposed to the weather)
- Non-loadbearing CMU: ASTM C129 or UBC 21-5, Type (I or II), weight (normal, medium, or light), unit size, color and texture (architectural block only), minimum compressive strength
- Concrete brick: ASTM C55 or UBC 21-3, Grade (N or S), Type (I or II), weight (normal or light), unit size, color and texture, manufacturer, minimum compressive strength

18.2.4 Construction

- Preconstruction conference
- Submittals, sample panels, mockups, testing
- Storage and protection of materials, hot and cold weather protection procedures
- Tolerances for placement and alignment of masonry

- Mortar mixing, retempering, placement, joint tooling, and pointing
- Wetting of brick with high initial rate of absorption (IRA), unit blending, unit placement
- Installation of flashing and weepholes, connectors, joint reinforcement, control joints, and/or expansion joints
- Placing reinforcement, grouting methods
- Temporary bracing and shoring, protections during construction, protection of finished work, moist curing

18.2.5 Quality Control Tests

Laboratory testing of materials and assemblages is usually limited to structural masonry rather than veneer systems. Mortar, grout, and masonry prisms may all be tested before construction to establish quality standards, and tested during construction to verify compliance. Tests may also be used as part of the material selection process.

When mortar is specified to have a certain minimum compressive strength for structural masonry, it is required to meet the property specification of ASTM C270 rather than the default proportion specification. To verify that the contractor's proposed mortar mix meets the strength requirements, a sample can be tested in accordance with ASTM C270, but the results will not be comparable for testing later field samples, because the methods of preparing the laboratory sample are not the same as those used in the field. If subsequent testing of field samples will also be required, both preconstruction and construction testing should be done in accordance with ASTM C780, *Standard Test Method for Preconstruction and Construction Evaluation of Mortars for Plain and Reinforced Unit Masonry.* The preconstruction test sets a quality standard against which field-tested samples may be compared. ASTM C780 actually includes several different types of tests, including compressive strength, board life, mortar aggregate ratio, water content, air content, and tensile strength. Specify only those tests which are needed.

Grout testing before and during construction can be done by a single test, ASTM C1019, *Standard Method of Sampling and Testing Grout,* which applies to both laboratory-prepared and field-prepared samples.

The compressive strength (f'_m) of structural masonry may be verified by the unit strength method or by the prism test method (refer to Chapter 12). If f'_m must be verified by the prism test method, an assemblage of the selected units and mortar must be constructed and tested in accordance with ASTM E447, *Test for Compressive Strength of Masonry Prisms,* or ASTM C1314, *Standard Test Method for Constructing and Testing Masonry Prisms Used to Determine Compliance with Specified Compressive Strength of Masonry.* These tests may be used both for preconstruction and construction evaluation of the masonry.

18.2.6 Sample Panels and Mockups

A sample panel is defined as a site-constructed panel of masonry to be used as a basis of judgment for *aesthetic* approval of the appearance of the materials and workmanship. Judging the appearance of masonry can be very subjective, but there are several basic things that should be considered.

- Compliance with allowable unit chippage and warpage

- Compliance with allowable size tolerances
- Unit placement
- Mortar joints and tooling
- Overall workmanship

Typical sample panels range in size from 4 ft×4 ft to 4 ft×6 ft or larger. The Masonry Standards Joint Committee (MSJC) *Building Code Specifications for Masonry Structures* (*ACI 530.1/ASCE 6/TMS 602*) requires a minimum sample panel size of 4 ft×4 ft.

A mockup panel goes a step beyond the sample panel because it includes other elements of the work not related to aesthetics. Mockups may be required instead of or in addition to sample panels. They may serve the dual purpose of setting criteria for both aesthetic and technical consideration, and they may also be built for testing purposes. Mockups should include all of the basic components of the masonry system, including reinforcement, connectors, shelf angles, flashing, weepholes, and expansion and control joints. If more elaborate mockups are required to show specific areas or details of the work such as window detailing, the panels should be delineated on the drawings or described adequately in the specifications to clearly identify the work required. Mockup panels are often larger than sample panels. The size will vary with complexity, but a basic panel without window element or other special components should be at least 4 ft×6 ft.

Sample panels and mockups can be built freestanding or as part of the permanent construction. If freestanding, they should be located where they will not interfere with subsequent construction or other job site activities because they must remain in place until the masonry work has been completed and accepted. Sample panels and mockups should be constructed early enough in the construction schedule to allow for rejection and reconstruction without delay to the work.

Since many of the items required in a mockup will be concealed, and since acceptance is based on procedure as well as appearance, the architect or engineer should be present during construction of the panel to observe the work. Documentation of concealed elements and procedural items may best be accomplished by photographing the work-in-progress. A cursory examination of a completed mockup panel will tell the observer nothing about what's inside the wall (or isn't inside the wall). Acceptance on such a basis does not give adequate criteria on which to accept or reject the project masonry.

Specifications typically say too little about sample panels and mockups. The construction documents should allow bidders to accurately estimate the cost of constructing the mockup. Size and number of panels required and all of the components to be included should be specified. Complex mock-ups which include various design elements should be illustrated on the drawings in plan, elevation, and section, and referenced to specific project details. The specifications should designate the accepted mockup as the project standard. They should also clearly establish the aesthetic and technical criteria on which acceptance or rejection of the panel will be based, as well as the person who will be responsible for evaluation (architect, engineer, construction manager, independent inspector, owner, etc.) Only specified products and materials or accepted substitutes should be used to construct the mockup. Units should be of the same production run that will be supplied for the project, and should represent the full range of color variation to be expected in the project. Mortar ingredients, including sand and water, should also be those which will be used for project construction since they have a significant effect on mortar color. The specification should also stipulate that the panel be built by a mason whose work is typical of that to be expected in the finished wall. A mason contractor should not assign the best bricklayer on the crew to build the sample, because if the rest of the crew cannot match that workmanship, there may be a basis for rejection of the finished work. Before construction of a sample panel or mockup begins, all project submittals should be reviewed for conformance to contract document requirements, and any required preconstruction testing should be complete.

18.3 SPECIFYING WITH *ACI 530.1*

The *Standard Building Code* (Southern Building Code Congress) and *National Building Code* (Building Officials and Code Administrators) both require use of the Masonry Standards Joint Committee (MSJC) *Building Code Requirements for Masonry Structures* (*ACI 530/ASCE 5/TMS 402*), which is jointly written by the American Concrete Institute (ACI), the American Society of Civil Engineers (ASCE) and The Masonry Society (TMS). The MSJC Code includes *Specifications for Masonry Structures* (*ACI 530.1/ASCE 6/TMS 602*). In all jurisdictions where either the *Standard Building Code* or *National Building Code* is enforced, using the Specifications is not an option. This mandate changes the way in which you must write masonry specs.

The MSJC *Specifications** establish a minimum quality standard for materials and construction, and attempt to ensure a level of testing and inspection commensurate with that required for concrete and steel structures. The document, however, must be coordinated with individual project specifications to avoid overlaps, duplications, conflicts, and omissions.

The MSJC Specifications are intended to be used "in its entirety by reference in the Project Specifications. Individual sections, articles, or paragraphs should not be copied into the Project Specifications since taking them out of context may change their meaning." A statement such as the following will serve to incorporate the Standard into the project spec.

> Masonry construction and materials shall conform to requirements of the Masonry Standards Joint Committee *Specifications for Masonry Structures* (*ACI 530.1-95/ASCE 6-95/TMS 602-95*) except as modified by this Section and Sections ____________.

The project specifications may stipulate more stringent requirements. They must supplement the MSJC Specifications in order to customize its application to each particular project and design.

The MSJC Specifications are not written as a guide specification with instructions or recommendations to the specifier. There is a Commentary published with the Code and the Specifications which gives some background information and suggestions on using the standards. A much more comprehensive handbook has been written by The Masonry Society for ACI, entitled the *Masonry Designers' Guide Based on Building Code Requirements for Masonry Structures* (*ACI 530-95/ASCE 5-95/TMS 402-95 and Specifications for Masonry Structures* (*ACI 530.1-95/ASCE 6-95/TMS 602-95*). The MSJC Specifications also include a checklist of mandatory items to which the specifier must respond, and a checklist of optional items where methods and materials other than the standard requirements may be specified. Items

*The title of this document is plural although there is only one specification. We treat it as plural to match the title.

required in addition to the MSJC Specifications must also be addressed in the project specifications. The MSJC Specifications must be well coordinated with the project specifications (including Division 1 requirements) to avoid overlaps, duplications, conflicts, and omissions.

Changes to the 1995 edition of the MSJC Specifications include a complete reorganization into Construction Specifications Institute (CSI) three-part SectionFormat™. While this is a welcome change, there are still many fundamental problems with the MSJC Specifications. The Code mandates use of the MSJC Specifications, and at the same time states that the MSJC Specifications do not govern where different provisions are specified. This allows the specifier to alter requirements through the project specifications. While the intent is to permit the project specification to impose more stringent requirements, it is equally possible that less stringent requirements could be specified, and these would take precedence over the MSJC Specifications.

The following discussion is intended to provide guidance on preparing project specifications which must be coordinated with the MSJC Specifications. Topics are listed in the order in which they appear in CSI's SectionFormat™.

18.3.1 General

References. The correct title, document number, issuing body and date of the MSJC Specifications should be given in the list of references. The MSJC Specifications include a list of ASTM references which should be checked for conflicts and omissions. The mandatory checklist then requires that sections, parts, and articles of the MSJC Specifications excluded from the project specifications be indicated, and articles at variance with the project specifications be listed. This list will vary for each project.

Quality Assurance. The checklists use the term "quality assurance" ambiguously to indicate both construction inspection and preconstruction testing. The mandatory checklist asks that the specifier (1) designate the owner's representatives (plural) for inspection and testing, implying that they are separate entities and functions and (2) specify the level of testing and inspection required. The testing laboratory information contained in the MSJC Specifications belongs in Division 1 and may conflict with other project requirements if not properly coordinated. Inspection of the work-in-progress (which is required) should be addressed in Part 3. Only pre-construction tests required to certify compliance should be placed under the Quality Assurance heading in Part 1.

The optional checklist includes three preconstruction tests that may be required. Prism tests may be required to verify the compressive strength of masonry on which the structural design is based (f'_m). If prism tests are not required in the project specifications, the unit strength method of determination (which is more conservative) is permitted by default, based on tables in the MSJC Specifications. The mandatory checklist requires that f'_m be given in the project specifications, but the MSJC Code also requires that it be shown on the structural drawings. Mortar tests are required only when mortar is specified to meet the property requirements of ASTM C270. The MSJC Specifications are otherwise based on the proportion requirements, and tests are not required. Compression testing of grout to verify the strength is also optional. For a detailed discussion of quality assurance and quality control in masonry construction, refer to Chapter 17. The project specifications should stipulate that the testing laboratory be certified in accordance with ASTM C1093, *Standard Practice for Accreditation of Testing Agencies for Unit Masonry,* to make sure that they have trained personnel, proper equipment, and are familiar with the specific requirements of masonry tests and inspections.

Submittals. The language of the 1995 MSJC Specifications leaves specific submittal requirements up to the project specifier. Certificates of compliance for masonry, mortar, and grout materials, and results of any preconstruction tests should be listed in the project specification, along with any samples, product data, or shop drawings needed. Compliance certificates should be required for reinforcing steel and accessories, as well as shop drawings showing all fabrication dimensions, locations, and details of the reinforcement Requirements for sample panels that are at variance with or in addition to the MSJC Specifications should be noted.

Cold weather construction requirements have been revised and reorganized in the 1995 Specifications, and now include provisions for glass unit masonry as well. Hot weather construction requirements have been revised and significantly expanded to reflect current industry practices, including moist curing by fog spray. Although these new requirements are more specific regarding the types of protection that are required, the contractor must submit detailed information about the *methods* that will be used to provide the required protection.

18.3.2 Products

Materials. The mandatory checklist contains a number of product-related items. The MSJC Specifications list all the ASTM clay, concrete, and stone masonry unit and material standards that are applicable to structural masonry systems. The specifier must indicate which units will be used and specify the required grade, type, size, and color. Mortar and grout ingredients must be specified, including any acceptable admixtures. The type and grade of reinforcement is required by the MSJC Code to be shown on the drawings, and by the MSJC Specifications to be given in the project specs. Wire fabric, if used, must be designated as either smooth or deformed. While the standard does list ASTM requirements for the materials used for anchors and ties, it does not specify the anchors and ties themselves. The exact types and sizes required for the project, including any proprietary products, must be given in the project specifications.

Although the 1995 MSJC Code now includes design requirements for masonry veneers, and passing reference is made to the need for flashing and weepholes, the MSJC Specifications do not include material or installation requirements for these items. Flashing and weepholes must still be covered in the project specifications. All required accessories, including flashing and weephole materials, must be specified, as well as the size and shape of joint fillers, and the size and spacing of pipes and conduits to be furnished and installed by the mason. If prefabricated masonry elements are used, specify any requirements supplemental to ASTM C901, *Standard Specification for Prefabricated Masonry Panels.*

The optional checklist includes requirements to specify mortar pigments, if used, and masonry cleaners other than water and detergents, which are permitted. If corrosion protection other than the listed types is required, all variances should be addressed, and items not requiring corrosion protection should be listed. Materials for shear keys used in control joints must also be selected. The standard lists acceptable materials for expansion joint fillers and joint sealants, but these items should be excluded here and covered in Division 7.

Mixes. Specify required mortar and grout mix types by proportion or property method in accordance with ASTM C270 and ASTM C476, respectively.

Testing. Testing and inspection requirements for source quality control and for verification of performance, when required, should be listed, including testing laboratory review of manufacturer's test reports and certificates of compliance.

18.3.3 Execution

Preparation. The optional checklist asks the specifier to note when wetting of the masonry units is required to ensure good bond between units and mortar. However, the wording in the standard itself prescribes these limits correctly as high-suction clay masonry units with initial rates of absorption in excess of 1 gram/minute/sq in., when tested in accordance with ASTM C67, *Standard Method of Sampling and Testing Brick and Structural Clay Tile.* The specifier should indicate, though, when tests are required, whether suction tests are to be laboratory- or field-conducted, and the method of wetting to be used when it is determined necessary. Units should not be wetted when the initial rate of absorption is acceptable, nor during winter construction.

Installation. There are several items on the optional checklist that apply to installation of the masonry. The specifier must indicate, first of all, if the pattern of units in the project is anything other than one half running bond, and if the joints are other than ⅜ in. Collar joints ¾ in. wide or less are to be solidly filled with mortar unless otherwise required by the project specs. Face shell bedding of hollow units also governs except in piers, columns, pilasters, starting courses at the foundation, and at grouted cells or cavities, where cross webs must also be mortared. If there are other locations that require full mortar bedding, these should be identified in the project specification. Variations from the standard full bedding requirements for solid units should also be noted, such as beveling to minimize mortar droppings in the cavity. The location of embedded sleeves for pipes and conduits should be shown on the drawings, and only the requirements for their installation covered in the specifications. And finally, requirements for the size, embedment, and spacing of both rigid and adjustable wall ties and anchors, if different from those in the standard, should be given.

The construction tolerances listed in the MSJC Specifications are structural tolerances intended to limit the eccentricity of applied loads. For veneer and other exposed masonry applications, tighter tolerances for aesthetic considerations may be included in the project specifications.

Inspection. The language of the 1995 MSJC Code and Specifications has been made more specific regarding the specification and implementation of a quality assurance program, but the Specifications remain vague and undefined. MSJC does not yet define "inspection," but proposed changes to the code may provide for different levels of inspection depending on the design procedure used (empirical or analytical) and the type of structure being built (essential or non-essential facility). In the 1995 edition of the

MSJC Code, the responsibility for deciding what tests should be required and how often inspections should be made still remains with the design professional. Since the design and complexity of the masonry will vary from project to project, this allows the engineer of a high-rise loadbearing masonry structure to require a higher level of quality assurance than the architect of a low-rise masonry veneer project would typically need. Refer to Chapter 17 for a discussion of quality assurance, quality control, testing, and inspection in masonry construction.

The MSJC Specifications stipulates that the contractor will pay for all inspections and tests needed to document submittals, certify product compliance, and establish mix designs. Inspection of foundations and work-in-progress, however, is paid by the owner, and may be the responsibility of the testing laboratory, the architect or engineer, or a third-party inspector, as the owner's designated representative. The intent of the MSJC Code and Specifications is to assure professional inspection at critical points in the construction. The size and complexity of the project may then influence whether this is continuous or periodic inspection.

Cleaning. The final item in the checklists relates to cleaning. If cleaning materials other than water and detergent have been included in the project specifications, the methods or limitations for their use should be covered here in Part 3.

There are many items not mentioned in the MSJC specifications which still must be covered in the project specifications. Among these are delivery, storage, handling and protection of materials, placement requirements for flashing and weepholes used in cavity wall construction, and protection of walls during construction. Coordinate your office master specifications with the requirements of the new standard to make sure that all aspects of material and workmanship requirements are covered.

18.4 FIELD OBSERVATION AND INSPECTION

Field observation and inspection have become increasingly important with the explosion of construction litigation. The intent of these site visits is to ensure that the finished work complies with the contract documents, and that the workmanship meets the required standards.

Good workmanship affects masonry performance, and is essential to high-quality construction. Masonry construction requires skilled craftsmen working cooperatively with the architect and engineer to execute design. The goal of quality workmanship is common to all concerned parties for various reasons of aesthetics, performance, and liability.

Responsibility for construction of a building project lies ultimately with the contractor. The A/E is not a party to the construction contract, but acts solely as the owner's representative in the field. As part of the team, the architect can assist the contractor and offer expertise in solving or avoiding potential problems. The architect must also act as interpreter of design intent, and safeguard the project quality by assuring proper execution of the work according to the requirements of the contract documents.

Independent inspection agencies or testing laboratories serve a different function. If required by the specifications, it is their responsibility to test various materials and assemblies to verify compliance with reference standards, design strengths, and performance criteria. Field observation and inspection procedures are necessary to assure the successful translation of the design, drawings, and specifications into a completed structure that functions as intended. An independent inspector's authority does not extend to supervision of the work, or to revision of details or methods without the written approval of the architect, owner, and contractor.

The following is intended as a guide to architectural field observation of masonry construction. It is not intended to serve for structural inspection of loadbearing masonry. For a discussion of structural inspection requirements, refer to Chapters 12 and 17.

18.4.1 Materials

An inspector must be familiar with the project specifications and must verify compliance of materials at the job site with the written requirements. Manufacturers must supply test certificates showing that the material properties meet or exceed the referenced standards as to ingredients, strength, dimensional tolerances, durability, and so on.

Unit masonry may be visually inspected for color, texture, and size, and compared to approved samples. Units delivered to the job site should be inspected for physical damage, and storage/protection provisions checked. Stone, brick, or concrete masonry that has become soiled, cracked, chipped, or broken in transit should be rejected. If the manufacturer does not supply test certificates, random samples should be selected and sent to the testing agency for laboratory verification of minimum standards. The inspector should also check the moisture condition of clay masonry at the time of laying, since initial rates of absorption affect the bond between unit and mortar, and the strength of the mortar itself. Visual inspection of a broken unit can indicate whether field tests of absorption rates should be performed (refer to *Fig. 15-2*). Concrete masonry units must be protected from wetting at the job site. CMU should be rejected if there is a question of excessive shrinkage potential or decreased bond.

Mortar and grout ingredients should be checked on delivery for damage or contamination, and to assure compliance with the specified requirements. Packaged materials should be sealed with the manufacturer's identifying labels legible and intact. Bagged ingredients that show signs of water absorption should be rejected. If material test certificates are required, check compliance with the specifications.

Acceptable mixing and batching procedures should be established at the preconstruction conference to assure quality and consistency throughout the job. If field testing of mortar prisms is required, laboratory samples should be prepared and tested sufficiently in advance of construction to allow changes or modifications of materials or procedures as indicated. Retempering time should be monitored to preclude the use of mortar or grout that has begun to set.

Accessories must also be checked for design compliance. The inspector must assure use of proper anchoring devices, ties, inserts, and reinforcement. Steel shelf angles and lintels should carry certification of yield strength and be properly bundled and identified for location within the structure.

18.4.2 Construction

Foundations, beams, floors, and other structural elements that will support the masonry should be checked for completion to proper line and grade before the work begins. Adequate structural support must be assured, and areas cleaned of dirt, grease, oil, laitance, or other materials that might impair bond of the mortar or grout. Overall dimensions and layout must be verified against the drawings and field adjustments made to correct discrepancies. Steel reinforcing dowels must be checked for proper location in relation to cores, joints, or cavities. The inspector should also keep a log of weather conditions affecting the progress or performance of the work. Inspectors should not interfere with the workers or attempt to direct their activities. If methods or procedures are observed that appear to conflict with the specifications or which might jeopardize the quality or performance of the work, they should be called to the attention of the contractor or foreman, and adjustments made as necessary.

18.4.3 Workmanship

Perhaps the single most important element in obtaining strong, weathertight masonry walls is full mortar joints. Partially filled head joints or furrowed mortar beds will produce voids which offer only minimal resistance to moisture infiltration. The first course of masonry must be carefully aligned vertically and horizontally, and fully bedded to assure that the remainder of the wall above will be plumb and level. Even if hollow CMU construction requires only face-shell bedding, this critical base course must have full mortar under face and web as well. Head joints must be fully buttered with mortar and shoved tight against the adjacent unit to avoid leaks (refer to *Fig. 15-14*). Units must not be moved, tapped, or realigned after initial placement, or the mortar bond will be destroyed. If a unit is displaced, all head and bed mortar must be removed and replaced with fresh material. Spot checks for proper bond can be made by lifting a fresh unit out of place to see if both faces are fully covered with adhered mortar.

The inspector should check for proper embedment and coverage of anchors, ties, and joint reinforcement, and should monitor vertical coursing and joint uniformity. Differential widths or thicknesses of mortar joints can misalign the modular coursing and interfere with proper location of openings, lintels, and embedded items. Storypoles, string lines, and tapes or templates should be used to check coursing between corner leads. Nail and line pinholes must be filled with mortar when string lines are removed to avoid water penetration through these voids. Work of other trades which penetrates the masonry should be incorporated during construction of the wall and not cut in later. Collar joints and cavities must be kept free of mortar fins and droppings to avoid plugging weepholes, damaging flashing, or interfering with grout pours. When they become thumbprint hard, joints should be tooled to compress the mortar surface.

The mason should place all vertical and horizontal reinforcement as the work progresses, holding the bars in correct alignment with spacers or wire. Minimum clearances should be maintained, and bar splices lapped and securely tied. The inspector should check to see that reinforcement is free of rust, loose scale, or other materials that could impair bond to the mortar. Care should be taken to avoid moving or jarring vertical steel that is already embedded in lower grouted courses.

Inspection should also include proper installation of flashing, control joints, expansion joints, lintels, sills, caps, copings, and frames. Door frames must be adequately braced until the mortar has set and the masonry work surrounding them is self-supporting.

Grouting is important to the structural integrity of reinforced masonry walls. Cavities and cores should be inspected before the grout is placed, and any remaining dirt, debris, mortar droppings, or protrusions removed before the work proceeds. Cleanout plugs left for high-lift pours allow visual inspection from below by use of a mirror inserted through the opening. Cleanout units should be fully mortared and shoved into place after inspection, then braced against blowout from the fluid pressure of the grout against the uncured mortar. The consistency of the grout should allow for easy pouring or pumping, and complete filling of the space. Rodding or vibrating to remove air bubbles and pockets also ensures that the grout covers fully around and between ties and reinforcement. Grout in contact with the masonry will harden more rapidly than grout in the center of the cavity. It is important that the rodding or vibrating be done immediately after the pour to avoid disturbance of any hardened grout. Timing of grout pours should be monitored to avoid excessive lateral pressure on uncured joints.

18.4.4 Protection and Cleaning

Throughout the construction period, both the masonry materials and the work must be protected from the weather. Materials must be stored off the ground under waterproof coverings to prevent wetting or staining. Exposed tops of unfinished walls must be covered each night to keep moisture out by draping waterproof plastic or canvas 2 ft down each side. Cold weather may require heating of materials and possibly the application of heat during the curing period. Hot, dry climates cause rapid evaporation, and mortar mixes may have to be wetter to compensate for premature drying. Both hot and cold weather may necessitate moist curing. The inspector must assure that these precautions are taken to avoid harmful effects, and must also see that completed work is protected from damage during other construction operations.

Suitable cleaning methods must be selected on the basis of the type of stain involved and the type of material to be cleaned. Improper use of cleaning agents can create more problems than are solved by their application. Mortar smears on the face of the masonry should be removed daily before they are fully hardened, and dry-brushed when powdery to prevent stains. Paints, textured coatings, or clear water repellents, if specified, should be applied carefully over clean, dry walls, and adjacent work protected against splatters and drips.

It is the inspector's job to see that the instructions and requirements of the drawings and specifications are carried out in the field. Safeguarding the quality of the work without impeding its progress is best achieved through cooperation with the contractor and workers. Good design and good intentions are not sufficient in themselves to assure quality of the finished product. The inspector can serve as interpreter and mediator in the proper execution of the work, ensuring masonry structures that are as durable and lasting as the materials of which they are made.

Appendix

GLOSSARY

A

Absorption The amount of water that a masonry unit absorbs when immersed in water under specified conditions for a specified length of time.

Absorption Rate The weight of water absorbed when a brick is partially immersed for 1 minute, usually expressed in either grams or ounces per minute. Also called suction or **initial rate of absorption.**

Abutment (1) That part of a pier or wall from which an arch springs, specifically the support at either end of an arch, beam, or bridge. (2) A **skewback** and the masonry that supports it.

Accelerator Ingredient added to mortar or grout to speed hydration of cementitious components to hasten set time.

ACI American Concrete Institute.

Adhered Attached by adhesion rather than mechanical anchorage, as in adhered veneer.

Adhesion-Type Ceramic Veneer Thin sections of ceramic veneer, held in place by adhesion of mortar to unit and to backing. No metal anchors are required.

Note: Boldface type within entries denotes terms for which there are main glossary entries.

Admixture A material other than water, aggregates, hydraulic cement, and fiber reinforcement used as an ingredient of grout or mortar and added to the batch immediately before or during its mixing.

Adobe Soil of diatomaceous content mixed with sufficient water so that plasticity can be developed for molding into masonry units.

Aggregate Granular mineral material such as natural sand, manufactured sand, gravel, crushed stone, and air-cooled blast furnace slag.

Air Drying The process of drying block or brick without any special equipment, simply by exposure to ambient air.

Air Entraining The capability of a material or process to develop a system of minute bubbles of air in cement paste, mortar, or grout during mixing.

AISC American Institute of Steel Construction.

Alumina The oxide of aluminum; an important constituent of the clays used in brick, tile, and refractories.

Anchor See **Connector.**

Anchored-Type Ceramic Veneer Thick sections of ceramic veneer held in place by grout and wire anchors connected to backing wall.

Angle Closer A special-shaped brick used to close the bond at the corner of a wall.

ANSI American National Standards Institute (formerly USAS).

Arch A vertically curved compressive structural member spanning openings or recesses; may also be built flat by using special masonry shapes or placed units.

Abutment The skewback of an arch and the masonry that supports it.

Arch Axis The median line of the arch ring.

Back Arch A concealed arch carrying the back of a wall where the exterior facing is carried by a lintel.

Camber The relatively small rise of a jack arch.

Constant-Cross-Section Arch An arch whose depth and thickness remain constant throughout the span.

Crown The apex of the arch ring. In symmetrical arches, the crown is at midspan.

Depth The depth (d) of any arch is the dimension that is perpendicular to the tangent of the axis. The depth of a jack arch is taken to be its greatest vertical dimension.

Extrados The exterior curve that bounds the upper extremities of the arch.

Fixed Arch An arch whose skewback is fixed in position and inclination. Plain masonry arches are, by nature of their construction, fixed arches.

Gothic or Pointed Arch An arch, with relatively high rise, whose sides consist of arcs or circles, the centers of which are at the level of the spring line. The Gothic arch is often referred to as a drop, equilateral, or lacent arch, depending on whether the spacings of the center are respectively less than, equal to, or more than the clear span.

Intrados The interior curve that bounds the lower extremities of the arch (see **soffit**). The distinction between soffit and intrados is that the intrados is linear, whereas the soffit is a surface.

Jack Arch An arch having horizontal or nearly horizontal upper and lower surfaces. Also called flat or straight arch.

Major Arch Arch with span greater than 6 ft and having equivalent uniform loads greater than 1000 lb/ft. Typically, a Tudor arch, semicircular arch, Gothic arch, or parabolic arch. Rise-to-span ratio greater than 0.15.

Minor Arch Arch with maximum span of 6 ft and loads not exceeding 1000 lb/ft. Typically, a jack arch, segmental arch, or multi-centered arch. Rise-to-span ratio less than or equal to 0.15.

Multi-centered Arch An arch whose curve consists of several arcs of circles which are normally tangent at their intersections.

Relieving Arch One built over a lintel, flat arch, or smaller arch to divert loads, thus relieving the lower member from excessive loading. Also known as discharging or safety arch.

Rise The rise (r) of a minor arch is the maximum height of arches off it above the level of its spring line. The rise (f) of a major parabolic arch is the maximum height of arch axis above its spring line.

Roman Arch A semicircular arch. If built of stone, all units are wedge-shaped.

Segmental Arch An arch whose curve is circular but less than a semicircle.

Semicircular Arch An arch whose curve is a semicircle.

Skewback The inclined surface on which the arch joins the supporting wall. For jack arches the skewback is indicated by a horizontal dimension (K).

Soffit The undersurface of the arch.

Span The horizontal distance between abutments. For minor arch calculations the clear span (S) of the opening is used. For a major parabolic arch the span (L) is the distance between the ends of the arch axis at the skewback.

Spring Line For minor arches the line where the skewback cuts the soffit. For major parabolic arches, the term commonly refers to the intersection of the arch axis with the skewback.

Trimmer Arch An arch (usually a low-rise arch of brick) used for supporting a fireplace hearth.

Tudor Arch A pointed four-centered arch of medium rise-to-span ratio.

Voussoir One of the wedge-shaped masonry units which forms an arch ring.

Arch Stone Voussoir stone.

Arching Action The distribution of loads in masonry over an opening. The load is usually assumed to occur in a triangular pattern above the opening extending from a maximum at the center of the span to zero at the supports.

Architectural Terra Cotta Hard-burned, glazed, or unglazed clay building units, plain or ornamental, machine-extruded or hand-molded, and generally larger in size than brick or facing tile. See **Ceramic Veneer.**

Arris A sharp edge of brick, stone, or other building element.

ASCE American Society of Civil Engineers.

ASHRAE American Society of Heating, Refrigeration, and Air Conditioning Engineers.

ASTM American Society for Testing and Materials.

Autoclave A type of curing system in the production of concrete masonry units which utilizes super heated steam under pressure to promote strength of units.

Axhammer An axe for spalling or dressing rough stone; has either one cutting edge and one hammer face, or two cutting edges. Also called **mason's hammer.**

B

Back Filling (1) Rough masonry built behind a facing or between two faces. (2) Filling over the extrados of an arch. (3) Brickwork in spaces between structural timbers, sometimes called brick nogging.

Backing Surface or assembly to which veneer is attached.

Backup That part of a masonry wall behind the exterior facing.

Band Course A continuous, horizontal band of masonry marking a division in the wall elevation. Sometimes called belt course, string course, or sill course.

Basalt Dark, fine-grained igneous rock used extensively as a paving stone and occasionally as a building stone.

Basketweave A checkerboard pattern of bricks or pavers, flat or on edge. Bricks or modular groups of bricks laid at right angles to those adjacent.

Bat A broken brick or piece of brick with one undamaged end. Also called a brickbat. Usually about one-half brick.

Batter Masonry that is recessed or sloping back in successive courses; the opposite of a corbel; to rack back.

Bearing Plate A plate placed under a truss, beam, or girder to distribute the load.

Bed (1) In masonry and bricklaying, the side of a masonry unit on which it lies in the course of the wall; the underside when placed horizontally. (2) The layer of mortar on which the masonry unit is set.

Bed Joint See **Joint.**

Bedding Course The first layer of mortar at the bottom of a masonry wall.

Belt Course A narrow horizontal course of masonry, sometimes slightly projected, such as window sills which are made continuous. Sometimes called **string course** or sill course.

BIA Brick Institute of America (formerly **SCPI**).

Blind Header A concealed header in the interior of a wall, not showing on the faces of the wall.

Block, Concrete See **Concrete Masonry Unit**

Blocking A method of bonding two adjoining or intersecting walls, not built at the same time, by means of offsets whose vertical dimensions are not less than 8 in.

Blockwork Masonry of concrete block and mortar.

Bolster A blocking chisel for masonry work. A broad-edged chisel made in a number of sizes, shapes, and weights.

Bond

Adhesion Bond The adhesion between masonry units and mortar or grout.

Masonry Bond Connection of masonry wythes with overlapping header units.

Metal Tie Bond Connection of masonry wythes with metal ties or joint reinforcement.

Mortar Bond or Grout Bond Adhesion between mortar or grout and masonry units, reinforcement, or connectors.

Pattern Bond Patterns formed by the exposed faces of the masonry units, for example, running bond or flemish bond.

American Bond Bond pattern in which every sixth course is a header course and the intervening courses are stretcher courses.

Basketweave Bond Modular groups of units laid at right angles to those adjacent to form a pattern.

Blind Bond Bond pattern to tie the front course to the wall where it is not desirable that any headers should be seen in the face work.

Common Bond Bond pattern in which five to seven stretcher courses are laid between headers.

Cross Bond Bond pattern in which the joints of the stretcher in the second course come in the middle of the stretcher in the first course composed of headers and stretchers intervening.

Dutch Cross Bond A bond having the courses made up alternately of headers and stretchers. Same as an *English cross bond.*

English Bond Bond pattern with alternating courses of headers and stretchers. The headers and stretchers are situated plumb over each other. The headers are divided evenly over the vertical joints between the stretchers.

English Cross Bond A variation of *English bond,* but with the stretchers in alternate courses centered on the stretchers above and below. Also called *Dutch cross bond.*

Flemish Bond Bond pattern in which each course consists of alternate stretchers and headers, with the headers in alternate courses centered over the stretchers in intervening courses.

Flemish Garden Bond Units laid so that each course has a header to every three to five stretchers.

Header Bond Bond pattern showing only headers on the face, each header divided evenly on the header under it.

Herringbone Bond The arrangement of units in a course in a zigzag fashion with the end of one unit laid at right angles against the side of a second unit.

Random Bond Masonry constructed without a regular pattern.

Running Bond The placement of masonry units such that head joints in successive courses are horizontally offset at least one quarter the unit length.

Stack Bond (1) The placement of units such that the head joints in successive courses are vertically aligned. (2) Units laid so no overlap occurs; head joints form a continuous vertical line. Also called plumb joint bond, straight stack, jack bond, jack on jack, and checkerboard bond.

Reinforcing Bond The adhesion between steel reinforcement and mortar or grout.

Bond Beam A course or courses of a masonry wall grouted and usually reinforced in the horizontal direction serving as an integral beam in the wall. May serve as a horizontal tie, bearing course for structural members, or flexural member itself.

Bond Beam Unit A hollow masonry unit with depressed sections forming a continuous channel in which reinforcing steel can be placed for embedment in grout.

Bond Breaker A material used to prevent adhesion between two surfaces.

Bond Course The course consisting of units that overlap more than one wythe of masonry.

Bond Strength The resistance of mortar or grout to separation from masonry units or reinforcement with which it is in contact.

Bonder A bonding unit. Also called a **header.**

Brick A manufactured masonry unit made from fired clay or shale, concrete, or sand-lime materials, which is usually formed in the shape of a rectangular prism, and typically placed with one hand.

Acid-Resistant Brick Brick suitable for use in contact with chemicals, usually in conjunction with acid-resistant mortars.

Adobe Brick An unfired, air-dried, roughly molded brick of earth or clay. When made with an emulsifier or fibers, called "stabilized adobe."

Angle Brick Any masonry unit shaped to an oblique angle to fit a salient corner.

Arch Brick Wedge-shaped masonry unit (usually a manufactured clay unit) for special use in an arch. Arch brick provide uniformity of mortar joint thicknesses as the arch is turned. Also refers to the hard-burned brick units found in the arch section of a scove kiln.

Building Brick Brick for building purposes not especially produced for texture or color (formerly called common brick).

Bullnose Brick A brick having one or more rounded corners.

Calcium Silicate Brick Brick made from sand and lime, with or without the inclusion of other materials.

Clay Brick A solid or hollow masonry unit of clay or shale, usually formed into a rectangular prism while plastic and burned or fired in a kiln.

Clinker Brick A very hard-burned brick whose shape is distorted because of nearly complete vitrification.

Common Brick See *building brick.*

Concrete Brick Brick made from portland cement, water, and suitable aggregates, with or without the inclusion of other materials.

Cored Brick A brick in which the holes consist of less than 25% of the section.

Dry Press Brick Brick formed in molds under high pressures from relatively dry clay (5 to 7% moisture content).

Economy Brick A brick whose nominal dimensions are 4×4×8 in.

Facing Brick Brick made especially for facing purposes.

Fire Brick (1) Any type of refractory brick, specifically fire clay brick. (2) Brick made of refractory ceramic material that will resist high temperatures.

Firebox Brick Brick manufactured from clay, fire clay, shale, or similar naturally occurring earthy substances and subjected to heat treatment at elevated temperatures.

Floor Brick Smooth, dense brick, highly resistant to abrasion, used as finished floor surfaces.

Gauged Brick (1) Brick that has been ground or otherwise produced according to accurate dimensions. (2) A tapered arch brick.

Glazed Brick A brick prepared by fusing on the surface a ceramic glazing material; brick having a glassy surface.

Hollow Brick Brick whose net cross-sectional area (solid area) in any plane parallel to the surface containing the cores, cells, or deep frogs is less than 75% of its gross cross-sectional area measured in the same plane.

Jumbo Brick A generic term indicating a brick larger in size than the standard. Some producers use this term to describe oversize brick of specific dimensions manufactured by them.

Norman Brick A brick whose nominal dimensions are $4 \times 2\frac{2}{3} \times 12$ in.

Paving Brick, Heavy Vehicular Brick intended for use in areas with a high volume of heavy vehicular traffic.

Paving Brick, Light Traffic Brick intended for use as paving material to support pedestrian and light vehicular traffic.

Roman Brick A brick whose nominal dimensions are $4 \times 2 \times 12$ in.

Salmon Brick Generic term for underburned brick that is more porous and lighter colored than hard-burned brick. Usually pinkish-orange in color.

Sand-Lime Brick See *Calcium silicate brick* in this list.

Sand-Struck Brick See *Soft-mud brick* in this list.

SCR Brick Brick whose nominal dimensions are $6 \times 2\frac{2}{3} \times 12$ in.

Sewer Brick Low-absorption, abrasion-resistant brick intended for use in drainage structures.

Soft-Mud Brick Brick produced by molding (often by a hand process) relatively wet clay (20 to 30% moisture).

Stiff-Mud Brick Brick produced by extruding a stiff but plastic clay (12 to 15% moisture) through a die.

Wire-Cut Brick Brick formed by forcing plastic clay through a rectangular opening designed to shape the clay into bars. Before burning, wires pressed through the plastic mass cut the bars into brick units.

Brick Ax Same as **brick hammer.**

Brick Facing See **Veneer.**

Brick Grade Designation denoting durability of clay brick expressed as SW (Severe Weathering), MW (Moderate Weathering), and NW (Negligible Weathering).

Brick Hammer A steel tool, one end of which has a flat, square surface used as a hammer, for breaking bricks, driving nails, and so on. The other end forms a chisel peen used for dressing bricks. (Also called **brick ax** or bricklayer's hammer.)

Brick Ledge A ledge on a footing or wall which supports a course of masonry.

Brick Trowel A trowel having a flat, trianglar steel blade in an offset handle used to pick up and spread mortar. The narrow end of the blade is called the "point"; the wide end, the "heel."

Brick Type Designation for clay brick that indicates qualities of appearance including tolerance, chippage, and distortion. Expressed as face brick standard (FBS), face brick extra (FBX), and face brick architectural (FBA) for solid brick; and hollow brick standard (HBS), hollow brick extra (HBX), hollow brick architectural (HBA), and hollow building brick (HBB) for hollow brick.

Brickbat See **Bat.**

Building Ecology The study of the dynamic interaction of a building with its external environment and its occupants, equipment, and processes.

Bullnose Unit A brick or concrete masonry unit having one or more rounded exterior corners.

Buttering Placing mortar on a masonry unit with a trowel.

Buttress A projecting structure built against a wall or building to give it greater strength and stability.

C

Carbonation A process of chemical weathering whereby minerals that contain sodium oxide, calcium oxide, potassium oxide, or other basic oxides are changed to carbonates by the action of carbonic acid derived from atmospheric carbon dioxide and water.

Cast Stone See **Stone.**

Cavity An unfilled space.

C/B Ratio The ratio of the weight of water absorbed by a masonry unit during immersion in cold water for 24 hours to weight absorbed during immersion in boiling water for 5 hours. An indication of the probable resistance of brick to freezing and thawing. Also called **saturation coefficient.** (See ASTM C67.)

Cell See **Core.**

Cellular Concrete Lightweight concrete product consisting of portland cement, cement-silica, cement-pozzolan, or lime-silica pastes, or pastes containing blends of these ingredients and having a homogeneous void or cell structure, attained with gas-forming chemicals or foaming agents. For cellular concretes containing binder ingredients other than, or in addition to, portland cement, autoclave curing is usually employed.

Cement A general term for an adhesive or binding material. See specific terms such as **portland cement** or **masonry cement.**

Cement Mortar A mixture of cement, sand, or other aggregates and water used for plastering over masonry or to lay masonry units.

Cement-Lime Mortar A mixture of cement, lime, sand, or other aggregates and water used for plastering over masonry or to lay masonry units.

Cementitious Material In proportioning masonry mortars, the following are considered cementitious material: portland cement, blended hydraulic cement, masonry cement, quicklime, and hydrated lime.

Cementitious Material, Hydraulic An inorganic material or a mixture of inorganic materials, which sets and develops strength by chemical reaction with water by formation of hydrates and is capable of doing so under water.

Centering Temporary formwork for the support of masonry arches, hearth extensions, or lintels during construction. Also called center.

Ceramic Broad term for products made from heat-resistant, nonmetallic, inorganic materials such as clay, bauxite, alumina, silicon carbide, etc., which have been fired to incipient fusion.

Ceramic Veneer A type of architectural terra cotta, characterized by large face dimensions and thin sections ranging from $1\frac{1}{8}$ to $2\frac{1}{4}$ in. in thickness.

Chimney Lining Fire clay or terra cotta material or refractory cement made to be built inside a chimney throat; that part of a chimney directly above the fireplace where the walls are brought close together.

Cinder An aggregate, sometimes used in the manufacture of concrete masonry units, made from burnt coal or volcanic lava.

Cinder Block See **concrete masonry unit.**

Clay An earthy or stony mineral aggregate consisting of hydrous silicates of alumina, plastic when sufficiently pulverized and wetted, rigid when dry, and vitreous when fired to a sufficiently high temperature.

Clay Mortar Finely ground clay used as a plasticizer for masonry mortars.

Cleanout An opening at the bottom of a grout space of sufficient size and spacing to allow the removal of debris.

Closer (1) The last masonry unit laid in a course. It may be whole or a portion of a unit. (2) A stone course running from one window sill to another (a variety of string course).

Closure Supplementary or short-length units used at corners or jambs to maintain bond patterns.

CMU Concrete masonry unit.

Cold Weather Construction Procedures used in constructing masonry when ambient air temperature or the temperature of the masonry units is below 40°F.

Collar Joint See **Joint.**

Color Pigment Inorganic matter used in mortar to vary the color.

Combustible Capable of undergoing combustion.

Composite Action Transfer of stress between components of a member designed so that in resisting loads, the combined components act together as a single member.

Concrete Block See **Concrete Masonry Unit.**

Concrete Brick See **Brick.**

Concrete Masonry Unit A manufactured masonry unit made from portland cement, mineral aggregates, and water, with or without the inclusion of other materials.

Concrete Block A hollow concrete masonry unit made from portland cement and suitable aggregates such as sand, crushed stone, cinders, burned clay or shale, pumice, scoria, and air-cooled or expanded blast furnace slag, with or without other materials.

A Block A cored masonry unit with one end closed by a cross web and the opposite end open or lacking an end cross web, typically forming two cells when laid in running bond. Also called *open end block.*

Cap Block A solid flat slab, usually 2¼ in. thick, used as a capping unit for parapet and garden walls. Also used for stepping stones, patios, veneering, etc.

Channel Block A hollow unit with web portions depressed less than 1¼ in. to form a continuous channel for reinforcing steel and grout.

Concrete Block A hollow or solid unit consisting of portland cement and suitable aggregates combined with water. Other materials such as lime, fly ash, air-entraining agent, or other admixtures may be permitted.

Ground Faced Block A concrete masonry unit in which the exposed surface is ground to a smooth finish.

Jamb Block A block specially formed for the jamb of windows or doors, generally with a vertical slot to receive window frames, etc.

Offset Block A concrete masonry unit that is not rectangular. Usually used as a corner block to maintain the masonry pattern on the exposed face of a single-wythe wall whose thickness is less than half the length of the unit.

Return Corner Block Concrete masonry unit that has one flat end for corner construction.

Screen Block Open-faced masonry units used for decorative purposes or to partially screen areas from the sun or outside viewers.

Sculptured Block Block with decorative formed or molded surfaces.

Shadow Block Block with a face formed in planes to develop surface shadow patterns.

Sill Block A solid concrete masonry unit used for the sills of a wall.

Slump Block Concrete masonry units produced so that they slump or sag in irregular fashion before they harden.

Split-Faced Block A concrete masonry unit with one or more faces purposely fractured to expose the rough aggregate texture to provide architectural effects in masonry wall construction.

Concrete Masonry Unit, Architectural Architectural concrete masonry units having textured or sculptured surfaces. Methods used to obtain different surface textures include splitting, grinding, forming vertical striations, and causing the units to "slump." Sculptured faces are obtained by forming projecting ribs or flutes, either rounded or angular, as well as vertical and horizontal scoring, recesses, and curved faces.

Concrete Masonry Unit, Moisture-Controlled A concrete masonry unit whose moisture content conforms to the requirements for Type I classification of ASTM Specification C55, C90, or C129.

Concrete Masonry Unit, Lightweight A unit whose oven dry density is less than 105 lb/cu ft.

Concrete Masonry Unit, Medium Weight A unit whose oven-dry density is at least 105 lb/cu ft and less than 125 lb/cu ft.

Concrete Masonry Unit, Normal Weight A unit whose oven-dry density is at least 125 lb/cu ft.

Concrete Masonry Unit, Pre-faced Concrete or calcium silicate masonry units with the exposed-to-view-in-place surfaces covered at the point of manufacture with resin, resin and inert filler, or cement and inert filler, to produce a smooth resinous tile facing.

Connector Mechanical devices, including anchors, wall ties, and fasteners, for securing two or more pieces, parts, or members together.

Connector, Anchor Metal rod, wire, or strap that secures masonry to its structural support.

Connector, Fastener Device used to attach non-masonry materials to masonry.

Connector, Wall Tie Metal connector which connects wythes of masonry walls together.

Control Joint See **Joint.**

Coping Masonry units laid on top of a finished wall. (1) A covering on top of a wall exposed to the weather, usually sloped to carry off water. (2) The materials or masonry units used to form a cap or a finish on top of a wall, pier, chimney, or pilaster to protect the masonry below from water penetration. Commonly extended beyond the wall face and cut with a drip.

Coping Unit A solid masonry unit for use as the top and finished course in wall construction.

Corbel (1) The projection of successive courses of masonry out from the face of the wall to increase the wall thickness or to form a shelf or ledge. (2) A shelf or ledge formed by successive courses of masonry projecting out from the face of a wall, pier, or column.

Corbeled Vault A masonry roof constructed from opposite walls or from a circular base, by shifting courses slightly and regularly inward until they meet. The resulting stepped surface may be smoothed or curved, but no true arch action is involved.

Core (1) The molded open space in a concrete masonry unit. (2) A hollow space within a concrete masonry unit formed by the face shells and webs. (3) The holes in clay units. Also called **cells.**

Corner Pole See **Story Pole.**

Cornerstone (1) Generally, a stone that forms a corner or angle in a structure. (2) More specifically, a stone prominently situated near the base of a corner in a building carrying information recording the dedicatory dates and other pertinent information. In some buildings, these stones contain or can cap a vault in which contemporary memorabilia are preserved.

Corrosion The chemical or electrochemical reaction between a material, usually a metal, and its environment that produces a deterioration of the material and its properties.

Corrosion Resistant Applied to a material that is inherently resistant to or treated or coated to retard harmful oxidation or other corrosive action.

Course A horizontal layer of units in masonry other than paving.

Crack A flaw consisting of complete or incomplete separation within a single element or between contiguous elements of constructions.

Crack Control Methods used to control the extent, size, and location of cracking in masonry, including reinforcing, providing movement joints, and ensuring dimensional stability of masonry materials.

Cramp Anchor A U-shaped metal connector used to hold adjacent units of masonry together, as in a parapet or wall coping, or to secure stone slab veneers together.

Crazing (1) The cracking that occurs in fired glazes or other ceramic coatings due to tensile stresses. (2) A network of apparent fine cracks at or beneath the surface of materials such as in transparent plastics, glazed ceramics, glass, or coatings.

Creep Time-dependent deformation due to sustained load.

Cross-Sectional Area, Gross (1) The area delineated by the out-to-out dimensions of masonry in the plane under consideration. (2) The total cross-sectional area of a specified section.

Cross-Sectional Area, Net (1) The area of masonry units, grout, and mortar crossed by the plane under consideration based on out-to-out dimensions. (2) The gross cross-sectional area minus the area of ungrouted cores, notches, cells, and unbedded areas. Net area is the actual surface area of a cross section of masonry.

Curing The maintenance of proper conditions of moisture and temperature during initial set to develop required strength and reduce shrinkage in concrete products and mortar.

Curtain Wall See **Wall.**

D

Dampcheck An impervious horizontal layer to prevent vertical penetration of water in a wall consisting of either a course of solid masonry, metal, or a thin layer of asphaltic or bituminous material. Generally near grade to prevent upward migration of moisture by capillary action. Also called damp course.

Damping Reduction of amplitude of vibrations due to energy loss (as in damping of vibrations from seismic shock).

Dampproofing The preparation of a wall to prevent moisture from penetrating through it.

Density The ratio of the mass of an object to its volume.

Density (of Masonry Units) Oven-dry weight per unit volume.

Diaphragm A roof or floor system designed to transmit lateral forces to shear walls or other vertical resisting elements.

Differential Movement Movement of two elements relative to one another that differs in rate or direction.

Dimensions of Masonry Units

Actual The measured dimensions of a masonry unit.

Height (1) The vertical dimension of the unit in the face of a wall. (2) Vertical dimension of masonry units or masonry, measured parallel to the intended face of the unit or units.

Length (1) The horizontal dimension of the unit in the face of the wall. (2) Horizontal dimension of masonry units or masonry, measured parallel to the intended face of the unit or units.

Nominal (1) A dimension greater than the specified (standard) dimension by the thickness of one joint, but not more than 13 mm or ½ in. (2) A dimension that may be greater than the specified masonry dimension by the thickness of a mortar joint .

Specified (1) the nominal dimension less the thickness of a standard mortar joint; that is, net dimension of the masonry unit. (2) The dimensions to which masonry units or constructions are required to conform. (3) The standard dimensions of a masonry unit, plus or minus any allowable size tolerances.

Thickness (1) That dimension designed to lie at right angles to the face of the wall, floor, or other assembly. (2) *Thickness* (*width*) Horizontal dimension of masonry units or masonry measured perpendicular to the intended face of the masonry unit or units.

Dimension Stone See **Stone.**

Dog's Tooth A brick laid with its corners projecting from the wall face.

Dolomite A sedimentary carbonate rock (a variety of limestone) that consists largely or entirely of the mineral dolomite.

Dolomitic Lime A trade term for high-magnesium lime; a misnomer as the product does not contain dolomite.

Dolomitic Limestone See **Limestone.**

Dovetail Anchor A splayed tenon that is shaped like a dove's tail, that is, broader at its base, that fits into the recess of a corresponding mortise.

Dowels Straight metal bars used to connect two sections of masonry.

Drip Groove or slot cut beneath and slightly behind the forward edge of a projecting stone member, such as a sill, lintel, or coping, to cause rainwater to drip off and prevent it from penetrating the wall.

Dry Stack Masonry Masonry work laid without mortar.

Durability The ability of a material to resist weathering action, chemical attack, abrasion, and other conditions of service.

Durability (Freeze-Thaw) The ability of masonry units to maintain integrity under the forces caused by the cyclic action of freezing and thawing in the presence of moisture.

E

Eccentricity (1) The distance between a vertical load reaction and a centroidal axis of masonry. (2) The normal distance between the centroidal axis of a member and the parallel resultant load.

Effective Height (1) Clear height of a braced member between lateral supports and used for calculating the slenderness ratio of a member. (2) The height of a member that is assumed when calculating the slenderness ratio.

Effective Thickness The thickness of a member that is assumed when calculating the slenderness ratio.

Effective Width That part of the width of a member taken into account in designing T- or L-beams.

Efflorescence, Water Soluble A crystalline deposit, usually white, of water-soluble compounds on the surface of masonry. Normally can be removed with water washing.

Efflorescence, Water Insoluble A crystalline deposit, usually white, of water soluble compounds which, on reaching the masonry surface, become water insoluble primarily through carbonation (sometimes called lime run or calcium carbonate stain); normally requires acid washing for removal.

Embodied Energy The amount of energy required to produce a given quantity of material, from raw material acquisition through processing, manufacture, and distribution.

Empirical Design Design based on application of physical limitations learned from experience or observations gained through experience, without structural analysis.

End-Construction Tile Structural clay tile units designed to be laid with the axis of the cells vertical.

Engineered Masonry Masonry which has been analyzed for vertical and lateral load resistance and whose members have been proportioned to resist design loads in accordance with working stress design or strength design principles.

Equivalent Thickness (1) Solid thickness to which a hollow unit would be reduced if the material in the unit were recast with the same face dimensions but without voids. (2) The percent solid volume times the actual width divided by 100. (3) E_T = net volume ÷ (specified unit length × specified unit height). (4) The average thickness of solid material in the unit.

Exfoliation Peeling or scaling of stone or clay brick surfaces caused by chemical or physical weathering.

Expansion Joint See **Joint.**

Extrados The exterior curve in an arch or vault.

F

Face (1) The exposed surface of a wall or masonry unit. (2) The surface of a wall or masonry unit. (3) The surface of a unit designed to be exposed in the finished masonry.

Face-Bedded Stone set with the stratification vertical.

Face Shell The side wall of a hollow concrete masonry or clay masonry unit.

Face Shell Bedding Application of mortar to only the horizontal surface of the face shells of hollow masonry units and in the head joints to a depth equal to the thickness of the face shell.

Facing Tile Structural clay tile for exterior and interior masonry with exposed faces. (Covered by ASTM C212 and ASTM C126.)

Field The expanse of wall between openings, corners, etc., principally composed of stretcher units.

Fire Box The interior of a fireplace furnace, serving as combustion space.

Fire Clay Sedimentary clay of low flux content.

Fire Resistance Property of a material or assemblage to withstand fire or give protection from it.

Fire Clay Mortar, Ground A refractory mortar consisting of finely ground raw fire clay.

Flagging (1) Collective term for flagstones. (2) A surface paved with flagstones. (3) The process of setting flagstones.

Flagstone A type of stone that splits easily into flags or slabs; also a term applied to irregular pieces of such stone split into slabs from 1 to 3 in. thick, and used for walks, patios, etc.

Flashing (1) Impermeable material placed in masonry to provide water drainage or prevent water penetration. (2) A technique of brick firing to produce a range of colors by controlling the atmospheric conditions in the kiln.

Flow A laboratory-measured property of mortar that indicates the percent increase in diameter of the base of the truncated cone of mortar when it is placed on a flow table and mechanically raised and dropped a specified number of times under specified conditions.

Flow after Suction Flow of mortar measured after it has been subjected to a vacuum to simulate suction of dry masonry units.

Flue Lining (1) A manufactured tubular non-loadbearing fired clay unit, normally used for conveying hot gases in chimneys. (2) A smooth, hollow clay or concrete tile unit used for the inner lining of masonry chimneys.

Fly Ash The finely divided residue resulting from the combustion of ground or powdered coal.

Freeze-Thaw Freezing and thawing of moisture in materials and the resultant effects on these materials and on structures of which they are a part or with which they are in contact.

Fretwork In masonry, any ornamental openwork or work in relief.

Frog An indentation in one bed surface of a brick manufactured by molding or pressing.

Furring A method of finishing the interior face of a masonry wall to provide space for insulation, to prevent moisture transmittance, or to provide a smooth or plane surface for finishing.

Furring Tile Tile for lining the inside of walls and carrying no superimposed loads.

Furring Units Thin masonry units used as furring.

Furrowing Striking a V-shaped trough in a mortar bed.

G

Glass Block Hollow or solid glass masonry unit.

Glaze A hard, glassy, fused ceramic coating which may have a matte or glossy finish.

Salt Glaze A lustrous glazed finish from the thermo-chemical reaction of the silicates of the clay body with vapors of salt or chemicals.

Clear Ceramic Glaze A ceramic glaze translucent or slightly tinted with a gloss finish.

Color Ceramic Glaze An opaque glaze of satin or gloss finish obtained by spraying the clay body with a compound of metallic oxides, chemicals, and clays.

Gneiss Coarse-grained metamorphic rock with discontinuous foliation caused by planar alignment of platy and lath-shaped minerals.

Gradation The particle size distribution of aggregate as determined by separation with standard screens. Gradation of aggregate is expressed in terms of the individual percentages passing standard screens. Sieve analysis, screen analysis, and mechanical analysis are terms used synonymously in referring to gradation of aggregate.

Granite A visibly granular, igneous rock generally ranging in color from pink to light or dark gray and consisting mostly of quartz and feldspars, accompanied by one or more dark minerals. The texture is typically homogeneous but may be gneissic or porphyritic. Some dark granular igneous rocks, though not geologically granite, are included in the definition.

"Green" Buildings Building structures that are designed, constructed, operated, and demolished in an environmentally enhanced manner.

"Green" Masonry A molded clay masonry unit before it has been fired in a kiln; an uncured concrete masonry unit.

Green Mortar Mortar that has set but not cured.

Greenstone A metamorphic rock, typically with poorly defined granularity, ranging in color from medium green or yellowish green to black.

Grid Pavers Open-type masonry units which allow grass to grow through openings and used for soil stabilization.

Ground Nailing strips placed in masonry walls as a means of attaching trim or furring.

Grout, Masonry A mixture of cementitious materials, aggregates, and water, with or without admixtures, used to fill voids in masonry; initially mixed to a consistency suitable for pouring or pumping without segregation of constituents.

Grout Lift (1) An increment of grout height within the total pour. A pour may consist of one or more lifts. (2) The height to which grout is placed in a cell, collar joint, or cavity without intermission.

Grout Pour The total height of a masonry wall to be grouted prior to erection of additional masonry. A grout pour will consist of one or more grout lifts.

Grout Pumping Method of installing masonry grout.

Grouted Hollow-Unit Masonry That form of grouted masonry construction in which certain designated cells of hollow units are continuously filled with grout.

Grouted Masonry Masonry construction composed of hollow units where designated hollow cells are filled with grout, or multi-wythe construction in which space between wythes is filled with grout.

Grouting, High Lift The technique of grouting where the masonry is constructed in excess of 5 ft high prior to grouting.

Grouting, Low Lift The technique of grouting as the wall is constructed, usually to scaffold or bond beam height, but not greater than 4 feet.

H

Hard-Burned Nearly vitrified clay products which have been fired at high temperatures.

Head Joint See **Joint.**

Header A masonry unit that overlaps two or more adjacent wythes of masonry to bind or tie them together. Also called a header bond. (1) A masonry unit that connects two or more adjacent wythes of masonry. (2) A masonry unit that overlaps two or more adjacent wythes of masonry to tie them together. Also called a bonder.

Header, Blind A concealed brick header in the interior of a wall, not showing on the faces.

Header, Clipped A bat that does not extend into the backup, placed to look like a header for purposes of establishing a pattern. Also called a false header.

Header Course A continuous bonding course of header brick. Also called a heading course.

Hearth (1) The masonry floor of a fireplace together with an adjacent area of fireproof material that may be a continuation of the flooring in the embrasure or some more decorative surfacing, as tile or marble. (2) An area permanently floored with fireproof material beneath and surrounding a stove.

Height of Wall The vertical distance from the foundation wall, or other similar intermediate support to the top of the wall, or the vertical distance between intermediate supports.

Height-Thickness (*H/T*) *Ratio* The height of a masonry wall divided by its nominal thickness. The thickness of cavity walls is taken as the overall thickness minus the width of the cavity.

Hot Weather Construction Procedures used in constructing masonry when ambient air temperature exceeds 100°F or 90°F with a wind velocity greater than 8 mph.

Hysteresis The irreversible expansion of marble building stone with cycles of heating and cooling.

I

Igneous Rock Rock formed by change of the molten material called magma to the solid state.

IMI International Masonry Institute.

Initial Rate of Absorption (IRA) A measure of the capillary suction of water into a dry masonry unit from a bed face during a specified length of time over a specified area.

Initial Set The beginning change from a plastic to a hardened state.

Inspection Observation to verify that construction meets the requirements of the contract documents.

Interlocking Block Pavers Solid masonry units capable of transferring loads and stresses laterally by arching or bridging action between units when subjected to vehicular traffic.

Intrados The curve that bounds the lower side of an arch.

J

Joint In building construction, the space or opening between two or more adjoining surfaces.

Mortar Joint In mortared masonry construction, the joints between units that are filled with mortar.

Bed Joint Horizontal layer of mortar on which a masonry unit is laid.

Collar Joint Vertical, longitudinal joint between wythes of masonry or between masonry wythe and backing.

Head Joint Vertical transverse mortar joint placed between masonry units within the wythe at the time the masonry units are laid.

Raked Joint A mortar joint where $\frac{1}{4}$ to $\frac{1}{2}$ in. of mortar is removed from the outside surface of the joint.

Shoved Joint Vertical joint filled by shoving a unit against the next unit when it is being laid in a bed of mortar.

Slushed Joint Head or collar joint constructed by "throwing" mortar in with the edge of a trowel.

Struck Joint A joint from which excess mortar has been removed by a stroke of the trowel, leaving an approximately flush joint.

Tooled Joint A mortar joint between two masonry units manually shaped or compressed with a jointing tool such as a concave or V-notched jointer.

Movement Joint In building construction, a joint designed to accommodate movement of adjacent elements.

Control Joint (1) In concrete masonry, a continuous joint or plane to accommodate unit shrinkage; may contain mortar or grout. (2) In building construction, a formed, sawed, tooled, or assembled joint acting to regulate the location and degree of cracking and separation resulting from the dimensional change of different elements of a structure. (3) In concrete, concrete masonry, stucco, or coating systems; a formed, sawed, or assembled joint acting to regulate the location of cracking, separation, and distress resulting from dimensional or positional change.

Expansion Joint (1) A continuous joint or plane to accommodate expansion, contraction, and differential movement; does not contain mortar, grout, reinforcement, or other hard materials. (2) In building construction, a structural separation between building elements that allows independent movement without damage to the assembly. (3) Discontinuity between two constructed elements or components, allowing for differential movement (such as expansion) between them without damage.

Saddle Joint A vertical joint along which the stone is lapped on either side to rise above the level of the wash on a coping or sill, thus diverting water from the joint.

Joint Reinforcement Metal bars or wires, usually prefabricated, to be placed in mortar bed joints.

Jointing The finishing of joints between courses of masonry units before the mortar has hardened.

K

Keystone Wedge-shaped stone at the center or summit of an arch or vault, binding the structure actually or symbolically.

Kiln A furnace, oven, or heated enclosure used for burning or firing brick or other clay material.

Kiln Run Bricks from the kiln that have not yet been sorted or graded for size or color variations.

King Closer A brick cut diagonally to have one 2-in. end and one full-width end.

L

Lap (1) The distance one masonry unit extends over another. (2) The distance one piece of flashing or reinforcement extends over another.

Lateral Support Bracing of walls either vertically or horizontally by columns, pilasters, cross walls, beams, floors, roofs, etc.

Lead The section of a wall built up and racked back on successive courses. A line is attached to leads as a guide for constructing a wall between them.

Leakage See **Water Leakage.**

Lewis Any of several metal devices for lifting stone blocks in the quarry or mill or for hoisting columns or other heavy masonry units in construction.

Lewis Bolt A bolt used to hang soffit stones or suspend the center part of lintels. It is conical or tapered and fits into slots cut from the back. It may also be leaded into stone, or be combined with expansion sleeves. Carries the weight on an I-beam or other supporting member above.

Lewis, Box An assembly of metal components, some or all of which are tapered upward, that is inserted into a downward-flaring hole (dovetail mortise) and cut into the tops of columns or other heavy masonry units for hoisting.

Lewis Hole An opening that is cut or drilled in stone blocks, columns, or other heavy masonry units to receive lewis hoisting devices. The shape and size of the hole varies with the lewis that is to be used.

Lewis Pin A metal peg, usually with its eye at the upper end, used for lifting stone blocks or masonry units. Lewis pins are used in pairs and are dependent on lever-action compression for gripping.

Lightweight Aggregate Aggregate of low density, such as expanded or sintered clay, shale, slate, diatomaceous shale, perlite, vermiculite, slag, natural pumice, scoria, volcanic cinders, tuff, diatomite, sintered fly ash, or industrial cinders.

Lime A general term for the various chemical and physical forms of quicklime, hydrated lime, and hydraulic hydrated lime.

Hydrated Lime Quicklime to which sufficient water has been added to convert the oxides to hydroxides.

Lime Mortar A lime putty mixed with an aggregate, suitable for masonry purposes.

Lime Putty The product obtained by slaking quicklime with water according to the directions of the manufacturer or by mixing hydrated lime and water to a desired consistency.

Quicklime A hot, unslaked lime. A calcined material, a major part of which is calcium oxide (or calcium oxide in natural association with lesser amounts of magnesium oxide) capable of slaking with water.

Slaked Lime Formed when quicklime is treated with water; same as hydrated lime.

Limestone (1) A rock of sedimentary origin composed principally of calcium carbonate (the mineral calcite), or the double carbonate of calcium and magnesium (the mineral dolomite), or some combination of these two minerals. (2) An initially sedimentary rock consisting chiefly of calcium carbonate or of the carbonates of calcium and magnesium. Limestone may be of high calcium, magnesian, or dolomitic.

Dolomitic Limestone Limestone containing from 35 to 46% magnesium carbonate ($MgCO_3$).

Magnesium Limestone Limestone containing from 5 to 35% $MgCO_3$.

High-Calcium Limestone Limestone containing from 0 to 5% $MgCO_3$.

Oolitic Limestone A limestone composed largely of the spherical or subspherical particles called oolites or ooliths.

Limestone Marble Compact, dense limestone that will take a polish is classified as marble in trade practice. Limestone marble may be sold as limestone or marble.

Line The string stretched taut from lead to lead as a guide for laying the top edge of a masonry course.

Line Pin A metal pin used to attach line used for alignment of masonry units.

Lintel A beam placed or constructed over an opening in a wall to carry the superimposed load and the masonry above the opening.

Loads, Allowable The permitted and projected safe load capacity through testing or calculated for a given structural element or combination of elements, including an acceptable safety factor for given material.

M

Marble Carbonate rock that has acquired a distinctive crystalline texture by recrystallization, most commonly by heat and pressure during metamorphism, and is composed principally of the carbonate minerals calcite and dolomite, singly or in combination.

Mason A worker skilled in laying brick, tile, stone, or block.

Masonry (1) Construction, usually set in mortar, of natural building stone or manufactured units such as brick, concrete block, adobe, glass block, or manufactured stone. (2) An assemblage of structural clay masonry units, concrete masonry units, stone, etc., or combination thereof, bonded with mortar or grout. (3) Construction of brick, block, or stone that is set in mortar, dry-stacked, or mechanically anchored.

Masonry Cement A hydraulic cement for use in mortars for masonry construction, containing one or more of the following materials: portland cement, portland blast-furnace slab cement, portland-pozzolan cement, natural cement, slag cement, or hydraulic lime; and in addition usually containing one or more materials such as hydrated lime, limestone, chalk, calcareous shell, talc, slag, or clay, as prepared for this purpose.

Masonry Cement Mortar Mortar produced using masonry cement.

Masonry Unit (1) Manufactured material, such as brick, concrete block, structural tile, or stone, suitable for the construction of masonry. (2) Natural or manufactured building unit of clay, concrete, stone, glass, or calcium silicate.

Masonry Unit, Clay Hollow or solid masonry unit of clay or shale, including clay brick, structural clay tile, and adobe brick.

Masonry Unit, Concrete A manufactured masonry unit made from portland cement, mineral aggregates, and water, with or without the inclusion of other materials.

Masonry Unit, Hollow A unit whose net cross sectional area in any plane parallel to the bearing surface is less than 75% of its gross cross-sectional area measured in the same plane.

Masonry Unit, Modular One whose nominal dimensions are based on the 4-in. module.

Masonry Unit, Solid A unit whose net cross-sectional area in any plane parallel to the bearing surface is 75% or more of its gross cross-sectional area measured in the same plane.

Mason's Hammer A hammer with a heavy steel head, one face of which is shaped like a chisel for trimming brick or stone.

Mason's Level Similar to a carpenter's level, but longer.

Mechanical Bond Tying masonry units together with metal ties, reinforcing steel, or keys.

Metamorphic Rock Rock altered in appearance, density, and crystalline structure, and in some cases mineral composition, by high temperature or high pressure or both.

Mix Design The proportions of ingredients to produce mortar, grout, or concrete.

Mixer A machine employed for blending the constituents of concrete, grout, mortar, or other mixtures.

Modular Coordination Dimensional coordination of masonry and other construction components through the use of standard incremental units.

Modulus of Elasticity Ratio of normal stress to corresponding strain for tensile or compressive stresses below proportional limit of material.

Moisture Content The amount of water contained, expressed as a percentage of the total absorption.

Mortar A mixture of cementitious materials, fine aggregate, and water, with or without admixtures, used to construct masonry.

Fat Mortar A mortar mixture containing a high ratio of binder to aggregate.

Harsh Mortar A mortar that is difficult to spread; not workable.

Lean Mortar Mortar deficient in cementitious components, and usually harsh and difficult to spread.

Ready-Mixed Mortar Mortar consisting of cementitious materials, aggregate, water, and set-control admixtures which are measured and mixed at a central location, using weight- or volume-control equipment. This mortar as delivered to a construction site shall be usable for a period in excess of $2\frac{1}{2}$ hours.

Surface Bonding Mortar A product containing hydraulic cement, glass fiber reinforcement with or without inorganic fillers, or organic modifiers in a prepackaged form requiring only the addition of water prior to application.

Mortar Bedding Construction of masonry assemblages with mortar.

Mortar Board A board approximately 3 feet square to hold mortar ready for the use by a mason.

Mortar Joint See **Joint.**

Movement Joint See **Joint.**

N

NBS National Bureau of Standards, now National Institute of Standards and Technology (**NIST**).

NCMA National Concrete Masonry Association.

Neat Cement In masonry, a pure cement undiluted by sand aggregate or admixtures.

Net Cross-Sectional Area The gross cross-sectional area of a unit minus the area of cores or cellular spaces.

NFPA National Fire Protection Association.

NIST National Institute of Standards and Technology (formerly National Bureau of Standards, **NBS**).

Noncombustible (1) Not combustible. (2) Any material that will neither ignite nor actively support combustion in air at a temperature of 1200°F when exposed to fire.

O

Overhand Work Masonry laid from the interior side a wall rather than the exterior side of a wall.

P

Parging The application of a coat of cement mortar to the back of the facing or the face of the backing in multi-wythe construction.

Paver A paving stone, brick, or concrete masonry unit.

Paving, Unit Vehicular or pedestrian traffic surfacing of unit masonry pavers.

Paving Brick See **Brick.**

Paving Stone Stone used in exterior pedestrian wearing surface as in patios, walkways, driveways, and the like.

PCA Portland Cement Association.

Pendentive A triangular segment of vaulting used to effect a transition at the angles from a square or polygon base to a dome above.

Perlite (1) Aggregate used in lightweight insulating concrete, concrete masonry units and plaster. (2) Insulation composed of natural perlite ore expanded to form a cellular structure.

Permeability Property of allowing passage of fluids.

Pick and Dip A method of laying brick whereby the bricklayer simultaneously picks up a brick with one hand and, with the other hand, enough mortar on a trowel to lay the brick. Sometimes called the Eastern, New England, or English method.

Pier An isolated column of masonry, or a bearing wall not bonded at the sides to associated masonry.

Pilaster A portion of a wall serving as an integral vertical column, and usually projecting from one or both wall faces. Sometimes called a pier.

Pilaster Block Concrete masonry units designed for use in construction of plain or reinforced concrete masonry pilasters and columns.

Plasticizer A substance incorporated into a material to increase its workability, flexibility, or extensibility.

Point (1) A wedge-shaped or pyramidal chisel. (2) To engage in the act of "pointing" mortar joints.

Pointing Troweling mortar into a joint after the masonry units are laid.

Repointing Filling in cut-out or defective mortar joints in masonry with fresh mortar.

Tuckpointing Decorative method of pointing masonry with a surface mortar that is different from the bedding mortar.

Portland Cement A hydraulic cement produced by pulverizing portland cement clinker, and usually containing calcium sulfate.

Prefabricated Masonry Masonry fabricated at a location other than its final location in the structure. Also known as preassembled, panelized, or sectionalized masonry.

Prefabrication To fabricate the parts at a factory or on site so that construction consists mainly of assembling and utilizing standard parts in a building structure.

Preservation, Building

Conservation Management of a natural resource, structure, or artifact to prevent misuse, destruction, or neglect. It may include detailed characterization and recording (technical or inventory) of provenance and history and application of measures.

Preservation The act or process of applying measures to sustain the existing form, integrity, or materials of a building, structure, or artifact and the existing form or vegetative cover of a site.

Protection The act or process of applying measures designed to affect the physical condition of a building, structure, or artifact by guarding it from deterioration, loss, or attack; or covering or shielding it from damage.

Rehabilitation, of a structure The act or process of returning a structure to a state of utility through repair or alteration which makes possible an efficient contemporary use.

Restoration The act or process of reestablishing accurately the form and details of a structure, site, or artifact as it appeared at a particular period in time, by means of removal of later work or by the reconstruction of missing earlier work.

Prism An assemblage of masonry units and mortar with or without grout used as a test specimen for determining properties of the masonry.

Prism Strength Maximum compressive strength (force) resisted per unit of net cross-sectional area of masonry, determined by testing masonry prisms.

Prism Testing Testing an assemblage of masonry units, mortar or grout to determine the compressive strength of masonry.

Productivity (1) Rate of production of masonry materials or assemblies. (2) Number of masonry units that a mason can install in a building structure during a given period of time.

Pumice Material of volcanic origin, being of cellular structure and highly porous, which is used as an aggregate for lightweight concrete or concrete masonry units.

Q

Quarry An open excavation at the earth's surface for the purpose of extracting usable stone.

Quarry Run Unselected materials of building stone within the ranges of color and texture available from the source quarry.

Quarry Sap Colloquial term for the natural moisture in stone as it comes from the quarry ledge. Varies in amount with the porosity of the structure.

Quartzite Geologically, metamorphic rock resulting from the annealing of quartz sandstone. See also **Sandstone.**

Queen Closer A cut brick having a nominal 2-in. horizontal face dimension.

Quoin Projecting courses of brick or stone at the corners and angles of buildings as ornamental features.

R

Racking Stepping back successive courses of masonry.

Racking Test Laboratory test for shear strength of masonry wall panels measured as diagonal tension.

Rain Penetration See **Water Penetration.**

RBM Reinforced brick masonry.

Reglet (1) A groove or recess to receive and secure metal flashing. (2) A groove in a wall or other surface adjoining a roof surface for the attachment of counterflashing.

Reinforced Masonry Masonry units and reinforcing steel bonded with mortar and/or grout in such a manner that the components act together in resisting forces.

Repointing See **Pointing.**

Retempering Adding more water to a hydraulic-setting compound such as masonry mortar after the initial mixing, but before partial set has occurred.

Retrofit In building, to add new materials or equipment not provided at the time of the original construction.

Riprap Irregularly broken and random-sized large pieces of rock.

Rowlock A brick laid on its face edge with the end surface visible in the wall face. Frequently spelled rolok.

Rustic (1) Masonry, generally of local stone, that is roughly hand-dressed, and intentionally laid with high relief in relatively modest structures of rural character. (2) A grade of building limestone characterized by coarse texture.

Rustic Joint A deeply sunk mortar joint that has been emphasized by having the edges of the adjacent stones chamfered or recessed below the surface of the face. Also called rusticated joint.

Rusticated Cut stone with strongly emphasized recessed joints and smooth or roughly textured block faces. The border of each block may be rebated, chamfered, or beveled on all four sides, at top and bottom only, or on two adjacent sides. The face of the block may be flat, pitched, or diamond point, and if smooth, may be hand- or machine-tooled.

S

Sand-Rubbed Finish Type of stone surface obtained by rubbing with a sand-and-water mixture under a block. Commonly applied with a rotary or belt sander.

Sandstone Sedimentary rock composed mostly of mineral and rock fragments within the sand size range (2 to 0.06 mm) and having a minimum of 60% free silica, cemented or bonded to a greater or lesser degree by various materials including silica, iron oxides, carbonates, or clay, and which fractures around (not through) the constituent grains.

Quartzite Highly indurated, typically metamorphosed sandstone containing at least 95% free silica, which fractures conchoidally *through* the grains.

Quartzitic Sandstone Sandstone containing at least 90% free silica (quartz grains plus siliceous cement), which may fracture around *or* through the constituent grains.

Saturation Coefficient The ratio of the weight of water absorbed by a masonry unit when immersed 24 hours in cold water to the weight of water absorbed after an additional immersion for 5 hours in boiling water. An indication of the probable resistance of brick to freezing and thawing. Also called C/B ratio.

SCPI Structural Clay Products Institute, now called Brick Institute of America (**BIA**).

SCR Trademark of the Brick Institute of America.

SCR Brick Brick whose nominal dimensions are $6 \times 2\frac{2}{3} \times 12$ in.

SCR Masonry Process A process providing greater efficiency, better workmanship, and increased production in masonry construction using story poles, marked lines, and adjustable scaffolding.

Screen Tile Clay tile manufactured for masonry screen wall construction.

Sedimentary Rock Rock formed from materials deposited as sediments in the sea, fresh water, or on land. The materials are transported to their site of deposition by running water, wind, moving ice, marine energy, or gravitational movements, and they may deposit as fragments or by precipitation from solution.

Serpentine Marble A rock consisting mostly or entirely of serpentine (hydrated magnesium silicate), green to greenish-black in color, commonly veined with calcite, and dolomite or magnesite, or both (magnesium carbonate).

Set A change in consistency from a plastic to a hardened state.

Shale (1) A thinly stratified, consolidated, sedimentary clay with well-marked cleavage parallel to the bedding. (2) A laminated sedimentary rock composed of clay minerals.

Shelf Angles Metal angles attached to a structural member used to support masonry.

Shrinkage Volume change due to loss of moisture or decrease in temperature.

Side-Construction Tile Structural clay tile intended for placement with the axis of the cells horizontal.

Skewback The inclined surface on which an arch joins the supporting wall.

Slate Microcrystalline metamorphic rock most commonly derived from shale and composed mostly of micas, chlorite, and quartz. The micaceous minerals have a subparallel orientation and thus impart strong cleavage to the rock which allows the latter to be split into thin but tough sheets.

Slenderness Ratio (1) The effective unsupported length of a uniform column divided by the least radius of gyration of the cross-sectional area. (2) The ratio of the effective height of a wall or column to its effective thickness; used as a means of assessing the stability of a masonry wall or column.

Smoke Chamber The space in a fireplace immediately above the throat where the smoke gathers before passing into the flue; narrowed by corbeling to the size of the flue lining above.

Soap A masonry unit of normal face dimension, having a nominal two-inch thickness.

Soapstone Massive soft rock that contains a high proportion of talc and that is cut into dimension stone.

Soffit The underside of a beam, lintel, or arch.

Solar Screen A perforated wall used as a sunshade.

Soldier A stretcher set on end with face showing on the wall surface.

Spall (1) To break away protrusions or edges on stone blocks with a sledge. (2) To flake or split away through frost action or pressure. (3) A chip or flake.

Specified Compressive Strength of Masonry f'_m Minimum compressive strength expressed as force per unit of net cross-sectional area required of the masonry used in construction by the project documents, and upon which the project design is based. Whenever the quantity f'_m is under the radical sign, the square root of numerical value only is intended and the result has units of pounds per square inch.

Squinch An arch or corbeling at the upper interior corners of a square tower for support of a circular or octagonal superstructure.

Stone

Building Stone Natural rock of adequate quality to be quarried and cut as dimension stone as it exists in nature, and used in the construction industry.

Cast Stone An architectural precast concrete building unit intended to simulate natural cut stone.

Cut Stone Stone fabricated to specific dimensions.

Dimension Stone Natural stone that has been selected, trimmed, or cut to specified or indicated shapes or sizes, with or without one or more mechanically dressed surfaces.

Fieldstone Natural building stone as found in the field.

Finished Stone Dimension stone with one or more mechanically dressed surfaces.

Flagstone A flat stone, thin in relation to its surface area, commonly used as a stepping-stone, for a terrace or patio, or for floor paving. Usually either naturally thin or split from rock that cleaves readily.

Stone Masonry Masonry composed of natural or cast stone.

Stone Masonry, Ashlar Stone masonry composed of rectangular units having sawed, dressed, or squared bed surfaces and bonded by mortar.

Ashlar Pattern A pattern bond of rectangular or square stone units, always of two or more sizes. If the pattern is repeated, it is patterned ashlar. If the pattern is not repeated, it is random ashlar.

Coursed Ashlar Ashlar masonry laid in courses of stone of equal height for each course, although different courses may be of varying height.

Random Ashlar Stone masonry pattern of rectangular stones set without continuous joints and laid up without drawn patterns. If composed of material cut to modular heights, discontinuous but aligned horizontal joints are discernible.

Stone Masonry, Rubble Stone masonry composed of irregular shaped units bonded by mortar.

Coursed Rubble Masonry composed of roughly shaped stones fitting on approximately level beds, well bonded, and brought at vertical intervals to continuous level beds or courses.

Random Rubble Masonry wall built of unsquared or rudely squared stones irregular in size and shape.

Squared Rubble Wall construction in which squared stones of various sizes are combined in patterns that make up courses as high as, or higher than, the tallest stones.

Stonemason A building craftsman skilled in constructing stone masonry.

Story Pole A marked pole for measuring masonry coursing during masonry construction.

Stretcher A masonry unit laid with its greatest dimension horizontal and its face parallel to the wall face.

Strike To remove excess mortar from the surface of a joint by cutting it flush with the unit surface using the edge of a trowel (see also **tooling**).

String Course A horizontal band of masonry, generally narrower than other courses, extending across the facade of a structure and in some structures encircling such decorative features as pillars or engaged columns.

Stringing Mortar The procedure of spreading enough mortar on a bed to lay several masonry units.

Structural Clay Tile Hollow masonry building units composed of burned clay, shale, fire clay, or combinations of these materials.

End-Construction Tile Tile designed to be laid with the axis of its cells vertical.

Facing Tile Tile for exterior and interior masonry with exposed faces.

Fireproofing Tile Tile designed for protecting steel structural members from fire.

Side-Construction Tile Tile intended for placement with the axis of the cells horizontal.

Suction See **initial rate of absorption.**

Surface Bonded Masonry Masonry units bonded by parging with a thin layer of fiber-reinforced mortar.

Surface Bonding Mortar A product containing hydraulic cement, glass fiber reinforcement with or without inorganic fillers, or organic modifiers in a prepackaged form requiring only the addition of water prior to application.

Sustainable Architecture/Design Building designs that fulfill the needs of the current generation without limiting the ability of future generations to produce what they need to sustain themselves.

T

Temper (1) In hydraulic setting compounds, to bring to a usable state by mixing in or adding water. (2) To moisten and mix mortar to a proper consistency.

Terra Cotta A fired clay product used for ornamental work.

Tie See **connector.**

Thermal Inertia Lag time required for a mass to heat or cool.

Thermal Mass Dense material capable of absorbing and storing heat.

Thermal Resistance The reciprocal of thermal transmittance (expressed by the notation R).

Tolerance Specified allowance of variation from a size specification.

Tooling Compressing and shaping the surface of a mortar joint (see also **strike**).

Toothing Constructing the temporary end of a wall with the end stretcher of every alternate course projecting. Projecting units are toothers.

Travertine A variety of crystalline or microcrystalline limestone distinguished by layered structure. Pores and cavities commonly are concentrated in some of the layers, giving rise to an open texture.

Travertine Marble Porous or cellularly layered, partly crystalline calcite of chemical origin.

Tuckpointing See **pointing.**

U

UBC Uniform Building Code.

Unit Masonry Construction of brick or block that is set in mortar, dry-stacked, or mechanically anchored.

Unreinforced Masonry (1) Masonry whose tensile resistance is considered in design and the resistance of any reinforcing steel present is neglected. (2) Masonry constructed without steel reinforcement, except that which may be used for bonding or reducing the effects of dimensional changes due to variations in moisture content or temperature.

V

Veneer A single facing wythe of masonry, anchored or adhered to a structural backing, but not designed to carry axial loads.

Vermiculite Micaceous aggregate used in lightweight insulating concrete, concrete masonry units, and plaster. (2) Insulation composed of natural vermiculite ore expanded to form an exfoliated structure.

Virtual Eccentricity The eccentricity of a resultant axial load required to produce axial and bending stresses equivalent to those produced by applied axial loads and moments. It is normally found by dividing the moment at a section by the summation of axial loads occurring at the section.

Vitrification Progressive reduction in porosity of a ceramic composition as a result of heat treatment, or the process involved.

Voussoir One of the truncated, wedge-shaped masonry units which forms an arch ring.

W

Wall (1) A vertical element with a horizontal length to thickness ratio greater than three, used to enclose space.

Bearing Wall A wall supporting a vertical load in addition to its own weight.

Bonded Wall A masonry wall in which two or more wythes are bonded to act as a structural unit.

Composite Wall A multi-wythe wall in which at least one of the wythes is dissimilar to the other wythe or wythes with respect to type of masonry unit.

Curtain Wall A nonbearing exterior wall, secured to and supported by the structural members of the building.

Masonry Bonded Hollow Wall Wall with an internal air space, the facing and backing wythes of which are connected with masonry headers.

Multi-Wythe Wall A masonry wall composed of two or more wythes.

Panel Wall An exterior non-loadbearing wall wholly supported at each story.

Parapet Wall That part of any wall entirely above the roof.

Partition Wall An interior wall one story or less in height. It is generally non-loadbearing. In Canada, partition means an interior wall of one-story or part-story height that is never loadbearing.

Pierced Wall See *screen wall.*

Perforated Wall See *screen wall.*

Prestressed Wall Reinforced concrete or masonry walls in which internal stresses have been introduced to reduce potential tensile stresses in the wall resulting from imposed loads.

Retaining Wall A wall not enclosing portions of a building, designed to resist the lateral displacement of soil or other material.

Screen Wall A masonry wall in which an ornamental pierced effect is achieved by alternating rectangular or shaped blocks with open spaces.

Serpentine Wall A wall that is a sine wave in plan.

Single-Wythe Wall A masonry wall only one unit in thickness.

Solid Masonry Wall A wall built of solid masonry units, laid contiguously, or with the collar joint between the units filled with mortar or grout.

Trombe Wall A masonry wall that is designed to absorb solar heat and release it into the interior of a building.

Water Absorption Process in which water enters a material or system through capillary pores and interstices and is retained without transmission.

Water Infiltration Process in which water passes through a material or system and reaches an area that is not directly or intentionally exposed to the water source.

Water Leakage Water infiltration that is unintended, uncontrolled, exceeds the resistance, retention, or discharge capacity of the system, or causes damage or accelerated deterioration.

Water Penetration Process in which water enters a material or system through an exposed surface, joint, or opening.

Water Permeation Process in which water enters, flows within, and spreads throughout a material or system.

Water Repellent A material or treatment for surfaces to provide resistance to penetration by water.

Water Retentivity That property of a mortar which prevents the rapid loss of water to masonry units of high suction. It prevents bleeding or water gain when mortar is in contact with relatively impervious units.

Water Saturation The maximum amount of water a material or system can retain without discharge or transmission.

Water Vapor Permeance Time rate of water vapor transmission through unit area of a flat material or construction induced by unit vapor pressure difference between two specified surfaces, under specified temperature and humidity conditions.

Waterproof Impervious to water.

Waterproofing (1) Treatment of a surface or structure to prevent the passage of water under hydrostatic pressure. (2) In building construction, treatment of a surface or structure to prevent the passage of liquid water under hydrostatic, dynamic, or static pressure.

Weephole (1) Opening in mortar joints or faces of masonry units to permit the escape of moisture. (2) A small hole allowing drainage of fluid.

Wind-Driven Rain Rain driven by the wind.

Workability Ability of mortar to be easily placed and spread.

Workmanship The art or skill of a worker; craftsmanship; the quality imparted to a thing in the process of creating it.

Wythe (1) Each continuous vertical section of a wall, one masonry unit in thickness. (2) The portion of a wall which is one masonry unit in thickness. A collar joint is not considered a wythe.

Appendix B

REFERENCE STANDARDS

The following reference standards of the American Society for Testing and Materials (ASTM)* and the International Conference of Building Officials (ICBO) *Uniform Building Code* are related to masonry products, testing, and construction.

Clay Masonry Units

ASTM C27, *Fire Clay and High Alumina Refractory Brick*
ASTM C32, *Sewer and Manhole Brick*
ASTM C34, *Structural Clay Loadbearing Wall Tile*
ASTM C43, *Terminology Relating to Structural Clay Products*
ASTM C56, *Structural Clay Non-Loadbearing Tile*
ASTM C62, *Building Brick*
ASTM C106, *Fire Brick Flue Lining for Refractories and Incinerators*
ASTM C126, *Ceramic Glazed Structural Clay Facing Tile, Facing Brick, and Solid Masonry Units*
ASTM C155, *Insulating Fire Brick*
ASTM C212, *Structural Clay Facing Tile*
ASTM C216, *Facing Brick*

*ASTM, formerly at 1916 Race St., Philadelphia, PA 19103, has moved to 100 Bar Harbor Drive, West Conshohocken, PA 19428.

ASTM C279, *Chemical Resistant Brick*
ASTM C315, *Clay Flue Linings*
ASTM C410, *Industrial Floor Brick*
ASTM C416, *Silica Refractory Brick*
ASTM C530, *Structural Clay Non-Loadbearing Screen Tile*
ASTM C652, *Hollow Brick*
ASTM C902, *Pedestrian and Light Traffic Paving Brick*
ASTM C1261, *Firebox Brick for Residential Fireplaces*
ASTM C1272, *Heavy Vehicular Paving Brick*
UBC 21-1, *Building Brick, Facing Brick and Hollow Brick (Made From Clay or Shale)*
UBC 21-9, *Unburned Clay Masonry Units and Standard Methods of Sampling and Testing Unburned Clay Masonry Units*

Cementitious Masonry Units

ASTM C55, *Concrete Building Brick*
ASTM C73, *Calcium Silicate Face Brick (Sand-Lime Brick)*
ASTM C90, *Loadbearing Concrete Masonry Units*
ASTM C129, *Non-Loadbearing Concrete Masonry Units*
ASTM C139, *Concrete Masonry Units for Construction of Catch Basins and Manholes*
ASTM C744, *Prefaced Concrete and Calcium Silicate Masonry Units*
ASTM C936, *Solid Concrete Interlocking Paving Units*
ASTM C1319, *Concrete Grid Paving Units*
UBC 21-2, *Calcium Silicate Face Brick (Sand-Lime Brick)*
UBC 21-3, *Concrete Building Brick*
UBC 21-4, *Hollow and Solid Loadbearing Concrete Masonry Units*
UBC 21-5, *Non-Loadbearing Concrete Masonry Units*

Natural Stone

ASTM C119, *Terminology Relating to Building Stone*
ASTM C503, *Marble Building Stone*
ASTM C568, *Limestone Building Stone*
ASTM C615, *Granite Building Stone*
ASTM C616, *Sandstone Building Stone*
ASTM C629, *Slate Building Stone*

Mortar and Grout

ASTM C5, *Quicklime for Structural Purposes*
ASTM C33, *Aggregates for Concrete*
ASTM C91, *Masonry Cement*
ASTM C144, *Aggregate for Masonry Mortar*
ASTM C150, *Portland Cement*

ASTM C199, *Pier Test for Refractory Mortar*
ASTM C207, *Hydrated Lime for Masonry Purposes*
ASTM C270, *Mortar for Unit Masonry*
ASTM C330, *Lightweight Aggregates for Structural Concrete*
ASTM C331, *Lightweight Aggregates for Concrete Masonry Units*
ASTM C404, *Aggregates for Masonry Grout*
ASTM C476, *Grout for Reinforced and Nonreinforced Masonry*
ASTM C658, *Chemical Resistant Resin Grouts*
ASTM C887, *Packaged, Dry, Combined Materials for Surface Bonding Mortar*
ASTM C1142, *Extended Life Mortar for Unit Masonry*
ASTM C1329, *Mortar Cement*
UBC 21-11, *Cement, Masonry*
UBC 21-12, *Quicklime for Structural Purposes*
UBC 21-13, *Hydrated Lime for Masonry Purposes*
UBC 21-14, *Mortar Cement*
UBC 21-15, *Mortar for Unit Masonry and Reinforced Masonry Other Than Gypsum*
UBC 21-19, *Grout for Masonry*

Reinforcement and Accessories

ASTM A82, *Cold Drawn Steel Wire for Concrete Reinforcement*
ASTM A153, *Zinc Coating (Hot-Dip) on Iron or Steel Hardware*
ASTM A165, *Electro-Deposited Coatings of Cadmium on Steel*
ASTM A167, *Stainless and Heat Resisting Chromium-Nickel Steel Plate, Sheet and Strip*
ASTM A185, *Welded Steel Wire Fabric for Concrete Reinforcement*
ASTM A496, *Deformed Steel Wire for Concrete Reinforcement*
ASTM A615, *Deformed and Plain Billet-Steel Bars for Concrete Reinforcement*
ASTM A616, *Rail-Steel Deformed and Plain Bars for Concrete Reinforcement*
ASTM A617, *Axle-Steel Deformed and Plain Bars for Concrete Reinforcement*
ASTM A641, *Zinc Coated (Galvanized) Carbon Steel Wire*
ASTM A951, *Joint Reinforcement*
ASTM B227, *Hard-Drawn Copper-Covered Steel Wire, Grade 30HS*
ASTM C1242, *Guide For Design, Selection and Installation of Exterior Dimension Stone Anchors and Anchoring Systems*
UBC 21-10, *Joint Reinforcement for Masonry*

Sampling and Testing

ASTM C67, *Sampling and Testing Brick and Structural Clay Tile*
ASTM C97, *Absorption and Bulk Specific Gravity of Natural Building Stone*

ASTM C109, *Compressive Strength of Hydraulic Cement Mortars*
ASTM C140, *Sampling and Testing Concrete Masonry Units*
ASTM C170, *Compressive Strength of Natural Building Stone*
ASTM C241, *Abrasion Resistance of Stone*
ASTM C267, *Chemical Resistance of Mortars, Grouts and Monolithic Surfacings*
ASTM C426, *Drying Shrinkage of Concrete Block*
ASTM C780, *Preconstruction and Construction Evaluation of Mortars for Plain and Reinforced Unit Masonry*
ASTM C880, *Flexural Strength of Natural Building Stone*
ASTM C952, *Bond Strength of Mortar to Masonry Units*
ASTM C1006, *Splitting Tensile Strength of Masonry Units*
ASTM C1019, *Sampling and Testing Grout*
ASTM C1072, *Method of Measurement of Masonry Flexural Bond Strength*
ASTM C1093, *Accreditation of Testing Agencies for Unit Masonry*
ASTM C1148, *Measuring the Drying Shrinkage of Masonry Mortar*
ASTM C1194, *Compressive Strength of Architectural Cast Stone*
ASTM C1195, *Absorption of Architectural Cast Stone*
ASTM C1196, *In Situ Compressive Stress Within Solid Unit Masonry Estimated Using Flatjack Method*
ASTM C1197, *In Situ Measurement of Masonry Deformability Using the Flatjack Method*
ASTM C1262, *Evaluating the Freeze-Thaw Durability of Manufactured Concrete Masonry and Related Concrete Units*
ASTM C1314, *Method of Constructing and Testing Masonry Prisms Used to Determine Compliance with Specified Compressive Strength of Masonry*
ASTM C1324, *Examination and Analysis of Hardened Masonry Mortar*
ASTM D75, *Sampling Aggregates*
ASTM E72, *Conducting Strength Tests of Panels for Building Construction*
ASTM E447, *Compressive Strength of Masonry Prisms*
ASTM E488, *Strength of Anchors in Concrete and Masonry Elements*
ASTM E514, *Water Permeance of Masonry*
ASTM E518, *Flexural Bond Strength of Masonry*
ASTM E519, *Diagonal Tension in Masonry Assemblages*
ASTM E754, *Pullout Resistance of Ties and Anchors Embedded in Masonry Mortar Joints*
UBC 21-6, *In-Place Masonry Shear Tests*
UBC 21-7, *Anchors in Unreinforced Masonry Walls*
UBC 21-16, *Field Test Specimens for Mortar*
UBC 21-17, *Compressive Strength of Masonry Prisms*
UBC 21-18, *Sampling and Testing Grout*
UBC 21-20, *Flexural Bond Strength of Mortar Cement*

Assemblages

ASTM C901, *Prefabricated Masonry Panels*

ASTM C946, *Construction of Dry Stacked, Surface Bonded Walls*

ASTM E835, *Guide for Dimensional Coordination of Structural Clay Units, Concrete, Masonry Units, and Clay Flue Linings*

ASTM C1283, *Practice for Installing Clay Flue Lining*

ASTM E1602, *Guide for the Construction of Solid Fuel-Burning Masonry Heaters*

UBC 21-8, *Pointing of Unreinforced Masonry Walls*

Appendix C

MASONRY ORGANIZATIONS

Technical guidance, design assistance, and general information about masonry materials and construction may be obtained through the following national and regional organizations.

National

American Concrete Institute
22400 W. Seven Mile Road
Detroit, Michigan 48219

Brick Institute of America
11490 Commerce Park Drive
Reston, Virginia 22091

Building Stone Institute
P. O. Box 5047
White Plains, New York 10602

Cast Stone Institute
2299 Brockett Rd.
Tucker, Georgia 30084

Concrete Paver Institute
2302 Horse Pen Road
Herndon, Virginia 22071

Expanded Shale, Clay and Slate Institute
2225 E. Murray Holladay Road, Suite 102
Salt Lake City, Utah 84117

Indiana Limestone Institute of America, Inc.
400 Stone City Bank
Bedford, Indiana 47421

International Masonry Institute
823 Fifteenth Street, N.W., Suite 1001
Washington, D.C. 20005

Marble Institute of America
30 Eden Alley, Suite 201
Columbus, Ohio 43215

Mason Contractors Association of America
1550 Spring Road, Suite 320
Oak Brook, Illinois 60521

The Masonry Society
3970 Broadway, Suite 201-D
Boulder, Colorado 80304

National Concrete Masonry Association
2302 Horse Pen Road
Herndon, Virginia 22071

National Lime Association
200 North Glebe Rd., Suite 800
Arlington, Virginia 22203

Portland Cement Association
5420 Old Orchard Road
Skokie, Illinois 60077

Regional

Alabama Masonry Institute
660 Adams Avenue, Suite 188
Montgomery, Alabama 36104

Arizona Masonry Guild
5225 North Central Avenue, Suite 115
Phoenix, Arizona 85012

Brick Institute of California
P.O. Box 879
Hermosa Beach, California 90254

Concrete Masonry Association of California and Nevada
6060 Sunrise Vista Drive, Suite 1875
Citrus Heights, California 95610

Masonry Institute of America
2550 Beverly Blvd.
Los Angeles, California 90057

Masonry Institute of Fresno
10452 N. State Hwy. 41
Madera, California 93638

Masonry Institute (Northern California)
62 Hacienda Circle
Orinda, California 94563

Masonry Resource of Southern California
P.O. Box 6337
Anaheim, California 92816

Western States Clay Products Association
62 Hacienda Circle
Orinda, California 94563

Rocky Mountain Masonry Institute
1780 South Bellaire, Suite 602
Denver, Colorado 80222

Masonry Institute of Connecticut
225 Grandview Drive
Glastonbury, Connecticut 06033

Florida Concrete and Products Association
649 Vassar Street
Orlando, Florida 32804

Masonry Association of Florida, Inc.
2064 Apricot Drive
Deltona, Florida 32725

Brick Institute of America, Region 9
5885 Glenridge Drive, Suite 200
Atlanta, Georgia 30328

Georgia Concrete and Products Association
100 Crescent Centre Parkway, Suite 110
Tucker, Georgia 30084

Cement and Concrete Products Industry of Hawaii
2828 Paa Street, Suite 1110
Honolulu, Hawaii 96819

Idaho Concrete Masonry Association
1300 East Franklin Road
Meridian, Idaho 83642

Illinois Masonry Institute
Masonry Advisory Council
1480 Renaissance Drive, Suite 401
Park Ridge, Illinois 60068

Mason Contractors Association of Greater Chicago
1480 Renaissance Drive, Suite 401
Park Ridge, Illinois 60068

Masonry Institute of Southern Illinois
Box 333
Lenzburg, Illinois 62255

Masonry Institute of Iowa
5665 Greendale Road, Suite C
Johnston, Iowa 50131

Kansas Masonry Industries Council
P.O. Box 15
Ottawa, Kansas 66067

Kentuckiana Masonry Institute
130 Fairfax Avenue, Suite 2A
Louisville, Kentucky 40207

Masonry Institute of Maryland
1200 Cowpens Avenue
Towson, Maryland 21286

The Masonry Institute, Inc.
4853 Cordell Avenue, Penthouse 1
Bethesda, Maryland 20814

International Masonry Institute New England Masonry Center
2 Park Plaza, Suite 315
Boston, Massachusetts 02116

New England Concrete Masonry Association
268 Main Street, Suite 241
North Reading, Massachusetts 01864

Masonry Institute of Michigan
32080 Schoolcraft, Suite 104
Livonia, Michigan 48150

Mason Contractors Association, Inc.
32080 Schoolcraft, Suite 104
Livonia, Michigan 48150

Brick Distributors of Minnesota
275 Market Street, Suite 409
Minneapolis, Minnesota 55405

Minnesota Concrete and Masonry Contractors Association
255 E. Kellogg Blvd.
St. Paul, Minnesota 55101

Minnesota Concrete Masonry Association
275 Market Street, Suite C-13
Minneapolis, Minnesota 55405

Minnesota Masonry Institute
275 Market Street, Suite 409
Minneapolis, Minnesota 55405

Mississippi-Louisiana Brick Manufacturers Association
812 N. President Street
Jackson, Mississippi 39202

Mason Contractors Association of St. Louis
1429 S. Big Bend Blvd.
St. Louis, Missouri 63117

Masonry Institute of St. Louis
1429 S. Big Bend Blvd.
St. Louis, Missouri 63117

Montana Contractors Association
P.O. Box 1510
Helena, Montana 59604

Nebraska Concrete Masonry Association
P.O. Box 7196
Omaha, Nebraska 68107

Nebraska Masonry Institute
11414 W. Center Road, Suite 211
Omaha, Nebraska 68144

Capital District Masonry Institute
6 Airline Drive
Albany, New York 12205

International Masonry Institute
Corporate Plaza West
286 Washington Avenue Extension, Suite 102
Albany, New York 12203

New York State Concrete Masonry Association, Inc.
2751-1/2 Lark Street, Suite 001
Albany, New York 12210

Brick Association of North Carolina
P.O. Box 13290
Greensboro, North Carolina 27415

Carolinas Concrete Masonry Association
1 Centerview Drive, Suite 112
Greensboro, North Carolina 27407

North Carolina Masonry Contractors Association
P.O. Drawer 40399
Raleigh, North Carolina 27629

North Dakota Ready-Mix and Concrete Products Association
P.O. Box 1070
Bismarck, North Dakota 58502

Northeastern Ohio Masonry Institute
21380 Lorain Road, Suite 200A
Fairview Park, Ohio 44126

Masonry Institute of Dayton
2077 Embury Park Road
Dayton, Ohio 45414

Masonry Institute of Northwest Ohio
136 North Summit Street, Suite 112
Toledo, Ohio 43604

Brick Institute of America, Mideast Region
P.O. Box 3050
North Canton, Ohio 44720

Masonry Institute of Columbus
1347 West 5th Avenue
Columbus, Ohio 43212

Masonry and Ceramic Tile Institute of Oregon
3609 S.W. Corbett, Suite 4
Portland, Oregon 97201

Delaware Valley Masonry Institute
Meetinghouse Business Center
140 West Germantown Pike, Suite 240
Plymouth Meeting, Pennsylvania 19462

Mason Contractors Association of Central Pennsylvania
P.O. Box 318
Elysburg, Pennsylvania 17824

Masonry Institute of Pennsylvania
2270 Novelstown Road
Pittsburgh, Pennsylvania 15205

Pennsylvania Concrete Masonry Association
P.O. Box 1230
Lebanon, Pennsylvania 17042

Brick Association of South Carolina
625-C Taylor Street
Columbia, South Carolina 29201

Masonry Institute of Tennessee
1136 Second Avenue, North
Nashville, Tennessee 37208

Masonry Institute of Texas
P.O. Box 34583
Houston, Texas 77075

Southwestern Brick Institute
P.O. Box 14667
Austin, Texas 78752

Concrete Masonry Manufacturers Association of Utah
1214 East Wilmington Avenue, Suite 301
Salt Lake City, Utah 84106

Virginia Masonry Council
P.O. Box 6386
Richmond, Virginia 23230

Masonry Institute of Washington
925 116th Avenue NE, Suite 201
Bellevue, Washington 98004

Masonry Industry Promotion Fund
East 102 Boone Avenue
Spokane, Washington 99202

Masonry Institute of Washington
925 116th Avenue, NE, Suite 201
Bellevue, Washington 98004

Northwest Concrete Masonry Association
40 Lake Bellevue Drive, Suite 100
Bellevue, Washington 98005

Washington State Conference of Mason Contractors
3101 Northup Way, Suite 105
Bellevue, Washington 98004

Brick Distributors of Wisconsin
P.O. Box 9282
Madison, Wisconsin 53715

Wisconsin Concrete Masonry Association
9501 S. Shore Drive
Valders, Wisconsin 54245

International

Masonry Council of Canada
4808 30th Street, SE
Calgary, Alberta T2B 2Z1
Canada

Masonry Institute of British Columbia
3636 East 4th Avenue
Vancouver, British Columbia V5M 1M3
Canada

Ontario Concrete Block Association
1013 Wilson Avenue, Suite 101
Downsview, Ontario M3K 1G1
Canada

Ontario Masonry Industry Promotion Fund
360 Superior Blvd.
Mississauga, Ontario L5T 2N7
Canada

British Masonry Society
c/o British Ceramic Research Ltd.
Queens Road, Penkhull
Stoke-on-Trent ST4 7LQ
England

National Association of Memorial Masons
Crown Buildings, High Street
Aylesbury, Buckinghamshire HP20 1SL
England

Concrete Masonry Association of Australia
25 Berry Street
North Sidney
New South Wales 2060
Australia

BIBLIOGRAPHY

ALLEN, EDWARD. *Stone Shelters.* Cambridge, Mass.: The MIT Press, 1969.

AMERICAN INSTITUTE OF ARCHITECTS. *General Conditions of the Contract for Construction,* AIA Document A201, 1987 ed. American Institute of Architects, Washington, D.C.

AMERICAN SOCIETY OF HEATING, REFRIGERATING AND AIR CONDITIONING ENGINEERS. *Fundamentals,* ASHRAE Handbook, Atlanta: 1989.

AMRHEIN, JAMES E. *Reinforced Masonry Engineering Handbook,* 4th ed. Los Angeles: Masonry Institute of America, 1983.

AMRHEIN, JAMES E., ET AL., *Masonry Design Manual,* 3d ed. Los Angeles: Masonry Industry Advancement Committee, 1979.

AMRHEIN, JAMES E. *Steel in Masonry.* Los Angeles: Masonry Institute of America, 1977.

AMRHEIN, JAMES E., AND MICHAEL W. MERRIGAN. *Reinforced Concrete Masonry Construction Inspector's Handbook,* 2d ed. Los Angeles: The Masonry Institute of America and the International Conference of Building Officials, 1989.

AMRHEIN, JAMES E., AND MICHAEL W. MERRIGAN. *Marble and Stone Slab Veneer,* 2d ed. Los Angeles: The Masonry Institute of America, 1989.

ARUMI, FRANCISCO N. *Thermal Inertia in Architectural Walls.* Herndon, Va.: National Concrete Masonry Association, 1977.

BALCOMB, J. DOUGLAS, ET AL. *Passive Solar Design Handbook.* Sacramento, Calif.: Solar Energy Information Services, 1981.

BELL, JOSEPH, ET AL., *From the Carriage Age...To the Space Age.* Herndon, Va.: National Concrete Masonry Association, 1970.

BERRYMAN, NANCY D., AND SUSAN M. TINDALL. *Terra Cotta: Preservation of an Historic Material.* Chicago: Landmarks Preservation Council of Illinois, 1984.

BORCHELT, J. GREGG, ED. *Masonry: Materials, Properties, and Performance, ASTM STP 778.* Philadelphia: American Society for Testing and Materials, 1982.

BOUDREAU, EUGENE H. *Making the Adobe Brick.* Berkeley: Fifth Street Press, 1971.

BRICK INSTITUTE OF AMERICA. *Technical Notes on Brick Construction,* Nos. 1–45. Reston, Va.: BIA.

BRICK INSTITUTE OF AMERICA. *Bricklaying Brick and Block Masonry.* Reston, Va.: BIA, 1988.

BRICK INSTITUTE OF AMERICA. *Principles of Brick Masonry.* Reston, Va.: BIA, 1989.

BUILDING OFFICIALS AND CODE ADMINISTRATORS INTERNATIONAL. *National Building Code.* Country Club Hills, Ill.: BOCA, 1990.

CAST STONE INSTITUTE. *Technical Manual with Case Histories.* Cast Stone Institute, Tucker, Ga.

CATANI, MARIO J., AND STANLEY E. GOODWIN. "Heavy Building Envelopes and Dynamic Thermal Response," *Journal of the American Concrete Institute,* February, 1976.

CLIFTON, JAMES R., ED. *Cleaning Stone and Masonry, ASTM STP 935.* Philadelphia: American Society for Testing and Materials, 1986.

COLLINS, GEORGE R. *Antonio Gaudi.* New York: George Braziller, Inc., 1960.

CONDIT, CARL W. *American Building.* Chicago: University of Chicago Press, 1968.

CONDIT, CARL W. *The Rise of the Skyscraper.* Chicago: University of Chicago Press, 1952.

CONSTRUCTION SPECIFICATIONS INSTITUTE AND ASSOCIATION FOR PRESERVATION TECHNOLOGY INTERNATIONAL. *Guide to Preparing Design and Construction Documents for Historic Projects* (CSI Document TD-2-8). Alexandria, Va.: CSI, 1996.

CONSTRUCTION SPECIFICATIONS INSTITUTE. *Manual of Practice.* Alexandria, Va.: CSI, 1992/1994.

CONSTRUCTION SPECIFICATIONS INSTITUTE. *MasterFormat Master List of Titles and Numbers for the Construction Industry,* Alexandria, Va.: Construction Specifications Institute and Construction Specifications Canada, 1988.

CONSTRUCTION SPECIFICATIONS INSTITUTE. *SpecGUIDE 04290 Adobe Unit Masonry.* Alexandria, Va.: CSI, 1994.

COX, WARREN J. *Brick Architectural Details.* Reston, Va.: Brick Institute of America, 1973.

DALZELL, J. RALPH. *Simplified Concrete Masonry Planning and Building.* New York: McGraw-Hill, 1972.

DAVEY, NORMAN A. *A History of Building Materials.* London: Phoenix House, 1961.

DAVIS, GERALD, ED. *Building Performance: Function, Preservation, and Rehabilitation, ASTM STP 901.* Philadelphia: American Society for Testing and Materials, 1986.

DETHIER, JEAN. *Down to Earth: Adobe Architecture.* New York: Facts on File, Inc., 1983.

DONALDSON, BARRY, ED. *New Stone Technology, Design and Construction for Exterior Wall Systems, ASTM STP 996.* Philadelphia: American Society for Testing and Materials, 1988.

DUNCAN, S. BLACKWELL. *The Complete Book of Outdoor Masonry.* Blue Ridge Summit, Pa.: TAB Books, Inc., 1978.

EGAN, M. DAVID. *Concepts in Building Fire Safety.* New York: John Wiley and Sons, 1978.

ELMINGER, A. *Architectural and Engineering Concrete Masonry Details for Building Construction.* Herndon, Va.: National Concrete Masonry Association, 1976.

FEININGER, ANDREAS. *Man and Stone.* New York: Crown Publishers, Inc., 1961.

FITCH, JAMES MARSTON. *Historic Preservation.* New York: McGraw-Hill, 1982.

GLOAG, JOHN. *The Architectural Interpretation of History.* New York: St. Martin's Press, 1975.

GOODWIN, STANLEY E., AND MARIO J. CATANI. *Simplified Thermal Design of Building Envelopes for Use with ASHRAE Standard 90-75.* Skokie, Ill.: Portland Cement Association, 1976.

GRIMM, CLAYFORD T. *Cleaning Masonry—A Review of the Literature.* Arlington, Texas: Construction Research Center of the University of Texas at Arlington, 1988.

GRIMM, CLAYFORD T. *Conventional Masonry Mortar—A Review of the Literature.* Arlington, Texas: Construction Research Center of the University of Texas at Arlington, 1994.

GRIMMER, ANNE E. *Keeping it Clean: Removing Exterior Dirt, Paint, Stains and Graffiti from Historic Masonry Buildings.* Washington, D.C.: U.S. Department of the Interior, National Parks Service, Preservation Assistance Division, Technical Preservation Services, 1988.

GROGAN, J. C., AND J. T. CONWAY, EDS. *Masonry: Research, Application, and Problems, ASTM STP 871.* Philadelphia: American Society for Testing and Materials, 1985.

HARRIS, HARRY A., ED. *Masonry: Materials, Design, Construction, and Maintenance, ASTM STP 992.* Philadelphia: American Society for Testing and Materials, 1988.

HEYMAN, JACQUES. *The Masonry Arch.* New York: John Wiley & Sons, 1982.

HUFF, DARREL. *How to Work With Concrete and Masonry.* New York: Popular Science Publishing Co., 1968.

INDIANA LIMESTONE INSTITUTE. *Indiana Limestone Handbook,* 17th ed. Bedford, Ind.: ILI.

INTERNATIONAL CONFERENCE OF BUILDING OFFICIALS. *Uniform Building Code.* Whittier, Calif.: ICBO, 1991.

INTERNATIONAL MASONRY INDUSTRY ALL-WEATHER COUNCIL. *Recommended Practices and Guide Specifications for Cold Weather Masonry Construction.* Washington, D.C.: International Masonry Institute, 1970.

INTERNATIONAL MASONRY INDUSTRY ALL-WEATHER COUNCIL TECHNICAL TASK COMMITTEE. *All-Weather Masonry Construction State-of-the-Art Report.* Washington, D.C.: International Masonry Institute, 1968.

JOHNSON, FRANKLIN B., ED. *Designing, Engineering and Constructing with Masonry Products.* Taken from the Proceedings of The International Conference on Masonry Structural Systems, The University of Texas at Austin, 1967. Houston: Gulf Publishing Co., 1969.

KREH, R. T., SR. *Masonry Skills.* New York: Van Nostrand Reinhold, Co., 1976.

LEBA, THEODORE, JR. *Design Manual—The Application of Non-reinforced Concrete Masonry Loadbearing Walls in Multistory Structures.* Herndon, Va.: National Concrete Masonry Association, 1969.

LEONTOVICH, VALERIAN. *Frames and Arches.* New York: McGraw-Hill, 1959.

LIEFF, M., AND H. R. TRECHSEL, EDS. *Moisture Migration in Buildings, ASTM STP 779.* Philadelphia: American Society for Testing and Materials, 1982.

LONDON, MARK. *Masonry: How to Care for Old and Historic Brick and Stone.* Washington, D.C.: The Preservation Press, 1988.

LOS ALAMOS SCIENTIFIC LABORATORY. *Passive Solar Design Handbook.* Washington, D.C.: U.S. Department of Energy, 1980.

MACKINTOSH, ALBIN. *Design Manual—The Application of Reinforced Concrete Masonry Loadbearing Walls in Multi-Story Structures.* Herndon, Va.: National Concrete Masonry Association, 1973.

MARS, G. C., ED. *Brickwork in Italy.* Chicago: American Face Brick Association, 1925.

MASONRY STANDARDS JOINT COMMITTEE. *Building Code Requirements for Masonry Structures (ACI 530/ASCE 5/TMS 402).* Detroit: American Concrete Institute, 1992.

MASONRY STANDARDS JOINT COMMITTEE. *Specifications for Masonry Structures (ACI 530.1/ASCE 6/TMS 602).* Detroit: American Concrete Institute, 1992.

MASONRY INSTITUTE OF AMERICA. *Masonry Veneer.* Los Angeles: MIA, 1974.

MASONRY INSTITUTE OF AMERICA. *Reinforcing Steel in Masonry.* Los Angeles: MIA, n.d.

MATTHYS, JOHN H., ED. *Masonry: Components to Assemblages, ASTM STP 1063.* Philadelphia: American Society for Testing and Materials, 1990.

MATTHYS, JOHN H., ED. *Masonry Designers' Guide Based on Building Code Requirements for Masonry Structures (ACI 530-92/ASCE 5-92/TMS 402-92 and Specifications for Masonry Structures) (ACI 530.1-92/ASCE 6-92/TMS 602-92).* The Masonry Society and The American Concrete Institute, 1993.

MAZRIA, EDWARD. *The Passive Solar Energy Book.* Emmaus, Pa.: Rodale Press, Inc., 1979.

MCHENRY, PAUL GRAHAM, JR. *Adobe and Rammed Earth Buildings: Design and Construction.* New York: John Wiley and Sons, 1984.

MCRAVEN, CHARLES. *Building with Stone.* Pownal, Vt.: Storey Communications, 1989.

MCKEE, HARLEY J. *Introduction to Early American Masonry—Stone, Brick, Mortar and Plaster.* Washington, D.C.: The Preservation Press, 1973.

MCLAUGHLIN, JACK. *Jefferson and Monticello: The Biography of a Builder.* New York: Henry Holt, 1988.

*NATIONAL BUREAU OF STANDARDS. *Fire-Resistance Classifications of Building Construction, Report BMS 92.* Washington, D.C.: U.S. Government Printing Office, 1942.

*NATIONAL BUREAU OF STANDARDS. *Fire-Resistance of Structural Clay Tile Partitions,* Report BMS 113. Washington, D.C.: U.S. Government Printing Office, 1948.

*NATIONAL BUREAU OF STANDARDS. *Fire Tests of Brick Walls, Report BMS 143.* Washington, D.C.: U.S. Government Printing Office, 1954.

*These documents, published by NBS, have not been reissued by NIST.

National Codes and Standards Council of the Concrete and Masonry Industries. *Thermal Mass Handbook: Concrete and Masonry Design Provisions Using ASHRAE / IES 90.1-1989.* Herndon, Va.: NCMA, 1995.

National Concrete Masonry Association. *Design Manual for Segmental Retaining Walls (Modular Concrete Block Retaining Wall Systems).* Simac, Michael R., et al. Herndon, Va.: NCMA, 1993.

National Concrete Masonry Association. *TEK Bulletins,* Nos. 1–164. Herndon, Va.: NCMA.

National Fire Protection Association. *Fire Protection Handbook,* 17th ed. Quincy, Mass.: NFPA, 1991.

National Fire Protection Association. *Life Safety Code Handbook.* Quincy, Mass.: NFPA, 1991.

National Lime Association. *Masonry Mortar Technical Notes Series.* Arlington, Va.: NLA.

National Park Service, U.S. Department of the Interior. *Preservation Briefs,* Nos. 1–23. Washington, D.C.: U.S. Government Printing Office.

Newman, Morton. *Structural Details for Masonry Construction.* New York: McGraw-Hill, 1986.

O'Brien, James J. *Construction Inspection Handbook,* 3d ed. New York: John Wiley and Sons, 1991.

O'Connor, Thomas F., ed. *Building Sealants: Materials, Properties and Performance, ASTM STP 1069.* Philadelphia: American Society for Testing and Materials, 1990.

O'Neill, Hugh. *Stone For Building.* London: William Heinemann, Ltd., 1965.

Olgyay, Aladar, and Victor Olgyay. *Solar Control and Shading Devices.* Princeton, N.J.: Princeton University Press, 1957.

Olgyay, Victor. *Design With Climate.* Princeton, N.J.: Princeton University Press, 1963.

Ontario Ministry of Citizenship and Culture. *Annotated Master Specifications for the Cleaning and Repointing of Historic Masonry.* Toronto: Heritage Branch of the Ontario Ministry of Citizenship and Culture, 1985.

Paulay, Tom, and M. J. N. Priestly. *Seismic Design of Reinforced Concrete and Masonry Buildings.* New York: John Wiley & Sons, 1992.

Plummer, Harry C. *Brick and Tile Engineering.* Reston, Va.: Brick Institute of America, 1962.

Przetak, Louis. *Standard Details for Fire-Resistive Building Construction.* New York: McGraw-Hill, 1977.

Proceedings: *Third Canadian Masonry Symposium, The University of Alberta at Edmonton,* 1983.

Proceedings: *Fourth Canadian Masonry Symposium, The University of New Brunswick at Fredericton,* 1986.

Proceedings: *Fifth Canadian Masonry Symposium, The University of British Columbia at Vancouver,* 1989.

Proceedings: *Second North American Masonry Conference, University of Maryland.* Boulder, Colo. The Masonry Society, 1982.

Proceedings: *Third North American Masonry Conference, University of Texas at Arlington.* Boulder, Colo.: The Masonry Society, 1985.

Proceedings: *Fourth North American Masonry Conference, University of California at Los Angeles.* Boulder, Colo.: The Masonry Society, 1988.

Proceedings: *Fifth North American Masonry Conference, University of Illinois at Urbana-Champaign.* Boulder, Colo.: The Masonry Society, 1991.

Ramsey, Charles G., and Harold S. Sleeper. *Architectural Graphic Standards,* 6th ed. New York: John Wiley & Sons, 1970.

Ramsey, Charles G., and Harold S. Sleeper. *Architectural Graphic Standards,* 8th ed. New York: John Wiley & Sons, 1988.

Randall, Frank A., and William C. Panarese. *Concrete Masonry Handbook,* 5th ed. Skokie, Ill.: Portland Cement Association, 1991.

Redstone, Louis G. *Masonry in Architecture.* New York: McGraw-Hill, 1984.

Sahlin, Sven. *Structural Masonry.* Englewood Cliffs, N.J.: Prentice-Hall, 1970.

Schneider, Robert R., and Walter L. Dickey. *Reinforced Masonry Design,* 2d ed. Englewood Cliffs, N.J.: Prentice-Hall, 1989.

Schwartz, Thomas A., ed. *Water in Exterior Building Walls: Problems and Solutions, ASTM STP 1107.* Philadelphia: American Society for Testing and Materials, 1991.

Smith, Baird M. *Moisture Problems in Historic Masonry Walls: Diagnosis and Treatment.* Washington, D.C.: U.S. Department of the Interior, National Park Service, Preservation Assistance Division, 1984.

Soles, Earl L., Jr. *The Colonial Williamsburg Historic Trades Annual,* Vol. II. Williamsburg, Va.: The Colonial Williamsburg Foundation, 1990.

Southern Building Code Congress. *Standard Building Code.* Birmingham, Ala.: SBCC, 1991.

Stahl, Frederick S. *A Guide to the Maintenance, Repair, and Alteration of Historic Buildings.* New York: Van Nostrand Reinhold, 1984.

Stokoe, James. *Decorative and Ornamental Brickwork.* New York: Dover Publications, 1982.

Szokolay, S. V. *Solar Energy and Building.* London: The Architectural Press, 1975.

The Underground Space Center, University of Minnesota. *Earth Sheltered Housing Design.* New York: Van Nostrand Reinhold, 1978.

Trechsel, Heinz R., and Mark Bomberg, eds., *Water Vapor Transmission Through Building Materials and Systems: Mechanisms and Measurement, ASTM STP 1039.* Philadelphia: American Society for Testing and Materials, 1989.

Watson, Donald, ed. *Energy Conservation Through Building Design.* New York: McGraw-Hill, 1979.

Woodforde, John. *Bricks to Build a House.* London: Routledge & Kegan Paul Ltd., 1976.

INDEX

A

B

G

H

I

J

K

L

M

ABOUT THE AUTHOR

Christine Beall is a consulting architect who has more than 25 years of experience in the design, specification, and construction of residential, commercial, institutional, and industrial buildings. She has published more than 90 articles and papers on masonry, sealants, glass, fire resistant materials, and related construction topics. Ms. Beall has also been a contributor to *Architectural Graphic Standards* and *Masonry Designer's Guide to the MSJC Masonry Code and Specifications*.